Concrete Buildings
in Seismic Regions
Second Edition

Concrete Buildings in Seismic Regions

Second Edition

George G. Penelis
Gregory G. Penelis

CRC Press
Taylor & Francis Group
Boca Raton London New York

CRC Press is an imprint of the
Taylor & Francis Group, an **informa** business

A SPON PRESS BOOK

CRC Press
Taylor & Francis Group
6000 Broken Sound Parkway NW, Suite 300
Boca Raton, FL 33487-2742

First issued in paperback 2021

ISBN 13: 978-1-03-209467-0 (pbk)
ISBN 13: 978-1-315-09799-2 (hbk)

Library of Congress Cataloging-in-Publication Data

Names: Penelis, George G., author. | Penelis, Gregory G., author.
Title: Concrete buildings in seismic regions / George Penelis and Gregory Penelis.
Description: 2nd ed. | Boca Raton : Taylor & Francis, CRC Press, 2018. |
Includes bibliographical references and index. |
Identifiers: LCCN 2018011311 (print) | LCCN 2018012074 (ebook) | ISBN
9781351578776 (Adobe PDF) | ISBN 9781351578769 (ePub) | ISBN 9781351578752
(Mobipocket) | ISBN 9781138106871 (hardback) | ISBN 9781315097992 (ebook)
Subjects: LCSH: Reinforced concrete construction. | Buildings, Reinforced
concrete--Earthquake effects. | Earthquake resistant design.
Classification: LCC TA683.2 (ebook) | LCC TA683.2 .P385 2018 (print) | DDC
693.8/52--dc23
LC record available at https://lccn.loc.gov/2018011311

Contents

11 Seismic pathology 561

Preface to the second edition

Reinforced concrete is one of the main building materials used worldwide, and an understanding of its structural performance under gravity and seismic loads, albeit complex, is crucial for the design of cost-effective and safe buildings.

Concrete Buildings in Seismic Regions comprehensively covers all of the analysis and design issues related to the design of reinforced concrete buildings under seismic action. It is suitable as a reference for the structural engineer dealing with specific problems during the design process and especially also for undergraduate and graduate structural, concrete and earthquake engineering courses.

This revised edition provides new and significantly developed coverage of seismic isolation and passive devices, and coverage of recent code modifications as well as notes on future developments of standards. It retains an overview of structural dynamics, the analysis and design of new R/C buildings in seismic regions, post-earthquake damage evaluation, pre-earthquake assessment of buildings and retrofitting procedures and several numerical examples.

The book outlines appropriate structural systems for many types of buildings, explores recent developments and covers the last two decades of analysis, design and earthquake engineering. It specifically addresses seismic demand issues and the basic issues of structural dynamics, considers the 'capacity' of structural systems to withstand seismic effects in terms of strength and deformation and highlights the assessment of existing R/C buildings under seismic action. All of the material has been developed to fit a modern seismic code and offers in-depth knowledge of the background upon which the code rules are based. It complies with the European Codes of Practice for R/C buildings in seismic regions, and includes references to current American standards for seismic design.

In closing, we express our thanks to our editor Tony Moore and his editorial assistant Gabriella Williams for their continuous support and confidence, as well as Mr Panagiotis Savvas and Mrs Ermina Damlamayan for drafting the manuscript. Also, we express our gratitude to the architect/photographer Yorgis Gerolympos for providing the cover picture of the Stavros Niarchos Foundation Cultural Center (SNFCC) during construction as well as the SNFCC organization for their permission to use the photograph.

Thessaloniki, June 2018

George G. Penelis
Gregory G. Penelis

Preface to the first edition

This book is addressed primarily to postgraduate students in earthquake engineering and to practicing structural engineers specializing in the design of R/C seismic-resistant buildings.

The basic aims that have guided the composition of the material of this book are the following:

1. The presentation of the content should be characterized by integrity, clarity and simplicity, particularly for the design procedure of new R/C buildings or assessment and retrofitting of existing ones. In this respect, it would constitute integrated knowledge for a student. We hope that the long experience of the first of the authors in teaching undergraduate and postgraduate students about R/C earthquake-resistant structures augurs well for the achievement of this aim.

2. The presentation of the scientific background of each subject should be made in a concise form with all the necessary but at the same time limited references to the sources so that enough of open field is available for a rigorous and systematic approach to the implementation of the scientific background in design procedure. In this context, this book would be valuable for a practicing engineer who wants to have in-depth knowledge of the background on which the Code rules are based. We hope that the extensive experience of both authors in the seismic design of new R/C buildings and in the assessment and retrofitting of existing ones, together with the wide experience of the first in posts of responsibility in seismic risk management in Greece, contribute to a balanced merging of the scientific background with practical design issues. At the same time, the numerical examples that are interspersed in the various chapters of the book also intend to serve this aim.

3. For furtherance of the above aims, all of the material of the book has been adjusted to fit a modern Seismic Code of broad application, namely EN1998/2004-5 (EC8/2004-5), so that quantitative values useful for the practice are presented. At the same time, comparative references are made to the American Standards in effect for seismic design. It is obvious that this choice covers the design requirements for Europe. At the same time, the comparative references to the American Standards enable an easy adjustment of the content to the American framework of codes.

In closing, we would like to express our thanks to the following collaborators for their contributions to the preparation of this book:

- Prof Andreas Kappos, co-author with G. Penelis of the book entitled *Earthquake Resistant Concrete Structures* (1997). We are grateful to him for his consent to reproduce from the above book a number of illustrations and some parts of the text as well.

- Dr Georgia Thermou, a lecturer in engineering at AUTH, and Mr John Papargyriou, MSc-DIC, for their contributions in the elaboration of the numerical examples.
- Dr Phil Holland for correction and improvement of the language of the text.
- Mr Panagiotis Savas for the elaboration of the figures.
- Ms Evaggelia Dara for the typewritten preparation of the text.
- Dr Kostas Paschalidis for the critical reading and correction of the text.

Thessaloniki, May 2013

George G. Penelis
Gregory G. Penelis

List of abbreviations

ACI	American Concrete Institute
ADRS	acceleration displacement response spectra
AFRP	aramid fibres
ASCE	American Society of Civil Engineers
ATC	Applied Technology Council
AUTH	Aristotle University of Thessaloniki
BI	bilinear form
CE	European Certification
CEB	Comité Euro-International du Béton
CEN	European Committee for Standardization
CF	confidence factor
CFRP	carbon fibres
CQC	complete quadratic combination
DBD	displacement-based design
DCH	ductility class high
DCL	ductility class low
DCM	ductility class medium
DDBD	direct displacement-based design
DIN	Deutsches Institut Für Normung
DL	damage limitation
DLS	damage limitation state
DSA	design seismic action
DT	destructive test
EERI	Earthquake Engineering Research Institute
ELOT	Greek Organization for Standardization
EMS-98 scale	European Macroseismic
EN	EuroNorm
EN 1990-1999	(Eurocode EC 1-EC 9)
ENV	E.N. (voluntarily)
EPP	elastic-perfectly plastic
EPS	Earthquake Protection Systems
EUROCODES (EC-CODES)	European Standards
FEM	finite element method
FEMA	Federal Emergency Management Agency
fib	International Federation for Structural Concrete
FRP	fibre reinforced plastic
GFRP	glass fibres

GMPW	Greek Ministry of Public Works
IDA	inelastic dynamic analysis
JBDPA	Japan Building Disaster Prevention Association
LS	limit state
MDOF	multi-degree-of-freedom
MEP	mechanical electrical plumping
MM scale	modified Mercalli scale
MRF	moment-resisting frames
MSK scale	Medvedev, Sponheuer, Karnik
NAD	National Application Document
NC	near collapse
NDT	non-destructive test
NIBS	National Institute of Building Sciences
NTU	National Technical University of Greece
NZS	New Zealand Code
OASP	Seismic Risk Management Agency of Greece
OCR	over consolidation ratio
PESH	potential earth-science hazard
PGA	peak ground acceleration
PGD	peak ground displacement
PGV	peak ground velocity
PSV	pseudovelocity spectra
PTFE	**polytetrafluoroethylene**
R/C	reinforced concrete
RILEM	International Union of Laboratories and Experts in construction materials
RVSP	rapid visual screening procedure
SD	significant damage
SDOF	single-degree-of-freedom
SEAOC	Structural Engineers Association of California
SI	seismic intensity
SNFCC	**Stavros Niarchos Foundation Cultural Center**
SIA	Schweizerische Normenvereinigung
SRSS	square root of the sum of the squares
THA	time–history analysis
TR	torsionally restrained system
TUR	torsionally unrestrained system
ULS	ultimate limit state
UNDP	United Nations Development Programme
UNIDO	United Nations Industrial Development Organization

Authors

George G. Penelis is an emeritus professor at the Aristotle University of Thessaloniki, Greece. He has served as a national representative on the drafting committee for Eurocode 2. He is an ordinary member of Academia Pontaniana, Italy. Penelis has published more than 250 technical papers and is a co-author of *Earthquake Resistant Concrete Structures* (Taylor & Francis, 2010).

Gregory G. Penelis is the CEO of Penelis Consulting Engineers S.A., and has been involved in the design/review of more than 100 buildings throughout Europe. He also has been involved in many research projects regarding the seismic assessment of listed and monumental buildings.

Chapter 1

Introduction

1.1 HISTORICAL NOTES

Earthquake engineering is an independent scientific discipline that has come into being along with engineering seismology over the past 100 years and is therefore still evolving, as happens in every new scientific field.

It started as a framework of codified rules for the seismic design of buildings at the beginning of the twentieth century, after the catastrophic Messina (Italy) earthquake of 1908. This design procedure was based, on one hand, on a framework of empirical rules for avoiding seismic damage observed in previous earthquakes and, on the other, on the simulation of the seismic action on a set of lateral forces equal to a percentage of the gravity loads of the building. This loading pattern constituted one additional load case, the 'seismic loading'. The above simulation was based on the fact that the acceleration of the masses of a building due to earthquake causes inertial lateral forces proportional to the masses of the building, and in this respect proportional to the gravity loads. The proportion of lateral seismic loads H_{seism} to gravity loads W resulted from the ratio of the peak ground acceleration (PGA) to the gravity acceleration g ($\varepsilon = PGA/g = H_{seism}/W$).

A critical point in the above procedure was the assessment of the PGA of an earthquake that should be taken into account for the determination of the seismic lateral loading, since for many years up to 1940 there were no instruments capable of determining the PGA (strong motion instruments). Therefore, for a period of almost 35 years after the San Francisco earthquake, qualitative observations were used, like the overturning of heavy objects located at ground level for the quantitative estimate of the PGA as a percentage of g (gravity acceleration). These values were introduced as seismic coefficients in the various Seismic Codes in effect and ranged usually from $\varepsilon = 0.04$ to 0.16, depending on the seismic hazard of each region.

At the same time, it was realised that the vibration of the superstructure of a building caused by seismic actions does not coincide with that of its foundation, particularly in the case of tall buildings, which are flexible. Thus, the first steps in structural dynamics were made early on, particularly for one degree of freedom systems, that is, single-storey buildings simulated by a concentrated mass and connected to their base by an elastic spring and a damper. The aim of these first steps was the determination of the time history of the mass motion in relation to its base for a given time history of motion at the base. The problem of the *resonance* of the vibrating mass was one of the basic points of interest, since such a resonance resulted in an *amplification* of the acceleration of the mass in relation to that of the base motion. These research activities were carried out for simple forms of vibration of the base, for example, sinusoidal or ramp excitations, since these were functions that could easily be integrated without computer facilities.

In 1940, the first accelerograph (strong motion instrument) installed at El Centro in California was activated by a strong earthquake. This was the first time-history acceleration record (Fintel and Derecho, 1974) known as the El Centro earthquake record. This was the first big step in earthquake engineering, since an objective method was invented for recording strong ground motions. It should be noted that the first measurement of the PGA of an earthquake had been preceded during the Long Beach earthquake in 1933. Nowadays, there are thousands of such instruments installed all over the world in seismically sensitive regions, enriching the collection of records of strong earthquakes.

The next big step was made 15 years later (in about 1955) with the development of computational techniques for obtaining *acceleration, velocity* and *displacement elastic response spectra* for a given time-history acceleration record. The use of acceleration response spectra enabled the easy and direct determination of the maximum acceleration that develops in the mass of a building of a given fundamental period for a seismic spectrum under consideration. The above step must be attributed mainly to the introduction of computer facilities in engineering (Housner et al., 1953).

The above research activities resulted in the conclusion that the *amplification factor* of the PGA in the case of resonance was very high for elastic structural systems. In this context, the peak accelerations of mass were many times higher than those for which the analysis of the buildings had been carried out for many decades according to the Codes then in effect. Therefore, in the case of strong earthquakes, extended collapses were to be expected even of earthquake-resistant buildings. However, while many buildings exhibited extensive damage, collapses were limited. Due to this fact, it was concluded that the inelastic behaviour of the structural system, which resulted in damages, led to a reduction of the peak acceleration of mass. So, in the 1960s, the energy balance theory of the structure under seismic action was born (Blume, 1960; Park and Paulay, 1975). According to this theory, a seismic action causes the transfer of a certain amount of kinetic energy from the foundation to the structural system. This energy is converted into potential energy, stored in the structural system in the form of strain energy and then partly dissipated during the successive cycles of vibration in the form of heat or other irrecoverable forms of energy. This cyclic conversion of energy may be accomplished either in the elastic range of a structure through viscous damping by the development of high internal forces accompanied by low strains (below yield), therefore producing no damage at all, or in the inelastic range, in the case of a structure with limited load-carrying capacity, by the development of inertial forces limited to its capacity and large strains beyond the yield point, therefore accompanied by damage or collapse, depending on the inelastic displacement capacities of the structural system. The above theory was the third big step in earthquake engineering.

The above three steps have provided the main core of modern seismic design and technology for the past 40 years. This may be summarised by the following concepts:

- Structural systems must resist low-intensity earthquakes without any structural damage.
- Structures should withstand a rare earthquake of moderate intensity (design earthquake) with light and repairable damage in the structural elements.
- Structures should withstand a very rare high-intensity earthquake without collapsing.

Although the acceptance of damages under the design earthquake meant an inelastic response of the structure, the core of the structural analysis procedure for seismic actions continued to be based on the linear theory. So, in order to solve this apparent discrepancy, two new concepts were introduced:

- The *ductility* coefficient
- The *capacity design* procedure

The former allows the reduction of seismic loading by a factor of about 1.5–5.0 times, having as prerequisite *the suitable reinforcing* of the structural elements so that the required inelastic deformation capacities are ensured. The latter readjusts the values of the internal forces due to seismic actions so that the risk of unexpected brittle local failures of structural elements is avoided. The above framework formulated the main body of the modern seismic philosophy known as 'force-based design' and constitutes the core of all modern Codes in effect. It should be noted here that while structures under the 'design earthquake' respond inelastically, the use of inelastic dynamic analysis was avoided, on purpose, by introducing the ductility coefficients and the capacity design due to computational difficulties, on one hand, and to difficulties for interpretation of the results on the other. So, the procedure of *inelastic dynamic analysis* remained a valuable tool for evaluating other simplified methods like the 'force-based design' and the 'displacement-based design' (see below).

While the force-based design procedure has been proven to be a reliable method for the design of new buildings, or structural systems in general, its implementation in the assessment and rehabilitation of existing buildings runs into many difficulties. Since the design of these buildings does not comply with the rules of the modern Codes in effect, *their capacities* for plastic deformation must be explicitly defined and compared to the *demands* of plastic deformations due to the design earthquake. Therefore, the force-based design is not proper anymore. Instead, the *displacement-based design* may be implemented. This procedure has been developed in the last 20 years and is based, on one hand, on the 'capacity curve' of the structural system generated by a static lateral loading of the structure up to failure (*pushover analysis*) and on the determination of the 'target displacement' of the structural system corresponding to the design earthquake, using *energy balance criteria*. Thus, the efficiency of an existing structural system is obtained by displacement verifications. In the last 15 years, successful efforts have been made for a *direct displacement-based design method for application in the case of new buildings*. At the same time, the development of the displacement-based design method enabled the introduction of 'the performance-based design' approach in the late 1990s, permitting the establishment of a clear relation matrix between well-defined up-to-collapse *damage levels* to the *return period* of the seismic action.

It should also be noted here that the basic theory of elastic or inelastic vibration had been previously developed for other scientific disciplines. However, its implementation in structural dynamics had encountered impassable obstacles due to the high degree of their redundancy and the random form of time-history seismic records. These two parameters made the numerical treatment of the relevant differential equations impossible. The rapid development of computers in the past 50 years enabled this obstacle to be overcome. It is important to note that until recently the above difficulties did not allow the adoption of even the 'modal response spectrum analysis', which is a method of dynamic analysis for linear structural systems, as the 'reference method' in Codes of Practice. Instead, Codes continued to have as a reference method the 'lateral force method of analysis', which, as was mentioned earlier, simulates the dynamic response of the structure with a laterally loaded static system.

Eurocode EC8-1/2004 is the first to have made this step forward to introduce the modal response spectrum analysis as a reference method.

In closing this short historical note on the development of earthquake engineering, reference should be made to the successful efforts for the development of alternative design techniques and devices for special types of buildings, namely the 'base isolation systems' and the 'dissipative devices' introducing additional dumping in the main structure.

1.2 STRUCTURE OF THE BOOK

The considerations of earthquake engineering briefly presented above will be adapted and applied in the forthcoming chapters to buildings with a structural system made of reinforced concrete, which constitute the major part of the building stock of the built environment in developed countries. The material of this book has been formulated into four main parts.

In the first part, seismic demand issues are examined (Chapters 2 through 6). More particularly, this part includes at the outset a short overview of basic issues of structural dynamics, which have been considered of special importance for the comprehension of the material of subsequent chapters. It also includes the procedure for the determination of the seismic actions and the ductility coefficients. The description of the acceptable methods for seismic analysis and the application of the capacity design rules to the seismic effects (internal forces) are included in this part as well. Finally, the conceptual design of building structural systems is also examined in detail, and guidelines are given for the proper structural system for various types of buildings.

In the second part of the book (Chapters 7 through 10), the capacity of structural systems to withstand seismic effects in terms of strength and deformation (local ductility) is examined in detail. At the same time, design methods are developed that ensure the safety verification of the structural members. More particularly, detailed reference is made to the behaviour of the basic materials (concrete-steel) and their bonding under cyclic loading. Further, a detailed presentation is made of the seismic capacity of the basic structural members of an R/C building, namely of beams, columns, joints, walls, diaphragms and foundations.

The third part is devoted to existing R/C buildings under seismic action (Chapters 11 through 14). More particularly, a detailed presentation is made of seismic pathology, post-earthquake emergency measures, assessment and rehabilitation procedures, the materials and techniques of repair and strengthening.

Finally, an overview on the alternative design techniques and devices for seismic protection is presented (Chapter 15), namely the 'base isolation systems' and the 'dissipative devices' introducing additional damping in the building.

It should be noted that the book has been adjusted to fit the European Codes and particularly to EN 1998 (Eurocode EC8/2004), with parallel extended references to the American Standards in cases that this has been considered necessary for comparisons.

From this point on, the reader him- or herself will judge if the structure of the extended material of this book fulfills its main aim of providing an integrated, comprehensible and clear presentation suitable for design practice.

Chapter 2

An overview of structural dynamics

2.1 GENERAL

The structural design of buildings was based for centuries, and even today for the greater part of the world, on the assumption that loading is not time-dependent, with only some special exceptions where cyclically moving masses cause vibrations. In this respect, for centuries the core of the education of a structural engineer has been based on statics and the strength of materials. Therefore, it can be easily understood why the first efforts for quantitative evaluation of the seismic actions on buildings, at the beginning of the twentieth century and for many decades since then, have been based on a simulation with static loads of the time-dependent inertial forces due to seismic vibrations. Even today, the American Codes of Practice for earthquake-resistant structures (ASCE 7-05) has *as a reference method of analysis* the 'lateral force method of analysis', where, of course, many of the consequences of dynamic behaviour of structures have been incorporated.

For many decades, the lateral inertial forces induced by an earthquake were calculated as the product of building mass times the maximum ground acceleration due to seismic action (Figure 2.1). In this respect, seismic actions were taken into account as additional static loading.

This approach was based implicitly on the assumption that the relative displacement of the structural system in relation to the foundation during the seismic action may be considered zero, and, therefore, the inertial forces on the storeys above depend only on the maximum ground acceleration $\ddot{x}_0$ and their mass. This assumption is acceptable to a degree for rigid buildings, or in other words low-rise buildings (1–3 storeys) with thick walls and small compartments, but not for the buildings that compose contemporary cities.

Since the 1950s, *structural dynamics has gradually become the core of the analysis and design of earthquake-resistant structures.* In the beginning, linear models with *viscous damping* were introduced, but soon after, steps were made towards models of inelastic behaviour with *hysteretic* damping, aiming at a more realistic approach in the response of structures to strong seismic motions, which cause damage.

This development was not easy at all, mainly due to computational difficulties. In spite of this, progress was accelerated by the need for higher seismic protection of modern cities with high-rise buildings located in seismic regions, where the consequences from a strong earthquake are many times greater than those of small towns. At the same time, the rapid development of computational means enabled the implementation in structural dynamics, knowledge already developed in the theory of vibration of simple systems for other scientific branches (e.g. electromagnetics, etc.).

Nowadays, structural dynamics is an autonomous branch of the 'theory of structures'. Therefore, a systematic approach to it is beyond the scope of this book. However, it was considered as a prerequisite to include an overview of the key issues of structural dynamics

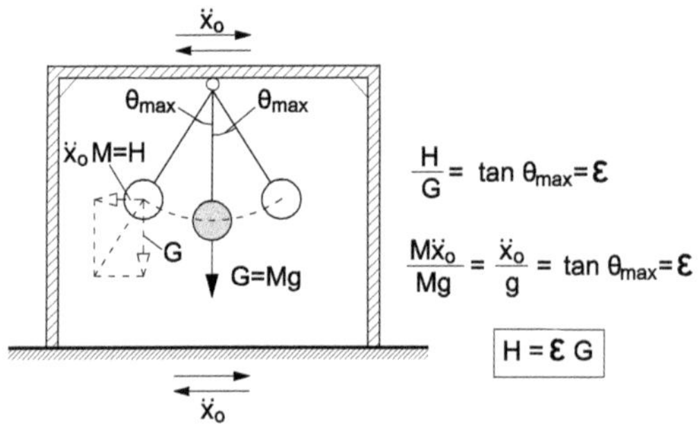

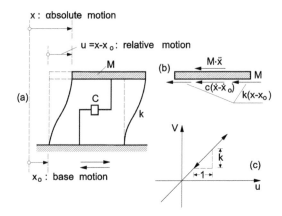

$$\frac{H}{G} = \tan\theta_{max} = \mathcal{E}$$

$$\frac{M\ddot{x}_0}{Mg} = \frac{\ddot{x}_0}{g} = \tan\theta_{max} = \mathcal{E}$$

$$\boxed{H = \mathcal{E}\,G}$$

Figure 2.1 Seismic loads of old seismic codes.

to which references will be made in subsequent chapters, since, as noted before, nowadays structural dynamics lies at the core of the analysis and design of earthquake-resistant structures.

2.2 DYNAMIC ANALYSIS OF ELASTIC SINGLE-DEGREE-OF-FREEDOM SYSTEMS

2.2.1 Equations of motion

The simplest structural system induced to seismic excitation of its base is the one shown in Figure 2.2. The understanding of the behaviour of this simple system, which has only one degree of freedom in the horizontal direction, is of substantial importance for structural dynamics in general, as will be presented next.

This system comprises one mass M on a spring in the form of two columns, which remain in the elastic range ($V_{el} = Ku$) during the vibration. The vibration is induced by the seismic excitation $x_0(t)$ of the base of the system. At the same time, the vibrating mass is connected

Figure 2.2 Single-degree-of freedom (SDOF) system excited by the base motion: (a) response of the system; (b) dynamic equilibrium condition; (c) shear force versus relative displacement diagram for the columns.

to the base of the system with a dash pot ($V_d = c\dot{u}$), which simulates the damping resistance of the oscillator to the vibration due to the internal friction of the structure.

The constant c will be defined later (see Section 2.3.2).

The motion parameters, which are the lateral displacement, velocity and acceleration of the mass of the oscillator, are given by the following expressions (functions of time):

$$x = x_o + u$$
$$\dot{x} = \dot{x}_o + \dot{u} \qquad \qquad (2.1a\text{–}c)$$
$$\ddot{x} = \ddot{x}_o + \ddot{u}$$

The notation of the above symbols is given in Figure 2.2.

The implementation of d'Alembert's concept for the equilibrium of the vibrating mass results in the following equation of vibration of the SDOF system:

$$M\ddot{x} + c\dot{u} + Ku = 0 \qquad \qquad (2.2)$$

or

$$\boxed{M\ddot{u} + c\dot{u} + K\ddot{x}_o(t)} \qquad \qquad (2.3)$$

Equation 2.3 shows that the most important parameter in relation to the seismic excitation for the description of the vibration response is the acceleration time history $\ddot{x}_o(t)$ of the base of the SDOF system. It is well known that since 1939, networks of special instruments known as *accelerographs* or *strong motion instruments* that record the time history of ground acceleration have been developed (Figure 2.3). These networks were developed first in the United States but they were later extended to seismic regions all over the world. These instruments are adjusted in advance so that they are triggered every time the ground acceleration exceeds the predefined limit. These instruments record the time history of the ground acceleration in a form that may be used as an input in Equation 2.3. In this respect, these records make possible the determination of the time-history response of the SDOF system to the recorded seismic action.

2.2.2 Free vibration

If during oscillation the exciting force becomes zero ($\ddot{x}_o(t) = 0$), the system continues to vibrate freely. In this case and for zero damping ($c = 0$), the equation of motion (Equation 2.3) takes the form: that is, it takes the form of a homogeneous differential equation of the second order. The general solution of this equation has the form

$$M\ddot{u} + Ku = 0 \qquad \qquad (2.3a)$$

which is a homogeneous differential equation of the second order. The general solution of this equation has the form

$$\boxed{u = u_o \cos \omega t + \frac{\dot{u}_o}{\omega} \sin \omega t} \qquad \qquad (2.4)$$

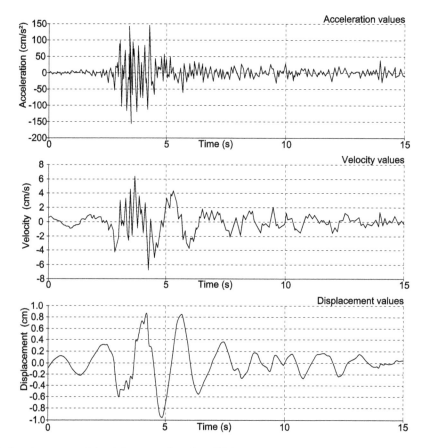

Figure 2.3 Accelerogram of the Athens earthquake of 7 September 1999 (N–S): components and derived diagrams of velocity and displacement. (Adapted from the Institute of Engineering Seismology and Earthquake Engineering [IESEE], Greece, 2003. *Strong Motion Data Base of Greece 1978–2003*, Thessaloniki, Greece.)

where

$$\omega = \frac{2\pi}{T_o} = \sqrt{\frac{K}{M}}$$ (2.4a)

is the *natural circular frequency* (in rad/s) and

$$T_o = 2\pi\sqrt{\frac{M}{K}}$$ (2.4b)

is the *natural period* of the system.

The constant coefficients in Equation 2.4 are determined by taking into account the following initial conditions:

For $t = 0$, the initial relative displacement u is supposed to be $u = u_o$, while the initial velocity, for $t = 0$, is supposed to be $\dot{u} = \dot{u}_o$. In other words, u_o and $\dot{u}_o$ are input data (Figure 2.4; Georganopoulou, 1982).

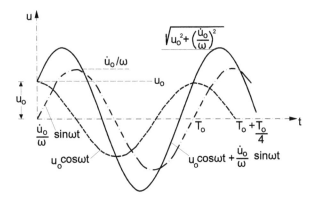

Figure 2.4 Diagram of the free vibration relative displacement of Equation 2.4 for input data: initial displacement $u = u_o$ and initial velocity $\dot{u} = \dot{u}_o$.

The natural period T_o is the dynamic constant of the system, the characteristics of which, the mass M and the spring stiffness K, have been incorporated into it. *Natural period T_o expresses the period of the free vibration of the system.*

When damping is not zero ($c \neq 0$), the homogeneous equation of motion (Equation 2.3a) takes the form

$$M\ddot{u} + c\dot{u} + Ku = 0 \tag{2.3b}$$

The general solution of this equation has the form

$$u = u_o e^{-\frac{ct}{2M}}(A \sin \gamma t + B \cos \gamma t) \tag{2.5}$$

where

$$\gamma = \sqrt{\frac{K}{M} - \left(\frac{c}{2M}\right)^2} \tag{2.6}$$

For $\gamma = 0$

$$\boxed{c = c_{cr} = 2\sqrt{MK}} \tag{2.6a}$$

The free vibration is transformed into an attenuation curve (Figure 2.5b). By setting $\zeta = c/c_{cr}$ (*critical damping ratio*), Equation 2.5 becomes

$$\boxed{u = u_o e^{-\zeta \omega t}(A \sin \omega_D t + B \cos \omega_D t)} \tag{2.7}$$

where

$$\omega_D = \omega\sqrt{1 - \zeta^2} = \gamma \tag{2.8}$$

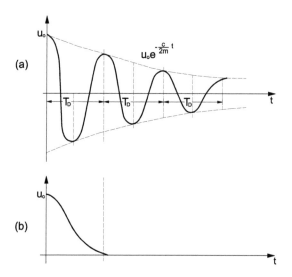

Figure 2.5 Free vibration with damping: (a) initial relative displacement u_o; (b) initial displacement in case of critical damping. (From Biggs, J.M. 1964. *Introduction to Structural Dynamics*. McGraw-Hill, New York. With permission.)

and

$$T_D = \frac{T}{(1-\zeta^2)^{1/2}}.$$ (2.9)

For values of $\zeta < 0.15$, the natural period T_D of the oscillating system is close to that of the undamped system T_o.

The constants A and B are defined by the initial conditions of the initial displacement u_o and the initial velocity $\dot{u}_o$, which gives

$$A = \frac{\dot{u}_o + \gamma u_o}{u_o w_o}, B = 1.0$$

2.2.3 Forced vibration

In case of a ground excitation $\left(\ddot{x}_o(t) \neq 0\right)$, the particular integral of Equation 2.3 must be added to the general integral (Equation 2.7).

In case of an artificial steady-state excitation in the form of a sinusoidal function (Figure 2.6; Georganopoulou, 1982), we get

$$\ddot{x}_o(t) = \ddot{x}_{o\max} \sin\Omega t.$$ (2.10)

The particular integral of the undamped forced vibration takes the form

$$u_p = \frac{\ddot{x}_{o\max}}{K/m} \frac{1}{1-\beta^2} \sin\Omega t$$ (2.11)

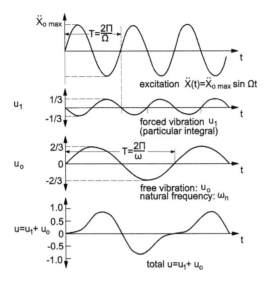

Figure 2.6 Forced vibration of an SDOF system excited by a sinusoidal base acceleration for $\Omega/\omega = 2$.

where

$$\beta = \frac{\Omega}{\omega} \tag{2.12}$$

In this respect, the damped forced vibration takes the form

$$u_{pd} = \frac{M\ddot{x}_{o\,max}/K}{\sqrt{(1-\beta^2)^2 + (2\zeta\beta)^2}} \sin\left(\Omega t - \arctan\frac{2\zeta\beta}{1-\beta^2}\right). \tag{2.13}$$

It is important to note that in case

$$\beta = \frac{\Omega}{\omega} = 1,$$

which means that the frequency Ω of the excitation coincides with the natural frequency (ω) of the SDOF system, the maximum value of u_p becomes (Figure 2.7)

$u_{p\,max} = \infty$ for an undamped system and

$$u_{pd\,max} = \frac{\ddot{x}_{o\,max} \cdot M}{2K\zeta} \text{ for a damped one} \tag{2.13a}$$

It is obvious that, in this case, *resonance* of the vibrating system takes place due to the coincidence of the natural period of the vibrating system and that of the exciting agent. As a result, the vibrating system *amplifies the ground maximum acceleration and the displacement.*

In case of a transient chaotic seismic excitation, the above process for the determination of the partial integral of Equation 2.3 must be modified radically since it is not possible to find

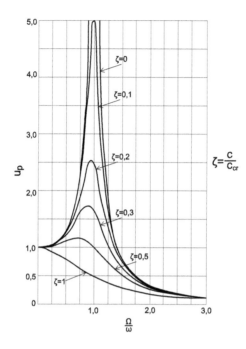

Figure 2.7 Maximum relative displacement u_p of an SDOF system activated by a sinusoidal forced vibration diagram of u_p versus natural frequencies ratio Ω/ω for various values of ζ (critical damping ratio). (From Biggs, J.M. 1964. *Introduction to Structural Dynamics*. McGraw-Hill, New York. With permission.)

a closed form of the partial integral. Thus, the function of the ground acceleration $\ddot{x}_o(t)$ is divided into successive, very short-duration impulses, as shown in Figure 2.8.

$$dP = \ddot{x}_o(\tau) \cdot M \cdot d\tau \tag{2.14}$$

Consider now one of these impulses, which end at time τ after the beginning of the ground motion (Equation 2.14). This impulse, in turn, causes an initial velocity on the vibrating system:

$$d\dot{u}_o(\tau) = \ddot{x}_o(\tau)d\tau \tag{2.15a}$$

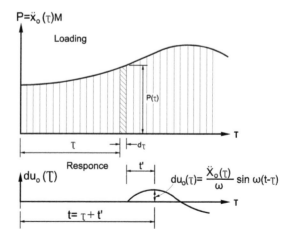

Figure 2.8 Response $u_o(t)$ of an SDOF system to a differential impulse.

while the initial displacement is

$$du_o(\tau) = 0 \qquad\qquad (2.15b)$$

Substituting Equations 2.15a and 2.15b into Equation 2.4 (undamped vibration) results to

$$du = \frac{\ddot{x}_o(\tau)}{\omega}\sin\omega(t-\tau)d\tau \qquad\qquad (2.16)$$

In the above expression, du is the displacement at time t due to the input ground acceleration $\ddot{x}_o(\tau)$ acting during the time differential $d\tau$. This results to

$$\boxed{u(t) = \frac{1}{\omega}\int_0^t \ddot{x}_o(\tau)\sin\omega(t-\tau)d\tau} \qquad\qquad (2.17)$$

This is the *Duhamel integral*, well known in structural dynamics. In case of damped vibration, Equation 2.17 takes the form

$$\boxed{u(t) = \frac{1}{\omega}\int_o^t \ddot{x}_o(\tau)e^{-\zeta\omega(t-\tau)}\sin\omega_D(t-\tau)d\tau} \qquad\qquad (2.17a)$$

The above Equations 2.17 and 2.17a give the displacement $u(t)$ during the ground motion at any moment t that, in other words, is the time history of the response of the system. It is obvious that the integral in the above equations can be quantified only by using numerical methods, and therefore only the introduction of computers in structural dynamics after 1949 made possible the application of Equations 2.17 and 2.17a in real case studies of seismic input. It should also be mentioned that in this case a strong amplification of the response is observed for natural periods of the vibrating system near to the predominant period of the excitation. This will be clarified in detail in Section 2.2.4.

2.2.4 Elastic response spectra

2.2.4.1 Definition: generation

The time history of the oscillation phenomenon is not always needed in practice in its entirety, as it is sufficient to know the maximum amplitude of the relative displacement, the relative velocity and the absolute acceleration developed during a seismic excitation. The reason is that, from these values, the maximum stress and strain state of the system can be determined. Thus, the concept of the *response spectrum* has been introduced. The response spectrum of an earthquake is a diagram in which ordinates present the maximum amplitude of one of the response parameters (e.g. relative displacement, relative velocity and acceleration) as a function of the natural period of the SDOF system.

For any seismic input (accelerogram $\ddot{x}_o(t)$), three elastic response spectra can be produced in order to generate ordinates that give the maximum amplitude of the relative displacement S_d, relative velocity S_v and absolute acceleration S_a, respectively, of an SDOF system versus its natural period T (Figure 2.9). This means that the elastic response spectrum of a seismic input reflects the behaviour of all elastic SDOF systems with natural period T between

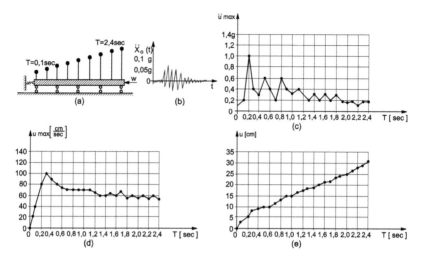

Figure 2.9 Response spectra of SDOF systems: (a) shaking table carrying SDOF systems with t = 0.1–2.4 s; (b) accelerogram; (c) acceleration spectrum; (d) velocity spectrum; (e) displacement spectrum.

Ø and ∞ during that specific excitation. The first elastic response spectra were produced experimentally by Biot in 1935 on a shaking table with SDOF oscillators and a natural period between 0 and 2.4 s (Figure 2.9).

After the introduction of computational means in structural dynamics, Housner and Kahn produced the first elastic response spectra analytically (Housner et al., 1953; Polyakov, 1974) using the Duhamel integral of (Equation 2.18), that is,

$$S_d = [u(t)]_{\max} = \frac{1}{\omega} \left[\int_0^t \ddot{x}_o(\tau) e^{-\zeta\omega(t-\tau)} \sin\omega_D(t-\tau)\,d\tau \right]_{\max} \qquad (2.18)$$

or Newmark methods for numerical integration.

On the other hand, the maximum velocity S_v can be approximated, assuming harmonic motion, by the expression

$$\boxed{S_v = \omega S_d = S_d\left(\frac{2\pi}{T}\right)} \qquad (2.19)$$

and the maximum acceleration by the expression

$$\boxed{S_\alpha = \omega S_v = \omega^2 S_d = S_d\left(\frac{4\pi^2}{T^2}\right).} \qquad (2.20)$$

The above assumption is acceptable for values of $\zeta < 0.30$. In reality, the maximum restoring force peaks display just before the point of the maximum displacement. Therefore, S_a in Equation 2.20 is slightly less than the true peak acceleration during an earthquake. Strictly speaking, S_a should be called 'pseudospectral acceleration' and S_v 'pseudospectral velocity'. However, for ζ up to 0.30, the above remark has no significance.

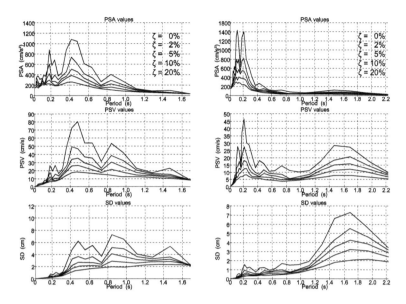

Figure 2.10 Elastic response spectra for 1999 Parnitha (Athens, Greece; right) and for 1978 Thessaloniki (Volvi, Greece; left) earthquakes for various damping values: acceleration (top), velocity (middle) and displacement (bottom). (Adapted from the Institute of Engineering Seismology and Earthquake Engineering [IESEE] Greece, 2003. *Strong Motion Data Base of Greece 1978–2003*, Thessaloniki, Greece.)

Thus, the velocity and acceleration spectra are derived by multiplying each ordinate of the displacement spectrum by ω, $(2\pi/T)$, and ω^2, $(4\pi^2/T^2)$, respectively.

From Equation 2.18, it can be seen that *as damping ζ increases, the spectral ordinates decrease* (Figure 2.10; I.E.S.E.E., 2003). For high values of damping, the spectra become smooth. In practice, for reinforced concrete structures, ζ is assumed equal to 0.05.

Practically, the computational procedure for the derivation of the elastic response spectra can be summarised as follows (Elnashai and Di Sarno, 2008):

1. Select a digitised form of the selected accelerogram from an available database.
2. For a given value of ζ and for successive values of T between 0.01 and 5.0 s, calculate successive values of S_d using one of the existing computing codes, which are based on the digitisation of the Duhamel integral (Equation 2.18) or other relevant methods.
3. Repeat the procedure of item 2 for successive values of ζ ranging between 0% and 20%.
4. Compute the velocity S_v and the acceleration S_a using Equations 2.19 and 2.20, respectively.
5. Plot the S_d, S_v, S_a spectra versus the natural period for the damping values selected.

2.2.4.2 Acceleration response spectra

The acceleration response spectra are of fundamental importance for the seismic design of buildings because they relate to the maximum inertial forces that develop during the earthquake. In this respect, they constitute the basis for the force-based design on which all modern Codes of Practice for a seismic design are based (see Section 3.4.3).

Therefore, a comprehensive examination of the acceleration response spectra will follow:

1. For $T = 0$, or in other words, for a completely rigid structure, the maximum acceleration of the system is equal to the *peak ground acceleration (PGA)* $\ddot{x}_o(t)_{max}$.
2. With the increase in the natural period T, the absolute acceleration S_a of the system increases as well, and, for a natural period near to the predominant period T_{predom} of the accelerogram $(\ddot{x}_o(t))$, S_a reaches its maximum value, which tends to be from *two* to *three* times the PGA for a 5% damping level. In this case, the system is *in resonance* with the seismic excitation (see Section 2.2.3).
3. For systems with natural period T longer than T_{predom}, S_a begins to decrease, since the system again goes out of phase. These are generally flexible systems.
4. Earthquakes occurring at small depths (up to 60 km) in the earth are the most frequent. These earthquakes have a predominant period on the order of 0.2–0.4 s; therefore, SDOF systems with a natural period within this range experience the highest acceleration. This natural period corresponds generally to the fundamental period of two- to four-storey buildings (see Section 5.6.4).
5. For deep, strong earthquakes (70–300 km in depth), which are rarer (1977 Bucharest earthquake, 1985 Mexico City earthquake), the high frequencies of motion are absorbed by the lithosphere, so only long-period motions reach at the ground surface. In this case, the predominant period appears to be 1.0–2.0 s; therefore, systems with a natural period within these limits experience the largest acceleration. Such a natural period generally corresponds to the fundamental period of 10- to 20-storey buildings (Figure 2.11).

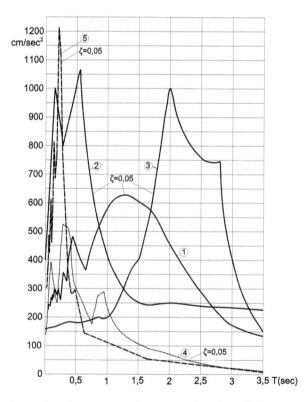

Figure 2.11 Characteristic acceleration spectra of strong earthquakes: (1) Bucharest earthquake 4.3.1977 N–S; (2) El Centro earthquake 18.5.1940 N–S; (3) Mexico City earthquake 19.9.1985 E–W SCl; (4) Thessaloniki earthquake 20.6.1978 E–W 'City' hotel; (5) Athens earthquake 07.09.1999.

6. Soft soils shift the maximum of the acceleration spectrum to the right of the diagram, as the predominant period of the soil is long. On the other hand, firm soils shift the maximum of the acceleration spectrum to the left (short predominant period). So flexible (high-rise) buildings with a long fundamental period are vulnerable to earthquakes when founded on soft soils, while stiff (low-rise) buildings appear to be vulnerable to earthquakes when they are founded on firm soil (Richart et al., 1970). A good example is the Mexico City earthquake (1985), where the accelerograms recorded in four different locations (Figure 2.12a) have given the acceleration response spectra depicted in Figure 2.12b (Penelis and Kappos, 1997).

7. Acceleration spectra allow an overall picture of the response of a large range of structures to an excitation. At the same time, they allow a static consideration of the seismic excitation, although it is a thoroughly dynamic phenomenon. This offers significant advantages to the structural engineer whose education is mainly focused on statics. Indeed, the relation between the inertial forces and the restoring ones (Figure 2.2b) has the following form:

$$V = K(x - x_o) = Ku = -(c\dot{u} + M\ddot{x}) \tag{2.21}$$

The maximum *base shear* develops when the relative displacement u reaches its maximum value, and therefore the relative velocity u takes a value equal to zero. So,

$$V_{max} = Ku_{max} = MS_a$$

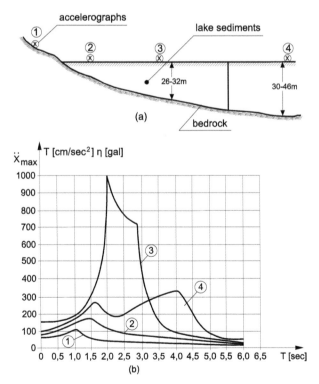

Figure 2.12 Mexico City earthquake 19.9.1985: (a) the ground stratification of Mexico City and (b) acceleration response spectra at points (1), (2), (3) and (4).

or

$$\boxed{V_{max} = KS_d = MS_a}$$

(2.22)

If we express V_{max} as a function of the weight W of the structure, the following expression results:

$$\boxed{\frac{V_{max}}{W} = \frac{MS_a}{W} = \frac{MS_a}{Mg} = \frac{S_a}{g} = \varepsilon}$$

(2.23)

The above relation means that in order for the maximum stress and strain of an SDOF system to be determined, the system can be loaded with a horizontal force V equal to the weight W multiplied by *the seismic coefficient ε*, resulting from the acceleration response spectrum of the specific earthquake scaled to g (Figure 2.13). It should be noted that the above conclusion is very close to the approach of the problem of base shear that was followed in the past, up until the early 1950s (see Figure 2.1). The only difference is that *ε is not a constant, but it is a value dependent on the natural period T_o of the structure.*

8. It should be noted that the acceleration response spectrum is the main tool for the 'modal response spectrum analysis' of multi-degree of freedom (MDOF) elastic systems (see Section 2.4.3 and Chapter 5.5.1).

9. Finally, reference should be made to *the vertical component of the ground motion*, which has generally been overlooked for many years. This has changed gradually in the last 20 years due to the increased number of near-source records combined with field observations confirming the destructive effects of high vertical vibrations (Papazoglou and Elnashai, 1996). The commonly used approach of taking the vertical spectrum as two-thirds of the horizontal one without a change in fundamental period content is

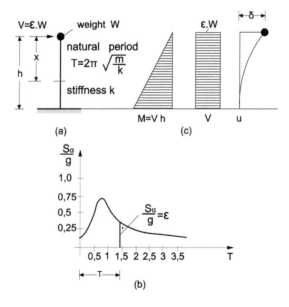

Figure 2.13 Transition from the dynamic to static response: (a) an SDOF system; (b) acceleration response spectrum; (c) maximum response of the structure.

now being abandoned (Elnashai and Papazoglou, 1997; Collier and Elnashai, 2001). For example, EC 8-1/2004 introduces forms for vertical elastic acceleration response spectra that are independent of those of the horizontal ones.

2.2.4.3 Displacement response spectra

These spectra have begun to gain interest in the last 15 years due to their direct application in the *displacement-based design (DBD) method*, which has been introduced recently for the assessment and redesign of existing buildings (EC8-1 and 3/2004-5).

As was explained in Section 2.2.4.1, there is a linear relation between S_a and Sd (Equation 2.20) for damping levels below 0.30. However, even for values of $\zeta = 0.05$ but for natural periods longer than 2.0 s, the resulting elastic displacement response spectra from Equation 2.20 have proven to be unreliable (Figure 2.14). The reason is that for many years, accelerograms were generated using analogue accelerographs with filters at 3.0 s to cut off noise frequency. Nevertheless, as various researchers have shown, the elastic response spectra are unreliable for fundamental periods longer than two-thirds of the filter cut-off period, meaning, longer than 2 s.

So, in the last few years, extended research has been carried out for the generation of reliable elastic (and inelastic) displacement response spectra based on digital accelerograms for

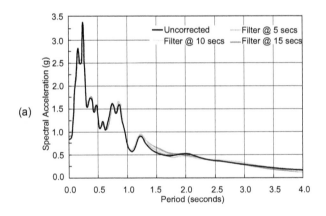

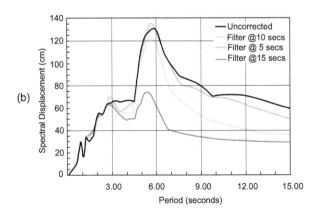

Figure 2.14 The 1978 Tabas (Iran) record filtered at 5, 10 and 15 s cut-off as well as baseline correction only: (a) acceleration spectra; (b) displacement spectra. (Adapted from Elnashai, A.S. and Di Sarno, L. 2008. *Fundamentals of Earthquake Engineering*, Wiley, West Sussex, UK.)

use in the DBD method, where periods T_{eq} longer than 2 s and damping coefficients higher than 15% appear due to the degradation of the structural system in the postelastic region (see also Section 13.5; Ambraseys et al., 1996; Bommer and Elnashai, 1999; Tolis and Faccioli, 1999). In conclusion, it is important to make the following remarks (Figure 2.14):

1. Spectral ordinates for all damping coefficients increase with the period from zero to a zone of maximum values and then decrease to converge at the peak ground displacement (PGD) for long periods.
2. This means that for long natural periods, much longer than the predominant period of the seismic motion, the mass of the structural system remains unaffected. The ground vibrates while the mass remains motionless, and so u_{max} (relative displacement) converges at the PGD (PGD = x_{omax}).

2.2.4.4 Velocity response spectra

Velocity response spectra are of great importance in seismic design because they are an index of the energy transmitted by the ground into the oscillator. Indeed, the energy transmitted into a vibrating mass is

$$E_{max} = \frac{1}{2}M\dot{u}_{max}^2 = \frac{1}{2}MS_v^2 \qquad (2.24)$$

In this respect, velocity spectra allow quantitative evaluation of the total seismic excitation, and, therefore, motions of different amplitude can be scaled (normalised) *to the same level of intensity*. This can be accomplished with the following integral, which was originally defined by Housner (Wiegel, 1970) *as spectrum intensity:*

$$SI = \int_{T_1}^{T_2} S_v(T,\zeta)dT \qquad (2.24a)$$

with integration limits $T_1 = 0.15$ and $T_2 = 2.55$. The above integral represents the area under the velocity spectrum for a given damping coefficient ζ, between the limits T_1 and T_2, and it is expressed in units of length. Table 2.1 gives some characteristic SI values (in mm) of several accelerograms, as well as the resulting normalisation factors with regard to the El Centro

Table 2.1 Values (in mm) and normalisation factors for the first 10 s of a series of accelerograms

Accelerogram		SI (mm)			
		$\zeta = 0\%$	$\zeta = 5\%$	$\zeta = 10\%$	Factor
El Centro	S00E	2334	1479	1230	1.00
Taft	N21E	1021	651	555	2.24
Taft	S29E	1176	749	605	2.00
Cal Tech	S90 W	634	414	338	3.60
Pacoima	S16E	5876	3910	3316	0.37
Thessaloniki	N30E	721	555	484	2.61
Thessaloniki	N60E	698	517	457	2.78

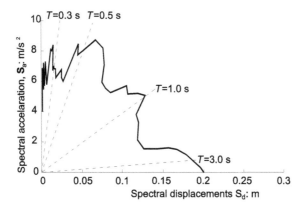

Figure 2.15 Capacity displacement spectrum (ADRS).

spectrum (Penelis and Kappos, 1997). The reliability of the scaling factors can be checked by studying the response of multi-storey structures for several accelerograms normalised to the same level of spectrum intensity (Kappos, 1990; Penelis and Kappos, 1997). It can be seen in such studies that although the accelerograms are normalised, the differences in response are significant. Recently, many efforts have been carried out by various researchers to improve the reliability of what is denoted as 'spectrum intensity SI' (Nau and Hall, 1984; Kappos, 1990; Martinez-Rueda, 1997; Elnashai, 1998). For more information, see Chapter 5.7.6.1.

2.2.4.5 Acceleration–displacement response spectra

These spectra are generated by plotting spectral acceleration directly against spectral displacement (Mahaney et al., 1993). This can be done by the implementation of Equation 2.20.

$$S_d = \frac{T^2}{4^2} S_a \tag{2.25}$$

Additionally, Equation 2.25 can be rearranged in the form

$$T = 2\pi \sqrt{\frac{S_d}{S_a}} \tag{2.26}$$

Figure 2.15 presents an ADRS spectrum. It is obvious that each curve refers to a predefined damping level. These spectra have been developed for the application of the capacity-spectrum technique in the DBD method (see Chapter 13.5.2).

2.3 DYNAMIC ANALYSIS OF INELASTIC SDOF SYSTEMS

2.3.1 Introduction

After the introduction of the concept of the response spectrum in earthquake engineering and the development of the first elastic spectra by Housner in 1949 (Housner et al., 1953), it was observed that the maximum acceleration of the vibrating masses in structures close to resonance with the seismic motion was two to three times higher than the PGA. Thus, for $\ddot{x}_{o\max}/g = 0.17$, that is, for a seismic motion of moderate destructiveness, the seismic coefficient ε reached the value of 0.35–0.50. On the other hand, all the existing engineered structures of the past were designed for E values between 0.04 and 0.16, according to seismic codes then in effect. However, the damage to these structures due to earthquakes that occurred in the meantime was not always destructive.

The difference was so great that it could not be attributed to the existing overstrength in the structures or to calculation errors. A more precise approach to the problem showed that this fact could be explained by taking into account the inelastic behaviour of the structures, which accompanied the structural damage. This behaviour led to the dissipation of a large percentage of the system's kinetic energy through damping.

2.3.2 Viscous damping

So far, the damping phenomenon of actual structures in the elastic range has been studied using the Kelvin–Voigt model (Figure 2.16). This model consists of a spring and a dashpot acting as a damper connected in parallel with the spring (Housner et al., 1953). Thus, the total force P required for the displacement u of the system is the sum of the restoring force P_e of the spring and the resistance P_d of the damper:

$$P = P_e + P_d = Ku + c\dot{u} \tag{2.27}$$

In this viscoelastic model of Kelvin–Voigt, the viscous damping is proportional to the *deformation velocity*. In real structural systems, however, springs and dampers are incorporated into elastic members that connect the masses to each other and to the ground. It is, therefore, important to evaluate at least qualitatively whether or not the introduced value for P_d in Equation 2.27 expresses with sufficient accuracy the viscoelastic damping in real structures.

It is known from material testing that the response of a material to an external loading depends on the rate of loading. The higher the rate of loading is, the larger the force required for the same deformation should be (Figure 2.17).

$$P = P_e + \Delta P = P_e + P_d \tag{2.28}$$

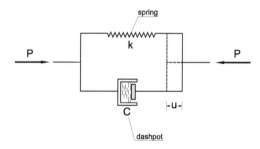

Figure 2.16 The Kelvin–Voigt viscoelastic model.

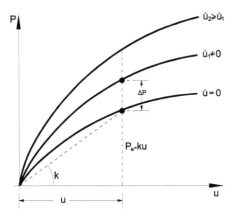

Figure 2.17 Qualitative diagrams of the axial force P versus normal strain u for various loading rates.

As Figure 2.17 shows, the viscous part of the loading is a complicated function of the deformation rate u. However, if this function is expanded in a polynomial series of u and only the first term is retained, P_d takes the approximate form

$$P_d = c\dot{u} \tag{2.29}$$

The above approximation, as we have already seen, has made possible the linearisation of the dynamic analysis problem of SDOF and MDOF systems, as we will see next.

It is of great interest to present graphically the total restoring force:

$$P = P_e + P_d = Ku + c\dot{u}. \tag{2.30}$$

If the system of Figure 2.2 is subjected to a vibration in the form

$$u = u_o \sin \omega t \tag{2.31}$$

then the relation 2.30 takes the form

$$P = Ku_o \sin \omega t + c\omega u_o \cos \omega t \tag{2.32}$$

Equations 2.31 and 2.32 define a P–u function, which describes an ellipse (Figure 2.18). The area of the shaded loop (area of ellipse) represents the dissipated energy due to the viscoelastic damping. The area of this ellipse is equal to

$$\Delta W = \int_{T}^{T+2\pi/\omega} P(t)\frac{du}{dt}\,dt = \pi c\omega u_o^2. \tag{2.33}$$

On the other hand, the maximum potential energy U of the system is equal to

$$U_e = \frac{1}{2}Ku_o^2. \tag{2.34}$$

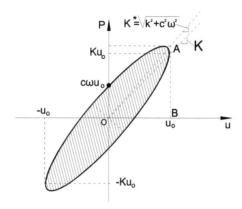

Figure 2.18 Diagram of the restoring force P versus relative displacement u for an excitation u = u$_o$ sin ωt of an SDOF with damping c.

Therefore,

$$\frac{\Delta W}{U_e} = \frac{2\pi c\omega}{K}.$$

(2.35)

Combining Equations 2.4a, 2.4b and 2.6a results in

$$c = \zeta_{el} \cdot c_{cr} = \zeta_{el}(2\sqrt{MK}) = \zeta_{el}(2\omega M).$$

(2.36)

Substituting 2.36 into 2.35, ζ_{el} takes the form

$$\boxed{\zeta_{el} = \frac{1}{4\pi}\frac{\Delta W}{U_e} = \frac{1}{4\pi}\frac{A_{ellipse}}{U_e}.}$$

(2.37)

This means that, the viscous damping ratio ζ_{el} is defined by the ratio of the loop area of Figure 2.18 divided by the maximum elastic strain energy equal to the area of the triangle OAB (Figure 2.18).

In the case of use of passive supplemental damping viscous dampers, these devices may be eventually designed to behave as nonlinear elements. This issue will be examined in detail in Chapter 15.

2.3.3 Hysteretic damping

In addition to the viscoelastic behaviour of materials, there are other factors that can lead to damping. The most significant, particularly in the case of inelastic deformations due to earthquakes, is 'the hysteretic behaviour of materials'. The stress–strain diagram of a material or a structural member under cyclic loading has the form of Figure 2.19, with several variations depending on special characteristics of the material or the member and the loading (see also Chapter 13.5.2). The area of the shaded loop in Figure 2.19 represents the energy that is dissipated in every loading cycle, due to the plastic behaviour of the *material*.

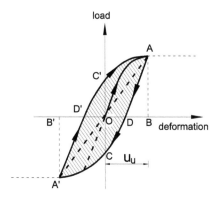

Figure 2.19 A typical hysteresis loop.

It is obvious that the larger the area of the hysteresis loop, more specifically the higher the plastic strain level of the material or of the structural member is, the larger the dissipated energy and the damping will be.

Comparing now Figures 2.18 and 2.19, it is evident that the hysteretic damping may be expressed approximately in the form of viscous damping using a hysteretic damping ratio (Jacobsen, 1960):

$$\zeta_{hyst} = \frac{1}{4\pi}\frac{\Delta W_{hyst}}{U_{el}} \qquad (2.38)$$

where

ΔW_{hyst} is the area of the hysteresis loop of Figure 2.19.
U_{el} is the area of the triangle OAB of Figure 2.19 equal to $(1/2)\ K_{ef}u_u^2$.
u_u is the ultimate inelastic deformation.

In this respect, the equivalent viscous damping is equal to

$$\zeta_{eq} = \zeta_{el} + \zeta_{hyst}. \qquad (2.39)$$

Equation 2.38 is a basis for the qualitative understanding of the phenomenon of hysteretic damping and a good indicator of dissipation per cycle of loading. Furthermore, it allows the application of the linear theory in structures exhibiting dissipative behaviour (see also Chapter 13.5.2) by introducing ζ_{eq}, provided it does not exceed values of 0.15–0.28. Last but not least, it is a very useful tool for DBD (see Chapter 13.5.2), as it makes possible the formulation of relations between *hysteretic damping* ζ_{hyst} and ductility μ ($\mu = (u_p/u_y)$) (see below).

2.3.3.1 Case study

Below we will determine the relation between ζ_{hyst} and μ in the case of a system with an elastic perfectly plastic (EPP) stress–strain diagram (Figure 2.20):

$$\Delta W_{hyst} = 2(u_p - u_y)2f_y = 4(u_p - u_y)f_y = 4f_y\left(\frac{u_p}{u_y} - 1\right)u_y \qquad (2.40)$$

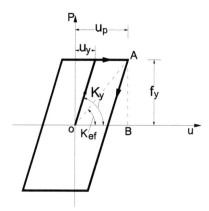

Figure 2.20 Relation between hysteretic damping ζ_{hyst} and ductility μ.

If we denote *ductility*, the ratio u_p/u_y,

$$\boxed{\mu = \frac{u_p}{u_y}} \tag{2.41}$$

and we introduce Equation 2.41 into Equation 2.40, this equation takes the form

$$\Delta W_{hyst} = 4f_y(\mu - 1)u_y. \tag{2.42}$$

The maximum potential energy of the system is given by the expression

$$U_e = \frac{1}{2}f_y u_p = \frac{1}{2}f_y \frac{u_p}{u_y}u_y = \frac{1}{2}f_y u_y \mu. \tag{2.43}$$

The introduction of Equations 2.42 and 2.43 into Equation 2.38 results in the following expression (Dwairi and Kowalsky, 2007):

$$\zeta_{hyst} = \frac{1}{4\pi} \frac{4f_y u_y(\mu - 1)}{(1/2)f_y u_y \mu}$$

$$\boxed{\zeta_{hyst} = \frac{2}{\pi}\left(1 - \frac{1}{\mu}\right)} \tag{2.44}$$

The above authors represented ζ_{hyst} in their study in the generalised form:

$$\zeta_{hyst} = \frac{C}{\pi}\left(1 - \frac{1}{\mu}\right) \tag{2.45}$$

where the coefficient C depends on the hysteresis rule (Priestley et al., 2007). In Figure 2.21, various hysteresis rules considered in inelastic analysis are depicted, while in Figure 2.22 (Penelis and Kappos, 1997; Priestley et al., 2007), the hysteretic component of equivalent

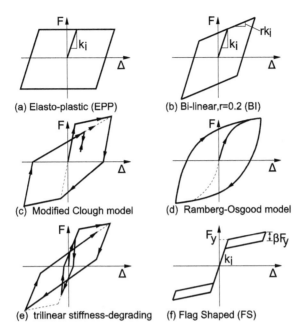

Figure 2.21 Hysteresis rules considered in inelastic time-history analysis. (a) Elastolastic (EPP); (b) bilinear, r = 0.2 (BI); (c) modified clough model; (d) Ramberg–Osgood model; (e) trilinear stiffness degrading; (f) flag shaped (FS).

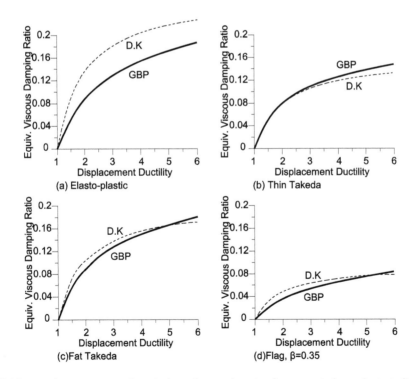

Figure 2.22 Hysteretic component of equivalent viscous damping from two independent studies. (DK indicates Dwairi and Kowalski, 2007; GBP indicates Grant et al., 2005.) (a) Elastoplastic; (b) thin takeda; (c) flat takeda; (d) flag, β = 0.35.

viscous damping versus displacement ductility μ of two independent studies is given (Grant et al., 2005; Dwairi et al., 2007).

In the case of use of hysteretic dampers in connection with seismic isolators, these devices demand a detailed design, which will be presented in Chapter 15.

2.3.4 Energy dissipation and ductility

The effects of inelastic behaviour on the response of structures to strong seismic motions may be clarified by studying an SDOF system.

Consider two SDOF systems with the same mass M and the same spring stiffness K, without damping (Figure 2.23). Suppose that both systems vibrate freely and that when they pass through their original equilibrium position, they both have the same velocity $\dot{u}_{max} = v_{max}$ (Park and Paulay, 1975). Suppose also that the first one has an elastic connection of ultimate strength V_{1u}, while the second has a strength V_{2u}, which is much lower than V_{1u} (Figure 2.24).

The mass of the first system is subjected to a displacement u_{o1} such that the potential energy stored in the form of strain energy, represented by the area of the triangle OBF, is equal to the kinetic energy of the system:

$$\frac{1}{2} M v_{max}^2 = \frac{1}{2} K u_{o1}^2.$$

(2.46)

Therefore,

$$u_{01} = \left(\frac{M}{K} \right)^{\frac{1}{2}} v_{max}.$$

(2.47)

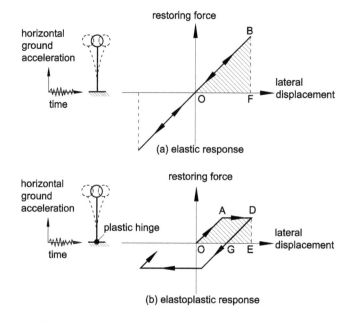

Figure 2.23 Response of SDOF systems to seismic action: (a) elastic response; (b) elastoplastic response.

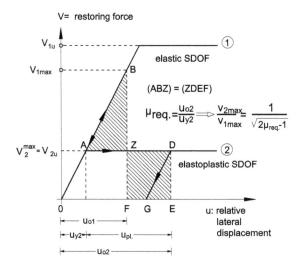

Figure 2.24 Quantitative relations between elastic and elastoplastic response of an SDOF system to an earthquake.

In this case, a maximum restoring force is developed in the elastic spring, equal to $V_{1max} < V_{1u}$, which coincides with the maximum inertial force $M\ddot{x}_{max}$. Given the fact that the velocity becomes zero, this restoring force starts to accelerate the system in the opposite direction, thus causing oscillations of constant amplitude, since viscous damping has been taken as zero.

The second system, unable to develop a restoring force equal to the first one, is led to the creation of a plastic hinge at the base, with maximum restoring force V_{2u} and maximum displacement u_{o2}, such that the area of the trapezoid OADE is equal to the kinetic energy of the system. Thus,

$$\frac{1}{2}Mv_{max}^2 = \frac{1}{2}V_{2u}u_{y2} + V_{2u}u_{pl}. \tag{2.48}$$

Therefore,

$$u_{pl} = \frac{1}{2V_{2u}}\left(Mv_{max}^2 - V_{2u}u_{y2}\right) \tag{2.49}$$

and the total displacement of the second system is equal to

$$u_{o2} = u_{y2} + u_{pl}. \tag{2.50}$$

For a displacement equal to u_{o2}, the system has consumed all its kinetic energy; therefore, under the influence of V_{2u}, it begins to move towards its original position. At the moment when V_2 becomes zero, the potential energy that has been transformed into kinetic energy is represented by the area of the triangle EDG because the energy represented by the parallelogram OADG has been dissipated by the plastic hinge in the form of heat and other irrecoverable forms of energy.

From the above it is obvious that, while in the elastic system there is a successive interchange between kinetic and potential energy, which results in a cumulative effect of the successive execution cycles without damping, in the elastoplastic system, only part of the kinetic energy is transformed into potential energy from cycle to cycle, a fact that results in a quick hysteretic damping. This means that the displacement u_{o2}, as defined above, is an upper limit for the elastoplastic system.

From the above discussion, it can be concluded that the seismic action on an oscillating system can be resisted either via large restoring forces and oscillation within the elastic range or smaller restoring forces and exploitation of the ability of the system to resist plastic deformations. This ability of the system is characterised as *ductility* and is a property of paramount importance for earthquake-resistant structures because it gives the designer the choice to design the structure for much lower forces than those of the elastic system.

Ductility may be defined either as *ductility demand* or *ductility supply*. The first refers to the ductility requirement for predefined yield strength of the structure under an earthquake action. Speaking in quantitative terms, ductility demand refers to the ductility requirement for a predefined reduction of the restoring force of an inelastic system under a seismic action. The second (ductility supply) refers to the maximum ductility that a structure can develop without collapse or other failure modes.

The ductility supply factor is defined as the ratio of the ultimate deformation at failure to the yield deformation δ_u/δ_y (Figure 2.25). The ultimate deformation at failure is defined for design purposes as the deformation for which the material or the structural element loses only a small predefined percentage of its ultimate strength (e.g. 15% for concrete). The larger the ductility supply factor of a structural element is, the larger the safety margins of the element against an earthquake in terms of displacements would be.

Of particular interest is the determination of the *ductility demand factor* μ_D of a structure for a given ratio of reduction R_μ of the elastic restoring force. Consider an elastoplastic system of mass M, stiffness coefficient K and damping ζ subjected to a seismic action of a given elastic response spectrum. Under the assumption of a fully elastic behaviour, the maximum restoring force V_{el}, which would act on the system, can be easily found from the above data. Now if the system has an ultimate strength V_u, smaller than V_{el}, then the ductility demand factor μ_D of the system, for the same seismic action, results from the equation of the potential energy in the two cases (Figure 2.24). Indeed, setting the area of the triangle ABZ equal to the area of the rectangle EDZF, the force reduction factor R_μ results in the following form:

$$\boxed{R_\mu = \frac{1}{q_\mu} = \frac{V_u}{V_{el}} = \frac{1}{\sqrt{2\mu_D - 1}}}$$

(2.51)

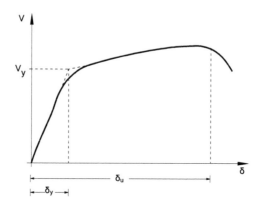

Figure 2.25 Definition of the ductility supply factor.

where R_μ is the reduction coefficient of the spectral value of force, and μ_D is the ductility demand for this force reduction.

In Equation 2.51, q_μ is called *the behaviour factor* of the structure and, as we will see later, it is the parameter in use in Eurocode 8/2004 in place of R_μ, which represents the reduction factor in use in the American Codes of Practice.

Up to now, in the preceding paragraph, the concept of energy dissipation and ductility was examined only in the first quarter of a loading cycle. Therefore, although the above presentation is very important for understanding the meaning of ductility as a concept, the results presented so far must be considered as just a first approximation. Indeed, detailed analytical studies on the dynamic response of inelastic SDOF systems have shown that, due to heavy damping, the *maximum displacement demands* of an elastic system and the corresponding inelastic one during the seismic excitation are approximately of the same magnitude (Figure 2.26).

Therefore,

$$\mu = \frac{u_{o2}}{u_{y2}} \text{ and } u_{o2} = \mu u_{y2} \tag{2.52a}$$

and

$$\boxed{q_\mu = \mu_D.} \tag{2.52b}$$

In any case, the analytical results so far have shown the following:

1. SDOF inelastic systems with short fundamental period, in other words, systems of high stiffness, present a reduction factor:

$$\boxed{R_\mu = \frac{1}{q_\mu} = 1.0} \quad \text{for } T < 0.05\text{s} \tag{2.53}$$

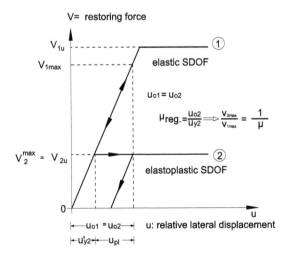

Figure 2.26 An alternative quantitative relation between elastic and elastoplastic response in an SDOF system.

2. SDOF systems with medium fundamental period present a reduction factor:

$$\boxed{R_\mu = \frac{1}{q_\mu} = \frac{1}{\sqrt{2\mu_D - 1}}} \quad \text{for } 0.125 < T < 0.5\text{s}} \tag{2.54}$$

3. SDOF inelastic systems with a long fundamental period present a reduction factor:

$$\boxed{R_\mu = \frac{1}{q_\mu} = \frac{1}{\mu_D}} \quad \text{for } T > 1.0\text{s}} \tag{2.55}$$

For intermediate periods, a linear interpolation is suggested (Newmark and Hall, 1982; Figure 2.27).

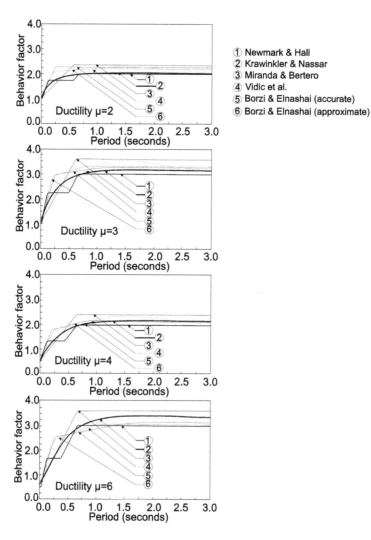

Figure 2.27 Comparison of (demand) derived by using different expressions for elastic—perfectly plastic systems on rock site; values of ductility μ: 2, 3, 4 and 6. (Elnashai, A.S. and Di Sarno, L.: *Fundamentals of Earthquake Engineering*. 2008. Copyright Wiley-VCH Verlag GmbH & Co. KGaA. Reproduced with permission.)

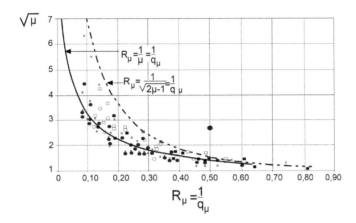

Figure 2.28 Force reduction factor R_μ. or $(1/q_\mu)$ versus $\sqrt{\mu}$.

The above is the first and simplest formulation used in practice and has been confirmed by other studies. Figure 2.28 shows the two curves described by Equations 2.51 and 2.52b as well as the results of an inelastic analysis of several SDOF systems for the 1940 El Centro earthquake, N–S component (Blume, 1960; Wiegel, 1970). Further improvements, however, have been accomplished by many studies since (Krawinkler and Nassar, 1992; Borzi and Elnashai, 2000). The main aim of the above efforts was to incorporate the fundamental period and the stress–strain law in the relation of q_μ and μ_D (Figure 2.27; Elnashai and Di Sarno, 2008).

Eurocode 8-1/2004 has adopted for the relation among behaviour factor q_o, displacement ductility μ_δ and fundamental period T the expressions presented by Vidic et al. (1994), as will be seen in detail in Equation 5.17 found in Chapter 5.4.4.1.

2.3.5 Physical meaning of the ability for energy absorption (damping)

It is useful to explain here what the term 'ability for energy absorption' means for reinforced concrete structures. Consider the cantilever of Figure 2.29, which is loaded at the top with a horizontal force V. By increasing V, the cantilever reaches an ultimate limit state. Failure can

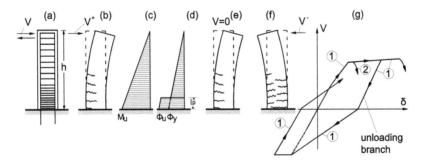

Figure 2.29 Inelastic response of a cantilever R/C beam under cyclic loading: (a) arrangement of cantilever loading; (b) loading with V + beyond yielding; (c) ultimate failure moment diagram; (d) distribution of curvature φ-plastic hinge zone l_p; (e) unloading V = 0; (f) loading with V − beyond yielding; (g) V–δ diagram for normally reinforced (I) and over-reinforced (2) cantilever.

occur in two ways, either by yielding of reinforcement in normally reinforced sections, or by the crushing of concrete in over-reinforced sections, where the strength of the compressive zone is lower than the yield strength of the reinforcement.

In the first case, for a very small increase of V above yielding, the displacement δ exhibits considerable values (Figure 2.29g); this is accomplished through opening of the cracks on the tensile side of concrete due to yielding of reinforcement. After successive steps and when large plastic deformations of steel in the tensile side have been developed, the width of the cracks increases, the depth of the compression zone is substantially decreased and the concrete at the compression side is crushed. Beyond this point, there is a rapid deterioration of the structural system and a steep descending branch on the V–δ diagram (Figure 2.29g, curve 1). In the second case (fracture of the compression zone without yielding of the reinforcement), there is a brittle failure (Figure 2.29g, curve 2) and a steep descending branch on the V–δ diagram, without the development of plastic deformations.

In the first case, there is a large amount of available ductility (supply); however, in order to make use of it, some yielding zones in the structure must be tolerated, which of course implies accepting *some degree of damage due to the appearance of wide cracks*. In the second case, the available ductility is very small. Therefore, over-reinforced systems present low ductility. The design and detailing of R/C structural elements and buildings as a whole will be discussed in detail in Chapters 8, 9 and 10.

At this point, it would be useful to approach the concept of the ductility factor (supply) and the way it is defined in some detail. So far, the ductility factor has been defined in terms of displacement for the SDOF system:

$$\mu_\delta = \frac{\delta_u}{\delta_y}.$$

However, the ductility factor may be defined in terms of plastic rotations of structural members at plastic hinges:

$$\mu_\theta = \frac{\theta_u}{\theta_y}$$

and at the end of yielding sections in terms of curvatures:

$$\mu_\varphi = \frac{\varphi_u}{\varphi_y}.$$

It is known from the theory of strength of materials that there are well-defined relations among $\mu_\varphi - \mu_\theta - \mu_\delta$. In Chapter 5.4.4, it will be presented in detail how one can go from the level of the section curvature to the level of the member rotations, and finally to the level of the displacement structural system as a whole.

At this point, it can only be noted that for a given ductility factor in terms of displacements, the ductility factor in terms of rotations is larger, while the ductility factor in terms of curvatures is much larger:

$$\mu_{\delta supl} \leq \mu_{\theta supl} < \mu_{\varphi supl} \qquad (2.56)$$

2.3.6 Inelastic response spectra

2.3.6.1 Inelastic acceleration response spectra

The main problem in the development of an inelastic response spectrum is the solution of the differential equation of motion of an SDOF inelastic system.

$$\boxed{M\ddot{u} + c\dot{u} + K(t)u = -M\ddot{x}_0(t)}$$ (2.57)

This equation is identical to Equation 2.3 except that the stiffness K is not a constant but a function of time, or better, a function of u, which is a function of time. Therefore, Equation 2.57 is not a linear differential equation anymore and, as a consequence, this equation can be integrated using only numerical methods. Various methods have been developed so far for this integration (e.g. Newmark methods), though the presentation of this integration is beyond the scope of this book (Biggs, 1964; Clough and Penzien, 1975; Chopra, 2001).

However, what should be noted here is that the integration of Equation 2.57 can be done only if the constitutive law (V–u diagram) of the SDOF inelastic system (e.g. Figure 2.30) is known. So, as the linear response spectrum gives the maximum response of the elastic system of a given viscous damping for a given accelerogram, in the same way the inelastic response spectrum gives the maximum response of the inelastic system of a given ductility demand, $\mu_D = (u_{max}/u_y)$ for a given accelerogram.

In Figure 2.31, the inelastic acceleration response spectra of the Kalamata, Greece earthquake of 1985, component N10°W are shown for various ductility demand factors μ_D (μ_D = 1, 2, 4, 6) and for the Clough–Riddel–Newmark hysteresis model (Clough and Johnston, 1966; Riddel and Newmark, 1979), which has been widely used for reinforced concrete.

It is evident that the development of inelastic acceleration response spectra directly from Equation 2.57 is a time-consuming and painful exercise. That is why extended studies have been carried out for years for the determination of relationships between R or q_μ and μ_D (see Section 2.3.4, Equations 2.51 and 2.52). So, *the inelastic acceleration response spectrum for a given ductility demand factor μ_D can easily be determined with a good approximation from the elastic one by reducing its ordinates by the reduction factors R of Equations 2.53, 2.54 and 2.55, or other similar ones proposed by other authors* (see Figure 2.27).

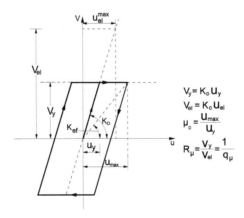

Figure 2.30 V–u diagram for an elastoplastic constitutive law of an SDOF system.

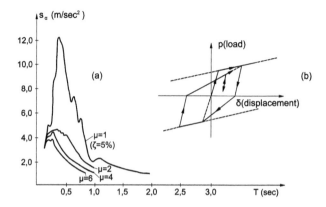

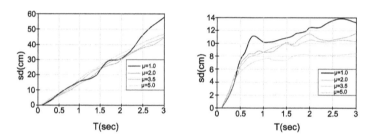

Figure 2.31 (a) Inelastic spectra of the Kalamata (1986), Greece, earthquake components N10°W, for various ductility factors μ; (b) Clough–Riddel–Newmark hysteresis model.

Figure 2.32 Inelastic displacement response spectra for mean alluvial ground layers in the United States (a) and in Greece (b). (Adapted from Kappos, A.J. and Kyriakakis, P. 2001. Soil Dynamics and Earthquake Engineering, 20, 111–123.)

The reduction of the elastic spectra by employing *R*-factors given above is the simplest and most popular approach to derive inelastic acceleration spectra and *is employed in most modern Codes of Practice for force-based design* (see Chapter 3.4.3).

2.3.6.2 Inelastic displacement response spectra

The inelastic displacement response spectra are similar to the elastic ones. Maximum displacements are obtained either as ordinates of equal damping coefficient curves or as ordinates of equal ductility demand versus natural periods, since these two options are equivalent according to Equations 2.44 and 2.45 (Figure 2.32). They have the same characteristics as the elastic ones with damping.

It is important to note that for periods up to 1.0 s, the maximum displacements are not much affected by the ductility factor.

2.4 DYNAMIC ANALYSIS OF MDOF ELASTIC SYSTEMS

2.4.1 Introduction

The scope of this book does not allow a detailed approach to the dynamic analysis of systems with more than one degree of freedom. Standard textbooks on structural dynamics (Biggs, 1964; Clough and Penzien, 1975; Warburton, 1976; Anastasiadis, 1989; Chopra, 2001) allow a detailed approach to the subject. However, since the elastic dynamic analysis of MDOF systems is the *reference method of analysis* for modern codes and particularly EC8-1/2004, it was decided that some major issues should also be presented here, for a better understanding of the main parts of this book.

2.4.2 Equations of motion of plane systems

The number of degrees of freedom of a lumped-mass system is determined by the minimum number of independent displacements and rotations of the lumped masses whereby their geometric position can be defined at a given moment. Thus, in a plane frame, with the mass concentrated in the beams of the storeys and with large axial stiffness of the beams, which are both very realistic assumptions for typical R/C structures, the degrees of freedom are determined by the number of storeys, while the independent variables of motion are their horizontal displacements u relative to the base (Figure 2.33).

The motion of the plane frame of Figure 2.33 is expressed by a system of n linear differential equations with constant coefficients, which can easily be generated. Indeed, the equation of motion of a storey (Figure 2.33) as in the case of the SDOF system (Section 2.2.1) can be generated using the dynamic equilibrium equation of d'Alembert:

$$P_{si}(t) + P_{di}(t) + m_i \ddot{x}_i(t) = 0 \qquad (2.58)$$

Keeping in mind that

$$P_{si}(t) = \kappa_{i1} u_1(t) + \kappa_{i2} u_2(t) + \cdots + \kappa_{ii} u_i(t) + \cdots + \kappa_{in} u_N(t)$$
$$P_{di}(t) = c_{i1} \dot{u}_1(t) + c_{i2} \dot{u}_2(t) + \cdots + c_{ii} \dot{u}_i(t) + \cdots + c_{in} \dot{u}_N(t)$$
$$m_i \ddot{x}(t) = m_i \ddot{x}_o(t) + m_i \ddot{u}_i(t)$$

the above equation of motion takes the form

$$m_i \ddot{u}_i(t) + \left(c_{i1} \dot{u}_1(t) + c_{i2} \dot{u}_2(t) + \cdots + c_{in} \dot{u}_{1N}(t) \right) +$$
$$+ \left(\kappa_{i1} u_1(t) + \kappa_{i2} u_2(t) + \cdots + K_{in} \dot{u}_{1N}(t) \right) = -m_i \ddot{x}_o(t). \qquad (2.58a)$$

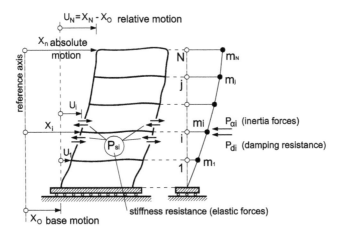

Figure 2.33 Dynamic equilibrium of elastic, damping and inertial forces at the level of a storey (d'Alembert equation).

So, the system of equations of motion takes the form

$$\mathbf{M}\ddot{u}(t) + \mathbf{C}\dot{u}(t) + \mathbf{K}u(t) = \mathbf{M}[\boldsymbol{\delta}]\ddot{x}_o(t) \tag{2.58b}$$

where

$$\mathbf{x}(t) = \mathbf{u}(t) + [\boldsymbol{\delta}]x_o(t)$$

$$\mathbf{M} = \begin{pmatrix} m_1 & & \\ & m_2 & \\ & & m_n \end{pmatrix}, \quad \mathbf{C} = \begin{pmatrix} c_{11} & c_{12} & \cdots & c_{1N} \\ c_{21} & c_{22} & \cdots & c_{2N} \\ & & \vdots & \\ c_{N1} & c_{N2} & \cdots & c_{NN} \end{pmatrix}$$

$$\mathbf{K} = \begin{pmatrix} K_{11} & K_{12} & \cdots & K_{1N} \\ K_{21} & K_{22} & \cdots & K_{2N} \\ & & \vdots & \\ K_{N1} & K_{N2} & \cdots & K_{NN} \end{pmatrix}, \quad \mathbf{u} = \begin{pmatrix} u_1 \\ u_2 \\ \vdots \\ u_N \end{pmatrix}, \quad \boldsymbol{\delta} = \begin{pmatrix} 1 \\ 1 \\ \vdots \\ 1 \end{pmatrix}$$

The notation of the above equations is given in Figures 2.33 and 2.34.

The free vibration behaviour of a plane frame is expressed by the above equation of motion (Equation 2.58a or 2.58b), taking into account that in this case the damping matrix $\mathbf{C}$ and the applied excitation vector $\mathbf{M}$ [1] $\ddot{x}_o(t)$ are both zero. So,

$$\mathbf{M}\ddot{u}(t) + \mathbf{K}u(t) = 0. \tag{2.59}$$

By comparing Equation 2.59 with Equation 2.3a of a free vibration of the SDOF system, one can easily assume that the displacement vector $\mathbf{u}$ may be expressed in the form

$$\mathbf{u} = \bar{u}\sin\omega t \tag{2.60}$$

and

$$\ddot{u} = -\omega^2\bar{u}\sin\omega t \tag{2.61}$$

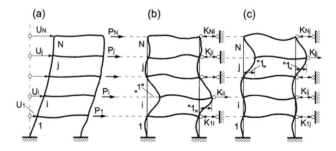

Figure 2.34 Determination of the vectors of the stiffness matrix K (K_i and K_j).

where $\bar{u}$ represents the amplitude of the vibratory motion and ω is the circular frequency. Introducing Equations 2.60 and 2.61 into Equation 2.59, we obtain

$$(-\omega^2 M\bar{u} + K\bar{u})\sin\omega t = 0 \qquad (2.62)$$

By taking into account that in Equation 2.62 the factor $\sin \omega t \neq 0$, it may be concluded that

$$\boxed{-\omega^2 M\bar{u} + K\bar{u} = 0.} \qquad (2.63a)$$

That is, a system of linear homogeneous equations for which it is known that in order for a solution for $\bar{u}$ other than zero to exist, the determinant must be equal to zero:

$$\mathrm{Det}\left|-\omega^2 M + K\right| = 0. \qquad (2.63b)$$

Equations 2.63a and 2.63b represent *a classic eigenvalue problem*. Its solution may be carried out by a variety of procedures. Standard digital computer codes are available for providing automatic solutions.

The solution of the eigenvalue equation of a plane frame system having n degrees of freedom provides the following output for each of its n modes of vibration, called *normal* or *natural* modes of vibration:

- The vibration frequency ω_i (or period $T_i = (2\pi/\omega_i)$)
- The vibration shape $\boldsymbol{\varphi}_i^T = [\varphi_{1i}; \varphi_{2i}, \varphi_{ii}\ \varphi_{ni}]$

The conclusions from the above analysis may be summarised as follows:

1. *The normal or natural modes* are the free, undamped periodic oscillations within which linear combinations represent the position of the system at every moment.
2. For every such normal mode, all the masses of the system oscillate *in phase*. This means that at each moment the ratio of the displacements of the vibrating masses remains constant. As a result, all masses go through rest position and reach maximum amplitude simultaneously.
3. The number n of normal modes is equal to the number of degrees of freedom. Every normal mode is related to a natural frequency or period of vibration known as the *natural period*. The normal mode with the longest natural period is by definition the *first or fundamental normal mode*.

Figure 2.35 shows the three normal modes of a three-storey frame. Note that the curves intersect the vertical axis at a number of points (including the one at the base), which coincide with the order of the natural mode. The amplitudes of each natural mode are normalised. Figure 2.35 shows the typical normal modes that correspond to the three-storey building of the figure. The eigen vectors of these modes are

$$\boldsymbol{\varphi}_1 = \begin{pmatrix} 1.00 \\ 0.78 \\ 0.48 \end{pmatrix}, \quad \boldsymbol{\varphi}_2 = \begin{pmatrix} -1.00 \\ 0.19 \\ 0.84 \end{pmatrix}, \quad \boldsymbol{\varphi}_3 = \begin{pmatrix} 0.53 \\ -1.00 \\ 0.88 \end{pmatrix}$$

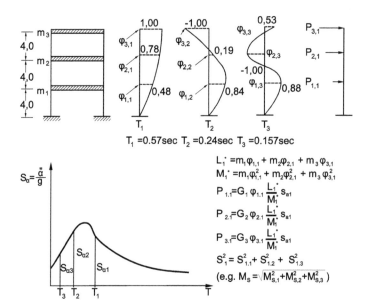

Figure 2.35 A three-storey plane frame analysed according to spectral modal analysis.

while the eigen periods are given below:

$$T_1 = 0.57\text{s}, T_2 = 0.24\text{s}, T_3 = 0.157\text{s}$$

It is important to note again that the ratio of the displacements at any moment is constant for each normal mode.

In the next paragraph, the *modal response spectrum analysis* for elastic MDOF plane systems will be presented in detail.

2.4.3 Modal response spectrum analysis

The analysis of multi-storey plane structures using this method is commonly performed with the aid of a proper computer code. The procedure is the following:

1. First, the natural periods and normal modes are determined:

$$\boldsymbol{\varphi}_1 = \begin{pmatrix} \varphi_{1,1} \\ \varphi_{2,1} \\ \vdots \\ \varphi_{i,1} \\ \vdots \\ \varphi_{n,1} \end{pmatrix}, \quad \boldsymbol{\varphi}_2 = \begin{pmatrix} \varphi_{1,2} \\ \varphi_{2,2} \\ \vdots \\ \varphi_{i,2} \\ \vdots \\ \varphi_{n,2} \end{pmatrix}, \quad \boldsymbol{\varphi}_i = \begin{pmatrix} \varphi_{1,i} \\ \varphi_{2,i} \\ \vdots \\ \varphi_{i,i} \\ \vdots \\ \varphi_{n,i} \end{pmatrix}, \quad \boldsymbol{\varphi}_n = \begin{pmatrix} \varphi_{1,n} \\ \varphi_{2,n} \\ \vdots \\ \varphi_{i,n} \\ \vdots \\ \varphi_{n,n} \end{pmatrix}$$

2. From each normal mode, the effective
   a. Modal mass M_i^*
   b. Modal excitation factor L_i^*
   c. Modal participation factor (L_i^* / M_i^*), and
   d. Modal maximum inertial forces P_{ij}
      are determined, as follows:

$$L_1^* = m_1\varphi_{1,1} + m_2\varphi_{2,1} + \cdots + m_i\varphi_{i,1} + \cdots + m_n\varphi_{n,1}$$

$$M_1^* = m_1\varphi_{1,1}^2 + m_2\varphi_{2,1}^2 + \cdots + m_i\varphi_{i,1}^2 + \cdots + m_n\varphi_{n,1}^2$$

participation factor: $\dfrac{L_1^*}{M_1^*}$

$$p_{1,1} = G_1\varphi_{1,1}\frac{L_1^*}{M_1^*}S_{a,1}$$

$$p_{2,1} = G_2\varphi_{2,1}\frac{L_1^*}{M_1^*}S_{a,1}$$

$$p_{i,1} = G_i\varphi_{i,1}\frac{L_1^*}{M_1^*}S_{a,1}$$

$$p_{n,1} = G_n\varphi_{n,1}\frac{L_1^*}{M_1^*}S_{a,1}$$

$\left.\right\}$ 1st mode

$$L_2^* = m_1\varphi_{1,2} + m_2\varphi_{2,2} + \cdots + m_i\varphi_{i,2} + \cdots + m_n\varphi_{n,2}$$

$$M_2^* = m_1\varphi_{1,2}^2 + m_2\varphi_{2,2}^2 + \cdots + m_i\varphi_{i,2}^2 + \cdots + m_n\varphi_{n,2}^2$$

participation factor : $\dfrac{L_2^*}{M_2^*}$

$$P_{1,2} = G_1\varphi_{1,2}\frac{L_2^*}{M_2^*}S_{a,2}$$

$$P_{2,2} = G_2\varphi_{2,2}\frac{L_2^*}{M_2^*}S_{a,2}$$

$$\vdots$$

$$P_{i,2} = G_i\varphi_{i,2}\frac{L_2^*}{M_2^*}S_{a,2}$$

$$\vdots$$

$$P_{n,2} = G_n\varphi_{n,2}\frac{L_2^*}{M_2^*}S_{a,2}$$

$\left.\right\}$ 2nd mode

$$L_1^* = m_1\varphi_{1,i} + m_2\varphi_{2,i} + \cdots + m_i\varphi_{i,i} + \cdots + m_n\varphi_{n,i}$$

$$M_1^* = m_1\varphi_{1,i}^2 + m_2\varphi_{2,i}^2 + \cdots + m_i\varphi_{i,i}^2 + \cdots + m_n\varphi_{n,i}^2$$

$$\text{participation factor}: \frac{L_i^*}{M_i^*}$$

$$P_{1,i} = G_1\varphi_{1,i} \frac{L_i^*}{M_i^*} S_{a,i}$$

$$P_{2,i} = G_2\varphi_{2,i} \frac{L_i^*}{M_i^*} S_{a,i}$$

$$\vdots$$

$$P_{i,i} = G_i\varphi_{i,i} \frac{L_i^*}{M_i^*} S_{a,i}$$

$$\vdots$$

$$P_{n,i} = G_n\varphi_{n,i} \frac{L_i^*}{M_i^*} S_{a,i}$$

ith mode

$$L_n^* = m_1\varphi_{1,n} + m_2\varphi_{2,n} + \cdots + m_i\varphi_{i,n} + \cdots + m_n\varphi_{n,n}$$

$$M_n^* = m_1\varphi_{1,n}^2 + m_2\varphi_{2,n}^2 + \cdots + m_i\varphi_{i,n}^2 + \cdots + m_n\varphi_{n,n}^2$$

$$\text{participation factor}: \frac{L_n^*}{M_n^*}$$

$$P_{1,n} = G_1\varphi_{1,n} \frac{L_n^*}{M_n^*} S_{a,n}$$

$$P_{2,n} = G_2\varphi_{2,n} \frac{L_n^*}{M_n^*} S_{a,n}$$

$$\vdots$$

$$P_{i,n} = G_i\varphi_{i,n} \frac{L_n^*}{M_n^*} S_{a,n}$$

$$\vdots$$

$$P_{n,n} = G_n\varphi_{n,n} \frac{L_n^*}{M_n^*} S_{a,n}$$

nth mode

In the above expressions
- G_i is the gravity load (in kN) or the respective mass ($G_i/10$) (kN/m/s^2).
- S_{ai} is the seismic motion pseudo-acceleration corresponding to the normal period T_i (Figure 2.35), expressed in the first case as a percentage of g ($g \cong 10$ m/s^2) or in the second one in natural value (m/s^2).

It should be noted that
a. The sum of all modal participation factors (L_i^*/M_i^*) is equal to unity.
b. The participation factors diminish in parallel with the shortening of the normal periods.

c. Consequently, for plane or symmetric pseudospatial systems, the first two or three eigenvalues are enough to cover the dynamic excitation of a very large percentage of the mass of the system (more than 90%).

3. For the maximum inertial forces of each normal mode, the probable maximum values of the response parameters (moments, shears, displacements, etc.) are determined through a classic static analysis. It should be noted that the superposition of SRSS should never be implemented in any case for the maximum inertial forces (e.g. $\sqrt{P_{1,2}^2 + P_{2,2}^2 + P_{2,3}^2}$ in Figure 2.35). If a procedure of this type is followed, *erroneous results* are derived for the various response parameters of the structure.

4. The above response quantities for the modes under consideration are superimposed by taking the square root of the sum of their squares (SRSS):

$$S_1 = \sqrt{S_{1,1}^2 + S_{1,2}^2 + S_{1,3}^2 + \cdots + S_{1,n}^2} \qquad (2.64)$$

Therefore, for the bending moments, for example, the above relation takes the form

$$M_s = \sqrt{M_{s,1}^2 + M_{s,2}^2 + M_{s,3}^2 + \cdots + M_{s,n}^2}. \qquad (2.65)$$

Thus, the superposition is based on the concept that all modes do not reach their maximum value simultaneously and that the responses in the vibration modes may be considered independent of each other. Therefore, according to probability theory (Clough, 1970), their most probable maximum value results through the SRSS.

It is evident that the results of this procedure may have either a plus or a minus sign and, therefore, in all following calculations they must be introduced with both signs. When the values of the natural period of any successive modes are very close, the above postulated independence does not apply. So, the concept of mode independence according to EC8-1 (CEN 1998-1, 2004) is considered to be fulfilled if

$$T_j \le 0.9 T_i \quad (i < j) \qquad (2.66)$$

where T_i, T_j are the natural periods of any two successive modes of vibration taken into account for the determination of seismic effects.

If Equation 2.66 is not satisfied, a more accurate procedure, the 'complete quadratic combination' (CQC) must be adopted (Wilson and Burton, 1982; Chopra, 2001):

$$S_1 \cong \left(\sum_{n=1}^{n} S_{1n}^2 + \underbrace{\sum_{i=1}^{n} \sum_{n=1}^{n}}_{i \ne n} \rho_{in} S_{1i} S_{1n} \right)^{1/2} \qquad (2.67)$$

and

$$\rho_{in} = \frac{8\zeta^2 (1 + \beta_{in}) \beta_{in}^{3/2}}{\left(1 - \beta_{in}^2\right)^2 + 4\zeta^2 \beta_{in} (1 + \beta_{in})^2} \qquad (2.68)$$

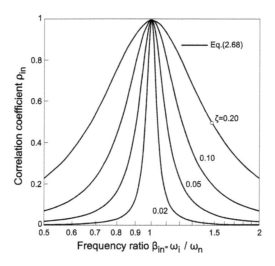

Figure 2.36 Variation of correlation coefficient ρ_{in} given by Equation 2.68. (Adapted from Chopra, A.K. 2001. *Dynamics of Structures: Theory and Application in Earthquake Engineering*, 2nd Edition. Prentice-Hall, New Jersey.)

where

$$\beta_{in} = \frac{\omega_i}{\omega_n} \tag{2.69}$$

This equation (Der Kiureghian, 1981) implies that

$$\begin{aligned} \rho_{in} &= \rho_{ni} \quad \text{for} \quad i \neq n \\ \rho_{in} &= 1 \quad \text{for} \quad i = n \end{aligned} \tag{2.70}$$

In Figure 2.36, the coefficient ρ_{in} is plotted as a function of $\beta_{in} = \omega_i/\omega_n$ according to Equation 2.68. It is apparent that this coefficient is reduced rapidly, as the two natural frequencies ω_i and ω_n move apart.

2.4.4 Pseudospatial structural single-storey system

2.4.4.1 General

The pseudospatial structural system is the most common model in use for the analysis and design of multi-storey buildings. Its extended use in recent decades must be attributed to the tremendous development in computer-aided structural analysis (Figure 2.37). This progress of the last 40 years has made it possible for the structural engineer to approach the structural system of buildings as a space structure without being obliged to analyse it in simpler plane systems, no matter if static or dynamic analysis is used. Even so, the assumption of considering the floors of buildings as rigid diaphragms in their plane simplifies the problem and diminishes the required computer capacity and computing time as well, because it leads to a drastic reduction in the number of unknown 'displacements' in application of the direct stiffness method.

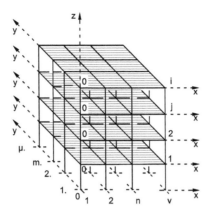

Figure 2.37 The geometry of a pseudospatial frame.

In R/C buildings, the in-plane stiffness of the floors is usually large enough in comparison to the lateral stiffness of the vertical structural elements that the displacement and rotation of any joint in-plane of the floor can be expressed as a linear function of two displacements, u_{mx} and u_{my}, of the centre of mass and a rotation ϕ_{mz} with respect to the z-axis ('displacements' of the rigid disc). Thus, the independent deformations of every joint i are limited to two rotations φ_{xi} and φ_{yi} with respect to x–x and y–y axes and a vertical displacement u_{zi}. This system is known as a 'pseudospatial structural system'.

The purpose of this section is the presentation of some important properties of the response of multi-storey 3D systems subjected to static or dynamic excitation. These properties, like 'centre of stiffness', 'torsional radius', 'radius of gyration' and so forth, will be met many times in subsequent chapters. So, it was decided to present a thorough *static* and *dynamic* analysis of the *single-storey pseudospatial system* here, where each of these properties will be introduced in the simplest form, since the single-storey system is the simplest form of this type of structure.

2.4.4.2 Static response of the single-storey 3D system

Consider the pseudospatial single-storey structure of Figure 2.38, which is loaded at the level of the diaphragm with two horizontal seismic forces H_x and H_y passing through the centre of mass m of the diaphragm and a torque M_z. According to what has been discussed in the previous paragraph, the displacement of every point of the diaphragm caused by the horizontal seismic forces and the torque may be determined by the displacement u_{mx} and u_{my} of the centre of mass M and the in-plane rotation of the diaphragm φ.

So, the displacement of the top of a column i (Figure 2.38) results from the relations

$$\left.\begin{aligned} u_{ix} &= u_{mx} - y_i\varphi \\ u_{iy} &= u_{my} + x_i\varphi \end{aligned}\right\} \tag{2.71}$$

If we call the stiffness of a frame i of x–x direction K_{ix} and the stiffness of a frame j of y–y direction K_{jy}, this stiffness expresses the required horizontal resistance (shear force) of the frame for the development of a relative displacement equal to '1' (Figure 2.39). As a result,

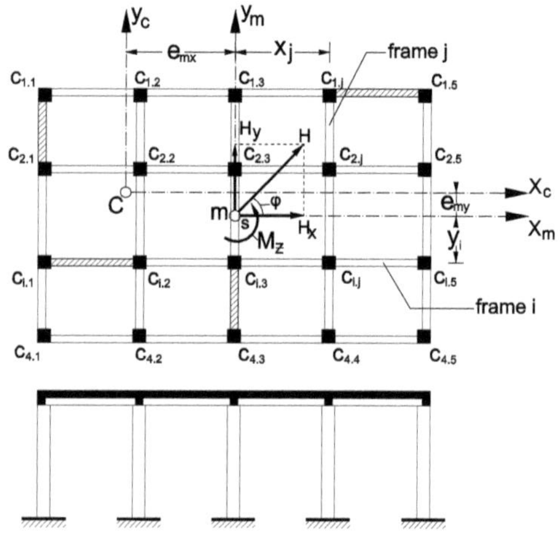

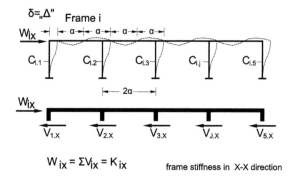

Figure 2.38 Plan and elevation of a single-storey pseudospatial structural system.

$$W_{ix} = \Sigma V_{ix} = K_{ix} \qquad \text{frame stiffness in X-X direction}$$

Figure 2.39 Definition of the stiffness K_{ix} of a frame i parallel to the x–x direction.

each of the frames i and j develops, respectively, the following horizontal shear forces (reactions) for the displacements of Equation 2.71:

$$\left. \begin{aligned} V_{ix} &= K_{ix}u_{mx} - y_i K_{ix}\varphi \\ V_{jy} &= K_{jy}u_{my} + x_j K_{jy}\varphi \end{aligned} \right\}. \tag{2.72}$$

The three equilibrium conditions for the floor result in the following equations:

$$\left. \begin{aligned} \sum_{i=1}^{n} V_{ix} &= H_x \\ \sum_{j=1}^{m} V_{jy} &= H_y \\ \sum_{j=1}^{m} x_j V_{jy} - \sum_{i=1}^{i=n} y_i V_{ix} + \sum T_i &= M_z \end{aligned} \right\} \tag{2.73}$$

where
ΣT_i the sum of the torsional moments developing at the columns due to φ $(T_i = K_{ip}\varphi)$.
If Equations 2.72 are incorporated into Equation 2.73, the following system results:

$$
\left.
\begin{aligned}
&\left(\sum_{i=1}^{n} K_{ix}\right) u_{mx} + \quad + 0 + \quad -\left(\sum_{i=1}^{n} y_i K_{ix}\right)\varphi = H_x \\
&0 + \quad +\left(\sum_{j=1}^{j=m} K_{jy}\right) u_{my} + \quad +\left(\sum_{j=1}^{m} x_j K_{jy}\right)\varphi = H_y \\
&-\left(\sum_{i=1}^{n} K_{ix}\right) u_{mx} + \left(\sum_{j=1}^{m} x_j K_{jy}\right) u_{my} \quad +\left(\sum_{j=1}^{m} K_{jy} x_j^2 + \sum_{i=1}^{n} K_{ix} y_i^2 + \sum K_{ip}\right)\varphi = M_z
\end{aligned}
\right\}
\tag{2.74}
$$

System 2.74 allows the determination of the unknown 'displacements' u_{mx}, u_{my} and φ. From this point on, for simplicity, the sum ΣT_i will be considered zero, since the torsional stiffness of the columns is very small compared to their bending stiffness.

System 2.74 may be reduced to a system of three uncoupled equations, if the reference point of the coordinate system is transferred from the *centre of mass* to another point C called *centre of stiffness* (see Chapter 5, Sections 5.2 and 5.3), such that the non-diagonal coefficients of system 2.74 take zero values, which are

$$
\left.
\begin{aligned}
&\sum_{i=1}^{i=n} y_{ic} K_{ix} = 0 \\
&\sum_{j=1}^{j=m} x_{jc} K_{jy} = 0
\end{aligned}
\right\}
\tag{2.75}
$$

It must be noted that point C, for which the above equations are in effect, is the centre of static moments of the fictitious forces K_{ix} and K_{iy}, respectively. So, the coordinates of C in relation to the centre of mass M are given by the following expressions (Figure 2.38):

$$
\left.
\begin{aligned}
e_{mx} &= \frac{\sum_{j=1}^{j=m} x_j K_{jy}}{\sum_{j=1}^{j=m} K_{jy}} \\
e_{my} &= \frac{\sum_{i=1}^{i=n} y_i K_{ix}}{\sum_{i=1}^{i=n} K_{ix}}
\end{aligned}
\right\}
\tag{2.76}
$$

Taking Equations 2.75 and 2.76 into account, system 2.74 is transformed as follows:

$$\left.\begin{array}{l} \left(\displaystyle\sum_{i=1}^{i=n} K_{ix}\right) u_{cx} + 0 + 0 = H_x \\[3mm] 0 + \left(\displaystyle\sum_{j=1}^{j=m} K_{jy}\right) u_{cy} + 0 = H_y \\[3mm] 0 + 0 + \left(\displaystyle\sum_{j=1}^{j=m} K_{jy} x_{jc}^2 + \displaystyle\sum_{i=1}^{i=n} K_{ix} y_{ic}^2\right) \varphi = M_c \end{array}\right\} \tag{2.77}$$

where

$$\left.\begin{array}{l} x_{jc} = x_{mja} - e_{mx} \\[2mm] y_{ic} = y_{mi} - e_{my} \\[2mm] M_c = H_y e_{mx} - H_x e_{my} + M_z \end{array}\right\} \tag{2.78}$$

It should be noted that according to Equation 2.78, frame coordinates and loading system (H_x, H_y, M_z) refer from that point on to the *centre of stiffness* instead of the *centre of mass*. The expression

$$\boxed{J_{TC} = \sum_{j=1}^{j=m} K_{jy} x_{jc}^2 + \sum_{i=1}^{n} K_{ix} y_{ic}^2} \tag{2.79}$$

is called *torsional stiffness* with respect to the *centre of stiffness*, and its meaning will be examined later (see Chapter 5.3).

By introducing the notation

$$\left.\begin{array}{l} K_x = \displaystyle\sum_{i=1}^{i=n} K_{ix} : \text{storey stiffness } x - x \\[4mm] K_y = \displaystyle\sum_{j=1}^{j=m} K_{jy} : \text{storey stiffness } y - y \\[4mm] J_{TC} = \displaystyle\sum_{j=1}^{j=m} K_{jy} x_{jc}^2 + \displaystyle\sum_{i=1}^{i=n} K_{ix} y_{ic}^2 : \text{torsional stiffness} \end{array}\right\} \tag{2.80}$$

where

ΣT_i the sum of the torsional moments developing at the columns due to φ $(T_i = K_{ip}\varphi)$.

If Equations 2.72 are incorporated into Equation 2.73, the following system results:

$$\left.\begin{array}{l} \left(\displaystyle\sum_{i=1}^{n} K_{ix}\right)u_{mx} + \quad +0+ \quad -\left(\displaystyle\sum_{i=1}^{n} y_i K_{ix}\right)\varphi = H_x \\[2em] 0+ \quad +\left(\displaystyle\sum_{j=1}^{j=m} K_{jy}\right)u_{my} + \quad +\left(\displaystyle\sum_{j=1}^{m} x_j K_{jy}\right)\varphi = H_y \\[2em] -\left(\displaystyle\sum_{i=1}^{n} K_{ix}\right)u_{mx} +\left(\displaystyle\sum_{j=1}^{m} x_j K_{jy}\right)u_{my} \quad +\left(\displaystyle\sum_{j=1}^{m} K_{jy} x_j^2 + \displaystyle\sum_{i=1}^{n} K_{ix} y_i^2 + \displaystyle\sum K_{ip}\right)\varphi = M_z \end{array}\right\} \tag{2.74}$$

System 2.74 allows the determination of the unknown 'displacements' u_{mx}, u_{my} and φ. From this point on, for simplicity, the sum ΣT_i will be considered zero, since the torsional stiffness of the columns is very small compared to their bending stiffness.

System 2.74 may be reduced to a system of three uncoupled equations, if the reference point of the coordinate system is transferred from the *centre of mass* to another point C called *centre of stiffness* (see Chapter 5, Sections 5.2 and 5.3), such that the non-diagonal coefficients of system 2.74 take zero values, which are

$$\left.\begin{array}{l} \displaystyle\sum_{i=1}^{i=n} y_{ic} K_{ix} = 0 \\[2em] \displaystyle\sum_{j=1}^{j=m} x_{jc} K_{jy} = 0 \end{array}\right\} \tag{2.75}$$

It must be noted that point C, for which the above equations are in effect, is the centre of static moments of the fictitious forces K_{ix} and K_{jy}, respectively. So, the coordinates of C in relation to the centre of mass M are given by the following expressions (Figure 2.38):

$$\left.\begin{array}{l} e_{mx} = \dfrac{\displaystyle\sum_{j=1}^{j=m} x_j K_{jy}}{\displaystyle\sum_{j=1}^{j=m} K_{jy}} \\[3em] e_{my} = \dfrac{\displaystyle\sum_{i=1}^{i=n} y_i K_{ix}}{\displaystyle\sum_{i=1}^{i=n} K_{ix}} \end{array}\right\} \tag{2.76}$$

Taking Equations 2.75 and 2.76 into account, system 2.74 is transformed as follows:

$$
\left.\begin{aligned}
\left(\sum_{i=1}^{i=n} K_{ix}\right) u_{cx} + 0 + 0 &= H_x \\
0 + \left(\sum_{j=1}^{j=m} K_{jy}\right) u_{cy} + 0 &= H_y \\
0 + 0 + \left(\sum_{j=1}^{j=m} K_{jy} x_{jc}^2 + \sum_{i=1}^{i=n} K_{ix} y_{ic}^2\right) \varphi &= M_c
\end{aligned}\right\}
\tag{2.77}
$$

where

$$
\left.\begin{aligned}
x_{jc} &= x_{mja} - e_{mx} \\
y_{ic} &= y_{mi} - e_{my} \\
M_c &= H_y e_{mx} - H_x e_{my} + M_z
\end{aligned}\right\}
\tag{2.78}
$$

It should be noted that according to Equation 2.78, frame coordinates and loading system (H_x, H_y, M_z) refer from that point on to the *centre of stiffness* instead of the *centre of mass*. The expression

$$
\boxed{J_{TC} = \sum_{j=1}^{j=m} K_{jy} x_{jc}^2 + \sum_{i=1}^{n} K_{ix} y_{ic}^2}
\tag{2.79}
$$

is called *torsional stiffness* with respect to the *centre of stiffness*, and its meaning will be examined later (see Chapter 5.3).

By introducing the notation

$$
\left.\begin{aligned}
K_x &= \sum_{i=1}^{i=n} K_{ix} : \text{storey stiffness } x - x \\
K_y &= \sum_{j=1}^{j=m} K_{jy} : \text{storey stiffness } y - y \\
J_{TC} &= \sum_{j=1}^{j=m} K_{jy} x_{jc}^2 + \sum_{i=1}^{i=n} K_{ix} y_{ic}^2 : \text{torsional stiffness}
\end{aligned}\right\}
\tag{2.80}
$$

in system 2.74 takes the following form:

$$\left.\begin{array}{l} K_x u_{mx} + 0 - e_{my} K_x \varphi = H_x \\ 0 + K_y u_{my} + e_{mx} K_y \varphi = H_y \\ -e_{my} K_x u_{mx} + e_{mx} K_y u_{my} + J_{TM} \varphi = M_z \end{array}\right\} \qquad (2.81)$$

where

$$J_{TM} = J_{TC} + e_x^2 K_y + e_y^2 K_x \qquad (2.82)$$

J_{TM} is called *torsional stiffness* with respect to the *centre of mass*. Its meaning will be discussed later in Chapter 5.3.

Now system 2.77 may be rearranged as follows:

$$\left.\begin{array}{l} K_x u_{cx} + 0 + 0 = H_x \\ 0 + K_y u_{cy} + 0 = H_y \\ 0 + 0 + J_{TC} \varphi = M_c \end{array}\right\} \qquad (2.83)$$

From the preceding discussion, the following conclusions may be drawn:

1. When $e_{mx} = e_{my} = 0$, that is, the centre of stiffness C coincides with the centre of mass and if torque M_z is zero, M_c in system 2.77 takes a zero value (Equations 2.78). Therefore, the pseudospatial structure exhibits only *translational* displacements, which means that when the centre of stiffness coincides with the centre of mass and the horizontal loading passes through the centre of mass (inertial forces), the system exhibits translational displacements.

2. If a structural system is symmetric about the x–x axis and is loaded only with H_y passing through the centre of mass, this system exhibits a translational motion u_{cy} parallel to the y–y axis and a torsional deformation φ:

$$u_{cy} = \frac{H_y}{\sum\limits_{j=1}^{m} K_{jy}} \qquad (2.84a)$$

$$\varphi = \frac{e_{mx} H_y}{J_{TC}}. \qquad (2.84b)$$

From Equation 2.84b, it can be seen that the torsional effect is proportional to the distance e_{mx} between M and C, and inversely proportional to the torsional stiffness of the system.

3. The relation

$$r_{xc} = \sqrt{\dfrac{J_{TC}}{\displaystyle\sum_{j=1}^{m} K_{jy}}} \tag{2.85}$$

is called the *torsional radius*, and its meaning will be examined later (Chapters 5.2 and 5.3).

4. In case of a pure *frame resisting* system or *pure shear wall system*, the stiffness K_{ix} or K_{iy} of the structural elements is approximately proportional to the moments of inertia of the cross section of the columns or the walls. In this case, the equations above may take the following approximate form:

   a. Coordinates of *centre of stiffness* C in relation to the *centre of mass* M:

$$\left. \begin{aligned} e_{mx} &= \dfrac{\displaystyle\sum_{j=1}^{j=m} x_j J_{jx}}{\displaystyle\sum_{j=1}^{j=m} J_{jx}} \\[2ex] e_{my} &= \dfrac{\displaystyle\sum_{i=1}^{i=n} y_i J_{iy}}{\displaystyle\sum_{i=1}^{i=n} J_{iy}} \end{aligned} \right\} \tag{2.86}$$

   b. *Torsional stiffness* with respect to the *centre of stiffness*:

$$\bar{J}_{TC} = \sum_{j=1}^{j=m} J_{jx} x_{jc}^2 + \sum_{i=1}^{n} J_{iy} y_{ic}^2 \tag{2.87}$$

   c. *Torsional stiffness* with respect to the *centre of mass*:

$$\bar{J}_{TM} = J_{TC} + e_{mx}^2 \left(\sum_{j=1}^{j=m} J_{jx} \right) + e_{my}^2 \left(\sum_{i=1}^{n} J_{iy} \right) \tag{2.88}$$

In case of a system symmetric to both axes x–x and y–y, the torsional deformation φ due to a torsional moment (torque) M_z is derived from Equation 2.84b as

$$\varphi = \dfrac{M_z}{J_{TC}}. \tag{2.89}$$

It is obvious that the value of J_{TC} depends mainly on the existence of elements with high flexural stiffness (i.e. structural walls) at the perimeter of the structure and parallel to it (Figure 2.40).

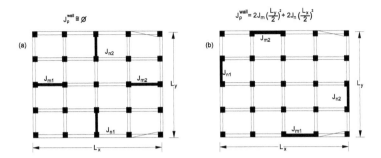

Figure 2.40 Effect of the arrangement of the structural walls on the torsional stiffness of the system: (a) system with low torsional stiffness; (b) system with high torsional stiffness.

d. *Torsional radius* with respect to the *centre of stiffness* C:

$$r_{xc} = \sqrt{\frac{\bar{J}_{TC}}{\sum_{j=1}^{m} J_{jx}}} \tag{2.90}$$

e. *Torsional radius* with respect to the *centre of mass*:

$$r_{xm} = \sqrt{\frac{\bar{J}_{TM}}{\sum_{j=1}^{m} J_{jx}}}. \tag{2.91}$$

5. It should be noted that in case of a space system consisting of *coupled frames* and *walls*, the above approximate equations cannot be used. In this case, taking into account that computer-aided analysis is available, the following procedure may be followed for the determination of the above values e_{mx}, e_{my}, J_{TC}, J_{TM}, r_{xc}, r_{xm}.
    a. The system is loaded with a torque M_z.
    b. Taking into account Equation 2.83, one may easily conclude that u_{cx} and u_{cy} are zero since H_x and H_y are zero too. Therefore, from the displacements of two counter-pairs of columns at the perimeter of the system, the position of the centre of stiffness can be determined geometrically, which means the values e_{mx} and e_{my} are defined (Figure 2.41). Likewise, the torsional deformation φ_z of the system is determined by the relations

$$\varphi_z = \frac{u_{x1} - u_{x2}}{l_y} = \frac{u_{y3} - u_{y4}}{l_x} \tag{2.92}$$

    c. From the third equation of Equation 2.83, the value J_{TC} can be easily determined. In fact,

$$J_{TC} = \sum_{j=1}^{j=m} K_{jy} x_{jc}^2 + \sum_{j=1}^{j=m} K_{ix} y_{ic}^2 = \frac{M_z}{\varphi_z} \tag{2.93}$$

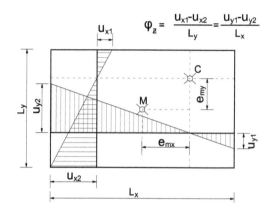

Figure 2.41 Determination of the centre of stiffness and the torsional rigidity with respect to the centre of stiffness, using a computer-aided procedure.

d. Then, the system is loaded with a horizontal force H in the x–x and y–y direction successively at its already determined centre of stiffness. The system will exhibit a translational displacement in the x–x and y–y directions, with u_{cx} and u_{cy} displacements respectively resulting from the analysis. By applying the first two equations of Equation 2.83, the stiffness of the system in the x–x and y–y direction becomes

$$K_x = \sum_{i=1}^{i=n} K_{ix} = \frac{H}{u_{cx}}$$

$$K_y = \sum_{j=1}^{j=m} K_{jy} = \frac{H}{u_{cy}}$$

(2.94)

e. Therefore, by combining Equations 2.93 and 2.94, the values of r_x, r_y and J_{TM} may be easily calculated.

2.4.4.3 Dynamic response of a single-storey 3D system

With regard that equations of motion of plane systems under seismic excitation (Equations 2.58b and 2.59) have been generated by the application of d'Alembert's equations for dynamic equilibrium, it is reasonable, also, to repeat the procedure here, although the motion of the system is not only *translational* but also *rotational*. In this respect, it should be remembered that Newton's second law of translational motion

$$F = m\ddot{x}$$

(2.95)

should be replaced, in case of rotational motion, by the equation (Figure 2.42)

$$T_{mz} = J_d \ddot{\varphi}$$

(2.96)

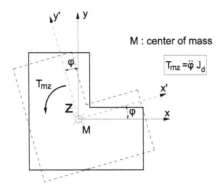

Figure 2.42 Torsional excitation of a pseudospatial single-storey system activated by a torque $T_{mz}(t)$.

where

F is the acting force.

m is the total mass of the floor.

$\ddot{x}$ is the absolute translational acceleration of the floor mass.

T_{mz} is the torsional moment of the rotating system with respect to the centre of mass of the floor.

$\ddot{\phi}$ is the angular acceleration of the floor with respect to the centre of mass.

J_d is the polar moment of inertia of the floor mass in plan with respect to the centre of mass of the floor. In case of a rectangular floor $b \times d$ with uniformly distributed mass, J_d is given by the expression:

$$J_d = m\frac{b^2 + d^2}{12}.$$

(2.97)

For simplicity, we assume that the system is damping-free, that is $c = 0$ (Equations 2.58b and 2.59).

Consider now the pseudospatial single-storey system of Figure 2.43, in which excitation at the ground level is determined by an accelerogram:

$$\ddot{x}_o(t) = \begin{bmatrix} \cos\theta \\ \sin\theta \end{bmatrix} \ddot{x}_o(t) = [\delta]\ddot{x}_o(t)$$

(2.98)

where θ is the angle of input motion to the x–x axis of the system. The motion of the system is expressed by two translational and one rotational motion of the centre of mass of the floor relative to the ground, which are

$u_{mx}(t)$ the displacement, parallel to the x-axis

$u_{my}(t)$ the displacement, parallel to the y-axis

$\phi(t)$ the in-plane rotation about the z-axis, passing through the *centre of mass*

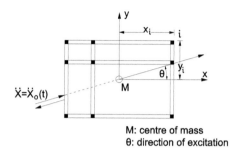

M: centre of mass
θ: direction of excitation

Figure 2.43 Pseudospatial single-storey system excited by an accelerogram $\ddot{x}_o(t)$ in the direction of angle θ with respect to the x–x axis.

The time-dependent motion of the top of a column i (Figure 2.43) results from the relations (see Equation 2.71)

$$\left.\begin{aligned} u_{ix}(t) &= u_{mx}(t) - y_i \varphi(t) \\ u_{iy}(t) &= u_{my}(t) + x_i \varphi(t) \end{aligned}\right\} \tag{2.99}$$

According to Equation 2.72, each of the frames i and j parallel to the x-axis and y-axis, respectively, develops the following horizontal restoring forces for the relative displacements of Equation 2.99:

$$\left.\begin{aligned} V_{ix}(t) &= K_{ix} u_{mx}(t) - y_i K_{ix} \varphi(t) \\ V_{jy}(t) &= K_{jy} u_{my}(t) + x_j K_{jy} \varphi(t) \end{aligned}\right\} \tag{2.100}$$

At the same time, due to the dynamic character of motion, the developing inertial forces are expressed as follows:

$$\left.\begin{aligned} f_{mx}(t) &= m\ddot{x}_m(t) = m\left[\ddot{x}_o(\cos\theta) + \ddot{u}_{mx}(t)\right] \\ f_{my}(t) &= m\ddot{y}_m(t) = m\left[\ddot{x}_o(\sin\theta) + \ddot{u}_{my}(t)\right] \\ T_{mz} &= J_d\left(\ddot{\varphi}(t) + \ddot{\varphi}_o(t)\right) = J_d\ddot{\varphi}(t) \end{aligned}\right\} \tag{2.101}$$

The value $\ddot{\varphi}_o(t)$ is taken to be equal to zero due to the fact that the usual types of accelerograms refer only to translational motion of the ground.

The three equilibrium equations of the floor (see Equation 2.73)

$$\left.\begin{aligned} \sum_{i=1}^{n} V_{ix}(t) + f_{mx}(t) &= 0 \\[2mm] \sum_{j=1}^{n} V_{jy}(t) + f_{my}(t) &= 0 \\[2mm] \sum_{j=1}^{n} x_j V_{jy}(t) - \sum_{i=1}^{n} y_i V_{ix}(t) + T_{mz} &= 0 \end{aligned}\right\} \tag{2.102}$$

result in the following linear system of differential equations:

$$\ddot{u}_{mx}(t)m + \left(\sum_{i=1}^{n} K_{ix}\right)u_{mx}(t) + \cdots + 0 + \cdots - \left(\sum_{i=1}^{n} y_i K_{ix}\right)\varphi(t) = -m\ddot{x}_o(t)(\cos\theta)$$

$$\ddot{u}_{my}(t)m + 0 + \left(\sum_{j=1}^{j=m} K_{jy}\right)u_{my}(t) + \cdots + \left(\sum_{j=1}^{m} x_j K_{jy}\right)\varphi(t) = -m\ddot{x}_o(t)(\sin\theta) \quad (2.103)$$

$$\ddot{\varphi}(t)J_d - \left(\sum_{i=1}^{n} y_i K_{ix}\right)u_{mx}(t) + \left(\sum_{j=1}^{m} x_j K_{jy}\right)u_{my}(t) + \left[\left(\sum_{j=1}^{m} K_{jy}x_j^2\right) + \left(\sum_{i-1}^{n} K_{ix}y_i^2\right)\right]\varphi(t) = 0$$

or in matrix form:

$$\begin{pmatrix} m & & \\ & m & \\ & & J_d \end{pmatrix}\begin{pmatrix} \ddot{u}_{mx} \\ \ddot{u}_{my} \\ \ddot{\varphi}_m \end{pmatrix} + \begin{pmatrix} K_x & 0 & -e_{my}K_x \\ 0 & K_y & e_{mx}K_y \\ -e_{my}K_x & e_{mx}K_y & K_{m\varphi} \end{pmatrix}\begin{pmatrix} u_{mx} \\ u_{my} \\ \varphi_m \end{pmatrix}$$

$$= -\begin{pmatrix} m & & \\ & m & \\ & & J_d \end{pmatrix}\begin{pmatrix} \cos\theta \\ \sin\theta \\ 0 \end{pmatrix}\ddot{x}_o(t) \quad (2.104)$$

where

$$\left.\begin{aligned} e_{mx} &= \frac{\displaystyle\sum_{j=1}^{m} x_j K_{jy}}{\displaystyle\sum_{j=1}^{m} K_{jy}} \\ e_{my} &= \frac{\displaystyle\sum_{i=1}^{n} y_i K_{ix}}{\displaystyle\sum_{i=1}^{m} K_{ix}} \end{aligned}\right\} \quad (2.105)$$

are the coordinates of the *centre of stiffness* (Equations 2.76) and

$$\left.\begin{aligned} K_x &= \sum_{i=1}^{n} K_{ix} \\ K_y &= \sum_{j=1}^{m} K_{jy} \\ K_{m\varphi} &= J_{Tm} \end{aligned}\right\} \quad (2.106)$$

System 2.104 may be rewritten in a more concise form, as follows:

$$\mathbf{M\ddot{u}}(t) + \mathbf{Ku}(t) = -\mathbf{M}[\boldsymbol{\delta}]\ddot{x}_{o}(t) \tag{2.107}$$

We see that the seismic motion of 3D structures obeys the same differential equations as with plane structures. The only difference is the meaning of the corresponding matrices and vectors. Therefore, from this point on, the procedure for integration is the same as the method developed in Sections 2.4.2 and 2.4.3. However, it is important to make some additional remarks here that will help us later on (Chapter 5.3) with the clarification of issues of 'torsional flexibility'.

2.4.4.4 Concluding remarks on the response of single-storey system

1. The three differential equations in Equation 2.103 or Equation 2.104 governing the three degrees of freedom – u_{mx}, u_{my} and φ – are coupled through the stiffness matrix because this matrix is not diagonal, since, in general, the *stiffness centre C* does not *coincide with the mass centre M*. Any thought of transferring the origin of coordinates from M to C must be abandoned because, in this case, u_x, u_y and φ would be uncoupled through the transformation of the coupling stiffness matrix to a diagonal one (see Equation 2.83), but at the same time, $\ddot{u}_x$, $\ddot{u}_y$ and $\ddot{\varphi}$ would be coupled through the transformation of the mass matrix from diagonal to a normal (coupling) one. This is quite reasonable, since the change of the origin of the coordinate system cannot change its dynamic properties. So, it can be concluded that each of the three eigen modes $[\phi_1]$, $[\phi_2]$, $[\phi_3]$ of free vibration represents combined motion of *translation* and *rotation*. It is well known from 'kinetics' that such a motion may be replaced by a rotational motion about three poles (Figure 2.44) – 0_1, 0_2, 0_3 – one for each mode. This problem will be analysed further in a next paragraph.

2. If the system is symmetrical in both directions, the *stiffness centre C* coincides with the *mass centre M*. In this case, e_{mx} and e_{my} in Equation 2.105 are zero and, therefore, the three equations of motion (Equation 2.103) become uncoupled and take the following form:

$$\left.\begin{aligned} m\ddot{u}_{mx} + K_x u_{mx} &= -m(\cos\theta)\ddot{x}_o \\ m\ddot{u}_{my} + K_y u_{my} &= -m(\sin\theta)\ddot{x}_o \\ J_d\ddot{\varphi}_m + J_{TC}\varphi_m &= 0 \end{aligned}\right\} \tag{2.108}$$

The three uncoupled eigen circular frequencies are given by the following expressions (see Equation 2.4a):

$$\left.\begin{aligned} \omega_{ux} &= \sqrt{\frac{K_x}{m}} \\ \omega_{uy} &= \sqrt{\frac{K_y}{m}} \\ \omega_{\varphi} &= \sqrt{\frac{J_{TC}}{J_d}} \end{aligned}\right\} \tag{2.109}$$

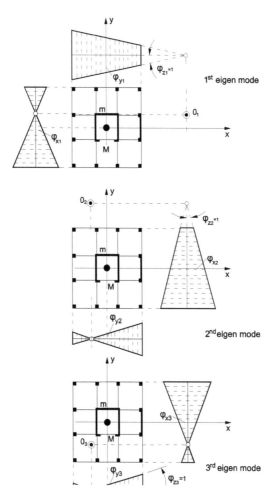

Figure 2.44 The three eigen modes of vibration of a single-storey pseudospatial structure.

If the rotational eigen frequency ω_φ is the shortest one, it means that the *rotational mode is the fundamental one*.
In this case,

$$\left.\begin{array}{c} \omega_\varphi < \omega_{ux} \\ \omega_\varphi < \omega_{uy} \end{array}\right\} \tag{2.110}$$

By introducing Equation 2.109 into Equation 2.110, the following expressions may be obtained:

$$\sqrt{\frac{J_{TC}}{J_d}} < \sqrt{\frac{K_x}{m}} \quad \text{and} \quad \sqrt{\frac{K_y}{m}} \tag{2.111}$$

or

$$\sqrt{\frac{J_d}{m}} > \sqrt{\frac{J_{TC}}{K_x}} \quad \text{and} \quad \sqrt{\frac{J_{TC}}{K_y}}.$$

(2.112)

Keeping in mind that

$$\sqrt{\frac{J_{TC}}{K_x}} \quad \text{and} \quad \sqrt{\frac{J_{TC}}{K_y}}$$

are the torsional radius r_{xc} and r_{yc} in the x–x and y–y directions and that

$$\sqrt{\frac{J_d}{m}} = l_s$$

(2.113)

is the polar radius of inertia, it may be concluded from Equation 2.112 that if

$$\boxed{l_s > r_{xc} \quad \text{and} \quad r_{yc}}$$

(2.114)

then the torsional eigen frequency is the shortest one, and in this respect *the rotational mode is the fundamental one*. In this context, the modal participation factor (Section 2.4.3) $(L_\varphi^*/M_\varphi^*)$ corresponding to torsional vibration is larger, and therefore the system must be considered 'torsionally flexible' with a lot of consequences for a building belonging to this category, as will be shown later (see Chapter 5.3).

At the same time, in the case of a symmetric system in two directions, since $e_{mx} = e_{my} = 0$, it may be concluded that

a. $J_{TC} = J_{TM}$.   (2.115)

b. Ground motion in the x-direction causes only u_x displacements. The same also holds true for the y-direction.

c. The system does not experience any torsional effect unless the base includes rotation about a vertical axis (torsional base acceleration $\ddot{\varphi}_o$) or the mass presents accidental eccentricities (Chopra, 2001).

3. For a structural system symmetric only about the x–x axis, the stiffness centre C lies on the x–x axis of symmetry, that is, $e_y = o$. Now, if the ground excitation acts only in the y–y direction, system 2.104 takes the form

$$
\begin{pmatrix} m & & \\ & m & \\ & & J_d \end{pmatrix}
\begin{pmatrix} \ddot{u}_{mx} \\ \ddot{u}_{my} \\ \ddot{\varphi}_m \end{pmatrix}
+
\begin{pmatrix} K_x & 0 & 0 \\ 0 & K_y & e_{mx}K_y \\ 0 & e_{mx}K_y & J_{TC} + e_{mx}^2 K_y \end{pmatrix}
\begin{pmatrix} u_{mx} \\ u_{my} \\ \varphi_m \end{pmatrix}
$$

$$
= - \begin{pmatrix} m & & \\ & m & \\ & & J_d \end{pmatrix}
\begin{pmatrix} 0 \\ 1 \\ 0 \end{pmatrix} \ddot{x}_o(t)
$$

(2.116)

This means that the motion u_x is uncoupled from the other two, and since there is no ground excitation in the x–x direction, there is no displacement in this direction. However, the eigen frequency is

$$\omega_{ux} = \sqrt{\frac{K_x}{m}} \tag{2.117}$$

and the corresponding eigen mode is a pure x–x displacement (Figure 2.45), meaning that

$$\varphi_{ux} = 1, \quad \varphi_{uy} = 0, \quad \varphi_{uz} = 0 \tag{2.118}$$

In case of free vibration, the other two equations take the following form:

$$\begin{pmatrix} m & \\ & J_d \end{pmatrix} \begin{pmatrix} \ddot{u}_{my} \\ \ddot{\varphi}_m \end{pmatrix} + \begin{pmatrix} K_{yy} & e_{mx}K_{yy} \\ e_{mx}K_{yy} & J_{TC} + e_{mx}^2 K_{yy} \end{pmatrix} \begin{pmatrix} u_{my} \\ \varphi_m \end{pmatrix} = 0 \tag{2.119}$$

Applying the procedure developed in Section 2.4.2, we find

$$\begin{pmatrix} K_{yy} - \omega^2 m & e_{mx}K_{yy} \\ e_{mx}K_{yy} & (J_{TC} + e_{mx}^2 K_{yy}) - \omega^2 J_d \end{pmatrix} \begin{pmatrix} \varphi_y \\ \varphi_z \end{pmatrix} = \begin{pmatrix} 0 \\ 0 \end{pmatrix} \tag{2.120}$$

In this respect, ω^2 results from the following equation:

$$\mathrm{Det} \begin{vmatrix} K_{yy} - \omega^2 m & e_{mx}K_{yy} \\ e_{mx}K_{yy} & (J_{TC} + e_x^2 K_{yy}) - \omega^2 J_d \end{vmatrix} = 0 \tag{2.121}$$

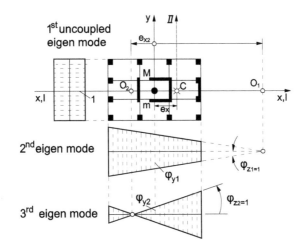

Figure 2.45 Eigen modes of vibration of a single-storey pseudospatial structure, symmetric about the x–x axis.

- It is evident that the two values of ω^2 (eigen frequencies) depend on the geometrical and dynamic values of the system, namely

$$m, J_d, K_{yy}, e_{mx} \text{ and } J_{TC}$$

- It is also evident that each of the two coupled eigen modes of motion has a coupled form of a displacement u_{my} and a rotation ϕ_m (Figure 2.45), which correspond to a rotation about two constant poles O_1 and O_2 on the x–x axis (Papapetrou, 1934). If the pole of the *shortest eigen frequency* lies within the boundaries of the floor (O_1 in Figure 2.45), the system is 'torsionally flexible' with all consequences resulting therefrom.
- The criterion for 'torsional flexibility' results from the analysis of Equation 2.121 (Anastasiadis, 1989; Anastasiadis and Athanatopoulou, 1996) and is

$$\boxed{r_{xc}^2 + e_{mx}^2 < l_s^2} \tag{2.122}$$

where

$$r_{xc} = \sqrt{\frac{J_{TC}}{K_{yy}}} \quad \text{the torsional radius}$$

$$l_s = \sqrt{\frac{J_d}{m}} \quad \text{the polar radius of inertia}$$

- In the case that $e_{mx} = 0$ (system symmetric in both directions), Equation 2.122 is reduced to Equation 2.114.

2.4.4.5 Static response of a pseudospatial multi-storey structural system

Single-storey pseudospatial systems have been analysed already for static and dynamic excitation in the previous paragraphs. The response of a multi-storey pseudospatial system to static lateral loading will be presented here, aiming at the integration of the picture of the qualitative response of 3D structural systems in use in seismic design of R/C buildings.

Consider the system of Figure 2.37 (Roussopoulos, 1956; Penelis, 1971). Under the action of lateral forces, each floor sustains a relative displacement with respect to the floor below, which can be described by three independent variables: the horizontal relative displacements u_{oj} and v_{oj} of the origin coordinate system and the rotation ω_j of the floor. Thus, the relative displacement of the frame m along the x-axis on the floor j is determined by the relationship

$$u_{jm} = u_{jo} - \omega_j y_m \tag{2.123}$$

while the relative displacement of a frame n along the y-axis on the same floor by the relationship

$$u_{jn} = u_{jo} - \omega_j x_m \tag{2.124}$$

The above relationships determine the displacements of the joint n, m of the floor j. In matrix form, they can be written as follows:

$$u_m = \mathbf{u}_o - \omega y_m$$

$$\mathbf{v}_n = \mathbf{v}_o + \omega x_n \qquad (2.125)$$

Next, the lateral stiffness of the plane frames will be defined by taking into account what has been discussed in Section 2.4.4.2. Consider the frame of Figure 2.46, which is loaded with horizontal forces H_j. *Storey shear* V_j is called the sum of the shears of the columns of storey j, that is,

$$V_j = \sum_{j=j}^{j=i} H_j \qquad (2.126)$$

If $u_1, u_2,..., u_j,..., u_i$ are the relative displacements of the floors due to the action of H_j, then the shear of the storey j is related to the u_j through the relationship

$$V_j = K_{j1}u_1 + K_{j2}u_2 + \cdots + K_{jj}u_j + \cdots + K_{ji}u_i \qquad (2.127)$$

or in matrix form

$$\mathbf{V} = \mathbf{K} \cdot u \qquad (2.128)$$

The above relationship, for $u_{oj} = 1$ and $u_1 = u_2 = \cdots = u_{j-1} = u_{j+1} = u_i = 0$ (Figure 2.47), results in

$$V_1 = K_{1j}, V_2 = K_{2j},... V_i = K_{ij} \qquad (2.129)$$

which means that the elements of the matrix K can be considered as the storey shears for a unit relative displacement of the storey. In the case of rigid girders and $s \neq i$, K_{sj} is zero, and the matrix K becomes diagonal.

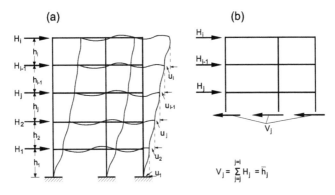

Figure 2.46 Displacement patterns of a plane frame under lateral loading (a) notation; (b) equilibrium of horizontal forces.

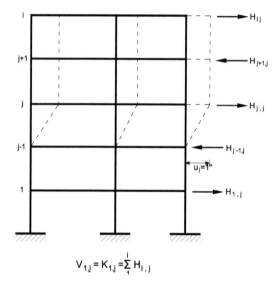

$$V_{1,j} = K_{1,j} = \sum_{1}^{i} H_{i,j}$$

Figure 2.47 Structural interpretation of the elements of matrix K.

From the equilibrium conditions of the shear forces of each storey towards the lateral forces that act on the floor under consideration, $3i$ equations with $3i$ unknowns derive, from which the relative displacements u, v and ω of the floors can be determined.

Indeed, for every storey, three equilibrium conditions are set forth. Those are

$$\left.\begin{array}{l} \sum V_{mj} = \sum H_{xj} = \overline{h}_{xj} \\[2mm] \sum V_{nj} = \sum H_{yi} = \overline{h}_{yi} \\[2mm] \sum V_{nj}x_n - \sum V_{mj}y_m = \overline{h}_{yj}x_G - \overline{h}_{xj}y_G \end{array}\right\} \qquad (2.130)$$

where x_G and y_G are the coordinates of the centre of mass, or in matrix form:

$$\left.\begin{array}{l} \sum V_m = \overline{h}_x \\[2mm] \sum V_n = \overline{h}_y \\[2mm] \sum V_n x_n - \sum V_m y_m = \overline{h}_y x_G - \overline{h}_x y_G \end{array}\right\} \qquad (2.131)$$

Substituting Equations 2.125 and 2.128 into Equation 2.131, we obtain

$$\left.\begin{array}{l} \left(\sum K_m\right)u_o + 0 \qquad -\left(\sum K_m y_m\right)\omega \qquad = \overline{h}_x \\[3mm] 0 + \left(\sum K_n\right)v_o + \left(\sum K_n x_n\right)\omega \qquad = \overline{h}_y \\[3mm] -\left(\sum K_m y_m\right)u_o + \left(\sum K_n y_n\right)v_o + \left(\sum K_n x_n^2 + \sum K_m y_m^2\right)\omega = \overline{h}_y x_G - \overline{h}_x y_G \end{array}\right\} \qquad (2.132)$$

These equations allow the calculation of the relative displacements and rotation of the floors in their plane, and, consequently, the load effects of the horizontal forces on the system. From the above presentation, it is obvious that for an efficient treatment of a pseudospatial system, even with the simplifications mentioned above under horizontal loadings, strong computational aid is needed.

If the origin of the coordinate system at every floor is replaced by a new point (see Section 2.4.4.2) such that the following relationships are fulfilled:

$$\sum_m K_m y_m = 0, \quad \sum_n K_n x_n = 0 \tag{2.133}$$

then Equations 2.132 take the following form:

$$\left. \begin{array}{l} \left(\sum K_m \right) \bar{u}_o = \bar{h}_x \\[2mm] \left(\sum K_n \right) \bar{v}_o = \bar{h}_y \\[2mm] \left(\sum K_n x_n^2 + \sum K_m y_m^2 \right) \bar{\omega} = h_y \bar{x}_G - h_x \bar{y}_G \end{array} \right\} \tag{2.134}$$

This means that the unknown $\bar{u}_o$, $\bar{v}_o$ $\bar{\omega}$ are not coupled anymore.

From now on, all coordinates refer to the new systems of coordinates, which, in general, are *different at each floor*. The coordinates of the origin of these new systems with respect to the original one derive from the following relationships:

$$x_{cj} = \frac{\sum K_n^j x_n}{\sum K_n^j}, \quad y_{cj} = \frac{\sum K_m^i y_m}{\sum K_m^i} \tag{2.135}$$

By keeping in mind the analysis of one-storey pseudospatial systems (Section 2.4.4.2), it may be concluded that the points with coordinates x_{cj} and y_{cj} with respect to the original system are the *centres of stiffness* of the successive storeys. These points in general *are not located on a vertical axis*. Only in case of a symmetric system with respect to both axes do the centres of stiffness of the successive storeys lie on a vertical axis passing through the centre of symmetry of the system.

In case of pseudospatial systems consisting only of frames or only of walls, the values of V_j in Equation 2.126 may be considered approximately as linear relations of the moments of inertia J_j of the columns or the walls under consideration. Therefore, in this case, the centre of stiffness for all storeys may be defined by the relations

$$x_c = \frac{\sum J_n^1 x_n}{\sum J_n^1}, \quad y_c = \frac{\sum J_m^1 y_m}{\sum J_m^1} \tag{2.136}$$

applied for the columns or the walls, say, at the ground floor. Thereafter, values of J_{TC}, J_{TM}, r_{xc}, r_{xm} may be defined by Equations 2.87, 2.88 and Equations 2.90, 2.91, which are valid for one-storey pseudospatial systems.

This simplification *cannot be applied in case of dual systems*, which are systems of coupled frames and ductile walls, since the response of such a system differs from the response of its constituent elements. In this case, which is the most common in practice, EC8-1/2004 allows the national authorities to adopt documented rules in their national annex that might provide computational procedures for the determination of a conventional centre of stiffness and of the torsional radius in multi-storey buildings.

In Greece, the Seismic Code in effect from 2000 until recently had adopted the following procedure (EAK 2000) (Figure 2.48):

- Load the successive storeys with a torque proportional to the storey height z:

$$T_z = G \cdot z$$

  G being the dead load of each storey

- Taking into account Equations 2.134, it may easily be concluded that $\bar{u}_o$ and $\bar{v}_o$ are zero since $\bar{h}_x$ and $\bar{h}_y$ are also zero. Therefore, from the values of the displacements of two counter pairs of columns on the perimeter of the system at each storey, the position of the centre of stiffness can be determined geometrically, meaning that the values x_{cj} and y_{cj} can be determined. From all these successive centres of stiffness, *only one is considered as the conventional centre of stiffness*, that is, the centre of stiffness of a storey that is nearest at a level $\cong 0.80h$, where h is the total height of the building. It has been proven (Makarios and Anastasiadis, 1997) that if the centres of stiffness of the successive storeys are considered as lying on a vertical axis passing through this point called the 'plasmatic axis of centres of stiffness', then the sum of the squares $\sum_j \omega_j^2$ of the twists of the successive floors for horizontal loads

$$H = G \cdot z$$

  passing through the centre of stiffness of each storey is minimum (least-squares method).

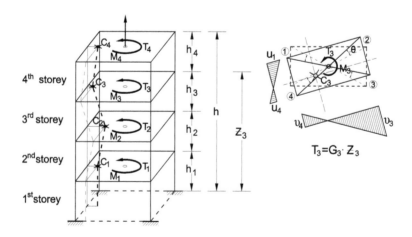

Figure 2.48 Determination of the 'plasmatic' axial of centre of stiffness C.

In parallel, the torsional deformation $\omega_{0.8H}$ at height $0.8h$ is determined by the relations

$$\omega_{0.8h} = \frac{u_1 - u_2}{l_y} = \frac{\upsilon_3 - \upsilon_4}{l_x} \qquad (2.137)$$

where u and v are the displacements of two counter pairs of columns at the perimeter of the storey near $0.80h$.

J_{TC} results from the relation

$$J_{TC} = \frac{\displaystyle\sum_{j=1}^{j=i} T_{jz}}{\omega_{0.8H}} \qquad (2.138)$$

where

$\sum_{j=1}^{j=i} T_{jz}$: the sum of the torques from storey 1 to storey i (last storey)

• Load the system with a pattern of horizontal forces in x–x and y–y directions successively. This pattern must have the form

$$H_{jx} = G \cdot z_j, H_{jy} = G \cdot z_j \qquad (2.139)$$

and must pass through the centres of stiffness lying on the 'plasmatic axis of centres of rigidity'. The system will exhibit translational – or almost translational – displacements in the x–x and y–y direction with $u_{c,0.80h}$ and $v_{c,0.80h}$, resulting from the analysis at storey with level near to $0.8h$. The stiffness of the system in the x–x and y–y direction results from the following equations:

$$\left.\begin{aligned} K_x &= \frac{\displaystyle\sum_{j=1}^{j=i} H_{jx}}{u_{c,0.8h}} \\[2em] K_y &= \frac{\displaystyle\sum_{j=1}^{j=i} H_{jy}}{\upsilon_{c,0.8h}} \end{aligned}\right\} \qquad (2.140)$$

Therefore, by combining Equations 2.138 and 2.140, the 'plasmatic' values of r_{xc}, r_{yc} and J_{TC}, J_{TM} may be easily determined.

In the case of a unisymmetric system (for example along the x-axis) both in geometry and loading, the stiffness centres are on the symmetry plane along which the loading $\bar{h}_x$ also acts. Consequently, Equation 2.134 takes the form

$$\left.\begin{aligned} \left(\sum K_m\right)\bar{u}_o &= \bar{h}_x \\[0.5em] \left(\sum K_n\right)\bar{\upsilon}_o &= 0 \\[0.5em] \omega\left[\sum K_n x_n^2 + \sum K_m y_m^2\right] &= 0 \end{aligned}\right\} \qquad (2.141)$$

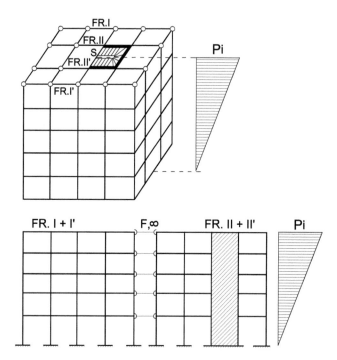

Figure 2.49 Analysis of a symmetric system.

This means that v_o and ω are equal to zero and the system is subjected to a translational displacement u_o only, so the problem can be simplified into a plane one (Figure 2.49; see Section 2.4.4.2).

In case the system is symmetric in both its main directions, the axis of centres of stiffness coincides with the axis of symmetry. Therefore, for loading in each of these two main directions passing through the centre of symmetry, uncoupled translational patterns of displacements appear in $x-x$ and $y-y$ directions. So, the system may be replaced by two plane systems in the $x-x$ and $y-y$ directions.

2.5 APPLICATION EXAMPLE

The response characteristics of three variations of a single-storey RC building are examined here, following both the static and the dynamic analysis approach described in Sections 2.4.4.2 and 2.4.4.3.

2.5.1 Building description

The buildings are rectangular in plan with dimensions b = 30 m in the horizontal x-axis and d = 25 m in the horizontal y-axis. They consist of six frames in both the $x-x$ and $y-y$ directions, each having five bays, with constant bay lengths equal to 6 and 5 m², respectively (Figure 2.50). The storey height is H = 5 m. The first building (Building A) is a frame building, whereas the other two buildings (Buildings B and C) are differentiated as per the

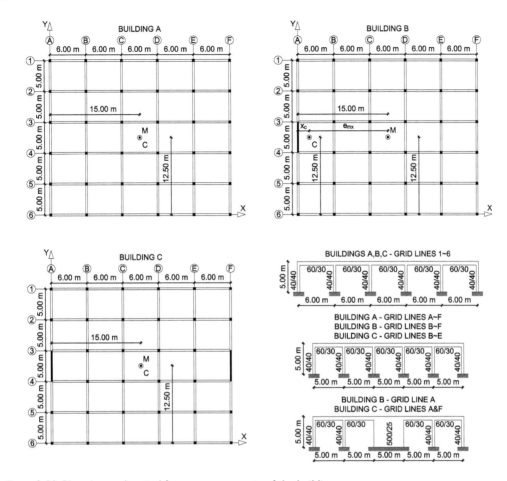

Figure 2.50 Plan view and typical frame arrangements of the buildings.

addition of RC walls on the perimeter of the building, along the y–y direction, as shown in Figure 2.50. All columns have a 40 cm square cross section. Both interior and exterior beams have a width of b_w = 30 cm and a height of b_b = 60 cm. The slab thickness is h_f = 18 cm. The walls have a thickness b_t = 25 cm and a length l_w = 5 m.

2.5.2 Design specifications

The factored distributed load at the floor area is taken to be 12 kN/m² for the $(g + \psi_2 q)$ load combination, ψ_2 = 0.30.

- Concrete quality is C25/30 with a modulus of elasticity E_c = 31 GPa. Steel quality is B500C.
- The PGA is a_g = 0.16 g.
- Seismic demand is defined for the EC8 Type I spectrum on ground C. The importance class is II.

2.5.3 Modelling assumptions

The stiffness of the vertical members is estimated by considering representative 2D frames (Section 2.4.4.2). The 3D systems are broken into three types of frames as depicted in Figure 2.50. A unit displacement is applied at the roof level, and the sum of the reactions of the vertical members corresponds to the stiffness of each frame.

- The stiffness of the concrete elements is taken to be equal to one-half (50%) of the corresponding stiffness of the uncracked elements, according to EC8 par. 4.3.1(7) (2004).
- Although the behaviour factor of the three building cases differs, for comparison reasons, it is taken to be equal to $q = 4$. Thus, a unique response spectrum is defined for all the buildings.
- The centre of mass lies at the intersection point of the symmetry axes of the building.

2.5.4 Static response

The stiffnesses of the three types of frames as depicted in Figure 2.50 are

1. For the frame in the x–x direction, $K_{f,x} = 17422.90$ kN/m.
2. For the frame in the y–y direction, $K_{f,y} = 17575.41$ kN/m.
3. For the frame with the wall at the mid-bay, $K_{wf,y} = 764001.30$ kN/m.

The individual frames at each direction are considered to function as a sequence of springs in parallel. The total stiffness in each direction is given as the sum of the stiffnesses of the individual frames:

Building A

$$x - x \text{ axis}: K_x = 6 \cdot K_{f,x} = 6 \cdot 17422.90 = 104537.40 \text{ kN/m}$$

$$y - y \text{ axis}: K_y = 6 \cdot K_{f,x} = 6 \cdot 17575.41 = 105452.46 \text{ kN/m}$$

Building B

$$x - x \text{ axis}: K_x = 6 \cdot K_{f,x} = 6 \cdot 17422.90 = 104537.40 \text{ kN/m}$$

$$y - y \text{ axis}: K_y = 5 \cdot K_{f,y} + K_{wf,y} = 5 \cdot 17575.41 + 764001.30 = 851878.35 \text{ kN/m}$$

Building C

$$x - x \text{ axis}: K_x = 6 \cdot K_{f,x} = 6 \cdot 17422.90 = 104537.40 \text{ kN/m}$$

$$y - y \text{ axis}: K_y = 4 \cdot K_{f,y} + K_{wf,y} = 4 \cdot 17575.41 + 2 \cdot 764001.30 = 1598304.24 \text{ kN/m}$$

2.5.5 Hand calculation for the centre of stiffness

The centre of stiffness for Buildings A and C coincides with the centre of mass, whereas in Building B, it is differentiated due to the existence of the 5 m RC wall in the mid-bay of the external frame along the y–y direction (Figure 2.50). The ordinates of the centre of stiffness are

$$x_C = \frac{K_{f,y} \cdot (6+12+18+24+30) + K_{wf,y} \cdot 0}{K_y} \Leftrightarrow$$

$$x_C = \frac{17575.41 \cdot (6+12+18+24+30) + 764001.30 \cdot 0}{851878.50} 35 = 1.86 \text{m}$$

$$y_C = \frac{K_{f,y} \cdot (0+5+10+15+20+25)}{K_x} \Leftrightarrow$$

$$y_C = \frac{17422.90 \cdot (0+5+10+15+20+25)}{104537.40} = 12.50 \text{m}$$

The eccentricities are estimated as

$$e_{mx} = x_M - x_C = 15.0 - 1.86 = 13.14 \text{m}$$
$$e_{my} = y_M - y_C = 12.5 - 12.5 = 0.00 \text{m}$$

2.5.6 Mass calculation

The total weight of each building is $W = 12\, b \cdot d = 12 \cdot 25 \cdot 30 = 9000$ kN.
The total mass of each building is $M = W/g = 9000/9.81 = 917.43$ t.

2.5.7 Base shear calculation

The design spectrum of EC8 (2004) for $S = 1.15$, $q = 4$ (Figure 2.51) is utilised for defining the base shear.
Building A

$$T_x = 2\pi \sqrt{\frac{M}{K_x}} = 2\pi \sqrt{\frac{917.43}{104537.40}} = 0.59\,\text{s} \rightarrow a_g = 0.115\text{g}, \quad V_{b,x} = 0.115 \cdot 9000 = 1035\,\text{kN}$$

$$T_y = 2\pi \sqrt{\frac{M}{K_y}} = 2\pi \sqrt{\frac{917.43}{105452.46}} = 0.59\,\text{s} \rightarrow a_g = 0.115\text{g}, \quad V_{by} = 0.115 \cdot 9000 = 1035\,\text{kN}$$

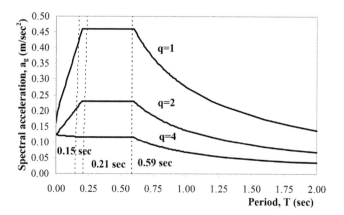

Figure 2.51 Design spectrum of EC8 (2004).

Building B

$$T_x = 2\pi\sqrt{\frac{M}{K_x}} = 2\pi\sqrt{\frac{917.43}{104537.40}} = 0.59\,\text{s} \rightarrow a_g = 0.115\text{g},\quad V_{b,x} = 0.115 \cdot 9000 = 1035\,\text{kN}$$

$$T_y = 2\pi\sqrt{\frac{M}{K_y}} = 2\pi\sqrt{\frac{917.43}{851878.35}} = 0.21\,\text{s} \rightarrow a_g = 0.116\text{g},\quad V_{b,x} = 0.116 \cdot 9000 = 1035\,\text{kN}$$

It may be noted that the estimated value of T_y for Building B is not accurate, since motion in the y–y direction is coupled.

Building C

$$T_x = 2\pi\sqrt{\frac{M}{K_x}} = 2\pi\sqrt{\frac{917.43}{104537.40}} = 0.59\,\text{s} \rightarrow a_g = 0.115\text{g},\quad V_{b,x} = 0.115 \cdot 9000 = 1035\,\text{kN}$$

$$T_y = 2\pi\sqrt{\frac{M}{K_y}} = 2\pi\sqrt{\frac{917.43}{1598304.24}} = 0.15\,\text{s} \rightarrow a_g = 0.117\text{g},\quad V_{b,x} = 0.117 \cdot 9000 = 1035\,\text{kN}$$

Despite the fact that the addition of walls substantially reduces the period in the y–y axis for Buildings B and C, as expected, this is not reflected in the estimated base shear values, since the estimated periods are either very close to or lie on the plateau of the spectrum.

The torsional stiffness with respect to the centre of stiffness, J_{TC}, and the torsional stiffness with respect to the centre of mass, J_{TM}, are calculated according to Equations 2.79 and 2.82, respectively, and presented in Table 2.2. In the same table, the torsional deformation ϕ_z, estimated according to Equation 2.89, as well as the horizontal displacements of the centre of mass and stiffness appear. Additionally, the torsional radii with respect to the centre of stiffness in both directions, r_x and r_y, are calculated (Equation 2.90). The analytical estimations are quoted for the case of Building B.

Table 2.2 Calculation of parameters J_{TC}, J_{TM}, ϕ_z, u_m, v_m, r_x, r_y

Building	$J_{TC}{}^a$ (kN m/rad)	$J_{TM}{}^a$ (kN m/rad)	$\phi_z{}^b$ (rad)	x–x u_{mx} (mm)	v_{mx} (mm)	y–y u_{my} (mm)	v_{my} (mm)	radii r_x (m)	r_y (m)
A	18.69	18.69	0	9.90	0	0	9.81	13.3	13.4
B	39.44	186.60	3.45	10.20	0	0	1.22	6.8	19.4
C	354.51	354.51	0	10.50	0	0	0.66	14.9	59.9

[a] X10⁶ → $\times 10^{6}$.
[b] X10⁻⁴ → $\times 10^{-4}$.

Building B
The torque is

$$M_z = V_{b,y} \cdot e_{mx} = 1035 \cdot 13.14 = 13603.19 \text{ kNm/rad}$$

The torsional stiffness with respect to the centre of stiffness is

$$J_{TC} = \sum_{j=1}^{j=m} K_{jy} x_{jc}^2 + n \sum_{i=1}^{i=m} K_{ix} y_{ic}^2 = K_{f,x} \cdot 2 \cdot (2.50^2 + 7.50^2 + 12.50^2)$$

$$+ K_{f,y} \cdot (4.14^2 + 10.14^2 + 16.14^2 + 22.14^2 + 28.14^2) + K_{wf,y} \cdot 1.86^2$$

$$= 39.44 \cdot 10^6 \text{ kN m/rad}$$

The torsional deformation is defined as

$$\varphi_z = \frac{M_z}{J_{TC}} = \frac{1035 \cdot 13.14}{39.44 \cdot 10^6} = 3.45 \times 10^{-4} \text{ rad}$$

The torsional stiffness with respect to the centre of mass is

$$J_{TM} = J_{TC} + e_x^2 \left(\sum_{i=1}^{i=n} K_{iy} \right) + e_y^2 \left(\sum_{j=1}^{j=m} K_{ix} \right) = 39.44 \times 10^6 + 13.14^2 \cdot 851878.35 + 0$$

$$= 186.60 \times 10^6 \text{ kN m/rad}$$

The torsional radii with respect to the centre of stiffness are (Equation 2.85)

$$r_{xc} = \sqrt{\frac{J_{TC}}{K_y}} = \sqrt{\frac{39.44 \times 10^6}{851878.35}} = 6.8 \text{ m}, \quad r_{yc} = \sqrt{\frac{J_{TC}}{K_x}} = \sqrt{\frac{39.44 \times 10^6}{104361.84}} = 19.4 \text{ m}$$

The displacements of the centre of mass are

$x - x$ axis: $u_{mx} = 10.2$ mm, $v_{mx} = 0$ m

$y - y$ axis: $u_{my} = 10.2$ mm, $v_{my} = 1.21$ mm

The maximum and minimum displacements of the corner columns are

$x - x$ axis: $u_x = 10.2$ mm, $v_x = 0$ m

$y - y$ axis: min: $u_{y,min} = 4.31$ mm, $v_{y,min} = 1.86$ mm

max: $u_{y,max} = 4.31$ mm, $v_{y,max} = 8.48$ mm

A comparison of the response parameters of Buildings A, B and C leads to the following conclusions:

- The addition of a single wall (Building B) increases J_{TC} by 2.1 times and the addition of the two walls (Building C) by 19.0 times compared to the stiffness of the planar frame building (Building A). This has a direct effect on the displacements of the system along the horizontal y–y axis, which are reduced substantially.
- In order to check the torsional rigidity of the three buildings, the radius of gyration of the floor mass in plan is estimated according to Equation 2.122 after substituting Equation 2.97:

$$l_s = \sqrt{\frac{b^2 + d^2}{12}} m = \sqrt{\frac{30^2 + 25^2}{12}} m = 11.27 \, m$$

Building A

$l_s = 11.27 m < r_x = 13.3 m; r_y = 13.4$ m

Building B (Figure 2.52)
x–x axis: Due to eccentricity along the x–x axis, Equation 2.122 applies:

$$l_s = 11.27 \, m < r_{mx} = \sqrt{r_x^2 + e_{mx}^2} = \sqrt{6.80^2 + 13.14^2} = 14.8 \, m$$

y–y axis:

$l_s = 11.27 m < r_y = 19.4$ m

Building C

$x - x$ axis: $l_s = 11.27 m < r_x = 14.9$ m

$y - y$ axis: $l_s = 11.27 m < r_y = 59.9$ m

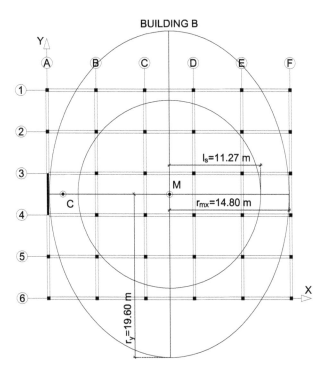

Figure 2.52 Definition of torsional rigidity in Building B.

From the above, it follows that all the buildings have sufficient torsional rigidity, and thus the fundamental translational periods in the two horizontal directions are longer than the torsional period (Tables 2.4 to 2.6).

2.5.8 Computer-aided calculation for the centre of stiffness

The methodology presented in Section 2.4.4.2 (conclusion 5) is implemented for the determination of parameters ϕ_z, J_{TC}, e_{mx}, K_x, K_y, r_x, r_y for Buildings A, B and C. The steps followed are as follows:

1. The centre of mass is loaded with a torque $M_z = 10,000$ kN m.
2. The displacements of two counterpairs of columns at the perimeter of the buildings are estimated in order to define ϕ_z and J_{TC} (Equations 2.92 and 2.93 apply):
   Building A

$$u_{x1} = 6.61\,\text{mm}, \quad u_{x2} = -6.61\,\text{mm}, \quad u_{y1} = 7.94\,\text{mm}, \quad u_{y2} = -7.94\,\text{mm}$$

$$\varphi_z = \frac{u_{x1} - u_{x2}}{d} = \frac{u_{y1} - u_{y2}}{b} = \frac{6.61 + 6.61}{25,000} = \frac{7.94 + 7.94}{30,000} = 5.29 \times 10^{-4}\,\text{rad}$$

$$J_{TC} = \frac{M_z}{\varphi_z} = \frac{10,000}{5.29 \times 10^{-4}} = 18.90 \times 10^{6}\,\text{kN m/rad}$$

Building B

$$u_{x1} = 3.16\,\text{mm}, \quad u_{x2} = -3.16\,\text{mm}, \quad u_{y1} = 0.47\,\text{mm}, \quad u_{y2} = -7.10\,\text{mm}$$

$$\varphi_z = \frac{u_{x1} - u_{x2}}{d} = \frac{u_{y1} - u_{y2}}{b} = \frac{3.16 + 3.16}{25,000} = \frac{0.47 + 7.10}{30,000} = 2.53 \times 10^{-4}\,\text{rad}$$

$$J_{TC} = \frac{M_z}{\varphi_z} = \frac{10,000}{2.53 \times 10^{-4}} = 39.60 \times 10^6\,\text{kN\,m/rad}$$

Building C

$$u_{x1} = 0.36\,\text{mm}, \quad u_{x2} = -0.36\,\text{mm}, \quad u_{y1} = 0.43\,\text{mm}, \quad u_{y2} = -0.43\,\text{mm}$$

$$\varphi_z = \frac{u_{x1} - u_{x2}}{d} = \frac{u_{y1} - u_{y2}}{b} = \frac{0.36 + 0.36}{25,000} = \frac{0.43 + 0.43}{30,000} = 2.85 \times 10^{-4}\,\text{rad}$$

$$J_{TC} = \frac{M_z}{\varphi_z} = \frac{10,000}{2.85 \times 10^{-4}} = 350.38 \times 10^6\,\text{kN\,m/rad}$$

3. The centre of stiffness is geometrically estimated:
   Buildings A, C: there is no eccentricity
   Building B

$$x_c = \frac{u_{y1} \cdot b}{(u_{y1} + u_{y2})} = \frac{0.47 \cdot 30,000}{(0.47 + 7.10)} = 1.86\,\text{m}$$

Hence,

$$e_{mx} = x_M - x_C = 15.0 - 1.86 = 13.14\text{m}$$

$$e_{my} = 0.00\text{m}$$

4. The centre of stiffness of each building is loaded by a horizontal force, $H = 10,000$ kN, in both the x- and y-axis. Having estimated the translational displacements u_{cx} and u_{cy} in the $x–x$ and $y–y$ directions, respectively, the corresponding stiffness is calculated according to Equation 2.94.
   Building A

$$x - x\,\text{axis:}\; u_{cx} = 0.0961\,\text{m} \rightarrow K_x = \frac{H}{u_{cx}} = \frac{10,000}{0.0961} = 104058.27\text{kN/m}$$

$$y - y\,\text{axis:}\; u_{cy} = 0.0952\,\text{m} \rightarrow K_y = \frac{H}{u_{cy}} = \frac{10.000}{0.0952} = 105042.02\text{kN/m}$$

Building B

$$x - x \text{ axis: } u_{cx} = 0.0991 \text{ m} \rightarrow K_x = \frac{H}{u_{cx}} = \frac{10,000}{0.0961} = 100908.17 \text{ kN/m}$$

$$y - y \text{ axis: } u_{cy} = 0.0119 \text{ m} \rightarrow K_y = \frac{H}{u_{cy}} = \frac{10,000}{0.0119} = 840336.13 \text{ kN/m}$$

Building C

$$x - x \text{ axis: } u_{cx} = 0.0969 \text{ m} \rightarrow K_x = \frac{H}{u_{cx}} = \frac{10,000}{0.0969} = 103199.17 \text{ kN/m}$$

$$y - y \text{ axis: } u_{cy} = 0.0063 \text{ m} \rightarrow K_y = \frac{H}{u_{cy}} = \frac{10,000}{0.0063} = 1587301.58 \text{ kN/m}$$

5. The torsional radii with respect to the centre of stiffness are estimated as (Equation 2.85) follows:
Building A

$$r_x = \sqrt{\frac{J_{TC}}{K_y}} = \sqrt{\frac{18.90 \times 10^6}{105042.02}} = 13.4 \text{ m}, \quad r_y = \sqrt{\frac{J_{TC}}{K_x}} = \sqrt{\frac{18.90 \times 10^6}{104058.27}} = 13.5 \text{ m}$$

Building B

$$r_x = \sqrt{\frac{J_{TC}}{K_y}} = \sqrt{\frac{39.60 \times 10^6}{840336.13}} = 6.9 \text{ m}, \quad r_y = \sqrt{\frac{J_{TC}}{K_x}} = \sqrt{\frac{39.60 \times 10^6}{100908.17}} = 19.8 \text{ m}$$

Building C

$$r_x = \sqrt{\frac{J_{TC}}{K_y}} = \sqrt{\frac{350.38 \times 10^6}{1587301.58}} = 14.9 \text{ m},$$

$$r_y = \sqrt{\frac{J_{TC}}{K_x}} = \sqrt{\frac{350.38 \times 10^6}{103199.17}} = 58.3 \text{ m}$$

Comparing the values of the response parameters (J_{TC}, e_{mx}, r_x, r_y) estimated by hand calculations to those derived with computer-aided analysis in Table 2.3, it may be concluded that the hand calculation procedure is sufficiently accurate for the type of buildings examined here.

Table 2.3 Comparison of parameters J_{TC}, e_{xm}, r_x, r_y estimated by hand calculations (static analysis) and by computer-aided analysis

	Static analysis				Computer-aided analysis			
Building	$J_{TC}{}^a$ (kN m/rad)	$e_{xm}(m)$	r_x (m)	r_y (m)	$J_{TC}{}^a$ (kN m/rad)	$e_{xm}(m)$	$r_x(m)$	r_y (m)
A	18.69	0	13.3	13.4	18.90	0	13.4	13.4
B	39.44	13.16	6.8	19.4	39.60	13.14	6.9	19.8
C	354.51	0	14.9	59.9	350.38	0	14.9	58.3

[a] $\subseteq 10^6$.

2.5.9 Dynamic response

Eigenvalue *analysis* is performed for all the buildings. The eigen periods and mass participation factors appear in Tables 2.4 to 2.6.

The displacements and rotations of the centre of mass and the corner columns of all the buildings for the first three modes are presented in Tables 2.7 and 2.8, respectively.

Table 2.4 Eigen periods and mass participation factors for building A

				Mass participation factors (%)								
	Eigen periods (s)			UX			UY			UZ		
Building	T_1	T_2	T_3	$\varepsilon_{1,UX}$	$\varepsilon_{2,UX}$	$\varepsilon_{3,UX}$	$\varepsilon_{1,UY}$	$\varepsilon_{2,UY}$	$\varepsilon_{3,UY}$	$\varepsilon_{1,UZ}$	$\varepsilon_{2,UZ}$	$\varepsilon_{3,UZ}$
A	0.69	0.69	0.33	100	0	0	0	100	0	0	0	0
					RX			RY			RZ	
				$\varepsilon_{1,RX}$	$\varepsilon_{2,RX}$	$\varepsilon_{3,RX}$	$\varepsilon_{1,RY}$	$\varepsilon_{2,RY}$	$\varepsilon_{3,RY}$	$\varepsilon_{1,RZ}$	$\varepsilon_{2,RZ}$	$\varepsilon_{3,RZ}$
				0	59	0	50	0	0	0	0	100

Table 2.5 Eigen periods and mass participation factors for building B

				Mass participation factors (%)								
	Eigen periods (s)			UX			UY			UZ		
Building	T_1	T_2	T_3	$\varepsilon_{1,UX}$	$\varepsilon_{2,UX}$	$\varepsilon_{3,UX}$	$\varepsilon_{1,UY}$	$\varepsilon_{2,UY}$	$\varepsilon_{3,UY}$	$\varepsilon_{1,UZ}$	$\varepsilon_{2,UZ}$	$\varepsilon_{3,UZ}$
B	0.69	0.56	0.10	100	0	0	0	86	14	0	0	0
					RX			RY			RZ	
				$\varepsilon_{1,RX}$	$\varepsilon_{2,RX}$	$\varepsilon_{3,RX}$	$\varepsilon_{1,RY}$	$\varepsilon_{2,RY}$	$\varepsilon_{3,RY}$	$\varepsilon_{1,RZ}$	$\varepsilon_{2,RZ}$	$\varepsilon_{3,RZ}$
				0	51	0	50	0	0	0	13	85

Table 2.6 Eigen periods and mass participation factors for building C

				Mass participation factors (%)								
	Eigen periods (s)			UX			UY			UZ		
Building	T_1	T_2	T_3	$\varepsilon_{1,UX}$	$\varepsilon_{2,UX}$	$\varepsilon_{3,UX}$	$\varepsilon_{1,UY}$	$\varepsilon_{2,UY}$	$\varepsilon_{3,UY}$	$\varepsilon_{1,UZ}$	$\varepsilon_{2,UZ}$	$\varepsilon_{3,UZ}$
C	0.69	0.18	0.08	100	0	0	0	100	0	0	0	0
					RX			RY			RZ	
				$\varepsilon_{1,RX}$	$\varepsilon_{2,RX}$	$\varepsilon_{3,RX}$	$\varepsilon_{1,RY}$	$\varepsilon_{2,RY}$	$\varepsilon_{3,RY}$	$\varepsilon_{1,RZ}$	$\varepsilon_{2,RZ}$	$\varepsilon_{3,RZ}$
				0	60	0	49	0	0	0	0	97

Table 2.7 Displacements and rotations of the centre of mass

	Building A			Building B			Building C		
Modes	u_{mx} (mm)	u_{my} (mm)	$\varphi_z{}^a$ (rad)	u_{mx} (mm)	u_{my} (mm)	$\varphi_z{}^a$ (rod)	u_{mx} (mm)	u_{my} (mm)	$\varphi_z{}^a$ (rod)
1	28.1	0	0	28.1	0	0	28.1	0	0
2	0	28.1	0	0	26.1	1.62	0	27.9	0
3	0	0	4.37	0	10.1	3.97	0	0	4.21

[a] φ_z value x 10^{-3}.

Table 2.8 Displacements and rotations of the corner columns

	Building A			Building B			Building C		
Modes	u_x (mm)	u_y (mm)	$\varphi_z{}^a$ (rad)	u_x (mm)	u_y (mm)	$\varphi_z{}^a$ (rad)	u_x (mm)	u_y (mm)	$\varphi_z{}^a$ (rad)
1	28.1	0	0	28.1	0	0	28.1	0	0
2	0	28.1	0	20.2	1.8 50.4	1.62	0	27.9	0
3	54.7	65.6	4.37	49.7	49.5 69.7	3.97	52.6	63.1	4.21

[a] φ_z value x 10^{-3}.

2.5.10 Estimation of poles of rotation for building B

Building B is symmetric along the x–x direction, and thus the centre of stiffness C and mass M lie on the x–x axis ($e_y = 0$). The motion is uncoupled in the x–x direction, and Equation 2.117 applies for calculating eigen frequency ω_x. The other two eigen modes are coupled (coupled form of a displacement along the x–x axis, u_{my} and rotation φ_z) and are estimated according to Equation 2.121. The poles of rotation O_1 and O_2 are estimated below (Figure 2.53):

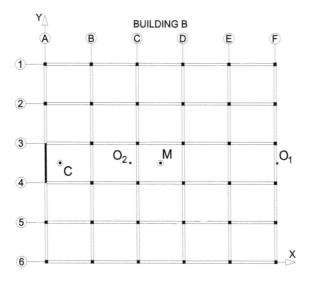

Figure 2.53 Poles of rotation in Building B.

The uncoupled eigen frequency in the $y-y$ direction is given as

$$\omega_y^2 = \frac{K_y}{m} = \frac{840336.13}{917.74} = 915.66\,\text{s}^{-2}$$

The eigen frequencies of the system are (coupled modes)

$$\text{Mode 2: } T_2 = 0.56\,\text{s} \rightarrow \omega_2^2 = \left(\frac{2\pi}{T_2}\right)^2 = \left(\frac{2\pi}{0.56}\right)^2 = 125.90\,\text{s}^{-2}$$

$$\text{Mode 3: } T_3 = 0.10\,\text{s} \rightarrow \omega_3^2 = \left(\frac{2\pi}{T_3}\right)^2 = \left(\frac{2\pi}{0.10}\right)^2 = 3947.84\,\text{s}^{-2}$$

The above data are used for the calculation of the ordinates of the poles of rotation relative to the centre of mass (Anastasiadis, 2001):

Coupled mode 2:

$$O_1 : e_{x2} = e_{xm}\frac{\omega_y^2}{\omega_y^2 - \omega_2^2} = 13.14\frac{915.66}{915.66 - 125.90} = 15.23\,\text{m}, e_{y2} = 0\,\text{m}$$

Coupled mode 3:

$$O_2 : e_{x3} = e_{xm}\frac{\omega_y^2}{\omega_y^2 - \omega_3^2} = 13.14\frac{915.66}{915.66 - 3947.84} = -3.96\,\text{m}, e_{y3} = 0\,\text{m}.$$

This may also be derived by geometrically plotting the eigen vector displacements as it may be seen in Chapter 5.9.6, Table 5.11.

Chapter 3

Design principles, seismic actions, performance requirements, compliance criteria

3.1 INTRODUCTION

The performance of a building in response to seismic actions is influenced mainly by three parameters:

1. The *seismic action*
2. The *level of damage* that is acceptable to society in the case of a strong earthquake and therefore is accepted by the legal framework in effect
3. The *quality* of the structural system of the building

The seismic action is a load case with the following characteristics, which differ drastically from all other loading types:

1. It is of high uncertainty in relation to the peak ground acceleration (PGA) $\ddot{x}_{\text{omax}}$, but also to its frequency content. This uncertainty is magnified by the fact that the existing strong motion records (accelerograms) cover only the period of the last 70 years.
2. Seismic actions in a region do not have a *reliable ceiling* as far as the expected PGA. This ceiling also depends on the *preselected return period of occurrence*, as will be shown later.
3. It has a short duration (a few seconds) in the form of vibration. In this respect, the induced 'strain energy' is limited in duration but causes high inelastic cyclic strains.

Limits of social acceptance of damages for seismic action of moderate and higher intensity are included in all modern Seismic Codes and were stated first in the SEAOC in 1978 (SEAOC, 1978) in the form of the following *principles*:

1. Structures should resist low-intensity earthquakes *without any structural damage*. Thus, during small and frequent earthquakes, all structural components forming the structure should remain in the elastic range.
2. Structures should resist an earthquake of moderate intensity ('design earthquake') with light and repairable damage at some structural members as well as at infill elements, which *do not put human life at risk*.
3. Structures should withstand high-intensity earthquakes with a return period much longer than their design life *without collapsing*.

In this respect, the transition of the structural system *to the inelastic range*, in case of a moderate or higher intensity earthquake, which entails the display of structural damage, must be considered *a basic social concession*.

Finally, the quality of the structural system is a parameter of controlled reliability, since it depends on the structural system layout, design, detailing and quality control of the construction.

In the sections that follow, a detailed presentation will be made of

- The principles of seismic design on the basis of *energy balance*
- The *seismic actions*
- The *performance-based design* and the relevant compliance criteria

The building and its design parameters will be examined in the next chapters.

3.2 CONCEPTUAL FRAMEWORK OF SEISMIC DESIGN: ENERGY BALANCE

3.2.1 General

From what has been presented so far, it is obvious that there is a substantial difference between conventional load cases (gravity loads, earth pressure, wind, hydrostatic pressure, etc.) and seismic action.

In the first category, if the *structural resistance* R_κ is bigger than the *action effects* E_κ (internal forces) by a specified percentage (safety factor), the structure is considered safe. This means that if

$$\frac{[R_\kappa]}{[E_\kappa]} \geq \gamma_f \cdot \gamma_m = \gamma_F \tag{3.1a}$$

and

$$\boxed{\frac{[R_\kappa]}{\gamma_m} \geq \gamma_f \cdot [E_k]} \tag{3.1b}$$

then the structure is covered by a safety factor:

$$\gamma_F = \gamma_f \cdot \gamma_m \tag{3.2}$$

In the above equations, the meaning of the notation in use is the following:
$[R_\kappa]$ is the characteristic strength of the structure.
$[E_\kappa]$ is the characteristic action effect on the structure.
γ_f, γ_m are partial safety factors for loads and materials (both greater than unity).

The above considerations are deemed to ensure the safety of the structure to an acceptable level of probability.

In the second category, the seismic action is introduced at the base of the structure in the form of a vibration of high frequency and short duration (a few seconds). Consequently, a limited amount of kinetic energy is introduced to the mass of the structure via the elastic or inelastic links of the mass to the foundation ground as long as the seismic motion lasts. Of course, if the mass of the structure is linked loosely to the ground via 'seismic isolators' (see Section 3.2.4), which at the same time have the ability to dissipate energy, the transfer of a considerable amount of kinetic energy to the mass of the structure is avoided. Therefore, in this case, the response of the structure to the seismic action is drastically reduced.

If the structure is in a position to transform the induced kinetic energy into potential strain energy without failure and then to dissipate a considerable percentage of this during its cyclic vibration in the form of hysteretic damping, then the structure must be considered safe.

Therefore, if

$$\frac{\left[W^u_{supply}\right]}{\left[W_{demand}\right]} \geq \gamma_w \cdot \gamma_m = \gamma_d \qquad (3.3a)$$

or

$$\frac{\left[W^u_{supply}\right]}{\gamma_m} \geq \gamma_w \cdot \left[W_{demand}\right] \qquad (3.3b)$$

in which

W^u_{supply} is the strain energy that the structure can absorb until failure
W_{demand} is the kinetic energy that is induced to the structure during the vibration
γ_w is the safety factor with respect to energy absorption

then the structure can undergo the earthquake motion with a safety factor:

$$\gamma_d = \gamma_w \cdot \gamma_m \qquad (3.4)$$

It is worth reviewing the preceding remarks on the diagrams of Figures 3.1 and 3.2.

Consider the building of Figure 3.1, which is loaded by permanent gravity loads and a constantly increasing lateral load p_i, which in turn causes a *base shear* V_i, which also is increasing continuously until *failure*.

This lateral loading results in a displacement of the top of the building relative to its base and equal to u. If the successive steps of $V_i - u_i$ are recorded in a diagram, a curve will be formed known as the *capacity curve of the structure* (Figure 3.2a).

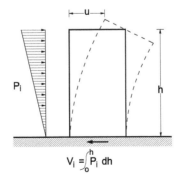

Figure 3.1 Lateral load P_i of a building developing base shear V_i and respective top horizontal displacement u.

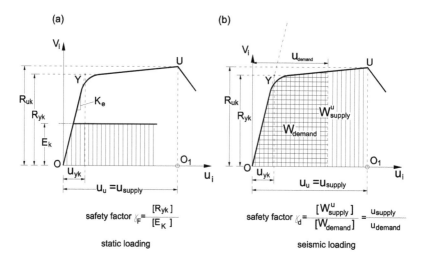

Figure 3.2 The main concept for safety verifications for (a) conventional loads and (b) seismic loads.

This curve has the following characteristic properties:

- It has a first *almost linear* branch of high slope (stiffness $K_e = \tan \theta$).
- At the end of this branch, there is the *yielding point* Y, where the slope of this branch is reduced drastically. The coordinates of this point are known as the *yield strength* ($R_{yк} = V_{yк}$) and the *yield strain* ($u_{yк}$) of the structure.
- The second branch of the diagram starts from the yield point Y with a very small slope (strain hardening) and continues up to a point U with coordinates $R_{uк} \cong (1.05 \div 1.20 R_{yк})$ known as strength at failure and u_u (strain at failure). After this point, *collapse takes place*, that is, an abrupt loss of strength capacity for small additional displacement. The area (W_{supply}) of the surface under the capacity curve of the structure ($OYUO_1$) represents the potential strain energy that can be accumulated in the structure for monotonous lateral loading up to collapse.

The above characteristics of the capacity curve have also been examined for the case of an RC cantilever beam in Section 2.3.5 (Figure 2.29).

For almost three centuries, the structural design for almost all types of loading, *except seismic actions*, has been based on the *basic requirement* that action effects ($E_к$) should not exceed a percentage of the yield strength ($R_{yк}$) of the structure (Equations 3.1a and 3.1b; Figure 3.2a). The reasoning behind this basic concept is that since loading is not instantaneous and acts permanently (static loadings), if it exceeds the yield capacity $R_{yк}$, the structure fails and collapses. Therefore, the safety margin must be based on strength capacity and inequalities 3.1a and 3.1b must be used.

In case of seismic action (Figure 3.2b), the vibration of the mass of the structure is the result of the kinetic energy W_{demand} that has passed from the foundation ground to the vibrating mass through the elastic or inelastic links between ground and mass. This energy (W_{demand}) *is limited*, due to the fact that the seismic duration is limited. Therefore, it is displayed in the form of a number of displacement reversals. In this context, it is accepted that in case that yield strength ($R_{yк}$) is exhausted, it would be enough that the displacement (u_{demand}) demand corresponding to the energy demand (W_{demand}) does not exceed a specified percentage of the displacement capacity at failure u_u, and therefore a percentage of the energy dissipation capacity W^u_{supply}:

$$u_u = u_{supply}$$
$$W_u = W_{supply}$$

The above reasoning is displayed in Equation 3.3.
By taking into account that

$$\left[W^u_{supply} \right] \cong u_{supply} \left[R_{y\kappa} \right] \qquad . \tag{3.5a}$$

and

$$\left[W_{demand} \right] \cong u_{demand} \left[R_{y\kappa} \right] \tag{3.5b}$$

it is concluded that

$$\frac{\left[W^u_{supply} \right]}{\left[W_{demand} \right]} \cong \frac{u_{supply}}{u_{demand}} \geq \gamma_w \cdot \gamma_m = \gamma_d \tag{3.6}$$

Now introducing Equation 2.52a, from Chapter 2.3.4, into Equation 3.6, the equivalent u_{supply} and u_{demand}, that is,

$$u_{supply} = \mu_{supply} u_y \tag{3.7a}$$

$$u_{demand} = \mu_{demand} u_y \tag{3.7b}$$

the following equation is obtained:

$$\frac{\left[W^u_{supply} \right]}{\left[W_{demand} \right]} \cong \frac{u_{supply}}{u_{demand}} = \frac{\mu_{supply}}{\mu_{demand}} \geq \gamma_w \cdot \gamma_m = \gamma_d \tag{3.8}$$

where γ_w is the safety factor referring to displacements.
From the above analysis, the following conclusions may be drawn:

1. For seismic actions, *and only for these*, it is accepted that during the earthquake the yield resistance $[R_{y\kappa}]$ of the structure may be completely exhausted. This means that it is accepted that due to the seismic action, the structure may pass into the inelastic range of the capacity curve.
2. For seismic actions, *and only for these*, it is accepted that the safety factor *may be satisfied on a displacement basis*, which means that the maximum displacement $[u_{demand}]$ must not exceed a specified percentage of $[u_{supply}]$ (Equations 3.6 and 3.8).
3. The above considerations in relation to those that have been presented in Chapter 2, Sections 2.3.4 and 2.3.6, constitute the core of the modern philosophy of seismic design, either as it is carried out in the form of *displacement-based design* or in the form of the *force-based design*.

3.2.2 Displacement-based design

3.2.2.1 Inelastic dynamic analysis and design

Taking into account the acceptance of inelastic response of the structure during an earthquake in the inelastic range, which means the acceptance of a specified level of damage, the direct approach to the problem of the analysis and design of seismic-resistant structures includes the following steps:

1. Determination of the capacity curve of the structural system.
2. Dynamic inelastic analysis of the structure for a series of normalised accelerograms to a code-specified intensity S.I.
3. Comparison of u_{supply} of the capacity curve with u_{demand}^{max} derived from the dynamic inelastic analysis of the structures (Figure 3.3).

The above procedure should include verifications for all crucial parameters. Those are

- Internal forces M, V, N
- Relative displacements
- Rotations θ of the nodes
- Inter-storey drift

As it will be explained in Chapter 5, Sections 5.5, 5.7.6 and 5.10.3, for the time being this procedure is used only in research and in special design cases (e.g. base-isolated systems).

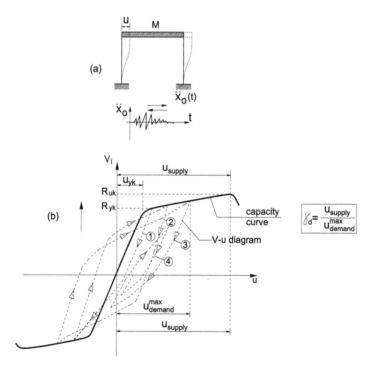

Figure 3.3 Displacement-based design based on inelastic dynamic analysis. (a) The structural system under seismic excitation; (b) $V_i - U_i$ curve for seismic loads and capacity curve of the structure.

3.2.2.2 Inelastic static analysis and design

This method has been developed in the last 20 years and is implemented basically for the assessment, verification and retrofitting of existing buildings.

A conceptual approach to this method will be given here for the SDOF system depicted in Figure 3.4a.

Step 1: Calculation and plotting of the capacity curve of the system (Figure 3.4b). This procedure assumes the implementation of an inelastic static step-by-step analysis of the system for an increasing lateral force ('pushover analysis').

Step 2: Determination of the fundamental period T_o, assuming linear elastic response $T_o = 2\pi\sqrt{M/K_e}$.

Step 3: Determination of the base shear of the system V^o_{demand}, assuming linear elastic response, using the elastic acceleration response spectrum (Figure 3.4c), that is,

$$V^o_{demand} = \ddot{\alpha}_{max(T_o)} \cdot M \tag{3.9}$$

Step 4: Determination of the behaviour factor q_{demand} equal to

$$\boxed{q_{demand} = \frac{V^o_{demand}}{R_{y\kappa}}} \tag{3.10}$$

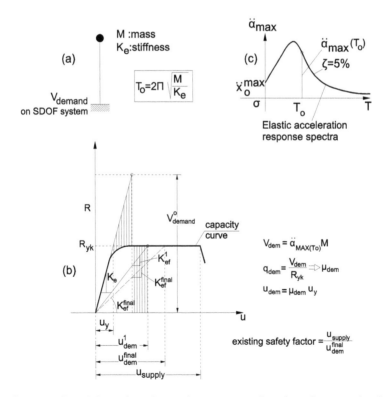

Figure 3.4 Displacement-based design based on inelastic static analysis (a pushover analysis). (a) Structural system; (b) capacity design curve and successive steps for the determination of *u* demand; (c) elastic acceleration response spectrum.

From q_{demand} using Equations 2.53 through 2.55 (see Chapter 2.3.4), μ^1_{demand} is determined.
Step 5: Using the relation

$$u^1_{demand} = \mu^1_{demand} \cdot u_y \tag{3.11}$$

u^1_{demand} is determined (Figure 3.4b).
    Step 6: Determination of the new effective stiffness K^1_{ef} equal to (Figure 3.4b)

$$K^1_{ef} = \frac{R_{yк}}{u^1_{demand}} \tag{3.12}$$

Step 7: For this new stiffness, K^1_{ef} steps 2 through 6 are repeated until successive iteration
    steps result in very close values for K_{ef}.
Step 8: For the final u^{final}_{demand} (Figure 3.4b), the existing safety factor is determined and
    compared with a codified safety factor:

$$\frac{u_{supply}}{u^{final}_{demand}} = \gamma_{existing} \geq \gamma_d \tag{3.13}$$

The above inequality constitutes the safety verification according to *displacement-based design*. Notation in use in the above presentation is given in Figure 3.4.

This method will be presented in detail in Chapters 5 and 13. It should be noted that in this method the inelastic static analysis requires the elaboration of the capacity curve of the system following a monotonous increasing in the lateral loading (*pushover analysis*).

It is obvious that for the elaboration of the capacity curve, it is necessary to know

- The material properties of the structure
- Its geometry
- The reinforcement detailing

This means that the structure has been previously designed, and in this respect, this method for the time being is used basically for the analysis, safety verification and retrofitting of *existing buildings*.

Recently, successful efforts have been in progress for the introduction of the above procedure in the analysis and design of new buildings in the form of a *direct displacement-based design* (DDBD) method (Priestley et al., 2007).

3.2.3 Force-based design

This method is implemented according to codes in effect *for the analysis and design of new buildings*, and it is considered nowadays to be the reference method for the seismic design.

In this method, the following procedure is followed:

1. The acceleration and, therefore, the *design base shear*, which results from the linear acceleration response spectrum, are reduced by the *behaviour factor q* ranging between 1.5 and 5.0. This *reduction factor* is specified by the Code of Practice,

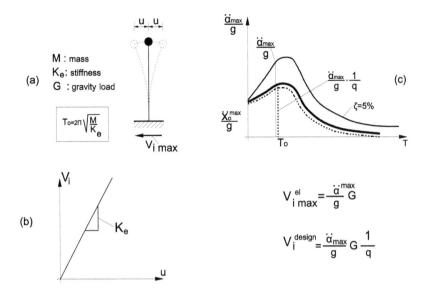

Figure 3.5 Force-based design based on 'model response spectrum analysis' or on 'lateral force method of analysis': (a) structural system under seismic action; (b) $V_i - u_i$ diagram; (c) relation between elastic and design acceleration response spectrum.

and its value depends on the structural type (Figure 3.5). For these reduced values of inertial forces, the structural system is analysed *under the assumption of linear behaviour.*

2. Based on the combination of load effects that have resulted from these inertial forces reduced by q and from the gravity loads wherein masses have been taken into account for the determination of the inertial forces, using *partial load safety factors* $\gamma_f = 1.0$, the dimensioning and safety verification of the structure is carried out for the ultimate limit state (ULS), that is,

$$[E_d] \leq \frac{[R_k]}{\gamma_m} \tag{3.14}$$

3. Based on the codified value of q and on the fundamental period of the structure T, ductility demand $_{\mu Dem}$ may be obtained from Equations 2.53 through 2.55 (Chapter 2.3.4).

Then from the following equation:

$$\mu_{supply} \geq \gamma_d \mu_{demand} \tag{3.15}$$

μ_{supply} is defined. Thereafter, the structural system is designed and detailed in such a way that its ductility μ_{supply} fulfils Equation 3.15. It is obvious that as the behaviour factor q increases, the value of μ_{demand} increases also, and therefore *the structure must be more ductile*. It should be noted that for the implementation of the procedure of this step, the capacity curve of the structure is needed, and in this respect the *inelastic static analysis* (pushover analysis) of the structure should be carried out.

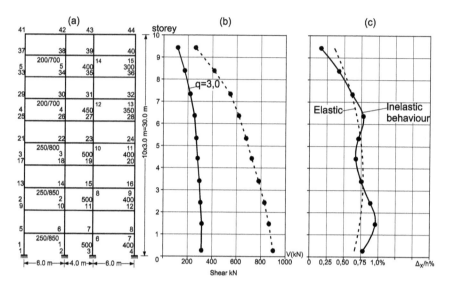

Figure 3.6 Comparison of shear forces and drift ratios (inter-storey deflections/storey height) of inelastic and elastic response at a typical ten-storey building: (a) structural system; (b) shear forces; (c) drift-ratios. (Adapted from Penelis, G.G. and Kappos, A.J. 1997. *Earthquake-Resistant Concrete Structures.* SPON E&FN, Chapman & Hall, London.)

4. In order to avoid inelastic static analysis, the safeguarding of the necessary ductility of the structure according to all modern Codes is covered by a series of *rules* that are characterised by increasing demands in parallel with the increase in behaviour factor *q*. In this respect, *the elaboration of the capacity curve of the structure as a whole and of its individual critical regions is avoided.* Only in specific critical regions and for high values of *q* is *local ductility* verified by calculations, according to rules explicitly stated in Codes.

5. However, it is not certain that the above procedure ensures the required ductility (Figure 3.6) of all structural members, since the structural analysis, either static or dynamic, is carried out in the elastic range. Therefore, a second series of rules is specified by Codes that ensure a desirable hierarchy in the sequence of the breakdown of the chain of resistance of the structure. In this way, the risk of early failure due to the formation of a collapse mechanism (e.g. inverted pendulum, etc.) is reduced and the structure may sustain catastrophic earthquakes much stronger than the 'design earthquake' (see Section 3.1), with extensive damages but without collapse. These rules constitute what is called *capacity design*, and they will be examined thoroughly in the following chapters (see Chapter 6).

Based on what has been presented above, it can easily be understood why

- The partial load safety factor for the seismic combination is $\gamma_f = 1.0$.
- At the same time, the inertial forces derived through a linear analysis are reduced by a *q*-factor ranging between 1.50 and 5.0.

The above considerations appear unreasonable to someone who uncritically implements a modern Seismic Code.

3.2.4 Concluding remarks

From what has been presented in this section so far, the following conclusions may be drawn.

1. The energy balance, or in other words the balance between the kinetic energy input (demand) in the mass and the potential strain energy storage (supply) in the structure, must be considered the core for seismic design, either for *force-based design* or for *displacement-based design*. Even for alternative design methods that are *base isolation systems* or *passive systems of energy dissipation*, the consideration of energy balance is of paramount importance.

2. *The force-based design method* will be the main issue in the major part of this book, since nowadays it constitutes the basis of all modern Codes for the analysis and design of new buildings. In this method, *strength* and *ductility* remain in a kind of counter-balance, which means that the *reduction* of the 'seismic design force' and, therefore, of the 'strength' R_{yk} of the structure due to a higher q-factor results in the need for a *ductility increase* in the structure so that the strength reduction can be counter-balanced.

3. *The displacement-based design method* will be examined in Chapter 5, where the calculation of the capacity curve will be presented, and in Chapter 13, where the assessment of the seismic capacity and retrofitting of existing R/C buildings will be presented. However, it should be noted that there is a high probability for the displacement-based design method to also be adopted by Codes for the design of new buildings in the near future (Priestley et al., 2007).

4. In the alternative design method of seismic base isolation, isolators are introduced between foundation and superstructure. This drastically reduces the kinetic energy input due to the increase in the fundamental period of the new system by removing the relevant ordinate on the elastic acceleration response spectrum to the right. At the same time, most of the kinetic energy input in this case is absorbed and dissipated at the isolators (Figure 3.7), leaving the superstructure almost free of strains.

5. The alternative design method of arranging *passive systems of energy dissipation* in some spans of frame systems protects the structure from damage due to drastic absorption and dissipation of the kinetic energy input by the dissipative devices (Figure 3.8).

3.3 EARTHQUAKE INPUT

3.3.1 Definitions

Earthquakes are ground vibrations that are caused mainly by the fracture of the crust of the earth or by sudden movement along an already *existing fault* (tectonic earthquakes), which is caused by the sudden release of elastic strain energy in the form of kinetic energy along the length of the fault ('elastic rebound theory'; Reid, 1911). Very rarely, earthquakes can be caused by volcanic eruptions. This energy accumulation can be explained by the theory of motion of *lithospheric plates*, into which the crust of the earth is divided (Figures 3.9 and 3.10; Strobach and Heck, 1980; Papazachos, 1986). Figure 3.11 shows some characteristic terms that are related to the phenomenon of earthquakes.

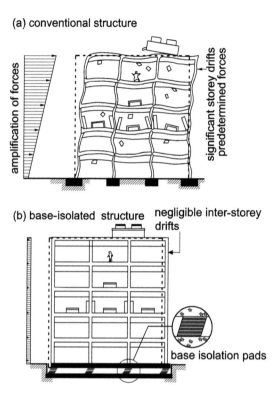

(a) conventional structure

amplification of forces

significant storey drifts
predetermined forces

(b) base-isolated structure　negligible inter-storey
drifts

base isolation pads

Figure 3.7 Comparative response to seismic action (a) of a conventional earthquake-resistant building; (b) of a building with seismic isolation.

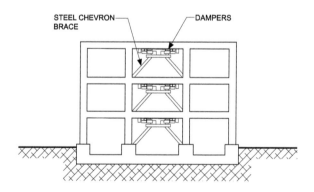

STEEL CHEVRON BRACE　　　　　　　DAMPERS

Figure 3.8 Passive energy dissipators (dampers) in a building.

The quantification of seismic motion is achieved with the use of two types of instruments, namely, *seismographs* and *accelerographs*. The first ones record displacements of the ground as a function of time and operate on a continuous real-time basis. Their recordings are of interest mainly to seismologists. The second ones record the acceleration of the ground as a function of time. They are adjusted to start operating whenever a certain ground acceleration is exceeded (strong motion instruments).

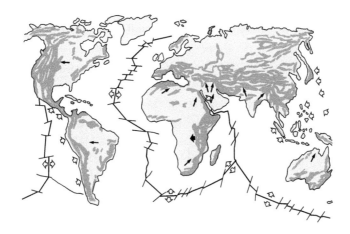

Figure 3.9 Motion of the lithospheric plates.

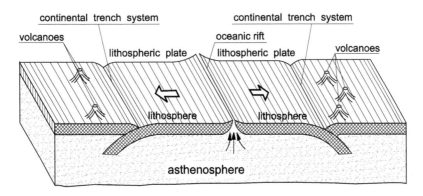

Figure 3.10 Motion system of lithospheric plates.

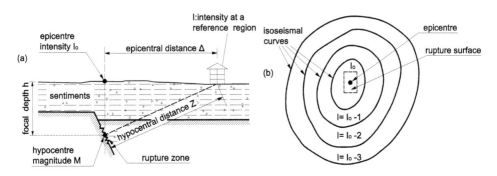

Figure 3.11 Terminology related to earthquakes (a) generation and propagation; (b) isoseismic curves.

They are of paramount importance in earthquake engineering since, as we have already seen, they provide *the seismic* input for static or dynamic analysis of structures.

The *magnitude* of an earthquake is a measure of this phenomenon in *terms of energy* released in the form of seismic waves *at the point of origin*. It is measured on various scales (Richter scale, M_L; body wave scale, m_b; surface wave scale, M_s; and moment scale, M_w).

The magnitude of any of the above scales is provided by using the relevant seismograms. Energy released at an activated fault may be expressed by the following semi-empirical relation (Richter, 1958):

$$\boxed{\log E = 11.80 + 1.50 M_s \, (\text{erg})} \tag{3.16}$$

As the magnitude increases by one unit, the energy release increases by a factor of $10^{1.5} = 31.6$.

The *intensity* of an earthquake is an index of the consequences this earthquake has on the population and the structures of a certain area. It is obvious that it is impossible to measure the damage due to an earthquake using a single-quantity system. Therefore, the damage is usually estimated qualitatively using *empirical intensity scales*. The most common macroseismic scales in use today are the modified Mercalli (MM) scale (Table 3.1), the Medvedev, Sponeur, Karnik (MSK) scale (Table 3.2) and the European Macroseismic (EMS) Scale. All have 12 intensity grades. Figure 3.12 shows the division of Greece into seismic zones (Papaioannou et al., 1994) according to the MM scale. It should be stressed that an earthquake *has only one magnitude* but different intensities from one place to the other.

The intensity generally *attenuates* as the distance from the epicentre increases. The soil conditions have a significant effect on the distribution of structural damage. If the points of equal intensity are connected on a map, the resulting curves are called *isoseismic contours* (Figure 3.11). From the design point of view, the intensity as it has been defined above is not of great interest.

The reason is that, on one hand, it does not provide any quantitative information about the parameters that are related to the ground motion (e.g. PGA, peak ground displacement, peak ground velocity, predominant period, duration). On the other hand, it is not an objective procedure since it evaluates the exciting force (the earthquake) using the response of the excited medium (structure), which depends on a series of variables such as strength, natural period and so forth, independent of the cause of damage.

However, considering that seismological records (from seismographs) do not exist for periods before mid-nineteenth century, that strong motion records do not exist for periods prior to 1939 and that the number of the latter in most seismic regions is limited even today, it is obvious that there is no other way but the one that combines the limited strong motion records with records based on qualitative intensity scales like that of the MM. Indeed, despite their subjective character, these macroseismic scales allow

- The use of the seismic history of a region
- The correlation of the maximum expected intensity in a certain period with existing records of strong motions in the same or even other areas and adoption of appropriate response spectra

Of course, it is not unusual for this kind of extrapolation to lead to serious mistakes, which make zoning revisions necessary after catastrophic earthquakes with unexpected spectral characteristics.

Table 3.1 The modified Mercalli scale

		Ground acceleration α	
		cm/s²	a/g
I	Not felt except by very few under especially favourable circumstances.		
II	Felt only by a few persons at rest, especially on upper floors of buildings. Delicately suspended objects may swing.	2	
		3	
III	Felt quite noticeably indoors, especially on upper floors of buildings, but many people do not recognise it as an earthquake.	4	
	Standing motor cars may rock slightly. Vibration like passing truck. Duration estimated.	5	0.005 g
		6	
IV	During the day, felt indoors by many, outdoors by few. At night some awakened. Dishes, windows, doors disturbed: walls make creaking sound. Sensation like heavy truck striking building. Standing motor cars rock noticeably.	7	
		8	
		9	
		10	0.01 g
V	Felt by nearly everyone, many awakened. Some dishes, windows, etc., broken: a few instances of cracked plaster, unstable objects overturned. Disturbances of trees, poles and other tall objects sometimes noticed. Pendulum clocks may stop.	20	
		30	
VI	Felt by all; many frightened and run outdoors. Some heavy furniture moved: a few instances of fallen plaster or damaged chimneys. Damage slight.	40	
		50	0.05 g
		60	
VII	Everybody runs outdoors. Damage negligible in buildings of good design and construction: slight to moderate in well-built ordinary structures: considerable in poorly built or badly designed structures: some chimneys broken. Noticed by persons driving motor cars.	70	
		80	
		90	
		100	0.1 g
VIII	Damage slight in specially designed structures: considerable in ordinary substantial buildings, with partial collapse: great in poorly built structures. Panel walls thrown out of frame structures. Fall of chimneys, factory stacks, columns, monuments, walls. Heavy furniture overturned. Sand and mud ejected in small amounts. Changes in well water. Disturbs persons driving motor cars.	200	
		300	
IX	Damage considerable in specially designed structures: well-designed frame structures thrown out of plumb: great in substantial buildings, with partial collapse. Buildings shifted off foundations. Ground cracked conspicuously. Underground pipes broken.	400	
		500	0.5 g
		600	
X	Some well-built, wooden structures destroyed: most masonry and frame structures destroyed with foundations: ground badly cracked. Rails bent. Landslides considerable from river banks and steep slopes. Shifted sand and mud. Water splashed over banks.	700	
		800	
		900	
		1000	1 g
XI	Few, if any, masonry structures remain standing. Bridges destroyed. Broad fissures in ground. Underground, pipelines completely out of service. Earth slumps and landslips in soft ground. Rails bent greatly.	2000	
		3000	
XII	Damage total. Waves seen on ground surfaces. Lines of sight and level distorted. Objects thrown upward into the air.	4000	
		5000	5 g
		6000	

Source: Derecho, A. and Fintel, M. 1974. Earthquake-resistant structures, *Handbook of Concrete Engineering*, Chapter 12, Van Nostrand Reinhold Co, New York.

Table 3.2 The MSK intensity scale

		Effect		
Degree	Intensity	*On people*	*On structures*	*On the environment*
1	Insignificant	Not felt		
2	Very light	Slightly felt		
3	Light	Felt mainly by people at rest		
4	Somewhat strong	Felt by people indoors	Trembling of glass Windows	
5	Almost strong	Felt indoors and outdoors, awakening of sleeping people	Oscillation of suspended objects, displacement of pictures on walls	
6	Strong	Many people are frightened	Light damage to structures, fine cracks in plaster	Very few cracks on wet soil
7	Very strong	Many people run outdoors	Considerable damage to structures, cracks in plaster, walls and chimneys	Landslides of steep slopes
8	Damaging	Everybody is frightened	Damage to buildings, large cracks in masonry, collapse of parapets and pediments	Changes in well-water Landslips of road Embankments
9	Very damaging	Panic	General damage to buildings, collapse of walls and roofs	Cracks on the ground, landslides
10	Extremely damaging	General panic	General destruction of buildings, collapse of many buildings	Changes on the surface of the ground, appearance of new water wells
11	Destructive	General panic	Serious damage to well-built structures	
12	General destruction	General panic	Total collapse of buildings and other civil engineering structures	Changes on the surface of the ground, appearance of new water wells

3.3.2 Seismicity and seismic hazard

For the seismic design of structures, it is essential to know the expected ground motion due to earthquakes. An earthquake, however, is a stochastic phenomenon with a random distribution of magnitude and intensity in time and space. Therefore, even for cases in which there are long-term seismic records, statistical processing of the latter is necessary for *the design earthquake* to be chosen with a preselected *probability of occurrence* in a certain period of time (e.g. 50 years, which is the design life of conventional buildings). For this reason, two concepts have been introduced: *seismicity* and *seismic hazard*.

3.3.2.1 Seismicity

Seismicity is a parameter that increases both with the magnitude and with the frequency of occurrence of earthquakes in an area. This parameter is expressed by the frequency of earthquakes (number of earthquakes per year), which exceed a predefined magnitude *M*.

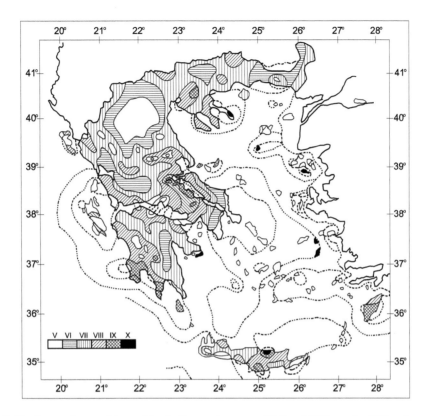

Figure 3.12 Maximum observed intensities in Greece between 1700 and 1981 on the MM scale.

Seismicity is expressed by the statistical law of Gutenberg and Richter (1956), as follows (Figure 3.13):

$$\ln N_m = a - bM$$

(3.17)

where N_m is the frequency per year of earthquakes with magnitude M or larger, and a and b are constants that are derived from statistical processing of the seismic records.

For example, for the area of Greece and for the period of 1901–1983, for a logarithmic basis of 10 instead of e, the values of a and b are (Papazachos, 1986)

$$a = 5.99, b = 0.94$$

It is evident that parameters a and b mainly describe the seismicity of an area.

Equation 3.17 is used in many instances for the statistical evaluations of many seismic parameters for everyday use in the seismic design of buildings. So, *the number of earthquakes per year* with a magnitude greater than M is deduced from Equation 3.17:

$$N = \frac{e^a}{e^{bM}}$$

(3.18)

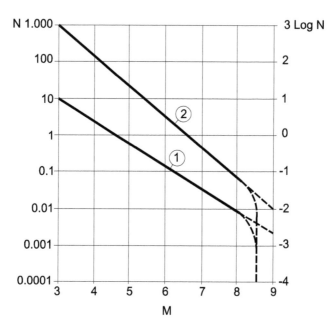

Figure 3.13 Cumulative function of earthquakes in (1) northern Greece; (2) Greece.

The corresponding *return period* T_r (the average time between two earthquakes of magnitude M or larger) is

$$T_r = \frac{1}{N} = e^{-(a-bM)} \tag{3.19}$$

Using Poisson's distribution in combination with Equation 3.17, it is easy to estimate *the probability of exceedance* $P_r(\%)$ during a period of t_r years (period of reference):

$$\ln(1 - P_r) = -\frac{t_r}{T_r} \tag{3.20a}$$

The period t_r is called the *reference period* and usually has to do with the estimated lifetime of the buildings.

For example, a 475-year return period (T_r) corresponds to a probability of exceedance in a 50-year building life (t_r):

$$\ln(1 - P_r) = -\frac{50}{475}$$
$$P_r = 10\%$$

The above example corresponds to *the seismic design input* for conventional buildings according to modern seismic codes (reference PGA a_{gR}; see Section 3.4.2).

3.3.2.2 Seismic hazard

According to the EERI glossary (EERI Committee on Seismic Risk, 1984; Dowric, 2005), 'Seismic hazard is any physical phenomenon associated with an earthquake that may produce adverse effects on human activities'.

The seismic hazard in an area is usually expressed quantitatively by the value of PGA (a_{gR}) or intensity I, for which the probability of exceedance of this value in a certain period of time (building lifetime) corresponds to a predefined value.

It has already been mentioned that the intensity I or the PGA (α_{gR}) generally decreases as the distance from the epicentre increases. The statistical evaluation of a large number of earthquakes has produced some *empirical attenuation laws*, which relate the maximum intensity I or PGA (a_{gR}) to the magnitude of the earthquake M and the distance Δ from the epicentre. For Europe, the following attenuation law for PGA (a_{gR}) has been proposed by Ambraseys and Bommer (1991):

$$\log a_{gR} = -0.87 + 0.217 M_s - \log r + 0.00117 r + 0.26 P \tag{3.21}$$

where

$$r = \sqrt{\Delta^2 + h^2}$$

with Δ the source distance and h the focal depth.

In Equation 3.21, P is zero for 50th percentile values and one for 84th percentile values. For Greece, a region of very high seismicity, the following attenuation laws for I and PGA (a_{gR}) have been proposed (Papazachos, 1986; Papaioannou et al., 1994):

$$I = 6.362 + 1.20 M_L - 4.402 \log (\Delta + 15) \tag{3.22}$$

$$\log a_{gR} = 3.775 + 0.38 M_L - 2.370 \log(\Delta + 13) \tag{3.23}$$

Over the last 20 years, extended research in engineering seismology has been in process all over the world, with interesting results in refining the above equations by introducing many other parameters (e.g. Takahashi et al., 2000; Ambraseys et al., 2005; Boore, 2005).

The scope of this book does not allow any further extension on the subject of attenuation. At this point, it is interesting to make use of Equations 3.18 through 3.20 for the probabilistic evaluation of PGA (a_{gR}) at various hazard levels (EC8-1/2004, par. 2.1(4), FEMA 356/2000: par. 1.6.1.3).

1. If a_{gR} is the PGA with a probability of exceedance P_r during a reference period of t_r years (reference lifetime of building), the following values may be defined:
   a. The corresponding return period T_r (the time between two earthquakes of equal or higher value than a_{gR}) results from Equation 3.20:

$$T_r = -\frac{t_r}{\ln(1 - P_r)} \tag{3.24}$$

b. The number of earthquakes per year of equal or higher value than a_{gR} results from Equation 3.19.

$$N_r = \frac{1}{T_r} \qquad (3.25)$$

2. If a_g is required for the same probability of exceedance as in item 1, but for a reference period of t_L years different from t_r (lifetime of a building), this value, based on Poisson's assumption, results from Equation 3.24:

$$\gamma_1 = \frac{a_g}{a_{gR}} \cong \left(\frac{t_L}{t_r} \right)^{\frac{1}{\kappa}} \qquad (3.26)$$

where
$\kappa \cong 3$ for $t_L > t_r$.
$\kappa \cong 2$ for $t_L < t_r$.
α_{gR} is the PGA for the reference period (life time of the building) t_r.
a_g is the PGA for the reference period t_L.

3. If a_g is required for the same reference period of t_r years as in item 1 (reference lifetime of the building), but for a given period of occurrence T longer or shorter than T_r, this value results for the same reason as in item 2 from the equations below:

$$\boxed{\ln(1 - P) = -\frac{t_r}{T} \Rightarrow P} \qquad (3.20b)$$

or

$$\boxed{\gamma_1 = \frac{a_g}{a_{gR}} \cong \left(\frac{P_r}{P} \right)^{\frac{1}{\kappa}}} \qquad (3.27)$$

In the above equations, γ_1 is called the *importance factor* (see Section 3.4.2.2).

The knowledge of the seismicity of a region together with attenuation laws enables the preparation of *seismic hazard tables and maps*, for example, Table 3.3 (Papaioannou et al., 1994) and Figure 3.14 (Drakopoulos and Makropoulos, 1983). Based on this information, *zonation maps* are issued by the national authorities of countries with high seismic hazard (Figure 3.15) and are incorporated in their seismic codes.

Such maps constitute for the time being the main contribution of engineering seismology to structural design as they provide, in effect, the seismic input.

However, the designer should not overlook the paramount importance of other seismic motion characteristics that are not included in the hazard maps, such as frequency content, earthquake duration and so on. These parameters are incorporated, to a degree, in *the design spectra*. These will be presented in Section 3.4.

Table 3.3 Values of maximum expected intensities
I and accelerations α_g in 10 Greek cities
for an 80-year reference period

Town	I (MM)	α_g/g
Rhodes	8.0	0.38
Larissa	7.8	0.37
Patra	7.6	0.37
Mitilini	7.6	0.30
Thessaloniki	7.3	0.26
Kalamata	7.2	0.24
Iraklion	7.1	0.23
Ioannina	7.1	0.20
Athens	6.7	0.17
Kavala	6.5	0.11

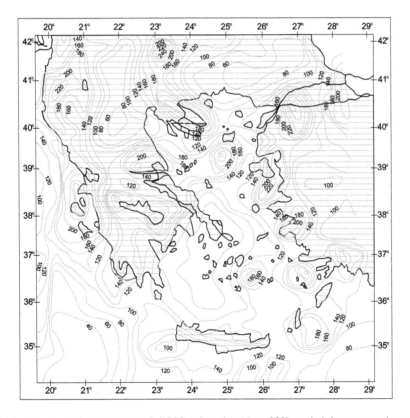

Figure 3.14 Maximum accelerations in gal (1000 gal = g) with a 90% probability to not be exceeded in 25 years reference period (Greece).

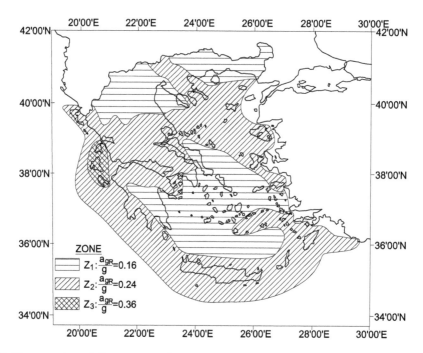

Figure 3.15 The zonation of Greece according to the National Annex (N.A.) of Greece attached to EC8-1/2004.

3.3.3 Concluding remarks

Summarising the material presented above, we should focus on the following points:

1. Earthquakes constitute a hazard primarily for human beings, for buildings and for structures in general.
2. Magnitude is a measure of the event in terms of energy release at the point of origin. Therefore, the destructiveness of an earthquake, although directly related to its magnitude, is also a function of many other parameters such as the focal depth, the distance from the epicentre, the soil conditions and the mechanical properties of the structures.
3. The intensity of an earthquake is an index of the consequences of an earthquake on the population and the buildings of a certain region. For many years only qualitative macroseismic intensity scales have been used for the damage estimate.
4. The assessment of the seismic hazard of a region is a complicated procedure based on the statistical analysis of existing strong motion records of the last few decades, together with the historical information on intensity that goes back to the past. Therefore, the result of this combination presented on hazard maps is information of limited credibility, but of crucial importance for design. That is why every now and then, major changes are introduced by the national authorities in the hazard maps.

3.4 GROUND CONDITIONS AND DESIGN SEISMIC ACTIONS

3.4.1 General

Having in mind (see Chapter 2.4.1) that the *usual method* of analysis in most modern Codes is the *elastic modal response spectrum analysis* and that the seismic motion input in this analysis is an *elastic acceleration response spectrum*, it is evident that Seismic Codes specify this spectrum in detail.

Elastic or inelastic response spectra (see Chapter 2, Sections 2.2.4 and 2.3.6) for a real seismic record give useful information for the assessment of the response of existing structures at the location where the seismic motion was recorded.

For design purposes, however, *the design spectrum* must be based on *a probabilistic prediction* based on various parameters. Those are

- Reference PGA a_{gR} with a given probability P_r of exceedance during a reference period t_r of years (reference lifetime of a building)
- Soil conditions
- Frequency content of the seismic motion
- The damping factor ζ
- The importance factor γ_1
- The behaviour factor q

So, the design spectra are *codified diagrams* based on a multi-functional evaluation of elastic and inelastic response spectra of past earthquakes combined with provisions that prevent non-conservative estimates for design actions. In this respect, *the design spectra do not correspond to any real earthquake of the past.* They are simply codified design tools with the following main characteristics:

- They are the averaged output, smoothed and normalised by a_{gR}, of real response spectra obtained from records of similar characteristics and a damping factor $\zeta = 5\%$.
- They have a generic form that is adjusted to the local conditions by introducing the proper values of the above-mentioned parameters.

In the next subsections, the formulation of seismic actions according to EC8-1/2004 will be explained in detail.

It is interesting to note that in the new version of EC8-1, which is under elaboration (EC8-1 draft 2, April 2017) and it seems to be in operation by 2022, the design seismic actions are going to be introduced in a way similar to that of the American Seismic Code (ASCE/SEI7-10), which means that instead of the PGA of the zoning maps, two parameters $S_{s,ref}$ and $S_{1,ref}$ are introduced. Those are, in detail,

- $S_{s,ref}$: the reference maximum specific acceleration, corresponding to the constant acceleration branch of the horizontal 5% (plateau) dumped elastic response spectrum, on site category A.
- $S_{1,ref}$: the reference spectral acceleration at the vibration period $T = 1$ s of the horizontal 5% dumped elastic response spectrum on site category A.

3.4.2 Ground conditions

3.4.2.1 Introduction

The seismic response of buildings is substantially influenced by the underlying soil conditions. It must be clear that there is an important distinction between the earthquake influence on ground cyclic motion and a series of other implications of the earthquake for the ground, like ground rupture, slope instability and permanent settlements caused by liquefaction or densification. This category of consequences may dramatically affect the building behaviour in response to the earthquake on a geological scale and, therefore, a geological and geotechnical investigation is required in order to make sure that the area is free of this kind of risk. If, however, the ground is susceptible to liquefaction or other types of failure, special geotechnical studies must be carried out for the proper design of the building. At the same time, depending on the importance of the structure, site investigation and laboratory tests must be carried out for the determination of the ground conditions in accordance with EC 8-1/2004.

3.4.2.2 Identification of ground types

Ground types are classified according to EC8-1/2004, in five categories labelled with letters A–E, and are presented in Table 3.4.

They are described by the stratigraphic profiles and are quantitatively characterised by the following parameters:

- Average shear wave velocity ($V_{s,30}$)
- Number of blows (N_{SPT})
- Undrained shear strength (c_u)

The characterisation of the ground type may be based on any one of the above parameters that is available.

In Table 3.4, two other ground types are included, namely, S_1 and S_2. These ground types are susceptible to liquefaction (S_2) and to anomalous site amplification (S_1). Therefore, a special geotechnical study should be carried out and special measures should be taken for soils of these categories so that the implications of ground failure for the building are diminished.

As we will see in the next section, the category in which the ground of a location is classified plays a paramount role for design seismic actions.

Table 3.4 Ground types

Ground types	A	B	C	D	E	S1	S2
$V_{s,30}$ (m/s)	>800	360–800	180–360	<180	–	<100 (indicative)	–
N_{SPT} blows/30 cm	–	>50	15–50	<15	–	–	–
c_u(kPa)	–	>250	70–250	<70	–	10–20	–

Description of stratigraphic profile:

A: Rock or other rock-like geological formation, including at most 5 m of weaker material at the surface.

B: Deposits of very dense sand, gravel or very stiff clay, at least several tens of metres in thickness, characterised by a gradual increase in mechanical properties with depth.

C: Deep deposits of dense or medium-dense sand, gravel or stiff clay with thickness from several tens to many hundreds of metres.

D: Deposits of loose-to-medium cohesionless soil (with or without some soft cohesive layers), or of predominantly soft-to-firm cohesive soil.

E: A soil profile consisting of a surface alluvium layer with v_s values of type C or D and thickness varying between about 5 and 20 m, underlain by stiffer material with v_s > 800 m/s.

S_1 Deposits consisting, or containing a layer at least 10 m thick, of soft clays/silts with a high plasticity index (PI > 40) and high water content.

S_2 Deposits of liquefiable soils, sensitive clays or any other soil profile not included in types A–E or S_1.

3.4.3 Seismic action in the form of response spectra

3.4.3.1 Seismic zones

As previously noted, it is necessary for the formation of the elastic response spectrum to know the reference PGA a_{gR}. In this context, national territories are divided by national authorities into *seismic zones* depending on the local hazard level, usually described in terms of the value a_{gR} *in rock or firm soil*. This acceleration corresponds to a reference return period of occurrence T_r of 475 years and coincides with the reference PGA a_{gR} with a 10% probability of exceedance during a reference period t_r of 50 years (lifetime of a building).

Then, the reference PGA a_{gR} is multiplied with an importance factor γ_1 *to produce the design ground acceleration*:

$$\boxed{a_g = \gamma_1 a_{gR}}$$

(3.28)

The importance factor γ_1 for conventional buildings is equal to 1.0, and it will be examined in more detail in Section 3.4.3.2.

Seismic zones with design ground acceleration not greater than 0.08 g are characterised as *low seismicity zones*, for which reduced or simplified seismic design procedures for certain types or categories of structures may be used. *The provisions of seismic codes need not be considered in seismic zones with design ground acceleration a_g not greater than 0.04 g.*

The selection of the categories of structures, ground types and seismic zones in a country for which low seismicity or very low seismicity characterisation is given is the responsibility of the national authorities, and they are included in the National Annexes of EC8-1/2004.

3.4.3.2 Importance factor

As clarified in the previous paragraph, the reference PGA a_{gR} included in seismic zones corresponds to a reference return period of occurrence T_r of 475 years and coincides with an a_{gR} with a 10% probability of exceedance during a reference period of 50 years, which is considered to be the design lifetime of normal buildings.

It is obvious that for different types of buildings, different hazard levels are established by the national authorities on the basis of the consequences of their failure. This reliability differentiation is implemented by classifying structures into different importance classes. *An importance factor γ_1 is assigned to each class category*, which reflects a higher or a lower value of the return period of occurrence T of the seismic event. Detailed guidance on the importance classes and the corresponding importance factors according to EC8-1/2004 is given below.

Buildings are classified into four importance classes depending on the size of the building, its value, its importance for public safety and the probability of human losses in case of collapse. The recommended values for the importance factor γ_1 are 0.8, 1.0, 1.2 and 1.4 for importance classes I, II, III and IV, respectively (Table 3.5).

It is important to note that in case a hazard level different from that assigned in Table 3.5 is specified, the importance factor γ_1 may be determined easily using Equations 3.26, 3.20 and 3.27.

Table 3.5 Importance classes of buildings and importance factors according to EC8-1/2004

Importance class	Buildings	Importance factor γ_1
I	Buildings of minor importance for public safety, for example, agricultural buildings, etc.	0.8
II	Ordinary buildings, not belonging in the other categories.	1.0
III	Buildings with a seismic resistance of importance in view of the consequences associated with a collapse, for example, schools, assembly halls, cultural institutions, etc.	1.2
IV	Buildings with integrity during earthquakes that is of vital importance for civil protection, for example, hospitals, fire stations, power plants, etc.	1.4

Note: Importance classes I, II and III or IV correspond roughly to consequences classes CC1, CC2 and CC3, respectively, defined in EN 1990:2002, Annex B.

EXAMPLE 3.1

If the design lifetime of an important building is specified as 100 years, then the importance factor γ_1 results from Equation 3.26, that is,

$$\gamma_1 = \frac{a_g}{a_{gR}} = \left(\frac{t_L}{t_r}\right)^{\frac{1}{\kappa}} \cong \left(\frac{100}{50}\right)^{\frac{1}{3}} = 1.25$$

For this building and for a probability of exceedance $P = 0.10$, the return period of occurrence results from Equation 3.20, that is,

$$l_n(1-P) = -\frac{t_r}{T}$$

$$l_n(1-0.10) = -\frac{100}{T}$$

$$T = \frac{100}{0.11} = 949 \, \text{years}$$

EXAMPLE 3.2

If the lifetime of the building is 50 years (reference lifetime, $t_L = t_r$) but the building must be designed for a return period of occurrence $T = 2.000$ years, then from Equation 3.20, the following may be obtained:

$$l_n(1-P) = -\frac{t_r}{T} = -\frac{50}{2000} = -0.025$$
$$1 - P = e^{-0.025} \Rightarrow P = 0.02$$

Introducing $P = 0.02$ into Equation 3.27, we obtain

$$\gamma_1 = \frac{a_g}{a_{gR}} \cong \left(\frac{0.10^{\frac{1}{3}}}{0.02} = 1.70 \right)$$

EXAMPLE 3.3

If the reference time for damage limitation is $t_L = 10$ years, then the importance factor γ_1 results from Equation 3.26, that is,

$$\nu = \frac{a_g}{a_{gR}} = \left(\frac{t_L}{t_r} \right)^{\frac{1}{\kappa}} \cong \left(\frac{10}{50} \right)^{\frac{1}{2}} \cong 0.44$$

For this hazard level and for probability of exceedance $P = 0.10$, the return period of occurrence results from Equation 3.20, that is,

$$l_n(1-P) = -\frac{t_r}{T}$$

$$l_n(1-0.10) = -\frac{10}{T}$$

$$T = \frac{10}{0.11} = 91 \text{ years}$$

3.4.3.3 Basic representation of seismic action in the form of a response spectrum

The *generic* form of the seismic action that is the generic acceleration response spectrum, henceforward called an 'elastic response spectrum', is presented in Figure 3.16. This form is the same for horizontal and vertical elastic response spectra as well. They are used for the ULS design seismic action and for the damage limitation state (see Section 3.5.3).

The abscissa of the corner points T_B, T_C, T_D (in seconds) and the parameter S for each ground type are given in Tables 3.6 and 3.7.

It must be noted that the *elastic response spectra* are grouped in two main categories. The first is characterised as Type 1 spectra and the second as Type 2 spectra. Values for Type 1 spectra are given in Table 3.6 and are recommended for regions affected mainly by earthquakes with a surface-wave magnitude M_s *greater than 5.5*. Values of Type 2 spectra are given in Table 3.7 and are recommended for regions affected mainly by earthquakes with a surface-wave magnitude M_s *not greater than 5.5*.

It should be noted that earthquakes of high magnitudes (>5.5) are deeper and excite a bigger earth mass. Therefore, the high-frequency content is filtered to a degree, so the maxima of the elastic spectra (the plateau area) are displaced to the right of the diagram (longer periods; Figure 3.17).

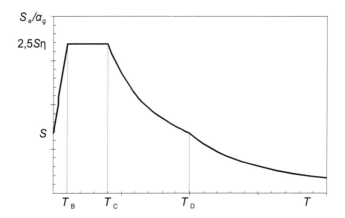

Figure 3.16 Generic shape of the elastic response spectrum. (From E.C.8-1/EN1998-1. 2004. Design of Structures for Earthquake Resistance: General Rules, Seismic Actions and Rules for Buildings. CEN, Brussels, Belgium. With permission of the British Standards Institution [BSI, CEN].)

Table 3.6 Values of the parameters describing the recommended Type 1 elastic response spectra

Ground type	S	$T_B(S)$	$T_C(S)$	$T_D(S)$
A	1.0	0.15	0.4	2.0
B	1.2	0.15	0.5	2.0
C	1.15	0.20	0.6	2.0
D	1.35	0.20	0.8	2.0
E	1.4	0.15	0.5	2.0

Table 3.7 Values of the parameters describing the recommended Type 2 elastic response spectra

Ground type	S	$T_B(S)$	$T_C(S)$	$T_D(S)$
A	1.0	0.05	0.25	1.2
B	1.35	0.05	0.25	1.2
C	1.5	0.10	0.25	1.2
D	1.8	0.10	0.30	1.2
E	1.6	0.05	0.25	1.2

Conversely, earthquakes of lower magnitude (<5.5) are shallow ones, and therefore the high-frequency content is not filtered. Therefore, the elastic spectra have their maxima (the plateau area) displaced to the left of the diagram (shorter periods; Figure 3.18).

It should also be noted that soft soils according to the diagrams of Figures 3.17 and 3.18 amplify the maxima of spectral diagrams more strongly than firm ones (Figure 3.19). This has been discussed in more detail in Chapter 2 (see Section 2.2.4).

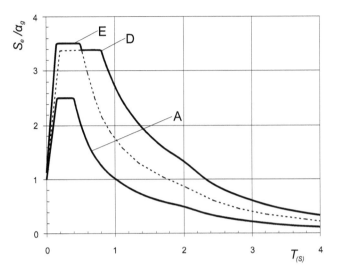

Figure 3.17 Recommended Type 1 elastic response spectra for ground types A–E (5% damping). (From E.C.8-1/EN1998-1. 2004. *Design of Structures for Earthquake Resistance: General Rules, Seismic Actions and Rules for Buildings.* CEN, Brussels, Belgium. With permission of the British Standards Institution [BSI, CEN].)

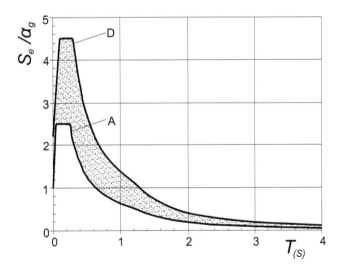

Figure 3.18 Recommended Type 2 elastic response spectra for ground A–E (5% damping). (From E.C.8-1/ EN1998-1. 2004. *Design of Structures for Earthquake Resistance: General Rules, Seismic Actions and Rules for Buildings.* CEN, Brussels, Belgium. With permission of the British Standards Institution [BSI, CEN].)

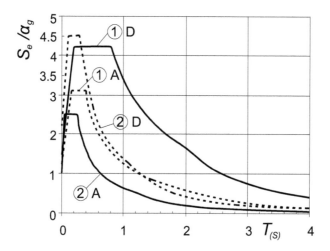

Figure 3.19 Soft soils (D) amplify the maxima of spectral diagrams more strongly than firm ones (A).

3.4.3.4 Horizontal elastic response spectrum

For the horizontal components of the seismic action, the elastic response spectrum $S_e(T)$ is defined by the following expressions (Figures 3.17 and 3.18):

$$0 \leq T \leq T_B : S_e(T) = a_g \cdot S \cdot \left[1 + \frac{T}{T_B} (\eta \cdot 2.5 - 1) \right] \tag{3.29}$$

$$T_B \leq T \leq T_C : S_e(T) = a_g \cdot S \cdot \eta \cdot 2.5 \tag{3.30}$$

$$T_C \leq T \leq T_D : S_e(T) = a_g \cdot S \cdot \eta \cdot 2.5 \left[\frac{T_C}{T} \right] \tag{3.31}$$

$$T_D \leq T \leq 4s : S_e(T) = a_g \cdot S \cdot \eta \cdot 2.5 \left[\frac{T_C T_D}{T^2} \right] \tag{3.32}$$

where

$S_e(T)$ is the elastic response spectrum.

T is the vibration period of a linear single-degree-of-freedom system.

a_g is the design ground acceleration on type A ground ($a_g = \gamma_1 \cdot a_{gR}$).

T_B is the lower limit of the period of the constant spectral acceleration branch (plateau).

T_C is the upper limit of the period of the constant spectral acceleration branch (plateau).

T_D is the value defining the beginning of the constant acceleration response range of the spectrum.

S is the soil factor.

η is the damping correction factor with a reference value of $\eta = 1$ for 5% viscous damping.

The value of the damping correction factor n may be determined by the expression

$$n = \sqrt{\frac{10}{5+\zeta}} \geq 0.55 \tag{3.33}$$

where ζ is the viscous damping ratio of the structure (see Chapter 2). From Equation 3.33, it can be seen that the elastic response spectra are credible for ζ values up to

$$\zeta \leq 28\% . \tag{3.34}$$

It should be noted that for *concrete structures* in the elastic range, ζ has a value of 5% and therefore $n = 1.0$.

The horizontal seismic actions are described by two orthogonal components considered independent and represented by the same response spectrum (Penzien and Watabe, 1974; Rosenblueth and Contreras, 1977).

3.4.3.5 Vertical elastic response spectrum

The commonly used approach in the past, of taking the vertical spectrum as two-thirds of the horizontal one without a change in frequency content, has been abandoned (Elnashai and Papazaglou, 1997; Collier and Elnashai, 2001). In this context, EC8-1/2004 has introduced the following equations for the vertical elastic response spectrum $S_{ve}(T)$ based again on the generic form of Section 3.4.3.3:

$$0 \leq T \leq T_B : S_{ve}(T) = a_{vg}\left[1 + \frac{T}{T_B}(\eta \cdot 3.0 - 1)\right] \tag{3.35}$$

$$T_B \leq T \leq T_C : S_{ve}(T) = a_{vg} \cdot \eta \cdot 3.0 \tag{3.36}$$

$$T_C \leq T \leq T_D : S_{ve}(T) = a_{vg} \cdot \eta \cdot 3.0 \left[\frac{T_C}{T}\right] \tag{3.37}$$

$$T_D \leq T \leq 4s : S_{ve}(T) = a_{vg} \cdot \eta \cdot 3.0 \left[\frac{T_C T_D}{T^2}\right] \tag{3.38}$$

The values of T_B, T_C, T_D and a_{vg} for each Type 1 or 2 of vertical spectra recommended by EC8-1/2004, as they have been defined in Section 3.4.3.3, are given in Table 3.8. It should be noted that these spectra are independent of the ground type ($S = 1$).

It is also important to note that the frequency content at high frequencies for the vertical component of the earthquake is higher than that of the horizontal ones (see Chapter 2.2.4). This explains the displacement of the plateau with the maximum values in the vertical response spectra diagram on the left, in relation to the corresponding diagram of horizontal response spectra.

Table 3.8 Recommended values of parameters describing the vertical elastic response spectra

Spectrum	α_{vg}/α_g	T_B (S)	T_C (S)	T_D (S)
Type 1	0.90	0.05	0.15	1.0
Type 2	0.45	0.05	0.15	1.0

It should also be noted here that usually only horizontal seismic actions are taken into account when designing a building. However, for the design of certain structures, the vertical component of the seismic action needs to be considered. According to EC8-1/2004, these structures are

- Pre-stressed beams
- Beams supporting columns
- Cantilever beams longer than 5.0 m
- Beams with spans over 20.0 m
- Base isolated structures

3.4.3.6 Elastic displacement response spectrum

The elastic displacement response spectrum $S_{De}(T)$ of the relative displacement u is obtained for a period T up to 4.0 s directly from the elastic acceleration response spectrum $S_e(T)$ using Equation 2.20 (Chapter 2).

$$S_{De}(T) = S_e(T)\left(\frac{T}{2\pi}\right)^2 \tag{3.39}$$

This part of $S_{De}(T)$ is codified by EC8-1/2004. For values of T longer than 4.0 s, the expressions 3.40 and 3.41 are proposed (Figure 3.20; EC8-1/2004, Annex A [informative]):

$$T_E \leq T \leq T_F : S_{De}(T) = 0.025 \cdot a_g \cdot S \cdot T_C \cdot T_D \cdot \left[2.5n + \left(\frac{T-T_E}{T_F-T_E}\right)(1-2.5n)\right] \tag{3.40}$$

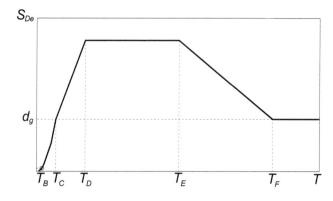

Figure 3.20 Elastic displacement response spectra. (From E.C.8-1/EN1998-1. 2004. *Design of Structures for Earthquake Resistance: General Rules, Seismic Actions and Rules for Buildings.* CEN, Brussels, Belgium. With permission of the British Standards Institution [BSI, CEN].)

Table 3.9 Additional control periods for
Type I displacement spectrum

Ground type	T_E (S)	T_F (S)
A	4.5	10.0
B	5.0	10.0
C	6.0	10.0
D	6.0	10.0
E	6.0	10.0

$$T \geq T_F : S_{De}(T) = 0.025 \cdot a_g \cdot S \cdot T_C \cdot T_D = d_g \tag{3.41}$$

The control periods T_E and T_F are presented in Table 3.9 (EC8-1/2004, Annex A).

Value d_g in Equation 3.41 represents the *design ground displacement* corresponding to the design ground acceleration, according to EC8-1/2004.

A reference to Chapter 2, Sections 2.2.4.3 and 2.3.6.2 should be made here in order to recall the scientific background of the codified expressions presented above.

3.4.3.7 Design spectrum for elastic analysis

As has already been discussed in detail (see Section 3.2), the capacity of structural systems of buildings to resist seismic actions in their nonlinear range permits their design for forces smaller than those corresponding to a linear elastic response.

To avoid explicit nonlinear analysis, the energy dissipation capacity of the structure is taken into account by performing a linear analysis based on a reduced response spectrum, henceforth called *design spectrum for elastic analysis*. This reduction is accomplished by introducing the *behaviour factor q*.

The design spectrum $S_d(T)$ for horizontal components and for the reference return period of 475 years, which is normalised by the gravity acceleration g, is defined (for $\zeta = 5\%$) in EC8-1/2004 by the following expressions:

$$0 \leq T \leq T_B : S_d(T) = a_g \cdot S \cdot \left[\frac{2}{3} + \frac{T}{T_B}\left(\frac{2.5}{q} - \frac{2}{3} \right) \right] \tag{3.42}$$

$$T_B \leq T \leq T_C : S_d(T) = a_g \cdot S \cdot \frac{2.5}{q} \tag{3.43}$$

$$T_C \leq T \leq T_D : S_d(T) \begin{cases} a_g \cdot S \cdot \dfrac{2.5}{q}\left[\dfrac{T_C}{T} \right] \\[2mm] \geq \beta \cdot a_g \end{cases} \tag{3.44}$$

$$T_D \leq T : S_d(T) \begin{cases} a_g \cdot S \cdot \dfrac{2.5}{q}\left[\dfrac{T_C T_D}{T^2} \right] \\[2mm] \geq \beta \cdot a_g \end{cases} \tag{3.45}$$

where

a_g, S, T_C, T_D are as defined in Section 3.4.3.4.

$S_d(T)$ is the design spectrum.

q is the behaviour factor.

β is the lower bound factor for the horizontal design spectrum.

The value to be ascribed to β for use in a country can be found in its National Annex. The recommended value for β is 0.2.

The values of q for horizontal components will be discussed in detail in Chapter 5.4.3, as they are related to the ductility and overstrength of the various R/C building types.

Comparing Equations 3.42 through 3.45 with the relevant ones of the horizontal elastic response spectrum, we can make the following remarks:

1. In design spectra, the value of n is equal to 1 as any additional damping is incorporated in the q factor (see Chapter 2).
2. In Equation 3.42 (first branch of design spectrum), there is a term equal to two-thirds instead of 1 (first branch of elastic response spectrum). This change covers the introduction of a q factor equal to 1.50 for structures of high stiffness (natural period equal to zero) due to the overstrength of structures of this type.
3. In design spectra, a minimum value has been introduced ($\beta = 0.20$) so that non-conservative estimates are prevented.

For the vertical component of the seismic action, the design action is given by the following expressions:

$$0 \leq T \leq T_B : S_{vd}(T) = a_{vg}\left[\frac{2}{3} + \frac{T}{T_B}\left(\frac{2.5}{q} - \frac{2}{3}\right)\right] \tag{3.46}$$

$$T_B \leq T \leq T_C : S_{vd}(T) = a_{vg} \cdot \frac{2.5}{q} \tag{3.47}$$

$$T_C \leq T \leq T_D : S_{vd}(T) = \begin{cases} a_{vg} \cdot \dfrac{2.5}{q}\left[\dfrac{T_C}{T}\right] \\ \geq \beta \cdot a_{vg} \end{cases} \tag{3.48}$$

$$T_D \leq T : S_{vd}(T) = \begin{cases} a_{vg} \cdot \dfrac{2.5}{q}\left[\dfrac{T_C T_D}{T^2}\right] \\ \geq \beta \cdot a_{vg} \end{cases} \tag{3.49}$$

where

a_{vg} is the design ground acceleration for the vertical spectral component.

β is the lower bound factor for the vertical design spectrum (recommended value $\beta = 0.20$).

q is the behaviour factor for the vertical spectrum component not greater than $q = 1.50$.

For the displacement-based design method where a design displacement response spectrum is needed, the ductility demand is introduced in the elastic displacement response spectrum in the form of equivalent damping ζ and therefore in the form of an equivalent n (see Chapter 13.5).

It is interesting to note that in the new version of EC8-1, which is under elaboration (EC8-1, draft 2, April 2017), the behaviour factor q is introduced in the form

$$q = q_R q_S q_D$$

aiming at the quantification of overstrength (q_R), deformation capacity (q_S) and energy dissipation (q_D) (see Chapter 5.7.3).

3.4.4 Alternative representation of the seismic action

3.4.4.1 General

It has already been noted that for the inelastic analysis of earthquakes, the time-history procedure is inevitable. In this context, the input seismic motion must be introduced in the form of a digitized accelerogram. EC8-1/2004, like all the other modern Codes of Practice (BSSC 2003, SEAOC 1999, ASCE 2007, etc.), foresees a well-defined procedure similar to those of other internationally known seismic codes for the generation of credible design accelerograms.

The specified procedures are the following:

- Generation of artificial accelerograms
- Recorded or simulated accelerograms

In case of a structural 3D model, the seismic motion must consist of three simultaneously acting accelerograms. The two horizontal ones may not be identical.

3.4.4.2 Artificial accelerograms

Artificial accelerograms are mathematical functions generated through random vibration theory. The most usual procedure is to generate a random signal similar to an accelerogram record, with an elastic response spectrum for 5% viscous damping that fits the codified elastic spectrum with a predefined accuracy, say 3–5% (Clough and Penzien, 1993). It must be noted that this method is an iterative one. Therefore, the output can fit as much as we like by employing more cycles of iteration. The scope of this book does not allow extended treatment of the subject. Much more information on the generation procedure may be found in more specific textbooks on engineering seismology (Elnashai and Di Sarno, 2008).

EC8-1/2004 specifies a series of requirements for such an accelerogram:

1. The duration of the accelerograms must comply with the earthquake magnitude and the relevant seismological information (e.g. distance and depth of the source) that were taken into account for the determination of a_{gR}.
2. The stationary part of the accelerogram, in the case that the seismological information on this issue is limited, must not be shorter than 10 s.
3. The suite of the artificial accelerograms should obey the following rules:
   a. At least three accelerograms are required for analysis.
   b. The mean value of PGA of these accelerograms must not be less than a_{gR} of the elastic design spectrum.

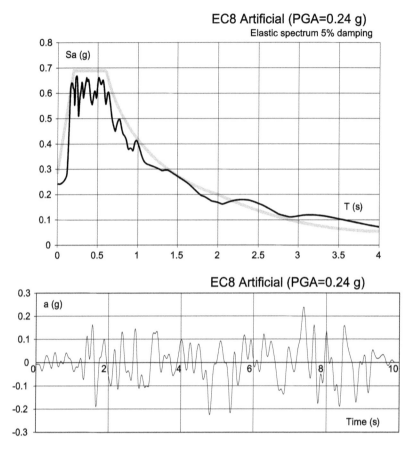

Figure 3.21 Artificial accelerogram for PGA = 0.24 g. (After Sextos, A., Pitilakis, K. and Kappos, A. 2003. *Earthquake Engineering and Structural Dynamics*, V, 32(4), 607–627.)

    c. The mean values of the 5% viscous damping elastic spectra of the above accelerograms must not be less than 90% of the elastic design spectrum in the period's region between $0.2T_1$ and $2T_1$, where T_1 is the fundamental period of the structure.

Various computer programs have been developed that are proper for the generation of artificial accelerograms, for example, the SIMQKE-1 platform (Gasparini and Vanmarcke, 1976) or ASING (Figure 3.21; Sextos et al., 2003).

The main disadvantage of the artificial accelerograms is that, many times, although all of them comply with the above-presented requirements, they present serious discrepancies among them as far as the response strains they cause to the structures. Additionally, very often they have a greater number of cycles of high amplitude than the natural accelerograms, a fact that leads to overly conservative response demands of the structure.

3.4.4.3 Recorded or simulated accelerograms

The use of natural records, and particularly those recorded at the reference region for which the accelerograms must be developed, constitutes theoretically the best procedure for obtaining this type of accelerogram by scaling them to the codified elastic acceleration spectrum. However, the requirements of EC8-1/2004 for the identical suite for both recorded and artificial ones make the procedure difficult. The reason is that many times the scaling to PGA leads to spectra not compatible with the codified ones and vice versa. So, very often, analysts have to run to various data banks and try many natural records of other regions with seismological characteristics similar to the reference location until the proper records are found that are compatible with the Code specifications (e.g. European Commission project site http://www.isesd.cv.ic.ac.uk/esd; Elnashai and Di Sarno, 2008).

The same holds for the simulated accelerograms that are generated through a physical simulation of the earthquake source, the wave path and the soil conditions. In conclusion, it should be mentioned that for many years extended research has been carried out for the generation of more reliable criteria based on quantitative intensity scales (see Chapter 2) for a comparative evaluation of the natural records instead of the requirements specified by modern Codes (Housner, 1953; Nau and Hall, 1984; Kappos, 1991; Martinez-Rueda, 1997; Elnashai, 1998).

3.4.5 Combination of seismic action with other actions

The design values E_d of the various action effects in the seismic design combination are determined according to EC8-1/2004 by combining the values of the relevant actions as follows:

$$\sum G_{\kappa j} + {}'A_{Ed} + {}'P_\kappa + {}' \sum \psi_{Ei} Q_{\kappa i}$$
(3.50)

where '+' implies 'combined with', Σ implies 'the combined effect of', $G_{\kappa j}$ is the characteristic value of permanent actions j (dead loads), A_{Ed} is the design value of the seismic action including the importance factor γ_1, P_κ is the characteristic value of eventual pre-stressing action, ψ_{Ei} is the combination coefficient for the variable action i and $Q_{\kappa i}$ is the characteristic value of variable action i.

The combination coefficient ψ_{Ei} is given according to EC8-1/2004 by the expression

$$\psi_{Ei} = \varphi \psi_{2i}$$

where
  ψ_{2i} is the combination coefficient for the quasi-permanent value of variable action $Q_{\kappa i}$.
  φ is a reduction factor ranging between 1.0 and 0.5 depending on the type of variable action.

The combination actions given in expression 3.50 are used for both the *ULS* and the *damage limitation state* (see Section 3.5.3).

The effects of the seismic action are defined by considering that all gravity loads and *consequently the relevant masses* appearing in the following combination of actions are present:

$$\sum G_{\kappa j} + {}' \sum \psi_{Ei} Q_{\kappa i}$$
(3.51)

where ψ_{Ei} is a combination coefficient for the variable action i presented above. Expressions 3.50 and 3.51 will be discussed in detail in Chapter 5.8.1.

3.5 PERFORMANCE REQUIREMENTS AND COMPLIANCE CRITERIA

3.5.1 Introduction

During the last 20 years, in the United States, there has been an increasing interest in defining design objectives based on the performance of the building (performance-based design). The term *performance levels* refers to *damage states* associated with the post-earthquake disposition of the buildings that are important to the building users.

It is important to note that the performance levels may be based on socioeconomic losses or nonstructural or structural building damage. Since the objective of this book is the structural design of R/C buildings in seismic regions, the discussion will be focused on structural performance levels.

In the document Vision 2.000 (SEAOC, 1995), which has exercised a strong influence on the formation of recent seismic design philosophy, four performance levels are defined. These levels, as they were later formulated by FEMA (FEMA 273, 274, 1997), are presented below:

- *Level 1: Fully operational* – The building continues to operate with insignificant damage.
- *Level 2: Immediate occupancy* – Damage is relatively limited. The structure retains a significant portion of its original stiffness and most or all of its original strength.
- *Level 3: Life safety* – Substantial damage has occurred to the structure and it may have lost a significant amount of its original stiffness. However, a substantial margin remains for additional lateral deformation before collapse would occur. In this respect, *life is protected*.
- *Level 4: Collapse prevention* – The building has experienced extreme damage. *Life is at risk*. If it is laterally deformed beyond this point due to post-earthquake action, the structure can experience instability and collapse.

It is obvious that the above descriptive definition of the performance levels must be given in an engineering form so that they may be expressed quantitatively. Figure 3.22a presents a typical capacity curve of a ductile building, where the various performance levels have been depicted, while Figure 3.22b presents the same curve for a 'brittle' building.

It should be noted that the structure should be designed and constructed to satisfy the specified performance levels. This constitutes one of the main objectives of the seismic design. It should also be noted that the compliance of the structural response with a specified performance level on the general capacity curve of the structure (Figure 3.22) does not ensure the satisfaction of the relevant performance level locally in the various elements of the structure. Seismic design has as a basic objective the assurance of the specified performance level for both *the structural elements and the structure as a whole*.

The above four performance levels are coupled through the matrix of Figure 3.23 with the following four levels of seismic excitation, which were introduced by FEMA (273; 274, 1997).

- EQ-I: 50% probability in 50 years (50% of EQ-III)
     Mean return period 72 years
- EQ-II: 20% probability in 50 years (70% of EQ-III)
     Mean return period 225 years

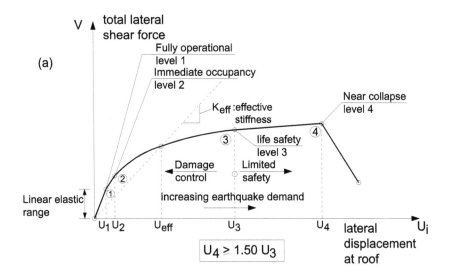

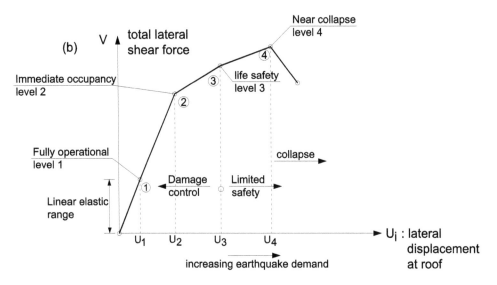

Figure 3.22 Typical capacity curve of an R/C building with the characteristic performance levels: (a) building with a ductile behaviour; (b) building with a brittle behaviour. (Adapted from FEMA 273, 274. 1997. NEHPR *Guidelines for the Seismic Rehabilitation of Buildings*, FEMA, Washington, DC.)

- EQ-III: 10% probability in 50 years (reference seismic action)
  Mean return period 475 years
- EQ-IV: 2% probability in 50 years (150% of EQ-III)
  Mean return period 2,475 years

The matrix diagonal corresponds to the basic objectives for normal buildings. This means that all diagonal combinations must be fulfilled.

It is evident that as the performance moves to higher levels, the relevant seismic action for which the structure must be analysed and designed is also higher.

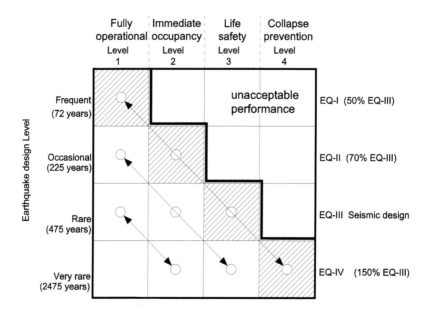

Figure 3.23 Matrix coupling of four performance levels with four levels of seismic excitation. The diagonal corresponds to the basic objectives for normal buildings. (Adapted from FEMA 273, 274. 1997. NEHPR *Guidelines for the Seismic Rehabilitation of Buildings*, FEMA, Washington, DC.)

3.5.2 Performance requirements according to EC 8-1/2004

In Europe, the performance levels have a long history, beginning in 1970 (CEB, 1970; Rowe, 1970) in the form of 'limit states'. These are states 'beyond which the structure no longer fulfils the relevant design criteria' (EN1990, 2002). The whole structure of Eurocodes is based on the concept of 'limit states' for all structural materials and all loadings. These are given below:

- Ultimate limit state (ULS)
- Serviceability limit state (SLS)

The first has to do with life safety and the second with comfort during operation. It is evident that the first is combined with higher loading (higher partial safety factor for loading) than the second one.

For seismic design, a proper transformation has been made by EC8-1/2004 in order for the system to comply with the philosophy of seismic design presented in Section 3.2. So, *the performance requirements* introduced are two, namely,

- *Non-(local) collapse requirement.* According to this requirement, the structure must be designed and constructed to withstand the design seismic action coupled to this requirement without *local* or *global collapse* retaining its structural integrity and a residual load-bearing capacity after the seismic event.
- *Damage limitation requirement.* The structure should be designed and constructed to withstand a seismic action that has a larger probability of occurrence than the design seismic action without sustaining damage that could impose any limitation on the use of the structure.

The above two performance requirements are coupled through the matrix of Figure 3.24 with the following two levels of seismic excitation:

- EQ-III: The reference seismic action a_{gR} associated with a reference probability of exceedance $P_{NCR} = 10\%$ in 50 years or a reference return period $T_{NCR} = 475$ years multiplied by the importance factor γ_1, that is, $a_g = \gamma_1 a_{gR}$ (recommended values for γ_1; see Section 3.4.3).
- EQ-I: The seismic action a_{gp} associated with a probability of exceedance $P_{DLR} = 10\%$ in 10 years or a reference return period $T_{DLR} = 95$ years, that is, $a_{gp} = v a_{gR}$, where $v = 0.5$ for importance classes I and II, and $v = 0.4$ for importance classes III and IV. The above values are recommended by EC8-1/2004. Different values for use in each E.U. member state may be found in its National Annex.

From what has been presented so far, it seems that the performance requirements specified by EC 8-1/2004 do not agree with those of FEMA (Figure 3.23), since they do not include the basic concept presented in Section 3.1, according to which in case of a high-intensity earthquake with a return period much longer than that of the design earthquake, the building should withstand it without collapse. This lack of agreement is rather superficial. Indeed, although no additional performance requirement for collapse prevention under a very rare seismic motion (e.g. mean return period of 2,475 years) is explicitly stated, it is considered that a series of provisions of EC 8-1/2004 for proper energy dissipation abilities of the structure together with the capacity design approach (see Chapter 6) implicitly ensure this third performance requirement (Figure 3.24, dashed part).

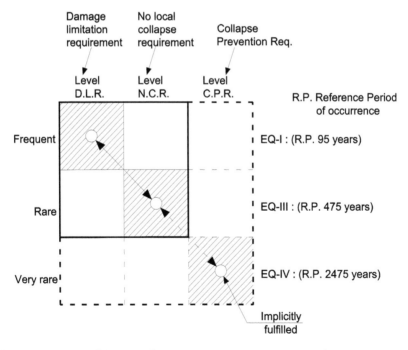

Figure 3.24 Matrix coupling of three performance levels with three levels of seismic excitation for new buildings according to EC8-1/2004.

It should be noted that in the new version of EC8-1, which is under elaboration (EC8-1, draft 2, April 2017), the four performance levels of FEMA 273, 274 (1997) are adopted with the following notation:

- LS Fully Operational (OP)
- LS of Damage Limitation (DL)
- LS of Significant Damage (SD)
- LS of Near Collapse (NC)

It is not known yet if this approach will be incorporated in the new version of EC8-1, which seems to be in effect by 2022.

3.5.3 Compliance criteria

3.5.3.1 General

For the satisfaction of the above-mentioned performance requirements, the following two limit states shall be checked:

- Ultimate Limitation State (ULS)
- Damage limitation (DLS)

The first is associated with local collapse or other forms of failure, which might *endanger life safety*. The second is associated with damage beyond which *the service ability* of the building is degraded.

In order to ensure the protection of the building against collapse under seismic actions much more severe than the design ones, a number of *specific measures* must be taken. This last principle of EC8-1/2004 is a preliminary answer to the remarks of the two last paragraphs of Section 3.5.2.

3.5.3.2 Ultimate limit state

Two parameters must be verified for the ULS:

- Sufficient strength
- Corresponding sufficient ductility

These two parameters are closely related, as there is a counter-balance between strength and ductility during seismic excitation. This issue has been discussed extensively in previous sections (see Section 3.2). Design for the ULS is *a force-based procedure*, as it was explained in Section 3.2, although the combination of strength and ductility is a matter of energy absorption and dissipation by the structure, which is finally expressed in the form of inelastic displacements. As it has already been noted over and over again, if this displacement demand is less than displacement supply scaled by a safety factor, the structure can withstand the seismic motion sufficiently. However, for the design of the ULS for new buildings, the *force-based design* method has been adopted in all modern Codes.

According to this procedure

1. The structure is analysed for gravity loads and seismic motion using *linear elastic–static* or linear *dynamic methods*. For the seismic motion, the codified elastic acceleration response spectrum is reduced by the *behaviour factor q*, ranging between 1.5

and 5.0 of its elastic values (design spectrum) depending on the ductility level of the structure (q-factor).

2. Then the structure is designed for a linear combination of internal forces (action effects) caused by gravity loads and reduced by q seismic actions. The dimensioning is carried out as for all other load cases implemented in the case of Eurocodes EC 2-1-1/2004, that is,

$$[E_d] < [R_d] \tag{3.52}$$

where

$[E_d]$ is the design internal forces (action effects) for load combinations of gravity loads and design seismic actions.

$[R_d]$ is the design resistance of each cross section calculated according to the rules of the relevant Codes in effect for all the other load cases.

3. In this way, it is obvious that the structure indirectly enters the post-elastic range, since it has been designed for reduced seismic actions by $q = 1.5$–5.0. Therefore, this reduced strength must be combined with measures for *sufficient ductility*. These measures are based on dimensioning and detailing rules. These rules are specified in Seismic Codes, and as the codified ductility demand becomes higher, these rules become stricter. All these rules will be presented, analysed, explained and discussed in Chapters 8 through 10. *The safety factor ensured in this way seems to be in terms of displacements between 1.50 and 2.00* (see Figure 3.22).

Summarising the procedure presented in the above three items, *it* can be said that the *benefit* due to seismic force reduction is partially balanced by the *cost* for higher ductility. Indeed, the results of extended cost–benefit analyses on this issue have shown that the choice to reduce forces and improve ductility is cost beneficial. What is more important, *it is much easier to improve ductility in order to withstand unexpected very rare severe earthquakes than to improve strength*.

In conclusion, it would be interesting to discuss the reasons for which the force-based design method has been adopted by modern Codes for the design of new buildings. This must be attributed to the following reasons:

1. The structure is analysed using linear methods that are simpler and much more reliable than the inelastic ones, which are necessary for a displacement-based design (generation of the capacity curve).
2. Under these conditions, *analysis and design* for seismic actions is carried out as for all other loadings, and therefore the usual superposition procedures can be adopted.
3. *Analysis and dimensioning are separated.* The analysis precedes and dimensioning follows based on the results of the analysis, while in the case of inelastic procedure, the dimensioning must proceed based on approximate methods. In other words, the procedure of the displacement-based design is a procedure for *assessment and verification*.

In the next chapters (Chapters 5, 8 and 9), a detailed discussion will be carried out on quantitative procedures for the relation between the overall ductility demand of the structure in the form of q-factor and ductility demand μ_D at the level of the structural elements (local ductility). In this respect, structural members will be either designed for ensuring local ductility supply in critical regions, for which there are relevant Code requirements, or will be detailed according to rules specified by Code.

Closing, it is interesting to note in the new version of EC8-1, which is under elaboration (EC8-1, draft 2, April 2017), the two methods of design namely

- The force-based design
- The displacement-based design

are introduced as equivalent ones with all consequences on the adopted methods of analysis and design.

At the same time, the compliance criteria necessary to satisfy the EC8-1 performance criteria are limited for the majority of the structures to the following levels:

- LS of Significant Damage (SD)
- LS of Damage Limitation (DL)

that is the same with the existing criteria in the Code EC8-1 in effect.

Of course, all other requirements referring to ductility, capacity design, etc. of the Code in effect continue to be included in the new version under elaboration.

3.5.3.3 Damage limitation state

This limit state protects the structure mainly from non-structural damage (in-fill walls, plasters, window glass, etc.) caused by frequent earthquakes. For these earthquakes (EQ-I, Section 3.5.2), the structure must present deformation (inter-storey drifts) that satisfies the deformation limits defined by the Code, so that the building is safe against unacceptable damage.

3.5.3.4 Specific measures

As it was noted at the beginning of this subsection, a series of specific measures is foreseen in all modern Codes for prevention of the collapse of the building in case of an unexpected, very rare and more severe earthquake than the design earthquake. These measures specified by EC8-1/2004 refer to

- Design
- Foundations
- Quality system plan

Particularly, for R/C buildings, the specific measures are imposed to cover

- General configuration of the building. The design measures specify rules and principles for regularity in plan and elevation, for ensuring a hierarchy in the loss of resistance of the structural members by means of *capacity design procedures*, so that premature failure may be avoided, and for taking into account soil deformability and adjacent structures in the formation of structural models.
- For the foundation, the Code imposes the basic principle for overstrength design of the foundation so that failure is limited to the super-structure.
- For the quality system plan, the Code specifies a series of obligations that should be taken into account during the preparation of drawings, technical reports and technical specifications for the design of the building.

- For resistance uncertainties in R/C buildings, the Code provides minimum dimensions and special detailing of the structural members.
- For ductility uncertainties in R/C buildings, minimum–maximum reinforcement limits are foreseen at the critical regions and limited normalised axial forces at the vertical R/C members (columns, structural walls).

In the chapters that follow, the implementation of the preceding principles presented in this section (Section 3.5) will be presented in detail.

Chapter 4

Configuration of earthquake-resistant R/C structural systems

Structural behaviour

4.1 GENERAL

One of the basic factors contributing to the proper seismic behaviour of a building is a rational conceptual design of the structural system in a way that lateral seismic actions (inertia-forces) are transferred to the ground without excessive rotations of the building and in a ductile manner. This cannot be achieved only through mandatory requirements of the Code. Therefore, there are also some general principles that can lead to the desirable result when they are followed. The guidelines that should govern a conceptual design against seismic hazard according to EC8-1/2004 are

- Structural simplicity
- Uniformity and symmetry
- Redundancy
- Bi-directional resistance and stiffness
- Torsional resistance and stiffness
- Diaphragmatic action at storey levels
- Adequate foundation

It should be mentioned that after a thorough examination of the 103 most badly damaged or collapsed R/C buildings in Athens after the earthquake of Parnitha (7 September 1999), it was found that 29 of these buildings had failed mainly due to their poor configuration (OASP, 2000). This was attributed mainly to the absence of collaboration between the architect and the structural engineer at the early stages of planning when a satisfactory compromise could have been reached. Of course, this is a usual situation only in small- and medium-sized buildings where there are not distinct stages of a preliminary design, predesign, and final design where this collaboration is consolidated.

It should also be remembered (see Chapter 3.4.2) that in case of big projects, geotechnical site investigations should be carried out and a geotechnical report should be prepared by an expert in soil mechanics to include all necessary information about

- The stratification profile
- The mechanical properties of the soil
- The ground type (see Chapter 3.4.2.2)
- The water table
- The presence of liquefiable soils
- The proposal on the type of foundation (shallow or deep)

It must be noted that even in case of small projects, it is necessary that soil investigation be carried out if there is no relevant information from buildings existing in the neighbourhood.

In the next section, a discussion will be presented on the guidelines stated above.

4.2 BASIC PRINCIPLES OF CONCEPTUAL DESIGN

4.2.1 Structural simplicity

The existence of simple structural systems with easily identified load paths for the transmission of gravity and seismic loads from the structural members to the foundation must be a basic objective of the conceptual design. It must be noted that in case of a simple structural system, the results of analysis and design are much more credible than those of a complicated one. In Figure 4.1, some simple structural systems are given in plan.

4.2.2 Structural regularity in plan and elevation

Buildings regular in plan and in elevation, without re-entrant corners and discontinuities in transferring the vertical loads to the ground, display good seismic behaviour. The presence of irregularities in plan leads to stress concentrations dangerous to the structure. In this case, if necessary, the entire building with re-entrant corners in plan may be subdivided by seismic joints into independent seismic compact parts (Figure 4.2).

Uniformity in elevation in mass and stiffness distribution is of essential importance for good seismic behaviour. Discontinuities in load transfer to the foundation with walls or columns 'planted on' beams and discontinued below, or discontinuities in deck diaphragms or building aspects with re-entrant corners, are bad signs for the behaviour of the building in case of a strong earthquake (Figures 4.4 and 4.5). Although the symmetrical arrangement

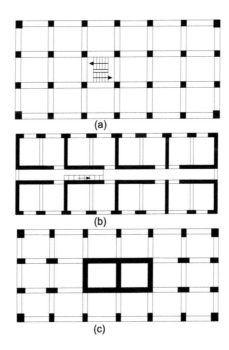

Figure 4.1 Structural systems characterised by simplicity: (a) a typical form of a frame system; (b) a typical configuration of an R/C shear wall system; (c) a dual system with an R/C core and frames.

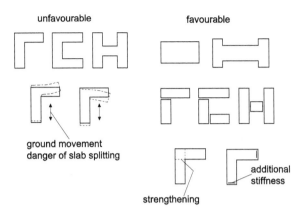

Figure 4.2 Unfavourable and favourable geometric configuration in plan.

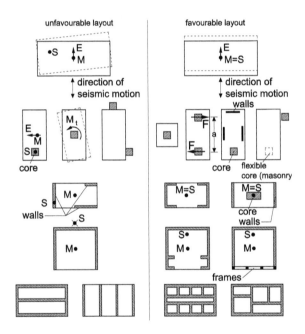

Figure 4.3 Distribution of mass and stiffness in plan.

of stiffness elements is not always possible due to architectural constraints, there should be a special concern in this direction so that torsionally flexible or asymmetric structures, which can cause failures to the corner columns and the walls at the perimeter, can be avoided (Figure 4.3; Baden Württemberg Innenministerium, 1985; see also Chapters 2.4 and 5.3).

4.2.3 Form of structural walls

In case R/C walls span voids between adjacent R/C columns, R/C structural walls should span the whole distance between them. In this way, the stiffness, strength and ductility of the structure are improved (Figure 4.6).

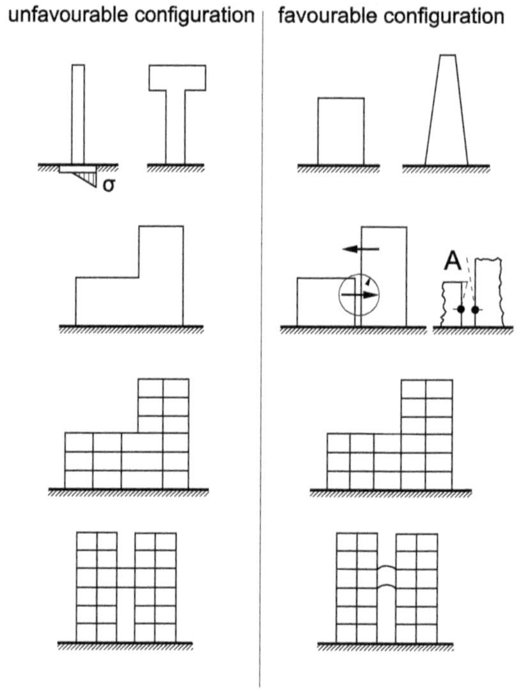

unfavourable configuration | favourable configuration

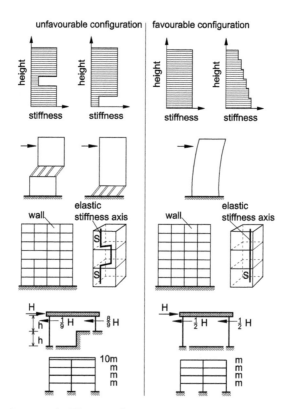

Figure 4.4 Unfavourable and favourable configuration in elevation.

unfavourable configuration | favourable configuration

Figure 4.5 Distribution of mass and stiffness in elevation.

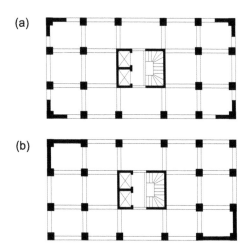

Figure 4.6 Layout of shear walls at the perimeter: (a) acceptable arrangement and (b) improved arrangement.

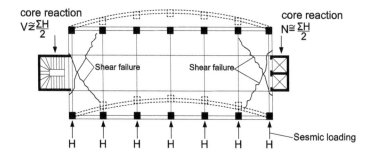

Figure 4.7 Unfavourable core arrangement; diaphragm at risk due to shear failure at the connections to the cores.

4.2.4 Structural redundancy

All the structural elements, including the foundation, should be well interconnected to build a monolithic, robust structure with high redundancy. High stiffness cores (staircases, shafts) lying in the perimeter of the building may be easily separated during an earthquake from the diaphragmatic system, leading the structure to unexpected response (Figure 4.7).

4.2.5 Avoidance of short columns

Short columns resulting from the presence of mezzanines or stiff masonry or R/C parapets below the windows should be avoided. If such arrangements cannot be avoided, their effect on the behaviour of the structure should be taken into account as far as the load effects, ductility and shear capacity are concerned (see Chapter 8.3.6; Figures 4.8 and 4.9).

4.2.6 Avoidance of using flat slab frames as main structural systems

Flat slab systems (Figure 4.10) without beams, although quite attractive in construction due to the low cost for formwork and the free space at storey for the arrangement of E/M ducts,

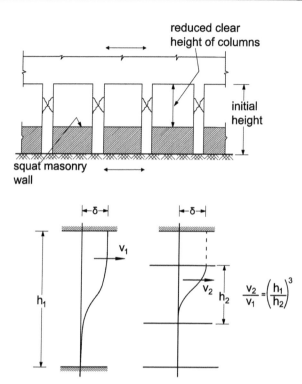

Figure 4.8 Concentration of large shear forces on short columns at the perimeter of the building.

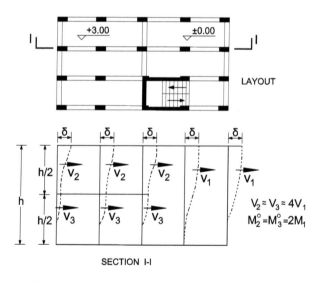

Figure 4.9 Concentration of large shear forces on short mezzanine columns.

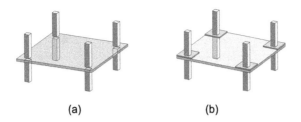

(a)   (b)

Figure 4.10 Flat slab systems; (a) slab directly on columns and (b) slab with drop panels.

should be avoided, as they are not covered completely by EC8-1/2004. *This does not mean that they cannot be combined with structural walls or cores* and frames capable of carrying the seismic actions (Figures 4.11 and 4.12).

4.2.7 Avoidance of a soft storey

Large discontinuities in the infill system in elevation (such as open-ground storeys) should be avoided (Figure 4.5). A stiffness discontinuity of this type generates a soft storey mechanism, which is very susceptible to collapse.

In case this type of structure cannot be avoided, as it happens in most Mediterranean countries where the General Building Code imposes an open storey at the ground level (Pilotis system), special measures should be taken in analysis and detailing of the structural walls and the columns.

4.2.8 Diaphragmatic behaviour

The system of the floors and roof of a multi-storey building constitutes the basic mechanism for transfer of inertial seismic forces from the slabs of the building, where the masses are distributed, to the vertical structural members (columns and structural walls) and thereby to the foundation. At the same time, the system of the slabs, particularly of cast in situ R/C buildings, ensures the behaviour of each storey deck as a rigid disc in plane, or in other words as a horizontal diaphragm, but one that is flexible in the vertical direction. In this way, the storey diaphragms contribute to increasing the system's redundancy. It is evident that the creation of this 3D structure with high redundancy can be generated very easily in case of R/C buildings cast in situ.

When an R/C building has a compact form in plan, it is obvious that there isn't any risk for structural failure of the diaphragms. However, when the structural system includes R/C cores of high stiffness at the limit of its perimeter (Figure 4.7) or in case of the existence of re-entrant corners in plan (Figure 4.2) or very large floor openings (Figures 4.13, 4.14 and 4.15), the diaphragmatic function may fail, and, therefore, special attention must be given to the analysis and design of the diaphragm itself (i.e. the analysis and design of the slab as a disc in-plane under the action of the inertial forces and the shear reactions of the vertical structural members on the disc).

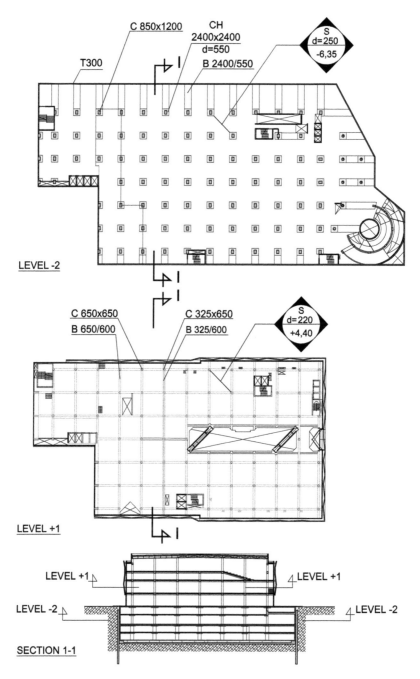

Figure 4.11 VRBANI, Zagreb mall; basements with flat slab frames and walls at the perimeter; frame system at the superstructure with limited shear walls.

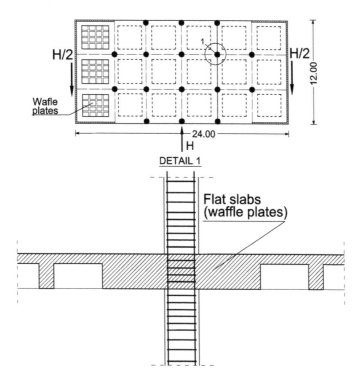

Figure 4.12 Flat slab frames combined with shear (ductile) walls.

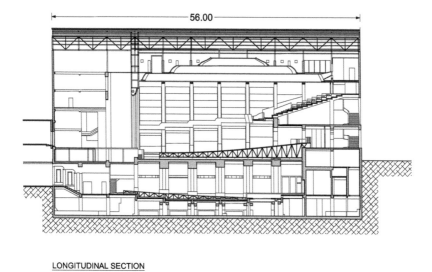

LONGITUDINAL SECTION

Figure 4.13 Palace – Aliki theatres, Athens, Greece; elevation.

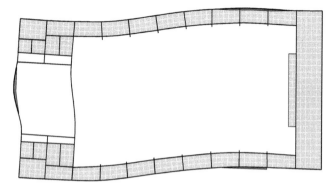

Figure 4.14 Palace – Aliki theatres, Athens, Greece third eigen mode of the structure at the level of balconies (FEM elastic modal spectrum analysis).

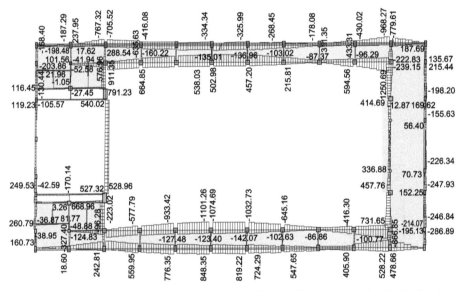

Figure 4.15 Palace – Aliki theatres, Athens, Greece. Axial forces of beams at the level of balconies due to seismic combination in the y–y direction.

4.2.9 Bi-directional resistance and stiffness

The structural elements should be arranged in an orthogonal in-plan structural pattern ensuring similar resistance, stiffness and ductility in both main directions (bi-directional function) since the seismic action may have any direction, and in this context the structure must be in a position to withstand any excitation with its two orthogonal components.

4.2.10 Strong columns–weak beams

Structures have to be composed of strong columns and weak beams for capacity design reasons. In Chapter 6, this recommendation will be discussed in detail.

4.2.11 Provision of a second line of defense

It is recommended to include in the structural system in parallel to shear walls a second line of defense formed by *ductile frames*. Thus, the dual system (structural walls combined with ductile frames) seems to be the most appropriate for resisting seismic action. ASCE 7-05 requires that, independently of the results of the analysis, 25% of the earthquake actions have to be carried by these frames. It should be noted that EC8-1/2004 does not impose such an obligation; instead, in case the frames resist for more than 35% of the base shear, the structural system is upgraded as far as its behaviour factor is concerned.

4.2.12 Adequate foundation system

The foundation plays a crucial role in the behaviour of the building in response to seismic actions. It should be noted that no matter what material is used for the superstructure, R/C is used almost exclusively for the foundation. The following recommendations should be kept in mind as far as the foundation is concerned:

- The site where the building will be constructed must be free of risks of soil rupture, slope instability and permanent settling caused by liquefaction or densification in the event of an earthquake (see also Chapters 3.4.2 and 10.2.3).
- In case of shallow foundations, the recommended system is a mat foundation or a grid of foundation beams or at least a grid of tie beams between the independent pads in case of firm soil or bedrock.
- In case of deep foundations (piles), the use of a foundation slab or tie beams between pile caps in both main directions is recommended.
- The existence of a basement with one or more underground storeys is a very good opportunity for the formation of an underground box structure. This box structure consists of a foundation mat or a grillage of foundation beams, a perimetric R/C wall system acting in parallel as a retaining wall for earth pressure, vertical structural members (columns, structural walls) in the basement space and the ground slab of the building. This box acts as a solid structure that safeguards the synchronous vibration of all vertical structural members at the level of the foundation (Figure 10.4), and in parallel it diminishes the overturning risk that exists in the case of eventual independent footings of the walls (Figure 4.16).

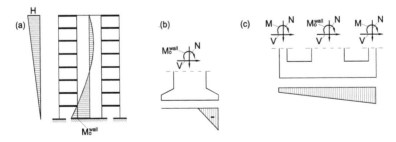

Figure 4.16 Foundation issues of shear (ductile) walls: (a) wall response in a dual system under lateral loading; its behaviour maybe simulated to a one end fixed cantilever supported by a spring at the other end; (b) wall foundation on a pad, overturning risk; (c) connections of the wall with the external columns with a foundation beam: diminishing of the risk of overturning due to enhancement of the total axial force and broadening of the foundation basis.

- It is evident that most of the recommendations for foundations so far have as their main objective making sure that all vertical elements have a synchronous excitation during a seismic motion, which is a basic design principle (see Chapter 2.4.4). Indeed, it should not be forgotten that the seismic actions reach the foundation in the form of waves (Figure 4.17). Thus, if footings are not well interconnected, each of them experiences an asynchronous vibration.
- Finally, it is recommended that all footings rest, if possible, on the same horizontal level (Figure 4.18).

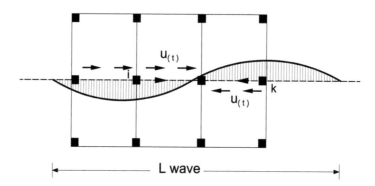

Figure 4.17 Relative displacement of the footings of columns *i* and *k* of a phase difference of the ground motion at points *i* and *k*.

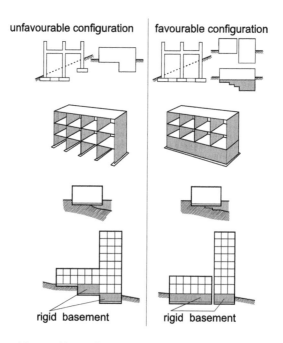

Figure 4.18 Unfavourable and favourable configuration of the foundation and the basement.

4.3 PRIMARY AND SECONDARY SEISMIC MEMBERS

EC8-1/2004 allows flexibility to the designer to characterise a number of structural members (beams and/or columns) as 'secondary' seismic members not forming part of the seismic action system of the building. This means that these elements will be introduced in the structural model as elements of zero bending stiffness. The reasoning behind this flexibility is that the designer might combine in the same building members that are not covered by the Code, like flat slab systems, together with others that are in accordance with the regulations, like ductile frames and/or walls. In this case, the ductile members constitute the *primary seismic system* while the secondary ones are used for carrying *only gravity loads*.

The Code specifies some restrictions in order to allow this procedure in analysis and design. These are the following:

1. The total contribution to lateral stiffness of all secondary seismic members should not exceed 15% of that of all primary seismic members. This can be checked by comparing the displacements of the centre of mass at the top storey for a horizontal loading in the two main orthogonal directions of the building, once taking into account the stiffness of all structural members, and another taking into account the stiffness of the primary members only. Particularly, in case of a flat slab system, a band of the slab (effective width) equal to the width of the columns plus two times the thickness of the slab should be taken into account (ACI 318 M-2011, EAK 2000) for the framing of the columns.
2. The design of the secondary R/C members should take into account, in addition to the action effects of gravity loads, the bending moments and shear forces that develop for the displacements at their ends (first- and second-order effects) due to the seismic action on the primary system.
3. It should be noted that the displacement of the primary systems resulting from the analysis for the design seismic action (see Chapter 3.4.3.7) should be increased by the behaviour factor q and by the P–Δ effect coefficient φ too (see Chapter 6.2.2.2). In this respect, the above requirement penalizes secondary members, as it requires them to remain *elastic* under the design seismic actions. This amounts to an over-strength factor in the range of q 'for secondary members', in contrast to the 'primary ones'.
4. Taking into account that the stiffness of the secondary members is not higher than 15% of the primary ones, and that the computer-aided analysis and design run very fast, it can be said that the above requirements for the secondary members can be fulfilled practically by analysing the system again with all its elements, primary and secondary ones. Next, the dimensioning of the secondary members proceeds for the combination of gravity loads and seismic actions, enhanced by q and φ but without any capacity considerations.
5. The dimensioning of the secondary seismic members is carried out as in non-seismic regions (e.g. EC2-1/2002), that is, without taking into account the regulations of EC8-1 for R/C primary seismic members.

4.4 STRUCTURAL R/C TYPES COVERED BY SEISMIC CODES

The structural system should preferably be composed of frames either alone or coupled with structural walls in two directions, so that a clearly defined flow of lateral forces is achieved. The structural walls in this case might be either independent plane members (discs) or combined plane members to form in plan L, T, C, Z sections or tubes. The structural systems covered by EC8-1/2004 should belong to one of the following structural types according to their behaviour under horizontal seismic actions:

- *Frame system*: A structural system in which both the vertical and lateral loads are mainly resisted by 3D ductile frames with a base shear that exceeds 65% of the total shear resistance of the whole structural system (Figure 4.1a).
- *Ductile wall system*: Structural systems in which vertical and lateral actions are mainly resisted by vertical structural ductile walls, either coupled or uncoupled, with a shear resistance at the building base that exceeds 65% of the total shear resistance of the whole structural system (Figure 4.1b). The percentage of shear resistance may be replaced approximately by the percentage of the shear action effects developed at the walls under lateral loading.
- *Dual system (frame or wall equivalent)*: A structural system in which support for the vertical loads is mainly provided by a 3D frame system and resistance to lateral loads is covered partly by the frame system and by coupled or uncoupled structural walls (Figure 4.1c). From the structural point of view, these systems must be classified in two different categories:

  Frame-equivalent dual systems, in which the shear resistance of the frame system at the building base is greater than 50% of the total shear resistance of the whole structural system

  Wall-equivalent dual systems, in which the shear resistance of the walls at the building base is higher than 50% of the total seismic resistance of the whole structural system

For the above classification, the percentage of the shear resistance may be replaced approximately by the percentage of the shear action effects developed at the walls under lateral loading.

The above three structural systems, which are

- Frame systems
- Wall systems
- Dual systems

are the main structural systems in use in earthquake-resistant R/C buildings. In Table 4.1, an overview is given of these systems together with their corresponding base shear resistance.

The quantitative classification of a structure in one of the above categories, according to the percentage of their shear resistance at the base of the building, in relation to the total shear resistance, is considered a safe index for the type of failure mode of the building that is likely to appear under a strong seismic action. Therefore, it is a good criterion for the evaluation of the 'behaviour factor' of the building, as we will see in Chapter 5.

Table 4.1 Main structural systems for earthquake resisting R/C buildings

Type of structural system	Percentage of base shear resisted by the system	
Frame system	>65%	Resisted by frames
Wall system	>65%	Resisted by walls
Dual system		
Frame equivalent	>50%	Resisted by frames
Wall equivalent	>50%	Resisted by walls

The simplification of the Code to allow the substitution of the 'shear capacity' of the vertical structural members at the base of the building by 'the demand' in estimating the percentage of shear participation of frames or walls *is a very useful approach for design practice.*

Indeed, for the determination of shear capacity of the structural members, the procedure of design must be integrated by dimensioning and detailing of the structural members. However, it is necessary to know the behaviour factor q at the beginning for the quantification of the design spectrum (see Chapter 3.4.3). This means that a time-consuming iterative procedure should be established until the correct q factor is determined. The introduction of the above 'simplification' allows the determination of the shear percentages of the structural elements in each main direction by applying a horizontal static loading at the rigidity centre of each storey (see Chapter 2, Sections 2.4.4 and 2.4.4.5) and determining the shear force at the base of the vertical structural elements. This can be done at the beginning after the preparation of the formwork drawings and the analytical model (geometry) of the structure before any analysis or design.

- *Torsionally flexible systems*: Another crucial point for the above structural systems is their *torsional rigidity*. According to the Code, there must be a distinction between 'torsionally flexible' and 'torsionally rigid' structural systems. Simply speaking, a 'torsionally flexible system' is a structural system wherein small eccentricities of the seismic horizontal forces cause large torsional deformations to the storey diaphragms and therefore excessive drifts at the perimeter columns of the system, disproportionate to those caused by the translational displacements (see Chapter 2.4.4; Figure 4.3).

This distinction will be discussed in detail in Chapter 5. For the time being, it must be noted only that in case the above systems fall into the category of the 'torsionally flexible', they continue to be accepted as systems covered by EC8-1/2004. However, a series of implications is activated. These implications have to do with the values of the q-factor of the system and its regularity in plan, which has an influence on the accepted methods of analysis for the system.

In addition to the four main structural systems presented above, two other less usual systems are covered by EC8-1/2004. These systems are the following:

- *System of large lightly reinforced walls*: This is a type of R/C building with the main structural system consisting of large R/C walls, which carry a large part of the gravity loads and the seismic action as well. According to EC8-1/2004, *a wall system* is classified as a system of large, lightly reinforced walls if in each of the two orthogonal directions it includes at least two walls with a horizontal dimension of not less than 4.00 m or $2h_w/3$ (h_w is the height of the wall), whichever is less. These walls in each main direction collectively must support at least 20% of the total gravity load, or, in other words, 40% of the gravity loads must be carried by walls in both orthogonal directions. Additionally, the fundamental period T_1 of the building, assuming fixed walls at the foundation, must be less than 0.5 s. These systems are considered to belong to 'wall family systems'. 'Behaviour factor' issues of this type of structural system will be discussed together with all other types in Chapter 5.

In case the system is torsionally flexible, it moves from the category of walls to the category of the torsionally flexible systems, like the main three systems described in the previous paragraphs.

- *Inverted pendulum system*: A system in which 50% or more of the mass is in the upper third of the height of the structure, such as with water or TV towers, is classified as an inverted pendulum system. In the same category, structural systems in which the dissipation of energy takes place mainly at the base of a single building element should be included, like a main shaft for elevators, stairs and E/M installations, combined with a few columns.

One-storey buildings – that is, buildings extended horizontally like industrial installations, covered stadia, cultural halls, auditoria and so on – *are excluded from the category of the inverted pendulum* if the normalised axial load v_d (see Chapter 8.3.4) does not exceed the value of 0.30.

It must be noted here again (see Section 4.2.6) that concrete buildings with *flat slab frames* used as primary seismic elements *are not covered* by EC8-1/2004.

4.5 STRUCTURAL CONFIGURATION OF MULTI-STOREY R/C BUILDINGS AND THEIR BEHAVIOUR TO EARTHQUAKE

4.5.1 General

The architectural form and the internal 'anatomy' of a building are governed by parameters of 'aesthetics' and 'functioning'. These parameters constitute for the owner the main criteria for his choice of the architectural design for his project, and, therefore, even of the team of designers to whom he will award the next steps of the project. Therefore, the structural system of the building must comply with the basic options for the above two parameters, and, for this reason, cooperation of the structural and MEP engineer with the architect of the project must be established from the early stages of design.

In this way, after mutual compromises, the requirements for

- The structural system
- The air conditioning
- The elevators for persons and goods
- The MEP storeys

will be incorporated in the final proposal for the project. Buildings may be classified into the following main categories:

1. *Multi-storey buildings.* In this category, the following types of buildings may be included: residential buildings, office buildings, malls, parking garages, hotels, etc. The majority of these buildings is constructed in the centre of cities and is subjected to numerous consequences because of their vicinity to existing buildings with foundations at various levels.
   The basic characteristics of these buildings may be summarised as follows:
   a. They usually have small (4.00–5.00 m) or medium size (7.00–8.00 m) spans and a high number of storeys.
   b. They have high gravity loads per plan unit at the foundation level due to the successive storeys.
   c. The centre of their mass is located higher as the number of storeys increases. The above three characteristics have a series of consequences for multi-storey buildings, given below.
   d. They present serious foundation problems due to the successive storeys and to their closeness to existing buildings in the case that the existing buildings have foundations at shallower level than the new ones.
   e. They need large column cross sections, particularly at the lower storeys. Therefore, a significant percentage of useful space on the ground floor and in basements is

lost. For this reason, in the last few years, high-quality concrete, up to C90, has been used in Europe for high-rise buildings.

f. The seismic horizontal loads are very high due to the large mass of the building. At the same time, the result of these forces is applied at a considerable height (at about two-thirds of the height of the building). Therefore, under this loading, the columns and the structural walls develop at their base high bending moments, shear forces, and antisymmetric axial forces (tension–compression), and, consequently, they are exposed to a high risk of overturning.

g. For buildings with increased number of storeys, the 'damage limitation state' from seismic action becomes critical for either the form of the structural system or its dimensioning.

2. *Buildings developed horizontally.* In this category, the following types of buildings may be included:

a. Industrial buildings, warehouses, malls, etc. These buildings are usually one-storey buildings with considerable storey height. Therefore, their gravity load per unit plan is limited. They usually have large spans (20.00–80.00 m) and are arranged in modular form. The above two characteristics have the following consequences for their structural behaviour.

b. Basically, the critical load combination is that of gravity loads. In the analysis and design of these buildings, the main concern is focused on girders (R/C or P/C beams, arches, shells, folded plates, etc.) under gravity loads.

c. The operational requirements for open plan spaces exclude almost completely the use of structural walls. So, the horizontal seismic forces are resisted usually by bending frames or cantilever columns in both directions.

d. Due to the small masses of the buildings and their low position in relation to the base of the building, the seismic action is not a critical issue for the design options. *It is often less important than wind loading.*

3. *Buildings for special use.* In this category, the following types of buildings may be included: stadiums, cultural centres, silos, water towers, water tanks, museums, etc. Each of these types of buildings exhibits special problems related to its basic function. Therefore, it is evident that the characteristics of these buildings cannot be classified easily into categories. Each of them is confronted by the structural engineer who is responsible for the design as a special case for which he mobilizes his knowledge, his experience and mainly his talent.

From what has been presented so far, it emerges that the most critical category for seismic design is that of multi-storey buildings, both in the stage of preliminary design and in the stage of the final design. Therefore, our interest will next be focused on the multi-storey building.

4.5.2 Historical overview of the development of R/C multi-storey buildings

Since its early steps in history, mankind has tried to express its magnificence by building mega-structures in height and extent (e.g. the Pyramids, the Tower of Babel, Hagia Sofia, Gothic cathedrals, etc.). The main materials that were used for centuries in the past were masonry and timber. These materials did not allow the design and construction of high-rise buildings with many storeys. At the same time, the operation of these buildings without the existence of elevators was out of the question for residential use.

The industrialisation of the nineteenth century and the subsequent explosion of the population of big cities have led to the extension of the cities upward. Steel, the main new industrial material of that period exhibiting high strength and ductility, has constituted the basic material for the structural system of multi-storey buildings and has dominated up to our days. At the same time, the development of elevators operated by electric power solved the problem of vertical transportation of humans and goods. Some buildings of the past with steel structural systems are mentioned below as examples (Fintel, 1974a; König and Liphardt, 1990):

- Home Insurance Building in Chicago, 1883: a 10-storey steel skeleton building
- Woolworth Building in Manhattan, 1913: a 60-storey steel skeleton building
- Chrysler Building in New York, 1920–1930: a 72-storey steel skeleton building
- The Empire State Building, New York, 1929: a 102-storey steel skeleton building

In many skyscrapers of this period, the foundations and the floor slabs were constructed of concrete (composite structures).

After World War I (1918), *multi-storey R/C buildings* from 10 to 12 storeys began to appear sporadically. The structural type in use was an imitation of the steel skeleton system, or in other words the traditional beam column frame system combined with R/C slabs, providing in this way a robust structure with diaphragmatic action at the level of the storeys (Figure 4.1a).

In the early 1950s, the introduction of *shear wall type of construction* extended the use of the R/C structural system to buildings of up to 30 storeys (Figure 4.1b).

In the same period, the combination of flat slab frames with shear wall structural systems, due to the low cost of construction and to the flexibility of decks without beams in passing E/M ducts, helped to spread the use of this R/C structural system to a large number of residential and office buildings of up to 20 storeys (Figures 4.11 and 4.12).

The development of the *frame tube* structural system in the 1960s, which is a perforated tube in the form of a moment-resisting frame at the perimeter of the building with closely spaced columns, allowed the extension of R/C buildings up to 40 to 50 storeys (e.g. the 30-storey CBS Building in New York, 1965; Figure 4.19).

The development of the *tube-in-tube system,* which consists of internal large cores of shear walls used as service shafts (staircases-lifts-air ducts) in combination with an external framed tube-type structure, increased resistance and stiffness and allowed the development of high-rise concrete buildings of up to 162 floors (Burj Dubai, 818 m high, 2009; Taranath, 2010; Figures 4.20 and 4.21).

The tube type and the tube-in-tube type structural systems fitted absolutely with the architectural style of the 1960s that *Mies van der Rohe* had imposed, which included high-rise buildings of prismatic form with an orthogonal, hexagonal or multicell plan (Figure 4.22). It can be said that these systems were the outcome of the close cooperation of architecture with structural engineering for confronting the aesthetic, operational and resistance problems of high-rise buildings.

Contemporary post-modern architecture after the Mies van der Rohe period has transformed the prismatic form of high-rise buildings into more plastic shapes. As the building is extended upward, its plan is transformed. Therefore, the tube-type structural systems can no longer fit post-modern plastic architectural forms. So, the external tube has been replaced by 'mega columns.' In Figure 4.23, the structural systems in use for R/C buildings with various numbers of storeys are depicted, while in Figure 4.24, the aspect of several of the world's tallest buildings is displayed.

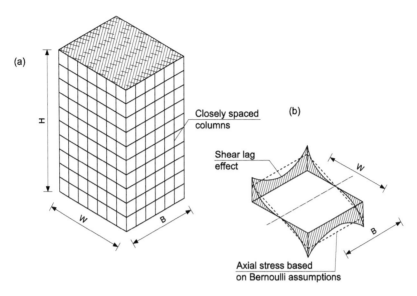

Figure 4.19 Framed system: (a) perspective view; (b) stress distribution under lateral loads. (Adapted from Elnashai, A.S. and Di Sarno, L. 2008. *Fundamentals of Earthquake Engineering*, Wiley, West Sussex, UK.)

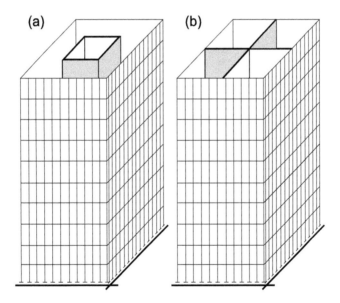

Figure 4.20 Dual system. (a) Tube-in-tube system; (b) external tube combined with shear walls. (König, G. and Liphardt, S.: *Hochhäuser aus Stahlbeton*. Beton Kalender, Teil II. 457–539. 1990. Copyright Wiley-VCH Verlag GmbH & Co. KGaA. Reproduced with permission.)

The development of reinforced concrete multi-storey building construction must be attributed mainly to the development of

- High-strength materials
- New design concepts
- New structural systems
- New construction methods

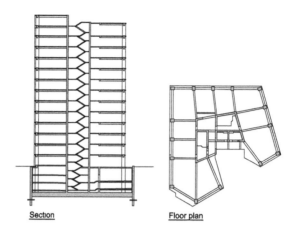

<u>Section</u>   <u>Floor plan</u>

Figure 4.21 Olympia tower, Bucharest, Romania; tube-in-tube system.

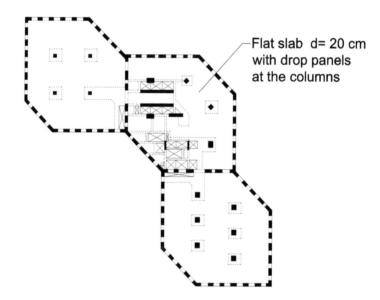

Flat slab d= 20 cm
with drop panels
at the columns

Figure 4.22 One Magnificent Mile; tube system with flat slabs and a hexagonal multicell floor plan. (König, G. and Liphardt, S.: *Hochhäuser aus Stahlbeton*, Beton Kalender, Teil II. 457–539. 1990. Copyright Wiley-VCH Verlag GmbH & Co. KGaA. Reproduced with permission.)

4.5.3 Structural systems and their response to earthquakes

4.5.3.1 General

The main structural systems in use in multi-storey buildings may be summarised as follows:

- Moment-resisting frames
- Shear wall systems
- Frame-wall or dual systems
- Flat slabs combined with shear walls and frames

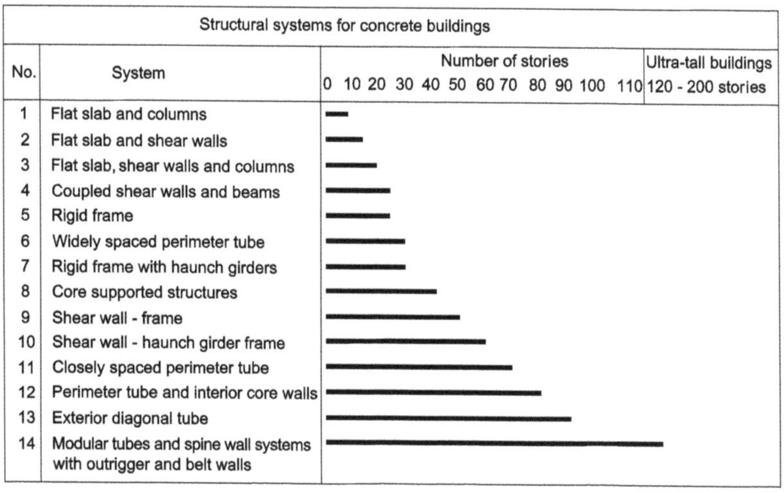

Structural systems for concrete buildings
No.
System
Number of stories
0 10 20 30 40 50 60 70 80 90 100 110
Ultra-tall buildings
120 - 200 stories
1 Flat slab and columns
2 Flat slab and shear walls
3 Flat slab, shear walls and columns
4 Coupled shear walls and beams
5 Rigid frame
6 Widely spaced perimeter tube
7 Rigid frame with haunch girders
8 Core supported structures
9 Shear wall - frame
10 Shear wall - haunch girder frame
11 Closely spaced perimeter tube
12 Perimeter tube and interior core walls
13 Exterior diagonal tube
14 Modular tubes and spine wall systems with outrigger and belt walls

Figure 4.23 Structural system categories. (Adapted from Taranath, B.S. *Reinforced Concrete Design of Tall Buildings*, CRC Press, Taylor & Francis Group, 2010.)

415 m/1362 ft
421 m/1380 ft
442 m/1451 ft
452 m/1483 ft
509 m/1671 ft
818 m/2684 ft
Hong Kong
Two international
Finance Centre
Shanghai
Jin Mao
Building
Chicago
Sears Tower
Kuala Lumpur
Petronas Towers
Taipei
Taipei 101
Dubai
Burj Dubai

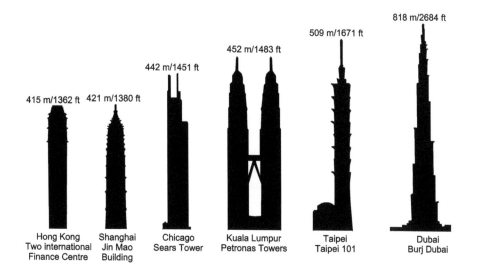

Figure 4.24 Comparative heights of some of the world's tallest buildings. (Adapted from Taranath, B.S. *Reinforced Concrete Design of Tall Buildings*, CRC Press, Taylor & Francis Group, 2010.)

- Tube systems
- Tube-in-tube systems
- Core-mega column systems

All the above systems are covered completely by EC8-1/2004.

From the discussion so far, it can easily be concluded that even a multimodal response spectrum analysis results in a series of inertial horizontal loadings, one for each mode (see Chapter 2.4.3). The final result is, in general, a load pattern of horizontal forces at the level of the storey decks parallel to the two main directions of the structure. Therefore, discussion on the behaviour of the structural systems in use for earthquake-resistant R/C buildings under horizontal loading is considered to be of major importance. In this respect, knowledge of the structural behaviour of pseudospatial systems, like R/C buildings, under horizontal loads turns out to be a useful tool for

- The conceptual design of the structure
- Qualitative evaluations of the computational output of the analysis of the seismic loading, no matter which method of analysis has been used

In the following, a short overview of the systems mentioned above and their response to lateral loading will be presented.

4.5.3.2 Buildings with moment-resisting frames

1. These systems consist of plane frames arranged in two orthogonal directions on a modular shaped plan (Figure 4.1a). The spans usually range from 4.00 to 8.00 m. Spanto-storey height ratio usually ranges from 1.50 to 2.50. Due to the gravity loads, larger spans result in greater beam depths and, therefore, in limited free storey height.
2. The system presents high redundancy, regularity, adequate torsional rigidity and structural ability to resist horizontal forces in any direction.
3. The values of moment, shear and axial force diagrams increase gradually from the top storeys to the fixed base (Figure 4.25). As a result, the demands for the cross sections of columns and beams increase at the lower storeys as the number of storeys increases.
4. The displacement at the top of the system results from the bending deformation of the beams, with the bending deformation of the columns, both constituting the shear racking of the rectangles formed by the beams and columns and the axial deformation of the columns (see Figures 4.25 and 4.26). Thus, the total displacement of a moment-resisting frame comprises
   a. Cantilever behaviour of the building (column axial forces): 15% to 20%
   b. Frame shear-type displacement (shear racking) due to bending of beams: 50% to 60%
   c. Frame shear-type displacement (shear racking) due to bending of columns: 20% to 30%

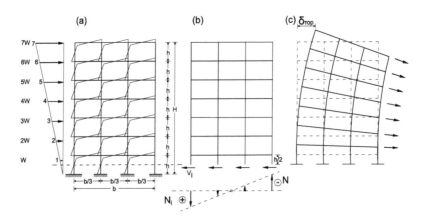

Figure 4.25 Frame system under lateral loading. (a) Moment diagram; (b) axial forces at the columns (opposite sign); (c) displacements due to axial strain of columns.

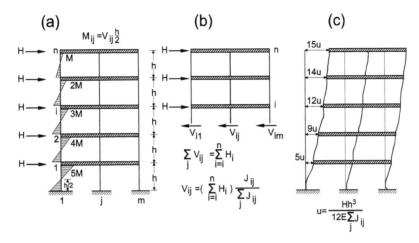

Figure 4.26 Action effects on shear frames under lateral loading; (a) M – diagram; (b) equilibrium between shear and lateral loads; (c) storey displacements.

5. Due to the low stiffness of moment-resisting frame systems, they present high inter-storey drifts. Therefore, after a number of storeys, the building design is governed by the 'damage limitation state'.
6. The high flexibility described above detunes the system from a short predominant period earthquake, particularly in the case of stiff soil.
7. For reasons discussed in the above item, as well as in item 5, this type of structural system is used in buildings of up to 20 storeys. Generally speaking, an acceptable aspect ratio (height to width) of these buildings is recommended to be less than 4.
8. Keeping in mind the M, V and N diagrams of a moment-resisting frame due to seismic action, it can easily be concluded that the critical regions that contribute mainly to seismic force resistance are the joints and a small region of the beams and columns joining there. Therefore, the strength and ductility of these regions are crucial for the seismic behaviour of the system, and special care should be taken for their design and detailing. Even so, if there is a feeling, due to the absence of skilled craftsmanship, that the construction of a frame system might be poor, this type of structure should be replaced by others (e.g. dual system or wall system), because in case of failure, soft storeys are formed with catastrophic consequences for the building, leading it many times to 'pancake' collapses (Booth and Key, 2006).

4.5.3.3 Buildings with wall systems

1. According to seismic codes in force, the vertical structural members with a ratio of cross-section dimensions greater than 4 are defined as 'walls'.
2. These members are characterised by their high stiffness in relation to columns with equivalent cross section. For example, the moment of inertia of a wall 0.20×2.5 m compared to that of a column 0.70×0.70 m with the same cross-section area is about 14 times more. Consequently, R/C walls constitute the most effective tool for the radical reduction of inter-storey drifts and the displacements of the storeys. At the same time, when walls and columns participate together in the resisting system of a building, walls with a total area equal to that of the columns undertake the main part of the base shear due to seismic motion. This justifies their definition as *structural walls*

or *shear walls* in various English-speaking countries. As already noted, these walls are characterised as *ductile walls* by EC 8-1/2004.

3. At the same time, shear walls appear to be more effective than columns in undertaking bending moments. In fact, it is known that the bending resistance of an R/C section results approximately from the relation

$$M_u = z f_{yd} A_s \cong 0.90 d f_{yd} A_s$$

(4.1)

where

-z is the lever arm of internal forces.
-d is the structural height of a section.
-A_s is the area of the reinforcement.
-f_{yd} is the design value of the yield strength of steel.

Therefore, given the same reinforcement, the ratio of bending resistance from the wall to the column, in the example above, is

$$\lambda = \frac{2.15}{0.65} \cong 3.30$$

(4.2)

In conclusion, it should be remembered that both walls and columns with equal cross-section area have almost the same shear capacity.

4. In this context, structural walls seem to respond very effectively to horizontal forces as far as strength and displacements are concerned. Of course, at first glance the above spectacularly better behaviour of walls in relation to columns is to a degree superficial, because it should not be forgotten that columns are not independent cantilevers in space; they are coupled with beams and slabs in frame structural systems. Columns with beams compose resisting frame systems with reduced bending moments at the columns compared to those of free cantilevers (shear system), due to their counter flexure deformation within the height of each storey. On the other hand, the walls respond like free cantilevers with coupled displacements at the levels of the floors (Figure 4.27).

So, the remarkable difference of behaviour between columns and walls observed above is reduced to a degree in conventional structures (Figure 4.28).

5. According to EC8-1/2004, the structural system of a building is characterised as a 'wall system' in case the walls carry at least 65% of the horizontal loading. Since the

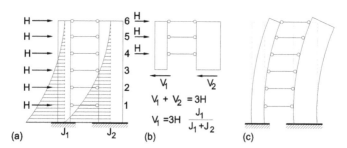

Figure 4.27 Action effects of shear (ductile) wall system under lateral loading. (a) M – diagram; (b) equilibrium between shear and lateral loads; (c) storey displacements.

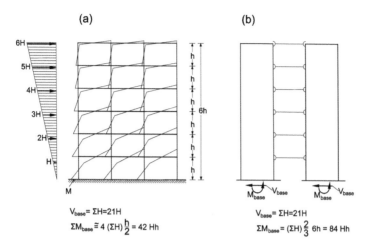

$$V_{base} = \Sigma H = 21H$$

$$\Sigma M_{base} \cong 4 \, (\Sigma H) \frac{h}{2} = 42 \, Hh$$

$$V_{base} = \Sigma H = 21H$$

$$\Sigma M_{base} = (\Sigma H) \frac{2}{3} \, 6h = 84 \, Hh$$

Figure 4.28 Comparison of the response of a frame to the response of a wall system under the same lateral loads. (a) Frame response; (b) wall system response.

walls are coupled at the level of the floors in relation to displacements only (diaphragmatic action), they respond like cantilevers. As a result, if these structural walls were reinforced adequately so that they failed first to bending and not to shear, this failure would happen at their base, where the bending moment caused by horizontal loading displays its maximum value. At this critical region, if the wall is adequately reinforced, plastic deformations may develop a plastic hinge. Consequently, the energy-dissipating zones in structural walls develop at their bases (Figure 9.14).

6. In case the coupling of the shear walls is strong enough with *spandrels*, these systems, according to EC8-1/2004, are classified as *coupled walls*, a fact that has serious consequences for their analysis and design, as it will become apparent in the following chapters (see Chapter 9.3). The bending moments at the base of these walls are reduced significantly due to the development of strong shear forces at their spandrels (Figure 4.29). At the same time, keeping in mind that the failure of the spandrels under horizontal loading precedes the failure at the base of the walls due to high shear, it may easily be concluded that these systems give an additional first line of defense to seismic action at their spandrels, where the failure precedes that of the plastic hinges at their bases. For this reason, EC8-1/2004 classifies coupled walls in a higher 'behaviour' category than independent walls (see Chapter 5.4.3).

7. In the category of wall systems, the systems of *large lightly reinforced walls* also belong, for which a first approach has been made in Section 4.4.

8. Of course, it should be noted that in order for a plastic hinge to develop at the base of a wall before its shear failure, the aspect ratio of the wall must be greater than 2 (see Chapter 9.4.4)

$$\frac{H}{l_w} \geq 2.0 \tag{4.3}$$

In case this relation is not fulfilled, as we will see in the next chapter, the 'behaviour factor' of the building, which is related directly to its structural ductility, is reduced drastically, according to EC8-1/2004.

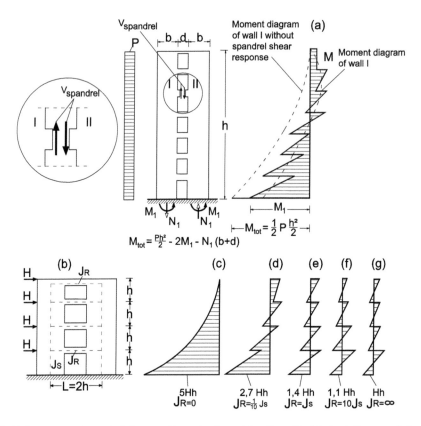

Figure 4.29 (a) Response of coupled shear walls under lateral loading; (b)–(g) the influence of the spandrel stiffness on the moment diagrams of the walls.

At the same time, in order for shear walls to fulfil their main objective, which is the reinforcement of buildings with high stiffness, according to design practice they must have an aspect ratio not higher than 7:

$$\boxed{\frac{H}{l_w} \leq 7.0}$$

(4.4)

From the two expressions above, it may be concluded that a reasonable length l_w of the wall must range between

$$\boxed{0.15H < l_w \leq 0.50H}$$

(4.5)

9. Wall systems are highly suitable for buildings up to *about 20 storeys*. For higher buildings, due to the quasicantilever behaviour of the structural system, the inter-storey drifts at the upper storeys of buildings begin to be critical at 'damage limitation state.' Therefore, the wall system must be assisted by moment-resisting frames coupled with the walls, which operate as retaining systems for the upper floors.

10. From what has been discussed so far, it can be concluded that the main advantages of wall systems may be summarised as follows:
    a. They provide high strength and stiffness at low cost.
    b. Their structural behaviour to past strong earthquakes has proven to be excellent. The creation of plastic hinges only at the base of the walls does not allow the formation of an inverted pendulum system as happens in frames, and therefore excludes collapses of the 'pancake' type that can flatten frames.
    c. The constructability of the walls is much better than that of frames where quality control might be easily lost at hundreds of joints, which are the critical regions of frames.
    d. They provide a very effective system for limited storey drifts in case of frequent small- or medium-sized earthquakes, and in this way they protect the non-structural elements that have a value much higher than that of the structural system.
11. At the same time, some disadvantages must be noted:
    a. The redundancy of the system is lower than that of a framed one.
    b. The walls present large moments at their base, which usually cause an uplift of independent footings even if the wall carries high axial loading. Therefore, their foundation must be well connected with the foundation of the rest of the structural elements, preferably in a box foundation system (Figures 4.16 and 10.4).
    c. Walls are, in general, undesirable elements in architectural design, since they put obstacles in the way of free communication from space to space at the floor level of the building.

4.5.3.4 Buildings with dual systems

1. In Sections 4.5.3.2 and 4.5.3.3, the characteristics of moment-resisting frame systems and of structural walls have been presented in detail. From the analysis so far, it can be concluded that the coupling of shear walls with frame systems can provide the building with the advantages of both and diminish their disadvantages. This combination of structural walls with moment-resisting frames is defined as a *dual system*.
2. In these systems, structural walls are usually arranged at the centre of the building around staircases, lifts, air ducts, etc., in the form of a core, while the frames are arranged at the perimeter of the building. So, a tube-in-tube system is generated. Not seldom, when the core is eccentric to the centre of mass of the building in plan, additional walls are arranged at the perimeter of the buildings so that the stiffness of the core is balanced. Beams are often arranged connecting the cores with the perimeter columns. Such an arrangement enhances the bearing capacity of slabs for the gravity loads and at the same time it increases the stiffness of the system by carrying a remarkable percentage of the overturning moment at the base of the building in the form of a pair of controversial axial forces acting at the columns of the perimeter. So, the displacements of the storeys due to seismic action diminish. However, it should be noted that the clear height of the storeys below the beam becomes shorter. Sometimes, in tube-in-tube buildings with more than 50 storeys, outrigger beams are arranged at some storeys, aiming at the diminishing of storey displacements, particularly in the 'damage limitation state'. This system (tube-in-tube with outriggers or tube-in-tube with outriggers and mega columns) constitutes the predominant one for buildings with a very great number of storeys (more than 100; Figure 4.30).

Outrigger beams   Perimeter frame

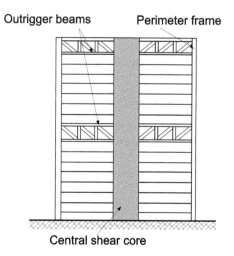

Central shear core

Figure 4.30 Outriggers in a shear core combined with perimeter frame (tube). (Adapted from Booth, E. and Key, D. 2006. *Earthquake Design Practice for Buildings*, Thomas Telford Ltd.)

3. In these systems, walls retain the frames at the lower storeys, while at the upper floors, the frames inhibit the large displacements and large inter-storey drifts of the walls (Figure 4.31).
4. The above response has the following consequences:
   a. For small intermediate and frequent earthquakes, the building exhibits limited inter-storey drifts, and consequently the damage limitation state is easily accomplished, even in very high buildings.
   b. In case of a strong earthquake, the first line of defence appears at the base of the walls in the form of plastic hinges, which begin to dissipate energy before the yield of the frames at their joints. Therefore, the risk for generation of a soft storey and for a 'pancake' collapse is drastically diminished.

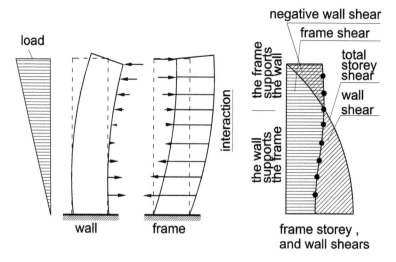

Figure 4.31 Interaction between frame and shear wall in a dual system under lateral loading.

    c. The moment-resisting frames constitute a second line of defence for a building, which is activated after the generation of the plastic hinges at the base of the walls. In this phase, the building has lost a part of its stiffness due to the plastic hinges of the walls and, therefore, is detuned with a subsequent diminishing of the inertial horizontal forces. ASCE 7-05 requires that the moment-resisting frames in dual systems should have a resisting capacity at least equal to 25% of the base shear of the system. EC8-1/2004 does not impose such an obligation. Instead, in case the frame resistance is higher than 35% of the base shear, the Code upgrades the behaviour factor of the system, which means that EC8-1/2004 gives incentives instead of imposing mandatory requirements.

5. The combination of moment-resisting frames with shear walls provides the building with a structural system of high redundancy.
6. In the case of outrigger beams, a special concern is necessary to exclude brittle failure of external columns due to axial overloading. In this context, the outriggers should be designed as ductile members that fail prior to column failure.

4.5.3.5 Buildings with flat slab frames, shear walls and moment-resisting frames

1. These systems constitute a very common and popular type of structural system for multi-storey buildings of up to 10 storeys. The structural system consists of flat two-way slabs on columns with or without column capitals, shear walls mainly at the core of the building and moment-resisting frames at the perimeter.
2. In this system, according to EC8-1/2004, the flat slab frames consisting of columns and slab strips in both directions are not considered suitable for earthquake loading, due to the low ductility of the column-slab joints. So, the flat slab frames in such a system should be considered as a secondary structural system, proper only for gravity loads (see Section 4.2.6; Figures 4.11 and 4.12).
3. These systems are very attractive in construction due to
    a. The simple and low-cost formwork
    b. The larger clear storey height
    c. The free passage for the E/M installations

4.5.3.6 Buildings with tube systems

1. In Section 4.5.3.2, it has already been made very clear that in the case of high-rise buildings, the beams of the bending resisting frame systems need to be deep enough so that the beams at the lower storeys are in position to carry moments and shears due to an earthquake. However, if the internal columns are omitted or if a flat slab system is arranged in the tube, the columns of the frames at the perimeter of the building may be arranged as closely as necessary, while the beams could take the form of deep spandrels. In this respect, the tube at the perimeter can be a stiff-framed system capable of carrying the lateral seismic loading.
2. In practice, these systems are used in buildings of up to 50 to 55 storeys.
3. The columns are spaced close enough (1.50–3.0 m), while spandrels have a depth of about 0.60 m.
4. It is important to note that this structural system has the behaviour of a perforated tube in response to horizontal actions. When the holes are circular, small and limited in number, it behaves like a closed tube and, if the Bernoulli concept holds (aspect ratio H/l greater than 2), the bending moment at the base is carried by normal forces with

almost linear distribution. If those holes take a rectangular form and their number increases, a large part of the horizontal forces is transferred to the base by the shear behaviour of the frames parallel to the load direction, that is, through 'shear lag effect' (Elnashai and Di Sarno, 2008; Figure 4.19).

5. The tubes are often hexagonal or multicell (Figure 4.22).
6. When the system described above is combined with a core in its centre, it is transformed to a *tube-in-tube system*, proper for very high-rise buildings (over 100 storeys), especially if it is combined with *outrigger beams* every after a certain number of storeys. From the structural point of view, as we have seen before, this system is also characterised as a dual system.

Chapter 5

Analysis of the structural system

5.1 GENERAL

For the analysis of a structural system for seismic actions, a set of properties of the system must be taken into account in advance, since according to EC8-1/2004 these properties influence

- Design actions
- Structural model
- Method of analysis
- Capacity design of columns

These properties are

- The structural regularity of the building
- Its torsional flexibility
- The ductility level of the structure

Most modern Codes for the above properties specify quantitative criteria with which the analysis must comply. The above issues will be examined in detail in the subsequent sections.

5.2 STRUCTURAL REGULARITY

5.2.1 Introduction

For many years up to 1988, the regularity of a building was a qualitative parameter that had to be taken into account in the preliminary structural design and was given as a recommendation of the Code. Since then, for the purpose of design, building structures in all modern Codes are separated into two categories:

- Regular buildings
- Non-regular buildings

Table 5.1 Consequences of structural regularity on seismic analysis

	Regularity		Allowed simplification		Behaviour factor
Plan	Elevation	Model	Linear-elastic analysis		For linear analysis
Yes	Yes	Planar	Lateral force		Reference value
Yes	No	Planar	Modal		Decreased value
No	Yes	Spatial	Lateral force		Reference value
No	No	Spatial	Modal		Decreased value

This distinction has implications for the structural model, the method of analysis and the value of the behaviour factor q, which is decreased for buildings that are non-regular in elevation. More particularly, according to EC8-1/2004

- The structural model can be either a simplified plane or a spatial one.
- The method of analysis can be either a simplified lateral force or a modal analysis.
- The reference value of the behaviour factor q_o given in Table 5.2 is decreased for buildings that are non-regular in elevation. Table 5.1 describes the implications for structural regularity on the design according to EC8-1/2004.

5.2.2 Criteria for regularity in plan

Buildings regular in plan must fulfil all the following requirements:

- The structural system of the building with respect to lateral stiffness and mass distribution must be approximately symmetric in plan in two orthogonal directions. It is obvious that in this case the *centre of stiffness* is very near to the *centre of mass*.
- The plan configuration must be compact, with re-entrant corners not affecting the area of the convex envelope of the floor more than 5% (Figure 5.1).

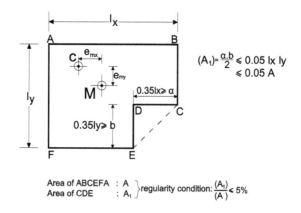

Area of ABCEFA : A ⎫ regularity condition: $\frac{(A_1)}{(A)} < 5\%$
Area of CDE : A_1 ⎭

Figure 5.1 Condition of regularity at the re-entrant corner.

- The floor must provide an efficient diaphragmatic action effect, that is, large in-plane stiffness compared to the stiffness of the vertical members.
- The slenderness

$$\lambda = L_{\max}/L_{\min}$$

of the building in plan must not be higher than 4, where $L_{\max}$ and $L_{\min}$ are the orthogonal dimensions in plan of the building.

At each level and for each direction x and y, the structural eccentricity e_m and the torsional radius r_c must be in accordance with the two conditions below:

$$\boxed{\begin{aligned} e_{mx} &\leq 0.30 r_{xc} \\ e_{my} &\leq 0.30 r_{yc} \end{aligned}}$$

(5.1a)

and

$$\boxed{\begin{aligned} r_{xc} &\geq l_{ms} \\ r_{yc} &\geq l_{ms} \end{aligned}}$$

(5.1b)

where

e_{mx}/e_{my} is the distance between the *centre of stiffness C* and the *centre of mass M* measured along the x/y direction, which is normal to the direction of analysis under consideration (Figure 5.1).

r_{xc}/r_{yc} is the *torsional radius* with respect to the centre of stiffness, which is the square root of the ratio of torsional stiffness with respect to the centre of stiffness to the lateral stiffness in the y/x direction, respectively.

l_{ms} is *the radius of gyration of the floor mass* in plan with respect to the *centre of mass M*.

The definition of the above parameters e_{mx}, e_{my}, r_{cx}, r_{cy}, l_{ms} together with r_{mx} and r_{my} has been given in Chapter 2.4.4. At the same time, in the same subsection, their structural meaning has been clarified in detail together with various simplifications acceptable to EC8-1/2004 for the calculation of their values. For the sake of convenience, the results of the analysis in the above subsection are also summarised below:

1. In the case of pseudospatial systems consisting only of *frames* or only of *walls*, the centres of stiffness C for all storeys lie approximately on a nearly perpendicular axis passing from a point at the plan of the first floor with coordinates with respect to the centre of mass given by Equation 2.86:

$$e_{mx} = \frac{\sum_{j=1}^{j=m} x_j J_{jx}}{\sum_{j=1}^{j=m} J_{jx}}, \quad e_{my} = \frac{\sum_{i=1}^{j=n} y_i J_{iy}}{\sum_{i=1}^{j=n} J_{iy}}$$

2. *Torsional stiffness* with respect to the *centre of stiffness* C is given by Equation 2.87:

$$\bar{J}_{TC} = \sum_{j=1}^{j=m} J_{jx} x_{jc}^2 + \sum_{i=1}^{j=m} J_{iy} y_{ic}^2$$

3. *Torsional stiffness* $\bar{J}_{TM}$ with respect to the *centre of mass* M is given by Equation 2.88:

$$\bar{J}_{TM} = \bar{J}_{TC} + e_{mx}^2 \left(\sum_{j=1}^{m} J_{jx} \right) + e_{my}^2 \left(\sum_{i=1}^{n} J_{iy} \right)$$

4. *Torsional radius* with respect to the *centre of stiffness* is given by Equation 2.90:

$$r_{xc} = \sqrt{\frac{\bar{J}_{TC}}{\sum_{j=1}^{m} J_{jx}}}$$

5. *Torsional radius* with respect to the *centre of mass* is given by Equation 2.91:

$$r_{xm} = \sqrt{\frac{\bar{J}_{TM}}{\sum_{j=1}^{m} J_{jx}}}$$

6. *Radius of gyration* with respect to the *centre of mass* (orthogonal plan) is given by Equation 2.113:

$$l_{sm} = \sqrt{\frac{J_d}{m}} = \sqrt{\frac{b^2 + d^2}{12}}$$

7. In the case of a pseudospatial system consisting of *dual systems*, the methodology of *plasmatic axis* of centres of stiffness may be followed (Chapter 2.4.4.5).

5.2.3 Criteria for regularity in elevation

Buildings regular in elevation must fulfil the following requirements:

- All lateral resisting systems must run without interruption from their foundation to the top of the building.
- In *framed buildings*, the ratio of the actual storey capacity to the demand required by the analysis should not vary disproportionately between adjacent storeys. Special concern must be given to masonry infilled frames (see Chapter 8.5.2.2).
- Both the lateral stiffness and the mass of the individual storeys must remain constant or diminish gradually from the base to the top of the building.
- Special consideration must be given in the case that setbacks are present.

Table 5.2 Basic values for the determination of q-factors

	Structural type	K_w	K_R^{el}		K_R^{over}		QSP		q_o	
			YES	NO	YES	NO	YES	NO	DCM	DCH
1	Frame system	1	1	0.8	$\dfrac{a_u}{a_1}$	$\dfrac{1+(a_u/a_1)}{2}$	1.2	1	3.0	4.5
2	Frame-equivalent dual system	1	1	0.8	$\dfrac{a_u}{a_1}$	$\dfrac{1+(a_u/a_1)}{2}$	1.2	1	3.0	4.5
3	Wall-equivalent dual system	$0.5<(1+a_o)/3\le1$	1	0.8	$\dfrac{a_u}{a_1}$	$\dfrac{1+(a_u/a_1)}{2}$	1.2	1	3.0	4.5
4	Coupled wall system	$0.5<(1+a_o)/3\le1$	1	0.8	$\dfrac{a_u}{a_1}$	$\dfrac{1+(a_u/a_1)}{2}$	1.2	1	3.0	4.5
5	Uncoupled wall system	$0.5<(1+a_o)/3\le1$	1	0.8	1	1	1.2	1	3.0	$4.0(a_u/a_1)$ or 4.0 $\dfrac{1+(a_u/a_1)}{2}$
6	Large lightly reinforced walls	$0.5<(1+a_o)/3\le1$	1	0.8	1	1	1.2	1	3.0	–
7	Torsionally flexible system	$0.5<(1+a_o)/3\le1$	1	0.8	1	1	1.2	1	2.0	3.0
8	Inverted pendulum	1	1	1.0	1	1	1.2	1	1.5	2.0

Note: $q = Kw \cdot K_R^{el} \cdot K_R^{over} \cdot K_{QSP} \cdot q_o$

In the case that any one of the above requirements is not fulfilled, the structural system is considered non-regular in elevation, and therefore the reference values of behaviour factors given in Table 5.2 *are decreased to 0.80 of their value.*

5.2.4 Conclusions

1. The criteria for regularity in plan are mainly qualitative and can be checked very easily at the beginning of the analytical procedure. Only the last one, the verification of Equations 5.1a and 5.1b, needs complicated calculations. However, acceptance by EC8-1/2004 of simplified procedures for the determination of the parameters included in Equations 5.1a and 5.1b, as was explained in Section 5.2.2, makes the whole procedure a little bit easier, and allows the decision making for regularity in plan at the beginning of the analysis.
2. Having in mind that non-regularity in plan does not have any computational implications for the structure except that of not allowing the introduction in the analysis of planar models together with some consequences of secondary importance for behaviour factor q_o (see Section 5.5), and that modern computational means have devalued the Code permit of using planar models and static seismic loadings, it seems that the requirements of Equations 5.1a and 5.1b are no longer of importance for the successive steps of analysis. Indeed, if the structural system is modelled as a pseudospatial system and the modal response spectrum analysis is used, which is the common practice

nowadays, the regularity criteria in plan and elevation become insignificant for the analytical procedure.

3. The only serious consequence for the successive steps of analysis is non-regularity in elevation, which imposes a decrease in the reference factor to 0.8 of its value. The check of regularity in elevation can be carried out very easily, without complicated calculations, at the preliminary stage of design.

4. The above requirement for decrease in the q_o factor is due to increased ductility demands at the level of soft storeys. Indeed, a standard modal response spectrum elastic analysis does not reveal the local ductility demands in case of a soft storey. The decrease in the value of the q_o factor, together with special measures taken at the areas of soft storey (capacity design, etc.), offer sufficient resistance in the critical areas (e.g. column joints) of the system.

5.3 TORSIONAL FLEXIBILITY

As was explained in Chapter 4.4, a 'torsionally flexible system' is a structural system where small eccentricities of the seismic horizontal forces cause large rotational deformations to the storey diaphragms and, therefore, excessive drifts at the columns of the perimeter, disproportional to those caused by the relevant translational displacements. This behaviour under special circumstances is amplified in case of dynamic excitation of the system, causing in many cases uncontrollable storey drifts. The above qualitative approach to the problem will be quantified in the next paragraph.

According to EC8-1/2004, a structural system is torsionally flexible if (see Section 5.2.2)

$$\boxed{\begin{aligned} r_{xc} &\leq l_{ms} \\ r_{yc} &\leq l_{ms} \end{aligned}}$$

(5.1c)

From the analysis of Chapter 2.4.4.4, it may be concluded (Equation 2.109) that *in the case of a double symmetric system*, the fulfilment of Equation 5.1c means that the three *uncoupled* eigen frequencies of a one-storey pseudospatial system are arranged as follows:

$$\boxed{\omega_{ux}, \omega_{ux} > \omega_{\varphi}}$$

(5.2)

that is, the greater eigen frequencies are translational ones and consequently the vibration of the system is dominated by the rotational vibration. In the above relation (Equation 5.1c), the right-side term expresses the lever arm of the inertial forces due to an eventual dynamic (seismic) excitation, while the term on the left side, r_{xc} or r_{yc} expresses the lever arm of the restoring forces. So, in the case that r_{xc} and r_{yc} are smaller than l_{ms} at every moment, the developing restoring forces are smaller than the inertial ones and the torsional mode prevails.

The above physical interpretation is expressed quantitatively by the fact that the first eigen mode is rotational, and in this respect, the greater part of the mass of the system is activated in a rotational motion.

Conversely, if

$$r_{xc} \text{ and } r_{yc} \geq l_{ms} \tag{5.3}$$

the translational excitations in the x and y direction are the predominant ones and therefore only limited drifts due to rotational vibration may develop.

In the case of a non-symmetric system, the fulfilment of Equation 5.1c as a criterion for torsional rigidity lies on the safe side. Indeed, from the analysis of Chapter 2.4.4.4 (Equation 2.122), it may be concluded that in the case of a non-symmetric system, Equations 5.1c are transformed to

$$\boxed{\begin{aligned} r_{xm} \leq l_{ms} \\ r_{ym} \leq l_{ms} \end{aligned}} \tag{5.1d}$$

where
r_{xm} and r_{ym} are given by Equation 2.91:

$$r_{xm} = \sqrt{\frac{\bar{J}_{TM}}{\sum_{j=1}^{m} J_{jx}}}$$

Since $\bar{J}_{TM} > \bar{J}_{TC}$ (see Equations 2.87 and 2.88), it is obvious that in the case that the centre of stiffness does not coincide with the centre of mass, the criterion for flexural flexibility established by Equation 5.1c is on the safe side.

In the case of a system with torsional flexibility, EC8-1/2004 specifies a considerable reduction in the values of the q_o-factors (Table 5.2) ranging from 0.66 to 0.75. At the same time, the structural system is not qualified for any release from joint capacity design, no matter if it is a wall system (more than 65% wall shear capacity) or a wall-equivalent dual system (more than 50% wall shear capacity).

The degree of protection offered to structures that abide by the criteria for structural regularity and torsional rigidity of EC8-1/2004, as expressed by Equations 5.1a and 5.1b, is clarified in Figure 5.2 as developed by Cosenza et al. (2000). In this figure, the behaviour of various one-storey pseudospatial systems designed according to various Codes and re-analysed as inelastic systems under dynamic excitation is presented by markers of various designation. Systems of good and satisfactory inelastic behaviour are designated as white and grey (increase of 10–20% in ductility demands), while systems of poor inelastic behaviour are designated as black (ductility demands greater than 50% compared to their torsionally balanced system). On this plot, the requirements for regularity and torsional rigidity of EC8-1/2004 are also plotted, that is, Equations 5.1a and 5.1b. Area ABCD corresponds to the area where the regularity criteria are fulfilled, while area AEFD corresponds to the area where the torsional rigidity criteria are fulfilled. From this figure, it can easily be concluded that most of the case studies that fulfil the above-mentioned criteria present a good to satisfactory inelastic behaviour. The same conclusion can be drawn from Figure 5.3, which corresponds to multi-storey pseudospatial structural systems.

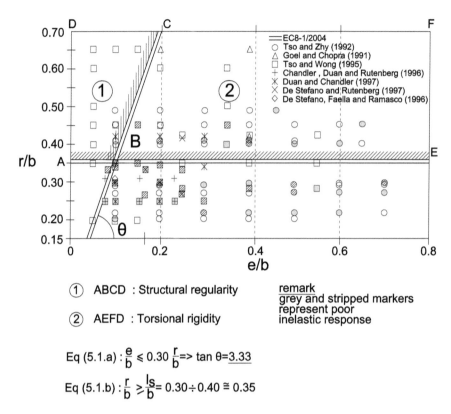

(1) ABCD : Structural regularity

(2) AEFD : Torsional rigidity

remark
grey and stripped markers
represent poor
inelastic response

Eq (5.1.a) : $\dfrac{e}{b} \leqslant 0.30\ \dfrac{r}{b}$ => tan θ=$\underline{3.33}$

Eq (5.1.b) : $\dfrac{r}{b} \geqslant \dfrac{l_s}{b}$= 0.30÷0.40 ≅ 0.35

Figure 5.2 Evaluation of the structural regularity and torsional rigidity criteria in a one-storey structure. (Adapted from Cosenza, E., Munfredi, G. and Realfonzo, R. 2000. Torsional effects and regularity conditions in R/C buildings. *Proceedings of the 12th World Conference on Earthquake* Engineering, Auckland.)

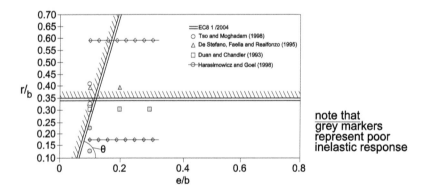

note that
grey markers
represent poor
inelastic response

Figure 5.3 Results for regularity and torsional rigidity of multi-storey building models. (Adapted from Cosenza, E., Munfredi, G. and Realfonzo, R. 2000. Torsional effects and regularity conditions in R/C buildings. *Proceedings of the 12th World Conference on Earthquake Engineering*, Auckland.)

From the discussion above, it may be concluded that the torsional flexibility of the structure must be determined in advance, at the beginning of the analysis procedure, so that a decision can be made on *q-factor* and *capacity* design issues from the beginning. For this decision to be made, the following procedure can be followed:

1. In the case of a frame system or wall system for the definition of e_{mx}, e_{my}, $\bar{J}_{TC}$, r_{cx}, r_{yc}, l_{ms}, Equations 2.86, 2.87, 2.88 and 2.113 may be used.
2. In the case of a dual system, the method of the 'plasmatic axis' of centres of rigidity may be followed (Chapter 2.4.4.5).
3. If the lower few eigen modes are determined together with their poles of rotation on each storey in case of a response spectrum 3-D analysis, these eigen modes may be used directly for checking the torsional rigidity. In fact, if the pole of the shortest eigen frequency lies in the boundaries of the floor, at any floor of the system, this system is 'torsionally flexible' (see Chapter 2.4.4.3), with all consequences resulting therein. It should be noted here that this procedure precedes the analysis of the system. In fact, after the elaboration of the analytical model and geometric, material and mass properties, the eigen modal analysis is executed, and thereof the decision on torsional flexibility may be taken.

5.4 DUCTILITY CLASSES AND BEHAVIOUR FACTORS

5.4.1 General

As was explained in Chapters 2.3.4, 3.2.1 and 3.4.3.7, the action effects (internal forces) due to seismic actions are defined in the force-based design method by taking into account that the structural system is in a position to dissipate seismic energy. Therefore, seismic actions are reduced by a factor q, which was called there the 'behaviour factor'. This factor is directly related to the ductility demand of the structure (Chapter 3.2). The reduced load effects resulting in this way from the analysis, *taking into account the behaviour factor* that has been introduced in the seismic actions, constitute the *demand* for the structural type under consideration and its structural members.

This *demand* must be covered by the capacity of the structure as a whole and of its structural members in strength and ductility (*supply*). Strength requirements are considered to be fulfilled if for all critical regions of the structural members the following relation is verified:

$$\boxed{E_d \leq R_d} \tag{5.4}$$

where
  E_d is the action effect due to reduced seismic action combined with gravity loads.
  R_d is the strength calculated by applying R/C mechanics supplemented by some additional rules, which will be presented in the next chapters.

On the other hand, ductility requirements are covered mainly by a series of rules specified by the Code in each case and in special cases by local ductility analytical verifications also specified by the Code. These rules and computational procedures for local ductility verifications will also be examined in detail in subsequent chapters.

5.4.2 Ductility classes

EC8-1/2004 classifies concrete structures into three ductility classes:

1. *Ductility class 'L'* (DCL – low ductility): corresponds to structures designed according to EC2-1-1/2004 (Eurocode for R/C structures) supplemented by rules enhancing available ductility. For all types of structures of this class, the specified value of the q-factor is

$$\boxed{q = 1.5}$$ (5.5)

2. *Ductility class 'M'* (DCM – medium ductility): corresponds to structures designed, dimensioned and detailed according to specific earthquake-resistance provisions, enabling the structure to enter well into the inelastic range under repeated reversal loading without suffering considerable loss of strength or brittle failures endangering the local or overall stability of the structure.
3. *Ductility class 'H'* (DCH – high ductility): corresponds to structures for which the design, dimensioning and detailing provisions are such as to ensure, in response to the seismic excitation, the development of a stable mechanism associated with large hysteretic energy dissipation, which has been chosen in advance and is under control during the excitation (capacity design).

In principle, the designer might choose any of the above three ductility classes for an earthquake-resistant structural system, since there is a trade-off between design seismic loading (relevant q-factor) and required ductility (μ_D) (Figure 5.4). However, modern Codes (e.g. EC8-1/2004 ASCE 7-05), taking into account the fact that it is easier to attain high ductility than high strength, put restrictions on the choice of ductility class, relating this choice to the seismicity of the region where the building is located. So, EC8-1/2004 allows

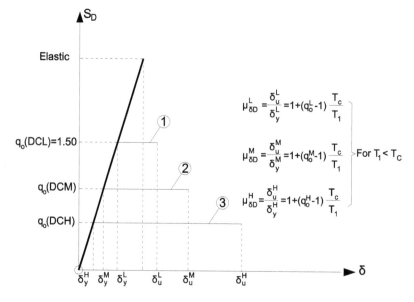

Figure 5.4 Relation between design force and ductility demand at the plateau of the design spectrum (see Chapter 3.2) for fundamental period $T_1 < T_c$ (Equation 5.17).

the design according to DCL class *only* for buildings in regions of *low seismicity*. The selection of the categories of structure, ground types and seismic zones in a country for which provisions of low seismicity apply is found in the National Annex of each country. EC8-1/2004 recommends considering as low seismicity cases either those in which the design ground acceleration of type A soil, a_g, is not greater than 0.08 g, or those where the product a_gS is not greater than 0.1 g. It can easily be concluded that buildings of DCL are designed without any particular provision for energy dissipation and ductility. Indeed, these buildings are designed according to EC2-1/2004 for load safety factors in earthquakes equal to 1.0, since earthquake action is considered accidental loading. The introduction *of behaviour factor q = 1.5 is attributable to the overstrength of the structure due to material (steel-concrete) overstrength, structural redundancy, minimum reinforcement requirements and so on (see Section 5.7.3)*.

On the other hand, in the case of *very low seismicity*, EC8-1/2004 exempts the structures from any obligation for earthquake design. The cases of very low seismicity are also defined by the National Annex of each country. EC8-1/2004 recommends considering as very low seismicity cases either those in which the design ground acceleration on type A soil, a_g, is not greater than 0.04 g or those where the product a_gS is not greater than 0.05 g.

Coming now to the other two ductility classes, DCM and DCH, it should be noted that according to EC8-1/2004, these two classes must be considered equivalent. However, Eurocode gives the option to country members of the European Union to specify geographical limitations on the use of ductility classes M and H in their relevant National Annexes.

5.4.3 Behaviour factors for horizontal seismic actions

The upper-limit value of the behaviour factor q introduced (see Chapter 3.2.3) to account for energy dissipation capacity must be derived for each design direction, according to EC8-1/2004, as the result of a multi-functional relationship of the following form:

$$\boxed{q = K_w \cdot K_R^{el} \cdot K_R^{over} \cdot K_Q \cdot q_o \geq 1.50} \tag{5.6}$$

where

q_o is the basic value of the behaviour factor, dependent on the type of the structural system, related with its redundancy, its ability to dissipate energy, the number of regions where energy can be dissipated and so on.

K_R^{over} is the factor reflecting the *overstrength and regularity in plan* of the building.

K_w is the factor reflecting the prevailing failure mode in structural systems with walls.

K_R^{el} is the factor reflecting the *regularity* or *irregularity* of the building in *elevation*.

K_Q is the factor reflecting the application or not of a Quality System Plan to the design, procurement and construction.

The values of the above factors are summarised in Table 5.2 and are examined in detail in the following items.

1. For buildings irregular in elevation, factor K_{reg}^{el} is reduced to 0.80, while for buildings regular in elevation, its value is 1.0.
2. The factor K_w reflects the prevailing failure mode in structural systems with walls. If the failure mode is of flexural ductile type with formation of plastic hinges in critical regions,

$$K_w = 1.0 \tag{5.7a}$$

This is the case of frames, frame-equivalent dual systems and flexural structural walls, that is, walls with an aspect ratio a_o (h_{wi}/l_{wi}: height/length) greater than 2. In the case that the aspect ratio of walls, wall-equivalent systems and torsionally flexible systems are less than 2, it may be concluded that the shear mode of failure (x-type failure) prevails and precedes that of a flexural failure. Therefore, ductility and the q-factor must be reduced. So, K_w takes the following values (see Table 5.2):

$$K_w = \begin{cases} 1.00, \text{ for frame and frame-equivalent dual systems} \\ (1+a_o)/3 \leq 1, \text{ but not less than } 0.5 \\ \text{for wall, wall-equivalent and torsionally} \\ \text{flexible systems} \end{cases} \tag{5.7b}$$

In the case of the existence of many walls in the structural system, the prevailing aspect ratio a_o may be determined from the following relation:

$$a_o = \sum h_{wi} \Big/ \sum l_{wi} \tag{5.8}$$

where
   h_{wi} is the height of wall i, and
   l_{wi} is the length of the same wall

3. The factor

$$\boxed{K_R^{over} = a_u/a_1} \tag{5.9}$$

in Table 5.2 reflects of the overstrength of a regular in-plan system on q-factor. Factors a_u and a_1 in Equation 5.9 are defined below:
a_1 is the value by which the horizontal seismic design action must be multiplied so that at least one member of the structure exhausts its flexural resistance while all other design actions participating in the 'seismic combination' remain constant.
a_u is the value by which the horizontal seismic design action must be multiplied so that plastic hinges are formed in a number of critical regions sufficient for the development of a mechanism, while all other design actions participating in the 'seismic combination' remain constant.

To clarify the above definitions and relations, consider two structural systems, one with high redundancy (Frame system) and the other with low redundancy (Wall system) loaded with constant gravity loads and design seismic actions H_{id} in the form of an inverted triangle (Figure 5.5).
   The V–δ diagram (pushover curve) of the first system is depicted in Figure 5.5a. As V_d increases, one of the structural elements reaches at its critical region its yielding strength in bending and, therefore, at this point the first plastic hinge is formed. The horizontal seismic action at this level is

$$V_{yd} = a_1 V_d \tag{5.10}$$

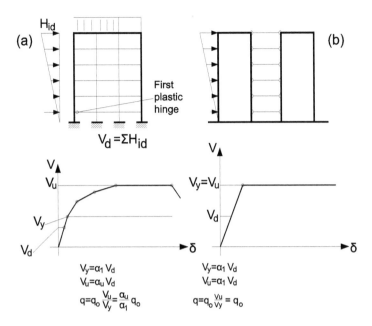

Figure 5.5 Justification of the magnifier factor a_u/a_i introduced in the q_o-factors of frame or frame-equivalent systems: (a) frame system; (b) shear wall system.

Therefore, from this point on, the stiffness of the system decreases and consequently the first knee appears in the diagram. Also, as V_d increases, new plastic hinges are generated, accompanied by new knees on the diagram until the structure collapses. The horizontal seismic action at this level is

$$V_u = a_u V_d \tag{5.11}$$

It is evident that the system presents an overstrength:

$$K^{ov} = \frac{V_u}{V_{yd}} = \frac{a_u}{a_1} \tag{5.12}$$

Therefore, since there is a trade-off between strength and ductility, the q_o factor must increase by the factor (a_u/a_1).

For the system of low redundancy (two uncoupled walls; Figure 5.5b), considering that the two walls have the same cross section, the plastic hinges are generated at both of them simultaneously.

Therefore,

$$V_{yd} = V_u \tag{5.13}$$

and

$$\frac{a_u}{a_1} = 1.00 \tag{5.14}$$

In other words, for structural systems of low redundancy, the factor K^{ov} is equal to 1.00 (see Table 5.2). In this respect, according to EC 8-1/2004, q_o is multiplied by K^{ov}, which is either 1.00 or a_u/a_i. In order to evaluate a_u/a_i, a static inelastic analysis for constant gravity loads and successively increasing horizontal seismic loads must be carried out (pushover analysis; Figure 5.5a) so that a_u and a_1 may be defined. It is obvious that this type of analysis assumes that the structural system has already been analysed elastically and designed properly. So, while a_u/a_i is included in the required data for the determination of the seismic actions used in the elastic initial analysis, it appears to be the output of an additional and extended inelastic analysis that should precede the elastic analysis and design of the system. To overcome this contradiction, which obviously implies an iterative process, EC 8-1/2004 for the determination of a_u/a_i gives the option to the designer to use either a pushover analysis or an approximate conservative value for a_u/a_i specified by the Code.

In the case of a pushover analysis, a_u/a_i must not exceed the value of 1.50 no matter what the output of the analysis. According to EC8-1/2004, the approximate values of a_u/a_i that might be introduced for the determination of the q-factor are the following:
a.   Frames or frame-equivalent dual system
   i. One-storey building: $a_u/a_i = 1.1$
   ii. Multi-storey one-bay frames: $a_u/a_i = 1.2$
   iii. Multi-storey, multi-bay frames or frame-equivalent dual systems $a_u/a_i = 1.3$
b.   Wall or wall-equivalent dual systems
   i. Wall systems with only two uncoupled walls per horizontal direction: $a_u/a_i = 1.0$
   ii. Other uncoupled wall systems: $a_u/a_i = 1.1$
   iii. Wall-equivalent or coupled wall systems: $a_u/a_i = 1.2$

The above values $K_R^{ov} = \alpha_u/\alpha_i$ correspond to buildings regular in plan. In the case of irregular buildings, in plan, the factor a_u/a_i is reduced to

$$K_R^{plan} = \left(1 + \frac{a_u}{a_1}\right) \Big/ 2 \qquad (5.15)$$

4. The factor K_Q reflects the application or not of a Quality System plan to the design, procurement and construction. The value of this factor is defined in the National Annex of the country members. This value is allowed to range between 1.0 and 1.20.
5. The basic value q_o of the behaviour factor is dependent on the type of structural system. Its values have been determined explicitly by the Code and are related to the redundancy of the system's capacity for energy dissipation, the degree of distribution of plastic hinges and so forth.

For inverted pendulum systems, q_o values for R/C buildings in EC8-1/2004 are in disagreement with relevant values of EC8-2/2004 for bridges with R/C single piers, which are in fact also inverted pendulum systems. In fact, q-factors for these piers are specified as equal to 3.50. To overcome this discrepancy, EC8-1/2004 allows the adoption of increased values for the q-factor for inverted pendulum systems, provided that it is shown that correspondingly higher energy dissipation is ensured at the base of the pendulum. This means that a higher value of q is allowed to be introduced in the elastic analysis if after the analysis and design has been finalised a 'pushover' inelastic analysis is carried out and from the capacity curve of the system a new q value is determined, taking into account also a displacement safety factor γ_d (see Chapter 3.2.3) of the order of 1.50. This new q-factor may be introduced in an elastic re-analysis and redesign of the pendulum.

It should also be recalled that the q-factor for vertical seismic excitation must be

$$\boxed{q_{vert} \leq 1.5}$$

(5.16)

unless a greater value is justified through an appropriate analysis.

6. Finally, it should be noted that in contrast to older European National Codes or even to ACI 318-14, the determination of the behaviour factor is related to a preliminary structural analysis at least for the characterization of the structural system (frame-equivalent/wall-equivalent), thus making the determination of the q-factor an 'iterative' procedure.

5.4.4 Quantitative relations between the q-factor and ductility

5.4.4.1 General

A crucial question emerging from the previous subsection concerns the method that has been followed for the definition of q_o-factors that were introduced by EC8-1/2004 in Table 5.2. It is also important to clarify how these values, which have been introduced into the Code, influence the rules of dimensioning and detailing of structural elements, so that these elements are in a position to grant the structural system the global ductility demand and the relevant q_o-factor that has been taken into account for the determination of seismic loading. In other words, it is necessary now to open the 'black box' of the relation between the q_o factor and the inelastic deformations (plastic curvatures ϕ_u) of the critical areas of the structural members that constitute the regions where the energy dissipation takes place.

According to what has been presented so far, the value of the q-factor is directly related to the global ductility factor of the structure *expressed in horizontal displacements* at the top of the structure. The relations between the q_o-factor and the displacement ductility μ_δ that have been adopted by EC8-1/2004 have the following form (Vidic et al., 1994; see Chapter 2.3.4):

$$\boxed{\begin{aligned} \mu_\delta &= q_o & \text{if } T \geq T_c \\ \mu_\delta &= 1+(q_o-1)\frac{T_c}{T} & \text{if } T < T_c \end{aligned}}$$

(5.17)

The value of μ_δ, as already known, is a multi-parametric function of geometry and of elastic as well inelastic properties of the structural members of the system. More precisely, inelastic displacements of a structure at its top are related to inelastic deformations of its structural members. At the structural member level, inelastic deformations are considered at two subsequent stages, that is, *as inelastic rotations* θ_{in} and *as inelastic curvatures* φ_{in}, which constitute the final (end) state, where plasticity is expressed.

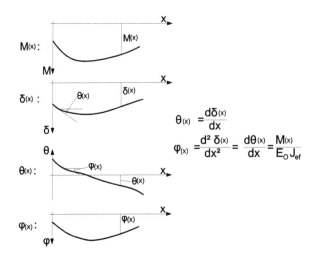

$$\theta_{(x)} = \frac{d\delta_{(x)}}{dx}$$

$$\varphi_{(x)} = \frac{d^2\delta_{(x)}}{dx^2} = \frac{d\theta_{(x)}}{dx} = \frac{M_{(x)}}{E_0 J_{ef}}$$

Figure 5.6 Differential relations between bending moment $M(x)$, curvature $\varphi(x)$, rotation $\theta(x)$ and displacement $\delta(x)$ established first by Mohr.

The aim of this subsection is the establishment of simple *generic* relationships among the q_o factor, displacement ductility μ_δ, the structure-rotational ductility μ_θ of its members and the curvature ductility μ_ϕ of member sections. Having established relations among

$$q_o - \mu_\delta - \mu_\theta - \mu_\varphi$$

it is obvious that

1. For given values of q_o, values of μ_ϕ may be defined in all critical member sections, and therefore simplified rules for design and detailing may also be established and codified.
2. For limit values of μ_ϕ (μ_ϕ-supply), values of q_o-factors (q_o-demand) may be defined and codified properly (Table 5.2).

It is important to remember here from Mechanics the differential relations among bending moment M, curvature φ, rotation θ and displacement δ established first by Mohr in the nineteenth century. These relations are summarised in Figure 5.6.

5.4.4.2 M–φ relation for R/C members under plain bending

5.4.4.2.1 Curvature φ_y at yield

Consider a rectangular beam section loaded by a bending moment M. Before steel yielding, the relation between curvature φ and bending moment M is (Nitsiotas, 1960)

$$\varphi = \frac{M}{E_b J_{ef}} \tag{5.18}$$

where
   $E_b J_{ef}$ is the effective stiffness of the member, which is considered constant up to steel yielding. Effective stiffness is defined below.

When bending moment increases at the level of steel yielding, curvature ϕ_y and bending moment M_y take the following form (Figure 5.7a):

$$\varphi_y = \frac{\varepsilon_c}{x_y} = \frac{\varepsilon_{sy}}{d - x_y} = \frac{\varepsilon_c + \varepsilon_{sy}}{d} \qquad (5.19)$$

$$x_y \cong \frac{D_{cy}}{0.8 f_{cu}} \cong \frac{A_s f_{sy}}{0.8 f_{cu}} \qquad (5.20)$$

$$M_y = A_s f_{sy} zy \cong 0.87 d A_s f_{sy} \qquad (5.21)$$

and

$$E_b J_{ef} = \frac{M_y}{\varphi_y} \qquad (5.22)$$

where
    ε_c is the concrete strain $\leq \varepsilon_{cu}$.
    ε_{cu} is the ultimate concrete strain.
    ε_{sy} is the steel strain at yield.
    f_{sy} is the steel strength at yield.
    D_{cy} is the concrete compression at the stage of steel yield.
    A_s is the steel area.

The rest of the symbols are depicted in Figure 5.7.

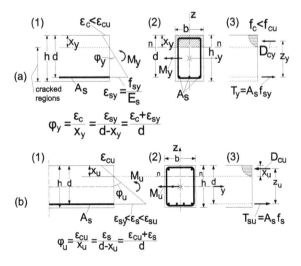

Figure 5.7 (a) Curvature φ_y at yield for R/C members under bending: (1) strain pattern at yield plane; (2) cross section at yield; (3) internal force pattern. (b) Curvature φ_u at failure for R/C members under plane bending: (1) strain pattern at failure; (2) cross section at failure; (3) internal forces at failure.

It is obvious that, if the geometry of the cross section and material properties are known, M_y, φ_y and $E_b J_{ef}$ may be easily calculated.

5.4.4.2.2 Curvature φ_u at failure

As is well known, for a regularly reinforced R/C member, for small steps of increase in the bending moment, steel strains increase disproportionally after yield. Compression zone x also decreases so that, for small moment increase beyond M_y, φ increases disproportionally and the member fails due to concrete crash caused mainly by the decrease in compressive zone x (see Chapter 2.3; Figure 5.7b). In this case,

$$\varphi_u = \frac{\varepsilon_{cu}}{x_u} = \frac{\varepsilon_s}{d - x_u} = \frac{\varepsilon_{cu} + \varepsilon_s}{d} \tag{5.23}$$

$$x_u \cong \frac{D_{cu}}{0.8 f_{cu}} \cong \frac{A_s f_s}{0.8 f_{cu}} \tag{5.24}$$

$$M_u = A_s f_s z_u \cong 0.9 d A_s f_s \tag{5.25}$$

and

$$E_b J_{inel} = \frac{M_u - M_y}{\varphi_u - \varphi_y} \tag{5.26}$$

where
   ε_s is the steel strain at failure $\varepsilon_{sy} < \varepsilon_s < \varepsilon_u$.
   ε_{su} is the ultimate elongation of reinforcing steel.
   D_{cu} is the concrete compression at failure (crush).
   $E_b J_{inel}$ is the mean stiffness in inelastic region.

It is obvious that for a given geometry of the cross section and material properties, M_u, φ_u and $E_b J_{inel}$ may be easily calculated. It should be noted that the relation M–φ from yield to failure stops being linear. However, in this region, it is also considered that the relation is approximately linear with inelastic stiffness that of Equation 5.26.
   So, taking these into account,

- For M values from zero to M_y, the relation M–φ is given by Equation 5.18:

$$\varphi = \frac{M}{E_b J_{ef}}$$

- For M values form M_y to M_u, the relation M–φ is given by the relation

$$\varphi = \frac{M - M_y}{E_b J_{inel}} \tag{5.27}$$

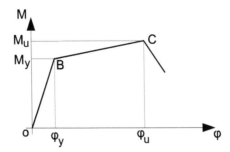

Figure 5.8 Moment–curvature diagram of a ductile R/C member under bending.

Based on the above considerations, the $M–\varphi$ diagram is depicted in Figure 5.8. Section curvature ductility is defined as follows:

$$\mu_\varphi = \frac{\varphi_u}{\varphi_y} = \frac{\varepsilon_{cu} + \varepsilon_s}{\varepsilon_c + \varepsilon_{sy}} \qquad (5.28)$$

A detailed quantitative procedure will be given in Chapters 8 and 9 for the determination of $\varphi_y – M_y$, $\varphi_u – M_u$ and μ_φ of beams, columns and walls.

5.4.4.3 Moment–curvature–displacement diagrams of R/C cantilever beams

Consider the cantilever R/C beam of Figure 5.9, loaded at the top with a horizontal load H, which is in position to cause steel yielding (H_y) at first step, and at second step failure (H_u). Curvatures developed along the length l of the cantilever due to loads H_y and H_u (Figure 5.9a), according to the presentation of the previous paragraph (Figure 5.8), have the form of Figure 5.9b. In order to simplify the calculations, the trapezium 00'2'2 in Figure 5.9b is replaced by an equivalent orthogonal. The area of this orthogonal

$$\theta_P = l_{pl} \cdot \varphi_u \qquad (5.29)$$

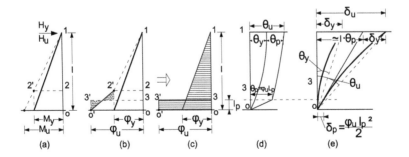

Figure 5.9 Moment–curvature displacement diagrams of an R/C cantilever beam: (a) loading pattern at yield and failure; (b) curvature diagram at yield and failure; (c) definition of the plastic rotation at the fixed end; (d) rotation of the plastic rotation at the fixed end; (d) rotation diagram at yield and failure; (e) displacement diagram at yield and failure.

represents the concentrated plastic rotation θ_p of the cantilever at a length l_{p1}, which is the *ideal length* of the *plastic hinge*.

Considering now successively, as for elastic loads, the curvatures ϕ_y and ϕ_u and implementing Mohr's theory, we define for δ_y and δ_u the following values:

$$\delta_y = \frac{\varphi_y}{3} l^2 \tag{5.30}$$

and

$$\delta_u = \frac{\varphi_y}{3} l^2 + (\varphi_u - \varphi_y) l_{pl} \left(l - \frac{l_{pl}}{2} \right) \tag{5.31}$$

Consequently,

$$\boxed{\mu_\delta = \frac{\delta_u}{\delta_y} = 1 + \frac{3 l_{pl}}{l} (\mu_\varphi - 1) \left(1 - 0.5 \frac{l_{pl}}{l} \right)} \tag{5.32}$$

The chord rotations are given by the following expressions:

$$\theta_y = \frac{\delta_y}{l}, \quad \theta_u = \frac{\delta_u}{l} \tag{5.33}$$

and therefore

$$\boxed{\mu_\theta = \frac{\theta_u}{\theta_y} = \mu_\delta} \tag{5.34}$$

The determination of the ideal length l_{p1} of the plastic hinge has been the object of extended research for many years. The value that has been adopted by EC8-3-ANNEX A/2005 has the following form:

$$\boxed{l_{pl} = 0.1l + 0.17h + 0.24 \frac{d_{bl} f_y (\text{MPa})}{\sqrt{f_{cu} (\text{MPa})}}} \tag{5.35}$$

where
    h is the depth of the member.
    d_{b1} is the mean diameter of the tension reinforcement.
    f_y is the steel stress at yield.
    f_{cu} is the concrete compressive strength.

For a range of usual values of the parameters included in Equation 5.35, the length of plastic hinges varies between the following limits (Fardis et al., 2005):

$$\left. \begin{array}{l} \text{Columns} \left\{ \begin{array}{l} 0.35l \\ 0.45l \end{array} \right. \quad \text{mean value: } 0.40l \\[3em] \text{Beams} \left\{ \begin{array}{l} 0.25l \\ 0.35l \end{array} \right. \quad \text{mean value: } 0.30l \\[3em] \text{Walls} \left\{ \begin{array}{l} 0.18l \\ 0.24l \end{array} \right. \quad \text{mean value: } 0.22l \end{array} \right\} \qquad (5.36)$$

where

l is the length of a cantilever loaded horizontally at the top or the distance of joints to zero-moment points of beams or columns of a frame loaded with horizontal loads at the storeys (shear span M/V). This length usually corresponds to the half-span or half-height of beams or columns, respectively.

5.4.4.4 Moment–curvature–displacement diagrams of R/C frames

Consider the closed frame of Figures 5.10 and 5.11, which is the basic unit of multi-storey multi-beam frames. Consider also a horizontal loading H at the top beam, which is in position to cause steel yielding (H_y) at a first step, and at a second step failure (H_u). For the mode of failure of this frame, two options exist:

1. In the case of strong beams – weak columns: formation of plastic hinges at the column ends (Figure 5.10c)
2. In the case of strong columns – weak beams: formation of plastic hinges at the beam ends (Figure 5.11b)

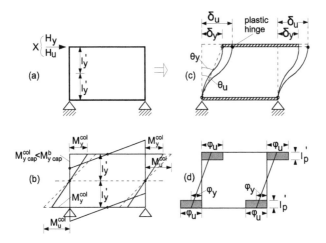

Figure 5.10 Moment–curvature displacement diagrams for R/C frames with strong beams–weak columns: (a) geometry; (b) moment diagrams at yield and failure; (c) displacement diagrams at yield and failure; (d) curvature diagrams at yield and failure.

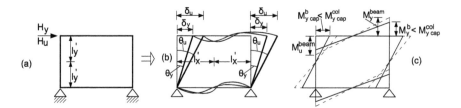

Figure 5.11 Moment–curvature displacement diagrams of R/C frames with strong columns–weak l_v beams: (a) geometry; (b) displacement diagrams at yield and failure; (c) moment diagrams at yield and failure.

If the procedure for the cantilever is repeated for the frame of Figure 5.10, the following results are derived:

- Frame with strong beams – weak columns (Figure 5.10a–d)

$$\delta_y = 2\frac{\varphi_y}{3}(l'_c)^2 \tag{5.37}$$

$$\delta_u = 2\frac{\varphi_y}{3}(l'_c)^2 + 2l'_{pl}(\varphi_u - \varphi_y)\left(l'_c - \frac{l'_{pl}}{2}\right) \tag{5.38}$$

$$\boxed{\mu_\delta = \frac{\delta_u}{\delta_y} = 1 + 3\frac{l'_{pl}}{l'_c}(\mu_\varphi - 1)\left(1 - 0.5\frac{l'_{pl}}{l'_c}\right)} \tag{5.39}$$

$$\theta_s = \frac{\delta_y}{h}, \quad \theta_u = \frac{\delta_u}{h} \tag{5.40}$$

$$\boxed{\mu_\theta = \frac{\theta_u}{\theta_y} = \mu_\delta} \tag{5.41}$$

- Frame with strong columns – weak beams (Figure 5.11a–d)

$$\theta_y^{beam} = 2\left(\frac{\varphi_y l'_b}{2}\right)\left(\frac{2}{3}l'_b\right)\frac{1}{2l'_b} = \frac{1}{3}\varphi_y l'_b \tag{5.42}$$

$$\delta_y = h\theta_y^{beam} = \frac{1}{3}\varphi_y l'_b h \tag{5.43}$$

$$\theta_u^{beam} = \frac{1}{3}\varphi_y l'_b + 2(\varphi_u - \varphi_y)l'_{pl}\left(l'_b - \frac{l'_{pl}}{2}\right)\frac{1}{2l'_b} \tag{5.44}$$

$$\delta_u = \theta_u^{beam} h$$

$$\mu_\delta = \frac{\delta_u}{\delta_y} = 1 + 3\frac{l'_{pl}}{l'_b}(\mu_\varphi - 1)\left(1 - 0.5\frac{l'_{pl}}{l'_b}\right) \quad (5.45)$$

$$\mu_\theta = \frac{\theta_u^{beam}}{\theta_y^{beam}} = \frac{\delta_u/h}{\delta_y/h} = \mu_\delta \quad (5.46)$$

5.4.4.5 Conclusions

From the above analysis, the following conclusions may be drawn:

1. For the basic forms of a structural system, that is, a cantilever (free standing wall) or a frame system, the relation between μ_θ and μ_ϕ has the same generic form, that is (Equations 5.32, 5.39, 5.45),

$$\mu_\delta = \mu_\theta = 1 + \frac{3l_{pl}}{l}(\mu_\varphi - 1)\left(1 - 0.5\frac{l_{pl}}{l}\right) \quad (5.47)$$

where
   l is the shear span (M/V), or, more simply, the length from the end of the structural member to the zero point of the bending moment diagram.
   l_{p1} is the ideal length of the plastic hinge (Equation 5.35).

Consequently, if Equation 5.47 is introduced in Equation 5.17, the following relationships are derived:

$$1 + \frac{3l_{pl}}{l}(\mu_\varphi - 1)\left(1 - 0.5\frac{l_{pl}}{l}\right) = q_o \quad \text{if} \quad T \ge T_c \quad (5.48)$$

and

$$1 + \frac{3l_{pl}}{l}(\mu_\varphi - 1)\left(1 - 0.5\frac{l_{pl}}{l}\right) = 1 + (q_o - 1)\frac{T}{T_c} \quad \text{if} \quad T < T_c \quad (5.49)$$

In this way, a direct relation *of generic type* is established between q_o and μ_ϕ for every structural type, that is, between the global q_o-factor of the structure and the curvature ductility at the ends of its structural members.

2. In EC8-1/2004, for reasons of simplicity and continuity with ENV edition (EN1998-1-3; Fardis et al., 2005), the following simpler expressions have been introduced in place of Equation 5.47, that is,

$$\mu_\delta = \mu_\theta = 1 + 0.5(\mu_\varphi - 1) \quad (5.50)$$

or

$$\boxed{\mu_\varphi = 2\mu_\theta - 1} \tag{5.51}$$

Consequently, introducing Equation 5.51 into Equation 5.17, the following relationships are drawn:

$$\boxed{\mu_\varphi = 2q_o - 1 \qquad \text{if} \quad T \ge T_c} \tag{5.52a}$$

and

$$\boxed{\mu_\varphi = 1 + 2(q_o - 1)\frac{T_c}{T} \qquad \text{if} \quad T < T_c} \tag{5.52b}$$

3. The simplified relation (Equation 5.51) is on the safe side. In fact, from the statistical analysis of μ_φ values, for the three types of structural members (beams, columns and walls) and for the full range of DCH and DCM buildings, taking into account usual l_{p1} values (Equation 5.36), the following mean safety factors have been derived:

for columns: 1.65
for beams: 1.35   (5.52c)
for walls: 1.10

4. In all above relations, according to EC8-1/2004 instead of final q-factors, *the basic ones, q_o,* are introduced, because q-factors generally present reduced values due to various penalties that are imposed on the procedure of analysis (design spectrum) to overcome various irregularities of the structure. It is not prudent to introduce this reduction of q_o-factors into Equations 5.48, 5.49 or 5.52 because it would result in a decrease of μ_φ-safety factors.
5. The above rationally concrete framework of relations between q_o–μ_δ–μ_θ–μ_φ has been verified through extensive prenormative analytical trials on real buildings. Indeed, a series of buildings was designed according to EC8-1/2004 in early prenormative stage, and then these buildings were evaluated through static or dynamic inelastic analysis. The conclusions of all these trials are that the conceptual basis of analysis and design established by EC8-1/2004 is sound and safe (Kappos, 1991; Kappos and Penelis, 1986, 1987).

5.4.5 Critical regions

From what has been presented in the previous subsection, the energy-dissipating zones of a structure are localised at the regions of the structural elements where plastic hinges might form. From the previous discussion, it can easily be concluded that these regions are located at the areas where the most adverse combinations of action effects occur (M, N, V, T). These regions of the primary seismic elements are defined as *critical regions*. Critical zones are located at the ends of beams and columns in R/C frames and at the bottom of R/C structural walls. The length of these regions is defined for each type of structural element by Codes. Having in mind that the plastic hinges undergo reversal cyclic rotations θ_{p1} during an earthquake, it is evident that the critical regions should be designed and reinforced properly. Length, reinforcement and detailing of critical regions will be discussed in detail in Chapters 8, 9 and 10 for each structural element separately.

5.5 ANALYSIS METHODS

5.5.1 Available methods of analysis for R/C buildings

Over the years, the computational capacity has increased (doubles every 18 months according to Moore's law; Moore, 1965) thus rendering the use of extremely advanced and detailed analysis of structures feasible. This, however, should not shift practicing structural engineers necessarily towards those methods, as their complexity creates several issues regarding input parameters that are not available, modelling approaches that are extremely complex and interpretation of results that is diverse and not straightforward. All these issues will be further discussed and tackled in the following paragraphs.

The most commonly used methods of analysis for R/C buildings under seismic actions are

- Equivalent static elastic analysis, called 'lateral force method of analysis'
- Modal spectral elastic analysis, called 'modal response spectrum analysis'
- Equivalent static inelastic (nonlinear) analysis, called 'pushover analysis'
- Time–history (*t-h*) inelastic (nonlinear) analysis (*t-h* linear analysis is not frequently used for buildings)

Of those, the reference method for EN1998-1/2004 is *the modal response spectrum analysis*, while for US codes (ACI318.14, IBC2012, SEAOC 09) the reference method *is the lateral force method of analysis*. This difference is significant as the two methods have very distinct advantages and disadvantages summarised in the following:

- *The modal response spectrum analysis* accurately depicts the dynamic behaviour of the structure by identifying the several important modes of vibration and using them in the calculation of the lateral forces at each level, but produces *unsigned results* for displacements, internal forces and stresses.
- *The lateral force method* uses the fundamental mode to calculate those forces but produces *signed results* for internal loads and stresses; the sign makes the results more comprehensive and clearer with regard to several aspects, such as the combination of dynamic and static load cases, the biaxial bending with axial load of columns, the integration of stresses of non-rectangular cross-section walls and cores and so forth.

Having in mind the above, there are elements of EN1998-1/2004 like the participation of walls shear in the total storey shear that actually require the execution of an equivalent static analysis (see Chapters 4.5 and 5.2), an issue solved by current analysis software by either using such an analysis or producing signed modal results (a notion theoretically wrong) according to a selected mode of vibration.

Of course all these issues are increased exponentially when dealing with the nonlinear approaches, which have the very serious distinction between the point hinge approach and the fibre approach (distributed plasticity over the cross section). These two approaches also have very distinct advantages and disadvantages, mainly summarised in the following:

- *The point hinge approach* is codified (through available M–θ diagrams) and easily applied in available commercial software, and produces controlled results ideal for performance base design/assessment. However, it is impossible to model shear walls and cores with 2-D shell elements, thus creating the need for a linear finite element approach, with uncertainties in the connectivity and the M–θ diagram of the complex section.
- *The fibre model provides* a more accurate solution on the element plasticity and also deals with the issue of complex section walls and cores, as their plasticity is inherently inserted in the modelling of the element. However, this accuracy requires a very deep knowledge of material properties of concrete, rebars and stirrups and their interface, which makes such a modelling approach extremely risky for practicing engineers not familiar with nonlinear analysis. It should also be noted that this modelling approach increases the instability of the nonlinear analysis and the possibility for errors.

Available commercial software provides solutions based on the point hinge approach with some fibre model capabilities for shell elements (walls) in order to overcome the obvious problems of the former method, and there is very reliable free academic software that provides solutions for the complete fibre model approach.

Finally, there is the distinction between *static* and *dynamic nonlinear analysis*, both of which are available tools in modern codes, and both of them also have advantages and disadvantages.

- *The static nonlinear analysis* provides a straightforward procedure for the input of the seismic forces as constantly increasing lateral loads (or displacements), requires 'simple' constitutive laws for point hinges or even fibre approaches and produces very comprehensive results as the capacity curve and the status of the elements of the building at each loading step. However, there are uncertainties on the lateral loads shape, the lateral loads point of application and the transformation of the building force–displacement curve (P–δ) to an equivalent SDOF capacity curve.
- *The dynamic t-h nonlinear analysis* is considered the most accurate and sophisticated analysis approach (for buildings) available, especially when the fibre approach is utilised. It produces results that have inherently taken into account the torsional effects, the seismic load distribution and the displacements, which do not require a transformation to an equivalent SDOF oscillator. On the other hand, this type of analysis is very sensitive to the selection of appropriate accellerograms (recorded, artificial, hybrid), the hysteretic behaviour of concrete and rebars, the distribution of masses and the interpretation of the results on an element (section) level (peak values, peak values within a time frame, effective values, etc.).

The current trend in academia is to try to combine the static nonlinear analysis with the dynamic modal analysis, either by producing a series of modal pushover analyses then combining them (Chopra and Goel, 2002), or by using the adaptive pushover approach, which detects the changes in the stiffness matrix, calculates new modes and accordingly modifies the lateral load value and shape at each step (Elnashai and Mwafy, 2000; Antoniou and Pinho, 2004a). This approach, although very promising, is neither codified nor suggested for practicing engineers at the moment. Instead, EUROCODE EC8-1/2004 introduces two different lateral load patterns for which the static inelastic analysis is carried out.

All the aforementioned issues are elaborated in the following sections.

5.6 ELASTIC ANALYSIS METHODS

5.6.1 General

As has been mentioned in previous chapters, there are several elastic analysis methods available that are usually called 'linear' although they might include geometric nonlinearity approaches (buckling), thus making the use of the term "elastic" more accurate. As explained previously, these methods are

- Equivalent static elastic analysis, called 'lateral force method of analysis'
- Modal spectral elastic analysis, called 'modal response spectrum analysis'
- Time–history (t-h) elastic analysis

5.6.2 Modelling of buildings for elastic analysis and BIM concepts

For either static or dynamic approaches, the modelling of three-dimensional structures in modern FE software packages has reached a level where all geometric features of a building may be accurately modelled. Especially with the introduction of building information modelling (BIM), the FE model of the building is an actual 3-D representation of the structure, which includes all structural and nonstructural elements, MEP installations, loads and boundary conditions, which are then automatically exported to a structural analysis software package that can perform the analysis, calculate the required reinforcement and feed it back to the BIM model. As an application example, the new opera in Athens in Figure 5.12 is shown.

It is, however, important for all structural engineers to be aware that these interfaces are only used, to date, for clash detection between architectural, structural and MEP design, while the structural modelling is done independently either from scratch or by using the BIM output as a reference model to be modified and verified by the structural engineer.

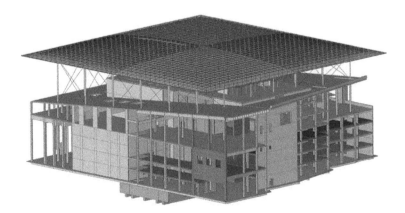

Figure 5.12 BIM application for the Athens Opera House (SNFCC) using Nemetschek Scia Engineer.

5.6.3 Specific modelling issues

The main problems of a full 3-D FE model of an actual building are the following:

- Simulation of walls, cores and openings
- Simulation of T- and Γ-shaped beams
- Diaphragm constraint
- Application of eccentricity of seismic loads

5.6.3.1 Walls and cores modelling

The modelling of walls and especially cores is available in most modern software packages through the use of planar (shell) finite elements, which reduce the requirements for modelling of rigid connections and equivalent stiffness, as was the case in the past for all types of analyses and remains to date for nonlinear analysis. The accurate modelling of cores in medium-rise and high-rise buildings is crucial, as it affects the total response and plays a significant role in the final architectural layout of these buildings.

However, another key issue is the design of these complex section elements, which, although modelled accurately for the analysis, are sometimes treated as individual rectangular sections (legs) for the design, thus underestimating their flexural capacity up to 25% (ECtools, 2013).

Finally, a very important issue for the design of walls is the accurate modelling of the openings, or, more precisely, the spandrels (connection beams) that are created. Unless properly modelled, these elements might elude the checking of the general concrete wall and remain poorly reinforced, while all the checks elaborated in the relevant chapters should be applied.

5.6.3.2 T- and Γ-shaped beams

As it is very well known, all modern codes require an effective flange length to be taken into account in the design of beams that are directly under slabs, thus resulting in a *T* section (where the slab is on both sides of the beam) or a Γ section (where the slab is on one side of the beam). There lies the risk of actually taking into account twice the part of slab that coincides with the part of the beam section, both regarding stiffness and self-weight.

Some modern commercial software packages allow the modelling of the hanging part of the beam (rib) separately while the modelling of the flange is done with the slab shell elements. Then the section of the beam is comprised by the two parts and the section forces are the combination of the integrated stresses of the required part of the slab and the section forces of the rib, as is shown in Figure 5.13.

However, other general-purpose software packages do not provide this option; hence, the engineer has to follow one of the two options below:

- Model the beams as T or Γ sections and reduce their self-weight and stiffness accordingly (through property modifiers) so that the total building self-weight and stiffness are correct.
- Model the beams as rectangular section, place them under the shell elements of the slab *by introducing an eccentricity* and then manually integrate the results (combining stresses from shell elements and section forces from the rectangular section) and calculate the required reinforcement for the actual T or Γ section.

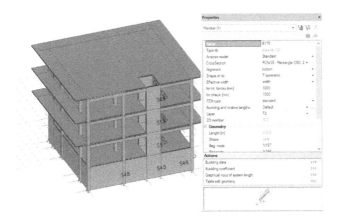

Figure 5.13 Modelling of ribs with one-dimension finite element (1-DFE) and of effective flange width from slab two-dimension finite element (2-DFE).

It is crucial to avoid modelling the flanges twice (as slab and as a beam section), as this seriously affects the stiffness, the mass and the resulting dynamic characteristics of the building.

5.6.3.3 Diaphragm constraint

All modern seismic codes for buildings have checks and verifications that have the inherent notion of a storey (storey forces, inter-storey drifts, etc.). In the past two decades, when computational power was significantly lower, it was important to seriously reduce the stiffness and mass matrices of finite element models of buildings, and for this main reason, the diaphragm constraint had been introduced, which essentially defines one master joint in the centre mass of a storey and connects all other nodes of the storey (as slave nodes) with rigid-body in-plane motion relative to that master joint. This allowed the use of a condensed stiffness and mass matrix, and all results were actually then extrapolated to all the connected slave nodes through shape functions (see Chapter 2.4.4).

It is obvious that nowadays, for normal buildings, the significance of the reduction of required computation force has been minimised; however, there are other issues that make the use of a diaphragm constraint a suggested option:

- Elimination of secondary eigen modes
- Application of the equivalent static analysis in a straightforward way (by applying the forces on the CM node)
- Application of accidental eccentricities in the equivalent static or spectral dynamic analysis
- Easily checked inter-storey drifts and storey displacements
- The philosophy of analysis and design of all modern Codes, which is based on the notion of diaphragm constraint

However, there are cases where the diaphragm constraint should or must be avoided:

- Buildings that have storeys with large openings (i.e. atriums)
- Buildings that, after a height, have twin towers
- Buildings with very flexible in-plane floors

All modern software packages foresee the following options:

- Rigid diaphragms
- Flexible diaphragms
- No diaphragms

It is clear that for most common buildings the use of the diaphragm constraint is suggested, while in complex buildings it is up to the structural engineer to decide upon using a flexible diaphragm or no diaphragm constraint at all. EN1998-1/2004 (EC8-1) notes that, 'the diaphragm is taken as being rigid, if, when it is modelled with its actual in-plane flexibility, its horizontal displacements nowhere exceed those resulting from the rigid diaphragm assumption by more than 10% of the corresponding absolute horizontal displacements in the seismic design situation'.

5.6.3.4 Eccentricity

The issue of the required eccentricity in the application of the seismic loads in either the equivalent static or the spectral analysis is elaborated in a following paragraph. However, the modelling of this eccentricity is a tricky issue, as one has the following options, depending on the modelling assumptions that have been adopted:

- Introduce the resulting torsion at the master joint when a diaphragm exists (for both static and spectral analysis).
- Divide the resulting torsion at all the nodes of the storey when a diaphragm does not exist (for both static and spectral analysis).
- Place either the lumped mass (for the case of modal spectral analysis) or the horizontal forces (for the case of equivalent static analysis) at four offset positions from the centre of mass, provided that a diaphragm constraint can be applied. This option, which theoretically is the most accurate, significantly increases the seismic lateral load cases from two (EX and EY) to four and the load combinations accordingly (exponentially). Some software codes perform this offset automatically and envelope the results, which, however, is not easily verified by the engineer.

It should be noted that according to EC8-1/2004, in order to account for uncertainties in the location of masses, the calculated centre of mass at each floor i shall be considered as being displaced in each direction by an accidental eccentricity of 5% the floor dimension perpendicular to the direction of the seismic action (L_i), $e_{ai} = \pm 0.05 L_i$.

5.6.3.5 Stiffness

As has already been elaborated, the stiffness in elastic analysis plays a significant role, especially in the modal spectral analysis as it affects the dynamic characteristics of the structure. Different codes provide different information of the effective stiffness of structural elements, using the basic notion that all elements, except columns that have a high axial load, are at stage II, that is, cracked, when an earthquake occurs; hence, beams and walls have a reduced effective stiffness of 50–70% of the uncracked one.

However, there are cases where this assumption creates misleading results for the gravity load combination, in the cases where this is critical. A very indicative example on the matter is for buildings with seismic isolation and the effect of the stiffness modifiers on the load distributions to each isolation unit.

5.6.4 Lateral force method of analysis

This method, for both main directions of the building, takes into account only the fundamental mode of vibration per direction. Based on the above modes of vibration, the respective fundamental periods T_{1x}, T_{1y}, and the relevant design spectrum, modified by the importance factor, the total inertia forces in the two main directions and their contribution along the height of the structure are defined (Chapter 3.2.3). For these loads, a static analysis of the structural system is carried out. In this context, this method might be characterised as an equivalent static analysis. From the above presentation, it is concluded that this type of analysis can only be applied to buildings in which the first two eigen modes are translational and response is not expected to have any essential contribution from higher modes of vibration.

EC8-1/2004 (par. 4.3.3.2.1) defines the following conditions for such an analysis to be used, in buildings that:

Condition 1
　　The fundamental periods of vibration T1 in two main directions are less than the following values:

$$T_1 \leq \begin{cases} 4T_C \\ 2.0s \end{cases} \tag{5.53}$$

　　where T_c is the upper limit of the period of the constant spectral acceleration branch (plateau).
Condition 2
　　Meet the criteria for regularity in elevation given in Chapters 4.2.2 and 5.2.3.

5.6.4.1 Base shear forces

The seismic base shear force V_B, for each horizontal direction in which the building is analysed, shall be determined using the following expression:

$$V_B = S_{d1}(T_1) \cdot m \cdot \lambda \tag{5.54}$$

where
　　$S_{d1}(T_1)$ is the ordinate of the design spectrum at period T_1.
　　T_1 is the fundamental period of vibration of the building for lateral motion in the direction considered.
　　m is the total mass of the building, above the foundation or above the top of a rigid basement.
　　λ is the correction factor, the value of which is equal to $\lambda = 0.85$ if $T_1 < 2\ T_C$ and the building has more than two storeys, or $\lambda = 1.0$ otherwise.

5.6.4.2 Distribution along the height

The base shear is distributed among the storeys in the same proportion as the inertia forces that correspond to the fundamental period of the structural system, which is homologous

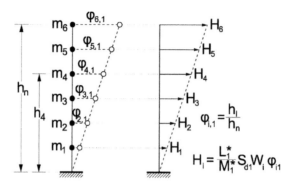

Figure 5.14 The fundamental mode of a multi-storey system and the corresponding inertial forces.

to the characteristic shape of the fundamental mode. Given the fact that the first mode of a multi-storey, multi-column system, with a limited number of storeys and sufficient lateral stiffness, appears to be linear (Figure 5.14; Biggs, 1964; Polyakov, 1974), the following relationships apply:

$$\varphi_{i1} = \frac{h_i}{h_n} \tag{5.55}$$

$$H_i = \lambda \gamma_1 m_i \varphi_{i1} \frac{L_1^*}{M_1^*} S_{d1} \tag{5.56}$$

(see Chapter 2.4.3)
and

$$\sum_i H_i = V_B = \lambda \gamma_1 \frac{L_1^*}{M_1^*} S_{d1} \sum_i m_i \varphi_{i1} \tag{5.57}$$

Therefore,

$$\lambda \gamma_1 \frac{L_1^*}{M_1^*} S_{d1} = V_B / \sum_i m_i \varphi_{i1} \tag{5.58}$$

Substituting Equations 5.55 and 5.58 into Equation 5.56, the following expressions are obtained:

$$H_i = \frac{m_i \varphi_{i1}}{\sum m_i \varphi_{i1}} V_B \tag{5.59}$$

or

$$H_i = \frac{m_i h_i}{\sum m_i h_i} V_B \qquad (5.60)$$

where
    h_i is the height of storey i from ground.

On the basis of this theoretical background, EC8 (par. 4.3.3.2.3) requires a distribution of forces as per the fundamental mode shape displacement at each storey level that is according to Equation 5.60.

This equation for constant values of m_i and storey heights h yields

$$H_i = \frac{m_i h_i}{m \sum h_i} V_B \qquad (5.61)$$

or

$$H_i = \left(n_i / \sum_i n_i \right) V_B \qquad (5.62)$$

where
    n_i is the number of the storey at the level of which the lateral force H_i is induced.

Equation 5.62 yields a *triangular distribution of seismic loading.*

It should be noted, however, that all advanced software packages provide the option to distribute the base shear linearly, uniformly or under any mode shape, thus allowing a different distribution of forces along their height for each direction of analysis.

5.6.4.3 Estimation of the fundamental period

The estimation of the fundamental period may be done either by approximate methods or by applying modal analysis to the building, as it is available to most engineering software packages. However, even in those cases, the following approximations provide a measure to check the accuracy of the modelling of the building.

For buildings with heights of up to 40 m, the value of T_1 (in s) may be approximated by the following expression:

$$T_1 = C_t \cdot H^{3/4} \qquad (5.63)$$

where
    C_t is 0.085 for moment-resistant space steel frames, 0.075 for moment-resistant space concrete frames and 0.050 for all other structures.

    H is the height of the building (in m) from the foundation or from the top of a rigid basement.

Alternatively, for structures with concrete shear walls, the value C_t may be taken as being

$$C_t = \frac{0.007}{\sqrt{A_c}} \tag{5.64}$$

where

$$A_c = \sum \left[A_i \cdot \left(0.2 + (l_{wi}/H)\right)^2 \right]$$

and

A_c is the total effective area of the shear walls in the first storey of the building (in m²).
A_i is the effective cross-sectional area of shear wall i in the direction considered in the first storey of the building (in m²).
L_{wi} is the length of the shear wall i in the first storey in the direction parallel to the applied forces (in m) with the restriction that l_{wi}/H should not exceed 0.9.

The estimation of T_1 (in s) may also be made by using the following expression:

$$T_1 = 2\pi \cdot \sqrt{d} \tag{5.65}$$

where

d is the lateral elastic displacement of the top of the building (in m) due to the gravity loads applied in the horizontal direction.

For the determination of the fundamental period of vibration period T_1 (in s) of the building, expressions based on methods of structural dynamics may be used. For example, the Rayleigh method may be properly adapted:

$$T_1 = 2\pi \left(\sqrt{\frac{1}{g} \frac{\Sigma W_i \delta_i^2}{\Sigma H_i \delta_i}} \right) \tag{5.66}$$

where H_i $(i = 1, 2,..., N)$ is a group of forces at the level of the floors, with a triangular distribution, δ_i $(i = 1, 2,..., N)$ is the corresponding displacements of the floors and W_i $(i = 1, 2,..., N)$ is the vertical loads at each storey i. For $N = 1$, from Equation 5.66 the previous Equation 5.65 can be derived.

It should be noted that all these approximate methods yield results that may differ from a modal analysis which considers the soil structure interaction, thus using equivalent springs instead of a fixed foundation.

5.6.4.4 Torsional effects

In the case of systems regular in plan and in elevation, as these properties have been defined in Section 5.2, the torsional effects taken into account in the 'simplified modal analysis' are only those related to an accidental eccentricity equal to (Section 5.6.3.4)

$$e_{ai} = \pm 0.05 \cdot L_i \tag{5.67}$$

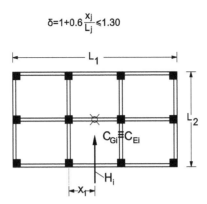

Figure 5.15 Evaluation of the torsional effects on a symmetric system with the aid of an amplification factor.

In this case, EC8 allows the torsional effects to be taken into account by amplifying the action effects on the individual load-resisting plane elements parallel to the seismic action using an amplification factor δ equal to (Figure 5.15)

$$\delta = 1 + 0.6\frac{x_i}{L_i} \leq 1.30 \tag{5.68}$$

If the analysis is performed using two planar models, one for each main horizontal direction, torsional effects may be determined by doubling the accidental eccentricity e_{ai}:

$$e_{ai} = \pm 0.10 L_i \tag{5.69}$$

which results in the following amplification factor δ:

$$\delta = 1 + 1.2\frac{x_i}{L_i} \leq 1.60 \tag{5.70}$$

It is obvious that such approaches are related to this type of simplified analysis, and whenever torsional sensitivity is considered an issue by the designer, at least a modal analysis should be executed.

It should be noted, again, that codes that promote (use as reference method of analysis) this type of static–elastic analysis suggest procedures that the dynamic characteristics of the building are determined through modal analysis. In that way, most of the drawbacks are eliminated while the main advantage, which is comprehensive and physically meaningful *signed results*, is preserved.

5.6.5 Modal response spectrum analysis

As already mentioned, this type of analysis is the *reference analysis method* according to EC8-1/2004. It may be applied to all types of buildings, even those that do not satisfy the conditions for applying the lateral force method of analysis.

For buildings complying with the criteria of regularity in plan but not in elevation, the analysis can be performed using two plane models, one for each main direction. Otherwise, the system must be analysed using a spatial model. Whenever a spatial model is used, the design seismic action will be applied along its two main directions determined by the resisting elements of the system. Otherwise, the design seismic action will be applied along all relevant horizontal directions and their orthogonal horizontal axes.

5.6.5.1 Modal participation

In a multi-modal analysis, the responses of all modes of vibration contributing significantly to the global response are taken into account (Clough and Penzien, 1975). This may be satisfied by either of the following:

- Demonstrating that the sum of the effective modal masses for the modes considered amounts to at least 90% of the total mass of the structure, that is,

$$\sum_{i=1}^{k} \frac{L_i^{*2}}{M_i^*} \geq 0.9 \sum_{i=1}^{n} M_i \tag{5.71}$$

where k is the number of modes considered and n the number of masses.
- Demonstrating that all modes with effective modal masses greater than 5% of the total mass are considered, that is,

$$\frac{L_j^{*2}}{M_j^*} \leq 0.05 \sum_{i=1}^{n} M_i \tag{5.72}$$

where j is the index of the modes not considered.

In the case of a spatial model, the above conditions must be verified for each main direction.

In buildings with a significant contribution from torsional modes, if the above conditions cannot be satisfied, the minimum number of modes k to be considered in a spatial analysis should satisfy the following condition (EN1998-1/2004):

$$k \geq 3 \cdot \sqrt{n}$$

and

$$T_k \leq 0.20\,s$$

where T_k is the period of vibration of mode k.

Here it should only be added that whenever a spatial model is used,

- The floor masses will be considered as either lumped masses concentrated at the centre of gravity of each floor or distributed depending on the diaphragm modelling approach adopted, as has been elaborated in Section 5.6.3.3.
- The accidental torsional effects may be determined using the appropriate modelling approach, which has been elaborated in Section 5.6.3.4.

5.6.5.2 Storey and wall shears

To evaluate the percentage of the storey shear that the walls receive (compared to the total), equivalent static analysis is utilised. This is necessary because the seismic modal analysis produces results that are unsigned, thus making them useless for such an evaluation. It should be noted that in dual systems, this percentage in the upper storeys can be negative, since the response of the wall differs from that of the frame.

This has been evident since the 1960s in the application of the McLeod method, as is shown in Figure 4.31.

5.6.5.3 Ritz vector analysis

Although eigen vectors have been used extensively in modal analysis, in recent years an approximate approach, called the Ritz vector analysis, has been utilised in buildings that have a very complex geometry. This analysis allows the identification of significant modes of vibration (excluding minor modes), as it uses as a starting point vectors defined by the engineer (the first vector is the displacement vector obtained from a static analysis) using the spatial distribution of the dynamic load vector as input. As has been demonstrated by Wilson (1985), dynamic analyses based on a unique set of Ritz vectors yield more accurate results than the use of the same number of exact mode shapes. It is considered beyond the scope of this chapter to further elaborate on this approach, which proves useful in dealing with complex spatial structures.

5.6.6 Time–history elastic analysis

The time–history elastic analysis is used only at specific cases of analysis where the exact response of the building through the duration of a set of excitations is required. Such requirements arise in the following types of buildings:

- Buildings with seismic isolation
- Buildings with damping devices
- Buildings with active mass systems
- High-rise buildings

In all the above cases, the modelling approaches analysed in the previous paragraphs are applicable, with the exception of the seismic devices which require modelling with additional data, such as

- Effective (elastic) stiffness
- Effective damping

These parameters are initially determined by the designing engineer and are then provided/confirmed by the selected manufacturer, as will be detailed in the relevant chapter.

The *t-h* analysis can provide, in such cases, information about the modification of the dynamic characteristics of the building, which are only approximated when using the previous types of analysis (i.e. torsional effects, participation of higher modes due to additional damping of first mode, etc.).

The selection of the excitation, scaling and interpretation of the results are issues that are elaborated in the relevant section for the nonlinear *t-h* analysis.

5.7 INELASTIC ANALYSIS METHODS

5.7.1 General

As is widely known, the design of new buildings is based on elastic methods (static or dynamic); however, buildings often sustain damage during an earthquake and thus develop some degree of nonlinear inelastic behaviour depending on the extent of damage. This behaviour is introduced into the design by modern codes by the use of a reduction factor for the seismic forces (q-factor, R-factor, etc.). On the other hand, for the assessment of the seismic behaviour of an existing building, many modern codes/guidelines suggest the use of a more 'accurate' approach – that of the nonlinear analysis.

Assessing the nonlinear behaviour of a building requires appropriate software, the most advanced case being the nonlinear step-by-step dynamic time–history analysis with a fibre model for the structural elements behaviour. This type of analysis, though very useful for the researcher/engineer as a benchmark, is extremely strenuous and non-design friendly as it requires a bundle of input data (accelerograms, damping values for each element, stress–strain laws for each material with a cyclic behaviour, etc.) and produces unclear results for design purposes (stress–strain over time steps, maximum–minimum values at different time steps, absorbed energy, etc.).

5.7.2 Modelling in nonlinear analysis

It is well known that the software available for nonlinear analysis can be generally divided into four different categories:

- Linear (1d) finite element point hinge models
- Linear (1d) finite element distributed nonlinearity (fibre) models
- Planar (2d shell) or spatial finite element nonlinear models (continuous or discrete cracking)
- Solid (3d) finite element nonlinear models (continuous or discrete cracking)

It is obvious that the methods are presented with increasing demand in modelling complexity and computational power. From these, only the first two are used for the analysis of complete structures while the latter are used mainly for substructures.

The point hinge approach concentrates the nonlinearity of the structural elements in nonlinear rotational springs at the ends of each element, the moment rotation curve of which corresponds to nonlinear behaviour of each element (see Chapter 2, Sections 2.3.5 and 2.5.2, and Section 5.4.4).

The fibre model utilises nonlinear material laws for concrete and reinforcement, and by dividing each element into sections and each section into regions (fibres), calculates, by section analysis, the moment–curvature for each load step for each section and then integrates along the length of the element, thus producing the nonlinear behaviour for each load step.

Modelling a building for nonlinear analysis requires different approaches than that of the modelling for linear analysis, specifically with regard to the following aspects:

- Slab and transfer of loads
- Diaphragm constraint
- Foundation
- R/C wall and cores
- Fibre or point hinge modelling
- Use of safety factors

5.7.2.1 Slab modelling and transfer of loads

Slabs in most modern commercial software packages are modelled by the use of some type of elastic shell element that transfers the gravity loads to the beams, facilitates the slab design itself and applies the diaphragm constraint. However, in nonlinear analysis, the shell elements for slab must not be present, as they will act elastically, thus interfering with the nonlinear behaviour of the T-beams. Therefore, two different problems arise: (i) the transfer of loads from slabs to beams; (ii) the modelling of the diaphragm constraint.

Regarding the first issue (i), some software packages include a macro element for slabs that only transfers the loads at an 'initial step' and then set the slab stiffness matrix to zero [0] in order to perform nonlinear analysis for gravity and lateral loads. In the cases where this option is not present, the loads are transferred manually using a triangulation approach, at the thirds of each beam, as shown in Figure 5.16 from a 16-storey actual building presented in the examples of this chapter.

It should also be noted, for the case of t-h nonlinear analysis, that the additional masses corresponding to the slab self-weight and gravity loads (mass combination) should be transferred directly to the nodes of each vertical element per storey to facilitate the nonlinear analysis. These concepts are shown graphically in Figure 5.16 for the typical storey of an actual building.

5.7.2.2 Diaphragm constraint

Regarding the issue of the diaphragm constraint, there are also nonlinear software packages that allow the use of a rigid diaphragm constraint in selected nodes; however, in cases where this option is not available, the engineer must manually add x-braces with numerically rigid elements (elastic material with $E_{Rigid} = 10 \cdot E_s$ and 1.0×1.0 m cross section) and end releases.

5.7.2.3 R/C walls and cores

As in the case with slabs, R/C walls and cores in most modern commercial software packages are modelled with elastic shell elements, an approach that has simplified the modelling of these elements to the average engineer and has essentially transformed the elastic structural modelling of a building to a 3-D BIM-type approach. Unfortunately, this simplification is not available, and not recommended even if available, for the nonlinear structural modelling as all elements should be simulated using 1-D elements either in point hinge or fibre model approaches. Therefore, the R/C walls and cores have to be modelled following the 'old' rules used in the 1990s for modelling such elements (then elastically).

Rectangular R/C walls should be modelled using one 1-D element corresponding to the dimensions, material properties and reinforcement of the wall itself. This element must be connected to the beams through the use of numerically rigid elements (elastic material with $E_{Rigid} = 10 \cdot E_s$ and 1.0×1.0 m cross section).

R/C cores of arbitrary geometry represent a higher challenge, as they have to follow different modelling approaches for point hinge models and fibre models. In the case of the point hinge model, the core must be modelled using one element placed at the centre of stiffness (shear centre) with the geometrical properties of the core section and a biaxial moment–rotation diagram (with axial force interaction) that corresponds to the nonlinear behaviour of the core. As this is not available by any standard, a detailed section analysis should be performed in order to define this set of curves. In the case of the fibre model, each leg of the core should be modelled as one 1-D element at the centre of stiffness of the leg

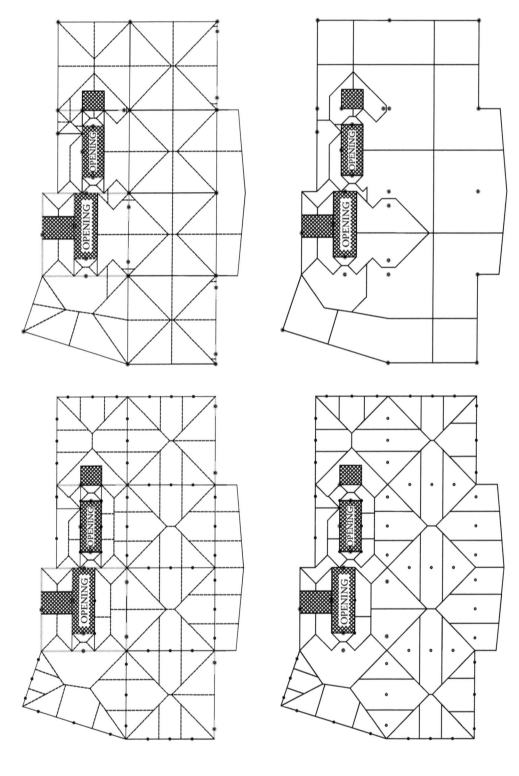

Figure 5.16 Gravity load distribution (left) and mass distribution (right).

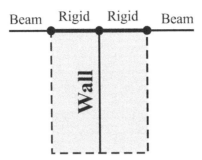

Figure 5.17 Wall modelling (elevation view).

with geometrical properties, material properties and reinforcement corresponding to this leg. In both cases, the 1-D elements are connected to the geometrical extremities of the core using numerically rigid elements (Figure 5.17).

It should be noted that if the proposed arrangement for the fibre model is applied in the point hinge model for the nonlinear simulation of cores, it will lead to a *significant unconservative error*, as the moment capacity of a pair of 1-D elements will be governed by their axial load–deformation curve, which in most approaches is elastic, thus giving the core *false infinite yield moment capacity.*

5.7.2.4 Foundation

It is suggested that the building is modelled as fixed at the level where it is laterally restrained (i.e. foundation box). This approach reduces the size of the nonlinear model and eliminates possibility of mistakes. Cases in which soil–structure interaction or foundation design is required should opt for simpler models of the superstructure and use geotechnical nonlinear software (e.g. Plaxis, 2012).

5.7.2.5 Point hinge versus fibre modelling

Two of the key issues in the point hinge modelling are the initial stiffness and the ultimate available rotation of each structural element. As the approach is approximate, these two assumptions seriously affect the results and their credibility. Usually the fibre model approach is used as a benchmark.

5.7.2.5.1 Initial stiffness

In U.S. guidelines such as FEMA356 or ATC40, the aforementioned modelling problems are dealt with in a, more or less, straightforward fashion. As far as reinforced concrete members elastic stiffness (second level) is concerned, the values defined in Table 6-5 of FEMA 356 where the beams are modelled with 50% of their uncracked flexural rigidity while columns with 50–70% of their uncracked flexural rigidity (Table 5.3) are proposed.

EC8-1/2004 suggests the use of cracked sections as defined in its Chapter 4.3.1, that is, 50% of the gross section for all elements. On the other hand, in text books (Leonhardt, 1977), the rigidity is modelled as a function of the flexural reinforcement ratio and varies from 25% to 50% of the uncracked rigidity (Figure 5.18).

Table 5.3 FEMA 356 cracked flexural and shear rigidity

Component	Flexural rigidity	Shear rigidity
Beams – nonprestressed	$0.5E_cI_g$	$0.4E_cA_w$
Beams – prestressed	E_cI_g	$0.4E_cA_w$
Columns with compression due to design gravity loads $\geq 0.5\ A_gf_c$	$0.7E_cI_g$	$0.4E_cA_w$
Columns with compression due to design gravity loads $\leq 0.3\ A_gf_c$ or with tension	$0.5E_cI_g$	$0.4E_cA_w$
Walls – uncracked (on inspection)	$0.8E_cI_g$	$0.4E_cA_w$
Walls – cracked	$0.5E_cI_g$	$0.4E_cA_w$
Flat slabs – nonprestressed	See FEMA 356/6.5.4.2	$0.4E_cA_g$
Flat slab – prestressed	See FEMA 356/6.5.4.2	$0.4E_cA_g$

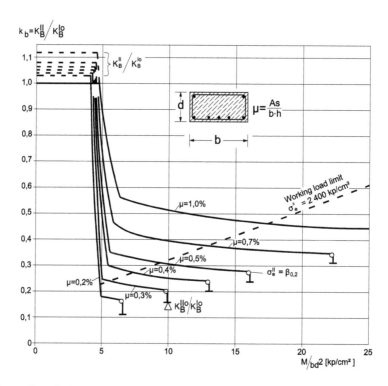

Figure 5.18 Ratio of cracked over uncracked rigidity as a function of reinforcement percentage.

One should be very careful when defining the initial stiffness in a point hinge model, as in most software packages, the initial elastic behaviour is defined by the elastic properties of the element while the point hinge is activated at the yield moment, thus requiring a rigid–plastic point hinge model, instead of an elastic–plastic one (that is prescribed in the codes), while the initial stiffness must be introduced in the general properties (by section modifiers or reduced E modulus) and not in the point hinge modelling.

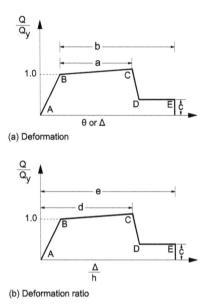

(a) Deformation

(b) Deformation ratio

Figure 5.19 Point hinge constitutive law suggested by U.S. guidelines. (a) Deformation; (b) deformation ratio.

5.7.2.5.2 Ultimate rotation

As far as member ductility is concerned, a very precise bilinear diagram is proposed, shown in principle in Figure 5.19 of the FEMA356, while the values for this diagram are derived from tables as a function of the axial and shear stress of the element (Table 5.4).

EN1998-3, on the other hand, provides very detailed equations for the derivation of these values (Annex A), and thus renders the whole approach very cumbersome and prone to errors.

The problem with all these data is that no analytical simulations of experimentally tested structures have been provided, making the user quite sceptic of their accuracy. This exercise, on the contrary, has been performed for some of the analytical nonlinear software packages used by academia, and the results have been very encouraging (ICONS project using Adaptic, 2012; Seismostruct, 2016; Elnashai et al. (ZEUS), 2002–2005).

5.7.2.6 Safety factors

Amongst the several parameters that affect the results of nonlinear analysis is the proper use of safety factors, local and global. More specifically of interest are, for local values, the material properties safety factors and the loads safety factors, as well as the global safety factor of the structure. Indicatively, FEMA356 uses the reliability of the material factor ($k = 0.75$–1.00) and characteristic values are used for gravity loads and mean values for deformation values without any additional safety factor. The safety factors are introduced in the end results, that is, the rotational capacity of structural elements that are clearly conservative (yet quantified), and the required over available ductility that ATC40 introduces a factor of 2.50.

For EC8-3/2005, the use of safety factors depends on several aspects, as is shown in Table 5.5 (where CF stands for confidence factor as defined in Chapter 13.8.3).

It is mainly up to the designing engineer to decide upon the safety factors introduced, and since the analysis is nonlinear and the analogy between input and output, which exists in

Table 5.4 Adaptation of the values for the point hinge diagram of columns (FEMA 356, Tables 5.6 through 5.8) to SI units and Eurocode notation

Conditions			Modelling parameters			Acceptance criteria				
			Plastic rotation angle, [rad]		Residual strength ratio	Plastic rotation angle, [rad]				
						Performance level				
						Component type				
						Primary			Secondary	
Columns controlled by flexure										
$\dfrac{N_{sd}}{b \cdot d \cdot f_{ck}}$	Stirrups	$\dfrac{V_{sd}}{b_w \cdot d \cdot \sqrt{f_{ck}}}$	a	b	c	IO	LS	CP	LS	CP
≤ 0.1	C	≤ 0.25	0.02	0.03	0.2	0.005	0.015	0.02	0.02	0.03
≤ 0.1	C	≥ 0.5	0.016	0.024	0.2	0.005	0.012	0.016	0.016	0.024
≥ 0.4	C	≤ 0.25	0.015	0.025	0.2	0.003	0.012	0.015	0.018	0.025
≥ 0.4	C	≥ 0.5	0.012	0.02	0.2	0.003	0.01	0.012	0.013	0.02
≤ 0.1	NC	≤ 0.25	0.006	0.015	0.2	0.005	0.005	0.006	0.01	0.015
≤ 0.1	NC	≥ 0.5	0.005	0.012	0.2	0.005	0.004	0.005	0.008	0.012
≥ 0.4	NC	≤ 0.25	0.003	0.01	0.2	0.002	0.002	0.003	0.006	0.01
≥ 0.4	NC	≥ 0.5	0.002	0.008	0.2	0.002	0.002	0.002	0.005	0.008

Table 5.5 Values of material properties and criteria for analysis and safety verifications

Type of element or mechanism (e/m)	Linear model (LM)		Nonlinear model		q-factor approach	
	Demand	Capacity	Demand	Capacity	Demand	Capacity
Ductile	Acceptability of Linear Model (for checking of $\rho_i=D_i/C_i$ values): From analysis. Use mean values of properties in model. Verification (if LM accepted): From analysis.	In terms of strength. Use mean values of properties. In terms of deformation. Use mean values of properties divided by CF.	From analysis. Use mean values of properties in model.	In terms of strength. Use mean values of properties divided by CF and by partial factor.	From analysis	In terms of strength. Use mean values of properties divided by CF and by partial factor.
Brittle	Verification (if LM accepted): If $\rho_i \leq 1$: from analysis. If $\rho_i > 1$: from equilibrium with strength of ductile e/m. Use mean values of properties multiplied by CF.	In terms of strength. Use mean values of properties divided by CF and by partial factor.		In terms of strength. Use mean values of properties divided by CF and by partial factor.	In accordance with the relevant Section of EN1998-1:2004.	

linear analysis, is not guaranteed, it indeed makes sense to use unfactored values and opt for a global safety factor in terms of ductility or displacements. However, it is suggested that in design applications (not assessment), *the use of design material properties is adopted* in order to safeguard against *possible legal or insurance implications,* thus obviously reducing the global safety factor opted for.

5.7.3 Pushover analysis

Taking into account all the issues raised previously regarding static and dynamic nonlinear analysis currently, the nonlinear method used for design purposes (FEMA 356, ATC40, EC8-1/2004) is the static nonlinear analysis, widely known as 'pushover'. According to EC8-1/2004, it is defined as 'a non-linear static analysis carried out under conditions of constant gravity loads and monotonically increasing horizontal loads'.

The static nonlinear analysis produces as a basic result the $P–\delta$ curve of the building, which demonstrates the capacity of the building to lateral loads (depicts the base shear capacity over lateral displacement). This curve corresponds to the multi-degree-of-freedom (MDOF) system, which with appropriate coefficients are transformed to the curve of the equivalent single-degree-of-freedom (SDOF) system oscillator called capacity spectrum (Figure 5.20).

Additionally, the displacement pattern and condition (level of developed nonlinearity) of each structural element is produced for each time step.

Three key points are defined on the capacity spectrum of a building:

- *Yield capacity*: Corresponds to the base shear which limits the linear-elastic behaviour of the building
- *Design capacity*: Correspond to the design base shear (V_d), which should be less than the yield base shear as safety factor for materials is used, as well as construction guidelines (minimum reinforcement, stirrups, capacity rules, etc.), which increase the actual base shear capacity.
- *Ultimate capacity*: Correspond to the maximum base shear when the building has been fully plasticised

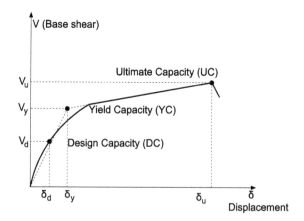

Figure 5.20 Capacity spectrum.

The following approximate relations connect those key points:

$$YC(V_y, \Delta_y): V_y = \gamma C_s \quad \Delta_y = \frac{V_y}{4\pi^2} T^2 \tag{5.73}$$

$$UC(V_u, \Delta_u): V_u = \lambda V_y = \lambda \gamma C_s \quad \Delta_u = \lambda \mu V_y = \lambda \mu \gamma C_s \frac{T^2}{4\pi^2} \tag{5.74}$$

where
  C_s is the seismic coefficient (percentage of building effective weight).
  T is the elastic fundamental period of the building.
  γ is the design overstrength factor (yield base shear/design base shear).
  λ is the actual overstrength factor (ultimate base shear/yield base shear).
  μ is the total ductility factor = δ_u/δ_y.

From the above relations, the building behaviour factor (q) may be defined as

$$\boxed{q = q_{os} \cdot q_\zeta \cdot q_\mu}$$

where
  q_{os} is the overstrength behaviour factor $q_{os} = 0.50(V_u + V_y)/V_d$.
  q_ζ is the damping behaviour factor = 1.0 for R/C buildings.

$$q_\mu : \begin{cases} \text{Ductility behaviour factor} = \mu & \text{for } T > 0.50\,\text{s} \\ \text{Ductility behaviour factor} = 1/\sqrt{2\mu - 1} & \text{for } T < 0.50\,\text{s} \end{cases}$$

as shown in Chapter 2.3.4.

5.7.4 Pros and cons of pushover analysis

Static nonlinear analysis facilitates the assessment of the behaviour of a structure by calculating the available strength capacity of the structural elements and the corresponding deformations (ductilities) and comparing them to the corresponding demands as defined for the design earthquake. The assessment is based on several parameters of structural behaviour as total drift, inter-storey drifts, ductility demand versus available ductility, node loads and element loads. The key benefit in using pushover analysis is that the designer has an estimate of the developed forces and deformations as the building enters the nonlinear range, taking into account the changes in the stiffness of the individual members and the redistribution of forces that take place.

The basic data that result from a pushover analysis, and not from an elastic static or dynamic analysis, are the following:

- Realistic values of forces on brittle elements such as short columns, coupling beams or high-depth beams and so on
- Estimate of the total (inelastic) deformations that must develop at critical parts of structural elements so that the structure can dissipate the seismic energy

- Effect of the yield or failure of a structural element on the overall behaviour of the structure and the redistribution of forces
- Pinpointing of the critical areas of elements that require high available ductility
- Highlighting of structural asymmetries in plan or along height, resulting from the plasticisation of critical elements that affect the structural and dynamic behaviour of the building (torsional effects, infill walls, soft storeys)
- Soft storey identification, either due to abrupt changes in stiffness, or more importantly due to abrupt changes in strength, along the height
- Capability of modelling and assessing the effect of infills, which when distributed unevenly seriously alter the desired behaviour of a building

The accuracy and validity of all of the aforementioned results depends highly on the assumptions and modelling approaches used during the analysis, especially regarding hysteretic behaviour of materials or whole elements, the load pattern and/or the corresponding displacement pattern and so on. Useful for the better understanding of the limits and capabilities of the pushover analysis is the following brief demonstration of principles upon which the method is based.

Pushover analysis and the assessment of buildings based upon it rely on the following two assumptions:

1. The response of an MDOF system that can be accurately represented by an equivalent SDOF system even in post-elastic range.
2. The behaviour is guided by a modeshape that remains constant during the duration of the excitation.

It is obvious that both assumptions are not entirely accurate; however, several parametric studies have shown that the estimate of the maximum seismic response of MDOF systems is acceptable provided that the behaviour is indeed guided by one modeshape (Figure 5.21; Fajfar and Fischinger, 1994; Saiidi and Sozen, 1981).

Regarding the second assumption, the works of Elnashai (2000) and Antoniou et al. (2002) demonstrate an extended pushover analysis, called adaptive pushover, which takes into account the following:

1. The changes (shifts) of the eigen modes, as several elements yield and cause a change in the stiffness matrix (K) of the structure, are taken into account by performing an eigen mode analysis at every 'significant' change of the stiffness matrix.
2. Higher significant eigen modes are taken into account corresponding to their participation mass ratio, as it results from the eigen analysis.
3. Use of the spectra of the specific excitations considered, from which the participating ratios are weighted.

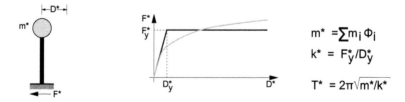

$$m^* = \sum m_i \, \Phi_i$$
$$k^* = F_y^*/D_y^*$$
$$T^* = 2\pi\sqrt{m^*/k^*}$$

Figure 5.21 Equivalent SDOF system. (From Fajfar, P. and Dolsek, M. 2000. A transparent nonlinear method for seismic performance evaluation. *3rd Workshop of the Japan-UK Seismic Risk Forum, Proceedings*, Imperial College Press, 2000.)

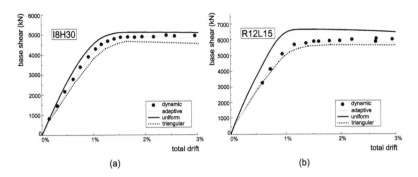

Figure 5.22 Adaptive pushover, pushover and nonlinear T-h: (a) asymmetric system of high ductility; (b) symmetric system of low ductility.

Results based on this type of analysis, of buildings sensitive to pushover analysis, are extremely close to ones by nonlinear time–history dynamic analysis for different levels of excitation. Indicative application of the method on an eight-storey non-symmetric moment-resisting frame (MRF), a 12-storey symmetric MRF and an 8-storey dual system has been presented in the work of Antoniou et al. (2002). Figure 5.22 shows the accuracy of the method compared to the nonlinear *t-h* analysis and the triangular and uniform distribution of loads for pushover analysis.

5.7.5 Equivalent SDOF systems

In modern earthquake engineering, the seismic excitation is either defined by design spectra or accelerograms; however, in codified design, the input is always defined, or at least correlated to spectra, which provide information about the acceleration and/or displacement of SDOF systems. Therefore, it is essential to correlate the results of the pushover analysis, which correspond the MDOF system to the properties of an equivalent SDOF system, taking into account the nonlinearity of the response both in terms of forces (base shear) and deformations. This is tackled by the use of an equivalent SDOF oscillator, a version of which is presented in the following two paragraphs, the first referring to the typical translational case, while the latter in the case of translational and rotational (torsional) behaviour.

5.7.5.1 Equivalent SDOF for torsionally restrained buildings

Several different approaches for the definition of the equivalent SDOF oscillator are available in literature, yet all start using the basic assumption that the deformation of the MDOF system may be described by a deformation vector [Φ], which remains constant through the loading time–history, regardless of the magnitude of the applied deformation.

In this paragraph, the well-known methodology for the definition of the SDOF oscillator for translational behaviour of spatial (3-D) structures, which was developed in the early work of Saiidi and Sozen (1981) (Krawinkler and Nassar, 1992), is presented, while in the next paragraph, an adaptation for translational and torsional deformation (asymmetric buildings) will be presented.

The equation describing the dynamic elastic response of the system (Figure 5.23) to excitation in vector form is

$$[M][\ddot{u}(t)] + [C][\dot{u}(t)] + [P(t)] = -[M][1]\ddot{u}_o(t) \tag{5.75}$$

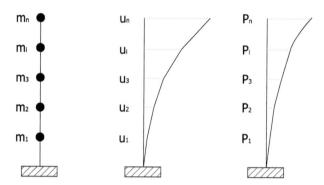

Figure 5.23 Modal deformation of a torsionally insensitive (translational) MDOF system.

By eliminating the damping terms from Equation 5.75, $[C][\dot{u}(t)]$, Equation 5.76 results:

$$[M][\ddot{u}(t)]+[P(t)]=-[M][1]\ddot{u}_o(t) \tag{5.76}$$

where

$$[M]=\begin{bmatrix} m_1 & 0 & - & 0 \\ 0 & m_2 & - & 0 \\ - & - & - & - \\ 0 & 0 & - & m_n \end{bmatrix}_n \quad \text{is the mass matrix}$$

$$[u]=\begin{bmatrix} u_1 \\ u_2 \\ - \\ u_n \end{bmatrix}, \text{ displacement vector}$$

$$[P]=\begin{bmatrix} P_1 \\ P_2 \\ - \\ P_n \end{bmatrix}, \text{ restoring force (internal) vector}$$

It is assumed that the displacement $[u]$ and the restoring force $[P]$ of the elastic MDOF system can be correlated to the corresponding parameters of the equivalent SDOF nonlinear oscillator u_n and P_n as a function of two vectors $[\Phi]$ and $[\Psi]$:

$$[u(t)]=[\Phi]u_n(t)=\begin{bmatrix} \varphi_1 \\ \varphi_2 \\ - \\ \varphi_n \end{bmatrix}u_n(t) \tag{5.77}$$

$$[P] = [\Psi]P_{\mathrm{n}}(t) = \begin{bmatrix} \Psi_1 \\ \Psi_2 \\ - \\ \Psi_{\mathrm{n}} \end{bmatrix} P_{\mathrm{n}}(t) \tag{5.78}$$

Therefore, using these transformations, the equation of vibration of the MDOF system becomes, in vector form

$$[M][\Phi]\ddot{u}_{\mathrm{n}}(t) + [\psi]P_{\mathrm{n}}(t) = -[M][1]\ddot{u}_{\mathrm{o}}(t) \tag{5.79a}$$

and in algebraic form

$$\left. \begin{aligned} m_1\varphi_1\ddot{u}_{\mathrm{n}}(t) + \psi_1 P_{\mathrm{n}} &= -m_1\ddot{u}_{\mathrm{o}}(t) \\ m_2\varphi_2\ddot{u}_{\mathrm{n}}(t) + \psi_2 P_{\mathrm{n}} &= -m_2\ddot{u}_{\mathrm{o}}(t) \\ \text{---------------------------------} \\ m_{\mathrm{n}}\varphi_{\mathrm{n}}\ddot{u}_{\mathrm{n}}(t) + \psi_{\mathrm{n}} P_{\mathrm{n}} &= -m_{\mathrm{n}}\ddot{u}_{\mathrm{o}}(t) \end{aligned} \right\} \tag{5.79b}$$

By multiplying Equation 5.79a times $[\Phi]^T$:

$$[\Phi]^{\mathrm{T}}[M][\Phi]\ddot{u}_{\mathrm{n}}(t) + [\Phi]^{\mathrm{T}}[\Psi]P_{\mathrm{n}}(t) = -[\Phi]^{\mathrm{T}}[M]\{1\}\ddot{u}_{\mathrm{o}}(t) \tag{5.80}$$

and by transforming the first term

$$[\Phi]^{\mathrm{T}}[M][\Phi] \frac{[\Phi]^{\mathrm{T}}[M]\{1\}}{[\Phi]^{\mathrm{T}}[M]\{1\}} \ddot{u}_{\mathrm{n}}(t) + [\Phi]^{\mathrm{T}}[\Psi]P_{\mathrm{n}}(t) = -[\Phi]^{\mathrm{T}}[M]\{1\}\ddot{u}_{\mathrm{o}}(t) \tag{5.81a}$$

or

$$[\Phi]^{\mathrm{T}}[M][1] \frac{[\Phi]^{\mathrm{T}}[M][\Phi]}{[\Phi]^{\mathrm{T}}[M]\{1\}} \ddot{u}_{\mathrm{n}}(t) + [\Phi]^{\mathrm{T}}[\Psi]P_{\mathrm{n}}(t) = -[\Phi]^{\mathrm{T}}[M]\{1\}\ddot{u}_{\mathrm{o}}(t) \tag{5.81b}$$

By defining

$$u^* = \frac{[\Phi]^{\mathrm{T}}[M][\Phi]}{[\Phi]^{\mathrm{T}}[M]\{1\}} u_{\mathrm{n}}(t) \tag{5.82a}$$

$$m^* = [\Phi]^{\mathrm{T}}[M]\{1\} \tag{5.82b}$$

Equation 5.81a becomes in vector form

$$m^* \ddot{u}^*(t) + [\Phi]^{\mathrm{T}}[\Psi]\left[P_{\mathrm{n}}(t)\right] = -m^* \ddot{u}_{\mathrm{o}}(t) \tag{5.83}$$

while Equations 5.82a and 5.82b are in algebraic form

$$m^* = m_1\varphi_1 + m_2\varphi_2 + \cdots + m_n\varphi_n = \sum_1^n m_i\varphi_i$$

$$[\Phi]^T[M][\Phi] = m_1\varphi_1^2 + m_2\varphi_2^2 + \cdots + m_n\varphi_n^2 = \sum_1^n m_i\varphi_i^2$$

$$[\Phi]^T[\Psi] = \sum_1^n \psi_i\varphi_i$$

meaning

$$m^* = \sum_1^n m_i\varphi_i \tag{5.84a}$$

$$u_n^*(t) = \frac{\sum_1^n m_i\varphi_i^2}{\sum_1^n m_i\varphi_i} u_n(t) \tag{5.84b}$$

$$[\Phi]^T[\Psi] = \sum_1^n \psi_i\varphi_i \tag{5.84c}$$

Equation 5.83 by introducing Equations 5.84a through 5.84c is transformed to

$$\left[\sum_1^n m_i\varphi_i\right]\ddot{u}^*(t) + \left[\sum_1^n m_i\psi_i\right]P_n(t) = -\left[\sum_1^n m_i\varphi_i\right]\ddot{u}_o(t) \tag{5.85a}$$

Taking into account that

$$\left[\sum_1^n m_i\psi_i\right]P_n(t) = V(t) \text{ or } P_n(t) = \frac{V(t)}{\left[\sum_1^n m_i\psi_i\right]} \tag{5.86}$$

where V is the base shear of the MDOF excited system.
Equation 5.85a may be re-written as

$$m^*\ddot{u}^*(t) + \frac{\sum_1^n \varphi_i\psi_i}{\sum_1^n \psi_i} V(t) = -m^*\ddot{u}_o(t) \tag{5.85b}$$

This corresponds to an SDOF system, which has defined the following properties:

$$m^* = [\Phi]^T[M]\{1\} = \sum_1^n m_i \varphi_i$$

(see Equation 5.84a)

$$V^*(t) = \frac{[\Phi]^T[\psi]}{\{1\}^T[\psi]} V(t) = \frac{\sum_1^n \varphi_i \psi_i}{\sum_1^n \psi_i} V(t)$$

(5.87)

where m^*, $u^*(t)$ and V^* are the mass, displacement and base shear of the equivalent SDOF oscillator, respectively.

The equation of vibration under excitation for this SDOF system is

$$m^* \ddot{u}^*(t) + V^*(t) = -m^* \ddot{u}_o(t)$$

(5.88)

having

$$u^*(t) = \Gamma_1 u(t)$$

(5.89a)

$$V^*(t) = \Gamma_2 V(t)$$

(5.89b)

$$\Gamma_1 = \frac{[\Phi]^T[M][\Phi]}{[\Phi]^T[M]\{1\}} = \frac{\sum_1^n m_i \varphi_i^2}{\sum_1^n m_i \varphi_i}$$

(5.89c)

$$\Gamma_2 = \frac{[\Phi]^T[\Psi]}{\{1\}^T[\Psi]} = \frac{\sum_1^n \varphi_i \psi_i}{\sum_1^n \psi_i}$$

(5.89d)

By introducing

$$\psi_i = \varphi_i$$

one gets

$$\Gamma_2 = \frac{\sum_1^n \varphi_i^2}{\sum_1^n \varphi_i}$$

(5.90a)

and for $m_i = m =$ constant

$$\Gamma_1 = \frac{\sum_1^n m_i \varphi_i^2}{\sum_1^n m_i \varphi_i} = \frac{m \sum_1^n \varphi_i^2}{m \sum_1^n \varphi_i} = \Gamma_2 \qquad (5.90b)$$

so in the end, one factor Γ is used, with

$$\Gamma = \frac{\sum_1^n \varphi_i^2}{\sum_1^n \varphi_i} \qquad (5.91)$$

This transformation factor is denoted as Γ in the Annex B of EC8-1/2004, Equation B3:

$$\Gamma = \frac{m^*}{\sum_1^n m_i \varphi_i^2} = \frac{\sum_1^n m_i \varphi_i}{\sum_1^n m_i \varphi_i^2} \qquad (5.92)$$

5.7.5.2 Equivalent SDOF for torsionally unrestrained buildings

This approach uses for the definition of the response quantities a generalised equivalent SDOF system with both translational and torsional response, by extending the methodology presented in the previous paragraph, wherein only the translational characteristics were accounted for.

Consider a multi-storey monosymmetric (i.e. stiffness symmetric with respect to one axis only, in the case of Figure 5.24, the horizontal axis) building (Paulay, 1996, 1997). The centre of mass is denoted as CM and the centre of resistance as CR. The equations

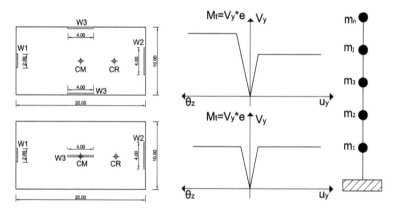

Figure 5.24 Torsionally restrained (TR-top) and torsionally unrestrained (TUR-bottom) multi-storey buildings.

describing the dynamic elastic response of the system to excitation in one direction (here, the y-direction) for an undamped system are (Chopra, 1995)

1st storey

$$m_1\ddot{u}_{y1}(t) + \sum_{j=1}^{n} u_{yj}(t) \cdot \sum_{1}^{i} k_{yi1j} + \sum_{j=1}^{n} \theta_{z1} \cdot \sum_{1}^{i} k_{yi1j}\alpha_{xi1} = -m_1\ddot{u}_{oy}(t)$$

$$m_1 r_1^2 \ddot{\theta}_{z1} + \sum_{j=1}^{n} u_y(t) \sum_{1}^{i} \alpha_{xi1} k_{yi1j} + \sum_{j=1}^{n} \theta_{zj}(t)\left(\sum_i k_{t1j} + \sum_i \alpha_{yi1}^2 k_{xi1j} + \sum_i \alpha_{xi1}^2 k_{yi1}\right) = 0$$

jth storey

$$m_j\ddot{u}_{yj}(t) + \sum_{j=1}^{n} u_{yj}(t) \cdot \sum_{1}^{i} k_{yijj} + \sum_{j=1}^{n} \theta_{zj} \cdot \sum_{1}^{i} k_{yijj}\alpha_{xij} = -m_j\ddot{u}_{oy}(t)$$

$$m_j r_j^2 \ddot{\theta}_{zj}(t) + \sum_{j=1}^{n} u_{yj}(t) \sum_{1}^{i} \alpha_{xij} k_{yijj} + \sum_{j=1}^{n} \theta_{zj}(t)\left(\sum_i k_{tjj} + \sum_i \alpha_{yij}^2 k_{xijj} + \sum_i \alpha_{xij}^2 k_{yij}\right) = 0$$

(5.93a)

nth storey

$$m_n\ddot{u}_{yn}(t) + \sum_{j=1}^{n} u_{yj}(t) \cdot \sum_{1}^{i} k_{yinj} + \sum_{j=1}^{n} \theta_{zn} \cdot \sum_{1}^{i} k_{yinj}\alpha_{xin} = -m_n\ddot{u}_{oy}(t)$$

$$m_n r_n^2 \ddot{\theta}_{zn}(t) + \sum_{j=1}^{n} u_y(t) \sum_{1}^{i} \alpha_{xin} k_{yinj} + \sum_{j=1}^{n} \theta_{zj}(t)\left(\sum_i k_{tnj} + \sum_i \alpha_{yin}^2 k_{xinj} + \sum_i \alpha_{xin}^2 k_{yin}\right) = 0$$

or, in vector notation,

$$\{M\}\{\ddot{u}\} + \{K\}\{u\} = \{M\}\{\delta\}\ddot{u}_{oy}(t)$$

(5.93b)

where

$$M = \begin{bmatrix} m_1 & 0 & \cdots & 0 & 0 & \cdots & 0 & 0 \\ 0 & m_1 r_1^2 & 0 & 0 & 0 & 0 & 0 & 0 \\ 0 & 0 & 0 & 0 & 0 & 0 & 0 & 0 \\ 0 & 0 & \cdots & m_j & 0 & \cdots & 0 & 0 \\ 0 & 0 & 0 & 0 & m_j r_j^2 & 0 & 0 & 0 \\ 0 & 0 & \cdots & 0 & 0 & \cdots & 0 & 0 \\ 0 & 0 & 0 & & 0 & 0 & m_n & \\ 0 & 0 & \cdots & 0 & 0 & \cdots & 0 & m_n r_n^2 \end{bmatrix}$$

$$K = \begin{bmatrix} K_{11} & K_{12} & \cdots & K_{1,2 \cdot j-1} & K_{1,2 \cdot j} & \cdots & K_{1,n-1} & K_{1,2 \cdot n} \\ K_{21} & K_{22} & \cdots & K_{2,2 \cdot j-1} & K_{2,2 \cdot j} & \cdots & K_{2,n-1} & K_{2,2 \cdot n} \\ \vdots & \vdots & \vdots & \vdots & \vdots & \vdots & \vdots & \vdots \\ K_{2 \cdot j-1,1} & K_{2 \cdot j-1,2} & \cdots & K_{2 \cdot j-1,2 \cdot j-1} & K_{2 \cdot j-1,2 \cdot j} & \cdots & K_{2 \cdot j-1,2 \cdot n-1} & K_{2 \cdot j-1,2 \cdot n} \\ K_{2 \cdot j,1} & K_{2,2 \cdot j} & \cdots & K_{2 \cdot j,2 \cdot j-1} & K_{2 \cdot j,2 \cdot j} & \cdots & K_{2 \cdot j,2 \cdot n-1} & K_{2 \cdot j,2 \cdot n} \\ \vdots & \vdots & \vdots & \vdots & \vdots & \vdots & \vdots & \vdots \\ K_{2 \cdot n-1,1} & K_{2,2 \cdot n-1} & \cdots & K_{2 \cdot n-1,2 \cdot j-1} & K_{2 \cdot n-1,2 \cdot j} & \cdots & K_{2 \cdot n-1,2 \cdot n-1} & K_{2 \cdot n-1,2 \cdot n} \\ K_{2 \cdot n,1} & K_{2 \cdot n,2} & \cdots & K_{2 \cdot n,2 \cdot j-1} & K_{2 \cdot n,2 \cdot j} & \cdots & K_{2 \cdot n,2 \cdot n-1} & K_{2 \cdot n,2 \cdot n} \end{bmatrix}$$

$$\{\delta\} = \begin{bmatrix} 1 \\ 0 \\ \vdots \\ 1 \\ 0 \\ \vdots \\ 1 \\ 0 \end{bmatrix} \quad \{\ddot{u}\} = \begin{bmatrix} \ddot{u}_{y1} \\ \ddot{\theta}_{z1} \\ \vdots \\ \ddot{u}_{yj} \\ \ddot{\theta}_{zj} \\ \vdots \\ \ddot{u}_{yn} \\ \ddot{\theta}_{zn} \end{bmatrix}, \{u\} = \begin{bmatrix} u_{y1} \\ \theta_{z1} \\ \vdots \\ u_{yj} \\ \theta_{zj} \\ \vdots \\ u_{yn} \\ \theta_{zn} \end{bmatrix}$$

and

$$K_{11} = \sum_1^i k_{yi1,1} \quad K_{12} = \sum_1^i k_{yi1,1} \alpha_{xi1}$$

$$K_{21} = \sum_1^i k_{yi1,1} \alpha_{xi1} \quad K_{22} = \sum_1^i k_{ti1,1} + \sum_1^i \alpha_{yi1}^2 k_{xi1,1} + \sum_1^i \alpha_{xi1}^2 k_{yi1,1}$$

....

$$K_{2j-1,2j-1} = \sum_1^i k_{yij,j} \quad K_{2j-1,2j} = \sum_1^i k_{yij,j} e_{xij}$$

$$K_{2j,2j-1} = \sum_1^i k_{yij,j} \alpha_{xij} \quad K_{2j,2j} = \sum_1^i k_{tij,j} + \sum_1^i \alpha_{yij}^2 k_{xij,j} + \sum_1^i \alpha_{xij}^2 k_{yij,j}$$

....

$$K_{2n-1,2n-1} = \sum_1^i k_{yin,n} \quad K_{2n-1,2n} = \sum_1^i k_{yin,n} \alpha_{xin}$$

$$K_{2n,2n-1} = \sum_1^i k_{yin,n} \alpha_{xin} \quad K_{2n,2n} = \sum_1^i k_{tin,n} + \sum_1^i \alpha_{yin}^2 k_{xin,n} + \sum_1^i \alpha_{xin}^2 k_{yin,n}$$

where

$m_1,\ldots m_j,\ldots m_n$ are the vibrating masses on storey 1, j and n, respectively.

k_{xi1j}, k_{yi1j} are the translational stiffnesses of individual resisting element (i) in the x- and y-directions, respectively, on storey 1 for unit displacement of storey j.

$k_{ti,1,j}$ are the rotational stiffnesses of individual resisting element (direction zz) on storey 1 for unit rotation of storey j.

α_{xij}, α_{yij} are the distances from individual resisting element i to the centre of mass (C_M) in x and y direction, respectively, on storey j.

$u_{yj}(t)$, $\theta_{zj}(t)$ are the displacement and rotation of the C_M as a function of time, on storey j.

$\ddot{u}_{oy}(t)$ is the excitation acceleration, as a function of time.

$\ddot{u}_{yj}(t)$, $\ddot{\theta}_{zj}(t)$ are the second derivatives of $u_y(t)$ and $\theta_z(t)$ with respect to time (translational and angular accelerations) on storey j.

In order to approximate the inelastic dynamic vibration of the system with an equivalent static one, by analogy to the case of translational multi-DOF system (presented in the previous paragraph), the following assumption is adopted:

$$\{u\} = \begin{bmatrix} u_{y1} \\ \theta_{z1} \\ \vdots \\ u_{yj} \\ \theta_{zj} \\ \vdots \\ u_{yn} \\ \theta_{zn} \end{bmatrix} = \begin{bmatrix} \varphi_{yo1} \\ \varphi_{zo2} \\ \vdots \\ \varphi_{yoj} \\ \varphi_{zoj} \\ \vdots \\ \varphi_{yon} \\ \varphi_{zon} \end{bmatrix} \cdot \bar{u}_y(t) \qquad \text{or} \qquad \{u\} = \{\varphi_0\}\bar{u}_y(t) \tag{5.94}$$

and

$$\{p\} = \begin{bmatrix} V_{y1} \\ M_{t1} \\ \vdots \\ V_{yj} \\ M_{tj} \\ \vdots \\ V_{yn} \\ M_{tn} \end{bmatrix} = \begin{bmatrix} \psi_{Po1} \\ \psi_{Mo1} \\ \vdots \\ \psi_{Poj} \\ \psi_{Moj} \\ \vdots \\ \psi_{Pon} \\ \psi_{Mon} \end{bmatrix} \cdot \bar{p}(t) \qquad \text{or} \qquad \{p\} = \{\psi_o\}\bar{p}(t) \tag{5.95}$$

where

V_{yj}, M_{tj} are the generalised lateral force and torque at the CM of storey j.

$\bar{u}_y(t), \bar{p}(t)$ are the 'dummy' time functions of displacement and force, respectively.

Hence, the assumption made is that the storey displacements (u_{yj}, θ_{zj}), as well as the storey forces, are expressed as a function of time using the preselected vectors φ_o (normalised spectral modal displacements u_{ymax}, θ_{ymax}) and ψ_o (normalised spectral modal loads V_{max}, M_{max}), and corresponding dummy functions that will be eliminated at the end of the procedure, hence need no explicit definition.

Referring to Equation 5.93a, the term

$$\sum_{j=1}^{n} u_{yj}(t) \cdot \sum_{1}^{i} k_{yijj} + \sum_{j=1}^{n} \theta_{zj} \cdot \sum_{1}^{i} k_{yijj} e_{xij}$$

expresses the generalised lateral restoring force for storey j, $\psi_{poj}\bar{p}(t)$,

$$\sum_{j=1}^{n} u_{yj}(t) \cdot \sum_{1}^{i} k_{yijj} + \sum_{j=1}^{n} \theta_{zj} \cdot \sum_{1}^{i} k_{yijj} e_{xij} = \psi_{poj}\bar{p}(t) \tag{5.96a}$$

while the term

$$\sum_{j=1}^{n} u_{y}(t) \sum_{1}^{i} e_{xi1} k_{yi1j} + \sum_{j=1}^{n} \theta_{zj}(t) \left(\sum_{i} k_{t1j} + \sum_{i} e_{yi1}^{2} k_{xi1j} + \sum_{i} e_{xi1}^{2} k_{yi1} \right)$$

expresses the generalised restoring torque for storey j, $\psi_{Moj}\bar{p}(t)$.

$$\sum_{j=1}^{n} u_{y}(t) \sum_{1}^{i} e_{xi1} k_{yi1j} + \sum_{j=1}^{n} \theta_{zj}(t) \left(\sum_{i} k_{t1j} + \sum_{i} e_{yi1}^{2} k_{xi1j} + \sum_{i} e_{xi1}^{2} k_{yi1} \right) = \psi_{Moj}\bar{p}(t) \tag{5.96b}$$

In the previous definitions, the terms expressing the restoring forces of the elastic MDOF system have been replaced by the forces calculated for the inelastic SDOF system.

Using these definitions, Equations 5.93a and 5.93b are reduced to the following:

1st storey

$$m_1 \varphi_{y01} \ddot{u}_y(t) + \psi_{P01}\bar{p}(t) = -m_1 \ddot{u}_{oy}(t)$$

$$m_1 r_1^2 \varphi_{z01} \ddot{u}_y(t) + \psi_{M01}\bar{p}(t) = 0$$

jth storey

$$m_j \varphi_{y0j} \ddot{u}_y(t) + \psi_{P0j}\bar{p}(t) = -m_j \ddot{u}_{oy}(t)$$

$$m_j r_j^2 \varphi_{z0j} \ddot{u}_y(t) + \psi_{M0j}\bar{p}(t) = 0 \tag{5.97a}$$

nth storey

$$m_n \varphi_{y0n} \ddot{u}_y(t) + \psi_{P0n}\bar{p}(t) = -m_n \ddot{u}_{oy}(t)$$

$$m_n r_n^2 \varphi_{z0n} \ddot{u}_y(t) + \psi_{M0n}\bar{p}(t) = 0$$

or, in vector notation,

$$\{M\}\{\varphi_0\}\ddot{u}_y + \{\psi_o\}\bar{p}(t) = -\{M\}\{\delta\}\ddot{u}_{oy}(t) \tag{5.97b}$$

Pre-multiplying Equation 5.97b by $\{\varphi_o\}^T$, the following equation is derived:

$$\{\varphi\}^T\{M\}\{\varphi_0\}\ddot{\bar{u}}_y + \{\varphi_0\}^T\{\psi_o\}\bar{p}(t) = -\{\varphi_0\}^T\{M\}\{\delta\}\ddot{u}_{oy}(t) \tag{5.98}$$

Taking into account that the lateral storey force (storey shear) can be defined as

$$V_y(t) = \{\delta\}\{b\} = \{\delta\}\{\psi_o\}\bar{p}(t) = \sum_{j=1}^{n}\psi_{P0j}\cdot\bar{p}(t) \tag{5.99a}$$

it follows that

$$\bar{p}(t) = \frac{V_y(t)}{\{\delta\}^T\{\psi_0\}} = \frac{V_y(t)}{\sum\limits_{j=1}^{n}\psi_{P0j}} \tag{5.99b}$$

Equation 5.98 can be transformed as follows:

$$\boxed{\{\varphi_0\}^T\{M\}\{\varphi_0\}\ddot{\bar{u}}_y + \frac{\{\varphi_0\}^T\{\psi_o\}\{V_y\}(t)}{\{\delta\}^T\{\psi_0\}} = -\{\varphi_0\}^T\{M\}\{\delta\}\ddot{\bar{u}}_{oy}(t)} \tag{5.100}$$

In order to proceed, the following notation is adopted:

$$m^* = \{\varphi_0\}^T\{M\}\{\delta\} = \sum_{j=1}^{n}m_j\varphi_{y0j} \tag{5.101a}$$

$$u_y^*(t) = \Gamma_1\bar{u}_y(t) \tag{5.101b}$$

$$V_y^*(t) = \Gamma_1 V_y(t) \tag{5.101c}$$

where

$$\Gamma_1 = \frac{\{\varphi_0\}^T\{M\}\{\varphi_0\}}{m^*} = \frac{\sum\limits_{j=1}^{n}m\left(\varphi_{yoj}^2 + r^2\varphi_{zoj}^2\right)}{\sum\limits_{j=1}^{n}m_j\varphi_{yoj}} \tag{5.102a}$$

$$\Gamma_2 = \frac{\{\varphi_0\}^T\{\psi_0\}}{\{\delta\}^T\{\psi_0\}} = \frac{\sum\limits_{j=1}^{n}\left(\varphi_{yoj}\psi_{poj} + \varphi_{zoj}\psi_{Moj}\right)}{\sum\limits_{j=1}^{n}\psi_{poj}} \tag{5.102b}$$

m^*, $u_y^*(t)$ and V_y^* are the mass, displacement and base shear of the equivalent SDOF oscillator, respectively.

Introducing the foregoing notation into Equation 5.100, the following expression is obtained:

$$m^* \ddot{u}_y^*(t) + V_y^*(t) = -m^* \ddot{u}_{oy}(t) \tag{5.103}$$

Equation 5.103 describes the inelastic response of the equivalent SDOF oscillator, while Equations 5.101 and 5.102 describe the relationship between the multi-storey 3-D building (modelled as an MDOF system) and the equivalent SDOF oscillator.

For the case of a single-storey building, Equations 5.101 and 5.102 are simplified as follows:

$$m^* = \{\varphi\}^T \{M\}\{\delta\} = m\varphi_{yo} \tag{5.101a'}$$

$$u_y^*(t) = \Gamma_1 \bar{u}_y(t) \tag{5.101b'}$$

$$V_y^*(t) = \Gamma_2 V_y(t) \tag{5.101c'}$$

where

$$\Gamma_1 = \frac{\{\varphi_o\}^T \{M\}\{\varphi_o\}}{m^*} = \frac{m\varphi_{yo}^2 + mr^2\varphi_{zo}^2}{m\varphi_{yo}} \tag{5.102a'}$$

$$\Gamma_2 = \frac{\{\varphi\}^T \{\psi_o\}}{\psi_{po}} = \frac{\varphi_{po}\psi_{po} + \varphi_{zo}\psi_{Mo}}{\psi_{po}} \tag{5.102b'}$$

For the application of the methodology, it is proposed that a single pushover analysis is used. The load vector is the set of storey forces causing the elastic spectral modal displacements of the building. As one may observe, this vector is slightly different from the spectral modal loads. The spectral modal displacements are a vector resulting by applying the SRSS rule at the displacement caused by the contribution of each mode, while the spectral modal loads are a vector resulting by applying the SRSS rule on each mode base shear contribution. The required vector may be correctly obtained as the storey shears, when applying the spectral modal displacement vector as a displacement constraint.

This methodology has been applied by Penelis and Kappos (2005) in single- and multi-storey buildings. Table 5.6 summarises the results for four different sets of accelerograms Q1 to Q4 (3–10 excitations each), where the target displacements and rotations vary around 10% from the corresponding dynamic *t-h* results for single-storey buildings.

The indicative graphical representation in Figure 5.25 shows the *t-h* results using inelastic dynamic analysis (IDA) versus the corresponding pushover for the case of a torsionally unrestrained (TUR) multi-storey building, and confirms the validity of the method.

It is useful to note here that, torsionally, unrestrained buildings demonstrated significant scatter in the resulting torsional rotations, even for IDA, as it is evident from Figure 5.25.

Table 5.6 Proposed pushover response versus *t–h* nonlinear analysis

	Dynamic			Static
Torsionally restrained single-storey building				
Q1 set	Mean	c.o.v.	Pushover	Diff.%
uy (cm)	0.792	12.86%	0.824	4.02%
Θz (rad)	4.96E–02	14.93%	5.30E–04	6.90%
Q2 set	Mean	c.o.v.	Pushover	Diff.%
uy (cm)	0.799	1.38%	0.783	2.06%
Θz (rad)	4.69E–04	0.73%	5.30E–04	4.94%
Q3 set	Mean	c.o.v.	Pushover	Diff.%
uy (cm)	1.034	32.54%	0.954	7.70%
Θz (rad)	4.90E–04	33.61%	4.41E–04	10.17%
Q4 set	Mean	c.o.v.	Pushover	Diff.%
uy (cm)	1.0709263	50.11%	1.041	2.78%
Θz (rad)	9.02E–04	60.96%	6.00E–04	33.48%
Q4 set pga	Mean	c.o.v.	Pushover	Diff.%
uy (cm)	2.255035	37.92%	1.822	19.20%
Θz (rad)	1.11E–03	43.88%	8.90E–04	19.84%
Torsionally unrestrained single-storey building				
Q1 set	Mean	c.o.v.	Pushover	Diff.%
uy (cm)	0.062	15.27%	0.765	13.61%
Θz (rad)	6.18E–04	23.68%	7.60E–04	23.05%
Q2 set	Mean	c.o.v.	Pushover	Diff.%
uy (cm)	0.128	24.82%	1.150	17.21%
Θz (rad)	1.28E–03	26.05%	1.10E–03	14.00%
Q3 set	Mean	c.o.v.	Pushover	Diff.%
uy (cm)	1.463	21.21%	1.531	4.66%
Θz (rad)	1.22E–03	22.54%	1.30E–03	6.95%
Q4 set	Mean	c.o.v.	Pushover	Diff.%
uy (cm)	5.834	66.11%	7.654	31.20%
Θz (rad)	5.15E–03	72.76%	4.30E–03	16.57%
Q4 set pga	Mean	c.o.v.	Pushover	Diff.%
uy (cm)	5.837	63.88%	5.640	3.37%
Θz (rad)	5.19E–03	69.34%	4.50E–03	13.27%

This high scatter may be explained in the case of torsionally unrestrained buildings (Figure 5.24, bottom) as yielding of one of the lateral resisting elements practically renders the building as pinned, for the torsional degree of freedom, rotating 'freely' around the remaining lateral resisting element (with the exception of the torsional rigidity of the element itself). This obviously represents an upper bound for the scatter of the torsional response and corresponds to a theoretical case, *as actual buildings always have some degree of torsional restraint provided by the existence of more than two lateral resisting elements in both principal directions.*

It should be noted that using the proposed methodology, the resulting pushover curve envelopes the nexus of dynamic response points, as is evident from Figure 5.25.

Obviously, one can also use other approaches, such as the modal pushover analysis presented by Chopra (2002) or the modified N2 method by Fajfar et al. (2005), as pushover analysis of torsionally sensitive buildings is currently an ongoing research issue.

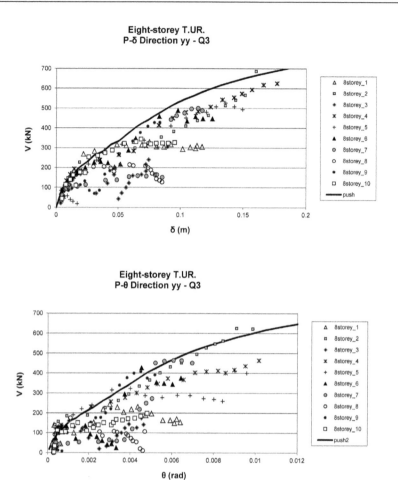

Figure 5.25 IDA of an eight-storey torsionally unrestrained MRF building (V–δ and V–θ charts).

5.7.6 Time–history nonlinear analysis

Time–history nonlinear dynamic analysis is considered to be the most accurate approach in the evaluation of the performance of a building under earthquake excitation, provided that all modelling approaches and input parameters have been introduced correctly. Its complexity has mainly limited this type of analysis to academic use, where it is often the benchmark approach for buildings, while it has application in the field by practicing engineers, on the modelling of seismically isolated buildings, where, however, the nonlinearity is limited to the isolator devices while the rest of the building is modelled elastically.

5.7.6.1 Input motion scaling of accelerograms

The input motion in the *t–h* nonlinear analysis is always a set of accelerograms. There are several types of accelerograms:

- Recorded accelerograms of previous earthquakes
- Artificial accelerograms compiled manually or using algorithms to match spectra
- Hybrid accelerograms, resulting from the modification of a recorded one in order to match a selected spectra type

Although intuitively an engineer would tend towards using actual recorded accelerograms, which include a more representative frequency content, the code specifications that require the use of a number of accelerograms with spectra close to the code elastic spectra render the use of artificial or hybrid accelerograms almost mandatory for practical applications. This has already been elaborated in Chapter 3.4.4.

The scaling of accelerograms has several different available techniques, the most common of which are

- Scaling according to maximum peak ground acceleration (PGA) target
- Scaling according to maximum velocity peak ground velocity
- Scaling according to Arias or Housner intensity

Although for unidirectional or bidirectional excitation in elastic analysis the aforementioned procedures do not result in serious differences, in the case of inelastic analysis, this selection results in significant differences.

The Housner intensity is described by the following expression, which in essence is the area under the Pseudovelocity Spectra PSV over the period (see Chapter 2.2.4.4):

$$\mathrm{SI} = \int_{0.1}^{2.5} PSV(t)\, dt \tag{5.104}$$

Proposals to modify this intensity by using a smaller period domain for the integration, that is, a time window close to the fundamental period of the building, have been considered and well documented for research purposes (Kappos and Kyriakakis, 2000).

The ARIAS intensity is described by the following integral:

$$\mathrm{AI} = \int_{t} a^2(t)\, dt \tag{5.105}$$

where $a(t)$ is the acceleration.

For the t-h inelastic analysis of a building, the scaling of the total energy of each excitation is critical; therefore, it is suggested that the ARIAS (or Housner) intensity is scaled for the sum of the energies of both directions, as can be shown in Table 5.7, which includes three accelerograms from the European Strong Motion Database (Ambraseys et al., 2000).

Table 5.7 Scaling of accelerograms

	Scaling factors for bidirectional t-h inelastic analysis				
Event	Direction	Arias I.	Total I.	Target intensity	Target/recorded (scale factor)
TABAS, IRAN (16/09/78 Ms = 7.3)	EW	11.21	23.17	9.71	0.65
	NS	11.97			
Friuli, IT (15/09/76 Ms = 6.0)	EW	1.09	1.82		2.31
	NS	0.73			
Gazli, UZB (17/05/76 Ms = 7.0)	EW	4.95	9.71		1.00
	NS	4.76			

5.7.6.2 Incremental dynamic analysis

The incremental dynamic analysis (IDA) is a procedure to define the capacity curve of a building by using a set of *t-h* inelastic analyses with an increasing intensity. Plotting the results of this analysis in the form of base shear Vstop displacement results in a trendline of points that includes the capacity curve of the building. An indicative IDA curve is shown in Figure 5.25, which refers to a 3-D eight-storey MRF both in terms of base shear displacement and base shear rotation (torsion).

This procedure has been extensively investigated by Vamvatsikos and Cornel (2002), and has also been used in 2-D research approaches.

In the case of 3-D analysis with torsional effects, the procedure requires additional attention to the selection of corresponding values for plotting this curve. There are several options that are summarised herein:

- Selecting the maximum responses for base shear and displacement (maxBS–maxD approach)
- Selecting the maximum top displacement and corresponding to this time (t_1) step base shear (maxD approach)
- Selecting the maximum base shear and corresponding to this time step (t_2) displacement (maxBS approach)
- Introducing a time window to the maxD and maxBS approaches as follows:

$$\text{maxD}(t_1), \text{maxBS}\{t_1 - \Delta t, t_1, t_1 + \Delta t\}$$

$$\text{maxBs}(t_2), \text{maxD}\{t_2 - \Delta t, t_2, t_2 + \Delta t\}$$

As has been theoretically explained by Penelis and Kappos (2005), the suggested approach is the maxD with a time window of one time step, which produces the most accurate results compared to the theoretical solution elaborated for the single-storey torsionally unrestrained system shown in Figure 5.26. In this figure, chart (a) shows the results of *t-h* inelastic analysis using all of the aforementioned matching pairs (rhombus for maxD, circles for maxD–maxBS and triangles for maxBS) and the resulting trendlines, while chart (b) shows the theoretical solution for torsionally restrained and torsionally unrestrained buildings of Figure 5.24.

5.8 COMBINATION OF THE COMPONENTS OF GRAVITY LOADS AND SEISMIC ACTION

5.8.1 General

Coming back to the linear methods of analysis, it should be noted that no matter which one of the two procedures presented above has been used, that is, 'the modal response spectrum analysis' or the lateral force method, the horizontal components of the seismic action should be considered, according to EC8-1/2004, as acting simultaneously in the two main directions. These two components may also be considered to have equal and uncorrelated intensities (Rosenblueth and Contreras, 1977).

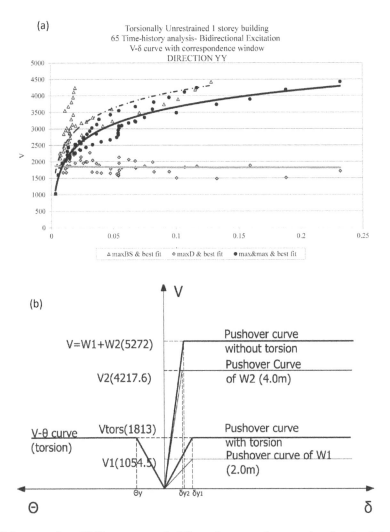

Figure 5.26 IDA curves for a TUR one-storey building using several approaches for the definitions versus the theoretical solution: (a) IDA curves for 65 time–history analysis; (b) the theoretical solution.

The combination of these two horizontal components for the determination of maximum seismic effects and, subsequently, their combination with the permanent gravity loads may be carried out as follows:

1. At first the structural response to each horizontal component shall be computed by means of the combination rules for modal responses given in Chapter 2.4.3 or by means of lateral force method of analysis (Section 5.6.4).
2. Then the maximum value of each action effect on the structure due to the two horizontal components of the seismic action may be estimated by the square root of the sum of the squared responses to each component of the seismic action, that is,

$$E_{max} = \pm\sqrt{E_x^2 + E_y^2}$$

(5.106)

where

E_{max} is the maximum action effect (M_x, M_y, M_z, V_x, V_y, N) due to the simultaneous action of the earthquake in both main directions.

E_x is the maximum action effect due to the application of the seismic action along the horizontal axis $x - x$ of the structure.

E_y is the maximum action effect due to the application of the seismic action along the horizontal axis $y - y$ of the structure.

In the case of a vertical element (column or wall) subjected to bending with axial force, the above are exemplified as follows:

a. For the $x - x$ earthquake component, the following internal forces are derived:

$$\pm M_{Ex\,x}, \quad \pm M_{Ey\,x}, \quad \pm N_{Ex}$$

b. For the $y - y$ earthquake component, the following internal forces are derived:

$$\pm M_{Ex\,y}, \quad \pm M_{Ey\,y}, \quad \pm N_{Ey}$$

Therefore, the extreme values of internal forces $\pm M_{Ex\,max}$, $\pm M_{Ey\,max}$, $\pm N_{Emax}$ will have the following form:

$$M_{Ex\,max} = \pm\sqrt{(M_{Exx})^2 + (M_{Exy})^2}$$
$$M_{Ey\,max} = \pm\sqrt{(M_{Eyy})^2 + (M_{Eyx})^2} \qquad (5.107a\text{--}c)$$
$$M_{E\,max} = \pm\sqrt{(N_{Ex})^2 + (N_{Ey})^2}$$

Since the alternative values for each load effect are two (±) and the number of the internal forces participating in the design of the element are three ($M_{Ex\,max}$, $M_{Ey\,max}$, N_E), it follows that the number of combinations necessary for the design is

$$\lambda = b^n = 2^3 = 8 \qquad (5.108)$$

where

n is the number of load effects participating in the design (in case of biaxial bending with axial load $n = 3$).

b is the number of alternative choices for each load effect ((±) $b = 2$).

3. The seismic action effects must be superimposed on the gravity load effects, that is, on (Chapter 3.4.5, Equations 3.50 and 3.51):

$$E_w = E(G + {}'\psi_{Ei}Q_i)$$

Therefore, the final action effects due to gravity loads and earthquake will have the form (Chapter 3.4.5)

$$\boxed{E_s = E(G' + {}'\psi_{Ei}Q_i)' + {}'E(\gamma_1 S_{dy}' + {}'\gamma_1 S_{dy})} \qquad (5.109)$$

where '+' implies 'to be combined with', G are the dead loads, Q_i is the characteristic value of variable action i, $S_{dx,y}$ is the design value of the seismic action parallel to $x - x$ and $y - y$ respectively, γ_1 is the importance factor and ψ_{Ei} is the combination coefficient for the variable action i (see Chapter 3.4.5).

It should be noted that the extreme values of seismic effects ($M_{x,ex}$, $M_{y,ex}$, $M_{z,ex}$, $V_{x,ex}$, $V_{y,ex}$, N_{ex}) determined above do not act simultaneously. Therefore, in the case that more than one load effect is needed for the safety verification at ultimate limit state (i.e. M_x, M_y, N for the cross-section of a column), the combination of the extreme values of all relevant load effects would be, at first glance, conservative.

In the following subsections, a theoretical approach to the problem will be presented, so that the reader may have a global view of the approximations involved in various procedures established in practice.

5.8.2 Theoretical background

According to the theoretical background for this issue, developed at the end of the 1970s (Rosenblueth and Contreras, 1977; Gupta, 1990), the response of a structural element (e.g. a column) to the combined simultaneous action of gravity loads and earthquake in the x and y axes is defined *in a three-dimensional response space* by an ellipsoid of the interacting load effects (M_x, M_y, N) with its centre at a vector $\bar{r}_0(M_{xo}, M_{yo}, N)$, where M_{xo}, M_{yo} and N represent the gravity load effect (Figure 5.27). In general, this ellipsoid has inclined axes, while the failure envelope for M_x, M_y, N is not susceptible to a simple description.

The design of the structural elements requires that the ellipsoid lies entirely within the safe domain of the failure envelope and, if possible, in contact with the failure surface. This is an extremely complicated procedure for routine design purposes, so various simplified procedures have been developed in which the ellipsoid has been approximated by a polyhedron, with coordinates of its vertices relating to the failure surface.

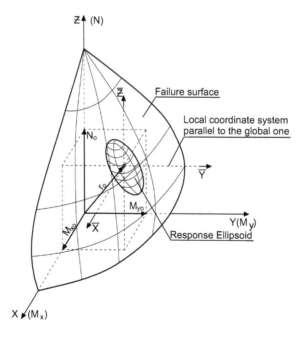

Figure 5.27 Response ellipsoid and failure envelope.

For a better understanding of the concept, a two-dimensional response plane is referenced (i.e. M and N on a cross section of an R/C wall). In this case, the 3d ellipsoid is collapsed to a 2d ellipse while the failure surface is reduced to a failure curve (Gupta and Singh, 1977; Panetsos and Anastasiadis, 1994). Below some methods developed so far are depicted in Figure 5.28, that is,

- The code reference method where the ellipse is approximated by the rectangle I, II, III, IV (EC8-1/2004)
- The Gupta and Chu (1977) method where the ellipse is approximated by a rectangle a, b, c, d.
- The Gupta and Singh (1977) method where the ellipse is approximated by a circumscribed octagon 1, 2, 3, 4, 1', 2', 3', 4'

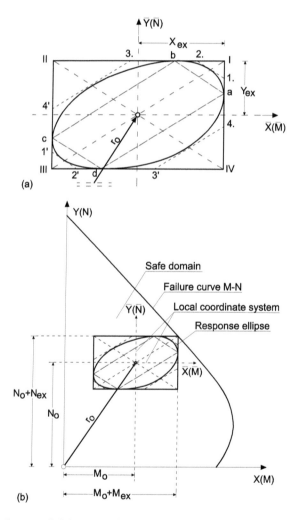

Figure 5.28 Response ellipse and failure envelope: (a) various simplified approaches to the problem; (b) response ellipse in the safe domain of the envelope (bending design).

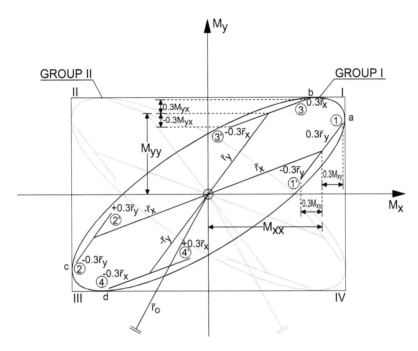

Figure 5.29 The Rosenblueth and Contreras procedure (alternative Code procedure).

Finally in Figure 5.29, the Rosenblueth and Contreras (1977) method is depicted where the ellipse is approximated by a vector

$$\bar{r}_c = \bar{r}_o + \alpha_x \bar{r}_x + \alpha_y \bar{r}_y \qquad (5.110)$$

where

$\bar{r}_c$ is the most probable extreme response vector due to the gravity loading $\bar{r}_o$ and the extreme earthquake actions $\bar{r}_x$ and $\bar{r}_y$ (Penelis and Penelis, 2014).

Through this last approach, it has been conclude that for

$$a_x = \begin{Bmatrix} 1.00 \\ 0.336 \end{Bmatrix} \text{ and } a_y = \begin{Bmatrix} 0.336 \\ 1.00 \end{Bmatrix} \qquad (5.111)$$

respectively, the maximum error is ±5.5%.

An alternative approach introduced by EC8-1/2004 has adopted for α_x and α_y the following values:

$$a_x = \begin{Bmatrix} \pm 1.00 \\ \pm 0.30 \end{Bmatrix} \text{ and } a_y = \begin{Bmatrix} \pm 0.30 \\ \pm 1.00 \end{Bmatrix} \qquad (5.112)$$

respectively. For these values, the maximum error is 4.4% on the safe side and 8.1% on the unsafe side.

5.8.3 Code provisions

5.8.3.1 Suggested procedure for the analysis

Bearing in mind, as already mentioned, that strong computational tools are available at low cost, it is suggested that, in the case that modal response spectrum analysis is applied, a spatial system with diaphragms at floor levels is used. At this case, the most convenient approach to the problem according to the authors' opinion is the following:

- The floor masses will be considered as lumped ones, concentrated at the centre of gravity of each floor.
- The accidental torsional effect ($e_i = \pm 0.05\ L_i$) may be determined as the envelope of the effects resulting from an analysis for static loading consisting of the torsional moments M_i about the vertical axis of the storey i.

$$M_i = e_{1i} \cdot H_i \qquad (5.113)$$

where M_i is the torsional moment of storey i about its vertical axis, e_{1i} is the accidental eccentricity of the storey mass i accounting for the two main directions and H_i is the horizontal force acting at storey i as derived from the application of lateral force method of analysis for the two main directions. The effect of the loading described above is considered, with alternating singes the same for all storeys. Therefore, the structural system will be analysed for the following actions:

$$\left. W = G' + {}'\Psi_{Ei} \cdot Q_i \right\} \text{ gravity load (see Chapter 3.4.5, Equations 3.50 and 3.51)} \qquad (5.114)$$

$$\left.\begin{array}{l} \text{Design spectrum } (S_{dx}) \text{ modified by} \\[4pt] \text{the importance factor for horizontal} \\[4pt] \text{excitation parallel to the } x-x \text{ axis and} \\[4pt] \text{masses at the centre of gravity of the floors} \\[4pt] M_i = e_{1y} \cdot H_{ix} \end{array}\right\} \begin{array}{l} x-x \text{ seismic action} \\[40pt] (\text{see Chapter } 5.6.3.4) \end{array} \qquad (5.115)$$

$$\left.\begin{array}{l} \text{Design spectrum } (S_{dy}) \text{ modified by} \\[4pt] \text{the importance factor for horizontal} \\[4pt] \text{excitation parallel to the } y-y \text{ axis and} \\[4pt] \text{masses at the centre of gravity of the floors} \\[4pt] M_i = \pm e_{1x} \cdot H_{iy} \end{array}\right\} \begin{array}{l} y-y \text{ seismic action} \\[40pt] (\text{see Chapter } 5.6.3.4) \end{array}$$

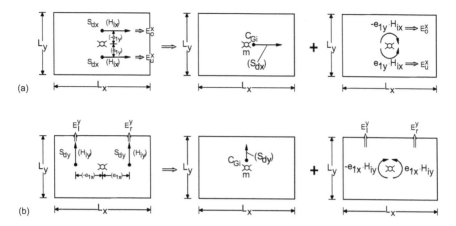

Figure 5.30 Seismic action effects E resulting from a modular response spectrum analysis: (a) seismic action in the $x-x$ direction; (b) seismic action in the $y-y$ direction.

The loading effects (E) from the above analysis, for each loading direction, will be obtained as follows (Figure 5.30):

$$E_w \Rightarrow G\text{'+'}\psi_{Ei} \cdot Q_i \qquad \big\}\text{gravity load}$$

$$\left.\begin{aligned} E_u^x &\Rightarrow S_{dx}\text{'+'}e_{1y} \cdot H_{ix} \\ E_o^x &\Rightarrow S_{dx}\text{'+'}(-e_{1y}) \cdot H_{ix} \end{aligned}\right\} x\text{--}x \text{ seismic action}$$

$$\left.\begin{aligned} E_r^y &\Rightarrow S_{dx}\text{'+'}e_{1x} \cdot H_{iy} \\ E_l^y &\Rightarrow S_{dy}\text{'+'}(-e_{1x}) \cdot H_{iy} \end{aligned}\right\} y\text{--}y \text{ seismic action}$$

(5.116)

where '+' means 'combined with'.

The notation above is given in Figure 5.30.

The above procedure seems to be the most convenient. All other options, for example, the use of two plane models combined with simplified torsional analysis, are proper only in the case that no efficient computational tools are available.

5.8.3.2 Implementation of the reference method adopted by EC8-I in case of horizontal seismic actions

The reference method of EC8-1/2004 for the calculation of the combinations of the seismic effects that should be taken into account in the design is based on the assumption that the horizontal components of the seismic action (see Section 5.8.1) are taken as acting simultaneously. This assumption introduces an error on the safe side, from 15% to 35%. In this case, the number of combinations of load effects that should be taken into account for the design of a column or a wall is eight.

In addition, it should be noted that due to the introduction of accidental eccentricities, four different centres of masses must be considered. Therefore, it follows that the total number of combinations for the design of the cross section of a column is 32 (4×2^3). In fact, substituting Equation 5.116 into Equation 5.106 and superimposing the gravity load effects, the following combinations result:

$$E = E_{\mathrm{w}} \pm \sqrt{\left(E_{\mathrm{u}}^x\right)^2 + \left(E_{\mathrm{r}}^y\right)^2} = \begin{cases} E_{1,\mathrm{I}} \\ \\ E_{1,\mathrm{II}} \end{cases}$$

$$E = E_{\mathrm{w}} \pm \sqrt{\left(E_{\mathrm{u}}^x\right)^2 + \left(E_{\mathrm{l}}^y\right)^2} = \begin{cases} E_{2,\mathrm{I}} \\ \\ E_{2,\mathrm{II}} \end{cases}$$

$$E = E_{\mathrm{w}} \pm \sqrt{\left(E_{0}^x\right)^2 + \left(E_{\mathrm{r}}^y\right)^2} = \begin{cases} E_{3,\mathrm{I}} \\ \\ E_{3,\mathrm{II}} \end{cases}$$

$$E = E_{\mathrm{w}} \pm \sqrt{\left(E_{0}^x\right)^2 + \left(E_{\mathrm{l}}^y\right)^2} = \begin{cases} E_{4,\mathrm{I}} \\ \\ E_{4,\mathrm{II}} \end{cases}$$

that is, eight different combinations for each load effect (M_x, M_y, M_z, V_x, V_y, N).

In the case of a column where three load effects must be considered, that is M_x, M_y and N, the number of combinations is 32:

$$\left.\begin{array}{lll} M_x = M_{1,\mathrm{I}x} & M_y = M_{1,\mathrm{I}y} & N = N_{1,\mathrm{I}} \\ M_x = M_{1,\mathrm{II}x} & M_y = M_{1,\mathrm{II}y} & N = N_{1,\mathrm{II}} \end{array}\right\} 8\,\text{combinations}$$

$$\left.\begin{array}{lll} M_x = M_{2,\mathrm{I}x} & M_y = M_{2,\mathrm{I}y} & N = N_{2,\mathrm{I}} \\ M_x = M_{2,\mathrm{II}x} & M_y = M_{2,\mathrm{II}y} & N = N_{2,\mathrm{II}} \end{array}\right\} 8\,\text{combinations}$$

$$\left.\begin{array}{lll} M_x = M_{3,\mathrm{I}x} & M_y = M_{3,\mathrm{I}y} & N = N_{3,\mathrm{I}} \\ M_x = M_{3,\mathrm{II}x} & M_y = M_{3,\mathrm{II}y} & N = N_{3,\mathrm{II}} \end{array}\right\} 8\,\text{combinations}$$

$$\left.\begin{array}{lll} M_x = M_{4,\mathrm{I}x} & M_y = M_{4,\mathrm{I}y} & N = N_{4,\mathrm{I}} \\ M_x = M_{4,\mathrm{II}x} & M_y = M_{4,\mathrm{II}y} & N = N_{4,\mathrm{II}} \end{array}\right\} 8\,\text{combinations}$$

Total   32 combinations

5.8.3.3 Implementation of the alternative method adopted by EC8-I in the case of horizontal seismic actions

As was already noted as an alternative to the above procedure, it is permitted according to EC8-1/2004 to compute the action effects due to both components using the following formulae (see Section 5.8.2):

$$\left.\begin{array}{l} E = E_x\text{'+'}0.30E_y \\ E = 0.30E_x\text{'+'}E_x \end{array}\right\} \tag{5.117}$$

The same formulae have been introduced in the United States by BSSC (2003) and SEAOC (1999).

This assumption introduces a maximum error of 4.4% on the safe side and 8.1% on the unsafe side. In this respect, this alternative results in steel savings in the vertical elements in comparison to the reference method (Penelis and Penelis, 2014).

However, the implementation of this method requires particular attention in the combination of the partial components of the seismic effects, especially in case that the modal response spectrum analysis is used. In this case, the partial components of the seismic effects result in absolute values for the output of the SRSS or complete quadratic combination (CQC) procedure (see Chapter 2.4.3).

For example, in the case of a vertical element (column or wall) subjected to bending with axial force, according to what has been presented in Section 5.8.1, the following load vectors develop.

For the $x - x$ earthquake component

$$\pm M_{Ex\,x}, \pm M_{Ey\,x}, \pm N_{Ex}$$

For the $y - y$ earthquake component

$$\pm M_{ex\,y}, \pm M_{Eyy}, \pm N_{Ey}$$

Therefore, the extreme values of the internal forces $M_{Ex\,max}$ with the corresponding M_{Ey} and N_E result from the following equations:

$$M_{Ex\,ext} = \pm M_{Ex\,x} \pm 0.30 M_{Ex\,y}, M_{Ey}^{cor} = \pm M_{Ey\,x} \pm 0.30 M_{Ey\,y}, \quad N_{ex}^{cor} = \pm N_{ex} \pm 0.30 N_{Ey} \quad (5.118)$$

while the extreme values of the internal forces $M_{Ey\,max}$ with the corresponding M_{Ex} and N_E result from the following ones:

$$M_{Ey\,extr} = \pm M_{Ey\,y} - 0.30 M_{Ey\,x}, M_{Ex}^{cor} = \pm M_{Ex\,y} \pm 0.30 M_{Ex\,x}, \quad N_{Ey}^{cor} = \pm N_{Ey} \pm 0.30 N_{Ex} \quad (5.119)$$

Since the alternative values for each load effect are four (± '+' ±) and the numbers of internal forces participating in the design of the element are three (M_x, M_y, N_E), it may be concluded that the number of combinations necessary for the design are

$$\lambda = b^n = 4^3 = 64$$

Additionally, bearing in mind that due to the introduction of accidental eccentricities four different centres of masses must be considered, it follows that the total number of combinations for the design of the cross section of a vertical element is

$$\lambda' = 4 \times 64 = 256$$

It is apparent that such an enormous number of load effect combinations cannot be afforded, despite the reinforcement savings if it were compared with the 32 load effect combinations required for the reference method. So, a comprehensive examination of Equations 5.118 and 5.119 should be carried out, combined with some approximations so that this large number of 256 combinations may be diminished.

Consider for a moment that N_E has a given value. Equations 5.118 and 5.119 may be arranged in two groups (Figure 5.29), that is,

Group (1)

Points $\boxed{1}$ 1.1': $M_{\text{Ex max}} = +M_{\text{Ex,x}} \pm 0.30 M_{\text{Ex,y}}$, $M_{\text{Ey}}^{\text{cor}} = +M_{\text{Ey,x}} \pm 0.30 M_{\text{Ey,y}}$, $N_{\text{ex}}^{\text{cor}} = N_E$

Points $\boxed{1}$ 2.2': $M_{\text{Ex min}} = -M_{\text{Ex,x}} \pm 0.30 M_{\text{Ex,y}}$, $M_{\text{Ey}}^{\text{cor}} = -M_{\text{Ey,x}} \pm 0.30 M_{\text{ey,y}}$, $N_{\text{Ex}}^{\text{cor}} = N_E$

Points $\boxed{1}$ 3.3': $M_{\text{Ey max}} = +M_{\text{Ey,y}} \pm 0.30 M_{\text{Ey,x}}$, $M_{\text{ex}}^{\text{cor}} = +M_{\text{ex,y}} \pm 0.30 M_{\text{Ex,x}}$, $N_{\text{Ey}}^{\text{cor}} = N_E$

Points $\boxed{1}$ 4.4': $M_{\text{Ey min}} = -M_{\text{ey,y}} \pm 0.30 M_{\text{Ey,x}}$, $M_{\text{ex}}^{\text{cor}} = -M_{\text{Ey,y}} \pm 0.30 M_{\text{Ex,x}}$, $N_{\text{Ey}}^{\text{cor}} = N_E$

$$(5.120)$$

Group (2)

Points $\boxed{2}$ 1.1': $M_{\text{Ex max}} = +M_{\text{Ex,x}} \pm 0.30 M_{\text{Ex,y}}$, $M_{\text{Ey}}^{\text{cor}} = -M_{\text{Ey,x}} \pm 0.30 M_{\text{Ey,y}}$, $N_{\text{ex}}^{\text{cor}} = N_E$

Points $\boxed{2}$ 2.2': $M_{\text{Ex min}} = -M_{\text{Ex,x}} \pm 0.30 M_{\text{Ex,y}}$, $M_{\text{Ey}}^{\text{cor}} = +M_{\text{Ey,x}} \pm 0.30 M_{\text{Ey,y}}$, $N_{\text{Ex}}^{\text{cor}} = N_E$

Points $\boxed{2}$ 3.3': $M_{\text{Ey min}} = -M_{\text{Ey,y}} \pm 0.30 M_{\text{Ey,x}}$, $M_{\text{Ex}}^{\text{cor}} = +M_{\text{Ex,y}} \pm 0.30 M_{\text{Ex,x}}$, $N_{\text{Ey}}^{\text{cor}} = N_E$

Points $\boxed{2}$ 4.4': $M_{\text{Ey max}} = +M_{\text{Ey,y}} \pm 0.30 M_{\text{Ey,x}}$, $M_{\text{Ex}}^{\text{cor}} = -M_{\text{Ex,y}} \pm 0.30 M_{\text{Ex,x}}$, $N_{\text{ey}}^{\text{cor}} = N_E$

$$(5.121)$$

Group (1) corresponds to the eight points 1–4 and 1'–4', which approximate according to Equation 5.117 ellipse (1) (Figure 5.29), while group (2) corresponds to the eight points 1–4 and 1'–4' which approximate ellipse (2), that is, the mirror of ellipse (1) (Figure 5.29).

Now, if the envelope of the design strength $M_{\text{xd}} - M_{\text{yd}}$ of a symmetric R/C cross section (e.g. column) for a given value of $N_{\text{Ed}} = N_E$ is plotted on the same plot with the ellipses (1) and (2) (Figure 5.31), the following cases may appear.

Case 1

The vector $\bar{r}_0 (M_{\text{ox}}, M_{\text{oy}})$ due to the gravity loads of seismic design may fall in the first or third quarter of the strength envelope, where $M_{\text{ox}} M_{\text{oy}}$ have the same sign, either (+, +) or (−, −). In this case, ellipse (1) becomes the crucial one for the dimensioning of the cross section, and in this respect, the combinations of group (1) are also the crucial ones for the design.

Case 2

The vector $\bar{r}_0 (M_{\text{ox}}, M_{\text{oy}})$ may fall in the second or fourth quarter of the strength envelope, where M_{ox}, M_{oy} have different sign, either (+, −) or (−, +). In this case, ellipse (2) becomes the crucial one for the dimensioning of the cross section, and in this respect, the combinations of group (2) are also crucial for the design.

Case 3

The vector $\bar{r}_0 (M_{\text{ox}}, M_{\text{oy}})$ may fall on one of the main axes $x - x$ or $y - y$, which means that the stress state due to gravity loads of seismic design is symmetric to one of the two main axes or to both of them. In this case, it is indifferent which of the two ellipses and, consequently, the two groups (1) and (2) will be used for the design.

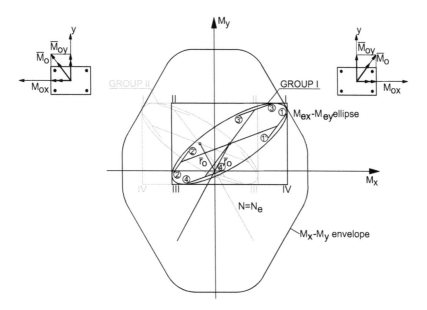

Figure 5.31 Choice of the proper ellipse (group I or group II) in regard to the position of the gravity load effects vector $\bar{r}_0$ in the quarters of the capacity envelope of a symmetric cross section.

The above considerations also hold for non-symmetric cross sections (e.g. a cross section with a U form; Figure 5.32).

From all of the above considerations, it may be easily concluded that the number of combinations in the case of use of the alternative method for seismic load combinations expressed by Equation 5.117 continues to be 32, as in the case of the reference method expressed by Equation 5.106.

It should be noted here that in the place of N_E, the following values are introduced: Group (1)

$$
\begin{aligned}
\text{Points}\boxed{1} \quad 1.1': N_{ex}^{cor} &= +N_E \pm 0.30 N_{Ey} \\
\text{Points}\boxed{1} \quad 2.2': N_{ex}^{cor} &= -N_{Ex} \pm 0.30 N_{Ey} \\
\text{Points}\boxed{1} \quad 3.3': N_{ey}^{cor} &= +N_{ey} \pm 0.30 N_{ex} \\
\text{Points}\boxed{1} \quad 4.4': N_{ey}^{cor} &= -N_{ey} \pm 0.30 N_{ex}
\end{aligned}
\tag{5.122}
$$

Group (2)

$$
\begin{aligned}
\text{Points}\boxed{2} \quad 1.1': N_{Ex}^{cor} &= +N_{Ex} \pm 0.30 N_{Ey} \\
\text{Points}\boxed{2} \quad 2.2': N_{Ex}^{cor} &= -N_{Ex} \pm 0.30 N_{Ey} \\
\text{Points}\boxed{2} \quad 3.3': N_{Ey}^{cor} &= -N_{Ey} \pm 0.30 N_{Ex} \\
\text{Points}\boxed{2} \quad 4.4': N_{ey}^{cor} &= +N_{Ey} \pm 0.30 N_{Ex}
\end{aligned}
\tag{5.123}
$$

The above procedure has been incorporated into the computer platform EC tools.

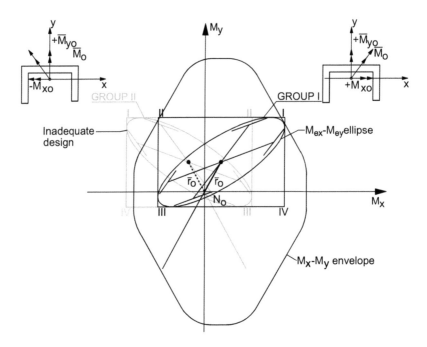

Figure 5.32 Choice of the proper ellipse (group I or group II) in regard to the position of the gravity load effects vector $\bar{r}_0$ in the quarters of the capacity envelope of a unisymmetric cross section.

5.8.3.4 Implementation of the alternative method for horizontal and vertical seismic action

It has been mentioned (Chapter 3.4.3) that the vertical component of the seismic action has to be considered only for certain structures (Luft, 1989). The effects of the vertical component according to EC8-1/2004 need only be taken into account for the elements under consideration and their directly associated supporting elements or substructures.

In the case that the horizontal components of the seismic action are also relevant for these elements, EC8 introduces the following combinations:

$$\left.\begin{array}{l} E = 0.30E_x \text{ '+' } 0.30E_y \text{ '+' } E_z \\ E = E_x \text{ '+' } 0.30E_y \text{ '+' } 0.30E_z \\ E = 0.30E_x \text{ '+' } E_y \text{ '+' } 0.30E_z \end{array}\right\} \tag{5.124}$$

where E_z is the action effect due to the application of the vertical component of the design seismic action.

The number of combinations needed for the design of R/C sections is equal to

$$\lambda = 4 \times 2 \times 2^3 = 64 \tag{5.125}$$

This number also includes the displacements of the centre of mass due to accidental eccentricities.

5.9 EXAMPLE: MODELLING AND ELASTIC ANALYSIS OF AN EIGHT-STOREY RC BUILDING

5.9.1 Building description

The building under investigation is an eight-storey RC building with one basement storey as shown in Figure 5.33. The total height of the building above the ground level is $H = 26$ m. The height of a typical storey is 3 m, whereas the height of the first storey and the basement are differentiated to 5 and 4 m, respectively (Figure 5.33). The building is rectangular in plan, with dimensions of 30 m on the horizontal x-axis and 25 m on the horizontal y-axis. The structural system consists of walls and frames. All columns have a 60 cm square cross section. Both interior and exterior beams have a width of $b_w = 30$ cm and a height of $h_b = 65$ cm. The slab thickness is $h_f = 18$ cm. Walls have a thickness of $b_t = 30$ cm, except for the perimeter basement walls, which have a thickness of $b_t = 25$ cm. Details relative to the length of the walls appear in Figure 5.33.

5.9.2 Material properties

- Concrete quality is C25/30, with a modulus of elasticity $E_c = 31$ GPa (EC2 Table 3.1).
- Steel quality is B500C.

5.9.3 Design specifications

- Permanent loads due to floor finishes and suspended ceilings: 2 kN/m².
- Permanent loads due to roof finishes: 3.5 kN/m².
- Permanent loads due to light partition walls: 0.5 kN/m².
- Permanent facade loads: 3.0 kN/m.
- Live load (whole building except roof): 5.0 kN/m².
- Live load on the roof: 2.0 kN/m².
- The PGA is $a_g = 0.24$ g.
- Seismic demand is defined for the EC8-Part I Type I spectrum on ground C. Importance class is II.

5.9.4 Definition of the design spectrum

5.9.4.1 Elastic response spectrum (5% damping)

Seismic actions are estimated according to the EC8-Part I elastic response spectrum, Type 1 (EC8-Part I, Section 3.2.2.2(1)P) for ground type C. The values of the periods T_B, T_C, T_D and of the soil factor S for ground type C are defined as follows: $T_B = 0.2$ s, $T_C = 0.6$ s, $T_D = 2.0$ s and $S = 1.15$. PGA is considered equal to $a_g = 0.24$ g. The building is classified in importance class II; hence the importance factor is $\gamma_I = 1$ (EC8-Part I, Section 4.2.5, Table 4.3). The elastic response spectrum is derived for 5% damping following the set of equations of EC8-Part I, Section 3.2.2.2(1)P.

5.9.4.2 Design response spectrum

For the definition of the design response spectrum, the knowledge of the behaviour factor, q, is necessary. Therefore, the building's structural system has to be classified according to its

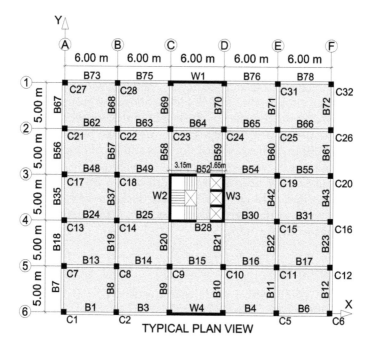

TYPICAL PLAN VIEW

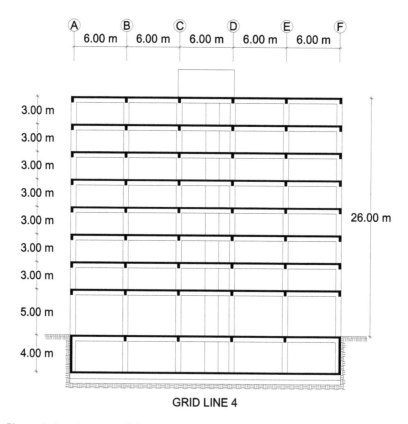

GRID LINE 4

Figure 5.33 Plan and elevation view of the eight-storey building.

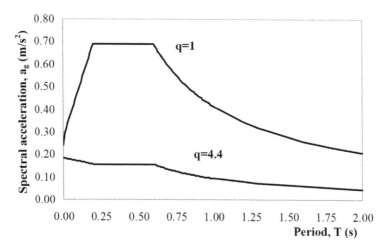

Figure 5.34 Elastic and design response spectrum of EC8-Part I (2004).

behaviour under horizontal seismic actions (EC8-Part I, Section 5.2.2.1). Moreover, the regularity in elevation and plan as well as the ductility class need to be taken into account for deriving the value of the behaviour factor. These criteria are examined in Sections 5.2.2 and 5.2.3, where more details appear for the extracted value of q. The design response spectrum (EC8-Part I, Section 3.2.2.5(4) P) utilised for the design of the building examined here is defined for $q = 4.4$ (the detailed procedure for estimation of the behaviour factor appears in Section 5.9.7). Both the elastic and the design response spectra are presented in Figure 5.34.

5.9.5 Estimation of mass and mass moment of inertia

The total weight of each floor is estimated by considering the combination $(G + 0.3Q)$ for permanent and live loads. The polar moment of inertia, J_d, is estimated as well (Equation 2.97):

$$J_d = m\frac{b^2 + d^2}{12} = m\frac{30^2 + 25^2}{12} = 127.08\,m$$

The results appear in Table 5.8.

Table 5.8 Storey weight, storey masses and moments of inertia

Level	Storey weight (kN)	Storey mass (ton)	Moment of inertia (ton · m²)
8th	7946.14	810.00	102,880.98
7th	8934.60	910.76	115,678.89
6th	8934.60	910.76	115,678.89
5th	8934.60	910.76	115,678.89
4th	8934.60	910.76	115,678.89
3rd	8934.60	910.76	115,678.89
2nd	8934.60	910.76	115,678.89
1st	9448.53	963.15	122,332.84
Total	71,002.29	7237.75	919,287.17

5.9.6 Structural regularity in plan and elevation

5.9.6.1 Criteria for regularity in plan

The regularity criteria in plan that need to be satisfied are (Section 5.2.2)

- The slenderness of the building in plan shall be $\lambda = L_{max}/L_{min} \leq 4$.
- At each floor, i, and for each direction of analysis, x and y, the structural eccentricity, e_{mx}, e_{my}, and the torsional radius, r_{xc}, r_{yc}, shall meet the following conditions (Equations 5.1):
  $x - x$ direction: $e_{mx,i} \leq 0.30 \cdot r_{xc,i}$, $r_{xc,i} \geq l_s$
  $y - y$ direction: $e_{my,i} \leq 0.30 \cdot r_{yc,i}$, $r_{yc,i} \geq l_s$
  The slenderness of the building is $\lambda = L_{max}/L_{min} = 30\ \text{m}/25\ \text{m} = 1.2 < 4$.

The determination of parameters e_{mx}, e_{my}, r_{xc}, r_{yc}, is carried out following the procedure described in Chapter 2.4.4.5.

1. Each floor is loaded at the centre of mass by $T_{zi} = G_i \cdot z_i$ and analysis is performed (G_i is the dead load of each storey, z_i is the storey height; Table 5.9).
2. The plasmatic axis of the centre of stiffness passes through the centre of rotation of floor ℓ (point P_o), being the one that lies closest to the level $z_o = 0.8 \cdot H = 0.8 \cdot 26 = 20.8$ m, which here is the sixth floor. The torsional deformation is (Equation 2.137)

$$\omega_{0.8H} = \frac{+0.1476 - (-0.1476)}{25} = \frac{+0.1792 - (-0.1751)}{30} = 0.0118\ \text{rad}$$

Torsional stiffness with respect to the centre of stiffness (Equation 2.138):

$$J_{TC} = \frac{\sum_{j=1}^{j=i} M_{jz}}{\omega_{0.8H}} = \frac{966428.58}{0.0118} = 81900727.12\ \text{kN m/rad}$$

3. The centre of stiffness of storey ℓ (point P_o) is geometrically estimated:

$$x_c = \frac{u_{y1} \cdot b}{(u_{y1} + u_{y2})} = \frac{0.175 \cdot 30000}{(0.175 + 0.179)} = 14.83\ \text{m}$$

Hence, $e_{mx} = x_M - x_c = 15.00 - 14.83 = 0.17$ m, $e_{my} = 0.00$ m.

Table 5.9 Definition of G_i, z_i and T_{zi} parameters

Storey	G_i	z_i	T_{zi}
8th	7509.49	26.00	195,246.80
7th	7842.98	23.00	180,388.57
6th	7842.98	20.00	156,859.62
5th	7842.98	17.00	133,330.68
4th	7842.98	14.00	109,801.74
3rd	7842.98	11.00	86,272.79
2nd	7842.98	8.00	62,743.85
1st	8356.91	5.00	41,784.53
Total	62,924.29		966,428.58

Table 5.10 Definition of parameters G_i, z_i, H_{jx}, H_{jy}

Storey	G_i	z_i	$H_{jx}= H_{jy}$
8th	7509.49	26.00	195,246.80
7th	7842.98	23.00	180,388.57
6th	7842.98	20.00	156,859.62
5th	7842.98	17.00	133,330.68
4th	7842.98	14.00	109,801.74
3rd	7842.98	11.00	86,272.79
2nd	7842.98	8.00	62,743.85
1st	8356.91	5.00	41,784.53
Total	62,924.29		966,428.58

4. The centre of stiffness of each storey is considered to coincide with the point where the plasmatic axis passes through the structure. A pattern of horizontal forces, $H_{jx} = H_{jy} = G_i z_i$, is applied to the centre of stiffness of every storey (Equation 2.139, Table 5.10). The translational displacements, $u_{c,0.080\,H}$ and $v_{c,0.80\,H}$, in the $x - x$ and $y - y$ directions, respectively, are estimated and the corresponding stiffnesses are calculated according to Equation 2.140.

   $x - x$ axis:

$$u_{c,0.80H} = 2.2583\,\text{m} \rightarrow K_x = \frac{\sum_{j=1}^{j=1} H_{jx}}{u_{c,0.8H}} = \frac{966428.58}{2.2583} = 427945.17\,\text{kN/m}$$

   $y - y$ axis:

$$v_{c,0.80H} = 3.0128\,\text{m} \rightarrow K_y = \frac{\sum_{j=1}^{j=1} H_{jy}}{v_{c,0.8H}} = \frac{966428.58}{3.0128} = 320774.22\,\text{kN/m}$$

5. The torsional radii with respect to the centre of stiffness are estimated as (Equation 2.85)

$$r_{xc} = \sqrt{\frac{J_{TC}}{K_y}} \sqrt{\frac{81900727.12}{320774.22}} = 15.97 \text{ m}, \ r_{yc} = \sqrt{\frac{J_{TC}}{K_x}} \sqrt{\frac{81900727.12}{427945.17}} = 13.83\,\text{m}$$

   The radius of gyration, l_s, is also estimated in order to apply the regularity criteria:

$$l_s = \sqrt{\frac{b^2 + d^2}{12}} = \sqrt{\frac{30^2 + 25^2}{12}} = 11.27\,\text{m}$$

   In all cases, $l_s < r_{xc}$, $l_s < r_{xy}$; hence, the system has sufficient torsional rigidity (Figure 5.35).

   The poles of rotation, O_1 and O_2, as estimated by ECtools (Penelis Software Ltd) for the first two eigen modes at each floor, appear in Table 5.11. The first two modes may

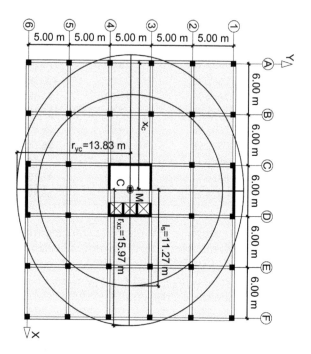

Figure 5.35 Definition of torsional rigidity.

Table 5.11 Poles of rotation at each storey

Storey	O_1 (y–y direction)		O_2 (x–x direction)	
	e_{xl} (m)	e_{yl} (m)	e_{xl}(m)	e_{yl} (m)
1st	728.57	0	0	1.00E+100
2nd	626.83	0	0	1.00E+100
3rd	593.81	0	0	1.00E+100
4th	589.17	0	0	1.00E+100
5th	601.44	0	0	1.00E+100
6th	626.64	0	0	1.00E+100
7th	662.83	0	0	1.00E+100
8th	705.56	0	0	1.00E+100

be considered uncoupled in both directions of loading, despite the small eccentricity presented along the $x - x$ axis ($e_{mx} = 0.17$ m corresponds to $0.56\%b$ where $b = 30$ m). Looking at the ordinates of O_1 and O_2, it may be seen that they are placed practically at an infinite distance from the centre of mass. Pole O_1 (it corresponds to loading in the y–y direction) is closer to the centre of mass compared to pole O_2 due to the small eccentricity presented along the x–x axis.

5.9.6.2 Criteria for regularity in elevation

All the conditions listed in Section 5.2.3 are satisfied. Thus, the building is characterised by regularity in elevation.

5.9.7 Determination of the behaviour factor q (Section 5.4.3)

The structural system of the building under investigation can be characterised as an *uncoupled wall system* in both directions. In this case, the shear resistance of the walls at the building base is higher than 65% of the total seismic resistance of the whole structural system. Taking into account the note provided by EC8-Part I (Section 5.1.2), where the fraction of shear resistance may be substituted by the fraction of shear forces, the base shear taken by walls is 88.9% and 82.4% in the x and y directions, respectively. The building is designed for high ductility class (DCH, Section 5.4.2).

The upper limit value of the behaviour factor is given by Equation 5.6:

$$q = K_{\mathrm{w}} \cdot K_{\mathrm{R}}^{\mathrm{el}} \cdot K_{\mathrm{R}}^{\mathrm{over}} \cdot K_{\mathrm{Q}} \cdot q_{\mathrm{o}}$$

with $K_{\mathrm{w}} = 1$, $K_{\mathrm{reg}}^{\mathrm{el}} = 1$, $K_{\mathrm{Q}} = 1$ and $q_{\mathrm{o}} = 4.0 \cdot a_{\mathrm{u}}/a_{1} = 4.4$ with $a_{\mathrm{u}}/a_{1} = 1.1$ for uncoupled wall systems.

5.9.8 Description of the structural model

The building is modelled as a spatial structural model in ETABS v9.7.4 (Computers and Structures Inc), comprising of beams, columns and walls (Figure 5.36). The following modelling assumptions are taken into consideration:

- Beams and columns are modelled with line elements using T and rectangular sections, respectively.
- The walls of the superstructure and the basement perimeter walls are modelled with shell elements.
- The slabs are modelled with shell elements and rigid diaphragm action is considered at each storey.
- The masses and moments of inertia are lumped at the centre of mass of each storey according to EC8-1/2004 (Paragraph 4.3.1(3)), and are evaluated by all gravity loads appearing in the combination of actions $\sum G_{\mathrm{k,i}} + \mathrm{E}\psi_{\mathrm{Ej}} \cdot Q_{\mathrm{k,j}}$ (EC8-Part I, Paragraph 3.2.4(2)).
- The foundation, in order to account for soil deformability, is modelled by means of foundation beams on elastic support and its parameter and modelling assumptions will be discussed in more detail in Chapter 10. Line and shell elements are used for this purpose.
- The bottom level of the first storey is considered as the base of the building, as far as the distribution of lateral seismic loads is concerned. First, response spectrum analysis is performed, and from the resulting shear forces at each storey, the horizontal seismic forces are calculated. These forces are applied to each storey both in the $x - x$ and the $y - y$ directions.
- Cracked concrete is assumed by multiplying the elastic stiffness parameters of the structural elements by 0.5 and the elastic torsional stiffness by 0.1.

Beams are modelled as line elements. The effective width is estimated according to EC2 (Paragraph 5.3.2.1). The code allows taking a constant width over the whole span in structural analysis (Paragraph 5.3.2.1(4)), which is the value of the span section. The effective width is estimated for two internal and two external beams (Figure 5.37). The following equations apply (Equation 5.7, EC8-Part I, Paragraph 5.3.2.1):

$$b_{\mathrm{eff}} = \sum b_{\mathrm{eff}} + b_{\mathrm{w}} \le b, b_{\mathrm{eff,i}} = 0.2 \cdot b_{\mathrm{i}} + 0.1 \cdot l_{0} \le \min(0.2 \cdot l_{0}; b_{\mathrm{i}}) \text{ for } l_{0} = 0.7 \cdot L$$

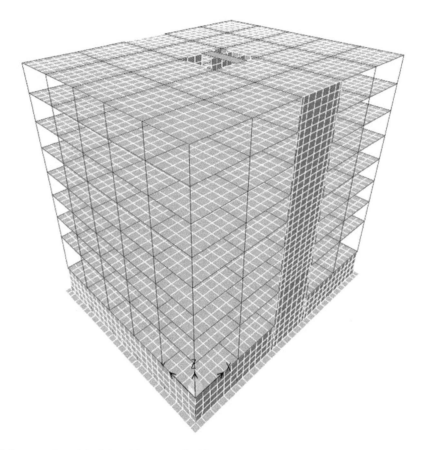

Figure 5.36 Structural model of the eight-storey building.

$x - x$ direction:

$$b_{\text{eff},1} = b_{\text{eff},2} = 0.2\frac{(5-0.30)}{2} + 0.1(0.7 \cdot 6.0) = 0.89\,m > \min\left(0.2 \cdot 0.7 \cdot 6.0; \frac{5}{2}\right) = 0.84\,\text{m}$$

Internal beams (BXIN): $b_{\text{eff,BXIN}} = 0.84 + 0.84 + 0.30 = 1.98$ m
External beams (BXOUT): $b_{\text{eff,BXOUT}} = 0.84 + 0.30 = 1.14$ m
$y - y$ direction:

$$b_{\text{eff},1} = b_{\text{eff},2} = 0.2\frac{(6-0.30)}{2} + 0.1(0.7 \cdot 5.0) = 0.92\,\text{m} > \min\left(0.2 \cdot 0.7 \cdot 5.0; \frac{6}{2}\right) = 0.70\,\text{m}$$

Internal beams (BYIN): $b_{\text{eff,BYIN}} = 0.70 + 0.70 + 0.30 = 1.70$ m
External beams (BYOUT): $b_{\text{eff,BYOUT}} = 0.70 + 0.30 = 1.00$ m

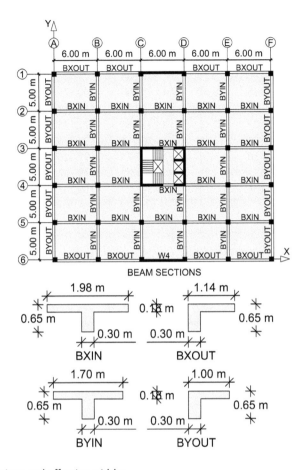

Figure 5.37 Beam sections and effective widths.

5.9.9 Modal response spectrum analysis

Response spectrum analysis is performed in two horizontal directions by considering the design response spectrum of Figure 5.34 ($q = 4.4$). The number of modes taken into account is defined by the sum of the effective modal masses, which has to be equal to at least 90% of the total mass of the structure. The CQC method is used to combine the results of the modes considered, in each direction. These results include the accidental torsional effects as described hereafter.

5.9.9.1 Accidental torsional effects

There are eight basic seismic load combinations which result from $G + 0.3\,Q \pm E_x \pm 0.3E_y$ and $G + 0.3\,Q \pm E_y \pm 0.3E_x$, where E_x and E_y are the seismic actions applied in the $x - x$ and $y - y$ directions, respectively. In the case that the centre of mass at each floor is considered to be displaced from each nominal location in each direction by an accidental eccentricity, then the resulting load combinations become 32 in number (8 basic seismic load combinations × 4 locations of mass).

Table 5.12 Response spectrum analysis for E_x

Storey	V_{xi} (kN)	V_{yi} (kN)	M_{xi} (kN m)	M_{yi} (kN m)	T_i (kN m) $e_{ai} = 0$	T_i (kN m) $e_{ai} = \pm 0.05\, L_i$
8th	1736.61	0.08	0.23	5209.84	21,708.37	23,874.54
7th	3278.27	0.15	0.67	14,943.00	40,979.46	45,197.69
6th	4592.75	0.22	1.32	28,520.13	57,410.58	63,448.52
5th	5693.97	0.28	2.15	45,310.68	71,175.77	78,884.20
4th	6626.00	0.34	3.15	64,765.64	82,826.13	92,057.79
3rd	7386.06	0.38	4.28	86,381.15	92,326.83	102,961.83
2nd	7980.14	0.42	5.51	109,671.82	99,752.88	111,634.49
1st	8422.90	0.44	7.70	150,705.27	105,287.36	118,341.19

Table 5.13 Response spectrum analysis for E_y

Storey	V_{xi} (kN)	V_{yi} (kN)	M_{xi} (kN m)	M_{yi} (kN m)	T_i (kN m) $e_{ai} = 0$	T_i (kN m) $e_{ai} = \pm 0.05\, L_i$
8th	0.09	1494.34	4483.03	0.27	22,929.52	25,166.25
7th	0.17	2808.99	12,816.05	0.79	42,944.08	47,292.45
6th	0.25	3915.30	24,383.14	1.52	59,715.70	65,921.25
5th	0.31	4842.32	38,629.07	2.43	73,742.95	81,696.95
4th	0.35	5633.68	55,112.33	3.48	85,735.69	95,294.71
3rd	0.39	6295.14	73,443.73	4.64	95,788.85	106,883.01
2nd	0.42	6830.78	93,264.90	5.89	103,972.78	116,457.88
1st	0.44	7252.29	128,392.70	8.07	110,455.47	124,279.18

The accidental eccentricity, according to EC8-Part I (Paragraph 4.3.3.3.3), is $e_{ai} = \pm 0.05 \cdot L_i$, where L_i is the floor dimension perpendicular to the direction of the seismic action (EC8-Part I, Equation 4.3).

In the example building studied here, a feature of ETABS is implemented, where the accidental eccentricity for each diaphragm is automatically taken into account for each of the seismic actions in directions $x - x$ and $y - y$. This eccentricity displaces the mass along the x- and y-axes, causing additional moments, the larger absolute value of which is then applied as torsion about the centre of mass, and the results are added to the response spectrum output. This is considered a conservative approach to the problem, but at the same time it eliminates the need to define a set of eight combinations for each mass location, since the effect of the eccentricity is already included in the results of the applied seismic actions and E_y, leaving in the end eight combinations.

In Tables 5.12 and 5.13, the influence of the accidental eccentricity at the response spectrum analysis results is shown for the $x - x$ and $y - y$ directions, respectively. As may be observed, the accidental eccentricity increases the value of the torsion T_i in both directions.

5.9.9.2 Periods, effective masses and modes of vibration

The modal properties are presented in Table 5.14 for the first 27 modes of vibration, which are considered in the response spectrum analysis, since the sum of the effective modal masses is over 90% of the total mass of the structure EC8-Part I (Paragraph 4.3.3.3.1).

Table 5.14 Periods and effective masses

Mode	T (s)	$M_{eff,x}$ (%)	$M_{eff,y}$ (%)	$M_{eff,Mz}$ (%)
1	0.82	0.00	75.36	0.02
2	0.71	73.98	0.00	0.00
3	0.60	0.00	0.02	70.57
4	0.20	0.00	10.76	0.06
...	...	...	...	...
27	0.02	0.00	0.00	13.79
ΣM_{eff} =		99.81	99.81	99.99
		1st mode, T = 0.82 s	2nd mode, T = 0.71 s	3rd mode, T = 0.60 s

It is noted that the criterion regarding the sum of the effective modal masses is satisfied in the $x - x$ and $y - y$ directions by considering the first 11 modes only. The need to comply with that criterion about the vertical axis too imposes the consideration of a larger number of modes. The 27th mode of vibration has a significant impact on the effective mass about the vertical axis.

The first three periods of vibration, considered to be the fundamental ones, are equal to 0.82, 0.71 and 0.60 s. According to the percentage of effective masses, it follows that the first two modes are predominantly translational in the $x - x$ and $y - y$ direction, whereas the third is predominantly torsional (Figure 5.38).

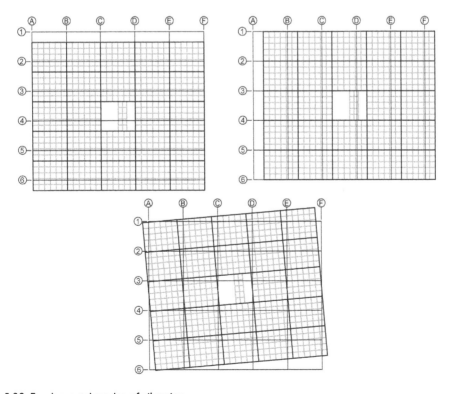

Figure 5.38 Fundamental modes of vibration.

5.9.9.3 Shear forces per storey

The base shear obtained by response spectrum analysis in the $x - x$ and $y - y$ directions is $V_{b,x} = 8422.90$ kN and $V_{b,y} = 7252.29$ kN in the first storey. The base shear distribution along the height of the building is shown in Table 5.15.

5.9.9.4 Displacements of the centres of masses

According to Chapter 6.2.3, the displacement d_s due to the design seismic action, could be the result of the elastic analysis of the structural system magnified by the behaviour factor q (Equation 6.43):

$$d_s = q \cdots d_e$$

The calculations are presented in Table 5.16 for $q = 4.4$. The drift at the eighth storey (roof level) is 0.40% (=0.1043/26) in the $x - x$ direction and 0.45% (=0.1175/26) in the $y - y$ direction.

Table 5.15 Storey shear forces

Storey	$V_{b,x}$	$V_{b,y}$
8th	1736.61	1494.34
7th	3278.27	2808.99
6th	4592.75	3915.30
5th	5693.97	4842.32
4th	6626.00	5633.68
3rd	7386.06	6295.14
2nd	7980.14	6830.78
1st	8422.90	7252.29

Table 5.16 Displacements at the centre of mass along the elevation in both the $x - x$ and $y - y$ directions

Storey	d_e (m)		$d_s = d_e^* q$ (m)	
	x–x	y–y	x–x	y–y
8th	0.0237	0.0267	0.1043	0.1175
7th	0.0215	0.0244	0.0946	0.1074
6th	0.0191	0.0218	0.0840	0.0959
5th	0.0165	0.019	0.0726	0.0836
4th	0.0137	0.0159	0.0603	0.0700
3rd	0.0107	0.0127	0.0471	0.0559
2nd	0.0077	0.0094	0.0339	0.0414
1st	0.0048	0.0061	0.0211	0.0268

5.9.9.5 Damage limitations

For buildings having non-structural elements of brittle materials attached to the structure (Equation 6.45)

$$d_{ri} \leq \frac{0.005 \cdot h_i}{r}$$

For buildings having ductile non-structural elements (Equation 6.46)

$$d_{ri} \leq \frac{0.0075 \cdot h_i}{r}$$

For buildings having non-structural elements fixed in a way so as not to interfere with structural deformation or without non-structural elements (Equation 6.47)

$$d_{ri} \leq \frac{0.010 \cdot h_i}{r}$$

The reduction factor r is taken to be equal to $r = 0.50$ for importance class II. The design inter-storey drift d_r is evaluated as the difference of the average lateral displacements ds at the top and bottom of the storey under consideration. The centre of mass is considered as the reference point for estimating the lateral displacements of the structural system. It may be noted that EC8-Part I (2004) does not provide a specific procedure for the estimation of the average lateral displacements. The calculations performed appear in Table 5.17. The ratio $(r\, d_{ri}/h_i)$ in all storeys is well below the strictest value of $d_{ri} = 0.005$. Hence, no damage is anticipated.

Table 5.17 Calculation of the damage limitation index

| Storey | d_{ri} (m) | | | $r \cdot d_{ri}/h_i$ | |
	x–x	y–y	h_i	x–x	y–y
8th	0.0097	0.0101	3	0.0016	0.0017
7th	0.0106	0.0115	3	0.0018	0.0019
6th	0.0114	0.0123	3	0.0019	0.0021
5th	0.0123	0.0136	3	0.0021	0.0023
4th	0.0132	0.0141	3	0.0022	0.0024
3rd	0.0132	0.0145	3	0.0022	0.0024
2nd	0.0128	0.0146	3	0.0021	0.0024
1st	0.0211	0.0268	5	0.0021	0.0027

5.9.9.6 Second-order effects

$P–\Delta$ effects need not be taken into account in the case that the inter-storey drift sensitivity coefficient θ_i is (Chapter 6.2.2, Equation 6.36):

$$\theta_i = \frac{\Delta_{el}^i \cdot q}{h_i} \cdot \frac{W_{tot}^i}{V_{tot}^i} \le 0.10$$

The term $(\Delta_{el}^i \cdot q)$ is equal to the design inter-storey drift d_{ri}. The sensitivity coefficients of all the storeys appear in Table 5.18. As may be observed, the second-order effects need not be taken into account, since $\theta_i \le 0.10$ in all storeys (Figure 5.39).

5.9.9.7 Internal forces

In Figures 5.40 through 5.45, the envelope diagram bending moments are presented for the seismic load combinations $(G + 0.3Q \pm E_x \pm 0.3E_y$ and $G + 0.3Q \pm E_y \pm 0.3E_x)$ for the elevation of grid axes B and C. The results at the foundation level are omitted.

Table 5.18 Estimation of the inter-storey drift sensitivity coefficients θ_i of each storey

Storey	$d_{ri}(=\Delta_{el}^i \cdot q)(m)$ x-x	y-y	W_{tot}(kN)	V_{tot} (kN) x-x	y-y	h_i (m)	θ_i (%) x-x	y-y
8th	0.0097	0.0101	7946.14	1736.61	1494.34	3	1.48	1.79
7th	0.0106	0.0115	16,880.74	3278.27	2808.99	3	1.82	2.30
6th	0.0114	0.0123	25,815.34	4592.75	3915.30	3	2.14	2.70
5th	0.0123	0.0136	34,749.94	5693.97	4842.32	3	2.50	3.25
4th	0.0132	0.0141	43,684.54	6626.00	5633.68	3	2.90	3.64
3rd	0.0132	0.0145	52,619.14	7386.06	6295.14	3	3.13	4.04
2nd	0.0128	0.0146	61,553.74	7980.14	6830.78	3	3.29	4.39
1st	0.0211	0.0268	71,002.27	8422.90	7252.29	5	3.56	5.25

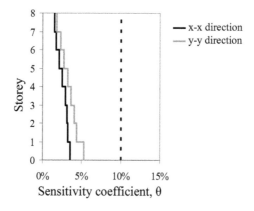

Figure 5.39 Comparison of the calculated inter-storey drift sensitivity coefficients θ_i with the Code upper limit $(\theta_i \le 0.10)$.

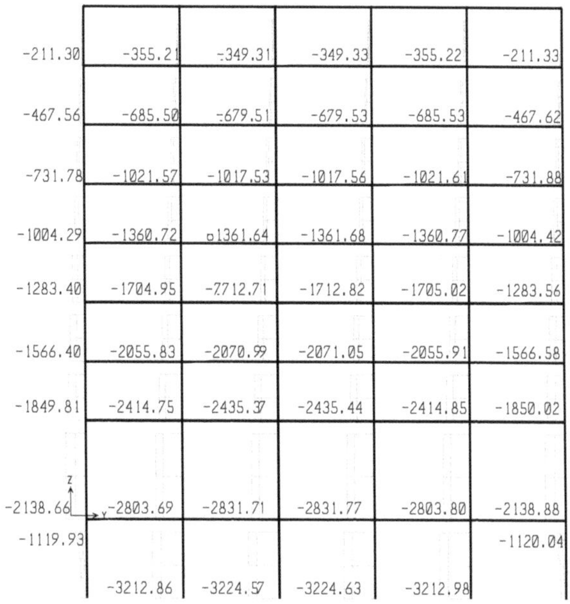

-211.30	-355.21	-349.31	-349.33	-355.22	-211.33
-467.56	-685.50	-679.51	-679.53	-685.53	-467.62
-731.78	-1021.57	-1017.53	-1017.56	-1021.61	-731.88
-1004.29	-1360.72	□1361.64	-1361.68	-1360.77	-1004.42
-1283.40	-1704.95	-7712.71	-1712.82	-1705.02	-1283.56
-1566.40	-2055.83	-2070.99	-2071.05	-2055.91	-1566.58
-1849.81	-2414.75	-2435.37	-2435.44	-2414.85	-1850.02
-2138.66	-2803.69	-2831.71	-2831.77	-2803.80	-2138.88
-1119.93					-1120.04
	-3212.86	-3224.57	-3224.63	-3212.98	

z
y

Figure 5.40 Envelope of axial forces – Grid axis B.

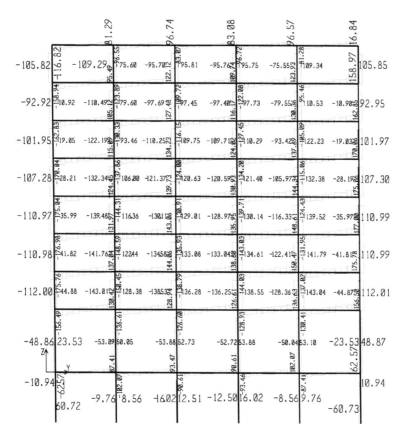

Figure 5.41 Envelope of shear forces – Grid axis B.

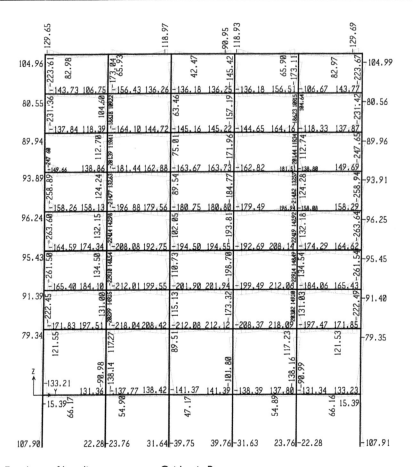

Figure 5.42 Envelope of bending moments – Grid axis B.

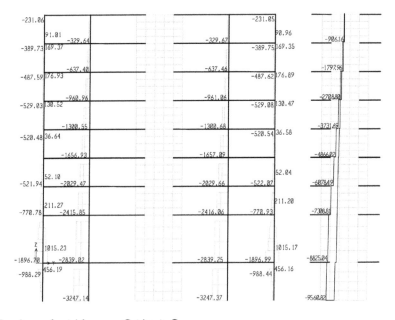

Figure 5.43 Envelope of axial forces – Grid axis C.

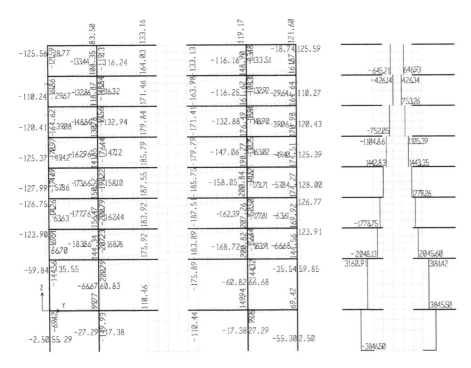

Figure 5.44 Envelope of shear forces – Grid axis C.

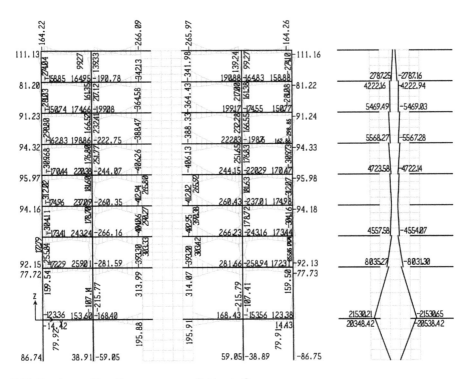

Figure 5.45 Envelope of bending moments – Grid axis C.

5.10 EXAMPLES: INELASTIC ANALYSIS OF A 16 STOREY BUILDING

5.10.1 Modelling approaches

Most of the modelling approaches mentioned this chapter, with regard to inelastic analysis, are represented in this application. This building is an actual structure in Bucharest, Romania, which has been designed according to EN1992:2004 and EN1998:2004 using the relevant Romanian National Annex. It is a dual R/C system with 16 storeys and four basements.

The plan view and section of the building is shown in Figure 5.46.

The building was initially analysed using Etabs (2013) and ECtools (2013), while for the nonlinear analyses, performed at a later stage and presented herein, the Zeus nonlinear software was used. More details on the modelling and the results may be found in the paper by Penelis and Papanikolaou (2009).

The modelling approach for the nonlinear analysis was the fibre model with linear elements for all structural elements. The finite element modelling is shown in Figures 5.47 and 5.48.

It is obvious that the suggestion to avoid the modelling of basements for the nonlinear analysis has been used.

A very useful tool to assess the accuracy of the NL model is to compare the eigen periods (elastic) to the ones of the elastic model used for design. In the case of the example, these are shown in Table 5.19.

Reinforced concrete sections (geometry and detailed reinforcement bar topology) were defined according to the original formwork drawings and were assigned to cubic elastoplastic frame elements, which have a tangent stiffness matrix that is integrated using second-order Gaussian quadrature (two Gauss points).

Structural walls were modelled using vertical frame elements along the wall mass centre and their horizontal kinematic constraint at storey levels with neighbouring beams was modelled explicitly, using numerically rigid elements (elastic material with $E_{\text{Rigid}} = 10\,E_s$ and 1.0×1.0 m cross section; Figure 5.17).

Rigid diaphragm action was considered at all storey levels. However, numerical treatment using master–slave joint constraints was not available in the current computational platform, and hence an explicit representation was necessary. This was realised using end-pinned, crossed diagonal rigid links on each quadrilateral slab region.

Short spandrels connecting walls with openings, which originally included embedded x-braces, were explicitly modelled by equivalent column sections and rigid link (Figure 5.49).

The total gravity load for each storey was calculated from the $G + \psi_2 Q$ seismic load combination of the initial design and was applied to the three beam inner nodes and wall end nodes, according to their geometrically derived tributary area.

For dynamic analysis, the total mass of each storey $m = (G + \psi_2 \cdot Q)/g$ was distributed to the end nodes of all vertical elements (columns and walls), according to their tributary area.

The horizontal loading pattern for inelastic static analysis was applied on the node coincident to the centre of mass (CM, Figure 5.47) of each storey level and, for dynamic analysis, biaxial excitation in both directions x and y (in the form of time–acceleration history) was applied on all base nodes.

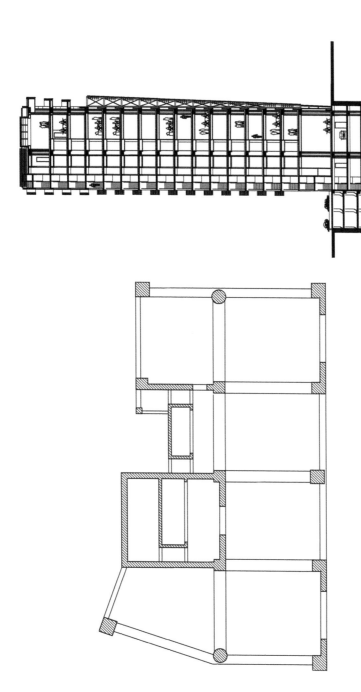

Figure 5.46 Plan view and elevation of a 16-storey building in Bucharest.

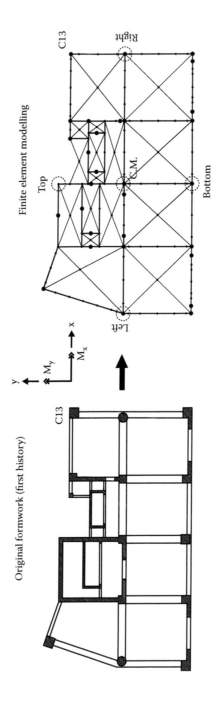

Figure 5.47 Typical storey plan view and modelling.

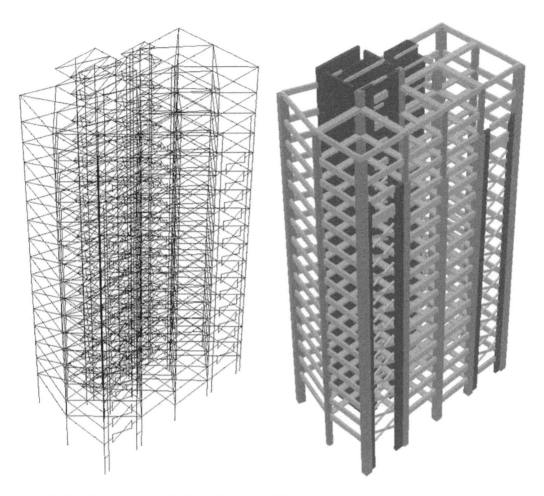

Figure 5.48 Finite element model of the 16-storey building.

Table 5.19 Eigen period from elastic and inelastic model

Mode	Zeus-NL (s)	ETABS (s)
1	1.03	0.98
2	0.81	0.73
3	0.53	0.56
4	0.25	0.22
5	0.21	0.21
6	0.16	0.16
7	0.14	0.12
8	0.13	0.09
9	0.12	0.08
10	0.11	0.08

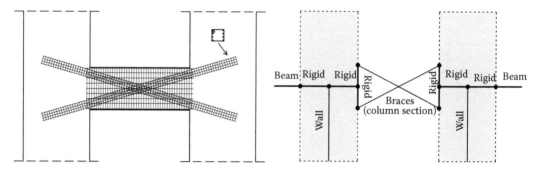

Figure 5.49 Finite element model for the short spandrels of the 16-storey building.

5.10.2 Nonlinear dynamic analysis

The excitations selected were three artificial 10 s strong motion records with a standard EC8-compatible frequency content for $a = 0.24$ g and soil C (Figure 5.50) and the 1977 Vrancea event (NS and EW components) unscaled.

Seven nonlinear time–history dynamic analyses were performed, six using the EC8 artificial records and one using the Vrancea 1977 event as follows:

1. EC8 #1 to #3 records: 100% in the x direction and 30% in the y direction $(x+0.30 \cdot y)$
2. EC8 #1 to #3 records: 30% in the x direction and 100% in the y direction $(0.30 \cdot x + y)$
3. Vrancea NS record in the x direction and EW record in the y direction $(x:\text{NS} + y:\text{EW})$

5.10.3 Nonlinear static analysis

The approach presented in Section 5.7.5.2 for torsionally sensitive buildings has been used for the application of the static nonlinear analysis.

Three inelastic static (pushover) analyses were performed, two using EC8-compatible static loads (for $a = 0.24$ g and soil C spectrum; see Figure 5.50) and one using the Vrancea event. The above analyses were performed in both positive and negative directions in order to capture the expected asymmetric structural response.

Specifically

1. EC8 spectrum: 100% in the x direction and 30% in the y direction $\pm(x+0.30 \, y)$
2. EC8 spectrum: 30% in the x direction and 100% in the y direction $\pm(0.30 \cdot x + y)$
3. Vrancea NS in the x direction and Vrancea EW in the y direction $\pm(x:\text{NS} + y:\text{EW})$

It is apparent that different loads (lateral and torsional) as well as different modification factors (c_1 and c_2) were derived for each static excitation, depending on the spectral shape. The capacity curves extracted from the analysis were converted into acceleration-displacement response spectrum (ADRS) and bilinearised using special software (Kappos, Panagopoulos and Penelis, 2008), employing the equal areas principle (see Chapter 13.5.2.2).

The target displacement of the SDOF oscillator was, in turn, determined iteratively using constant ductility capacity spectra, and finally the target displacement of the MDOF building was calculated. The above procedure is schematically depicted in Figure 5.51 (see Chapter 13.5.2.2).

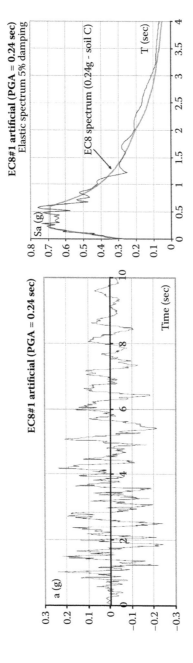

Figure 5.50 EC8 artificial strong motion records (0.24 g and soil C; left) and respective elastic spectra (right).

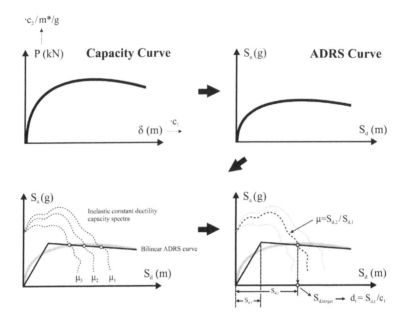

Figure 5.51 Derivation of target displacement from the capacity curve using the ADRS method.

5.10.4 Results: global response

The global response results presented hereinafter for each excitation and analysis type are the following:

1. Static and dynamic capacity curves $(P-\delta)$ (Figures 5.52 and 5.53), which demonstrate that there is a close correlation between static and dynamic nonlinear behaviour and that the building shows asymmetric response due to structural eccentricity. The maximum top displacement from nonlinear dynamic analysis for the EC8-compatible excitations is approximately 20 cm for the x direction and 30 cm for the y direction (Figures 5.52 and 5.53, respectively, and Table 5.20), which renders the

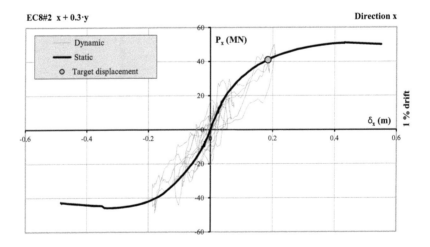

Figure 5.52 Global P–δ comparison between static and dynamic analysis for the EC8#2 $x + 0.3 \cdot y$ excitation.

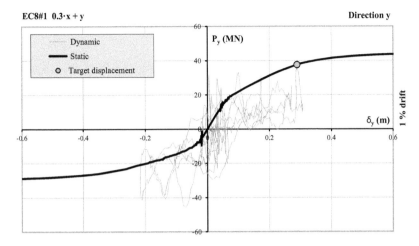

Figure 5.53 Global P–δ comparison between static and dynamic analysis for the EC8#1 0.3 · x + y excitation.

Table 5.20 Comparison of top displacements at centre of mass and opposite sides between static and dynamic analysis

Excitation	Dynamic (maximum)	Static (at target disp.)	Difference (%)
Response at centre of mass (CM)			
EC8#1 $x + 0.3 \cdot y$ (δ_x)	0.181	0.200	10.6
EC8#2 $x + 0.3 \cdot y$ (δ_x)	0.208	0.185	11.2
EC8#3 $x + 0.3 \cdot y$ (δ_x)	0.205	0.200	2.5
EC8#1 $0.3 \cdot x + y$ (δ_y)	0.308	0.290	6.0
EC8#2 $0.3 \cdot x + y$ (δ_y)	0.229	0.290	26.7
EC8#3 $0.3 \cdot x + y$ (δ_y)	0.251	0.270	7.6
Vrancea x: NS + y:EW (δ_x)	0.259	0.295	13.8
		Average	11.2
Response at opposite sides			
EC8#1 $x + 0.3 \cdot y$ (δ_x)–Top	0.141	0.151	7.3
EC8#1 $x + 0.3 \cdot y$ (δ_x)–Bottom	0.222	0.250	12.8
EC8#2 $x + 0.3 \cdot y$ (δ_x)–Top	0.177	0.138	22.2
EC8#2 $x + 0.3 \cdot y$ (δ_x)–Bottom	0.240	0.232	3.4
EC8#3 $x + 0.3 \cdot y$ (δ_x)–Top	0.175	0.151	13.7
EC8#3 $x + 0.3 \cdot y$ (δ_x)–Bottom	0.238	0.250	5.1
EC8#1 $0.3 \cdot x + y$ (δ_y)–Left	0.290	0.227	21.8
EC8#1 $0.3 \cdot x + y$ (δ_y)–Right	0.328	0.352	7.3
EC8#2 $0.3 \cdot x + y$ (δ_y)–Left	0.219	0.226	3.1
EC8#2 $0.3 \cdot x + y$ (δ_y)–Right	0.244	0.352	44.6
EC8#3 $0.3 \cdot x + y$ (δ_y)–Left	0.238	0.187	21.4
EC8#3 $0.3 \cdot x + y$ (δ_y)–Right	0.264	0.299	13.0
Vrancea x:NS + y:EW (δ_x)–Top	0.222	0.231	4.2
Vrancea x:NS + y:EW (δ_x)–Bottom	0.300	0.362	20.7
		Average	14.3

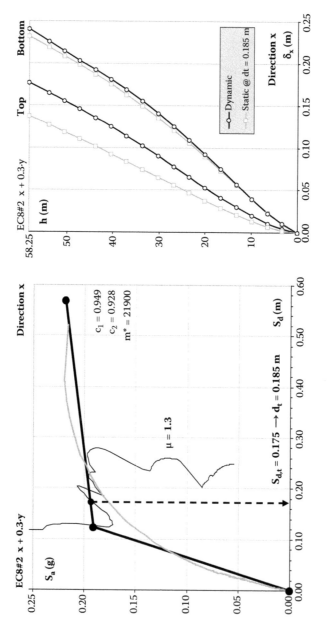

Figure 5.54 Derivation of target displacement and displacement profiles for the EC8#2 x+0.3 · y excitation.

estimation of 36 and 68 cm calculated from the elastic design of the building (EC8), respectively, a safe approach. Moreover, the base shear capacity of the building is approximately 50 MN for the x direction and 40 MN for the y direction, while the base shear design force is approximately 24 MN.

2. The storey displacements in the x and y directions for both sides of the building (x: top and bottom of plan, y: left and right of plan), which demonstrate not only the translational response, but also the rotational (Figure 5.54). In the same figure, the derivation of the target displacement for inelastic static analysis (using the ADRS method) is also shown. The differences between the dynamic and static nonlinear approach regarding the displacement profiles are presented in Table 5.20. It is important that the average difference regarding the response at the centre of mass is about 11%, while the average difference for the sides response (or rotational response) is about 14%.

5.10.5 Results: local response

The local response results are shown for the corner column C13 (top-right corner of plan) in order to investigate the biaxial nonlinear stress state of the section under a varying axial load. Figure 5.55 shows the moment-chord rotation curves (static and dynamic) as well as the maximum dynamic demand in comparison to the static demand calculated at the target displacement step of the inelastic static analysis. The large difference between moment capacities for the x and y directions (shown in Figure 5.55), even though the cross section is almost symmetrical, is justified if one takes into account the biaxial moment stress of the column with different axial load at different time steps.

The results show an acceptable agreement between dynamic and static approach, considering the different variation of the column axial load between static and dynamic nonlinear analysis. Furthermore, the nonlinear demands (column rotations) are well within the design requirements of EC8.

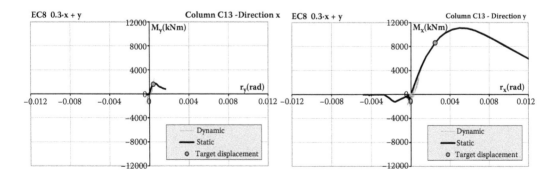

Figure 5.55 Local *M–r* comparison between static and dynamic analysis for the EC8#2 *x*+ 0.3 · *y* excitation.

Chapter 6

Capacity design – design action effects – safety verifications

6.1 IMPACT OF CAPACITY DESIGN ON DESIGN ACTION EFFECTS

6.1.1 General

Earthquakes belong to the category of accidental actions; therefore,

- They are not combined with other accidental actions, and
- Earthquake loading is combined with gravity loads with partial safety factors for actions equal to 1.0 (see Chapter 5.8.3).

According to the procedure described in detail in Chapters 3, 4 and 5, the action effects are defined in a deterministic way, as if the loads were statistically reliable and the response of the structure was in the elastic range. These two weak points in the calculation of the action effects make necessary a more reliable approach to the problem, which would ensure the existence of adequate strength and ductility in crucial regions of the structure, so that premature local or general collapse is excluded.

Indeed, since it is impossible to predict with accuracy the characteristics of the seismic motion due to an earthquake bigger than the design earthquake, it is impossible to estimate with accuracy the response of an R/C building to this earthquake. However, it is possible to provide the structure with those features that will ensure the resistance shifting of the building in displacement terms beyond life safety limits. In terms of ductility, energy dissipation, damage or failure pattern, this means that the sequence in the breakdown of the chain of resistance of the structure will follow a predefined desirable hierarchy (Park and Paulay, 1975; Penelis and Kappos, 1997; Elnashai and Di Sarno, 2008; Fardis, 2009). In order to ensure a certain sequence in the failure mechanism of the resistance chain, the resistance of every link should be known. This knowledge should not be based on assumptions of disputable reliability, but on quantified strength of the structural elements that will be subjected to very large deformations (due to formation of plastic hinges) during a catastrophic earthquake.

Although the nature of the design actions is probabilistic, the ability to have a semi-deterministic allocation of strength and ductility in the structural members provides an effective tool for ensuring a successful response and prevention of collapse during a catastrophic earthquake.

Such a response may be achieved if the successive regions of energy dissipation are rationally chosen and secured through a proper design procedure, so that the predecided energy dissipation mechanism would hold throughout the seismic action. This design concept is included in a procedure called *capacity design procedure*.

According to this procedure, the structural elements that are designated to dissipate the seismic energy via bending reversals are reinforced accordingly, with a special concern for avoiding brittle failure due to shear in regions out of the plastic hinges. In the same way, all other members are provided with adequate strength reserves, in order to ensure that the chosen dissipating mechanism will be activated and preserved during a strong earthquake without premature brittle failure of non-ductile regions or members.

This means that the action effects resulted from the analysis *serve only as a guide and must be properly modified* in order to accommodate the capacity design of the structure. It is obvious that this modification cannot be based solely on the knowledge and ingenuity of the designer, and it should be formulated in a disciplined procedure with reference to the Code. It is also obvious that this modification of the seismic effects should be a function of the chosen ductility class.

The aforementioned concepts have been incorporated into all modern Codes for earthquake design and since the early 1980s in EC 8-(ENV).

6.1.2 Design criteria influencing the design action effects

From the design criteria for R/C buildings included in EC8-1/2004, only

- Local resistance criteria
- Capacity design criteria

influence the determination of the design action effects. All the others refer to dimensioning and detailing of the R/C structural members and will, therefore, be addressed in Chapters 8 through 10. The design criteria influencing the design action effects are the following, in detail:

1. All critical regions of the structure must exhibit resistance adequately higher than action effects, developed in these regions under the seismic design situation (e.g. minimum reinforcement in tensile zones, minimum reinforcement in compressive zones, etc.)
2. Brittle or other undesirable failure modes, such as
   a. Shear failure of the structural members
   b. Failure of beam–column joints
   c. Yielding of foundations
   d. Yielding of any other element intended to remain elastic (e.g. structural walls beyond critical zones) must be excluded. This can be ensured if the design action effects of purposely selected regions are derived *from equilibrium conditions* when flexural plastic hinges with their potential overstrengths have occurred in adjacent areas.
3. Extensive distribution of plastic hinges should be ensured, avoiding their concentration in any single storey (*'soft storey' mechanism*) and particularly at both ends of a number of columns in the same storey. This can be achieved with sufficient reliability, if it is ensured that plastic hinges develop only in beams and not on columns, except for the unavoidable formation of plastic hinges at the base of the building. (Figure 6.1).

The implementation of these criteria for the determination of the design action effects of the various structural elements of a building is given below.

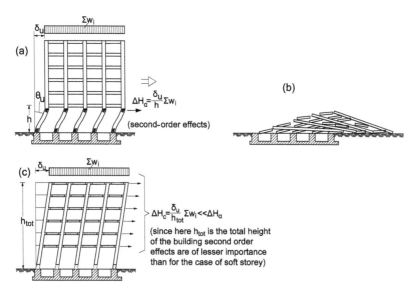

Figure 6.1 Plastic mechanism of a frame system with strong beams–weak columns. (a) Generation of a soft storey. The second-order effects lead to an overturning of the formed inverted pendulum. (b) 'Pancake' collapse pattern. (c) A frame system with strong columns–weak beams. This ensures the existence of a strong vertical spine preventing collapse.

6.1.3 Capacity design procedure for beams

According to EC8-1/2004, CEB/MC-SD 185 (CEB 1985), ASCE/SEI7-10, NEHRP 2003 and SEAOC 1999, the design values of the bending moments of a beam for all ductility classes are obtained from the analysis of the structure for the seismic loading combinations, as described in detail in Chapter 5.8.3, without any modifications except for a possible redistribution.

However, according to all relevant Codes, beams need an additional compression reinforcement at their support equal to at least 50% of the corresponding tension reinforcement in order to ensure an adequate local ductility level (Chapter 8.2.5). Based on the capacity design concept, these reinforcement bars are appropriately anchored in concrete, so that they can operate as tension reinforcement in case of moment reversal due to an unexpected severe earthquake. Therefore, the moment resistance envelope of the beams is considerably improved at low cost (the cost of anchorages of the compression reinforcement) no matter what the values are of the design action effects, which have been derived from the analysis (Figure 6.2a).

This means that the beam, as it is designed, can carry much larger moment fluctuations generated by an earthquake than the design action moments. However, in order to ensure this behaviour, the structural element has to be protected from a premature shear failure, because it is well known that shear failure does not present a ductile mode (see also Chapter 8.2.5). Therefore, the design shear for DCM and DCH R/C buildings should not be the one resulting from the analysis, but the shear corresponding to the equilibrium of the beam under the appropriate gravity load and a rational adverse

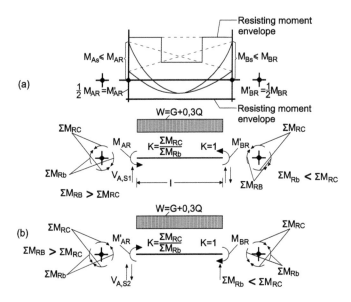

Figure 6.2 Capacity design values of shear forces acting on beams; (a) resisting moment envelope; (b) equilibrium conditions for the determination of shear forces.

combination of the actual bending resistances of the cross sections at the ends of the beam (Figure 6.2b):

$$
\begin{aligned}
V_{AS1} &= \frac{wl}{2} + \gamma_{Rd} \cdot \frac{\kappa_A M_{AR} + \kappa_B M'_{BR}}{l} \\
V_{AS2} &= \frac{wl}{2} - \gamma_{Rd} \frac{\kappa_B M_{BR} + \kappa_A M'_{AR}}{l} \\
V_{BS1} &= -\frac{wl}{2} + \gamma_{Rd} \frac{\kappa_A M_{AR} + \kappa_B M'_{BR}}{l} \\
V_{BS2} &= -\frac{wl}{2} - \gamma_{Rd} \frac{\kappa_B M_{BR} + M'_{AB}}{l}
\end{aligned}
\tag{6.1}
$$

where $M_{AR}, M'_{AR}, M_{BR}, M'_{BR}$ are the actual resisting moments at the ends of the beam, accounting for the actual area of the reinforcing steel (all moments positive), and γ_{Rd} is an amplification factor due to materials overstrength but also taking into account the reduced probability that all end moments exhaust simultaneously all strength reserves. This γ_{Rd} factor counterbalances the partial safety factor of steel that has been introduced for the fundamental combination (see Section 6.2.2), as well as for the seismic one, and covers the hardening effects as well. In the absence of more reliable data, according to EC8-1/2004, γ_{Rd} may be taken as

$$
\begin{aligned}
&\bullet \quad \text{Ductility class medium}: \gamma_{Rd} = 1.00 \\
&\bullet \quad \text{Ductility class high}: \gamma_{Rd} = 1.20
\end{aligned}
\tag{6.2}
$$

Coefficients κ_A and κ_B have been introduced to respond to the case of strong beams and weak columns. Indeed, in the case that the columns are strong and the beams weak, which is the prevailing condition for earthquake-resistant R/C buildings, the sum of the resisting design bending moments of the columns at the joints, taking into account their axial load effects, is larger than the sum of the resisting design bending moments of the beams framing into the joints, that is, $\Sigma M_{Rc} > \Sigma M_{Rb}$. In this case, plastic hinges are formed at the beam ends. Therefore, coefficient κ is equal to 1.00,

$$\kappa = 1.00 \tag{6.3}$$

since at the beam ends their resisting design bending moments may develop to their highest value.

Conversely, in the case of strong beams and weak columns, the sum of the resisting design bending moments of the beams framing the joints is larger than the sum of the resisting design bending moments of the columns running to the joints, taking into account, of course, their axial load effects, that is, $\Sigma M_{Rb} > \Sigma M_{Rc}$. In this case, plastic hinges are likely to form at column ends at the joints. Therefore, coefficient κ must be less than 1.00, because at the beam ends their resisting design bending moments cannot develop their highest values, as these are limited by the lower resisting moments of the adjacent columns. In this case, κ should have a reduced value equal to

$$\kappa = \frac{\Sigma M_{Rc}}{\Sigma M_{Rb}} \leq 1 \tag{6.4}$$

In this respect, according to EC8-1/2004, κ should be equal to (Figure 6.2b)

$$\kappa = \min\left(1, \frac{\Sigma M_{Rc}}{\Sigma M_{Rb}}\right) \tag{6.5}$$

The sign of the ratio

$$\zeta = \frac{V_{AS2}}{V_{AS1}} \text{ or } \frac{V_{BS2}}{V_{BS1}} \tag{6.6}$$

has a considerable effect on the shear design of the beams, as it will be explained in Chapter 8. The above capacity design procedure, according to EC8-1, applies to DCM and DCH buildings. For DCL buildings, the design values of the acting shear forces are obtained from the analysis of the structure for the seismic load combination, as described in detail in Chapter 5.8.3.

6.1.4 Capacity design of columns

6.1.4.1 General

The basic concept in the capacity design of frame or frame-equivalent dual systems is that the failure mechanism must include plastic hinges only at the base of the columns, while all other plastic hinges of the mechanism are distributed at the beams (Figure 6.1c). In this context, a prevailing strong vertical spine is ensured in the building, which could potentially protect the building against the formation of a 'soft storey' (inverted pendulum) and a collapse in a 'pancake' pattern in the case of a strong unexpected earthquake (Figure 6.1b,c; Chapter 11, Figure 11.49). Of course, this assumes that bending failure, which is dissipative, precedes that of shear, which is brittle.

6.1.4.2 Bending

It has already been noted that the formation of plastic hinges in the columns during the earthquake should be avoided so that the energy is dissipated by the beams only (Park, 1986). The reasons for this requirement, which are very clearly stated in EC8-1/2004 (see Section 6.1.2), are the following:

1. Due to axial compression, columns have less available ductility than beams, as it will be clarified in Chapter 8.3.2. On the other hand, for the same displacement at the top of the frame (Figure 6.3), or in other words for the same global ductility expressed in

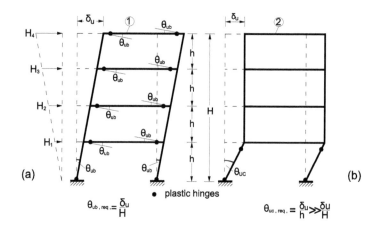

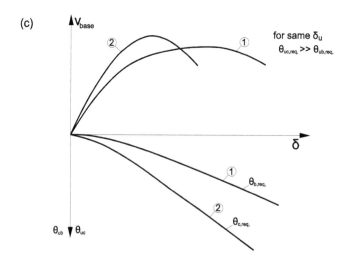

Figure 6.3 Failure mechanism of a frame: (a) beam mechanism; (b) storey mechanism; (c) interpretation of the vulnerability of a storey mechanism via $V_{base} - \delta - \theta$ diagram.

terms of displacements (see Chapter 5.4.4), much larger plastic column rotations are required than beam rotations. Indeed, in the case of plastic hinges at columns, the total plastic rotations required are developed at the top and bottom of the columns of one storey (soft storey), while in the case of plastic hinges, they develop at the beams and are spread to all storeys of the frame (Figure 6.3). Therefore, for the same global ductility, a larger local column ductility in rotations is required (Figure 6.3a) over local beam ductility. Thus, while (Park and Paulay, 1975)

$$\theta_{uc}^{avail} \leq \theta_{ub}^{avail} \tag{6.7a}$$

for the same δ_{ureq} (i.e. the same μ_{ureq})

$$\theta_{uc}^{required} = \frac{\delta_{ureq}}{h} \geq \theta_{ub}^{required} = \frac{\delta_{ureq}}{H} \tag{6.7b}$$

2. While beam failure exhibits extended cracking only in the tension zones, due to the yielding of the reinforcement, column failure mode presents, in successive steps close to one another, spalling of concrete, breaking of ties, crushing of concrete core and buckling of the longitudinal rebars. This process leads to the creation of a collapse mechanism due to the inability of the columns to carry the axial gravity loads after their failure. Therefore, avoiding column failure is much more crucial for the overall safety of the structure than avoiding beam failure in bending.
3. The formation of plastic hinges in the columns leads to significant inter-storey drifts, so that the relevant second-order effects may lead to a premature collapse of the structure.

In order to decrease the probability of plastic hinge formation in the columns, *frames or frame-equivalent dual systems* must be designed to have 'strong columns and weak beams' (Park, 1986; Paulay et al., 1990; Priestley and Calvi, 1991; Penelis and Kappos, 1997). This concept is adopted in the requirements of EC8-1 and other relevant Codes. They state that the sum of the resisting design moments of the columns at a joint, taking into account the action of normal force, should be greater than the sum of the resisting design moments of all beams framing the joint for each one of the two orthogonal directions of the building and for both positive and negative actions of the seismic motion (Figure 6.4), that is,

$$\boxed{\begin{aligned} \left|M_{R,1}^{c,o}\right| + \left|M_{R,1}^{c,u}\right| &\geq 1.30\left(\left|M_{R,1}^{b,l}\right| + \left|M_{R,1}^{b,r}\right|\right) \\ \left|M_{R,2}^{c,o}\right| + \left|M_{R,2}^{c,u}\right| &\geq 1.30\left(\left|M_{R,2}^{b,l}\right| + \left|M_{R,2}^{b,r}\right|\right) \end{aligned}} \tag{6.8}$$

Factor 1.30 has been introduced in order to take into account the variability of the yield stress f_y of the reinforcement and the probability of strain-hardening effects (overstrength factor).

Therefore, the capacity design is satisfied if the columns are designed for the following moments:

$$\left.\begin{aligned} M_{S1CD} &= a_{CD1}M_{S1} \\ M_{S2CD} &= a_{CD2}M_{S2} \end{aligned}\right\} \tag{6.9}$$

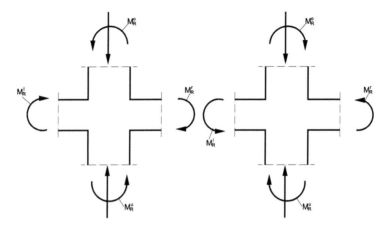

Figure 6.4 Strong columns–weak beams.

where

$$a_{CD1} = 1.30 \frac{\left|M_{R,1}^{b,l}\right| + \left|M_{R,1}^{b,r}\right|}{\left|M_{S,1}^{c,o}\right| + \left|M_{S,1}^{c,u}\right|} \Bigg\}$$

$$a_{CD2} = 1.30 \frac{\left|M_{R,2}^{b,l}\right| + \left|M_{R,2}^{b,r}\right|}{\left|M_{S,2}^{c,o}\right| + \left|M_{S,2}^{c,u}\right|} \Bigg\}$$

(6.10)

In the above relationships,

- M_S^c are the action effects (bending moments) of the columns derived from the analysis for the seismic combination.
- M_R^b are the design resisting moments of the beam derived from the design of the beams, which has already preceded column design.

EC8-1 allows a relaxation of the above capacity design criterion whenever the probability of full reversal of beam end-moment is relatively low (wall-equivalent dual systems, uncoupled wall systems). The following cases are also exempted from the requirements of the above procedure:

- In single-storey R/C buildings and in the top storey of multi-storey buildings
- In one quarter of the columns of each storey in plane R/C frames with four or more columns
- In two-storey R/C buildings if the value of the normalised axial load vd at the bottom storey does not exceed 0.3 in any column

In EC8-1/2004, it is clearly stated that the capacity design procedure for columns is implemented *in frame* and *frame-equivalent dual* systems. No reference is made to these systems in case they are 'torsionally flexible'. It is the author's opinion that *the torsionally flexible frame or frame-equivalent systems should also comply with the capacity design procedure* because of their additional vulnerability, which is attributable to the torsional behaviour of the system.

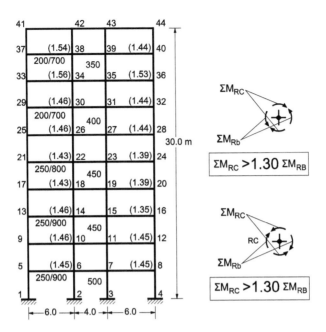

Figure 6.5 Values of the magnification factors α_{CD} for a 10-storey, 4-column R/C frame.

Finally, for DCL buildings, the design bending moments of columns are determined from analysis of the structure for the seismic load combination without any application of the capacity design criterion.

The magnification factor a_{CD} (Equations 6.9 and 6.10) takes rather high values. In the example of Figure 6.5, where a plane frame has been analysed for gravity loads '+' seismic actions, the values of a_{CD} for DC M range from 1.35 to 1.56.

6.1.4.3 Shear

According to the capacity design criterion, and following the rationale developed for beams (Section 6.1.3), shear forces are determined by considering the equilibrium of the column under the actual resisting design moments at its ends (Figure 6.6):

$$V_{sd.CD} = \gamma_{Rd} \frac{\kappa_A M_{AR} + \kappa_B M_{BR}}{l_c} \qquad (6.11)$$

where M_{AR} and M_{BR} are the actual resisting moments at the ends of the column, taking into account the axial existing forces and accounting for the actual area of the reinforcing steel (all moments positive), and γ_{Rd} is an amplification factor due to the materials overstraining but also taking into account the reduced probability that all end moments exhaust simultaneously all strength reserves. In the absence of more reliable data, according to EC8-1/2004, γ_{Rd} may be taken as

$$\text{Ductility class medium} : \gamma_{Rd} = 1.10$$
$$\text{Ductility class high} : \gamma_{Rd} = 1.30 \qquad (6.12)$$

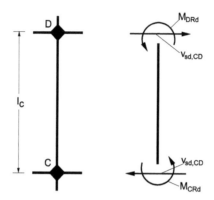

Figure 6.6 Capacity design of shear forces acting on columns derived by equilibrium condition of the column under shear forces and resisting moments M_{Rd}.

Coefficients κ_A and κ_B have been introduced to account for the case of strong columns and weak beams. Indeed, in the case that the beams are strong and the columns weak, the sum of the resisting design bending moments of the beams framing into the joints is larger than the sum of the resisting design bending moments of the columns running to the joints. Taking into account their axial load effects, this means $\Sigma M_{Rc} < \Sigma M_{Rb}$. In this case (Figure 6.6), plastic hinges are formed at column ends. Therefore, the coefficient κ is equal to 1.0,

$$\kappa = 1.00 \tag{6.13}$$

since at the column ends their resisting design bending moments may develop to their highest value.

Conversely, in the case of strong columns and weak beams, the sum of the resisting design moments of the beams framing into the joints are smaller than the sum of the resisting design bending moments of the columns running to the joints, taking into account, of course, their axial load, that is, $\Sigma M_{Rc} > \Sigma M_{Rb}$. In this case (Figure 6.6), plastic hinges are likely to form at the beam ends at the joints. Therefore, the coefficient κ must be less than 1.00 because at the column ends their resisting design bending moments cannot develop their highest values, as these are limited by the lower resisting moments of the adjacent beams. In this case, κ should have a reduced value equal to

$$\kappa = \frac{\sum M_{Rb}}{\sum M_{Rc}} \leq 1.0 \tag{6.14}$$

In this respect, according to EC8-1/2004, κ should be equal to (Figure 6.6)

$$\boxed{\kappa = \min\left(1, \frac{\sum M_{Rb}}{\sum M_{Rc}}\right)} \tag{6.15}$$

Finally, for DCL, the design action shear forces are determined by the analysis of the structure for the seismic load combination without any application of the capacity design criterion.

6.1.5 Capacity design procedure for slender ductile walls

6.1.5.1 General

The basic concept in the capacity design of wall or wall-equivalent dual systems is that the failure mechanism must include plastic hinges only at the base of the walls and of the eventual columns of the system, while all other plastic hinges of the mechanism are distributed at the beams. At the same time, bending failure must precede any shear failure, which is brittle, and in this way it interrupts the dissipative procedure. Likewise, as in the case of a frame or frame-equivalent dual system with strong columns and weak beams, a vertical spine is formed for the building, which shifts its resistance in terms of displacements beyond the life safety limits (Figure 6.7).

6.1.5.2 Bending

The moment diagrams of slender ductile (shear) walls ($h_w/l_w > 2$) under static seismic action have the form of Figure 6.8. However, a dynamic response analysis results in moment diagrams with approximate linear variation (Paulay and Priestley, 1992). Thus, the design moment diagram introduced by EC8-1 has the form of a trapezoid covering the saw-like M-diagram (Paulay, 1986).

The introduction of this M-diagram ensures that the plastic hinge will be generated at the base of the shear wall, while the rest of the wall will remain in the elastic region during a strong seismic motion. In this respect, the shear wall will have ductile behaviour so that it may be considered synonymous with a 'ductile wall'.

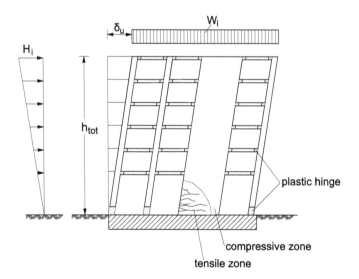

Figure 6.7 Plastic mechanism of a wall-equivalent dual system designed according to capacity design requirements (plastic hinges at beam ends and at the base of the vertical elements).

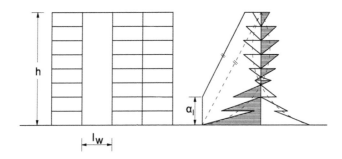

Figure 6.8 Slender ductile wall moment diagram and the capacity design envelope.

The value a_1 expresses the tension shift equal to (Figures 6.8 and 6.9)

$$a_1 = z \cot\theta \cong 0.9l_w \cot\theta \tag{6.16}$$

For the usual case where

$$\theta = 45°$$

$$\boxed{a_1 \cong 0.9l_w} \tag{6.17}$$

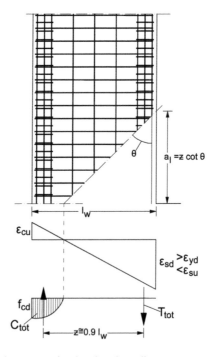

Figure 6.9 Tension T_{tot} shift mechanism in slender ductile walls.

In addition to the above changes of moment diagrams of ductile walls, the Code allows the following redistributions in order to account for uncertainties in analysis and post-elastic dynamic effects:

1. Redistribution of seismic action effects (bending moments, shear forces) up to 30% in all types of ductile walls.
2. In coupled walls, the above redistribution of 30% should be redistributed from the walls, which are under low compression to those which are under high compression. Such redistribution alleviates the shear requirements, as will be seen in the next section. At the same time, redistribution of seismic action effects between coupling beams of different storeys up to 20% is allowed under the assumption that the axial force at the base of the walls is not modified (see Chapter 9.3.2).

6.1.5.3 Shear

It is well known that the available moment resistance M_{rd} at the base of a ductile wall is usually larger than the moment demand M_{Ed}. Therefore, according to the capacity design criterion, design shear demand derived from the analysis must be magnified by a magnification factor ε so that the above-defined possible moment increase, together with various uncertainties in the analysis and post-elastic dynamic effects, is taken into account.

- DCM buildings
- For ductile walls, the magnification factor ε may be taken as equal to

$$\boxed{\varepsilon = 1.50} \tag{6.18}$$

- For ductile walls in dual systems, a modified design envelope of shear forces is adopted (Figure 6.10). The design envelope of the shear forces along the height of the wall is derived as follows:
  - For $z < 1/3\, h_w$

$$\boxed{V_{Sd} = 1.50 V'_{Sd}} \tag{6.19}$$

  - For $1/3\, h_w < z < h_w$
  V_{Sd} at the top is equal to

$$\boxed{V_{Sd} = \frac{1}{2} V_{Sd}^{base}} \tag{6.20}$$

Variation between $z < 1/3\, h_w$ and $z = h_w$ must be considered linear (see Figure 6.10). The notation of the above equations is given below:

- V'_{Sd} is the shear force along the height of the wall obtained from the analysis.
- V_{Sd} is the capacity design value.
- h_w is the height of the ductile wall.
- z is the reference height measured from the base of the wall.

The adoption by EC8-1/2004 of the above envelope of design shear forces for ductile walls in dual systems accounts for the uncertainties of contribution of higher modes.

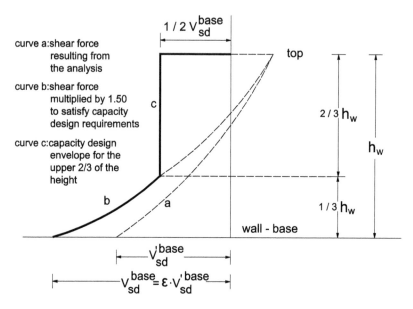

Figure 6.10 Design envelope of the shear forces in the walls of a dual system. Curve a: Shear forces resulting from the analysis. Curve b: Shear forces multiplied by 1.50 to satisfy capacity design requirements. Curve c: Capacity design envelope for the upper 2/3 of the height.

In addition, inelastic analyses performed so far (Eibl and Keintzel, 1989) have shown that the resulting shears are much higher than the shears derived from an elastic response analysis, due to changes of frame reactions on the walls in the post-elastic range.

- DCH buildings

    For this class of buildings, in the case of slender ($h_w/l_w > 2.0$) ductile walls, the magnification factor ε may be estimated as follows:

$$\varepsilon = q \left[\left(\frac{\gamma_{Rd}}{q} \frac{M_{Rd}}{M_{Ed}} \right)^2 + 0.1 \left(\frac{S_e(T_c)}{S_e(T_1)} \right)^2 \right]^{1/2} \leq q \qquad (6.21)$$

where
- q is the behaviour factor used in the design.
- M_{Ed} is the design bending moment at the base of the wall.
- M_{Rd} is the design flexural resistance at the base of the wall.
- γ_{Rd} is the factor to account for overstrength due to the steel-hardenings recommended value of 1.20.
- T_1 is the fundamental period of vibration of the building in the direction of consideration.
- T_c is the upper-limit period of the constant spectral acceleration branch (see Chapter 3.4.3).
- $S_e (T)$ is the ordinate of the elastic response spectrum.

In any case, the magnification factor ε must be limited between

$$1.5 \leq \varepsilon \leq q \qquad (6.22)$$

It should be clarified that the design envelope of the shear forces along the height of the wall in the case of a dual-wall system in DCH buildings has the same form as that of DCM ones (Figure 6.10).

It is recommended that both design moments (M_{Ed}, M_{Rd}) in Equation 6.21 should be calculated for the same design axial force. Otherwise, erroneous results are derived, particularly in the case of coupled walls, where the axial force of gravity loads may even change sign due to the coupled action under horizontal seismic loading (see Chapters 9.3.2 and 4.5.3.3, Figure 4.29). The erroneous results mentioned above tend to be smoothened if moment redistribution is implemented from the walls, which are under low compression to those which are under high compression.

6.1.6 Capacity design procedure for squat walls

As was mentioned before, walls with an aspect ratio $h_w/l_w < 2.0$ are defined as 'squat walls'. Capacity design rules for bending and shear for ductility classes M and H are given below.

6.1.6.1 DCH buildings

- In regard to flexure, there is no need to modify the M-diagrams derived from the analysis.
- The shear force V'_{Ed} from the analysis should be increased as follows:

$$\boxed{1.50V'_{Ed} < V_{Ed} = \gamma_{Rd}\left(\frac{M_{Rd}}{M_{Ed}}\right)V'_{Ed} \le qV'_{Ed}}$$

(6.23)

The definition and variable values are given in the paragraph above.

6.1.6.2 DCM buildings

- Referring to flexure, there is no need to modify the M-diagrams derived from the analysis.
- The shear force V'_{Ed} from the analysis should be increased as follows:

$$\boxed{V_{Ed} = 1.50V'_{Ed}}$$

(6.24)

6.1.7 Capacity design of large lightly reinforced walls

- In Chapters 4 and 9, detailed reference is made to this type of structural system and to its structural behaviour, particularly to the procedure of energy dissipation through wall uplifting from the soil or through opening and closing of horizontal cracks.
- The additional dynamic axial forces developed in large walls due to the dissipating mechanisms described above should be taken as being ±50% of the axial force of the wall due to gravity loads, unless they are evaluated using a more precise calculation. Therefore, while bending moments derived by the analysis are not modified, the corresponding axial forces must be modified as follows:

$$\boxed{N_{Ed} = (1.0 \pm 0.50)N'_{Ed}}$$

(6.25)

where
- N'_{Ed} is the axial force from the analysis corresponding to the design bending moment.
- N_{Ed} is the modified capacity design value.

    In the case that the behaviour factor q of the wall does not exceed 2.0, meaning in the case of an aspect ratio less than 1.0 or a torsionally flexible system, the effect of the dynamic axial force may be neglected.
- In order to ensure that flexural failure precedes shear failure of the wall, the shear force V'_{Ed} from the analysis must be increased at every storey in accordance with the following expression:

$$\boxed{V_{Ed} = V'_{Ed}\, \frac{q+1}{2}}$$

(6.26)

where q is the behaviour factor of the building.
- It should be noted here that, since the large lightly reinforced walls are characterised as DCM structures, the maximum value of the q-factor can be $q = 3.0$. In this respect, the magnification factor for V'_{Ed} takes a value equal to 2.0, that is, higher than the magnification factor ε imposed on DCM ductile walls.

6.1.8 Capacity design of foundation

It should be remembered here that the exclusive material for the foundation members (pads, tie beams, foundation beams, rafts, piles) is reinforced concrete, no matter if the superstructure is an R/C, steel or composite structural system. It should also be noted from the beginning that by the term 'foundation', reference is made not only to the structural components (pads, tie beams, foundation beams, rafts, piles) but also to the soil. These two components constitute a composite system that might be used for energy dissipation in case of an earthquake (Figures 6.1 and 6.7). However, keeping in mind that the mechanical properties of soil are of limited reliability, particularly in the inelastic range, it would not be reasonable for somebody to rely upon a dissipative mechanism that might extend to the soil.

At the same time, the foundation is a part of the structure buried in the ground, and, therefore, it cannot be easily examined for eventual damage after an earthquake; it is more difficult to intervene for retrofitting.

Consequently, it is reasonable for the foundation, specifically the structural members and the soil, to be kept in elastic range for the design seismic action, and, therefore, to be designed not for the seismic effects resulting from the analysis for the seismic combination but for the resistance of the vertical members of the superstructure at their joints with foundation members.

The above considerations have formulated the *capacity design* procedure for foundations (structural members and soil) adopted by EC8-1/2004. This procedure is given below:

1. The action effects for the foundation elements shall be derived on the basis of capacity design considerations, taking into account the development of possible overstrength. Of course, this procedure must not result in values of action with an effect greater than those resulting for $q = 1$ (elastic behaviour).
2. If the action effects for the foundation have been determined using the value of the behaviour factor q

$$\boxed{q = 1.5}$$

applicable to DCL structures, no capacity design considerations in accordance with (1.) are required. This might be a convenient simplification in computational procedure. However, for individually founded vertical members or even for a common foundation of all of them (e.g. a raft foundation) in the case of large building aspect ratios (h_w/l_w), such a simplification, due to the decreased value of q, leads to big tensile action effects at the perimeter or at the corner points of the foundation, or to overturning or even sliding instabilities of the whole building considered as a rigid body (Figure 6.11). Therefore, this simplification should be used with much concern for its consequences.

3. For foundations of individual walls or columns, the capacity design procedure is considered to be satisfied if the design values of the action effects E_{Fd} on the foundations are derived as follows:

$$\boxed{E_{Fd} = E_{F,G} + \gamma_{Rd}\Omega E_{F,E}} \tag{6.27}$$

where

a. γ_{Rd} is the overstrength factor, taken as being equal to

$$\left.\begin{array}{ll} \gamma_{Rd} = 1 & \text{for } q \le 3 \\ \gamma_{Rd} = 1.2 & \text{for } 3.0 \le q \end{array}\right\} \tag{6.28}$$

b. $E_{F,G}$ is the action effect due to the non-seismic actions (gravity load effects) included in the seismic combination.

c. E_{FE} is the action effect from the analysis of the design seismic action.

d. Ω is a magnification factor of the element i of the structure, which has the highest influence on the effect E_F under consideration.

4. For foundations of structural walls or columns of moment-resisting frames on individual footings, Ω is the *minimum value* of the ratio M_{Rd}/M_{Ed} in the two orthogonal principal directions at the lowest cross section where a plastic hinge can form in the vertical element in the seismic design situation. It is self-evident that M_{Rd} is defined by taking into account the axial force of the corresponding seismic combination that has resulted M_{Ed}. So,

$$\Omega = \min\left[(M_{Rd,x}/M_{Ed,x}); (M_{Rd,y}/M_{Ed,y})\right] \tag{6.29}$$

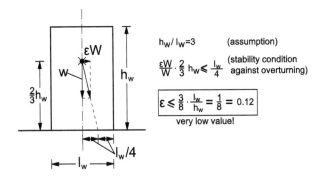

$h_w/l_w = 3$ (assumption)

$\dfrac{\varepsilon W}{W} \cdot \dfrac{2}{3}\, h_w \le \dfrac{l_w}{4}$ (stability condition against overturning)

$$\varepsilon \le \frac{3}{8} \cdot \frac{l_w}{h_w} = \frac{1}{8} = 0.12$$

very low value!

Figure 6.11 Stability condition against overturning.

It should be remembered (see Chapter 5) that M_{Ed} may have eight different values for each orthogonal direction. So, the computation of 16 Ω values for each individual vertical member and therefore for each independent footing tends to be a tiresome procedure.

5. For common foundations of more than one vertical element (foundation beams, rafts, etc.), Ω used in expression 6.27 is derived from the vertical element with the largest horizontal shear force in the design seismic situation or, alternatively, by introducing for $\gamma_{Rd}\Omega$ in Equation 6.27 the value

$$\boxed{\gamma_{Rd}\Omega = 1.40} \tag{6.30}$$

6. Under the above considerations, dimensioning and design of foundation members are carried out according to EN1992-1-1:2004 and to the limited additional rules in use for DCL buildings, since the procedure developed above ensures that the dissipative zones of the building are limited only to the superstructure.

Keeping in mind that the common foundations for almost all vertical elements are the usual cases for earthquake-resistant buildings, it can easily be concluded, according to Equation 6.30, that seismic actions for the foundation are increased by 40% in comparison to those for the superstructure of the building. At this point, Eurocode is very conservative compared to US codes (SEAOC, 1999; Booth and Key, 2006), which by contrast reduce the overturning moments at the base from the analysis by 10% to 25% depending on the method in use for the elastic analysis, taking implicitly into account the ability of the foundation soil to dissipate energy to a degree (cf. Figure 6.11).

Finally, it should be noted that EC8-1/2004 allows as an alternative the design action effects for foundation elements to be derived on the basis of the analysis for the seismic design situation without the capacity design consideration presented above. In this case, the design of these elements follows the corresponding rules for elements of the superstructure, which are the rules for DCM or DCH building. The reasoning behind the above alternative approach is based on the assumption that plastic hinges might also be acceptable in the foundation elements (tie beams, foundation beams, etc.; Figure 10.3). *However, the above alternative refers only to the structural element of the foundation and not to the soil.* The above alternative simplifies the analysis and design of common base foundations and results in more cost-effective solutions.

All above considerations presented here in principle will be examined in detail in Chapter 10.

6.2 SAFETY VERIFICATIONS

6.2.1 General

As already mentioned in Chapter 3.5.3, in order to satisfy the fundamental requirements of *no (local) collapse* and *damage limitation, three compliance criteria* should be considered, namely,

- Ultimate limit state
- Damage limitation
- Specific measures

For buildings of importance classes other than IV (see Chapter 3.4.3.2), the requirements of earthquake-resistant design may be considered satisfied if the total base shear due to seismic action combination, calculated with a behaviour factor $q = 1.5$ (DCL), is less than that due to other relevant combinations (e.g. wind combination) for which the building is designed on the basis of a linear elastic analysis. This means that in regions of low seismic hazard, where wind or other horizontal loading subjects the structure to base shear higher than that caused by the seismic actions (for $q = 1.5$), the design of the structure should be carried out without taking into account the seismic actions. In seismic regions where the preceding conditions are not fulfilled, the following verifications must be considered.

6.2.2 Ultimate limit state

Safety against (local) collapse is considered to be satisfied if the following conditions are met:

- Resistance condition
- Second-order effects
- Ductility condition
- Equilibrium condition
- Resistance of horizontal diaphragms
- Resistance of foundations
- Seismic joint conditions

In the following paragraphs, the above conditions will be examined in detail.

6.2.2.1 Resistance condition

The resistance condition is satisfied if, for every structural element, the following relations are fulfilled:

$$\boxed{\begin{aligned} E_{d1} &\leq R_d \\ E_{d2} &\leq R_d \end{aligned}}$$

(6.31a–b)

where
- E_{d1} are the design action effects for the gravity combination (e.g. $E(1.35G \text{ '+' } 1.50Q)$).
- E_{d2} are the design action effects on the structural element for gravity loads and seismic actions, also taking into consideration the capacity design rules on the seismic action effects.
- R_d is the corresponding design resistance of the same element. For the determination of R_d, the Code for R/C structures is implemented (EC2-1-1/2004) unless additional rules are imposed by the Seismic Code in particular cases (see Chapters 8 through 10).

For the determination of R_d, the partial safety factor – γ_s for steel and γ_c for concrete – may be taken from Eurocode EC2-1-1/2004 for the fundamental load combinations:

$$\boxed{\begin{aligned} \gamma_s &= 1.15 \\ \gamma_c &= 1.50 \end{aligned}}$$

(6.32a–b)

The choice of the above values is based on the assumption that, due to the local ductility provisions, the ratio between the residual strength after the seismic degradation of the R/C member and its initial strength is roughly equal to the ratio between γ_m values of accidental and fundamental load combinations. From the above consideration, it can be seen that in any case the partial material safety factors for seismic loading should not be those introduced in EC2-1-1/2004 for accidental loading, although seismic action is considered accidental. This must be attributed to the fact that an inelastic cyclic loading results in degradation of the member resistance (see EN 1990/2002) and, therefore, such a reduction of safety factors would be risky.

At the same time, the introduction of the safety factors of the fundamental load combination allows the same value for the design resistance R_d to be used for the gravity combination and for the seismic design situation.

6.2.2.2 Second-order effects

Due to their inelastic response, most structural systems under the action of seismic forces sustain large horizontal displacements, resulting in the creation of large secondary effects (Wilson and Habibullah, 1987; Luft, 1989; Paulay et al., 1990).

Consider the frame of Figure 6.12. When this frame, for some external reasons (an earthquake, in this case), is displaced by Δ, each of the two $W/2$ column loads can be analysed into an axial force on the column with a value approximately $W/2$ and a horizontal one:

$$\Delta H_{1,2} = \frac{\Delta}{h} \cdot \frac{W}{2}$$

Thus, the floor is loaded additionally (second-order effect) by a horizontal force equal to

$$\boxed{\Delta H = \frac{\Delta}{h} W} \tag{6.33}$$

In the case of a seismic action, the displacement Δ according to what has been explained in the chapter about ductility for a fundamental period longer than T_c (corner point C of the

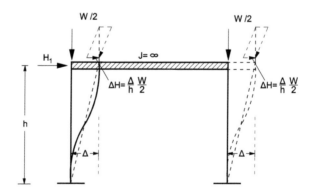

Figure 6.12 Second-order effect on a one-storey, two-column frame.

acceleration spectrum) is equal to Δ_{el}, which results from the design seismic loading of the Code multiplied by the behaviour factor q of the structure:

$$\Delta = \Delta_{el} \cdot q \tag{6.34}$$

For simplicity's sake, Eurocode EC8-1/2004 also allows the use of the above expression in the case where $T < T_c$ (see Section 6.2.3). Therefore, the additional shear force of the storey due to the second-order effect is equal to

$$\boxed{\Delta V = \frac{\Delta_{el} q}{h} W} \tag{6.35}$$

Eurocode EC8-1/2004 specifies the following procedure for the verification of the influence of second-order effects:

1. For

$$\boxed{\theta_i = \frac{\Delta V}{V} = \frac{\Delta_{el}^i q}{h} \frac{W_{tot}^i}{V_{tot}^i} \leq 0.10} \tag{6.36}$$

$P–\Delta$ effects need not to be taken into account. In the above relation, the various symbols have the following meaning:
- θ_i is the inter-storey drift sensitivity coefficient of each storey.
- W_{tot}^i is the total gravity load at and above the storey under consideration in the seismic design situation.
- Δ_{el}^i is the elastic displacement inter-storey drift (difference of displacements between the top and the bottom of the storey under consideration) at the centre of mass for design seismic loading.
- q is the behaviour factor of the building.
- V_{tot}^i is the total seismic storey shear.
- h is the inter-storey height.

2. For

$$\boxed{0.10 \leq \theta_i \leq 0.20} \tag{6.37}$$

the $P–\Delta$ effects must be taken into account. In this case, an acceptable approximation could be to increase the relevant seismic action effects by a factor equal to

$$\boxed{\varphi = \frac{1}{(1-\theta)}} \tag{6.38}$$

3. For

$$\boxed{0.20 \leq \theta_i \leq 0.30} \tag{6.39}$$

it is necessary for the $P–\Delta$ effects to be taken into account in analysis; otherwise, the stiffness of the system must be increased, and finally

4. For

$$0.30 \leq \theta_i$$

(6.40)

the stiffness of the system must be increased, since EC8-1/2004 does not allow such large inter-storey drift sensitivities.

In any case, it is recommended that a high degree of lateral stiffness is provided for the structural system so that at least second-order effects are prevented. This can be easily achieved in R/C buildings, especially in case of dual structural systems.

Finally, it should be noted that P-Δ effect verifications should be carried out *at the beginning of the analysis and design*, because at that moment, decisions must be made for the type of elastic analysis, in other words for the consideration of P-Δ effects or not, and for the eventual increase of load effects (in the case that $0.10 < \theta < 0.20$). Therefore, a preliminary static analysis for seismic loads is usually carried out (see Chapter 5.5.1) so that all the parameters in use in Equation 6.36 are defined in advance.

6.2.2.3 Global and local ductility condition

The ductility condition is satisfied by means of

- Specific material-related requirements
- Maximum–minimum requirements for the reinforcement for flexure in critical regions of the structural members
- Minimum reinforcement requirements for shear in critical regions
- Minimum reinforcement requirements for confinement of critical regions
- Calculation of the required confinement reinforcement in critical regions of members of major importance for the seismic resistance of the structure (columns-structural walls)
- Appropriate reinforcement detailing
- Application of capacity design procedure

The last requirement has already been examined in detail in the previous section. It is obvious that it is taken into account for the determination of the load effects E_{d2} and, therefore, in the verification of resistance condition. All the other requirements are closely related to the selected ductility class of the building (ductility demand; see Chapter 5.4.2).

These requirements will be presented and discussed in detail in Chapters 8 through 10 for every type of structural component, together with all necessary experimental and analytical evidence.

6.2.2.4 Equilibrium condition

Equilibrium condition refers to the stability of the building under the set of actions given by the combination rules described in Chapter 5.8.4. It should be noted that *the seismic actions even for the equilibrium condition are introduced with their reduced values due to the behaviour factor q of the building, but also by taking into account the capacity design magnification that has been introduced in Section 6.1.8.* This is a reasonable consequence of the fact that due to the dissipative mechanisms generated in the critical regions of the structure, satisfying the selected q-factor, the internal forces that develop in the structural members cannot overcome their resistance, and, therefore, those of the elastic analysis carried out for the design seismic actions multiplied by the capacity design magnification factors.

In equilibrium conditions, mainly the overturning and sliding of the building as a rigid body is included. It is obvious that these types of verifications refer mainly to buildings with large aspect ratios h_w/l_w (total height to building/small dimension in plan), where overturning problems are critical. They also refer to buildings susceptible to sliding (see Chapter 10).

6.2.2.5 Resistance of horizontal diaphragms

The resistance of horizontal diaphragms refers to their ability to transmit with sufficient over-strength the design seismic action effects to the various lateral load-resisting systems to which they are connected. As already discussed in Chapters 4 and 5, R/C slab diaphragms without re-entrant corners or discontinuities in plan present a high level of diaphragmatic resistance, and, therefore, safety verification is not needed. In the case, however, a resistance verification of the diaphragm has been decided, the forces or stresses obtained from the analysis as acting on the diaphragm must be multiplied by an overstrength factor γ_d greater than 1.0. The recommended values by EC8-1/2004 for the relevant resistance verifications are

- For brittle failure mode (shear)

$$\boxed{\gamma_d = 1.30} \tag{6.41}$$

- For ductile failure mode

$$\boxed{\gamma_d = 1.10} \tag{6.42}$$

Design provisions for R/C diaphragms (analysis–dimensioning–detailing) will be given in Chapter 9.7.

6.2.2.6 Resistance of foundations

Action effects for foundation design, taking into account capacity design rules, have been discussed in detail in Section 6.1.8. Detailed analysis and design issues for foundations will be discussed and presented in Chapter 10.

6.2.2.7 Seismic joint condition

Seismic joint condition is satisfied by means of joints between adjacent buildings or between adjacent statically independent units of the same building. These joints must be wide enough to protect them from earthquake-induced pounding.

For buildings that do not belong to the same property, the distance of the building from the property line must be at least equal to the maximum horizontal displacement $d$smax resulting from Equation 6.43 (see Section 6.2.3).

For adjacent independent units of the same building, the distance between them must not be less than the square root of the sum of the squares (SRSS) of the maximum horizontal displacements of the two units also resulting from Equation 6.43.

Finally, in the case that the floor elevations of the adjacent independent units under design are the same, the minimum distance mentioned above may be reduced by a factor of 0.7.

6.2.3 Damage limitation

The task of damage limitation is to ensure the protection of non-structural elements (masonry, glass panels, tiles, etc.) from premature failure for seismic actions of a higher probability of occurrence than that of the design. The main parameter of seismic action that influences the behaviour of these elements is the ratio of the inter-storey drift d_r to the inter-storey height h, which is induced in them by the adjacent R/C members of the building (Figure 6.13; Uang and Bertero, 1991; Stylianidis, 2012).

The reason for this must be attributed to the fact that masonry and other low-strength brittle materials fail at much lower inter-storey drift d_{ri}/h_i than the surrounding R/C frames (Figure 6.14).

From the above, it may be concluded that for the verification of the damage limitation criterion (serviceability limit state), the displacement due to the design seismic actions must be calculated and then reduced so that a shorter return period is taken into account.

The displacement d_s due to the design seismic action could be the result of the elastic analysis of the structural system magnified by the behaviour factor q, since the ordinates of the design response spectrum corresponding to the analysis include a reduction factor q. Consequently,

$$d_s = qd_e \tag{6.43}$$

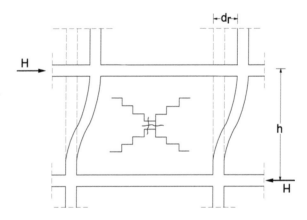

Figure 6.13 Masonry failure with x-shaped cracks due to the R/C frame inter-storey drift d_r.

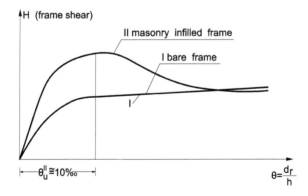

Figure 6.14 H − θ diagram of a masonry-infilled frame.

where
- d_s is the displacement of a point of the structural system induced by the design seismic action.
- q is the q-factor introduced into the analysis of the system.
- d_e is the displacement of the same point of the structural system as determined by the linear analysis based on the design response spectrum.

In this respect, the relevant inter-storey drift d_r/h is equal to

$$\frac{d_{ri}}{h_i} = \frac{d_{si} - d_{s(i-1)}}{h_i}$$
(6.44)

where
- d_{si} and $d_s(i-1)$ are the displacements of storeys i and $i-1$, respectively, at the reference point.
- h_i is the storey height.

To ensure the damage limitation according to EC8-1/2004, the inter-storey displacement d_{ri} must fulfil the following limits:

1. For buildings having non-structural elements of brittle materials attached to the structure:

$$d_{ri}r \leq 0.005\, h_i$$
(6.45)

2. For buildings having ductile non-structural elements:

$$d_{ri}r \leq 0.0075\, h_i$$
(6.46)

3. For buildings having non-structural elements fixed in a way so as not to interfere with structural deformations, or without non-structural elements (Figure 6.15),

$$d_{ri}r \leq 0.010\, h_i$$
(6.47)

The reduction factor r accounts for the consideration of the shorter return period that should be taken into account for the damage limitation in relation to the return period for ULS.

Recommended values by EC8-1/2004 for r are given below:

- For buildings of importance classes

  III and IV: $r = 0.40$

- For buildings of importance classes

  I and II: $r = 0.50$

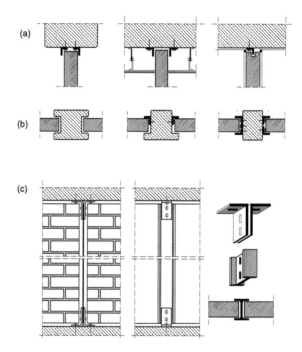

Figure 6.15 Detailing of separation joints between partition walls and R/C frame: (a) upper boundary; (b) lateral boundary; (c) reinforcement for out-of-plane overturning.

The above reduction factors correspond to a probability of exceedance $P_{\text{DLR}} = 10\%$ in $t_{\text{L}} = 10$ years.

For this hazard level, the reference return period of occurrence is reduced from 475 years (ultimate limit state) to 95 years (see Chapter 3.5.2). The exact value for the above conditions results in

$$r = 0.44 \tag{6.48}$$

The small increase of r for classes III and IV in contrast to the small decrease of r for classes I and II must be attributed to the fact that the importance factor γ_1 introduced for classes III and IV has already increased the reference period of occurrence of 475 years to a higher level for these classes, while the importance factor γ_1 introduced into class I has decreased it.

Finally, it should be noted that the damage limitation verification is implicitly based on the assumption that the ordinates of the elastic response spectrum for the 90-year reference period of occurrence *are proportional* to those of the elastic response spectrum for 475 years, which has been used for the ultimate limit state, which means that the same design response spectrum is used in both cases.

6.2.4 Specific measures

It has been noted already in Chapter 3.5.3 that a series of specific measures is foreseen in all modern Codes for the collapse prevention of a building in the case of a very rare earthquake and, therefore, more severe than the design earthquake. Some of these measures have already been presented in detail in the previous chapters, while others will be presented in Chapters 8 through 10. However, it would be useful to summarise these measures here, so that a global view of the three *compliance criteria* mentioned at the beginning of this section can be obtained. These measures specified by EC8-1/2004, particularly for R/C buildings, refer to

- Design of the superstructure
- Foundations
- Quality system plan
- Resistance uncertainties
- Ductility uncertainties

6.2.4.1 Design

- Structures should have a simple and regular form in plan and elevation (Chapters 4.2.2, 5.2.2 and 5.2.3).
- In order to ensure an overall ductile behaviour, brittle failure or premature formation of unstable mechanisms must be avoided. For this reason, the capacity design procedure should be adopted (Section 6.1).
- Since the seismic performance of the structure depends on the behaviour of its critical regions, the detailing of these regions must be such that they maintain under cyclic conditions the ability to transmit the necessary forces and to dissipate energy. For this reason, the detailing of connections between structural elements and regions where nonlinear behaviour is foreseeable deserves special care in design (Chapters 8 through 10).
- The analysis must be based on an appropriate structural model (Chapter 5).

6.2.4.2 Foundations

- The stiffness of the foundation must be adequate in terms of transmitting to the ground as uniformly as possible the actions received from the superstructure (Section 6.1.8 and Chapter 10).
- Only one foundation type must, in general, be used for the same structure (Chapters 4 and 10).

6.2.4.3 Quality system plan

- The choice of materials and construction techniques must be in compliance with the design assumptions. The design documents must indicate structural details, sizes and quality provisions.
- Elements of special structural importance requiring special checking during construction should be identified on the design drawings. In such cases, the checking methods to be used should also be specified.

6.2.4.4 Resistance uncertainties

Important resistance uncertainties could be produced by geometric errors. To avoid these uncertainties, rules referring to the following items should be applied:

- Certain minimum dimensions of structural elements (Chapters 8 through 10).
- Appropriate limitations of column drifts must be provided (Section 6.2.2).
- Special detailing rules should be applied in reinforcing R/C elements, so that unpredictable moment reversals and uncertainties related to the position of the inflection point are taken into account (Chapters 8 through 10).

6.2.4.5 Ductility uncertainties

In order to minimise ductility uncertainties, the following rules must be applied:

- An appropriate minimum local ductility is needed in every seismic-resistant part of the structure (Chapters 8 through 10); thus, by enhancing the redistribution capacity of the structure, some of the model uncertainties are alleviated.
- Minimum–maximum reinforcement percentages in all critical regions are specified to take into account ductility requirements and to avoid brittle failure upon cracking (Chapters 8 through 10).
- The normalised design axial force values are kept at a low level to avoid decrease in local ductility at the top and bottom of the columns (Chapters 8 through 10).

6.2.5 Concluding remarks

In conclusion, it should be noted that until recently there was no Code providing a direct computational procedure for quantifying the safety level of a new structure designed according to its principles against collapse in the case of a very rare and high-intensity earthquake (e.g. collapse prevention requirement under a seismic motion of a mean return period of 2,475 years). Even today, it is considered that the series of provisions for the design of a structure that have been already presented, such as appropriate configuration in plan and in elevation, correct elastic analysis based on a reliable design spectrum, conformation of the seismic design effects to the capacity design considerations and design verifications according to the code provisions and so forth, form a reliable set of conditions that ensure *qualitatively* the safety of a structure against collapse under severe unexpected seismic actions.

During the last 20–25 years, numerous attempts have been made to control the collapse requirement quantitatively, using inelastic seismic analysis. In the beginning, inelastic dynamic seismic analysis was used mainly in research and particularly for the evaluation of the codified design procedure by means of extended parametric analyses (Kanaan and Powel, 1973; Park et al., 1985; Kappos and Penelis, 1986, 1989; Kappos et al., 1991; Michailidis et al., 1995). Many of the principles and rules introduced in modern Codes are owed to the results of this research.

Later, the development of displacement-based design, which has as a core the inelastic static analysis for horizontal loading (push-over analysis), made possible the quantitative evaluation of the structural safety against collapse in everyday practice. Indeed, the development of the computational power of desktops in the last 15 years and the parallel development of commercial computational codes for push-over analysis enabled the use of displacement-based design for the quantitative follow-up of the post-elastic behaviour of a structure as a whole and of its structural elements up to collapse (see Chapters 5.7 and 13.5). 5EC8-1/2004 is the first Code that has introduced, in case of uncertainties, the redesign of a new building by the push-over analysis for detailed evaluation of the post-elastic performance of the structural system, that is, the early development of plastic hinges and the eventual creation of a soft storey, the brittle failure of structural elements and, most important, the safety factor of the structure against collapse *in terms of displacements*.

EC8-1/2004 goes one step further, allowing the country members of the EU to make use of a push-over analysis as the prime method (displacement-based design) under well-defined prerequisites instead of the force-based design that has been presented so far and that constitutes the reference method for the design of buildings in seismic regions for all modern Codes.

Finally, it should be noted that for the assessment and retrofitting of existing buildings, EC 8-3/2005 introduces displacement-based design as a method of reference. The above brief remarks on displacement-based design will be elaborated in greater detail in Chapter 13.

Chapter 7

Reinforced concrete materials under seismic actions

7.1 INTRODUCTION

It is well known that reinforced concrete is a composite material consisting of *concrete* and *steel* reinforcement. These two materials, *bonded together* due to their physicochemical and mechanical properties, have supplied one of the basic materials for the construction of building structural systems for the past 120 years. The behaviour of reinforced concrete is, therefore, quite well known under static loading.

Here, reference will be made mainly to its behaviour under seismic loading and only brief reference will be made to its behaviour under static loading, when it is necessary for a comprehensive understanding of its behaviour under cyclic loading.

Static loading is imposed on materials or structural elements in the lab at slow rates, increasing from zero to failure in two different procedures, which are

- Load-controlled procedure
- Displacement-controlled procedure

According to the first one, which was the traditional method for more than one century up to early 1970s, for each successively increasing loading step, the corresponding displacement or strain is measured at a reference point of the specimen. So, a diagram of P–δ or σ–ε is plotted, which is known as the *constitutive law of material of the structural element* under consideration. This procedure extends up to P_{max} or σ_{max} and the corresponding displacement or strain. From that point on, the experimental procedure 'collapses', since the test set-up is load controlled.

Since the late 1960s, new types of testing machines have been developed for the study of the post-elastic behaviour of materials or structural members, as well as study of the descending branch of the diagram P–δ or σ–ε. The main characteristic of these testing machines is that displacement or deformation is controlled. In this case, for each increasing deformation or displacement step imposed, the loading is determined (measured) until the complete disintegration of the material of the structural element. The function of these machines is controlled by servo-electronic valves and is based on a computerised programming of the *deformation* or *displacement path*, which is transformed by means of the servo-electronic valves to an analogue signal for the proper operation of the hydraulic jacks of loading.

Seismic action can be simulated by a reversed cyclic loading with a limited number of cycles. The loading frequency is also low, ranging between 0.5 and 10 Hz. Last but not least, it should be noted that according to what has been presented in previous chapters, these cyclic loading reversals are extended in the post-elastic region of the P–δ or σ–ε diagram. In other words, in case of seismic excitation, *low cycle fatigue* phenomena prevail in contrast to what happens in the case of dynamic loaded works in their elastic region as bridges, industrial installations and so forth, where a high-cycle fatigue problem may appear.

For the study of a reversed cyclic loading of a material or a structural element, displacement controlled machines are used, equipped with servo-electronic valves and double-acting loading jacks. A computerised program of the displacement or strain time history is given as input and, as the displacement or the strain is imposed, the corresponding loading is determined (measured) and depicted on a P–δ or σ–ε constitutive diagram.

Main parameters that influence the capacity of a specimen for energy absorption and dissipation are the following:

1. Plastic deformation that may be developed under monotonic loading without serious strength degradation
2. Relation of the monotonic curve of a load–deformation diagram with that of the envelope of the cyclic loading
3. Loop area of each cycle of reversed loading
4. Low cycle fatigue limit
5. Influence of the strain rate on the envelope of the cyclic loading

In the past 50 years, extended research has been carried out either in the lab or analytically for the experimental investigation and analytical modelling of the above parameters, as far as material and structural elements are concerned. The basic conclusions of this extended research may be summarised as follows:

1. The envelope of recycling loading does not substantially deviate from the monotonous one. Low cycle fatigue, even if the load level exceeds 85–90% of the strength under monotonic action, does not result in failure for a low number of reversals, in other words for seismic action. The cases of *shear* and *bonding slip* of reinforcement should be exempted, since in these cases the degradation of the strength of structural elements is serious even for a low number of cycles. For this reason, a great effort has been made for these types of strains in structural systems to be kept at a low level by means of a capacity design approach.
2. For low cyclic loading strain rates similar to those of earthquake excitation, strength does not substantially exceed the strength of monotonic loading.
3. From all of the above, it may be concluded that the energy dissipation capacity of a material or a structural element depends on the following characteristics:
   a. The available inelastic deformation (ductility supply) under monotonic loading
   b. The area of the energy dissipation loop of each cycle, since it is known (see Chapter 2.3.3) that this area expresses the dissipated seismic energy for every cycle reversal

In this context, the basic aim of the seismic design of critical regions of a structural system is the provision of satisfactory ductility and 'full' dissipation recycling loops and more specifically the avoidance of strength degradation due to *shear* and *bond slip*.

In this chapter and in Chapters 8 through 10, a comprehensive presentation will be made of the techniques and analytical tools that are used in the seismic design of materials and structural elements for qualitative as well as quantitative safeguarding of the ductility capacity of the structure.

7.2 PLAIN (UNCONFINED) CONCRETE

7.2.1 General

It is known that concrete presents a rather high compressive strength but a limited and unreliable tensile strength, so that in the design of R/C structures, the tensile concrete strength is ignored (cracked tensile zones). It is also a brittle material, like all other stone or stone-like materials. In this context, it is necessary that its use in a seismic design be combined with special measures that mitigate the effects of this property.

7.2.2 Monotonic compressive stress–strain diagrams

The form of this diagram is depicted in Figure 7.1 (Park and Paulay, 1975). It should be noted that as the strength of the material increases, its brittleness also increases. In any case, the maximum compressive strength corresponds to a strain ranging from 2‰ to 1.7‰. Various analytical models have been developed for the simulation of the σ_c–ε_c curve (fib, 2012).

The best-known analytical expression for the ascending branch of the diagram is Hognestad's expression (Park and Paulay, 1975):

$$\sigma_c = f_c \left[\frac{2\varepsilon_c}{\varepsilon_{c1}} - \left(\frac{\varepsilon_c}{\varepsilon_{c1}} \right)^2 \right] \qquad (7.1)$$

where

- $\varepsilon_{c1} = \dfrac{2f_c}{E_{co}}$: (see diagram of Figure 7.2).
- E_{co} is the initial tangent modulus of elasticity.
- $\varepsilon_{c1} = 2.0‰$.

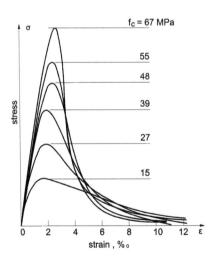

Figure 7.1 Stress–strain curves for various concrete classes under uniaxial compression. (Adapted from CEB. 1993. CEB-FIP Model Code 1990, *Bulletin d'Information*, CEB, 213/214, Lausanne.)

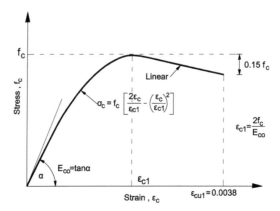

Figure 7.2 Idealised stress–strain curve for concrete under uniaxial compression. (Park, R. and Paulay, T.: *Reinforced Concrete Structures.* 1975. Copyright Wiley-VCH Verlag GmbH & Co. KGaA. Reproduced with permission.)

For the descending branch, the model of Kent and Park (1971) (Park and Paulay, 1975) may be considered the simplest, as this branch is simulated by a descending straight line presented by the following expression:

$$\sigma_c = f_c[1 - z(\varepsilon_c - \varepsilon_{c1})] \tag{7.2}$$

where

$$z = \frac{0.50}{\varepsilon_{c50} - \varepsilon_{c1}}$$

and

$$\varepsilon_{c50} = \frac{3 + 0.29f_c}{145f_c - 1000} \, (f_c \text{ in MPa}) \tag{7.3}$$

which is an expression obtained by fitting to experimental results.

For large strains, a residual strength equal to $0.20 \, f_c$ was introduced by Kent and Park in the above model. Later in this chapter, the provisions of EC2-1-1/2004b will be presented for the simulation of the σ_c–ε_c diagram.

7.2.3 Cyclic compressive stress–strain diagram

Such a diagram is depicted in Figure 7.3 (Karsan and Jirsa, 1969). It should be noted that the successive loops are almost in contact with the curve of monotonic loading. From extended experimental work, the following results may be drawn:

1. For high values of strain rate, strength and stiffness are strongly influenced positively (Figure 7.4; Priestley and Wood, 1977). However, in the case of seismic action, which has a strain rate that is rather low, at a range of $\dot{\varepsilon} = 4$–$5 \, 10^{-2}$/s, the positive influence is rather small, ranging from 10% to 15%.

   Therefore, this overstrength may be ignored, as it is on the safe side.

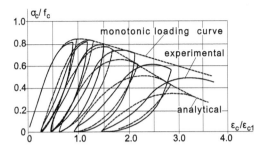

Figure 7.3 Stress–strain diagrams for concrete subjected to repeated uniaxial compression. (Adapted from Karsan, I.D. and Jirsa, J.O. 1969. *Journal of the Structural Division*, ASCE 95(ST 12), 2543–2563.)

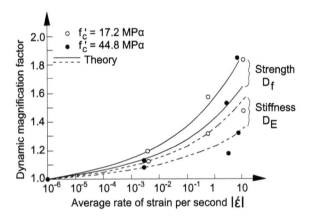

Figure 7.4 Dynamic magnification factors D_f and D_E to allow for strain rate effects on strength and stiffness. (Priestley, M.J.N. and Wood, J.H.: Behaviour of a complex prototype box girder bridge. *Proceedings of the RILEM International Symposium on Testing In-Situ of Concrete Structures*, Budapest, I, 140–153. 1977. Copyright Wiley-VCH Verlag GmbH & Co. KGaA. Reproduced with permission.)

2. For successive loading cycles below 85% of the axial compressive strength of monotonic loading and for a large number of reversals (about 200), specimens do not display any sign of failure (low cycle fatigue). If loading exceeds 90% of the compressive strength, 19 to 20 cycles are enough for the strength degradation and failure of the specimen (Figure 7.5; Karsan and Jirsa, 1969; Fardis, 2009). However, keeping in mind that the number n of cycles during a seismic event is limited ($n \cong t/T$, where t is the duration of the event and T the predominant period of the seismic motion) and that only a small number of cycles exhaust the strength limits, it may be concluded that low cycle fatigue does not seriously influence concrete strength during a seismic motion. In any case, in order to take measures against this strength degradation, EC8-1/2004 recommends the use of material partial factors for seismic design equal to those in force for static loading, although seismic action is considered to be accidental (see Chapter 6).

The above considerations have influenced the formulation of programs for displacement-controlled tests in the laboratory. In fact, two such programs are mainly in use. The first includes one cycle of displacement or strain per step, while the second one includes two to four cycles (Figure 7.6; Stylianidis, 2012).

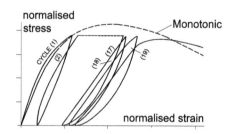

Figure 7.5 σ–ε diagram of concrete under cyclic uniaxial compression. (Adapted from Karsan, I.D. and Jirsa, J.O. 1969. *Journal of the Structural.* Division, ASCE 95(ST 12), 2543–2563.)

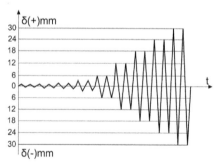

Figure 7.6 Loading program (two cycles per step) for testing of infilled R/C frames. (Adapted from Stylianidis, K. 2012. *The Open Construction and Building Technology Journal,* 6, 194–212.)

7.2.4 Provisions of Eurocodes for plain (not confined) concrete

According to EC8-1/2004, all properties that are specified by EC2-1-1/2004 (R/C structures) for plain concrete are accepted also for earthquake-resistant R/C buildings. A summary of the most important properties is given below.

- Strength
  - For buildings of ductility class medium (DCM): Concrete class may range from C16/20 to C90/105
  - For buildings of ductility class high (DCH): Concrete class may range from C20/25 to C90/105

- Stress–strain diagram (σ_c–ε_c) of monotonic compressive loading for inelastic analysis
  The constitutive law of σ_c–ε_c is depicted in Figure 7.7 and results from the expression

$$\boxed{\frac{\sigma_c}{f_{cm}} = \frac{\kappa n - n^2}{1 + (\kappa - 2)n}}$$

(7.4)

where

$$n = \frac{\varepsilon_c}{\varepsilon_{c1}}$$

(7.5)

$$\kappa = 1.05 E_{cm} \, |\varepsilon_{c1}| / f_{cm}$$

(7.6)

Expression 7.4 is valid for $0 < |\varepsilon_c| < |\varepsilon_{cu1}|$.
The meaning of the notation in the above expressions is the following:
- f_{cm} is the mean compressive strength (MPa).
- E_{cm} is the mean modulus of elasticity.
- ε_{c1} is strain at maximum stress.

Values for the various concrete classes may be found in Table 3.1 of EC2-1-1/2004 (Table 7.1 in this chapter). Equation 7.4 is a generalised form of Hognestad's expression 7.1. In expression 7.1, κ has a constant value equal to 2, while in Code's expression 7.4, κ varies according to concrete class.
- Stress–strain diagram (σ_c–ε_c) for section design
  The constitutive law σ_c–ε_c for the design of R/C sections in bending with or without axial load is given in Figure 7.8 and results from the following expressions:
- For $0 \leq \varepsilon_c \leq \varepsilon_{c2}$

$$\boxed{\sigma_c = f_{cd}\left[1 - \left(1 - \frac{\varepsilon_c}{\varepsilon_{c2}}\right)^n\right]}$$

(7.7)

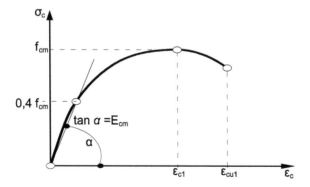

Figure 7.7 Schematic representation of the stress–strain relation of concrete for structural analysis. (Adapted from EC2-1-1/EN1992-1-1. 2004. *Design of Concrete Structures – Part 1-1: General Rules and Rules for Buildings*. BSI, CEN, Brussels, Belgium.)

Table 7.1 Strength and deformation characteristics for concrete

	Strength classes for concrete									
f_{ck} (MPa)	12	16	20	25	30	35	40	45	50	55
$f_{ck,cube}$ (MPa)	15	20	25	30	37	45	50	55	60	67
f_{cm} (MPa)	20	24	28	33	38	43	48	53	58	63
f_{ctm} (MPa)	1.6	1.9	2.2	2.6	2.9	3.2	3.5	3.8	4.1	4.2
$f_{ctk,0.05}$ (MPa)	1.1	1.3	1.5	1.8	2.0	2.2	2.5	2.7	2.9	3.0
$f_{ctk,0.95}$ (MPa)	2.0	2.5	2.9	3.3	3.8	4.2	4.6	4.9	5.3	5.5
ε_{cm} (GPa)	27	29	30	31	33	34	35	36	37	38
ε_{c1} (‰)	1.8	1.9	2.0	2.1	2.2	2.25	2.3	2.4	2.45	2.5
ε_{cu1} (‰)					3.5					3.2
ε_{c2} (‰)					2.0					2.2
ε_{cu2} (‰)					3.5					3.1
n					2.0					1.75
ε_{c3} (‰)					1.75					1.8
ε_{cu3} (‰)					3.5					3.1

Source: EC2-1-1/EN1992-1-1. 2004. *Design of Concrete Structures – Part 1-1: General Rules and Rules for Buildings*. BSI, CEN, Brussels, Belgium. With permission.

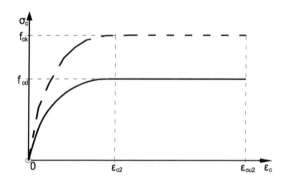

Figure 7.8 Parabola–rectangle diagram for concrete under compression for design. (Adapted from EC2-1-1/EN1992-1-1. 2004. *Design of Concrete Structures – Part 1-1: General Rules and Rules for Buildings*. BSI, CEN, Brussels, Belgium.)

- For $\varepsilon_{c2} \leq \varepsilon_c \leq \varepsilon_{cu2}$

$$\boxed{\sigma_c = f_{cd}}$$

(7.8)

where
- n is an exponent taken from Table 3.1 of EC2-1-1/2004 (Table 7.1).
- ε_{c2} is the strain corresponding to the maximum strength according to Table 7.1.
- ε_{cu2} is the strain at failure according also to Table 7.1.

For simplicity's sake, the compressive strength and the corresponding strain are introduced in the above expressions with a positive sign.
- Remarks
  - The exclusion of some low classes of concrete from the earthquake-resistant R/C building aims at an increase in the ductility of the structural system, as we will see later.

- For concrete classes from C16/20 to C50/60, the values of the above parameters are given below:
  - ε_{c1}: 1.9‰–2.45‰
  - ε_{cu1}: 3.5‰ (constant)
  - ε_{c2}: 2.0‰ (constant)
  - ε_{cu2}: 3.5‰ (constant)
  - n: 2

  It should be noted that ε_{c2}, ε_{cu1}, ε_{cu2}, n for concrete classes in common use *are constant*.
- For concrete classes from C55/67 to C90/105, the values of ε_{c1}, ε_{c2} increase successively, while the values of ε_{cu1}, ε_{cu2} decrease with a parallel decrease in the curvature of the ascending branch of the σ_c–ε_c diagram. In other words, this diagram tends to be linear without a descending or horizontal branch. In effect, as the strength of concrete increases above C50/60, its behaviour tends to be more brittle (Figure 7.1).
- In EC8-1/2004, a bold step was made by introducing high concrete classes above C50/60 and up to C90/105. This step was based on extended research carried out in the last few years, and the objective was to make R/C competitive with steel in the tall building industry. In order to limit the use of high-strength concrete to high-rise buildings, EC2-1-1/2004 recommends that member states decide about such limitations in their National Annexes. Some countries already specify the obligation of the user to obtain a special permit from the relevant national authorities for the use of a concrete class above C50/60.

7.3 STEEL

7.3.1 General

It is well known to every structural engineer that steel is a material with high and reliable tensile strength as well as high ductility. The same properties hold also in compression of steel rebars if they are *secured against buckling*. Therefore, steel is the material used in the form of bars for reinforcing all R/C structural members in zones under tension. Similarly, particularly in R/C seismic structures, it is used to confine and to strengthen the structural member or zones under compression because, as will be seen in subsequent chapters, in this way the ductility of the critical regions is increased.

7.3.2 Monotonic stress–strain diagrams

Stress–strain diagrams σ_s–ε_s of various steel grades are given in Figure 7.9, while the forms of hot-rolled and cold-deformed steel are given in Figure 7.10a and b. It should be noted that for hot-rolled steel in use in R/C structures, the stress and strain at failure σ_{uk} and ε_{uk} are in the range of 550 MPa and 75‰, respectively (Table 7.2; see Table C.1 EC2-1-1/2004, Annex C).

Keeping in mind that the area of the σ_s–ε_s diagram (Figure 7.11) expresses the energy absorbed up to failure per unit of material volume, and comparing the corresponding areas for steel and concrete, we see that the ratio of energy absorption W_s to W_c of steel to concrete for typical classes of materials is in the order of

$$\frac{W_s}{W_c} = 1000 - 1200 \tag{7.9}$$

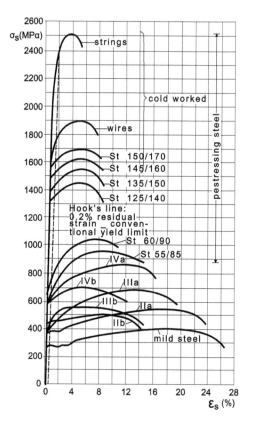

Figure 7.9 Stress–strain diagrams for steel rebars of various grades. (Adapted from Penelis, G.G. and Kappos, A.J. 1997. *Earthquake-Resistant Concrete Structures*. SPON E&FN, Chapman & Hall, London.)

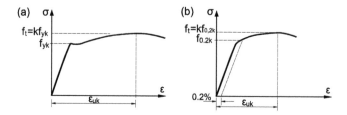

Figure 7.10 Stress–strain diagrams of typical reinforcing steel: (a) hot-rolled steel; (b) cold-rolled steel. (Adapted from EC2-1-1/EN1992-1-1. 2004. *Design of Concrete Structures – Part 1-1: General Rules and Rules for Buildings*. BSI, CEN, Brussels, Belgium.)

So, even in the case that the volumetric steel percentage of R/C members is in the order of 2% (140–150 kg/m³), the capacity of steel for seismic energy absorption and dissipation is about 20–25 times higher than for that of concrete. From what has been presented above, it may be concluded that the capacity for ductile behaviour of R/C members relies basically on steel.

Table 7.2 Properties of reinforcement

Product form	Bars and de-coiled rods			Wire fabrics		
Class	A	B	C	A	B	C
Characteristic yield strength f_{yk} or $f_{0.2k}$ (MPa)	400–600					
Minimum value of $k = (f_t/f_y)_k$	≥1.05	≥1.08	≥1.15 <1.35	≥1.05	≥1.08	≥1.15 <1.35
Characteristic strain at maximum force, ε_{uk} (%)	≥2.5	≥5.0	≥7.5	≥2.5	≥5.0	≥ 7.5
Bendability	Bend/Rebend test			–		
Shear strength	–			0.3 A f_{yk} (A is area of wire)		

Source: EC2-1-1/EN1992-1-1. 2004. *Design of Concrete Structures – Part 1-1: General Rules and Rules for Buildings.* BSI, CEN, Brussels, Belgium. With permission.

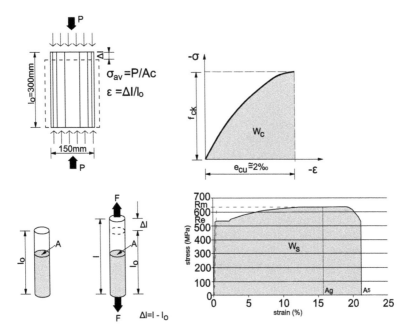

Figure 7.11 Determination of the ratio of strain energy absorbed up to failure per volume unit of steel to concrete.

7.3.3 Stress–strain diagram for repeated tensile loading

A typical stress–strain σ_s–ε_s diagram for repeated tensile loading is depicted in Figure 7.12 (Blakeley, 1971; Park, 1972). It is important to note that the successive loops are in contact with the curve of the monotonic tensile loading diagram. It is also important to note that the following useful conclusions may be drawn from extended laboratory research:

1. For low values of strain rates ($\cong 1.01 \times 10^{-2}$/s), the yield limit increases 10–20%, with the higher values of increase corresponding to steels of low strength.
2. In the case of a repeated tensile cyclic loading (about 90% of the yield strength) and for a large number of loading cycles (above 1000) under high strain rates (above 1.01×10^{-2}/s), the failure load is limited very little in comparison to the failure static loading. In other

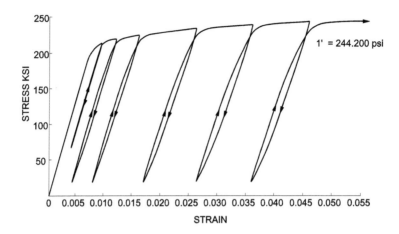

Figure 7.12 Stress–strain curve for steel wire under repeated tensile loading. (Adapted from Sinha, B.P., Gerstle, K.H. and Tulin, L.C. 1964. *Journal of the American Concrete Institute*, Proceedings, 61(8), 1021–1038; Park, R. 1973. Theorization of structural behaviour with a view to defining resistance and ultimate deformability. *Symposium on Resistance and Ultimate Deformability of Structures Acted on by Well-Defined Repeated Loads.* IABSE, Lisboa, 1973. With permission of IABSE.)

words, the increase in the yield strength due to the high strain rate is counter-balanced by the fatigue symptom. All the above characteristics are also in effect for compressive axial loading under the assumption that buckling is prohibited.

7.3.4 Stress–strain diagram for reversed cyclic loading

Under reversed cyclic loading, the stress–strain properties of steel become quite different from those of purely tensile or compressive stress. This is known as the *Bauschinger effect* and results in a lowering of the reversed yield stress (Figure 7.13; Park, 1972). Once a full loop is integrated, the linear relation between stress–strain (constant E_s) is disrupted for all

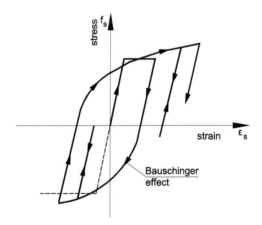

Figure 7.13 Stress–strain diagram for steel under reversed loading (Bauschinger effect). (From Park, R. 1973. Theorization of structural behaviour with a view to defining resistance and ultimate deformability. *Symposium on Resistance and Ultimate Deformability of Structures Acted on by Well-Defined Repeated Loads.* IABSE, Lisboa, 1973. With permission of IABSE.)

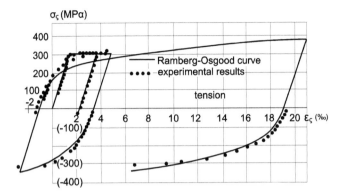

Figure 7.14 Modelling of the cyclic loading of steel bars and comparisons with relevant test data. (Park, R. and Paulay, T.: *Reinforced Concrete Structures*. 1975. Copyright Wiley-VCH Verlag GmbH & Co. KGaA. Reproduced with permission.)

successive cycles. Many attempts have been made to elaborate reliable analytical models for simulating the stress–strain constitutive law for reversed cyclic loading (Popov, 1977; CEB, 1991b). One of the most reliable expressions is that of Ramberg–Osgood, based on curve fitting to available test data (Figure 7.14; Park and Paulay, 1975; Popov, 1977).

7.3.5 Provisions of codes for reinforcement steel

Generally, reinforcement steel categories in Europe are specified by EN1992-1-1:2004 and by EN10080 (Table 7.2; Figure 7.15). The various categories are classified into three classes, namely

- Steel class B (400–600) A
- Steel class B (400–600) B
- Steel class B (400–600) C

Class A refers to cold deformed steel bars. *This class is considered to be unsuitable for primary structural members* of seismic-resistant buildings, even for DCL buildings, because of their low strain at failure ε_{su}. For buildings of low or moderate ductility (DCL/DCM buildings), the other two steel classes B and C may be used, while for buildings of behaviour class DCH (high ductility), only steel rebars of class C may be used.

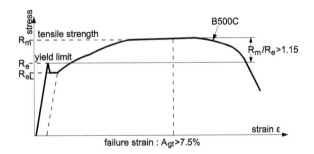

Figure 7.15 Stress–strain diagram of steel B500C (EN 10080).

Table 7.3 Requirements for steel reinforcement bars according to EN1998-1/2004

Magnitude	EN 1998-1/2004				ELOT: EN 1420-2-3	
	DCL/DCM structures		DCH structures		Steel class B500 A	Steel class B500 C
	Steel class B	Steel class C	Steel class B	Steel class C		
R_m/R_e	≥1.08	≥1.15 <1.35	Ø	≥1.15 <1.35	≥1.05	≥1.15 <1.35
Agt (%)	>_5.0	>_7.5	Ø	≥7.5	>2.5	≥7.5
$R_{e\,act}/R_{e\,nom}$	<1.30	<1.30	Ø	<1.25	Ø	≤1.25
	EC 1992-1/2004					

With the exceptions of closed stirrups and ties, *only ribbed bars* may be used as reinforcing steel in primary structural members.

It should be mentioned here that the majority of steel factories in southern Europe, where seismicity is high, produce mainly steel rebars and weldable meshes of class B500C (weldable tempcore steel) in diameters from $d = 6$ to $d = 40$ mm. This steel category is suitable for buildings of all ductility classes. At the same time, they produce steel of class B500A in diameters of $d = 5$ to $d = 8$ mm in the form of welded meshes for non-primary seismic structural members (e.g. in Greece ELOT: EN 1420-2-3).

In Table 7.3, the steel classes that may be used in each ductility class are given together with their main mechanical properties.

7.3.6 Concluding remarks

1. Basic requirements for steel rebars in use for the primary seismic-resistant members are the following:
   a. High strength
   b. High ductility
   c. High bond to concrete
   In this respect,
   a. Yield stress is about 500 MPa (almost twice the yield stress of classic mild steel).
   b. Steel class A with a low value of ε_{su} (2.5%) is excluded from primary seismic members of all ductility classes.
   c. Steel of class B with a medium value of ε_{su} (≥5.0%) is allowed for use only in buildings of low and moderate ductility (DCL and DCM).
   d. Only steel of class C with a high value of ε_{su} (≥7.5%) is allowed for use in buildings of ductility class DCH.
   e. Ribbed steel rebars are obligatory for ensuring a high bond between rebars, and concrete, hoops and ties are exempted from the rule. A detailed examination of this issue will be presented in a subsequent paragraph, because rebar slip due to bond disintegration causes a quick degradation of rebar anchorage in concrete.
2. In addition to the above requirements, steel rebars for seismic-resistant primary members must fulfil a requirement for a minimum ratio $\kappa = f_t/f_y$, namely

$$\kappa = f_t/f_y \left. \begin{cases} \geq 1.05, \text{ for steel of class A} \\ \geq 1.08, \text{ for steel of class B} \\ \geq 1.15, \text{ for steel of class C} \end{cases} \right\} \qquad (7.10a\text{–}c)$$

where

- f_y is the yield strength.
- f_t is the tensile strength.

The above requirement imposes a minimum level of strain hardening with increasing value from steel class A to steel class C. The reasoning behind this is that the length of the plastic hinge that might be formed in the critical zone of a structural member (Figure 7.16) is closely related to the strain hardening of the reinforcement. Indeed, the plastic rotation capacity θ_u^{avail} of a joint is expressed by the dashed area of the curvature diagram in Figure 7.16. Therefore, the plastic curvatures must extend to a sufficient length of the member l_{p1}, known as 'the length of the plastic hinge'. In fact, the yield and failure moments M_y and M_u, respectively, at the base of the cantilever in Figure 7.16 result with a good degree of approximation from the relations

$$
\left.
\begin{aligned}
M_y &\cong z_y \cdot f_y \cdot A_s \\
M_u &\cong z_u \cdot f_u \cdot A_s
\end{aligned}
\right\}
\tag{7.11a,b}
$$

where

- f_y and f_u are the yield and the tensile strength.
- z_y and z_u are the lever arms of internal forces at yield and failure and that they might approximately be considered equal.

Therefore, the plastic hinge length may be given from the expression below:

$$
l_{pl} = l\left(1 - \frac{M_y}{M_u}\right) \cong l\left(1 - \frac{f_y}{f_t}\right)
\tag{7.12}
$$

From the above expression, it may be concluded that as the ratio f_y/f_t tends to 1.0, the length of the plastic hinge l_{p1} decreases and therefore the capacity of plastic rotation at the hinge also decreases.

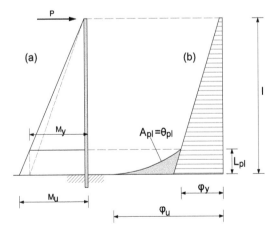

Figure 7.16 Schematic presentation of the reasoning behind the requirement that $f_t/f_y > 1.0$. (a) Moment diagrams yield and failure; (b) curvature diagram: plastic rotation $\theta_{pl} = A_{pl}$. In conclusion, if there is no degree of strain hardening ($f_t > f_y$), there is no plastic hinge.

3. Finally, according to European Codes, steel of class C, which is suitable for use in high-ductility R/C buildings (DCH), must also have a guaranteed upper limit of tensile strength, that is,

$$\left. \begin{aligned} \kappa &= \frac{f_t}{f_y} \le 1.35 \\[2mm] &\text{and} \\[2mm] \frac{f_{y\kappa0.95}}{f_{ynom}} &\le 1.25 \end{aligned} \right\} \tag{7.13a,b}$$

where
- $f_{y\kappa0.95}$ is the upper characteristic (95% fractile) of the actual yield strength.
- f_{ynom} is the nominal yield strength.

The above requirement for high-ductility steel (class C) allows the introduction of reasonable and reliable values of overstrength γ_{Rd} in the determination of capacity design action effects (see Chapter 6).

7.4 CONFINED CONCRETE

7.4.1 General

Since the beginning of the twentieth century, it has been clear that concrete under triaxial compressive loading, which is a predominant axial stress σ_1 and a hydrostatic lateral pressure ($\sigma_2 = \sigma_3 = p$) (Figure 7.17), presents an increase in its axial strength and its ability for plastic deformation. Lateral pressure p is accomplished mainly by using circular spirals or narrow spaced hoops. When axial stress exceeds 70% of the axial compressive strength of plain concrete, the lateral reinforcement is activated due to the volume increase in the concrete core laterally. At the same time, the member under compression begins *to spall* little by little. So, finally, the concrete core inside the spiral reinforcement continues to be uncracked, confined by the steel reinforcement, while the cover has been spalled completely. When the axial stress exceeds a critical value, *confinement steel yields first*, while at the same time, longitudinal cracks appear in the core parallel to the axial stress, which leads to the failure of the core. From what has been presented so far, the lateral reinforcement, due to its tensile strain, exerts pressure on the concrete core, which confines its lateral extension. For this reason, the term 'confined concrete' has been introduced. In Figure 7.18, σ_1–ε diagrams are presented, showing the influence of the degree of confinement on concrete (Scott et al., 1982).

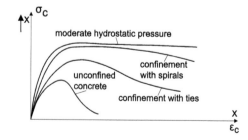

Figure 7.17 Stress–strain diagrams for various types of confinement.

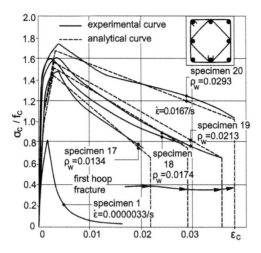

Figure 7.18 Stress–strain diagrams for concrete confined by different numbers of hoops. (Adapted from Scott, B.D., Park, R. and Priestley, M.J.N. 1982. *ACI Structural Journal*, Vol. 79(1), 13–27.)

Recently, FRP warps have been used for the confinement of existing R/C structures. A detailed treatment of this issue will be made in Chapter 14.5 dealing with the repair and strengthening of R/C members.

In closing, it should be noted that confinement is the most effective tool for increasing the strength of concrete and *mainly for its transformation from a brittle material into a material with sufficient ductility.*

7.4.2 Factors influencing confinement

Confinement is influenced by the following factors:

1. *The volumetric coefficient of transverse reinforcement* ρ_w. This coefficient is defined as the ratio of the volume of spiral reinforcement or of hoops to the volume of relevant concrete that is confined by the spiral or hoop reinforcement.
2. *The yield strength of transverse reinforcement* f_{yw}. It is obvious that this and the previous factor define the maximum possible value of confinement.
3. *The compressive strength* f_c *of concrete.* The influence of confinement is more effective on low-strength concrete.
4. *The spacing of hoops or the spiral pitch.* It is obvious that widely spaced spirals or hoops leave zones of the compressed member unconfined.
5. *The form and the arrangement of the spirals and hoops.* Cyclic hoops are more effective than orthogonal ones (Figure 7.19). Hoops or ties that penetrate the concrete core are also more effective than orthogonal ones since in this case unconfined zones are limited.
6. *The strain rate* $\dot{\varepsilon}$ *of axial deformation.* A high strain rate $\dot{\varepsilon}$ influences the increase in strength and the decrease in plastic deformation, as in the case of plain concrete.
7. *The strain gradient in the cross section.* R/C members under compression are likewise affected by bending moments, causing a strain gradient on the affected section. This eccentric loading does not have a serious influence on the confinement of the structural member.

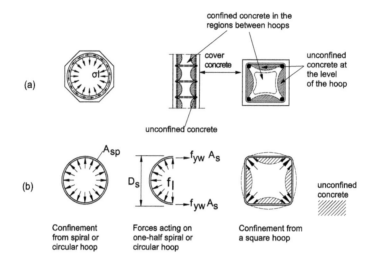

Figure 7.19 Confinement of concrete by circular and square hoops: (a) region confined by hoops; (b) stress pattern due to confinement by hoops. (Adapted from Penelis, G.G. and Kappos, A.J. 1997. *Earthquake-Resistant Concrete Structures.* SPON E&FN, Chapman & Hall, London.)

The above factors have been investigated for many years and have been simulated by analytical models of varying reliability. The scope of this book does not allow an extended treatment of this research. In any case, the results of this research have influenced the Codes of Practice in the dimensioning of members under compression.

A detailed examination of this issue will be made in the following subsections. Here, the presentation will be limited to the changes of the σ_c–ε_c diagrams that have been adopted by Codes in effect (EC2-1-1/2004 and EC8-1/2004) in the case of transverse hydrostatic stresses independently of the type of confinement.

7.4.3 Provisions of Eurocodes for confined concrete

7.4.3.1 Form of the diagram σ_c–ε_c

The diagram σ_c–ε_c of confined concrete that has been adopted by EC2-1-1 and EC8-1 is given in Figure 7.20. The values of $f_{ck,c}$, $\varepsilon_{c2,c}$ and $\varepsilon_{cu2,c}$ are given below:

For $\sigma_2 \leq 0.05 f_{ck}$:

$$f_{ck,c} = f_{ck}(1.0 + 5.0\sigma_2/f_{ck})$$

(7.14a)

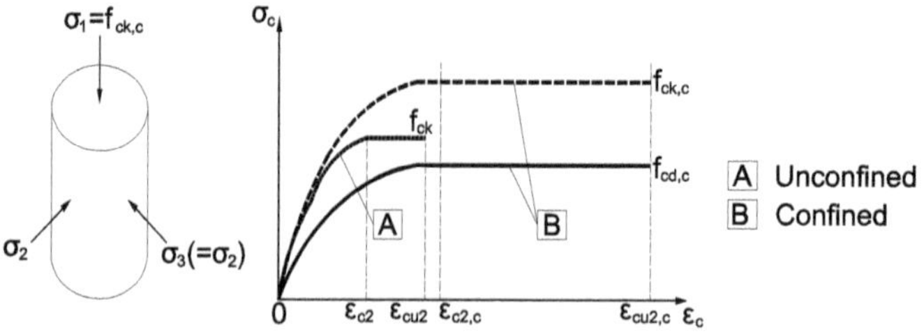

Figure 7.20 Stress–strain diagrams of confined concrete. (Adapted from EC2-1-1/EN1992-1-1. 2004. *Design of Concrete Structures – Part 1-1: General Rules and Rules for Buildings.* BSI, CEN, Brussels, Belgium.)

For $0.05 f_{ck} \leq \sigma_2$:

$$f_{k,c} = f_{ck}(1.125 + 2.5\sigma_2/f_{ck}) \tag{7.14b}$$

$$\varepsilon_{c2,c} = \varepsilon_{c2}(f_{ck,c}/f_{ck})^2 \tag{7.15}$$

$$\varepsilon_{cu2,c} = \varepsilon_{cu2} + 0.2\sigma_2/f_{ck} \tag{7.16}$$

Symbols in use are depicted in Figure 7.20. The above relations, which have been adopted first by Model Code 90/CEB/FIP and then by EC2-1-1/2004, underestimate the consequences of confinement in relation to other models of higher reliability, like that of Newman and Newman (1971).

According to this model (Figure 7.21; Fardis, 2009)

$$f_{ck,c} = f_{ck}(1 + \kappa) \tag{7.17}$$

$$\varepsilon_{c2,c} = \varepsilon_{c2}(1 + 5\kappa) \tag{7.18}$$

$$\varepsilon_{cu2,c} = 0.004 + 0.5\frac{\sigma_2}{f_{ckc}} \tag{7.19}$$

where

$$\kappa \cong 3.7\left(\frac{\sigma_2}{f_{ck}}\right)^{0.86} \tag{7.20}$$

This model has been adopted by EC8-3/2005 for the assessment and strengthening of existing structures.

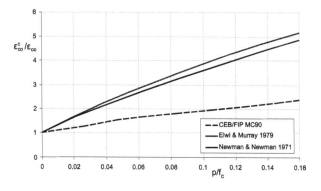

Figure 7.21 Comparison of the predictions of various confinement models for the enhancement of concrete strain at maximum strength; p, confinement stress; ε_{co}, strain at failure of unconfined concrete; ε_{co}^c, strain at failure of confined concrete. (Adapted from Newman, K. and Newman, J.B. 1971. *Failure Theories and Design Criteria for Plain Concrete. Solid Mechanics and Engineering Design.* J. Wiley-Interscience, New York. With kind permission from Springer Science+Business Media: Fardis, M.N. 2009. *Seismic Design, Assessment and Retrofitting of Concrete Buildings.*)

7.4.3.2 Influence of confinement

According to what has been presented in the previous paragraph, the determination of σ_2 in Equations 7.14 through 7.20 is the basic parameter for confinement consequences. In the case of confinement by means of spirals, hoops, stirrups and ties, the confined concrete core *fails just after the yield of confining reinforcement*. Therefore, the crucial point in the design is the determination of σ_2 at the stage of yielding of the reinforcement of confinement.

Consider first a circular cross section confined by means of circular spiral reinforcement with a core diameter D_c and a pitch of the spiral s (Figure 7.22). At an advanced stage of axial loading, after spalling of the cover, lateral stresses σ_2 are developed in the form of a hydrostatic pressure. If the spacing of spirals is very small, these stresses may be considered uniformly distributed along the member height. Therefore, according to known expressions from the theory of strength of materials (Figure 7.22)

$$2A_s f_{yw} = \sigma_2 D_c s \tag{7.21}$$

and

$$\sigma_2 = \frac{2A_s f_{yw}}{D_c s} \tag{7.22}$$

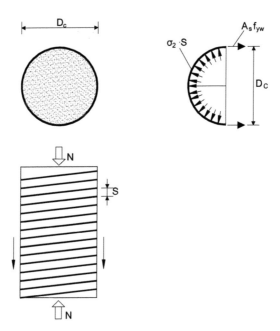

Figure 7.22 Stress pattern for the determination of the confinement degree (σ_2) as a function of the mechanical percentage (ω_w) of spiral confinement on concrete strength (f_{ck}).

Now, taking into account that the mechanical percentage of lateral confining reinforcement is equal to

$$\omega_w = \rho_w \frac{f_{yw}}{f_{ck}} = \frac{A_s \pi D_c}{\pi D_C^2 s / 4} \frac{f_{yw}}{f_{ck}} = \frac{4A_s}{D_C s} \frac{f_{yw}}{f_{ck}} \tag{7.23}$$

and introducing Equation 7.23 into Equation 7.22, we obtain

$$\sigma_2 = 0.5\omega_w f_{ck} \tag{7.24}$$

In the above equations, the various symbols have the following meaning:

- D_c is the core diameter (Figure 7.22).
- s is the pitch of the spiral.
- A_s is the area of the cross section of a spiral.
- f_{ck} is the characteristic concrete strength.
- f_{yw} is the yield strength of the spiral reinforcement.
- $\rho_w = \dfrac{4A_s}{D_c s}$ is the volumetric percentage of spiral reinforcement.
- $\omega_w = \rho_w \dfrac{f_{yw}}{f_{ck}}$ is the mechanical percentage of spiral reinforcement.

In practice, members under compression are usually neither circular nor laterally reinforced with spirals or hoops in very narrow spacing. Therefore, expression 7.24 must be modified so that the real stress condition is defined with sufficient accuracy. So, in Equation 7.24, a coefficient of reduction factor α must be introduced for the estimation (see Section 7.4.2) of

- Spacing of the spirals, hoops or stirrups
- Arrangement of the stirrups and ties in the cross section
- Longitudinal reinforcement of the member

that is,

$$\boxed{\sigma_2 = 0.5\alpha\omega_w f_{ck}} \tag{7.25}$$

In this respect, Equations 7.14b and 7.16 take the following form:

$$\boxed{f_{ck,c} = f_{ck}(1.0 + 2.5\alpha\omega_w)} \tag{7.26}$$

$$\boxed{f_{ck,c} = f_{ck}(1.125 + 1.125\alpha\omega_w)} \tag{7.27}$$

$$\boxed{\varepsilon_{cu2,c} = \varepsilon_{cu2} + 0.10\alpha\omega_w} \tag{7.28}$$

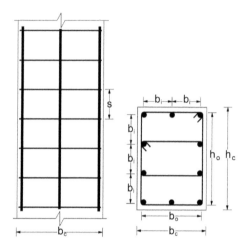

Figure 7.23 Confinement of concrete orthogonal cross section with hoops.

The mechanical models that have been used for the evaluation of α have taken into account the attenuations of the core confinement in longitudinal and transversal directions as well as (Figure 7.23) the form of hoops, stirrups and ties (circular or orthogonal). In this respect, for each type of lateral confinement, α is expressed as the product of two factors, α_s and α_n, that is,

$$\alpha = \alpha_s \cdot \alpha_n \qquad (7.29)$$

In the above relation, α_s expresses the influence of spiral hoops or stirrups spacing (s), while α_n is the influence of the distance of longitudinal rebars that are encaged by stirrups or ties.

- For orthogonal members under compression and orthogonal stirrups (Figure 7.23), which is the most common case in practice, EC8-1/2004 specifies α_s and α_n as follows:

$$\alpha_s = (1 - s/2b_o)(1 - s/2h_o) \qquad (7.30)$$

and

$$\alpha_n = 1 - \sum_n b_i^2 / 6 b_o h_o \qquad (7.31)$$

where
- n is the number of longitudinal rebars with lateral constraint by stirrups or ties.
- b_i is the distance between laterally engaged longitudinal bars by means of stirrups and ties.

The rest of the symbols may be found in Figure 7.23.

- For circular cross sections reinforced with hoops,

$$\boxed{\alpha_n = 1} \tag{7.32}$$

$$\boxed{\alpha_s = (1 - s/2D_c)^2} \tag{7.33}$$

where
D_c is the core diameter measured between the axes of hoop rebars.
s is the spacing of hoops.
- In the case of eccentric axial load of an orthogonal cross section (Figure 7.24; Fardis, 2009),

$$a_n = 1 - \frac{\sum b_i^2}{6x_o b_{yo}} \tag{7.34}$$

$$\alpha_s = \left(1 - \frac{s}{2b_{xo}}\right)\left(1 - \frac{s}{4x_o}\right) \tag{7.35}$$

The symbols of Equations 7.34 and 7.35 may be found in Figure 7.24.

Many researchers have contributed to the development of the above mechanical models in the last 40 years (Park et al., 1982; Sheikh and Uzumeri, 1982; Mander et al., 1988; Kappos, 1991b).

At the same time, the output of analytical models has been cross-checked by means of extended laboratory evidence (see reviews in Park and Paulay, 1975; Aoyama and Noguchi, 1979; Sakai and Sheikh, 1989).

The procedure for determination of α_s will be presented as an example in the case of circular hoops. It is assumed that the load path of confinement stresses from hoops to the core follows the form depicted in Figure 7.19. In this respect, the envelope of the confined zone

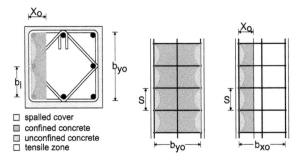

Figure 7.24 Calculation of confinement effectiveness in the compression zone of the confined core of a member in flexure. (With kind permission from Springer Science+Business Media: Fardis, M.N. 2009. *Seismic Design, Assessment and Retrofitting of Concrete Buildings.*)

between two hoops is an axisymmetric parabolic surface with a minimum diameter $D_{c\,min}$ at the mid-height of distance s of two successive hoops equal to

$$D_{c\,min} = D_c - \frac{s}{4}2 = D_c - \frac{s}{2} \tag{7.36}$$

Therefore, the area $A_{c\,min}$ of the minimum cross section of the confined core is

$$A_{c\,min} = \frac{\pi}{4}\left(D_c - \frac{s}{2}\right)^2 \tag{7.37}$$

and finally

$$\alpha_s = \frac{A_{c\,min}}{A_c} = \frac{\left(D_c - \dfrac{s}{2}\right)^2}{D_c^2} = \left(1 - \frac{s}{2D_c}\right)^2 \tag{7.38}$$

Equation 7.38 is the same expression as that in Equation 7.33, which was adopted in EC8-1/2004.

7.5 BONDING BETWEEN STEEL AND CONCRETE

7.5.1 General

It is well known from the design of R/C structures under static loading that *bonding* between steel rebars and concrete is of paramount importance for the mechanical behaviour of reinforced concrete. In fact, bonding is a prerequisite for the transfer of any stress of a steel rebar to concrete in the form of shear bond stresses.

Consider Figure 7.25. Equilibrium conditions result in the following expression:

$$\frac{\pi d_s^2}{4}d\sigma_s = \pi d_s \tau_b dx \tag{7.39}$$

or

$$d\sigma_s = \frac{4}{d_s}\tau_b dx \tag{7.40}$$

Integrating Equation 7.40 between two points S_0 and $S_\perp$ in a distance l_b, the following expression is obtained:

$$\sigma_o = \sigma_{S0} - \sigma_{S1} = \frac{4}{d_s}\int_{S_2}^{S_1} \tau_b\, dx = \frac{4}{d_s}\tau_{b\,mean}l_b \tag{7.41}$$

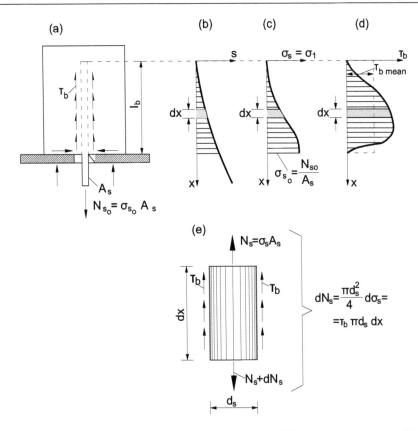

Figure 7.25 Bond shear stress distribution at the pull-out test: (a) experimental layout; (b) rebar slip; (c) tensile stresses at rebar; (d) bond stresses; (e) stress pattern at a differential element dx of the rebar.

The meaning of the above symbols may be found in Figure 7.25.

It may be easily concluded from Equation 7.40 that any change of axial stresses along the rebar results in the development of bond shear stresses τ_b at the interface of steel and concrete, and conversely, *it is impossible for tensile stresses to develop in rebars* without bonding between concrete and reinforcement, in other words if $\tau_{bmean} = 0$.

These bond shear stresses are caused by three different transfer mechanisms. These mechanisms may be briefly described as follows:

1. *Chemical adhesion* of the cement paste on the surface of the steel bar. The degree of adhesion depends mainly on the roughness and the cleanness of the steel rebar surface. Its value τ_o ranges from 0.5 to 1.0 MPa (ACI Committee 408, 1991) and is exhausted before any slip δ between steel and concrete.
2. *Bond due to friction.* When the loading of the bar causes a bond shear stress

$$\tau > \tau_o$$

adhesion breaks down and bonding is developed *by friction* shear stresses, even *for a very small slip* of the rebar in the concrete. It is known that for the development of shear stresses due to friction, a transverse pressure on the steel surface is needed. These pressures may be caused by transverse confinement of concrete, transverse loading and microscopic anomalies (pitting) of the rebar surface. The friction coefficient μ ranges between 0.30 and 0.60, depending on the roughness of the steel surface. This type of bond is distinct only in the case of smooth rebar surfaces.

3. *Mechanical interlock.* Mechanical interlock is developed in the case of rebars with ribs, and it leads to an imposing enhancement of the total bond shear. In this case, the slip of the rebar in relation to the surrounding concrete is accompanied by internal cracks of concrete at the points of interlock of the ribs (Figure 7.26; Leonhardt, 1973). These cracks display vertically with respect to the main tensile stresses. This means that a slip at failure is accompanied by widening of the concrete circumference around the bar. As a consequence, strong circumferential stresses are developed, which cause splitting failure at the free surface of the concrete (Figure 7.27). At this point, the bond, to a large degree, disintegrates. In this context, bonding is enhanced if there is adequate confinement. Bond strength in the case of deformed bars, according to EC2-1-1/2004, is estimated to be 3.0–7.0 times higher than that of smooth rebars, depending on the position of the bar in the concrete member. This can explain why in earthquake-resistant R/C structures the main rebars must be deformed.

For the study of bonding, the three load transfer mechanisms presented above are examined together as one, in the form of bond shear f_b. In this context, bond shear f_b is defined experimentally using specimens and loading in the form depicted in Figure 7.28 (RILEM/CEB/FIP). The mean values of τ_b for the successive loading steps and the corresponding relative slip between steel and concrete are depicted in a τ–s diagram, which represents the constitutive law of bonding. The conventional value of f_{bd} is defined as the mean bond shear stress for which a slip of 0.1 mm is displayed.

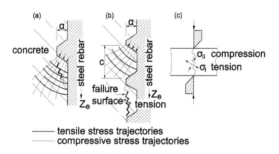

Figure 7.26 Qualitative stress and failure surface distribution of concrete at the ribs of deformed bars: (a) for large rib spacing; (b) for small rib spacing; (c) failure surface at shear (according to E. Morsh). (With kind permission from Springer Science+Business Media: Leonhardt, F. and Mönnig, E. 1973. *Vorlesungen uber Massivbau, Erster Teil.*)

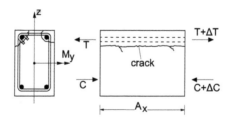

Figure 7.27 Longitudinal cracks due to bonding failure in the case of deformed steel rebars.

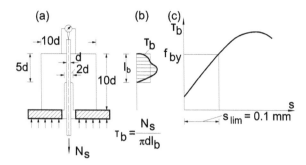

Figure 7.28 Pull-out test: (a) experimental layout; (b) normalised bond stress τ_b; (c) τ_b–s diagram. (RILEM/ CEB/FIB. With kind permission from Springer Science+Business Media: Leonhardt, F. and Mönnig, E. 1973. *Vorlesungen über Massivbau, I. Teil.*)

The scope of this book does not allow an extended examination of bond issues (anchorage lengths, reinforcement splices, reinforcement overlapping, etc.) for static loading, since all the above belong to the common practice for R/C structures in general. Here, we will focus our interest mainly on the following issues:

    a. Monotonic loading up to failure of the bonding mechanism, including the descending branch of the τ_b–s diagram
    b. Reversed cyclic loading up to failure

The above two parameters, together with those in effect for static loading, govern the bond mechanism in seismic structures and specifically at critical regions like joints, where significant bond stresses develop and sometimes lead to bond failure.

7.5.2 Bond–slip diagram under monotonic loading

Bonding between steel and concrete has been a subject of extended research since the beginning of the twentieth century due to its importance for the behaviour of reinforced concrete. However, the complete diagram of τ_b–s, including also its descending branch up to complete failure of the bond, has been studied in the last 40 years, mainly in relation to the study of seismic-resistant R/C structures in the post-elastic stage.

In Figure 7.29, the idealised form of the diagram τ_b–s of *deformed bars* is depicted, based on laboratory tests carried out by Eligehausen et al. (1983) for unconfined and confined concrete. Simultaneously, the mechanical model adopted by Model Code 1990/CEB/FIB, 1993, is given in Figure 7.30. The main characteristics of the diagrams for bonding of deformed bars under monotonic loading are the following:

    1. *Specimens without lateral confinement*
      a. For loading up to a certain level $\tau_b = \tau_o$, no slip s is displayed because bonding of steel and concrete is based on the chemical adhesion. The value of τ_o ranges between 0.5 and 1.0 MPa (ACI Committee 408, 1991).
      b. For $\tau_o < \tau < \tau_{max}^{un}$, after the rupture of adhesion, friction begins to be activated together with the mechanical interlock of ribs of deformed bars in concrete.
      c. For the next loading steps, internal micro cracks appear in concrete at its interface with rebars. As a result, the slope of the τ_b–s diagram begins to decrease.

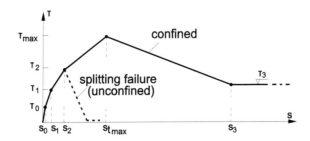

Figure 7.29 Idealised diagram of τ_b–s. (Adapted from Eligehausen, R., Popov, E.P. and Bertero, V.V. 1983. *Local bond stress–slip relationships of deformed bars under generalized excitations. Report EERC-83/23*, Univ. of California, Berkeley; Penelis, G.G. and Kappos, A.J. 1997. *Earthquake-Resistant Concrete Structures*. SPON E&FN, Chapman & Hall, London.)

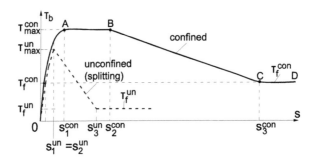

Figure 7.30 Idealised diagram of τ_b–s according to Model Code 1990. (Adapted from CEB 1993. *CEB-FIP Model Code 1990, Bulletin d'Information*, CEB, 213/214, Lausanne.)

    d. For a value $\tau_b = \tau_{max}^{un}$, splitting is displayed at the concrete surface, accompanied by an abrupt decreasing of τ_b (descending branch in Figure 7.30 of unconfined bonding).

    e. The characteristic values of τ_b and s accompanied by analytical expressions for the ascending branch in Figure 7.30 are given in Model Code 1990 of CEB/FIB (1993) and are presented below:

        i. $s_1 = s_2 = 0.6$ mm

        ii. $s_3 = 1.0$ and 2.5 mm: for 'good' and 'poor' bond conditions, respectively

        iii. $\tau_{max}^{un} = 2.0\ \sqrt{f_{ck}}$ and $1.0\ \sqrt{f_{ck}}$: for 'good' and 'poor' bond conditions, respectively

        iv. $\tau_f^{un} = 0.15\tau_{max}$

        v. $\tau = \tau_{max}^{un} \left(\dfrac{s}{s_1} \right)^{0.40}$: for the ascending branch

  2. *Specimens with lateral confinement*

    a. In the case of lateral confinement caused either by lateral reinforcement or by transversal external pressure, the diagram τ_b–s (Figure 7.30) has the same form as that of the unconfined specimen with the higher value of τ_{max} as its main characteristic, and a strength plateau at τ_{max}.

b. After the concrete splitting, a horizontal plateau is displayed in the τ_b–s diagram due to lateral confinement, and then the descending branch follows, which corresponds to the concrete pulverising between successive ribs.

c. The characteristic values of τ_b and s accompanied by an analytical expression for the ascending branch in Figure 7.30 are given in Model Code 1990/CEB/FIB, 1993, and are presented below:

   i. $s_1^{con} = 1.0$ mm $= s_f$

   ii. $s_2^{con} = 3.0$ mm

   iii. $s_3^{con} =$ rib spacing

   iv. $\tau_{max} = 2.5 \sqrt{f_{cк}}$ and $1.25 \sqrt{f_{cк}}$: for 'good' and 'poor' bond conditions, respectively

   v. $\tau_f = 0.4 \tau_{max}$

   vi. $\tau = \tau_{max} \left(\dfrac{s}{s_f} \right)^{0.40}$: for the ascending branch

The above parameters are in effect for lateral confinement defined below:

$$
\left.
\begin{array}{l}
p \geq 7.5\,\text{MPa} \qquad\qquad 7.42a \\[2mm]
\text{or} \\[2mm]
\dfrac{\sum A_{sw}}{(nA_s)} \geq 7.5 \qquad 7.42b
\end{array}
\right\} \qquad (7.42)
$$

where
- A_s is the area of the cross section of each bonded rebar.
- n is the number of bonded bars.
- ΣA_{sw} is the area of the cross section of the confining stirrups along the bond length under consideration.

For

$$
\left.
\begin{array}{l}
0 \leq p \leq 7.5\,\text{MPa} \qquad\qquad 7.43a \\[2mm]
\text{or} \\[2mm]
0.25 \leq \sum A_{sw}/(nA_s) \leq 1 \qquad 7.43b
\end{array}
\right\} \qquad (7.43)
$$

a linear interpolation is recommended between the respective values of unconfined and confined concrete.

3. From extensive laboratory tests, it has been verified that the τ_b–s diagram for bars under compression is identical to that for rebars under tension in the case of lateral confinement.

4. Finally, in the case that steel bars under tension exhaust their yield strength, their respective lateral contraction causes a decrease in bond. The opposite happens in case of rebars under compression.

7.5.3 Bond–slip diagram under cyclic loading

A bond–slip diagram under repeated loading of an embedded bar has as an envelope the monotonic diagram of bond–slip presented in the previous paragraph.

Results of a reversed cyclic loading in conceptual form are depicted in the diagram of Figure 7.31 (Balázs, 1989; Fardis, 2009) and Figure 7.32. In this case, it is important to note the following:

1. Reversed loading leads to a degradation of high order of bonding after a number of cycles.
2. The main characteristic of bond hysteretic loops is that for bond stresses above τ_f (see Section 7.5.2), the loading reversals are accompanied by horizontal branches at the loops caused by almost free slip of the rebars in the surrounding concrete, due to the disintegration and pulverising of the cement paste between the successive ribs (Figure 7.32). This form of loops results in the loss of a big percentage of the capacity for energy dissipation at the region of bonding. In other words, before the yielding of a rebar by means of which the hysteretic energy dissipation mechanism is activated, the disintegration of the bonding of the rebar with the surrounding concrete leads to the drastic degradation of the ductility of the structure. Consequently, it is of major importance that special measures are taken at the critical regions, for example, joints of seismic-resistant frames, so that this risk may be eliminated or at least be minimised (see Chapter 8.4.3). The most important of these measures in Codes of Practice is the introduction of very low values for f_{db}. In this way, bond zones remain in the elastic

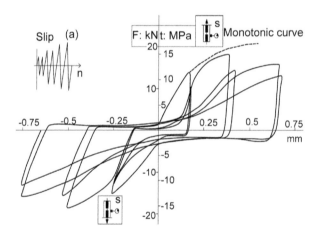

Figure 7.31 Bond stress–slip diagram under cyclic loading with f_c = 25 MPa. (Adapted from Balázs, G.L. 1989. Bond softening under reversed load cycles. *Studie Ricerche, Politecnico di Milano,* No. 11, Milano, 503–524.)

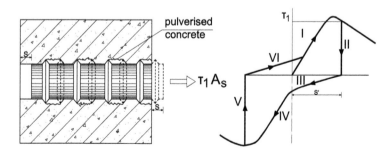

Figure 7.32 Bond stress–slip diagram under cyclic loading in conceptual form (I–VI: diagram branches of one cycle).

domain of the τ_b–s curve, and the degradation of the anchorage action is not significant. However, this requirement for low values for f_{db} leads to considerable lengths of anchorage or splices.

3. The formation of hooks at the end of embedded rebars, usually at right angles, seems to be a very effective measure for improving anchorage. From the extensive laboratory research of Eligehausen et al. (1983), the following remarks may be drawn:

   a. In the τ–s diagram under monotonic loading after the maximum value of τ has been reached, the descending branch continues to be very stable with a very smooth slope, which means that the bond continues to be stable against large slips. This must be attributed to the resistance of the hook to a pull-out effect, which contributes substantially to the internal forces of the bond.

   b. The τ–s curve for reversed cyclic loading continues to be very close to the curve of monotonic loading for a large number of reversals.

   c. The above remarks are taken into account very seriously for bar anchorage in joints of frames, particularly at the external columns (see Chapter 8.4.3).

In the last 30 years, many analytical models have been formulated that simulate in one way or another the behaviour of bond under reversed cyclic loading, as has been presented in experimental research (Tassios, 1979; Eligehausen et al., 1983; Filippou et al., 1983; ACI Committee 408, 1991; CEB Task Group 22 [CEB 1991b]; Soroushian et al., 1991; Darwin et al., 2002a,b; Cairns, 2006; Eligehausen and Lettow, 2007).

7.5.4 Provisions of Eurocodes for bond of steel to concrete

7.5.4.1 Static loading

It should be noted again that, as was explained in Section 7.5.1, the basic provisions for bond anchorage lengths, splices and overlaps are specified by Codes of Practice for conventional R/C structures. In Europe, EC2-1-1/2004 covers this issue in detail.

A summary of the main points of these provisions will be presented below:

1. The synergy of the three bonding mechanisms presented in Section 7.5.1 is approached by the assumption that constant design bond shear strength f_{db} develops along the interface of a rebar and concrete. This is defined by the following expressions (Figure 7.25).

   a. For deformed steel rebars:

   $$\boxed{f_{bd} = 2.25 f_{ctк}/\gamma_c \cong 0.32 f_{ck}^{2/3}} \tag{7.44}$$

   where
   – $f_{ctк}$ is the characteristic tensile strength of concrete (5% fractile).
   – γ_c is the partial safety factor for concrete equal to 1.5.
   – f_{ck} is the characteristic compressive strength of concrete (5% fractile).

   b. Eurocode EC2-1-1/2004 does not make any reference to smooth steel rebars, since this type of reinforcement is excluded from the design of R/C structures.

   It should be remembered that the above value of f_{bd} corresponds to a slip s relative to concrete equal to 0.1 mm, in contrast to $s_f = 0.6$ mm, which is considered the limit of s, for which τ_b max is developed (see Section 7.5.2) in an anchorage without confinement. Therefore, it may be concluded that Codes are very conservative with respect to bond values since they are formulated to exclude bond failure at anchorage or splicing zones.

2. As far as the influence of confinement on f_{bd} is concerned, an enhancement coefficient equal to the following is specified:

$$\boxed{1/(1-0.04p) \leq 1.4}$$

(7.45)

or

$$1/(1 - \kappa\lambda) \leq 1.4$$

where
- p is the confinement due to external pressure along the anchorage length.
- $\lambda = (\Sigma A_{st} - \Sigma A_{st\ min})/A_s$.
- ΣA_{st} is the area of transverse reinforcement along the anchorage length.
- A_s is the area of the cross section of the anchored rebar with the bigger diameter.
- ΣA_{stmin} is the area of the minimum transverse reinforcement (0.25 As for beams and zero for slabs).
- κ is a coefficient ranging from 0 to 0.1 depending on the arrangement of the transverse reinforcement.

3. The above values of f_{bd} refer to a 'good' position of rebars in concrete, for example,
   a. At the bottom of the beams
   b. At least 300 mm below the upper surface of the concrete
   c. At an angle greater than 45° in relation to the horizontal line
      In the case of 'poor' position, the above values of f_{bd} should be reduced to 0.70 of those for 'good' position.

4. As far as the basic anchorage or splicing length l_b, this is defined under the assumption that rebar yields before bond failure. Taking into account Equation 7.41 and introducing there, in place of σ_{so}, the design yield stress f_{yd} of steel, and in place of $\tau_{b\ mean}$ the value f_{bd} that results

$$f_{yd} \leq \frac{4}{d_s} f_{bd} l_b$$

(7.46)

or

$$\boxed{l_b \geq \frac{f_{yd}}{f_{bd}} \frac{d_s}{4}}$$

(7.47)

where
- f_{yd}, f_{bd} are design stresses for steel and bond, respectively.
- d_s is the rebar diameter.
- l_b is the basic anchorage or splicing length.

The above basic length is modified in order to comply with special conditions, for example, the position of the anchorage zone, the existence or not of hooks and so on.

7.5.4.2 Seismic loading

Keeping in mind that the provisions for design anchorage and splice lengths for static loads are based on the assumption that rebar design yield precedes the design bond failure (bond slip, <0.1 mm), it can easily be concluded that these provisions also cover, in general, the requirements for seismic design of bonding, since the provisions for static load *are based on a capacity design concept for bonding*.

So, seismic codes for R/C structures include only some additional requirements, which are summarised below:

1. Only deformed rebars are accepted for main reinforcement due to their high bond strength.
2. Measurement of anchorage length should start some diameters inside the frame joints, so that even if a limited length of bond anchorage is degraded at the beginning of the joint because of load reversals, its influence would be of no significance for the overall behaviour of the anchorage. In fact, the rebar tensile stress diminishes rapidly at a distance of some diameters of the rebar inside the limit of the joint to the beam or the column and, therefore, bond degradation is limited only in this additional anchorage length, which is not taken into account. The above concept imposes special rules on the design of anchorages or splices of R/C members. These rules will be examined in detail in Chapters 8 and 9.
3. Determination of anchorage of splice length should take into account that rebars may be affected in tension or compression, and, therefore, the most unfavourable situation must be considered.
4. Special measures are foreseen in critical zones. There, due to concrete cracking by bending, bonding is vulnerable, and therefore special rules should be implemented in case of steel splicing.
5. All of the above, together with a series of special rules, will be presented in detail in subsequent chapters in the examination of the behaviour of seismic-resistant R/C members.

7.6 BASIC CONCLUSIONS FOR MATERIALS AND THEIR SYNERGY

From what has been presented in the previous subsections, the following conclusions may be drawn.

1. The basic material on which structural ductility is based is steel. In fact, steel has tremendous capacity for absorbing and dissipating seismic energy.
2. Concrete is a brittle material with very limited capacity for absorbing and dissipating seismic energy. However, thanks to its high compressive strength, it allows R/C members under bending to develop high curvature at failure stage, and, therefore, it allows the steel, which is embedded in the flexural zone, to develop high post-elastic tensile strains and consequently to absorb and dissipate a considerable amount of seismic energy.
3. The strength and ductility of concrete are substantially enhanced by a proper confinement by means of spirals, hoops or stirrups.
4. The bonding of steel and concrete is a crucial parameter for the seismic response of R/C structures. In the case that the slippage of a steel rebars exceeds the design limit

(0.1 mm) due to reversal loading, a disintegration of the bond takes place and the hysteretic loops at the post-elastic stage of steel are replaced by the s-shaped loops of bond, which are unstable and 'thin'. In fact, in this case, bond degradation takes place before steel yield, and, therefore, the bond loop prevails.

5. Consequently, R/C structures have the capacity for developing adequate plastic deformations and, therefore, of absorbing and dissipating seismic energy, under the assumption that they are properly reinforced and designed (capacity design). In the following chapters, the design of the main structural members of R/C buildings will be presented in detail.

6. It is important to note the paramount importance of the development of analytical models for monotonic or cyclic loading for the basic materials and their bonds. As we will see later, these models will prove very useful for the design of R/C members. At the same time, they constitute the basis for the formulation of constitutive laws for reversed cyclic loading of R/C members, which are one of the necessary inputs for the nonlinear analysis of R/C structures (see Chapter 5.7.2).

7. Last but not least, it should be noted that both materials, concrete and steel, display a stress–strain diagram under monotonic loading, which is with a very good approximation the envelope of cyclic loading for both materials.

The above fact allows the reasonable consideration that in cross sections of R/C members where axial stresses prevail (bending with axial force), M–φ diagrams defined for monotonic loading are also the envelopes for cyclic loading caused by seismic action. In this respect, this curve may be used either in elastic or inelastic dynamic or static analysis for the control of demands in relation to the capacities. Extensive testing of R/C members under prevailing bending has confirmed the above consideration, as we will see in the next chapter.

Chapter 8

Seismic-resistant R/C frames

8.1 GENERAL

It has already been noted in Chapter 4 that seismic-resistant R/C *frames* as well as R/C *walls* and *dual systems* constitute the three basic seismic-resistant systems in use for earthquake-resistant buildings. Frame systems are usually 3-D structures consisting of plane frames arranged in two orthogonal directions, usually in the form of pseudo 3-D structure. It is obvious that each of these frames is affected biaxially, at least at its columns, due to the simultaneous seismic actions in both main directions. However, for a better understanding, the plane frame will be examined first and then reference will be made to biaxial loading.

It should be noted from the beginning that the elements of a frame are beams, columns and joints. These elements will be examined in detail in the following sections.

It has been noted already that seismic design is carried out for seismic actions in combination with the gravity loads for which masses have been taken into account for the determination of the inertial seismic forces, no matter if the analysis has been carried out using a linear static or linear dynamic method.

The form of moment diagrams M_{wd} of gravity loads W_{dE}, which participate in seismic load combinations, is given in Figure 8.1. It should be remembered that these diagrams correspond to the combination

$$W_{dE} = \sum G_{kj}{'}{+}{'}\sum \varphi\psi_{2i}Q_i \qquad (8.1)$$

(see Chapter 3.4.5). On the other hand, the relevant forms of moment diagrams of seismic action effects (32 combination cases) E_d are given in Figure 8.2.

It is obvious that in Figure 8.2, for simplicity, only 1 of the 32 combinations is depicted. It should also be noted that for a reversed seismic action, these diagrams change sign.

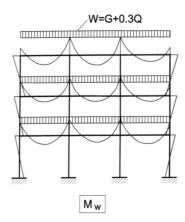

Figure 8.1 Bending moment diagram for gravity loads.

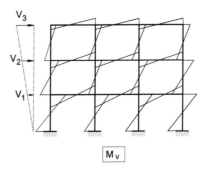

Figure 8.2 Bending moment diagram for horizontal seismic actions.

The combinations of gravity and seismic action effects for beams are given in Figure 8.3, while the relevant diagrams for columns are given in Figures 8.4 and 8.5. The main remarks for the above diagrams are the following:

1. Beams are affected mainly in bending and shear.
2. Columns are affected by normal forces, moments and shear.
3. Action effects in beams and columns increase from the upper to the lower storeys.
4. Axial forces of the external columns *are strongly affected* by the seismic action to such a degree that in the case that the aspect ratio H/L of the frame is high, they may change sign (Figure 8.6).
5. The moments of the beams due to seismic action adjacent to a joint have the opposite sign and are counterbalanced by the relevant moments of the adjacent columns framing to the same joint. This means that in the core of the joint, very strong changes of moments and shears take place both in the beam as well as in the column direction. As a result, very strong shear and bond stresses develop in the limited area of the joint core. For this reason, as we will see later, *these areas are very vulnerable* to seismic action and need special concern (Figure 8.7).

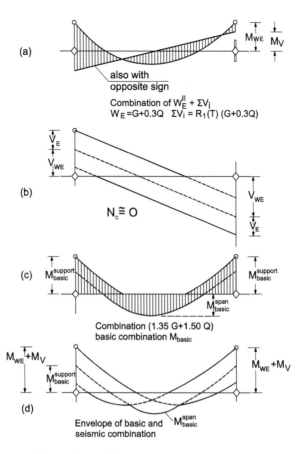

Figure 8.3 Moment diagram of frame beams for various load cases: (a) seismic combination with the corresponding gravity loads; (b) shear diagrams corresponding to load case combination (a); (c) basic load combination (1.35G+1.50Q); (d) bending moment envelope of basic load combination (c) with seismic load combination (a).

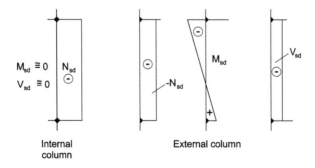

Figure 8.4 M, N, V column diagrams under gravity loads.

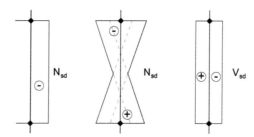

Figure 8.5 M, N, V column diagrams under horizontal seismic and corresponding gravity loads.

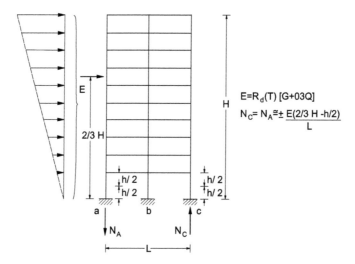

Figure 8.6 Axial reactions at the base of the external columns of a frame due to horizontal seismic action.

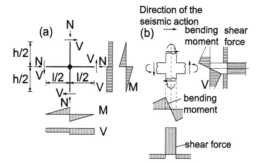

Figure 8.7 (a) M–V diagrams of a frame sub-assemblage due to seismic action (sway); (b) magnification of M–V diagrams in the region of the joint.

8.2 DESIGN OF BEAMS

8.2.I General

For a thorough examination of the action effects of a beam as a component of a ductile frame, the following diagrams of a typical internal beam are depicted in detail.

- In Figure 8.3a and b, typical M_d, V_d diagrams are given for the seismic combination W_{dE} '+' E_d.
- In Figure 8.3c, typical M_d, V_d diagrams are given for the 'basic' combination (i.e. $1.35G$ '+' $1.50Q$).
- In Figure 8.3d, an envelope for the above two cases has been elaborated.

From the examination of the above diagrams, the following remarks should be noted:

- Bending and shear prevail while axial forces are of secondary importance for beams.
- The influence of seismic actions increases from the top storeys to the base so that the combination of moment diagrams of seismic action and corresponding gravity loads present reversed sign at the critical zones near joints of lower storeys.
- The same also holds for shear diagrams. However, it should be remembered that design shear forces are determined on the basis of the capacity design procedure developed in Chapter 6.1.
- Moment diagrams of seismic actions induce contra-flexural deformation of beams (Figure 8.8). In this context, each beam behaves like two cantilevers, each with a span equal to about the half of the length of the beam under a reversed loading V_{Ed} at their free end. This behaviour explains why in almost all laboratory tests for beams under seismic cyclic excitation, the model in use is a cantilever imposed on a cyclic loading at its free end.
- The envelope of moment diagrams for 'the basic combination' (gravity design loads) and for the seismic combinations (Figure 8.3d) shows that a beam may be found under tension in almost its entire length at both its flanges (top and bottom) during load reversals, depending on the direction of seismic action and the level of the storey where the beam is located. This is taken into account very seriously by Codes in beam detailing, as will be presented later.

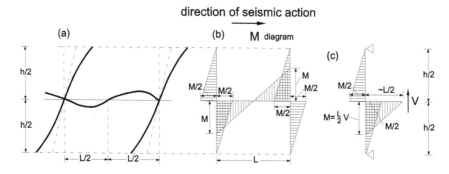

Figure 8.8 Flexural behaviour of a frame sub-assemblage: (a) deformations; (b) moment diagrams; (c) substitute sub-assemblage.

8.2.2 Beams under bending

8.2.2.1 Main assumptions

The assumptions made for the design of R/C beam sections under bending and for the elaboration of M–φ diagrams are the same as those in use for R/C members under static loading. These assumptions are the following:

1. Equilibrium conditions are considered for a structure without taking into account its deformations (first-order theory).
2. Concrete cannot carry tensile stresses (stage II).
3. Plane cross sections vertical to the axis of a beam under bending continue to be plane and vertical to the deformed axis up to failure (Bernoulli concept). In this respect, strain distribution over a cross section continues to be linear up to failure.
4. Stress–strain diagrams of steel and concrete are those given by the Code for stress analysis (see Chapter 7, Sections 7.2.4 and 7.3.5).
5. There is perfect bonding between steel rebars and concrete, which means that strains of concrete and steel at any point of a cross section are identical, that is,

$$\boxed{\varepsilon_{si} \equiv \varepsilon_{ci}} \tag{8.2}$$

In this context, for low strains where σ_c may be considered as a linear function of the relevant strain, meaning that for $\sigma_c < 0.7\,f_c$, the following expressions are in effect:

$$\varepsilon_{si} = \frac{\sigma_{si}}{E_s}, \varepsilon_{ci} = \frac{\sigma_{ci}}{E_c}$$

and, therefore,

$$\boxed{\sigma_{si} = \sigma_{ci}\frac{E_s}{E_c} = n\sigma_{ci}} \tag{8.3}$$

where
- E_s is the modulus of elasticity of steel.
- E_c is the modulus of elasticity of concrete.
- n is the ratio (E_s/E_c).

6. Failure criterion for bending is the ultimate strain of concrete (Figure 8.9).

8.2.2.2 Characteristic levels of loading to failure (limit states)

Consider the cantilever beam of Figure 8.10, which is loaded by a force V vertically to its axis at its free end. By increasing V from zero to a limit value V_{crack}, tensile and compressive stresses develop at a cross section near the fixed end, wherein extreme values are given by the expression (Figure 8.10a)

$$\left.\begin{array}{c}\sigma_o \\ \sigma_u\end{array}\right\} \cong \pm\frac{M_A}{bh^2/6} = \pm\frac{lV}{bh^2/6} \leq f_{ctm,fl} \tag{8.4}$$

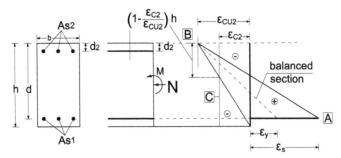

A   Ultimate strain of steel

B   Ultimate strain of concrete

C   Ultimate strain of concrete under plain compression

Figure 8.9 Failure criteria in bending with axial force.

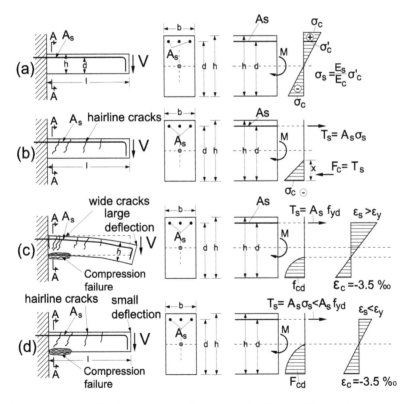

Figure 8.10 Successive steps of loading to failure of a beam under flexure: (a) stage I (uncracked); (b) stage II (cracked tensile zone); (c) failure due to steel yield (normally reinforced beam); (d) failure due to compression failure of concrete before steel yield (over-reinforced beam).

where
- $f_{\text{ctm,fl}}$ is the mean tensile strength of concrete to flexure; all other symbols may be found in Figure 8.10.

Up to this level, an R/C cross section does not display any cracks, and, therefore, the whole concrete section participates in carrying axial stresses, compressive and tensile. In this context, the curvature of the cross section is given by the well-known expression

$$\varphi = \frac{M_A}{E_c J} = \frac{lV}{E_c bh^3/12} \tag{8.5}$$

where
E_c is the modulus of elasticity of concrete.

As V gradually increases, tensile stresses σ_o approach $f_{\text{ctm,fl}}$; at this stage, when

$$\sigma_o = f_{\text{ctm,fl}} \tag{8.6}$$

concrete in the tension zone cracks. Therefore, for

$$M_{cr} = V_{cr} \cdot l = f_{\text{ctm,fl}} \frac{bh^2}{6}, \tag{8.7}$$

cracks display at the base of the beam vertically to its axis. At this stage, the existing reinforcement in the tensile zone must be in position to substitute the tensile forces of concrete that has cracked; otherwise, *a brittle collapse will take place* (Figure 8.10b).
Therefore,

$$A_{s\,\min} f_{yk} z \geq M_{cr} = f_{\text{ctm,fl}} bh^2/6 \tag{8.8}$$

Taking into account that

$$z \cong 0.87 d = 0.87 \cdot 0.93\, h = 0.81\, h$$

Equation 8.8 takes the form

$$\boxed{\rho_{\min} = \frac{A_{s\,\min}}{bh} = 0.205 \frac{f_{\text{ctm,fl}}}{f_{yk}}} \tag{8.9a}$$

At the same time, the relation between $f_{\text{ctm,fl}}$ and f_{ctm} according to EC2-1-1/2004 is given by the expression

$$f_{\text{ctm,fl}} = \max\ \{(1.6 - h/100)f_{\text{ctm}}; f_{\text{ctm}}\}$$

where
- h is the total member depth in mm.
- f_{ctm} is the mean axial tensile strength of concrete.

For a depth of 350–450 mm, the above relation results in

$$f_{ctm,fl} = (1.25 \div 1.15)f_{ctm}$$

Introducing this value in Equation 8.9a, we obtain

$$\rho_{min} \cong (0.26 \div 0.24)\frac{f_{ctm}}{f_{yк}} \tag{8.9b}$$

It should be noted that the recommended value of EC2-1-1/2004 for ρ_{min} is the same as the above, with a coefficient equal to 0.26.

Thereafter, as V continues to increase above V_{cr}, cracks slowly extend deeper, exhibiting a stable situation. Of course, the slope of the moment–curvature diagram becomes smoother (Figure 8.12) due to the fact that the tensile zone of concrete has cracked, and, therefore, the stiffness of the beam gradually diminishes. This procedure continues until one of the two materials, steel or concrete, fails.

- In the case that the beam is *normally reinforced,* steel yields first, before concrete failure occurs at the compressive zone (Figure 8.10c). As a consequence, for a small additional load, cracks expand, the height of the compression zone decreases, due to the deepening of the cracks, and, finally, the compression zone fails due to crushing. At this stage, the *beam collapses.* However, between steel yielding and concrete crushing in the compression zone, large plastic deformations take place, accompanied by a low-strain 'hardening' of the relevant branch of the moment–curvature diagram (Figure 8.12). Therefore, the available ductility of the beam in terms of moment–curvature diagrams is significant.
- In the case that the beam is over-reinforced before steel yielding (Figure 8.10d), concrete at the compression zone fails due to crushing before the appearance of wide and deep cracks at the tension zone. In fact, since steel has not yielded yet, the width of cracks at the tension zone continues to be small, while the compression zone, if it is not confined, fails in *a brittle mode* without any warning. It is obvious that, in this case, there are not any plastic deformations in terms of curvature (Chapter 2, Figure 2.29).
- The critical point between these two types of failure is the *balanced failure.* In this case, both materials, steel and concrete, fail simultaneously. This means that at this stage, steel yields while concrete exceeds its extreme deformation $\varepsilon_c = \varepsilon_{cu2}$ (Figure 8.11). For example, in the case of unconfined concrete of class ≤C50/60 and steel of class B500c, a cross section under balanced failure displays the deformation pattern of Figure 8.11:

$$x \cong 0.58d \tag{8.10a}$$

In this context, for an orthogonal cross section, which is simply reinforced, the reinforcement percentage ρ_b at balanced failure is equal to

$$\rho_b \cong 0.47\frac{f_{ck}}{f_{yк}} \tag{8.10b}$$

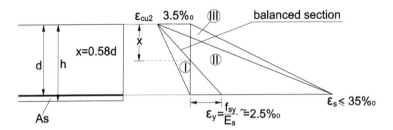

Figure 8.11 Deformation pattern of cross section. Region I: brittle failure mode due to over-reinforcement of the section. Region II: ductile failure mode of normally reinforced sections. Region III: brittle failure mode due to under-reinforcement of the section.

or in case of a cross section reinforced also with reinforcement $\rho_2 = A_{s2}/bd$ in its compression zone, ρ_{max}, in the tension zone takes the form

$$\rho_{max} - \rho_2 = 0.47 \frac{f_{cк}}{f_{yк}} \tag{8.10c}$$

For a ductile type of failure, ρ must be less than ρ_b. *In this case, the deformation pattern of the cross section is located in region II of Figure 8.11.* For ρ greater than ρ_b, a compression brittle failure prevails. In this case, the deformation pattern of the cross section is located in region I of Figure 8.11. Finally, it should be noted that region III of Figure 8.11 corresponds to $\rho < \rho_{min}$, in other words, to tension brittle failure due to inefficient steel reinforcement capable of substituting tensile stresses of concrete at the cracking stage.

8.2.2.3 Determination of the characteristic points of M–φ diagram and ductility in terms of curvature for orthogonal cross section

From the presentation so far, three characteristic points have been defined (Figure 8.12):

- Point A: corresponding to cracking stage (M_{cr}–φ_{cr})
- Point B: corresponding to steel yield stage (M_y–φ_y)
- Point C: corresponding to failure stage (M_u–φ_u)

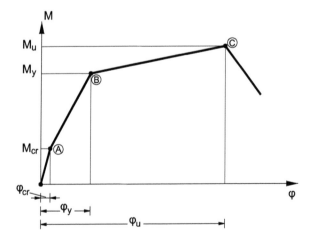

Figure 8.12 M–φ diagram for normally reinforced cross section.

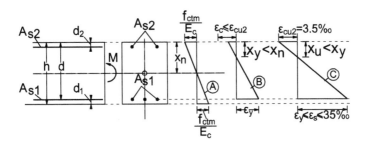

Figure 8.13 Deformation pattern of a cross section: (a) cracking; (b) yield; (c) failure.

The cross-section deformation patterns for these three points are given in Figure 8.13.

In the following paragraph, closed expressions will be derived for orthogonal cross sections using some approximations that simplify the procedure. For cross sections of a complex form (e.g. T, U, Z, Γ, etc.), computer platforms have been developed to allow the determination of points A, B and C (e.g. NOUS (2005), RCCOLA-90, etc.).

1. *Point A: $M_{cr}-\varphi_{cr}$*
   From Equations 8.4 and 8.5, it follows that

$$M_{cr} = f_{ctm,fl}bh^2/6$$ (8.11a)

$$\varphi_{cr} = \frac{M_{cr}}{E_c bh^3/12}$$ (8.11b)

2. *Point B: $M_y-\varphi_y$ (Figure 8.14)*
   Taking into account that the deformation pattern of the cross section for this point of the diagram is depicted in Figure 8.14, it may be concluded that

$$\varphi_y = \frac{\varepsilon_c}{x} = \frac{\varepsilon_{s1y}}{d-x} = \frac{\varepsilon_c + \varepsilon_{s1y}}{d} = \frac{\varepsilon_{s2}}{x-d_2}$$ (8.12a)

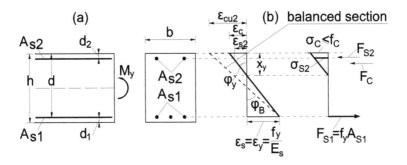

Figure 8.14 Cross section at yield: (a) geometry; (b) deformation pattern; (c) internal forces.

or

$$\boxed{\varphi_y = \frac{\varepsilon_c}{\xi d} = \frac{\varepsilon_{s1}}{d - \xi d} = \frac{\varepsilon_{s2}}{\xi d - d_2} = \frac{\varepsilon_c + \varepsilon_{s1}}{d}}$$

(8.12b)

where

$$\xi = \frac{x}{d}$$

Assuming that the stress distribution in the compression concrete zone is linear, since at this stage usually $\sigma_c < 0.7 f_c$ (Chapter 7.2.2), it can be seen that

$$\varepsilon_c = \frac{\sigma_c}{E_c}, \varepsilon_{s1} = \frac{\sigma_{s1}}{E_s}, \varepsilon_{s2} = \frac{\sigma_{s2}}{E_s}$$

(8.13)

Substituting the above expressions in Equation 8.12b, the following values of σ_{s1} and σ_{s2} are determined:

$$\sigma_{s1} = \frac{1 - \xi}{\xi} n\sigma_c$$

(8.14a)

$$\sigma_{s2} = \frac{\xi d - d_2}{\xi d} n\sigma_c$$

(8.14b)

The equilibrium condition of internal forces of cross section (Fc, $Fs1$ and $Fs2$) gives

$$\boxed{F_c + F_{s2} - F_{s1} = 0}$$

(8.15)

Substituting into Equation 8.15 the values of σ_{s1} and σ_{s2} of Equation 8.14, and taking into account that the distribution of the compressive stresses of concrete has been assumed to be linear, Equation 8.15 takes the following form:

$$\frac{1}{2} b\xi d\sigma_c + \sigma_{s2} A_{s2} - \sigma_{s1} A_{s1} = 0$$

(8.16a)

$$\frac{1}{2} b\xi d\sigma_c + \frac{\xi d - d_2}{\xi d} n A_{s2}\sigma_c - \frac{1 - \xi}{\xi} n A_{s1}\sigma_c = 0$$

(8.16b)

or

$$\xi^2 + 2n(\rho_1 + \rho_2)\xi - 2n\left(\rho_1 + \frac{d_2}{d}\rho_2\right) = 0$$

(8.16c)

where

$$n = \frac{E_s}{E_c}, \rho_1 = \frac{A_{s1}}{bd}, \rho_2 = \frac{A_{s2}}{bd} \tag{8.17}$$

The solution of Equation 8.16c gives the following results:

$$\boxed{\xi_y = \sqrt{n^2(\rho_1 + \rho_2)^2 + 2n\left(\rho_1 + \frac{d_2}{d}\rho_2\right)} - n(\rho_1 + \rho_2)} \tag{8.18}$$

If $\xi_y = (x/d) \geq 0.58$ (for steel class B500c), brittle collapse prevails due to concrete failure under compression (see Section 8.2.2.2) and the procedure ends; otherwise the procedure continues as follows:

If ξ is introduced from Equation 8.18, in Equation 8.12b, the following equation is obtained:

$$\boxed{\varphi_y = \frac{\varepsilon_{s1}}{d(1-\xi_y)} = \frac{f_y/E_s}{d(1-\xi_y)}} \tag{8.19}$$

The corresponding value of M_y is equal to the moments of internal forces F_{s2} and F_{s1} about the centroid of the triangle of compression stresses of concrete:

$$M_y = A_{s1}f_y d\left(1 - \frac{\xi}{3}\right) + A_{s2}\sigma_{s2}d\left(\frac{\xi}{3} - \frac{d_2}{d}\right) \tag{8.20}$$

Introducing in Equation 8.14a the value of f_y in place of σ_{s1} and substituting $n\sigma_c$ into Equation 8.14b from Equation 8.14a, we get the following expression for σ_{s2}:

$$\sigma_{s2} = \frac{(\xi_y - d_2/d)}{(1-\xi_y)}f_y \tag{8.21}$$

Introducing now Equation 8.21 into Equation 8.20, we finally get

$$\boxed{M_y = f_y d\left[A_{s1}\left(1 - \frac{\xi_y}{3}\right) + A_{s2}\frac{(\xi_y - d_2/d)}{(1-\xi_y)}\left(\frac{\xi_y}{3} - \frac{d_2}{d}\right)\right]} \tag{8.22}$$

where ξ_y is taken from Equation 8.18. So, Equations 8.18, 8.19 and 8.22 allow the determination of ξ_y, M_y and φ_y.

3. *Point C: M_u–φ_u (Figure 8.15)*

Taking into account that the strain pattern of the cross section at this point of the diagram M–φ is depicted in Figure 8.15, it may be concluded that

$$\varphi_u = \frac{\varepsilon_{cu2}}{x} = \frac{\varepsilon_{s1u}}{d-x} = \frac{\varepsilon_{cu2} + \varepsilon_{s1u}}{d} = \frac{\varepsilon_{s2}}{x_u - d_2} \tag{8.23a}$$

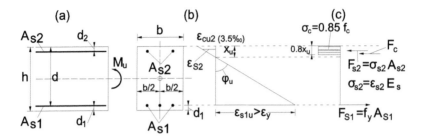

Figure 8.15 Cross section at failure: (a) geometry; (b) deformation pattern; (c) internal forces.

or

$$\varphi_u = \frac{\varepsilon_{cu2}}{\xi_u d} = \frac{\varepsilon_{s1u}}{d - \xi_u d} = \frac{\varepsilon_{s2}}{\xi_u d - d_2} = \frac{\varepsilon_{cu2} + \varepsilon_{s1u}}{d} \qquad (8.23b)$$

where

$$\xi_u = \frac{x_u}{d}$$

In this case (Figure 8.15), the equilibrium condition (Equation 8.15) may be written as follows:

$$0.8 x_u \, b f_c + A_{s2} f_y - A_{s1} f_y = 0 \qquad (8.24)$$

Therefore,

$$x_u = \frac{(A_{s1} - A_{s2}) f_y}{0.8 f_c b} \qquad (8.25)$$

$$\xi_u = \frac{x_u}{d} = \frac{(\rho_1 - \rho_2)}{0.8} \cdot \frac{f_y}{f_c} \qquad (8.26)$$

In Equations 8.24 through 8.26, it has been assumed that σ_{s2} is equal to yield stress, and consequently that

$$\varepsilon_{s2} \geq \frac{\sigma_{s2}}{E_s} = \frac{f_y}{E_s} \qquad (8.27)$$

According to Figure 8.16, this assumption is fulfilled if

$$\varepsilon_{s2} = \varepsilon_{cu2} \frac{x_u - d_2}{x_u} \geq \frac{f_y}{E_s} \qquad (8.28)$$

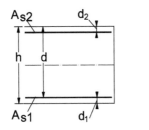
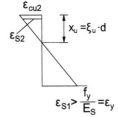

Figure 8.16 Strain pattern for $\varepsilon_{S2} \geq (f_y/E_S) = \varepsilon_y$.

or

$$\overline{\xi}_u \geq \frac{\varepsilon_{cu2} d_2/d}{\varepsilon_{cu2} - f_y/E_s}$$

(8.29)

Introducing the value of $\overline{\xi}_u$ from Equation 8.29 into Equation 8.26, we find

$$\rho_1 - \rho_2 \geq 0.8\overline{\xi}_u \frac{f_c}{f_y}$$

(8.30)

EXAMPLE

For concrete class C20, steel class B500c and $d_2/d = 0.07$ (cross sections with moderate height), according to EC2-1-1/2004, we have

$$\varepsilon_{cu2} = 3.5\text{‰}, f_c = 20 \text{ MPa}, f_y = 500 \text{ MPa}, E_s = 200 \text{ GPa}, \frac{f_y}{E_s} = 2.5\text{‰}$$

Therefore,

$$\overline{\xi}_u = \frac{3.5 \cdot 0.07}{3.5 - 2.5} = 0.245$$

$$\rho_1 - \rho_2 \geq 0.8\overline{\xi}_u \frac{f_c}{f_y} = 0.8 \cdot 0.245 \cdot \frac{20}{500} = 0.008$$

which means that if ρ_1 and ρ_2 differ by less than 8‰, Equation 8.30 gives erroneous results.

In this context, if ξ_u is introduced from Equation 8.26, φ_u results from Equation 8.23b:

$$\varphi_u = \frac{\varepsilon_{cu2}}{\xi_u d}$$

(8.31)

The corresponding value of M_u is equal to the moments of internal forces F_{s2} and F_{s1} about the centroid of the orthogonal of compression stresses of concrete:

$$M_u = A_{s1}f_{y1}(d - 0.4x_u) + A_{s2}f_{y2}(0.4x_u - d_2) \tag{8.32}$$

Introducing Equation 8.26 into Equation 8.32, we get

$$M_u = A_{s1}f_y d\left(1 - \frac{0.4}{0.8}(\rho_1 - \rho_2)\frac{f_y}{f_c}\right) + A_{s2}f_y d\left(\frac{0.4}{0.8}(\rho_1 - \rho_2)\frac{f_y}{f_c} - \frac{d_2}{d}\right) \tag{8.33a}$$

$$M_u = A_{s1}f_y d\left(1 - \frac{(\rho_1 - \rho_2)}{2}\frac{f_y}{f_c}\right) + A_{s2}f_y d\left(\frac{(\rho_1 - \rho_2)}{2}\frac{f_y}{f_c} - \frac{d_2}{d}\right) \tag{8.33b}$$

The above equations are valid under the following assumptions:

- $\sigma-\varepsilon$ diagram of steel is elastoplastic without strain hardening.
- A_{s2} has yielded.
- A_{s2} has not buckled.

4. Ductility μ_φ in terms of curvature

Ductility μ_φ in terms of curvature is determined by the following expression:

$$\boxed{\mu_\varphi = \frac{\varphi_u}{\varphi_y}} \tag{8.34a}$$

By introducing φ_u and φ_y in the above expression from Equations 8.19 and 8.31 and replacing ξ_u from Equation 8.26, we get

$$\boxed{\mu_\varphi = \frac{0.80\varepsilon_{cu2} \cdot f_{ck} \cdot E_s}{(\rho_1 - \rho_2)f_{yk}^2}(1 - \xi_y)} \tag{8.34b}$$

It must be noted again that the above expression is in effect in the case that steel under compression yields.

From the above expression, the following conclusions may be drawn:

- High concrete strength f_c increases μ_φ since it is included in the numerator of Equation 8.34b.
- Confinement also increases μ_φ since in case of confinement, ε_{cu2} is enhanced.
- High-yield strength f_y of steel decreases μ_φ since its square is included in the denominator of Equation 8.34b.
- A high percentage ρ_1 of steel in the tensile zone decreases μ_φ since ρ_1 is included in the denominator. At the same time, ρ_1 contributes to the enhancement of ξ_y (Equation 8.18) and therefore to a decrease in μ_φ.
- Steel in the compressive zone increases μ_φ since its value is subtracted from ρ_1 in the denominator of expression 8.34b.

In the case that $\varepsilon_{s2} < \varepsilon_y = (f_y/E_s)$, μ_φ may be determined from the following expression (Park and Paulay, 1975):

$$\mu_\varphi = 0.8 E_s \varepsilon_{cu2}(1-\xi_y)\left(f_y \left\{ \left[\left(\frac{\rho_2 \varepsilon_{cu2} E_s - \rho_1 f_y}{1.7 f_c} \right)^2 + \frac{d_2}{d} \cdot \frac{\rho_2 \varepsilon_{cu2} E_s}{1.25 f_c} \right]^{1/2} - \left(\frac{\rho_2 \varepsilon_{cu2} E_s - \rho_1 f_y}{1.7 f_c} \right) \right\} \right)^{-1}$$

(8.35)

The ratio M_u/M_y for simply reinforced cross sections averages around 1.05, while for double reinforced ones, this ratio averages a little bit higher (1.07–1.10).

8.2.2.4 Determination of the characteristic points of M–φ diagram and ductility in terms of curvature for a generalised cross section

As mentioned earlier, points A, B and C of M–φ diagrams of Figure 8.12 may be determined for any form of the cross section using proper computer codes. In the following paragraphs, the method that is in the core of the computer program NOUS will be outlined in brief (Penelis, 1969a; EC Tools, 2013). It should be noted that although in the following paragraphs the method is presented for a symmetric cross section, the program can also be used for non-symmetric cross sections either for the design or for the determination of M–φ diagrams (see Chapter 9.2.2.3).

1. *Point A:* M_{cr}–φ_{cr}

    Consider the cross section of Figure 8.17 symmetric to the z–z axis. For the determination of M_{cr} and φ_{cr}, the corresponding deformation pattern is defined by two points, the extreme concrete point at tension on the z–z axis, where the strain is

$$\varepsilon_{ctm} = \frac{f_{ctm}}{E_c}$$

(8.36)

and the centroid of the composite cross section $\left(A_c + n\sum A_{si} \quad \text{where } n = E_s/E_c \right)$ where the strain is zero. In fact, since

$$\sigma_{ct} \leq f_{ctm}$$

(8.37)

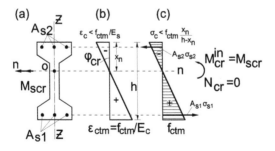

Figure 8.17 Cross section at crack stage: (a) cross section; (b) deformation position at crack stage; (c) corresponding stresses and internal forces.

where
- σ_{ct} is the extreme tension stress of concrete.
- f_{ctm} is the mean tension strength of concrete at flexure.

the cross section is not yet cracked, for this deformation pattern.
In this context, φ_{cr} is equal to

$$\varphi_{cr} = \frac{\varepsilon_{ctm}}{h - x_n} = \frac{f_{ctm}}{(h - x_n)E_c} \tag{8.38}$$

where
x_n is the distance of neutral axis $n-n$ from the external point of concrete in compression on the $z-z$ axis.
At the same time, the moment of all internal forces for this deformation pattern about the centroid of the extreme rebar in tension has the moment M_{cr} as an output at the crack.

2. *Point C:* $M_u-\varphi_u$

Consider the cross section of Figure 8.18 symmetric to the $z-z$ axis. For the determination of M_u and φ_u, the deformation pattern of the cross section must be defined.

This pattern must have an ordinate at the extreme point of concrete under compression equal to ε_{cu2} (Figure 8.18b). Therefore, one point of this straight line is explicitly defined. If this line now intersects the $z-z$ axis at a distance x_n from the extreme point of concrete under compression, then φ_u is determined by the relation

$$\varphi_u = \frac{\varepsilon_{cu2}}{x_n} \tag{8.39}$$

and M_u results from the sum of the moments of all internal forces corresponding to this deformation pattern about the centroid of the extreme steel bars under tension.

Therefore, the crucial point of the whole procedure is the determination of the position of the neutral axis, in other words the value x_n.

The determination of x_n is based on the condition that the sum of the internal forces corresponding to this deformation pattern is equal to zero ($\Sigma N_{in} = 0$), since the cross section is under plain bending.

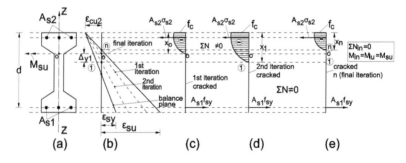

Figure 8.18 Cross section at concrete failure under compression: (a) cross section; (b) deformation pattern at successive iteration cycles 0.1, n; (c) internal forces of iteration (1); (d) internal forces of iteration (2); (e) internal forces of final iteration (n).

It should be noted that the extreme steel strain equal to

$$\varepsilon_{se} = \frac{\varepsilon_{cu2}}{x_n}(d - x_n) \tag{8.40a}$$

must be greater than ε_y, in order for a plastic deformation to exist, that is,

$$\varepsilon_{se} = \frac{\varepsilon_{cu2}}{x_n}(d - x_n) > \varepsilon_{sy} = \frac{f_y}{E_s} \tag{8.40b}$$

Otherwise, the bending member displays a brittle failure before the yielding of steel. In this case, the procedure for determining M_y and φ_y has no meaning.

In any case, the following procedure is followed for the determination of x_n:

Step 1: For the determination of a second point of the deformation pattern, we suppose that at first position of deformation plane intersects the z–z axis at the centroid 0 of the composite uncracked cross section $(A_c + n\Sigma A_{si})$. This is an arbitrary position, which is considered to be the first iteration.

Step 2: For this position ΣN_{io} is determined as the sum of all internal forces corresponding to the assumed deformation pattern. Obviously this sum is not zero since x_o has been selected arbitrarily.

Step 3: For a small displacement Δy_1 of the neutral axis parallel to its initial position, ΣN_{io} becomes equal to

$$\sum N_{io} + \Delta N_{io} \tag{8.41a}$$

If this displacement is the proper one, the above relation must be equal to zero:

$$\sum N_{io} + \Delta N_{io} = 0 \tag{8.42a}$$

The value of ΔN_{io} is approximately a linear relation of Δy_1 and is equal to

$$\Delta N_{io} = F_o \Delta y_1 \tag{8.43a}$$

where
F_o is a function of

- The first arbitrary position of the neutral axis n–n (x_o)
- A series of geometrical parameters of the cracked cross section (concrete-steel) related to x_o
- The characteristic strength and deformation of concrete $(f_{ck}, \varepsilon_{cu2})$ and steel $(f_{yk}, \varepsilon_{yk} = f_{yk}/E_s)$

It should be noted that for relatively small displacements Δy, the values of F_o tend to be independent of Δy (Newton–Raphson approach).

Therefore, Equation 8.42a takes the form

$$\sum N_{io} + F_o \Delta y_1 = 0 \tag{8.44a}$$

or

$$\Delta y_1 = -\frac{\sum N_{io}}{F_o} \tag{8.45a}$$

Step 4: For this new position of the neutral axis ($x_1 = x_o + \Delta y_1$), the new ΣN_{i1} is determined, which continues to be other than zero. Consequently, a new small displacement Δy_2 of the neutral axis $n\text{–}n$ is required so that

$$\sum N_{i1} + \Delta N_{i1} = 0 \tag{8.41b}$$

$$\sum N_{i1} + F_1 \Delta y_2 = 0 \tag{8.43b}$$

or

$$\Delta y_2 = -\frac{\sum N_{i1}}{F_1} \tag{8.45b}$$

Step 5: After a small number of iterations, the neutral axis converges very quickly to its final position, for which

$$\sum N_n \cong 0 \tag{8.46}$$

Step 6: For the final position of the neutral axis ($x = x_n$), the moments of all internal forces, corresponding to the final deformation pattern about the extreme rebar under tension, result in the bending moment M_u at flexure.

The ultimate curvature at section level is given as in paragraph (ii) by the expression

$$\boxed{\varphi_u = \frac{\varepsilon_{cu2}}{x_n}} \tag{8.47}$$

In the case that

$$\boxed{\varepsilon_{su} = \frac{\varepsilon_{cu2}}{x_n}(d - x_n) < \varepsilon_y = \frac{f_y}{E_s}} \tag{8.48}$$

the failure of concrete to compression precedes steel yield at tension zone.

Therefore, the procedure stops here and what is presented in the next paragraph is in this case meaningless for the procedure. Otherwise, the procedure continues as follows:

3. *Point B: $M_y\text{–}\varphi_y$*

Consider the cross section of Figure 8.19 symmetric to the $z\text{–}z$ axis. For the determination of M_y and φ_y, the deformation pattern of the cross section must be defined. This pattern must have an ordinate at the extreme rebar under tension equal to κ_y (Figure 8.19b). Therefore, one point of this straight line is explicitly defined. For the time

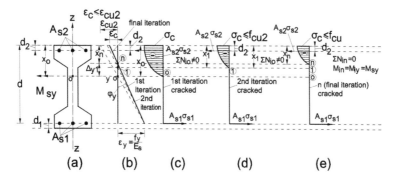

Figure 8.19 Cross section at yield stage (a) cross section; (b) deformation pattern at successive iteration steps 0.1, n; (c) internal forces of step (1); (d) internal forces of step (2); (e) internal forces of final step (n).

being, we consider that this line intersects z–z at a distance x_n from the extreme point of concrete under compression. We recall that this point is located on the neutral axis n–n. The condition for the determination of x_n (the position of the neutral axis n–n) is based again on the requirement that the sum of the internal forces corresponding to this deformation pattern be equal to zero ($\Sigma N_{in} = 0$), since the cross section is under plain bending. If x_n is known, φ_y is determined by the relation

$$\varphi_y = \frac{\varepsilon_y}{d - x_n} \tag{8.49}$$

while M_y results from the sum of the moments of all internal forces corresponding to this deformation pattern about the centroid of the extreme steel rebar under tension. It should be remembered that the case of having a final extreme strain ε_{cextr} of concrete under compression greater than ε_{cu2} has already been excluded by means of the procedure developed in paragraph (ii).

For the determination of x_n, the following procedure is followed:

Step 1: For the determination of the position of the strain pattern, we suppose first that the straight line of its boundary intersects the z–z axis at the centroid 0 of the composite uncracked cross section. This is an arbitrary position of the strain pattern considered a first iteration to the final position. It should be noted that in case the corresponding extreme strain deformation ε_{cext} of concrete under compression exceeds ε_{cu2}, the deformation plane for the first iteration is replaced by the balanced situation ($\varepsilon_{cu2}, \varepsilon_{sy}$).

Steps 2, 3, 4, 5: The same procedure is followed as that of the corresponding steps in paragraph (ii).

Step 6: For this final position of x_n of the neutral axis, values M_y and φ_y are determined according to what has been presented at the beginning of this paragraph.

8.2.3 Load–deformation diagrams for bending under cyclic loading

8.2.3.1 General

As already noted (Chapter 2.3.3), the notion of ductility capacity should be accompanied by that of energy dissipation, related to moment–curvature or force–displacement loops under

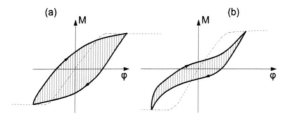

Figure 8.20 M–φ diagrams: (a) spindle-shaped loops when flexure prevails; (b) pinched loops when shear prevails.

cyclic loading. In fact, it has been explained in detail in Chapter 2.3.3 that the capacity of a member to dissipate energy under cyclic loading is expressed by the area of the loops of cyclic loading. The moment–curvature or force–displacement diagram of monotonic loading is assumed to be the envelope of cyclic loading, and this has to be proven by experimental evidence. Figure 8.20 makes apparent the above considerations. Although both diagrams have as a backbone the same M–φ curve, the diagram in Figure 8.20a with spindle-shaped loops displays a much higher capacity for energy dissipation than the diagram of Figure 8.20b with pinched loops. Therefore, it is of paramount importance to comment on moment–curvature or force–deflection diagrams of beams under cyclic loading.

From a review of extensive experimental data, it may be seen that the behaviour of critical regions of beams under cyclic loading should be classified as follows:

- Critical regions where inelastic behaviour is controlled by bending (flexural critical regions)
- Critical regions where inelastic behaviour is controlled by high shear while bending is of secondary importance

In the next paragraph, flexural behaviour under cyclic loading will be examined, while in Section 8.2.4, behaviour under prevailing shear will be examined.

8.2.3.2 Flexural behaviour of beams under cyclic loading

From the abundant experimental data, two characteristic cases will be examined below:

1. The case of a cantilever with an orthogonal cross section 406 × 203 mm reinforced with a reinforcement percentage $\rho_1 = 1.4\%$ and $\rho_2 = 0.74\%$ and a shear span $M/Vd = 4.5$. Nominal shear stress is rather low, $\tau_n = 0.26\sqrt{f_c}$ (Figure 8.21; Bertero and Popov, 1977; Penelis and Kappos, 1997).
2. The case of a cantilever beam T cross section with dimensions also depicted in Figure 8.21, with top reinforcement $\rho_1 = 1.4\%$ while bottom reinforcement $\rho_2 = 0.74\%$. In both cases, the bottom flange reinforcement is 50% of the reinforcement at the top, which is in accordance with the Seismic Code provisions for the reinforcement of critical regions of T and orthogonal beams. Shear span M/Vd continues to be equal to 4.5, while the nominal shear stress is the same as that of case 1.

From detailed examination of the above diagrams, the following remarks may be made:

1. Loops of flexural behaviour are *spindle-shaped* and, therefore, absorb and dissipate adequate amounts of seismic energy.
2. Cyclic loading *is stable,* having as its envelope the corresponding M–φ or P–δ curve of monotonic loading.

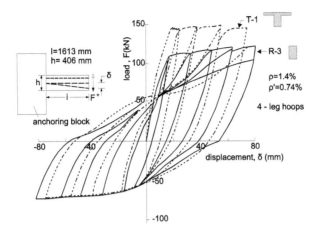

Figure 8.21 Hysteresis loops of R/C members subjected to predominantly flexural cyclic loading.

3. No significant stiffness degradation occurs at service levels, in contrast to the behaviour at post-elastic levels, where stiffness is reduced instantaneously after each reversal of moment in which the peak curvature is increased beyond its previous value. This may be mainly attributed to the Bauschinger effect, which influences the steel stress–strain diagram under reversals of loading (see Chapter 7.3.4).

4. Repeated reversals up to the same post-elastic level of deformation show a remarkable strength stiffness and energy dissipation stability.

5. Finally, it should be noted that failure is usually due to buckling of the main reinforcement under compression (Figure 8.22). Factors controlling buckling are

- Concrete cover
- The spacing size and detailing of hoops
- Strain history of the rebars that may cause rupture of one or more of them

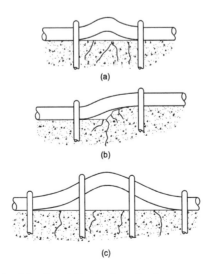

Figure 8.22 Different modes of buckling in reinforcing bars (a–c).

8.2.4 Strength and deformation of beams under prevailing shear

8.2.4.1 Static loading

Dimensioning of R/C beams against shear under-static loading is one of the main issues in the design of R/C conventional structures. Therefore, a detailed presentation of this subject is beyond the scope of this work. Here, an overview of the main issues of the problem will be made so that a subsequently easy transition to cyclic loading may be made. Failure under shear of a reinforced beam with *longitudinal rebars and stirrups* in the web may occur by *diagonal tension* (Figure 8.23a; diagonal concrete cracking accompanied by stirrup yield) or *by diagonal compression* (diagonal concrete crush; Figure 8.23b).

8.2.4.1.1 Failure caused by diagonal tension

This type of failure is resisted by two mechanisms, that is,

- Beam action in the shear span without web reinforcement
- Truss mechanism of web reinforcement (stirrups, bent-up rebars)

8.2.4.1.1.1 BEAM ACTION IN THE SHEAR SPAN WITHOUT WEB REINFORCEMENT

This type of failure is governed mainly by concrete tensile strength and the existing tensile flexural reinforcement in the shear span (Figure 8.24).

Extended experimental evidence (Leonhardt and Walther, 1962; Leonhardt and Mönnig, 1973) has shown that shear failure in the form of diagonal cracking exhibits before failure to flexure in the case that the shear span-to-depth ratio

$$\boxed{\frac{a}{d} = \frac{M_{Eu}}{V_E d}}$$

(8.50)

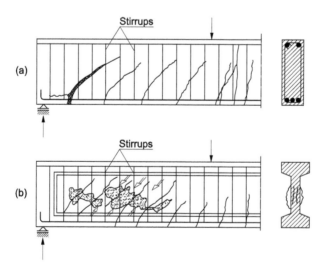

Figure 8.23 Failure of an R/C beam to shear: (a) failure due to diagonal tension; (b) failure due to crushing of the compressive diagonal struts.

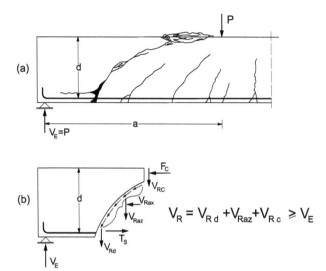

Figure 8.24 Shear failure of a beam without shear reinforcement: (a) failure mode; (b) factors contributing to shear resistance.

ranges between the following limits (Figure 8.25):

$$2.0 \leq \frac{a}{d} \leq 7.0$$

(8.51)

The characteristic picture of this failure mode displays diagonal cracks that start up from existing already flexural ones at the shear span. These inclined cracks develop rapidly up to the compression zone, leading to a *brittle* failure.

- Shear strength of compressive zone: V_{Rc}
- Aggregate interlock: V_{Ra}
- Dowel effect: V_{Rd}

Therefore, in order to avoid this type of failure, minimum web reinforcement is foreseen by Codes, even if the dimensioning of the beam to shear does not require it.

More specifically, parameters that contribute to shear strength in this case are (Figure 8.24b)

- The shear strength of the compression zone
- Aggregate interlock
- Dowel effect of tensile longitudinal reinforcement

Shear strength of the uncracked compression zone depends on concrete strength. On the other hand, aggregate interlock and the dowel effect are strongly influenced by the percentage of reinforcement in the tension zone of the beam, since the width of the diagonal cracks is strongly affected by the tensile stress of longitudinal bars and, therefore, by its percentage. Based on the above remarks, it can easily be concluded that the shear strength of a beam without web reinforcement is determined mainly by

1. The tensile strength of concrete
2. The percentage of tensile reinforcement in the tension zone of the shear span

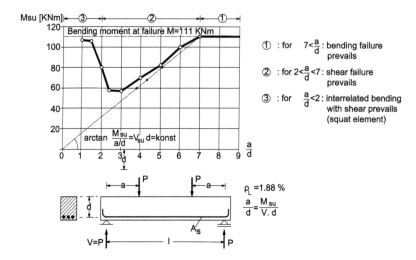

Figure 8.25 The influence of the ratio *a/d* on the shear strength of beams without shear reinforcement. Results of Stuttgart test. Beam with constant cross section and reinforcement ratio at bottom flange $\rho_L = 1.88\%$ $V_{su} = (M_{su}/a)$. (Adapted from Leonhardt, F. and Walther, R. 1962. *Schubversuche an einfeldrigen Stahlbetonbalken mit und ohne Schubbewehrung*. DA für Stahlbeton Heft 151, W. Ernst und Sohn, Berlin. With kind permission from Springer Science+Business Media: Leonhardt, F. 1973. *Vorlesungen uber Massivbau, I. teil.*)

Eurocode EC2-1/1991 ENV (CEN, 1992) had quantified shear strength in this case by the expression

$$V_{Rd,c} = \tau_{Rd} \cdot \kappa \cdot (1.2 + 40\rho_1)b_w \cdot d \tag{8.52}$$

where
- b_w is the smallest web width.
- d is the effective height of the cross section.
- κ is $1.6 - d \not< 1.0$ (in meters) expressing the size effect of the cross section.
- ρ_1 is the longitudinal reinforcement ratio at shear span.
- $\tau_{Rd} = (f_{ctk}/\gamma_d)\xi$ is the reduced tensile strength of concrete to the height x of the compression zone.
- $\xi = x/d$ (estimated to $\xi \cong 0.25$).

The influence of the above-mentioned two parameters (concrete tensile strength and percentage of tensile reinforcement at the shear span), together with the size effect of the beams, is displayed with scientific transparency in Equation 8.52. At the same time, this approach complies with the modifications necessary for seismic action effects as we will see in the next paragraph. A similar expression has been adopted for many years by the American Codes (ACI 318-2005, 2008, 2011) as well as by the Codes of many other countries.

Eurocode EC2-1-1/2004 has adopted the following expression for the determination of V_{Rdc}:

$$V_{Rd,c} = C_{Rd,c} \cdot \kappa \cdot (100\rho_1 f_{cκ})^{1/3} b_w \cdot d \tag{8.53}$$

This value should not be *less than* ($V_{Rd,cmin}$):

$$\boxed{V_{Rd,c\,min} = 0.035\kappa^{3/2}f_{c\kappa}^{1/2} \cdot b_w d}$$ (8.54)

where
- $C_{Rd,c}$ is the coefficient derived from tests (recommended value 0.12).
- κ is the size effect factor $\kappa = 1 + \sqrt{\dfrac{200}{d}}$ with d in mm.
- ρ_l is the longitudinal reinforcement ratio in tension zone of the shear span (≤ 0.02).
- $f_{c\kappa}$ is the characteristic concrete compression strength (MPa).
- b_w is the smallest web width (mm).
- d is the effective height of the cross section (mm).
- $V_{Rd,c}$ (N).

8.2.4.1.1.2 TRUSS MECHANISM OF WEB REINFORCEMENT. ALTERNATIVE (A)

Web reinforcement (stirrups, bent-up rebars) generates for the beam an additional resisting mechanism to shear. This mechanism, according to the well-known Mörsh truss analogy, consists of (Figure 8.26)

- A chord of longitudinal reinforcement under tension at the bottom of the beam.
- A chord of the compressive stresses of concrete and of the compressive forces of reinforcement under compression concentrated at the centroid of the compressive zone.
- Stirrups and bent-up rebars in the web acting as tension members.
- Inclined concrete struts under compression between the successive inclined shear cracks, which are developed after V_{Ed} has exceeded $V_{Rd,c}$. For almost one century and even nowadays, according to the ACI318-2011, the inclination of these struts was considered to be 45°.

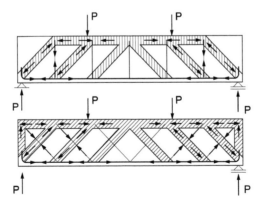

Figure 8.26 Mörsh truss mechanism: (a) stirrups as shear reinforcement; (b) bent-up bars as shear reinforcement.

In this context, shear force V_{wd} resisted by web reinforcement is equal to

$$V_{wd} = \frac{A_{sw}}{s} 0.9d \cdot f_{ywd} \qquad (8.55)$$

where
- A_{sw} is the cross-sectional area of web reinforcement.
- s is the spacing of the stirrups.
- f_{ywd} is the design yield strength of the shear reinforcement.
- $0.9d$ is a good approximation for the lever arm z of the internal forces of the beam in bending.

The same procedure was followed in EC2-1-1/1991 ENV.
It should be noted here that

1. According to what has been presented so far, a beam with web reinforcement in the form of stirrups comprises two shear-resisting mechanisms to shear failure caused by diagonal tension. At ultimate state, both mechanisms must be exhausted. Therefore, the shear strength of the beam to diagonal tension is equal to

$$V_{Rd} = V_{Rd,c} + V_{Wd} \qquad (8.56)$$

2. It is equally important to note that while the beam mechanism without stirrups is *brittle*, the arrangement of web reinforcement stabilises the inclined cracks and their width, until stirrups yield. Therefore, web reinforcement together with the enhancement of the shear strength of the beam improves radically its ductility.
3. Beams should be reinforced at least with a minimum web reinforcement in the form of stirrups even if

$$V_{Ed} \leq V_{Rd,c\,min} \qquad (8.57)$$

The minimum web reinforcement ratio that is specified by Codes

$$\rho_w = \frac{A_{sw}}{b_w s} \qquad (8.58)$$

covers the concrete (unreinforced) shear resistance $V_{Rd,c}$.

8.2.4.1.1.3 TRUSS MECHANISM OF WEB REINFORCEMENT – ALTERNATIVE (B)

A different procedure was followed by EC2-1-1/2004 for failure caused by diagonal tension. In case V_{Ed} exceeds the value of $V_{Rd,c}$, given by Equations 8.53 and 8.54, the *truss model analogy with concrete struts of variable inclination is applied* (Figure 8.27).

Approach "Variable inclination struts":

Stage 1: web uncracked in shear
Stage 2: inclined cracks occur
Stage 3: stabilised inclined cracks
Stage 4: yielding of stirrups,
            further rotation, finally
            web crushing

Strut rotation as measured in tests
(TU Delft)

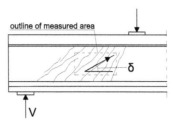

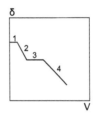

Figure 8.27 The principle of variable truss action. (Adapted from Walraven, J.C. 2002. Delft, University of Technology, Background document for EN1992-1-1. Eurocode 2: Design of Concrete Structures-Chapter 6.2: Shear, Delft.)

The only difference of this model from the 'Mörsh truss model' is that the inclination of the cracks and therefore of the compressed concrete struts is considered to be of variable inclination δ, varying between

$$21.8° \leq \delta \leq 45°$$ (8.59a)

or

$$2.5 \geq \cot \delta \geq 1.0$$ (8.59b)

In this context, shear force $V_{\omega d}$ resisted by web reinforcement is equal to

$$V_{wd} = V_{Rd} = \frac{A_{sw}}{s} 0.9 df_{ywd} \cot \delta$$ (8.60)

The actual value of δ that should be introduced in Equation 8.60 is defined by the strut inclination for which shear failure due to diagonal tension and diagonal compression occur simultaneously (see next paragraph).

It should be noted here that in contrast to the ENV edition of EC2-1-1/1991 and to ACI318-2011 in the above expression (Equation 8.60), both mechanisms of Equations 8.52 and 8.55 are incorporated into one, characterised *the variable inclination strut*.

The above mechanism is based on extended experimental evidence (Walraven, 2002), according to which shear failure exhibits the following stages:

*   For $V_{Ed} \leq V_{Rd,c}$, the web remains uncracked to shear.
*   For $V_{Rd,c} \leq V_{Ed}$, inclined cracks occur at the beginning at an angle in relation to the beam axis at about 45°.

- These cracks remain stable while V_{Ed} increases and while stirrups begin to develop tensile stresses below yielding stress.
- Shear failure occurs when stirrups yield, followed by strut rotation, until finally the web crushes (Figures 8.23b and 8.27).

8.2.4.1.2 Failure caused by diagonal compression

It is apparent that this type of failure does not refer to the beam action without web reinforcement, since in this case tensile diagonal failure prevails due to the very low tensile strength of concrete. Instead, this type of failure may be critical for beams with web reinforcement.

For the determination of shear resistance in the case of *diagonal concrete crushing*, the Mörsh truss model already presented above is used.

According to this model (Figure 8.28a,b),

$$\boxed{V_{Rd,max} = \alpha_{cw}b_w 0.9 \cdot d(v_1 f_{cd})/2}$$
(8.61)

for $\delta = 45°$ (ACI318-2011), or

$$\boxed{V_{Rd,max} = \alpha_{cw}b_w 0.9d(v_1 f_{cd}) \cdot \cot\delta/(1+\cot^2\delta)}$$
(8.62)

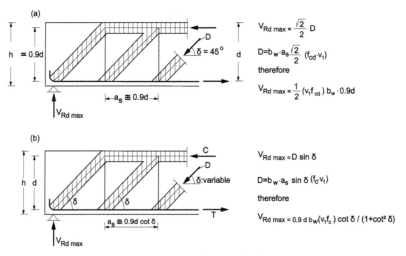

Figure 8.28 Determination of the shear resistance $V_{Rd\ max}$: (a) diagonal compression failure at $\delta = 45°$; (b) variable diagonal compression failure at $21.8° \le \delta \le 45°$. (Adapted from Penelis et al. 1995; Walraven, J.C. 2002. Delft, University of Technology, Background document for EN1992-1-1. Eurocode 2: Design of Concrete Structures-Chapter 6.2: Shear, Delft.)

for δ variable (21.8°≤ δ ≤ 45°) (EC2-1-1/2004),
where
- δ is a variable ranging between (21.8° ≤ δ ≤ 45°) (EC2-1-1/2004).
- v_1 is a factor for concrete compressive strength reduction of concrete already cracked by shear:

$$v_1 = 0.6\left[1 - \frac{f_{cк}}{250}\right] \tag{8.63}$$

($f_{cк}$ in MPa)
- α_{cw} is a coefficient equal to 1.0 for beams without axial loading.

8.2.4.1.3 Design procedure for shear under static loading according to codes

Based on the above presentation, an overview of the design procedure for shear under static loading according to Codes in use would be useful.

- *Step 1*: Determine the span-to-depth ratio:

$$\frac{a}{d} = \frac{M}{V \cdot d}$$

if $(a/d) \geq 2.0$, the following procedure will be followed:
- *Step 2*: Shear resistance $V_{Rd,c}$ of the beam without web reinforcement is determined using Equation 8.52 (ENV EC2-1/1991) or using Equations 8.53 and 8.54 (EC2-1-1/2004). If

$$V_{Ed} \leq V_{Rd,c}$$

web reinforcement is not required. However, *minimum web reinforcement* in the form of stirrups must be foreseen to ensure the beam against a potential brittle shear failure.
- *Step 3*: In the case that

$$V_{Ed} \geq V_{Rd,c}$$

two different procedures are foreseen, depending on the Code under consideration:
- In the case that ACI 318-2011 is in effect, design shear for V_{Ed} must be less than the sum of the shear contribution of the concrete $V_{Rd,c}$ and the shear contribution of the web reinforcement V_{wd}, under the assumption that the concrete struts exhibit a constant indication δ = 45°:

$$\boxed{V_{Ed} \leq V_{Rd,c} + V_{wd}} \tag{8.64}$$

This expression allows the determination of the required web shear reinforcement. At the same time, the concrete struts under compression must have a sufficient safety factor from crushing. Therefore, V_{Ed} must also satisfy the following expression:

$$\boxed{V_{Ed} \le V_{Rd,max} \quad \text{for } \delta = 45°} \tag{8.65}$$

Usually, Equation 8.65 is satisfied without any difficulty unless webs are very thin. In this case, web width or the nominal depth of the cross section must be increased.

- In the case that EC2-1-1/2004 is in effect, design shear force must be covered by the *truss model analogy* with concrete struts of *variable inclination* simultaneously for both the web reinforcement yield and concrete inclined struts trussing:

$$\boxed{V_{Ed} \le V_{Rd} = \frac{A_{sw}}{s} 0.9d \cdot f_{ywd} \cot \delta} \tag{8.66a}$$

and

$$\boxed{V_{Ed} \le V_{Rd,max} = \alpha_{cw} b_w 0.9d(v_1 f_{cd}) \cdot \cot \delta/(1 + \cot^2 \delta)} \tag{8.66b}$$

with $21.8° \le \delta \le 45°$.

The accomplishment of both expressions 8.66a and 8.66b is satisfied by successive trials for various values of δ.

- *Step* 4: In the case that the shear span-to-depth ratio (shear-span ratio) is less than or equal to 2.0:

$$\frac{a}{d} \leqq 2.0$$

a special procedure must be followed, since in this case the Bernoulli concept for a plane strain pattern over the cross section of the member does not prevail, and therefore a two-dimensional stress state prevails where strong relation among shear and normal stresses exists. This case refers mainly to spandrels between shear walls, to joints of frames and to low shear walls. Therefore, detailed reference will be made to this issue in relevant chapters (see Section 8.3.6 and Chapter 9.4).

8.2.4.2 Cyclic loading

It has been noted many times thus far that beams of ductile R/C frames display potential plastic hinges at their ends. During a strong seismic action, these plastic hinges yield first in one direction and then in the other, as the frames sway to the right and left successively. Under these conditions of cyclic loading of combined bending and shear, if shear prevails

$(2 \leq (M/V \cdot d) \leq 7.0)$, diagonal cracks progressively develop crosswise in the region of the plastic hinge. These cracks widen from cycle to cycle, since plastic strains of flexural and of shear reinforcement are accumulated (Park and Paulay, 1975; Bertero, 1979; Paulay and Bull, 1979; Scribner and Wight, 1980; Paulay and Priestley, 1992; Penelis and Kappos, 1997; Booth and Key, 2006; Fardis, 2009). This tends to eliminate the aggregate interlock, the dowel effect and the shear strength of the compressive zone, which are the basic parameters that contribute to the beam shear action together with the truss mechanism of web reinforcement. For this reason, Codes of Practice specify that the above contribution to shear (beam shear strength) should be eliminated unless the ductility demand of the building is of a limited level. In this context, shear action effect should be resisted *only by the web reinforcement* determined by the 'Mörsh' truss analogy mechanism.

Furthermore, the widening of diagonal cracks in the hinge plastic region leads to a 'pitching effect' on the V–δ diagram. In fact, for a diagonal strut under compression, which has been cracked transversally in the previous step to be activated, the transverse cracks must close first. This leads to a situation where there is very little resistance to shear, and, therefore, stiffness around the midpoint of the loading cycle diminishes gradually from cycle to cycle (Figure 8.29). In this context, the capacity of the beam for energy absorption and dissipation in the region of plastic hinge is radically reduced in the case of diagonal shear cracks.

In closing, it should be noted that the cyclic widening and closing of diagonal cracks, in combination with the corresponding widening and closing of the flexural cracks, leads to the degradation of the plastic hinge region and finally to failure in a mode of *sliding shear* (Figure 8.30).

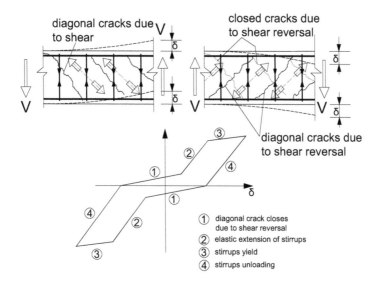

Figure 8.29 Shear deformation in reversing hinge zones: (a) crack pattern; (b) deformation of truss; (c) shear versus deformation in reversing hinge. (Adapted from Booth, E. and Key, D. 2006. *Earthquake Design Practice for Buildings.* Thomas Telford Ltd. With permission.)

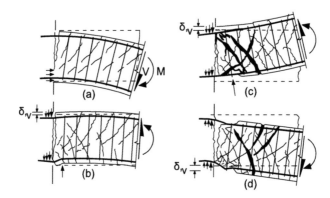

Figure 8.30 Characteristic phases of the response of an R/C beam to reverse cyclic loading: (a) loading downwards; (b) loading upwards up to original position; (c) continuation of the upward loading; (d) reverse loading downwards accompanied with sliding shear.

8.2.4.3 Concluding remarks on shear resistance

From what has been presented so far regarding the behaviour of plastic joints in response to shear, the following remarks may be made:

1. The design of the beams for seismic shear effects may be carried out basically like that of conventional static loading.
2. In particular, additional special measures should be taken at the regions of plastic hinges (critical regions). These measures may be summarised as follows:
    a. Enhancement of shear effects based on the capacity design concept (Chapter 6.1.3), so that yield of shear reinforcement at the plastic hinges does not develop during seismic load reversals
    b. Elimination of the contribution of the beam resistance to shear in case of high-ductility demand
    c. Arrangement of diagonal reinforcement in the region of plastic hinges in the case that the design shear effects present reversed sign, so that sliding is avoided
3. In case that shear reinforcement yields, adverse consequences are evident at the region of a plastic hinge, the most significant of which are the following:
    a. Quick strength degradation of the hinge region
    b. Pinching effects on the P–δ diagram
    c. Shear sliding failure at a position where wide flexural cracks have been developed
4. It is the author's opinion that the design method for shear of the variable compression strut inclination model, adopted by EC2-1-1/2004, presents some weak points in comparison to that of the constant inclination model adopted by ACI 318-2011.
    a. There is inadequate transparency of the mechanical models behind the design method of EC2-1-1/2004.
    b. In fact, for values of $V_{Ed} \leq V_{Rd,c}$, a beam model is used, where the two factors (concrete shear strength and flexural reinforcement percentage) are not clearly distinct.
    c. Furthermore, in the case that $V_{Ed} \geq V_{Rd,c}$, the beam model is completely abandoned and it is incorporated in the truss model of variable angle strut, without an apparent engineering concept behind it, except the 'best-fitting' statistical concept of extensive experimental results.

d. Last but not least is the case of buildings of high-ductility demand. A return to the truss model of constant angle (45°) in order to eliminate the beam mechanism contribution appears to be unjustified from the engineering point of view.

5. For joints, spandrels and deep beams, a special design procedure should be followed for both bending and shear (see Section 8.4 and Chapter 9.4), since bending and shear can no longer be considered two independent strain states.

8.2.5 Code provisions for beams under prevailing seismic action

8.2.5.1 General

The design rules of beams under seismic actions according to EC8-1/2004 are given below, together with the required justification in terms of the theory developed in the previous subsections. It should be noted that the design specifications are directly related to the ductility demand under consideration, DCL, DCM and DCH. For higher ductility class, specifications tend to be stricter in order for a higher local ductility in terms of curvature to be obtained.

Specifications for the design effects, taking into account capacity design procedure, have already been presented in Chapter 6.1.3.

Material issues have also been presented in Chapter 7, Sections 7.2.4, 7.3.5 and 7.5.4 for all ductility classes.

It should be mentioned here that for the design of beams of buildings of class DCL, Code specifications for conventional R/C structures (EC2-1-1/2004) are adopted, except where concrete and steel reinforcement qualities have been specified by EC8-1/2004, as has already been presented in Chapter 7. For this reason, design of beams for DCL buildings will not further occupy our attention in this book. Therefore, in the following paragraphs, only the design of beams for DCM and DCH buildings will be presented.

In closing, it should be noted that all rules and specifications of the corresponding Code for conventional R/C structures (EC2-1-1/2004) continue to be in effect, unless they contradict what will be presented below.

8.2.5.2 Design of beams for DCM buildings

1. Geometrical constraints
   a. The eccentricity of a beam axis relative to that of the column with which it is connected should be limited to $b_c/4$, where b_c is the largest cross-sectional dimension of the column normal to the longitudinal axis of the beam (Figure 8.31):

$$e \leq b_c/4 \tag{8.67}$$

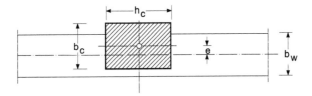

Figure 8.31 Eccentric arranged beam in relation to the supporting column.

In this way, cyclic moments can be transferred safely from the columns to the beams.
  b. The beam width b_w should be (Figure 8.31)

$$b_w \leq \min (b_c + h_w, 2b_c) \tag{8.68}$$

where h_w is the depth of the beam.

   In this way, a large percentage of the horizontal bars of the beam pass at the joints through the column and, therefore, take advantage in their bond of the favourable effect of column compression.

 2. Resistance in bending and shear
   a. Bending and shear resistance of beams for DCM are determined in accordance with the Code in effect for conventional R/C structures, i.e. EC2-1-1/2004. It should also be remembered that the design for shear is not carried out for the design shear effects that result from the analysis but for those from the capacity design procedure (Chapter 6.1.3).
   b. The top reinforcement of the end cross sections with T- or Γ-shaped sections should be placed mainly within the width of the web, and only a part of these rebars might be placed outside the web but within the effective flange width b_{eff}. The effective flange width is depicted in Figure 8.32 for various types of column–beam joints.

 3. Detailing for local ductility
   a. The *critical regions* of the beam are defined as the regions of a length $l_{cr} = h_w$ (where h_w is the depth of the beam) measured from the two ends of the beam to its span. As already explained, these are regions of potential yield of steel reinforcement and, therefore, the positions of potential plastic hinges (Figure 8.33).

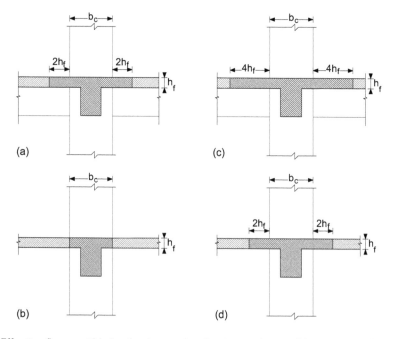

Figure 8.32 Effective flange width b_{eff} for beams framing into columns: (a) exterior column with transverse beam; (b) exterior column without transverse beam; (c) interior column with transverse beam; (d) interior column without transverse beam. (Adapted from E.C.8-I/EN1998-I. 2004. *Design of Structures for Earthquake Resistance: General Rules, Seismic Actions and Rules for Buildings.* CEN, Brussels, Belgium.)

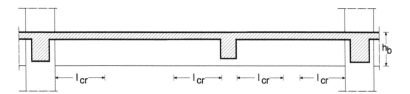

Figure 8.33 Critical regions of beams.

b. In beams supporting discontinued (cut-off) vertical elements (Figure 8.33), the regions up to a distance of $2h_w$ on each side of the supported vertical element should also be considered as being critical regions.

c. The value of the curvature ductility demand μ_φ at the critical regions must satisfy Equations 5.52a and 5.52b, that is,

$$\boxed{\mu_\varphi = 2q_o - 1 \qquad\qquad \text{if}\quad T_1 \geq T_c} \tag{5.52a}$$

$$\boxed{\mu_\varphi = 1 + 2(q_o - 1)\frac{T_c}{T_1} \qquad \text{if}\quad T_1 < T_c} \tag{5.52b}$$

This is deemed to be satisfied if the following conditions are met at both flanges of the beam:

1. At the compression zone, reinforcement equal or greater than half of the reinforcement provided at the tension zone ρ_1 should be placed, in addition to any compression reinforcement computed for the ULS design.
2. The reinforcement ratio ρ_1 in the tension zone should not exceed a value of

$$\boxed{\rho_{1\max} \leq \rho_2 + \frac{0.0018}{\mu_\varphi \cdot \varepsilon_{sy,d}}\frac{f_{cd}}{f_{yd}}} \tag{8.69}$$

where (Figure 8.34)

$$\rho_1 = \frac{A_{s1}}{bd} \tag{8.70a}$$

$$\rho_2 = \frac{A_{s2}}{bd} \tag{8.70b}$$

The above expression results from Equation 8.34b. In fact, if in Equation 8.34b f_{ck}, f_{yk} are replaced by f_{cd}, f_{yd}, and
a. ε_{cu2} is replaced by the conventional value of 0.0035 corresponding to concrete classes C16–C50.

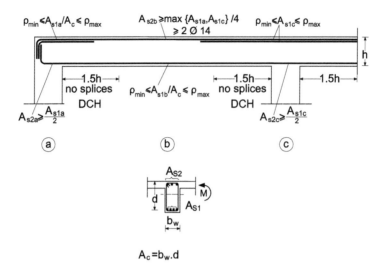

Figure 8.34 Arrangement of longitudinal reinforcement in earthquake-resistant R/C beams.

b. ξ_y is replaced by its mean value equal to 0.34, resulting from Equation 8.18. Equation 8.34b takes the form

$$\rho_1 - \rho_2 = \frac{0.8 \cdot 0.0035 \cdot f_{cd}}{\mu_\varphi \cdot f_{yd} \cdot f_{yd}/E_s} 0.66$$

or

$$\rho_{1max} \leq \rho_2 + \frac{0.0018}{\mu_\varphi \cdot \varepsilon_{sy,d}} \frac{f_{cd}}{f_{yd}} \qquad (8.69)$$

Keeping in mind that $\mu_{\varphi demand}$ values resulting from Equation 5.52a or 5.52b exhibit a mean safety factor for beams on the order of 1.35 (see Chapter 5.4.4.5) and that the ratio

$$\frac{f_{cd}}{f_{yd} \cdot \varepsilon_{sy,d}} = \frac{f_{ck} \cdot 1.15^2}{1.5 \cdot f_{yк} \cdot \varepsilon_{sy}} = \frac{1}{1.14} \frac{f_{cк}}{f_{yк} \cdot \varepsilon_{sy}}$$

it may easily be concluded that a safety factor on the order of $1.35 \cdot 1.14 \cong 1.54$ is introduced in terms of curvature ductility by implementing expression 8.69 (see Chapter 3.2) for determination of ρ_{1max}.

Equation 8.69 is very restrictive for the top reinforcement ρ_1 at beam supports, particularly for DCH buildings for which high values of $\mu_{\varphi demand}$ result (on the order of 10–12). Therefore, the only way to satisfy both requirements
i. Strength verification

$$\rho_{1d} \cong \frac{M_{Ed}}{0.9d^2 b f_{yd}} \qquad (8.71)$$

ii. Ductility requirement

$$\rho_{1max} \le \rho_2 + \frac{0.0018}{\mu_{\varphi\,dem} \cdot \varepsilon_{sy,d}} \frac{f_{cd}}{f_{yd}}, \tag{8.72}$$

is either to change the cross section of the member or to increase the reinforcement percentage ρ_2 at the compression zone, which is the easy way, since the change of cross sections of the members has consequences for the analysis (change of member stiffness) and mainly for the building operation.

Finally, it should be noted that in the case of use of reinforcement steel class B, the curvature ductility factor μ_φ in expression 8.69 should be magnified by a factor 1.5. This requirement results in the enhancement of steel reinforcement ρ_2 under compression, and in this respect in the dissuasion of the designer from using class B steel reinforcement instead of class C.

In Table 8.1, the maximum reinforcement ratio ρ_{1max} (‰) may be found for various concrete classes, for steel reinforcement class B500c, for $\rho_2 = (1/2)\,\rho_1$ and for ductility classes DCM and DCH (Ignatakis, 2011).

3. The reinforcement ratio of the tension zone at any point along the beam must not be less than the following minimum value ρ_{min}:

$$\boxed{\rho_{min} = 0.50 \left(\frac{f_{ctm}}{f_{y\kappa}} \right)} \tag{8.73}$$

so that pre-emptive brittle failure due to concrete tension cracks is avoided.

This quantity is almost twice as high as that recommended by EC2-1-1/2004 for conventional R/C structures (see Equation 8.9). This discrepancy must be attributed to the fact that Equation 8.73, although based on the same concept as Equation 8.9, refers to T-shaped beams, which are the usual case in R/C buildings. In fact, the part of the web under tension of a T beam (in uncracked stage) is much larger – almost double – than that of the orthogonal section where the tension zone is limited to the mid-height of the web. The output of Equation 8.73 is much closer to those of the relevant requirements of the American Code ACI 318-2011. In any case, it is the author's opinion that in the next revision of Eurocodes, this discrepancy should be clarified.

Finally, it should be noted that in Equation 8.73, a safety factor of at least 1.15 is included, since steel overstrength for steel class C is $f_{ut}/f_{y\kappa} \ge 1.15$.

In Table 8.2, the minimum reinforcement ratio ρ_{min} (‰) may be found for various concrete classes and for steel reinforcement class B500c (Ignatakis, 2011).

4. Within the critical regions, hoops satisfying the following conditions should be provided:
   a. The diameter d_{bw} of the hoops should not be less than 6 mm.
   b. The spacing, s, of hoops (in mm) should not exceed

$$\boxed{s = \min\,(h_w/4; 24d_{bw}; 225\ \text{mm}; 8d_{bL})} \tag{8.74}$$

Table 8.1 Maximum per mil (‰) of tensile reinforcement at I_{cr}: $\rho_{L,max} = A_{s,max}/(b \cdot d)$ (Steel Class B500c)

Materials		C16	C20	C25	C30	C35	C40	C45	C50	C55	C60
$P_{L,max}$ (‰)	DCM	5.97	7.47	9.34	11.20	13.07	14.94	16.80	18.67	20.54	22.40
	DCH	3.80	4.75	5.93	7.12	8.31	9.49	10.68	11.87	13.05	14.24

Assumption: $\rho' = \rho_{L,max}/2$, $\varepsilon_{sy,d} \approx 2.174$‰, $\mu_\varphi = 6.8$(DCM) or 10.7(DCH), $\mu_\varphi =$ Steel Class B500c: C($f_{yd} = 500/1.15$ MPa).

Table 8.2 Minimum per mil (‰) of tensile reinforcement at tensile zones: $\rho_{L,min} = A_{s,min}/(b \cdot d)$ (Steel Class B500c)

Materials	C16	C20	C25	C30	C35	C40	C45	C50	C55	C60	C70	C80	C90
$\rho_{L,min}$ (‰)	1.90	2.20	2.60	2.90	3.20	3.50	3.80	4.10	4.20	4.40	4.60	4.80	5.00

Steel Class B 500c: f_{yk} = 500 MPa.

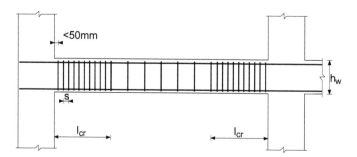

Figure 8.35 Transverse reinforcement in critical regions of beams. (Adapted from E.C.8-1/EN1998-1. 2004. *Design of Structures for Earthquake Resistance: General Rules, Seismic Actions and Rules for Buildings.* CEN, Brussels, Belgium.)

where
- d_{bL} is the minimum diameter of longitudinal bars (in mm).
- h_w is the beam depth (in mm).

c. The first hoop should be placed at a distance not more than 50 mm from the adjacent face of the column (Figure 8.35).

8.2.5.3 Design of beams for DCH buildings

8.2.5.3.1 Geometrical constraints

In addition to the rules for DCM buildings, the following rules should be implemented for DCH buildings:

1. The width of beams must not be less than 200 mm.
2. The height-to-width ratio of the web should satisfy the following expressions:

$$\frac{l_{ot}}{b} \leq \frac{70}{(h/b)^{1/3}} \quad \text{and} \quad \frac{h}{b} \leq 3.5 \tag{8.75}$$

where
- l_{ot} is the distance between torsional constraints. In the case of beams, l_{ot} may be defined as the clear span between the adjacent columns where the beam is jointed.
- h is the beam depth at mid-span, including also the depth of the slab.
- b is the width of the flange under compression. For a beam under seismic action, b is the width of the web.

Beam safety against lateral stability is deemed to be satisfied by the above expression (Equation 8.75).

8.2.5.3.2 Resistance to bending

The rules in effect for beams of DCM buildings in bending also apply for DCH buildings without any exceptions.

8.2.5.3.3 Resistance to shear

- For shear resistance, the Code for conventional R/C buildings (i.e. EC2-1-1/2004) applies, unless otherwise specified below.
- Shear action effects, as mentioned before, will derive from capacity design rules (see Chapter 6.1.3), so that brittle shear failure may be prevented.
- In the critical regions, *the strut inclination δ in the truss model should be 45°*. This provision aims indirectly at eliminating concrete shear resistance (see Section 8.2.4).
- With regard to the arrangement of shear reinforcement within the critical region, the following cases should be distinguished, depending on the algebraic value of the ratio:

$$\boxed{\zeta = V_{Ed,min} / V_{Ed,max}} \tag{8.76}$$

between the minimum and maximum acting shear forces as derived by application of the capacity design rule (see Chapter 6.1.3).

1. If $\zeta \geq -0.5$, the shear resistance provided by the reinforcement should be computed in accordance with the Code for conventional R/C structures, *but for strut inclination δ equal to 45°*.
2. If $\zeta \leq -0.5$, that is, when a reversal of shear forces prevails, then
   a. If

$$\left. V_E \right|_{max} \leq (2 + \zeta) \cdot f_{ctd} \cdot b_w \cdot d \tag{8.77}$$

where
   - f_{ctd} is the design value of concrete tensile strength specified by EC2-1-1/2004
       then the rule of paragraph 1 also applies in this case.
   b. If

$$\left. V_E \right|_{max} \geq (2 + \zeta) \cdot f_{ctd} \cdot b_w \cdot d \tag{8.78}$$

inclined reinforcement should be provided in two directions, usually at ±45° to the beam axis. X-shaped reinforcement should be capable of resisting half of $|V_E|_{max}$, while the other half should be resisted by stirrups. In this case, the verification is carried out by means of the truss analogy (Figure 8.36), that is,

$$\boxed{0.5 V_{E\,max} \leq 2 A_s \cdot f_{yd} \cdot \frac{\sqrt{2}}{2}} \tag{8.79}$$

where
   - A_s is the area of the inclined reinforcement in one direction, crossing the potential sliding plane (usually the beam end section).

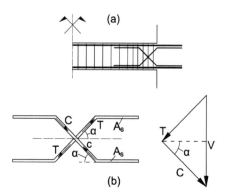

Figure 8.36 X-shaped type of reinforcement for beams with high shear: (a) arrangement of shear reinforcement; (b) design of the x-shaped diagonal reinforcement.

If the angle of the inclined reinforcement differs from 45°, which is the case of short beams (type of spandrel) where the inclined reinforcement follows the two diagonals of the beam in place of $\left(\sqrt{2}/2\right)$ in Equation 8.79, the value of sin α is introduced, where α is the angle of the diagonals of the beam to its axis (Figure 8.36). The provision for inclined shear reinforcement at the critical regions aims at the avoidance of failure in sliding shear mode (Figure 8.30) to which detailed reference has been made in Section 8.2.4.2.

8.2.5.3.4 Detailing for local ductility

- Rules for the local ductility of DCM buildings also apply for DCH buildings, unless they contradict the following provisions referring to such buildings.
- The critical regions of a beam of a DCH building are defined as the regions of a length $l_{cr} = 1.50h_w$ (where h_w is the depth of the beam).
- At least two ribbed bars with $d_{bL} = 14$ mm must be provided, both at the top and the bottom of the beam.
- One-fourth of the maximum top reinforcement at the supports should run along the entire beam length (Figure 8.34). This requirement, in combination with the requirement for a minimum reinforcement in compression zones at the end of the beams, is deemed to secure upper and bottom flanges of the beam from the risk of being found under tension over almost all their length at successive load reversals, depending on the level of the storey where the beam is located.
- The spacing, s, of hoops (in mm) within the critical region in the case of DCH buildings should not exceed

$$s = \min\left(h_w/4; 24d_{bw}; 175 \text{ mm}; 6d_{bL}\right)$$

(8.80)

8.2.5.4 Anchorage of beam reinforcement in joints

1. For the anchorage of beam reinforcement, all rules specified by the Code for conventional R/C structures (EC2-1-1/2004) are in effect unless otherwise specified in the following paragraphs.
2. The additional rules specified below by EC8-1/2004 are in effect for both DCM and DCH buildings.

3. The part of longitudinal beam reinforcement bent in joints of the frame for anchorage should always be placed inside the corresponding column hoops.
4. The diameter of beam longitudinal reinforcement passing through beam column joints must be in accordance with the following limitations.
   - For interior beam–column joints,

$$\frac{d_{bl}}{h_c} \leq \frac{7.5}{\gamma_{Rd}} \cdot \frac{f_{ctm}}{f_{yd}} \cdot \frac{1+0.8v_d}{1+0.75\kappa_D \cdot \rho_2/\rho_{1max}} \qquad (8.81)$$

   - For exterior beam–column joints,

$$\frac{d_{bl}}{h_c} \leq \frac{7.5 \cdot f_{ctm}}{\gamma_{Rd} \cdot f_{yd}} \cdot (1+0.8v_d) \qquad (8.82)$$

where
- h_c is the width of the column parallel to the beam.
- f_{ctm} is the mean value of the tensile strength of concrete.
- f_{yd} is the design value of the yield strength of steel.
- v_d is the normalised minimum design axial force in the column for the seismic design combination ($v_d = N_{Ed}/f_{cd} \cdot A_c$).
- κ_D is the factor reflecting the ductility class.
- $\kappa_D = 1.0$ for DCH buildings.
- $\kappa_D = 0.66$ for DCM buildings.
- ρ_{1max} is the maximum allowed tension steel ratio in critical regions.
- ρ_2 is the compression steel ratio.
- γ_{Rd} is the model uncertainty factor of the design value of resistances, taken as being equal to 1.2 or 1.0, respectively, for DCH or DCM.

In Table 8.3, the maximum allowed rebar diameters may be found for various concrete classes, for internal and external joints, for steel class B500c, for $\rho_2/\rho_{1max} = 0.50$ and for $v_d = 0.40$ (Ignatakis, 2011).

From the above table, it may be concluded that the ratio h_c/d_{bl}, for internal joints, ranges from 38.4 (C16) to 17.8 (C50) for DCH buildings. This means that the requirements imposed by expressions 8.81 and 8.82 are crucial at the prestudy stage for the choice of the dimensions of the columns and the concrete class in relation to the diameters of the rebars that are going to be used in beams. In fact, if the limits of the diameters are small, it will be necessary to have a large number of longitudinal rebars in the beam, closely spaced hoops to avoid buckling and, therefore, high construction cost together with many problems in concrete casting. In this respect, these requirements

Table 8.3 Maximum diameter sizes for a column width h_c = 500 mm

Concrete				C16	C20	C25	C30	C35	C40	C45	C50	C55	C60
$\emptyset_{L,max}$ (mm)	DCM	$k_D = 2/3$	+	17	20	24	26	29	32	35	37	38	40
		$\gamma_{Rd} = 1.0$	⊣	22	25	30	33	36	40	43	47	48	50
	DCH	$k_D = 1.0$	+	13	15	18	20	22	24	26	28	29	30
		$\gamma_{Rd} = 1.2$	⊣	18	21	25	28	30	33	36	39	40	42

Steel Class B500c: f_{yd} = 500/1.15 MPa, v_d = 0,40, ρ'/ρ_{max} = 0.50.

appear to have many implications for the design and construction of R/C buildings. However, they have been justified by extensive laboratory tests on interior joints under cyclic loading (Kaku and Asakusa, 1991; Kitayama et al., 1991).

The derivation on the above expressions 8.81 and 8.82 is based on analytical considerations of steel bonding in the joint (Fardis, 2009). These considerations are based mainly on the following assumptions (Figure 8.37):

- Steel bars yield at one face of the joint in tension, while at the other face, they yield simultaneously in compression (capacity design concept).
- The joint core is confined due to the axial forces of the column.
- Bars are bonded to a length equal to $0.8h_c$.
- In DCH structures, the anchorage length of beam or column bars anchored within column–beam joints is measured from a point on the bar at a distance $5d_{bL}$ inside the face of the joint, to take into account the yield penetration due to cyclic inelastic deformations (Figure 8.38a; see Chapter 7.5.4).

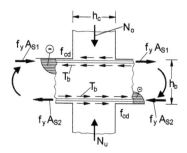

Figure 8.37 Internal force pattern on the bonded rebars in a joint.

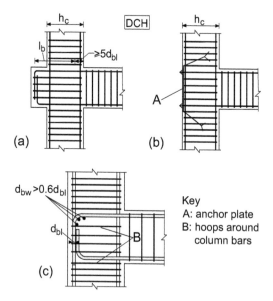

Figure 8.38 Additional measures for anchorage in exterior beam–column joints: (a) beam extension; (b) anchorage plate; (c) bonds and transverse reinforcement. (Adapted from E.C.8-1/EN1998-1. 2004. *Design of Structures for Earthquake Resistance: General Rules, Seismic Actions and Rules for Buildings.* CEN, Brussels, Belgium.)

5. If the requirements of expression 8.82 cannot be satisfied in exterior beam–column joints, because of the limited depth h_c of the column, the following additional measures may be taken to ensure the anchorage of the longitudinal rebars.
   • The beam may be extended horizontally in the form of exterior stubs (Figure 8.38a) if this may be accepted by the architectural design.
   • Headed bars or anchorage plates welded to the end of the bars may be used (Figure 8.38b).
   • Bends with a minimum length of 10 d_{bl} and transverse reinforcement placed tightly inside the bend of the bars may be added (Figure 8.38c). In this case, h_c/d_{bl} may be reduced to 70% of results from expression 8.82. The risk of bond failure for bars under compression in the joint is eliminated due to the closely spaced hoops of the column that encloses the bends.
6. Top and bottom bars passing through joints should terminate in the adjacent beams not less than l_{cr} from the external face of the joint (Figure 8.34).

8.2.5.5 Splicing of bars

1. For splicing of bars, all rules imposed by the Code for conventional R/C structures (EC2-1-1/2004) are in effect, unless it is otherwise specified in the following paragraphs.
2. The additional rules specified below by EC8-1/2004 are in effect for both DCM and DCH buildings.
3. Lap splicing by welding is prohibited in the critical zones.
4. The following requirements should apply for the transverse reinforcement within the lap length, in addition to the provisions of EC 2-1-1/2004.
   a. If the anchored and the continuing bar are arranged in a plane parallel to the transverse reinforcement, the sum of the area of all spliced bars ΣA_{SL} will be used in the calculation of the transverse reinforcement (Figure 8.39a).
   b. If the splicing is arranged within a plane normal to the transverse reinforcement, the area of the transverse reinforcement will be calculated on the basis of the area of the larger-lapped longitudinal bar A_{SL} (Figure 8.39b).
   c. The spacing s of the transverse reinforcement in the lap zone will not exceed

$$s = \min\{h/4; 100\}(\text{mm})$$ (8.83)

where h is the minimum cross-sectional dimension.

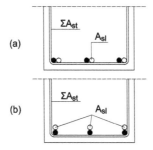

Figure 8.39 Arrangement of splicing bars in relation to the transverse reinforcement: (a) in a plane parallel to the bottom flange of the beams; (b) in planes vertical to the bottom flange.

8.3 DESIGN OF COLUMNS

8.3.1 General

For a thorough examination of the action effects of a column as a component of a seismic-resistant frame, the following diagrams of load effects of a typical internal and a typical external column have been already depicted.

In Figure 8.4, typical M_{Ed}, N_{Ed}, V_{Ed} diagrams are given for the 'basic' combination (i.e. 1.35 G '+' 1.50 Q).

In Figure 8.5, typical M_{Ed}, N_{Ed}, V_{Ed} diagrams are given for the seismic combination W_{dE} '+' E_d.

Based on the examination of the above diagrams, the following remarks should be made:

1. 'Basic' combination $1.35G$ '+' $1.50Q$
   a. Internal columns exhibit
      i. Large axial compressive forces N_{Ed}
      ii. Insignificant bending moments M_{Ed}
      iii. Insignificant shear forces V_{Ed}
   b. External columns exhibit
      i. Large axial compressive forces N_{Ed} but generally smaller than those of the internal columns
      ii. Intermediate bending moments M_{Ed}
      iii. Intermediate shear forces V_{Ed}
2. 'Seismic' combination W_{dE} '+' E_d
   a. For both internal as well as external columns, axial load effects N_{Ed} exhibit high values, still lower, however, than those corresponding to the 'basic' combination $1.35G$ '+' $1.50Q$, since the partial load coefficients for 'seismic' combination are for gravity loads equal to 1 and for live loads usually equal to $\psi = 0.30$.
   b. N_{Ed} exhibits strong variations due to reversals of seismic effect (Figure 8.6), particularly at the external columns of frames with a large aspect ratio H/L.
   c. M_{Ed} and V_{Ed} exhibit high values with reversed sign due to the seismic effect reversals.
   d. In frame systems, all action effects M_{Ed}, N_{Ed}, V_{Ed} increase from the top to the base of the building (see Chapter 4.5.3.2).
   e. In dual systems, while N_{Ed} of columns increase from the top to the base of the building, M_{Ed} and V_{Ed} do not present serious changes from the top to the base (see Chapter 4.5.3.4).
3. Concluding remarks
   a. Columns are affected by strong axial forces increasing from the top to the base of the building. At the same time, these axial forces exhibit significant variations due to the seismic action reversal.
   b. Seismic action also causes large moments and shears increasing from the top to the base of the building, with reversals in sign due to the cyclic character of seismic action.
   c. Moment diagrams exhibit their extreme values at the ends of columns, while at their mid-height they are about zero. In this context, each column behaves like two cantilevers, each with a span equal to about half of the height of the column, under a cyclic loading $\pm V_{Ed}$ at their free ends and an axial load equal to N_{Ed}.
   d. Finally, it should be noted that each column in a frame system belongs to two plane frames orthogonally arranged in the building since the building is a 3-D structure, and in this context, columns are induced to biaxial bending with axial loading.

8.3.2 Columns under bending with axial force

8.3.2.1 General

According to most modern Codes of Practice, a linear R/C structural member is characterised as a column when the normalised axial load

$$v_d = \frac{N_{Ed}}{A_c f_{cd}} \lessgtr 0.10$$

(8.84)

where
- N_{Ed} is the normal action effect (compression as positive).
- A_c is the area of its cross section.
- f_{cd} is the design strength of concrete in compression.

1. The assumptions for the design of R/C column sections under axial load with bending
   and for the generation of $M{-}\varphi$ diagrams are the same as those of R/C beams already
   presented in Section 8.2.2.1.
2. Some critical conclusions from the design of orthogonal cross sections under M_{Ed} and
   N_{Ed} would be useful for the next steps of this chapter. Therefore, they are summarised
   below.
3. $M{-}N$ interaction diagram for an orthogonal symmetrically reinforced cross section
   under uniaxial bending with axial force in qualitative form is given in Figure 8.40.
4. In the same figure, the strain pattern of the cross section is depicted for the three
   branches (a), (b), (c) of the interaction curve (Figure 8.40a), described below:
   - Branch A–B
     Corresponds to the region (a) of the strain diagram, which means that this
     branch corresponds to post-yielding strains of steel in tension, and in this context,
     the cross section may develop a post-elastic deformation in terms of curvature.

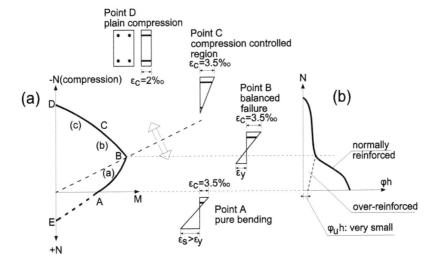

Figure 8.40 (a) *M–N* interaction diagram for an orthogonal symmetrically reinforced cross section of a
column and relevant strain patterns; (b) qualitative presentation of ductility capacity in relation to N. (Adapted from Zararis, P. 2002. *Design Methods of Reinforced Concrete* (in Greek).
Kyriakides Bros. Ltd, Thessaloniki.)

- Point B: Corresponds to *the balanced failure* of the cross section (see Section 8.2.2(b)), that is, $\varepsilon_c = \varepsilon_{cu2}$ and $\varepsilon_{s1} = \varepsilon_{y1}$.
- Branch B–C: Corresponds to the region (b) of the strain diagram, which means that this branch corresponds to tension steel strains lower than yield.
- Branch C–D: Corresponds to region (c) of the strain diagram; in other words, this branch corresponds only to compression concrete strains over the entire cross section.
- The balanced failure (point B) corresponds to a normalised axial load ν ranging between 0.30 and 0.45 (Figure 8.41). This means that for ν greater than or around this value, post-elastic deformations may be achieved either by increasing the strength of concrete so that ν is diminished or by increasing concrete strain at failure through confinement.

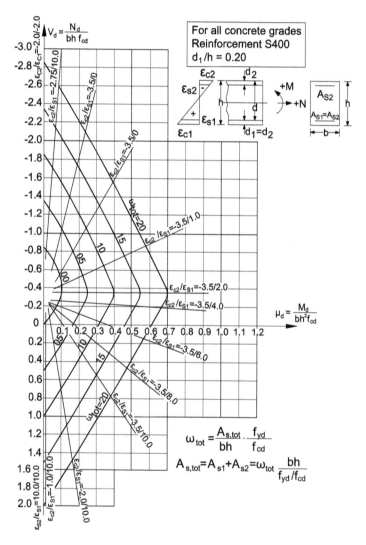

Figure 8.41 Interaction M–N diagrams and the relevant strain patterns at characteristic points. (Adapted from Zararis, P. 2002. *Design Methods of Reinforced Concrete* (in Greek). Kyriakides Bros. Ltd, Thessaloniki.)

- Keeping in mind that for column design the load effects result from a capacity design procedure, according to which potential plastic hinges around the joint develop only at the end of the beams, it is reasonable to consider that ductility supply for the columns is not essential. However, for various reasons, the probability of ductility demand at the ends of columns cannot be excluded. Some of these reasons are listed below:
  - The strain hardening of the longitudinal rebars of the beams ranging between 10% and 25% may cause overloading of the beams in terms of bending moments, leading to unacceptable deformations of the ends of the columns in case of non-existence of a controlled capability for post-elastic rotations of their ends (Paulay, 1986a).
  - As shown in Figure 8.40, the flexural strength of a column is strongly influenced by the axial loading, which is not constant during the seismic action, particularly for columns at the perimeter of the building. This variation of the axial load may easily exceed the values predicted by the analysis, particularly because of unexpected vertical seismic accelerations (Papazoglou and Elnashai, 1996). In this respect, if the column is not in a position to carry the load effects, it is necessary to be in a position of entering into an inelastic range. Therefore, the availability of a quantified ductility capacity should be sought for the columns.
  - Dynamic inelastic analysis of multi-storey R/C frames has shown that in order for columns to remain in elastic range all along their height except at their fixed ends at the foundation, beam overstrength factor γ_{Rd} for the capacity design of the joints should be in the range of 2–2.5, while, as we have seen in Chapter 6.1.4, their codified values, for cost reasons, range between $\gamma_{Rd} \approx 1.10$ and 1.30. Therefore, the need for ductility at the end of columns is again obvious.
  - The 3-D functioning of the columns of the frames in a 3-D structure reduces the reliability of the results of the capacity design procedure for the columns (see Section 8.3.5).

For these reasons and some others of secondary importance, columns must be designed for axial load and bending as ductile members. In this respect, the inelastic moment–curvature (M–φ) diagram up to failure must be examined as it has been done for beams, but taking into account the influence of the axial load, which is based on uniaxial bending considerations.

8.3.2.2 Determination of characteristic points of M–φ diagram and ductility in terms of curvature under axial load for an orthogonal cross section

From the presentation of beams, it is known that three characteristic points must be defined (Figure 8.12):

1. Point A corresponding to cracking stage (M_{cr}–φ_{cr})
2. Point B corresponding to steel yield stage (M_y–φ_y)
3. Point C corresponding to failure stage (M_u–φ_u)

It is apparent that these three points for the same R/C cross section have different coordinates each time, depending on the axial load.

In the following paragraph, closed expressions will be developed for an orthogonal cross section using some approximations that simplify the whole procedure (Park and Paulay, 1975; Penelis and Kappos, 1997; Fardis, 2009). For cross sections of any complex form and reinforcement, computer platforms have been developed to allow the determination of points A, B and C (e.g. RCCOLA-90, NOUS 3P, 2005; Papanikolaou, 2012). The theoretical background of NOUS (2005) has already been presented for beams and can also be used for columns.

8.3.2.2.1 Point A: $M_{cr}-\phi_{cr}$

Keeping in mind from classic strength of materials that (Figure 8.42)

$$\sigma_{M+N}^{max} = \frac{M}{bh^2/6} - \frac{N}{bh} \leq f_{ctm,fl} \tag{8.85}$$

($+N$ for compression)
the following expressions may be obtained:

$$M_{cr} = f_{ctm,fl}\frac{bh^2}{6} + N\frac{h}{6}$$

$$\phi_{cr} = \frac{M_{cr}}{E_c bh^3/12} \tag{8.86a–b}$$

where
- $f_{ctm,fl}$ is the mean tensile strength of concrete to flexure.
- E_c is the modulus of elasticity of concrete.

All other symbols are given in Figure 8.42.
It is obvious that for the same cross section, the coordinates of point A increase with an increase of axial force N.

8.3.2.2.2 Point B: $M_y-\phi_y$

Consider a column with orthogonal cross section, double-reinforced (Figure 8.43) and loaded by a constant compressive axial load N and a uniaxial bending moment M increasing from M_{cr} to yield of the reinforcement in the tensile zone.
The following normalised parameters will be introduced in the following presentation.

1. Normalised axial force:

$$\nu = \frac{N}{dhf_c} \tag{8.87a}$$

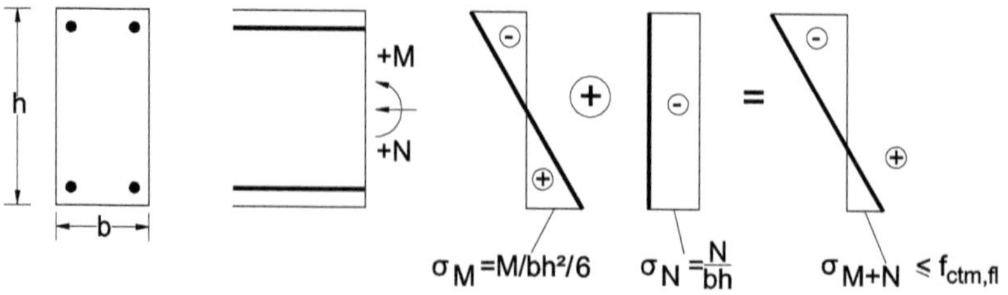

Figure 8.42 Stress pattern of a column under bending M with axial force N before cracking.

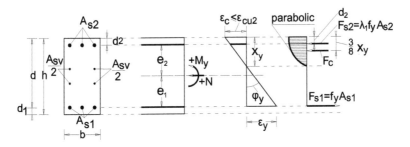

Figure 8.43 Strain and stress pattern of an R/C column at yield.

2. Percentage of reinforcement under tension:

$$\rho_1 = \frac{A_{s1}}{bh} \tag{8.87b}$$

3. Percentage of reinforcement under compression:

$$\rho_2 = \frac{A_{s2}}{bh} \tag{8.87c}$$

4. Percentage of web longitudinal reinforcement:

$$\rho_v = \frac{A_{sv}}{bh} \tag{8.87d}$$

The following assumptions are taken also into account for the derivation of the main expressions:

- $d \cong 0.9h$.
- $\sigma_c \cong f_c$, which is a reasonable approximation for bending with axial load at yield.
- The stress pattern of concrete is parabolic at yield.
- $\sigma_{s2} = \lambda_1 f_y$, where $\lambda_1 \leq 1$ (stress of the reinforcement A_{s2} in the compression zone).

From the strain pattern at yield, depicted in Figure 8.43 and under the above assumptions, the equilibrium between external loading and internal forces gives

$$N = F_c + F_{s2} + F_{sv} - F_{s1} \tag{8.88}$$

or

$$\nu f_c b \cdot h = \frac{2}{3} f_c b \cdot x_y + \lambda_1 f_y \rho_2 b \cdot h + \rho_v f_y b \cdot h - f_y \rho_1 b \cdot h \tag{8.89}$$

$$\nu f_c = \frac{2}{3} f_c \xi_y \frac{d}{h} + \lambda_1 f_y \rho_2 - f_y \rho_1 + \rho_v f_y \tag{8.90}$$

or

$$\xi_y = \frac{1}{0.6}\left(v + (\rho_1 - \lambda_1\rho_2 + \rho_v)\frac{f_y}{f_c}\right)$$

(8.91)

Parametric investigations (Tassios, 1989) have led to the following values for λ_1:

- For

  $v < 0.1$ ('*beams*'): $\lambda_1 = 0.5 + 18\rho_1$

  (8.92a)

- For

  $v = 0.1$: $\lambda_1 \cong 2/3$

  (8.92b)

- For

  $v = 0.2$: $\lambda_1 \cong 0.9$

  (8.92c)

- For

  $v > 0.2$: $\lambda_1 \cong 1.0$

  (8.92d)

It should be noted that v usually ranges between 0.20 and 0.60.

The value of ξ_y determined above must be compared to $\xi_b = \frac{x_b}{d} = 0.58$ (for steel class B500c, see Section 8.2.2.2), which corresponds to the *balanced failure* of a cross section.

If $\xi_b < 0.58$, failure at yield prevails and the procedure continues as in the following paragraph, or else brittle failure of concrete precedes steel yielding and the procedure ends here.

In case $\xi_b < 0.58$, the corresponding bending moment M_y at yield may be determined using the moment equilibrium equation between external loading and internal forces with respect to the reinforcement in tension:

$$M_y = f_{s2}0.9d + \frac{2}{3}f_cbd^2\xi_y\left(1 - \frac{3}{8}\xi_y\right) - Ne_1$$

(8.93)

or

$$M_y = bd^2\left[\lambda_1 f_y\rho_2 + \frac{2}{3}f_c\xi_y\left(1 - \frac{3}{8}\xi_y\right)\right] - Ne_1$$

(8.94)

Remark

The moments of web reinforcement A_v are not taken into account because their influence is very small.

The curvature of the cross section at yield results from the relation

$$\varphi_y = \frac{\varepsilon_y}{d - x_y} = \frac{\varepsilon_y}{d(1 - \xi_y)} \tag{8.95}$$

Equations 8.91, 8.94 and 8.95 allow the determination of point B of the M–φ diagram.

8.3.2.2.3 Point C: M_u–φ_u

Taking into account that the strain pattern of the cross section for this point of the diagram is depicted in Figure 8.44, it may be concluded that

$$x_u = \frac{\varepsilon_{cu2}}{\varepsilon_{cu2} + \varepsilon_s} d \tag{8.96}$$

Assuming that the stress pattern of concrete is orthogonal (Figure 8.44) at failure and that the same expression is in effect for σ_{s2} as at yield, that is, $\sigma_{s2} = \lambda_1 f_y$, the equilibrium equation between internal forces and external loading gives

$$N = F_{cut} + F_{s2u} + F_{svu} - F_{s1u} \tag{8.97}$$

or

$$v f_c bh = 0.8 f_c b x_u + \lambda_1 f_y \rho_2 bh + \rho_v f_y bh - f_y \rho_1 bh \tag{8.98a}$$

$$v f_c = 0.8 f_c \xi_u \frac{d}{h} + \lambda_1 f_y \rho_2 + \rho_v f_y - f_y \rho_1 \tag{8.98b}$$

$$\xi_u = \frac{1}{0.72} \left[v + (\rho_1 - \lambda_1 \rho_2 + \rho_v) \frac{f_y}{f_c} \right] \tag{8.99a}$$

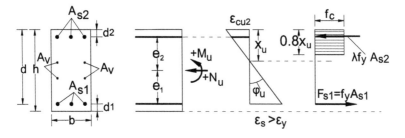

Figure 8.44 Strain and stress pattern of an R/C column at failure.

Taking into account that for columns under seismic action $\rho_1 = \rho_2$ and that v is usually greater than 0.20, the above expression takes the following form:

$$\xi_u = \frac{1}{0.72}\left(v + \rho_v \frac{f_y}{f_c}\right) = \frac{1}{0.72}(v + \omega_v)$$

(8.99b)

where

- ω_v is the mechanical ratio of web reinforcement $\left(\omega_v = \frac{A_{sv}}{bh}\frac{f_y}{f_c}\right)$.

The corresponding bending moment M_u at failure may be determined using the moment equilibrium equation between external loading and internal forces with respect to the reinforcement at tension, that is,

$$M_u = F_{s2} 0.9d + 0.8x_u f_c b (d - 0.4x_u) - N \cdot e$$

or

$$M_u = bd^2[\lambda_1 f_y \rho_2 + 0.8f_c \xi_u (1 - 0.4\xi_u)] - N \cdot e$$

(8.100)

The curvature of the cross section at failure has the following form:

$$\varphi_u = \frac{\varepsilon_{cu2}}{x_u} = \frac{\varepsilon_{cu2}}{\xi_u d}$$

(8.101)

Equations 8.99b, 8.100 and 8.101 allow the determination of point C of the M–φ diagram. The above equations are in effect under the following assumptions:

- σ–ε diagram of steel is elastoplastic without strain hardening.
- A_{s2} has not buckled.

8.3.2.2.4 Ductility μ_φ of the column

Ductility μ_φ of the column is obtained by the following expression:

$$\mu_\varphi = \frac{\varphi_u}{\varphi_y}$$

(8.102)

Introducing φ_u and φ_y into the above expression from Equations 8.95 and 8.101 results in

$$\mu_\varphi = \frac{\varepsilon_{cu2}}{\varepsilon_y}\frac{1 - \xi_y}{\xi_u}$$

(8.103)

or

$$\mu_\varphi = 1.2 \frac{\varepsilon_{cu2}}{\varepsilon_y} \left[\frac{0.60}{\nu + (\rho_1 - \lambda_1 \rho_2 + \rho_v)} - 1 \right] \qquad (8.104a)$$

Taking into account that $\lambda_1 = 1$ and $\rho_1 = \rho_2$, the above expression takes the following form:

$$\mu_\varphi = 1.2 \frac{\varepsilon_{cu2}}{\varepsilon_y} \left[\frac{0.60}{\nu + \omega_v} - 1 \right] \qquad (8.104b)$$

Equations 8.104a and 8.104b, although approximate, offer an effective tool for the study of the parameters that influence the value of μ_φ. In fact, if it is taken into account that

- For columns under seismic action $\rho_1 = \rho_2$
- ν is usually greater than 0.20 and therefore $\lambda_1 = 1$
- For concrete classes between C20 and C50 $\varepsilon_{cu2} = 3.5‰$
- For conventional R/C structures, the reinforcement class is B500c, that is,

$$\varepsilon_y = \frac{500}{200.10^3} = 2.5 \cdot 10^{-3}$$

μ_φ takes the following form:

$$\mu_\varphi = 1.2 \frac{3.5}{2.5} \left(\frac{0.60}{\nu + \omega_v} - 1 \right) \qquad (8.104c)$$

where

- $\omega_v = \rho_v \dfrac{f_c}{f_y}$ is the mechanical percentage of longitudinal steel in web.

Expression 8.104c for $\nu \geq 0.33$ and $\omega_v \cong 0.00$ results in μ_φ having a value $\mu_\varphi \leq 1$. This conclusion is in agreement with what has been mentioned in Section 8.3.2.1. Even for $\nu \cong 0.20$, which is a rather low value for ν in the design of columns, the above expression results in

$$\mu_{\varphi supl} \cong 1.2 \frac{3.2}{2.5} \left(\frac{0.60}{0.20} - 1 \right) \cong 3.36$$

This is a very low value for $\mu_{\varphi supl}$, even for this low normalised axial load, if it is taken into account that according to Equations 5.52a and 5.52b, $\mu_{\varphi demand}$ must be higher than

$$\mu_\varphi = 2q_o - 1 \qquad \text{if} \quad T_1 \geq T_c$$

$$\mu_\varphi = 1 + 2(q_o - 1)\frac{T_c}{T_1} \qquad \text{if} \quad T_1 < T_c$$

In this respect, in order for the expression

$$\mu_\varphi^{demand} \leq \mu_\varphi^{supply} \tag{8.105}$$

to be fulfilled, *the only way is the confinement of the critical zones of the columns with hoops* so that ε_{cu2} is increased in Equation 8.104b to the required degree so that the above inequality 8.105 is accomplished.

It should be added here that the expression of μ_φ^{supply} adopted by EC 8-1/2004 has a different starting point. It is based on the assumption that φ_y may be expressed by the following semi-empirical equation:

$$\boxed{\varphi_y = \frac{\kappa\varepsilon_y}{h}} \tag{8.106}$$

where κ has the following values:
- $\kappa = 1.75$ for rectangular beams and columns
- $\kappa = 1.44$ for rectangular walls
- $\kappa = 1.57$ for T, U or hollow rectangular section

The above semi-empirical expression 8.106 is based on procedures best fitting with experimental results (Biskinis, 2007; Fardis, 2009).

Taking into account expressions 8.101 and 8.106, μ_φ takes the form

$$\mu_\varphi^{supply} = \frac{\varphi_u}{\varphi_y} = \frac{\varepsilon_{cu2}h}{\xi_u d \cdot \kappa \cdot \varepsilon_y} = \frac{\varepsilon_{cu2}}{\varepsilon_y}\frac{h}{d}\frac{0.72}{(\nu+\omega_v)\cdot\kappa}$$

or

$$\boxed{\mu_\varphi^{supply} = \frac{\varepsilon_{cu}}{\varepsilon_y}\frac{0.72}{0.9\cdot\kappa(\nu+\omega_v)} \cong 0.80\frac{\varepsilon_{cu}}{\kappa\cdot\varepsilon_y(\nu+\omega_v)}} \tag{8.107}$$

This equation will be used in Section 8.3.4 and Chapter 9.2.4 for the determination of the required confinement reinforcement of the critical regions of columns and ductile walls, according to their ductility class.

8.3.2.3 Behaviour of columns under cyclic loading

1. As has already been noted in Chapter 4.5.3.2, column failure in frame or frame-equivalent systems has destructive consequences for the building because of the loss of support for storeys above, and, therefore, of the risk of a 'pancake' type collapse. So, a special concern should be given to column protection from flexural or shear failure.
2. The basic difference between beam and column behaviour under cyclic loading is attributable to the axial compressive load of the column. Extensive laboratory research

has focused the influence of the axial loading to column behaviour under cycling bending on the following consequences:

- Failure to flexure depends significantly on the value of the compressive force N (Figures 8.40 and 8.45). For increasing values of N from zero to N-balance, bending moments also increase with parallel decrease in the capacity of the column for post-elastic deformation. For a further increase in N beyond the balance point, the moment at failure decreases with a parallel drastic diminishing of the post-elastic deformation of the column.
- The axial compressive strain caused by axial loading is added to the compressive strain due to bending at the compression zone at each cycle. As a result, concrete degradation accompanied by spalling and crushing in the compressive zone, and, therefore, also steel rebar buckling, is almost inevitable, unless adequate *confinement* by means of closely spaced hoops in the critical regions is specified.
- Flexural and shear cracks formed in each loading semi-cycle tend to close during the reversing semi-cycle that follows due to axial load. This has as a consequence the diminishing of the pinching effect on the hysteresis loops of cycling loading, which are present in the case of beams. Additionally, concrete disintegration due to shear failure does not prevail, and, therefore, it is reasonable for axial load favourable contribution to shear to be taken into account in the shear design.
- The change of the axial force N during a cyclic seismic action, caused mainly by the overturning moment of the building (see Chapter 4.5.3.2, Figure 4.25) to the columns of the perimeter, leads to asymmetric hysteresis loops (see Figure 8.46) due to the variation of the axial load of the column.
- Axial tension is not impossible in external columns, particularly in buildings with a high aspect ratio, H/L (see Figure 8.6). In this case, the hysteresis loops tend to have zero area, and, therefore, the capacity of the column for energy dissipation is eliminated.
- Finally, it should be noted that in the case of large inter-storey drifts, second-order effects influence in an adverse way the bending moment at failure.

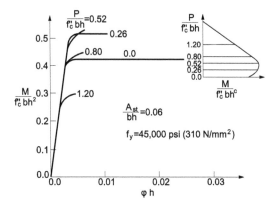

Figure 8.45 Moment–curvature curves for column sections at various levels of axial load. (Park, R. and Paulay, T.: *Reinforced Concrete Structures*. 1975. Copyright Wiley-VCH Verlag GmbH & Co. KGaA. Reproduced with permission.)

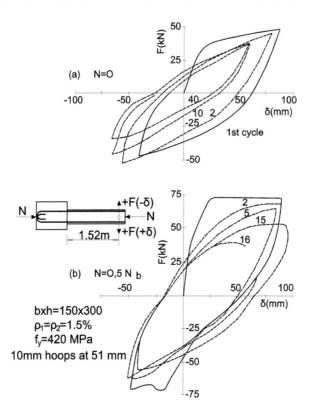

Figure 8.46 Hysteresis loops of elements under flexure and axial loading at various levels: (a) axial load N = 0; (b) axial load N = 0.5 N_b. (Adapted from Jirsa, J.O. 1974. Factors influencing behaviour of reinforced concrete members under cyclic overloads. *Proceed of 5th World Conf. on Earthq. Eng.* June 1973, Rome, Italy, 2, pp. 1198–1204.)

8.3.3 Strength and deformation of columns under prevailing shear

8.3.3.1 General

Column design to shear follows the same procedure under either static or dynamic loading, as in the case of beams (Section 8.2.4). Additionally, in the case of columns, the beneficial influence of compressive axial load should be taken into account. *This allows the concrete contribution to shear to be taken into account* for both ductility classes (DCM and MCH). In this respect, x-shaped reinforcement in the critical regions is not obligatory for columns under seismic action.

Reference should be made here again to the following remarks from the presentation of beams:

- For shear span-to-depth ratio

$$\frac{\alpha}{d} = \frac{M_{Eu}}{V_{Eu}d} \geq 7.0,$$ (8.108)

bending prevails. In this case bending failure occurs before any shear failure, no matter if shear reinforcement exists or not.

- For shear span-to-depth ratio

$$2.0 \le \frac{\alpha}{d} \le 7.0, \tag{8.109}$$

the failure mode depends on the shear reinforcement of the web. For columns of conventional R/C structures, the main concept of the design is that both shear and bending strengths satisfy the strength inequalities (see Chapter 3.2.1, Equations 3.1a and b) for the corresponding load effects resulting from the analysis. In this respect, bending failure is likely to precede or be simultaneous with shear failure. For columns of earthquake-resistant R/C structures, special concern is given to the requirement that a column must yield at both of its ends before shear failure. Therefore, shear effects derive from the capacity design procedure (see Chapter 6.1.4).

- For shear span-to-depth ratio

$$\frac{\alpha}{d} \le 2.0, \tag{8.110}$$

i.e. in the case of short R/C columns, a special design procedure must be followed so that an explosive cleavage failure of the short column is avoided (see Section 8.3.6).

From the above presentation, it may be concluded that in this section only the design of regular R/C columns will be dealt with, meaning columns with shear-span ratio

$$\frac{\alpha}{d} = \frac{M_{Ed}}{V_{Ed}} \ge 2.0$$

Short columns will be discussed in detail in Section 8.3.6.

8.3.3.2 Shear design of rectangular R/C columns

Failure under shear of a reinforced column with longitudinal rebars on both sides and transverse hoops may occur, as in the case of a beam failing to *diagonal tension* (Figure 8.23a) or *diagonal compression* (Figure 8.23b).

8.3.3.2.1 Failure caused by diagonal tension

As was noted in Section 8.2.4 for beams, this type of failure is resisted by two mechanisms:

- Beam action in the shear span without web reinforcement
- Truss mechanism of web reinforcement (stirrups–bent-up rebars)

8.3.3.2.1.1 SHEAR RESISTANCE IN THE SHEAR SPAN WITHOUT REINFORCEMENT

According to EC2-1/1991 ENV, in order for the beneficial influence of a compressive axial force N (N compressive is introduced with positive sign) to be taken into account, an additional term is introduced in Equation 8.52:

$$\boxed{V_{Rdc} = \tau_{Rd} \cdot \kappa (1.20 + 40\rho_1 + 0.15\sigma_{cp}) b_w d} \tag{8.111}$$

where

$$\sigma_{cp} = N_{Ed}/A_c \tag{8.112}$$

- N_{Ed} is the axial force due to loading or prestress.

The meaning of all other symbols may be found in Section 8.2.4.1 for beams.

A similar expression has also been adopted by the American Codes (ACI 318-2005, 2008, 2011), as well as by the Codes of many other countries.

Eurocode EC2-1-1/2004 has adopted for the determination of V_{Rdc} the following expression in the case of existence of an axial load:

$$\boxed{V_{Rdc} = \left[C_{Rd,c} \cdot \kappa (100\rho_1 f_{ck})^{1/3} + \kappa_1 \sigma_{cp} \right] b_w d} \tag{8.113}$$

and not less than ($V_{Rd,c}$ min)

$$\boxed{V_{Rd,c\ min} = \left(0.035\kappa^{3/2} f_{ck}^{1/2} + \kappa_1 \sigma_{cp} \right) b_w d} \tag{8.114}$$

where
- κ_1 is a coefficient with a recommended value of 0.15.
- N_{Ed} is the axial force in section caused by loading or prestress (N).
- A_c is the area of the concrete section (mm²).

$$\boxed{\sigma_{cp} = N_{Ed}/A_c \leq 0.2 f_{cd}\ (\text{MPa})} \tag{8.115}$$

$V_{Rd,c}$ is in N.

8.3.3.2.1.2 TRUSS MECHANISM OF WEB REINFORCEMENT – ALTERNATIVE (A)

As was explained in Section 8.2.4.1.1.2 in the beam case, the truss mechanism, according to ACI 318-2005-2011, is based on the concept of concrete strut diagonals with a constant indication of 45°. The same assumption was also in effect in EC2-1-1/1991 ENV. According to this assumption, even in the case of the existence of an axial compressive load, the shear resistance by the column web reinforcement is given by the expression

$$\boxed{V_{wd} = \frac{A_{sw}}{s} 0.9d \cdot f_{ywd}} \tag{8.116}$$

The meaning of all symbols above has been defined for the relevant equation for beams (Equation 8.55).

Taking all the above into account, it may be concluded that, as in the case of columns, shear strength to diagonal tension is equal to

$$V_{Rd} = V_{Rdc} + V_{wd} \tag{8.117}$$

The beneficial influence of the compressive axial force has been incorporated above in V_{Rdc}.

8.3.3.2.1.3 TRUSS MECHANISM OF WEB REINFORCEMENT – ALTERNATIVE (B)

A different procedure has been followed by EC2-1-1/2004 for failure caused by diagonal tension for both beams and columns. In the case that V_{Ed} exceeds the value of V_{Rdc} given by Equation 8.113 and 8.114, *the truss model analogy with concrete struts of variable inclination is applied.*

In this context, as was explained in the case of beams,

$$\boxed{V_{wd} = V_{Rd} = \frac{A_{sw}}{s} 0.9d \cdot f_{ywd} \cot \delta} \tag{8.118}$$

where (Equation 8.59b)

$$2.5 \geq \cot \delta \geq 1.0$$

It should be noted that in expression 8.118, both mechanisms of Equations 8.113 and 8.116 *are incorporated into one* that is *the variable inclination strut.*

The beneficial influence of compressive axial force in this procedure is incorporated into the capacity of cot δ to take values closer to 2.5 (more inclined struts), since, as we will see later, $V_{Rd\,max}$ increases due to N_{Ed}.

8.3.3.2.2 Failure caused by diagonal compression

As explained in the case of beams, for the determination of *diagonal concrete crushing*, 'the truss analogy' model is used, taking into account the influence of the axial load:

- For $\delta = 45°$ (ACI 318-2011, EC2-1-1/1997 ENV)

$$\boxed{V_{Rd\,max} = \alpha_{cw} b_w 0.9d (\nu_1 f_{cd})/2} \tag{8.119}$$

- For $2.5 \geq \cot \delta \geq 1.0$ (EC2-1-1/2004)

$$\boxed{V_{Rd\,max} = \alpha_{cw} b_w 0.9d (\nu_1 f_{cd}) \cot \delta / (1 + \cot^2 \delta)} \tag{8.120}$$

Particularly for columns or prestressed members, according to EC2-1-1/2004, the following values are recommended for α_{cw} (Figure 8.47):

$$\alpha_{cw} = 1 + \sigma_{cp}/f_{cd} \quad \text{for} \quad 0 < \sigma_{cp} \leq 0.25 f_{cd} \tag{8.121a}$$

$$\alpha_{cw} = 1.25 \quad \text{for} \quad 0.25 f_{cd} < \sigma_{cp} \leq 0.55 f_{cd} \tag{8.121b}$$

$$\alpha_{cw} = 2.50 (1 - \sigma_{cp}/f_{cd}) \quad \text{for} \quad 0.5 f_{cd} < \sigma_{cp} \leq 1.0 f_{cd} \tag{8.121c}$$

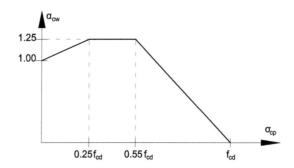

Figure 8.47 Increase of web crushing capacity to shear by axial compression or prestressing. (Adapted from Walraven, J.C. 2002. Delft, University of Technology, Background document for EN1992-1-1. Eurocode 2: Design of Concrete Structures-Chapter 6.2: Shear, Delft.)

The meaning of the above symbols has been clarified in the relevant Equations 8.61 and 8.62 for beams.

In this context, in case of compressive axial force, $V_{Rd\,max}$ increases and, therefore, the relevant value of cot δ in Equation 8.118 may be higher, leading to a lower requirement for web shear reinforcement.

8.3.3.2.3 Design procedure for columns under shear

Based on the above presentation, an overview of the design procedure for columns under shear according to Codes in use would be useful.

- *Step 1*: Determine the span-to-depth ratio:

$$\frac{\alpha}{d} = \frac{M}{V \cdot d}$$

  If $\frac{\alpha}{d} \geq 2.0$, the following procedure will be followed:
- *Step 2*: Shear resistance V_{Rdc} of the column without web reinforcement is determined using Equation 8.111 (EC2-1/1991 ENV) or using Equation 8.113 (EC2-1-1/2004). If $V_{Ed} \leq V_{Rdc}$, web reinforcement is not required.

  However, minimum web reinforcement in the form of hoops must be foreseen to secure the column from a potential brittle failure to shear.
- *Step 3*: In case that

$$V_{Ed} \geq V_{Rdc}$$

  two different procedures are foreseen depending on the Code under consideration:
  - In the case that ACI 318-2011 is in effect, design shear force V_{Ed} must be less than the sum of the shear contribution of the concrete V_{Rdc} (Equation 8.111; beam action) and the shear contribution of the web reinforcement V_{wd} (Equation 8.116), under the assumption that concrete struts exhibit a constant indication δ = 45° (Equation 8.64):

$$\boxed{V_{Ed} \leq V_{Rdc} + V_{wd}}$$

This expression allows the determination of the required web shear reinforcement. It is obvious that the beneficial influence of a compression axial force is included in V_{Rdc}.

At the same time, the concrete struts must have a safety factor sufficient to resist crushing. Therefore, V_{Ed} must also satisfy the following expression (Equation 8.65):

$$\boxed{V_{Ed} \leq V_{Rd\,max}}$$

for $\delta = 45°$

$V_{Rd\,max}$ is given in Equation 8.119.

- In the case that EC2-1-1/2004 is in effect, design shear force must be covered by the *truss model analogy* with concrete struts of *variable inclination* simultaneously for both the web reinforcement yield and concrete inclined struts crushing:

$$\boxed{\begin{aligned} V_{Ed} \leq V_{Rd} &= \frac{A_{sw}}{s} 0.9 d f_{ywd} \cot\delta \\ V_{Ed} \leq V_{Rd\,max} &= \alpha_{cw} b_w 0.9 d (v_1 f_{cd}) \cot\delta/(1+\cot^2\delta) \end{aligned}}$$

(8.122a–b)

where

$21.8° \leq \delta \leq 45°$

The accomplishment of both expressions 8.122a and 8.122b is satisfied by successive trials. It should be noted here that the beneficial influence of N_{Ed} is included in Equation 8.122b and particularly in the coefficient α_{cw}.

- *Step 4*: In the case that the span-to-depth ratio (shear-span ratio) is

$$\frac{\alpha}{d} \leq 2.0$$

a special procedure must be followed (see Section 8.3.6).

In closing, it would be worthwhile to repeat here that, thanks to compressive axial load, columns do not exhibit crosswise widening bending and shear cracks from cycle to cycle of seismic loading. This allows the shear strength contribution of concrete to be taken into account in the shear design of columns under seismic action. In this context, in the procedure imposed by EC2-1-1/2004, the inclination of concrete struts *also continues to be considered variable* in the case of seismic design for all ductility classes. At the same time, x-shaped reinforcement in the critical regions of the column is not obligatory.

8.3.4 Code provisions for columns under seismic action

8.3.4.1 General

The design rules for columns under seismic actions according to EC8-1/2004 are given below, together with the required justification in terms of the theory developed in the previous subsections. It should be noted that the design specifications, as in the case of beams, are directly related with the class of ductility demand, DCL, DCM and DCH. For higher ductility classes, they tend to be stricter in order to obtain a higher local ductility in terms of curvature.

Specifications for the design effects, taking into account capacity design procedure, have already been presented for columns in Chapter 6.1.4.

Material issues have also been presented in Chapter 7, Sections 7.2.4, 7.3.5 and 7.5.4 for all ductility classes. It should also be mentioned here that for the design of columns of buildings of class DCL, Code specifications for conventional R/C structures are applied (EC2-1-1/2004), except that concrete and steel reinforcement qualities are specified by EC 8-1/2004 and have already been presented in Chapter 7. For this reason, design of columns for DCL buildings will not be further examined.

In closing, it should be noted that all rules and specifications of the corresponding Code for conventional R/C structures (EC2-1-1/2004) continue to be in effect unless they contradict what will be presented below.

8.3.4.2 Design of columns for DCM buildings

8.3.4.2.1 Geometrical constraints

Unless $\theta \leq 0.1$ (see Chapter 6.2.2.2), the cross-sectional dimensions of columns should not be smaller than 1/10 of the larger distance between the point of contra-flexure and the ends of the column.

8.3.4.2.2 Resistance to flexure with axial load and shear

Flexural and shear resistance will be determined in accordance with EC2-1-1/2004, as in the case of conventional R/C buildings, by using the value of the axial force from the analysis of the seismic design combination.

- Biaxial bending, which is the most common case, may be simplified by substituting two uniaxial bending situations for it, one for every main direction, introducing a reduction of 30% in the uniaxial bending resistance of the cross section:

$$\boxed{0.7 M_{\mathrm{Rid}}(N_{\mathrm{Ed}}) \geq M_{\mathrm{Eid}}(i = x, y)} \tag{8.123}$$

where
- M_{Eid} is the acting bending moment, in $i = x, y$ directions.
- N_{Ed} is the most unfavourable axial load resulting from the seismic combination.
- M_{Rid} is the bending strength under N_{Eid} in $i = x, y$ direction.

  The above simplification is insignificant, if it is taken into account that today there are computer platforms that can carry out the biaxial verification very easily and quickly (e.g. NOUS, 2005; RCCOLA-90; FAGUS/Cubus, 2011.
- In columns of seismic-resistant frames, the value of the normalised axial force ν_{d} should comply with the following expression:

$$\boxed{\nu_{\mathrm{d}} = \frac{N_{\mathrm{Ed}}}{A_{\mathrm{c}} f_{\mathrm{cd}}} \leq 0.65} \tag{8.124a}$$

where
- N_{Ed} is the design axial force.
- A_c is the area of the cross section of the column.
- f_{cd} is the design concrete strength.

Taking into account that $f_{cd} = \dfrac{f_{ck}}{1.50}$, it may be concluded that

$$\boxed{\nu_d = \frac{N_{Ed}}{A_c f_{ck}/1.50} = 1.50\nu_k}$$

and therefore

$$\boxed{\nu_k \leq \frac{0.65}{1.50} \cong 0.43}.$$

(8.124b)

This requirement ensures that ν_k is very near to $\nu_{balance}$ (see Section 8.3.2.1).

8.3.4.2.3 *Detailing of columns for local ductility*

- *The critical regions* of a column are defined as the regions of a length l_{cr} quantified below and measured from the two ends of the column to its height. As already explained, these are regions of potential plastic hinges (Figure 8.48).

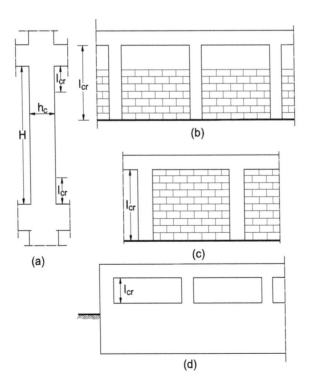

Figure 8.48 Column critical regions.

- *The length of the critical region l_{cr}* (in meters) may be determined from the following expressions:

$$l_{cr} = \max\{h_c; l_c/6; 0.45\}$$ (8.125)

where
- h_c is the largest cross-sectional dimension of the column (in meters).
- l_c is the clear length of the column (in meters).

- If $l_c/h_c < 3$ (short column), the entire height of the column should be considered a critical region. This type of column will be examined separately and in detail in the next section.
- *The total longitudinal reinforcement* ρ_i should not be less than 0.01 and not more than 0.04, that is,

$$0.01 \leq \rho_l \leq 0.04$$ (8.126)

and should be arranged symmetrically in symmetric cross sections ($\rho_1 = \rho_2$).
- At least one intermediate bar should be provided between corner bars along each column side to ensure the integrity of the beam–column joints.
- In the critical region *at the base* of the columns, at the bottom storey, a value of the curvature ductility factor μ_φ should be provided, equal to that specified for the critical region of beams (see Section 8.2.5.2.3; Equations 5.52a and 5.52b). In all other critical regions of the columns, the required local ductility is deemed to be covered by the confinement specified in subsequent paragraphs *without quantitative verification*.
- Particularly in the critical region at the base of a column, the above-mentioned requirement will be satisfied by a proper confinement of the critical region by means of hoops, which will allow the enhancement of the strain capacity of concrete from –0.0035 to a higher value, ensuring the development of a quantified post-elastic deformation without concrete spalling. This requirement is deemed to be satisfied if

$$\alpha\omega_{wd} \geq 30\mu_\varphi v_d \varepsilon_{syd} \frac{b_c}{b_o} - 0.035$$ (8.127)

where (Figure 8.49)
- ω_{wd} is the mechanical volumetric ratio of confining hoops within the critical region.

$$\left[\omega_{wd} = \frac{\text{volume of confining hoops}}{\text{volume of concrete core}} \cdot \frac{f_{yd}}{f_{cd}} \right]$$ (8.128)

- μ_φ is the required value of the curvature ductility factor.
- v_d is the normalised design axial force ($v_d = N_{Ed}/A_c f_{cd}$).
- h_c is the cross-sectional depth.
- h_o is the depth of the confined core (to the centreline of the hoops).

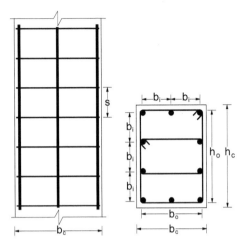

Figure 8.49 Confinement of concrete core.

- ε_{syd} is the design value of tension steel strain at yield.
- b_c is the cross-sectional width.
- b_o is the width of the confined core (to the centreline of the hoops).
- α is the confinement effectiveness factor, equal to

$$\alpha = \alpha_n \cdot \alpha_s \qquad (8.129)$$

In the above expression, α_n and α_s take the following values.

1. For rectangular cross sections:

$$\alpha_n = 1 - \sum_n b_i^2 / 6 b_o h_o \qquad (8.130a)$$

$$\alpha_s = (1 - s/2b_o)(1 - s/2h_o) \qquad (8.131a)$$

where
- n is the number of longitudinal bars laterally encaged by hoops or cross ties.
- b_i is the distance between consecutive encaged bars (see also Figure 8.49 for b_o, h_o).

2. For circular cross sections with circular hoops and diameter of confined core D_o (to the centreline of hoops):

$$\alpha_n = 1 \qquad (8.130b)$$

$$\alpha_s = \left(1 - \frac{s}{2D_o}\right)^2 \qquad (8.131b)$$

3. For circular cross sections with spiral hoops:

$$\alpha_n = 1 \tag{8.130c}$$

$$\alpha_s = \left(1 - \frac{s}{2D_o}\right) \tag{8.131c}$$

4. For cross sections of L, T, C and so on, detailed reference will be made in the next chapter, where local ductility issues of shear wall cores will be dealt with.

The derivation of $\alpha = \alpha_n \, \alpha_s$ was presented in Chapter 7.4.3.2 for concrete confinement by means of hoops.

The justification of Equation 8.127 was presented in details in the first edition of this book (Penelis and Penelis, 2014). From this justification, it has resulted that Equation 8.127 ensures a safety factor for the local ductility of columns of the order of $\gamma_{\mu\varphi} \approx 3.00$.

It should be noted here that the relevant American Code of Practice (ACI 318-2008, Chapter 21) specifies semi-empirical formulas for the quantification of confinement reinforcement in all critical regions of columns, justifying this choice by the fact that μ_φ and ν_d are not known with sufficient accuracy. In this way, the design of the local ductility reinforcement in the critical regions is greatly simplified, and the design is carried out either manually or is computer-aided.

- A minimum value of ω_{wd} equal to 0.08 should be provided within the critical region at the base of the columns. This value corresponds in the case of concrete class C25 and steel class B500c to a volumetric percentage of

$$\rho_{w,min} = 0.08 \frac{25 \cdot 1.15}{500 \cdot 1.5} = 0.003 \quad (3\text{\textperthousand})$$

- Within the critical regions of columns, hoops and cross ties with a diameter $\not< 6.00$ mm will be provided at spacing such that a minimum ductility is ensured and local buckling of longitudinal bars is prevented. These requirements are satisfied if the following conditions are taken into account (Figures 8.50 and 8.51):

1. The spacing s of the hoops (in mm) does not exceed

$$s = \min\{b_o/2; 175 \text{ mm}; 8d_{bl}\} \tag{8.132}$$

where
- b_o is the minimum dimension of the concrete core (in mm).
- d_{bl} is the minimum diameter of the longitudinal bars (in mm).

2. The distance between consecutive longitudinal bars encaged by hoops does not exceed 200 mm.

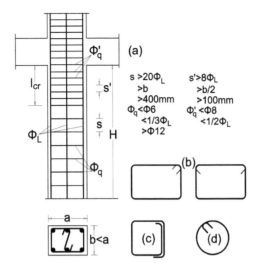

Figure 8.50 Arrangement of column reinforcement: (a) arrangement of hoops; (b) closing of external hoops; (c) closing of internal hoops; (d) closing of cyclic hoops (DCM buildings).

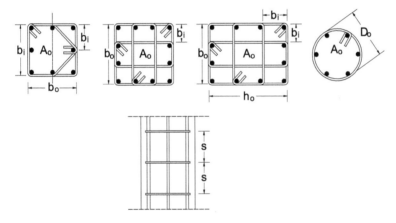

Figure 8.51 Various types of hoop arrangement in orthogonal and circular cross sections of columns, with the required notation for the implementation of expressions 8.130a, 8.131a and 8.131b.

8.3.4.3 Design of columns for DCH buildings

8.3.4.3.1 Geometrical constraints

In addition to the requirements for columns of DCM buildings, in the case of DCH, the minimum cross-sectional dimension should not be less than 250 mm.

8.3.4.3.2 Resistance to flexure with axial load and shear

The rules for DCM are also here in use. The only difference is that

$$\boxed{v_\mathrm{d} = \frac{N_\mathrm{Ed}}{A_\mathrm{c} f_\mathrm{cd}} \leq 0.55}$$

(8.133)

instead of 0.65, which is imposed on DCM buildings. This value of ν_d corresponds to a value for ν_k equal to

$$\boxed{\nu_k = \frac{0.55}{1.50} \cong 0.36}.$$
(8.134)

This requirement ensures that ν_k is below $\nu_{balance}$ (see Section 8.3.2.1).

8.3.4.3.3 Detailing of columns for local ductility

The rules for DCM are also in effect here, unless they contradict what is specified below.

1. The length of the critical region l_{cr} may be determined as follows (in meters):

$$\boxed{l_{cr} = \max\{1.5h_c; l_{cl}/6; 0.6\}}$$
(8.135)

where
- h_c is the largest cross-sectional dimension of the column (in meters).
- l_{cl} is the clear length (in meters).

2. The detailing of critical regions above the base of the columns, at the bottom storey, should be designed for a minimum value of the curvature ductility factor μ_φ derived from Equations 5.52a and 5.52b for reduced values of q_o to two-thirds of their values that have been taken into account for the determination of the q-factor of the building. In this case, $\alpha\omega_{wd}$ should again be determined by applying Equation 8.127.
3. In any case, the minimum value of ω_{wd} should be
   - In the critical region of the base at the bottom storey

$$\boxed{\omega_{wd} \geq 0.12}$$
(8.136)

   - In all other critical regions of the column

$$\boxed{\omega_{wd} \geq 0.08}$$
(8.137)

   - Within the critical regions of columns, hoops and cross ties with a diameter

$$\boxed{d_{bw} \geq 0.4 d_{bL\,max} \sqrt{f_{ydL}/f_{ydw}}}$$
(8.138)

   should be provided at a spacing such that a minimum ductility is ensured and local buckling of longitudinal bars is prevented. These requirements are satisfied if the following conditions are fulfilled.
   - The spacing s of the hoops (in mm) does not exceed

$$\boxed{s = \min\{b_o/3; 125 \text{ mm}; 6d_{bl}\}}$$
(8.139)

where
- b_o is the minimum dimension of the concrete core (in mm).
- d_{bl} is the minimum diameter of longitudinal bars (in mm).

- The distance between consecutive longitudinal bars encaged by hoops does not exceed 150 mm.
- The critical regions in the columns of the two lower storeys should be increased in length by 50%.
- The amount of longitudinal reinforcement at the base of the columns of the bottom storey should not be less, in any case, than that provided at the top of the columns of this storey.

8.3.4.4 Anchorage of column reinforcement

1. For the anchorage of column reinforcement, all rules specified by the Code for conventional R/C structures (EC2-1-1/2004) are in effect unless otherwise specified in the following paragraphs.
2. The additional rules specified below by EC8-1/2004 are in effect for both DCM and DCH buildings.
3. The anchorage length l_{anc} of column bars within critical regions should be calculated for a ratio of the required area of reinforcement $A_{s,req}$ to the available area $A_{s,avail}$ equal to 1.0, that is

$$\boxed{\frac{A_{s,req}}{A_{s,avail}} = 1.0} \tag{8.140}$$

no matter if $A_{s,avail}$ is greater than $A_{s,required}$.
4. If under the seismic design combinations, the axial force in the column is tensile, the anchorage lengths should be increased by 50% relative to those specified by EN1992-1-1/2004.

8.3.4.5 Splicing of bars

- In addition to what has been developed in the relevant paragraph for beams (Section 8.2.5.5), the following rules are specified for columns.
- Mechanical couplers may be used in columns if these devices are covered by appropriate testing documents.
- The required area of transverse reinforcement of columns spliced at the same location could be calculated from the following expression:

$$\boxed{A_{st} = s(d_{bl}/50)(f_{yld}/f_{ywd})} \tag{8.141}$$

where
- A_{st} is the area of one *leg* of the transverse reinforcement.
- d_{bl} is the diameter of the spliced bar.
- s is the spacing of the transverse reinforcement.
- f_{yld} and f_{ywd} are the design values of the yield strength of the longitudinal and transverse reinforcements, respectively.

8.3.5 Columns under axial load and biaxial bending

8.3.5.1 General

So far, column design has been approached as if columns were members of plane frames. Even for the verification at ultimate limit state, Code allows the substitution of biaxial bending by two independent uniaxial states reducing in parallel the design resistance to bending in the direction under consideration by 30%.

For ensuring column ductility, the confinement degree has been determined separately in each main orthogonal direction, in other words under a uniaxial consideration in two main directions.

$M-\theta$ diagrams under monotonic or cyclic loading, their stability under cyclic loading and the form of their hysteresis loops have also been examined under uniaxial bending with axial force.

However, as noted at the beginning of this section, columns in buildings are subjected to a biaxial loading due to seismic action. Therefore, it is important to examine the degree of reliability of all these simplifications that have been introduced by Codes in the analysis and design of columns as 3-D frame members. Unfortunately, the existing test results on axially loaded members under biaxial bending are limited, due to the practical difficulties of such testing (Fardis, 2009).

At the same time, the bidirectional loading history of the column has a very important role for the results and the relevant conclusions (Figures 8.52 and 8.53).

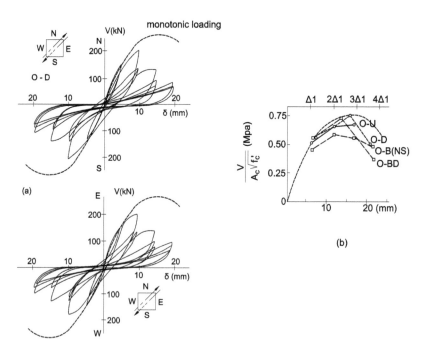

Figure 8.52 Diagonal cyclic loading of columns: (a) hysteresis loops in each direction; (b) envelope of the resultant action. (Adapted from Jirsa, J.O., Maruyama, K. and Ramirez, H. 1980. The influence of load history on the shear behaviour of short RG columns. *Proceedings of the 7th World Conference on Earthquake Engineering*, Istanbul, 6, 339–146.)

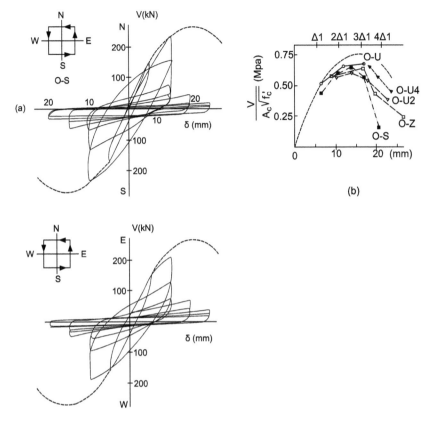

Figure 8.53 Square-like cyclic loading of columns: (a) hysteresis loops in each direction; (b) envelope of the resultant action. (Adapted from Jirsa, J.O., Maruyama, K. and Ramirez, H. 1980. The influence of load history on the shear behaviour of short RG columns. *Proceedings of the 7th World Conference on Earthquake Engineering*, Istanbul, 6, 339–146.)

In any case, taking into account the existing state of the art on the behaviour of columns under biaxial bending, the following issues will be examined:

- Strength
- Yield and failure stages
- Ductility μ_φ at curvature level
- Stability of M–θ curve under cyclic loading
- Form of hysteresis loops under biaxial loading

8.3.5.2 Biaxial strength in bending and shear

It is well known from the design of conventional R/C structures that biaxial bending with axial load is depicted by the 3-D envelope of Figure 8.54. In case of an orthogonal, symmetrically reinforced cross section and for a given axial force N (compressive is considered positive), the safe region is enclosed by the curve depicted in Figure 8.55a. The form of this curve is related to the percentage of steel reinforcement $\rho_{tot} = \dfrac{A_s}{bh}$ and its class.

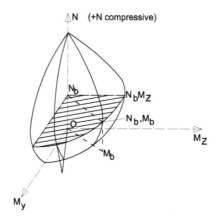

Figure 8.54 Interaction diagram of biaxial bending with axial compressive force.

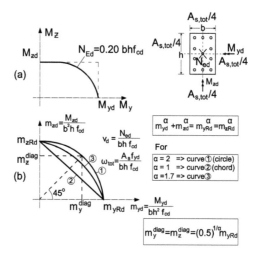

Figure 8.55 M_y–M_z interaction diagrams for a given axial force N. (a) M_y–M_z–N; (b) normalised diagram of M_y–M_z–N.

If the parameters of the problem are normalised,

$$m_{yRd} = \frac{M_{yRd}}{bh^2 f_{cd}} \tag{8.142a}$$

$$m_{zRd} = \frac{M_{zRd}}{b^2 h \cdot f_{cd}} \tag{8.142b}$$

$$\omega_{tot} = \frac{A_{s,tot}}{bh} \frac{f_{yd}}{f_{cd}} = \rho_{tot} \frac{f_{yd}}{f_{cd}} \tag{8.143}$$

$$\nu_d = \frac{N_{Ed}}{bh f_{cd}}, \tag{8.144}$$

then the curve depicted in Figure 8.55a is transformed into the curve depicted in Figure 8.55b (the various symbols are defined in Figure 8.55), which is expressed by the following equation:

$$m_{yd}^{\alpha} + m_{zd}^{\alpha} = m_{Rd}^{\alpha}$$ (8.145)

where

$$m_{Rd} = m_{yRd} = m_{zRd}$$ (8.146)

The value of α, for reasonable values of ω_{tot}, ν_d and steel classes, ranges between

$$\alpha = 1.33 - 2.00,$$ (8.147)

which means that the capacity envelope lies between a circle and an inflated rhombus (Figure 8.55b). This is the basic form of the curves of all charts for the design of R/C orthogonal cross sections under biaxial bending with axial load (Hassoun and Al-Manaseer, 2008).

From the above chart, it is concluded that biaxial loading leads to a strength decrease, since both normalised moments m_{yd} and m_{zd} corresponding to the m_{zd}–m_{yd} envelope are less than m_{Rd} (Figure 8.55). Furthermore, taking into account that the codified loading combination is given by an expression in the form

$$m_y '+' 0.30 m_z \quad \text{or} \quad m_z '+' 0.30 m_y,$$ (8.148)

it may be concluded that the corresponding maximum reduction of m_{Rd} that should be introduced in the verification of the substitute uniaxial bending at ULS is derived from the following expression:

$$m_{yd}^{1.33} + m_{zd}^{1.33} \leq m_{Rd}^{1.33}$$ (8.149)

where

$$\frac{m_{zd}}{m_{yd}} \cong 0.30 \quad \text{or} \quad m_{zd} \cong 0.30 m_{yd}$$

Therefore,

$$m_{yd} \leq 0.78 m_{Rd}$$ (8.150)

Expression 8.150 explains the adoption of a value equal to 0.70 for this reduction in the Code (see Section 8.3.4.2). It should be pointed out again that the above codified simplification is insignificant nowadays, when commercial computer platforms are in use for the design of any form of R/C cross section under biaxial bending with axial force.

Taking into account that capacity design for the joints of frame or frame-equivalent build-ings is carried out for the two main orthogonal plane frame directions independently, while strength verification of the columns is carried out in one direction under the capacity design moment values and in the other under the M values resulting from the seismic analysis, it can easily be concluded that in case of an earthquake with effects exceeding the design seis-mic action, it is probable that in both directions the capacity moments might be developed at the ends of a column. In this case, and taking into account what has been presented earlier, columns might be overloaded and fail. Therefore, the case mentioned above constitutes an additional reason for the confinement of the critical regions of the columns, so that plastic deformations are developed in the case of overloading due to unexpected malfunctioning of the main concept for 'strong columns–weak beams'.

In conclusion, it should be pointed out that although test results of columns under cyclic biaxial bending with axial load are very limited (Fardis, 2009), the results on a series of 35 specimens (Bousias et al., 2002; Biskinis, 2007) have given experimental results very close to analytical ones, that is, $M_{yy,\,exp}/M_{yy\,an}$ and $M_{zy,\,exp}/M_{zy\,an}$ are very close to 1.0.

8.3.5.3 Chord rotation at yield and failure stage: skew ductility μ_φ in terms of curvature

From the laboratory tests on 35 specimens to which reference was made in the previous paragraph, the following conclusions can be drawn:

- The chord rotation at the yield stage fulfils with reasonable accuracy (7–10% scatter-ing) the following expression:

$$\left(\frac{\theta_{yy\,exp}}{\theta_{yy\,uniax}}\right)^2 + \left(\frac{\theta_{yz\,exp}}{\theta_{yz\,uniax}}\right)^2 = 1 \tag{8.151}$$

  which is a relation similar to that of Equation 8.145.
- A similar expression is in effect for the failure stage,

$$\left(\frac{\theta_{uy\,exp}}{\theta_{uy\,uniax}}\right)^2 + \left(\frac{\theta_{uz\,exp}}{\theta_{uz\,uniax}}\right)^2 = 1 \tag{8.152}$$

  with a scattering up to 17%.
- Finally, a similar expression is obtained analytically for the skew ductility μ_ϕ in terms of curvature:

$$\frac{(\mu_{\varphi y})^2}{(\mu_{\varphi 1})^2} + \frac{(\mu_{\varphi z})^2}{(\mu_{\varphi 2})^2} = 1 \tag{8.153}$$

This means that for given values of $\mu_{\varphi 1}$ and $\mu_{\varphi 2}$ (capacities of ductility in terms of uniaxial bending), when the seismic action is skew, the available capacity in the main directions diminishes. For example, in the case of an angle of 45° ($\mu_{\varphi y} = \mu_{\varphi z}$), the available capacity in the main directions takes the form

$$\mu_{\varphi y} = \mu_{\varphi z} = \sqrt{\frac{\mu_{\varphi 1}^2 \mu_{\varphi 2}^2}{\mu_{\varphi 1}^2 + \mu_{\varphi 2}^2}} \tag{8.154}$$

It is apparent that these values of $\mu_{\varphi y}$ and $\mu_{\varphi z}$ are smaller than the relevant values of $\mu_{\varphi 1}$ and $\mu_{\varphi 2}$. However, this negative consequence of the skew bending on the ductility capacity of the columns is covered by the existing safety factor for their local ductility, which has been estimated to be about 3.00 (see Section 8.3.4.2).

8.3.5.4 Stability of M–θ diagrams under cyclic loading: form of the hysteresis loops

According to test results of Jirsa et al. (1980) and Umehara and Jirsa (1984) (Figures 8.52 and 8.53), the following conclusions have been drawn:

- For cyclic loading into inelastic range, the biaxial response appears to be inferior to the uniaxial one. Even less favourable conditions arise if a bidirectional loading history such as the square loading history is applied. In this case, the reduction in strength is more pronounced.
- The hysteresis loops of square columns reinforced with closely spaced hoops have shown that their behaviour under biaxial cyclic loading along the diagonal of the cross section is quite satisfactory, characterised by stable hysteresis loops even at ductility levels $\mu_f > 8$ (Priestley and Park, 1987). The same conclusions have been drawn from the laboratory tests on a series of the 35 specimens mentioned before.
- From the above contradictory results to those presented in the diagrams of Figures 8.52 and 8.53, it may be concluded that reliable conclusions on the above issues cannot yet be drawn, as the number of tests is limited.

8.3.5.5 Conclusions

From the presentation above, it is concluded that code provisions for strength, ductility and capacity design, although based on the concept of a plane frame, cover completely the 3-D function of the frame system and particularly its columns, which are subjected to 3-D loading. This is accomplished by the introduction of various safety keys like greater safety factors, local ductility demands and so forth.

8.3.6 Short columns under seismic action

8.3.6.1 General

As clarified in previous sections of this chapter, the design procedure for columns presented thus far has referred to regular columns with slenderness (Figure 8.56)

$$\lambda = \frac{L}{h} \geq 4.00 \tag{8.155}$$

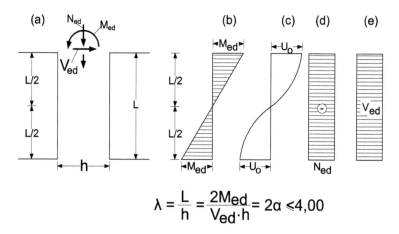

$$\lambda = \frac{L}{h} = \frac{2M_{ed}}{V_{ed} \cdot h} = 2\alpha < 4{,}00$$

Figure 8.56 Squat column: M, N, V deflection diagrams; (a) geometry; (b) bending moments; (c) deflections; (d) axial forces; (e) shear forces.

or to antisymmetrically loaded columns with a shear span-to-depth ratio

$$\alpha = \frac{M_{Ed}}{V_E \cdot h} = 0.5\lambda \geq 2.0 \tag{8.156}$$

Members with $\alpha \leq 2.0$, reinforced conventionally, have substantially different behaviour under cyclic loading, characterised by a high vulnerability to brittle failure in a mode of x-shaped diagonal splitting of concrete due to a diagonal compressive field leading to a premature *explosive cleavage shear fracture* (Figure 8.57; Minami and Wakabayashi, 1980; Tegos and Penelis, 1988). The inability of the conventional design procedure for columns of regular slenderness to give a reliable solution also for short columns must be attributed to the fact that in the case of short columns, the plane strain distribution concept over the

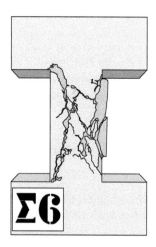

Figure 8.57 Premature explosive cleavage shear fracture of a short column. (Adapted from Tegos, I. 1984. Contribution to the study and improvement of earthquake-resistant mechanical properties of low slenderness structural elements. PhD thesis, Aristotle University of Thessaloniki (in Greek).)

cross section of the member stops prevailing. Therefore, a two-dimensional stress field is developed with strong interrelation between shear and normal stresses.

Short columns are often found in industrial or school buildings where continuous masonry infills are used to create openings extending along full spans of R/C frames at the facade of the building (Figure 8.48). They are also found at the perimeter of elevated underground storeys, with their deck slab 1 to 1 1/2 m above the surrounding ground level. Finally, in case of multi-storey buildings with a regular storey height of 3.00–3.20 m, the loads of the columns at the lower storeys often require dimensions for the columns more than 0.80 × 0.80 m, which corresponds to a low slenderness value of $\lambda = \dfrac{3.20}{0.80} = 4.0$. For all of these buildings, there are examples of impressive collapses due to a premature explosive cleavage shear fracture exhibited at short columns.

For this reason, extensive laboratory and analytical research has been conducted since the early 1980s aiming at the examination of the behaviour of short columns under cycling horizontal loading with axial load.

The arrangement of the reinforcement that has been examined refers to (Penelis and Kappos, 1997)

- Conventional reinforcement consisting of longitudinal rebars with transverse hoops (Figure 8.58)
- Bidiagonal reinforcement (Figure 8.59)
- Rhombic truss reinforcement (Figure 8.60)

The use of cross-inclined diagonal bars (bidiagonal reinforcement) has been proposed by Minami and Wakabayashi (1980), while the use of multiple cross-inclined bars forming a rhombic truss has been proposed by Tegos and Penelis (1988).

Test results have indicated that the use of either bidiagonal or rhombic reinforcement leads to an increase in shear capacity, as well as in stiffness and energy dissipation of short columns.

As shown in Figures 8.59 and 8.60, the use of non-conventional reinforcement patterns prevents brittle modes of failure in the form of diagonal splitting and leads to a shear type of failure, similar to the one observed in normal slenderness columns.

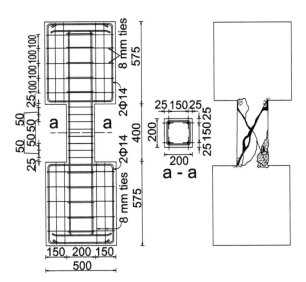

Figure 8.58 Conventional reinforcement of short columns against shear (close spaced ties).

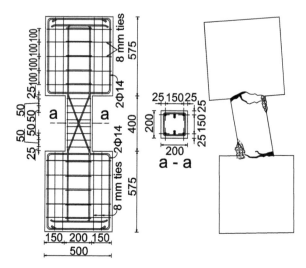

Figure 8.59 Bidiagonal reinforcement and ties of short columns against shear.

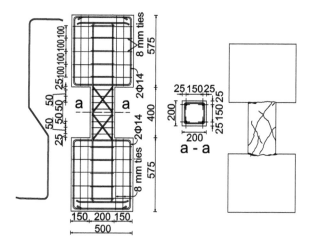

Figure 8.60 Rhombic reinforcement and ties of short columns against shear.

Bidiagonal reinforcement is practically preferable for slenderness $L/h < 1.5$, while rhombic reinforcement is preferable for $1.5 \leq L/h \leq 4.0$.

Detailed design procedure for the reinforcing of short columns may be found in the first edition of this book (Penelis and Penelis, 2014).

Concluding, it should be noted that the best protection from premature explosive cleavage shear is to avoid the existence of short columns by foreseeing joints between columns and the adjacent walls, which causes the formation of the short column (Figure 8.48).

8.3.6.2 Shear strength of short columns with inclined bars

According to an ultimate strength model developed by Tegos and Penelis (1988), the total shear capacity of a short column ($\alpha \leq 2$) reinforced with inclined bars, longitudinal bars and hoops may be estimated by superposition of the following three partial mechanisms:

- The well-known truss mechanism, wherein hoops are the web elements in tension and concrete struts between inclined shear cracks are the elements in compression, the longitudinal steel is the tension cord and the compression flexural zone is the compression cord; this mechanism carries a shear force

$$V_{R1} = (\tan\theta - \sin\theta)\, A_s f_y \tag{8.157}$$

where A_s is the area of longitudinal reinforcement in the tension cord and θ is the angle of incline concrete struts with respect to the column axis
- The rhombic truss mechanism of the inclined bars, which is able to carry a shear force

$$V_{R2} = 2A_s f_y \sin\theta \tag{8.158}$$

where A_s is the area of inclined bars, which coincides with the area of tension reinforcement (to be entered in Equation 8.157 if no extra [straight] longitudinal bars are used).
- A compression parallelogram formed by the compressive stress path in concrete, as compression is transmitted from one end of the member to the other through a double-arch action; this mechanism carries a shear

$$V_{R3} = 0.5N \tan\theta \tag{8.159}$$

where N is the axial load of the column. The total shear capacity is given by the relationship

$$V_R = V_{R1} + V_{R2} + V_{R3} \tag{8.160}$$

Transverse reinforcement (hoops) is required both for the truss mechanism and for balancing the splitting force of the compression parallelogram. The required amount of hoops may be calculated from the relationship (Tegos and Penelis, 1988)

$$\rho_w = (V_R - V_{R2})/2\alpha bh f_{yw} \tag{8.161}$$

which implies that no transverse reinforcement is required for the development of the rhombic truss mechanism. Nevertheless, in spite of the amount of shear carried by the inclined bars (V_{R2}), it is essential that a certain amount of hoops is present to provide the necessary confinement to concrete.

8.3.6.3 Code provisions for short columns

Despite the extended experimental and analytical research and the proposed design procedures so far (Shohara and Kato, 1981; AIJ, 1994; Tegos, 1984; Tegos and Penelis, 1988),
Eurocodes EC2-1/2004 and EC8-1/2004 do not include any special provision for short columns. The only provision is that the critical region of the column should be extended to the
total length of its height in the case that $\lambda = L/h$ is less than 3. ACI 318-11 does not include
any requirement for short columns either.

   However, it should be noted that various authors (Wight and Sozen, 1975) have come to
the following conclusions, based on statistical analyses of experimental results:

1. If $\alpha = \dfrac{M}{Vh} \geq 2$, the column does not exhibit diagonal explosive shear failure (splitting).
2. A reasonable minimum value for transverse reinforcement in the form of hoops is ρ_w
   min $\cong 6‰$ or ω_w min $\cong 0.16$.

   In conclusion, it should be mentioned that some EU Member States, in the framework of
their National Annexes for EC8-1/2004 (e.g. Greece), have introduced special rules for short
columns, which are based on a simplified approach of existing analytical methods (National
Annex of Greece, 2010).

8.4 BEAM–COLUMN JOINTS

8.4.1 General

Beam–column joints are the most crucial parts of seismic-resistant R/C frames because in
their limited volume, the stress state field that develops due to seismic action is very high.
   Indeed, keeping in mind that the joints are considered for the structural analysis as solid
elements, it may be concluded that they must be in a position to sustain in their core

- High shear stresses acting horizontally and vertically, as well (Figure 8.7)
- High bond stresses acting at the interface of concrete and longitudinal steel bars passing through the joint because of bending at the ends of the beams and the columns
  framing at the joint (Figure 8.37)

   The determination of the stress field in joints is a very difficult and questionable procedure, even in the case of a plane frame, due to the existence of the longitudinal rebars in
the joint subjected to high stresses, and also of the contribution of horizontal hoops to the
shear resistance of the core of the joint and its confinement. The whole situation becomes
more complicated in the case of 3-D joints, which constitute the most usual case (Figure
8.61) due to development of a triaxial stress state. Finally, the existence of a slab acting as
a diaphragm makes the stress state even more complicated, due to the confinement that it
causes to the joint. Consequently, it is not surprising that there is a strong disagreement
among various investigators about the design procedure that should be followed for R/C
joints. This disagreement is also reflected in the various Seismic Codes for R/C structures.
In any case, the main concept of all codes is that in the design procedure, *failure should be
excluded from the joint region* and be limited to the ends of the beams or at least of the
columns framing at the joint.

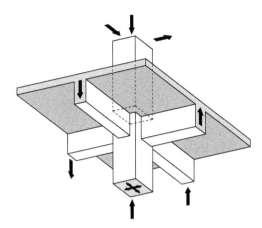

Figure 8.61 Alternative detailing of interior joints. (Paulay, T. and Priestley, M.J.N.: *Seismic Design of Reinforced Concrete and Masonry Buildings*. 1992. Copyright Wiley-VCH Verlag GmbH & Co. KGaA. Reproduced with permission.)

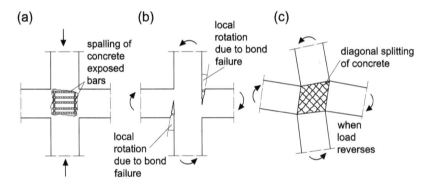

Figure 8.62 Types of failures at beam–column joints: (a) spalling of joint core; (b) anchorage failure of beam bars; (c) shear failure of the joint core.

The failure modes of some basic types of joints are depicted in Figures 11.23, 11.24 and 11.25 (Chapter 11.1.5). The usual types of failure of a joint may be classified as follows:

- *Spalling* of cover concrete at the faces of the joint core (Figure 8.62a)
- *Bond failure* of the longitudinal bars of the beam, which leads to strength deterioration and to a significant stiffness degradation due to fixed-end rotations (Figure 8.62b)
- *Diagonal splitting* caused by shear (Figure 8.62c)

8.4.2 Design of joints under seismic action

In the following paragraphs, a short overview will be presented on the two methods developed for the design of joints under seismic action:

- The design of joints to shear developed by Park and Paulay (1975) and Paulay and Priestley (1992)
- The design method adopted by EC 8-1/2004

8.4.2.1 Demand for the shear design of joints

The forces acting on the core of an interior joint of a plane frame under seismic action are shown in Figure 8.63. The horizontal joint shear V_{jh} is counterbalanced by $T_{b1} + C_{b2} - V_{col}$, i.e.

$$\boxed{V_{jh} = T_{b1} + C_{b2} - V_{col}}$$

(8.162)

where

- T_{b1} is the tensile force of the reinforcement at the top of beam 1 (to the left of the joint).
- C_{b2} is the resultant of the compressive stresses at the top of beam 2 (to the right of the joint).
- T_{b2} is the tensile force of the reinforcement at the bottom of beam 2 (to the right of the joint).

Taking into account that the axial force in the beam is very small, it may be concluded that

$$C_{b2} = T_{b2}$$

(8.163)

Therefore, Equation 8.162 takes the form

$$\boxed{V_{jh} = T_{b1} + T_{b2} - V_{col}}$$

(8.164)

Assuming that both top and bottom reinforcements of the beam are at yield point, Equation 8.164 takes the following form:

$$\boxed{V_{jh} = \gamma_{Rd} \cdot f_{yd}(A_{s1} + A_{s2}) - V_{col}}$$

(8.165)

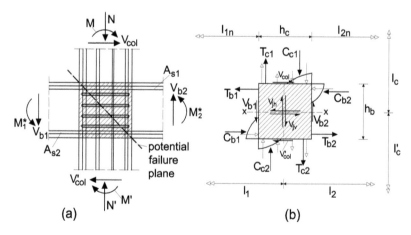

(a)   (b)

Figure 8.63 (a) Seismic actions in the joint core; (b) internal forces in the joint core (T results from tensile forces and C results from compressive forces at each face of the joint).

where

> γ_{Rd} is the overstrength due to steel strain hardening. For DCH buildings, γ_{Rd} should not be less than 1.20 ($\gamma_{Rd} \not< 1.20$).

For the determination of V_{col}, the values of bending moments at the top and bottom end of the column above the joint should be known at the yield stage of the beams framing to the joint. These values are difficult to define due to the permanent change of member stiffness in post-elastic stage. Given this, Paulay et al. (1978) suggested a capacity relationship for estimating the column shear V_{col} on the basis of the beam moments at the joint faces (Figure 8.63):

$$V_{col} = \frac{(l_1/l_{1n})M_1^* + (l_2/l_{2n})M_2^*}{(l_c + l_c')/2} \tag{8.166}$$

where

- l_1 and l_2 are the beam spans measured from the column centre lines.
- l_{1n} and l_{2n} are the corresponding clear spans of the beams.
- l_c and l_c' are the column heights.
- M_1^* and M_2^* are the capacity design moments of the beams at the joint faces.

Expression 8.166 has been adopted by the New Zealand Code (NZS 3101, 1995) but not by EC8 (see Section 8.4.3).

The vertical joint shear V_{jv} can be easily derived by applying the Caushy theorem, according to which $\upsilon_{jh} = \upsilon_{jv}$. Therefore,

$$\upsilon_{jh} = \frac{V_{jh}}{b_j h_c} = \frac{V_{jv}}{b_j h_b} = \upsilon_{jv} \tag{8.167}$$

where

> b_j is the thickness of the joint

or

$$\boxed{V_{jv} = \frac{h_b}{h_c} V_{jh}} \tag{8.168}$$

The same procedure may be applied for the determination of V_{jh} at an exterior joint. Taking into account that there is only one beam, Equation 8.165 takes the form

$$\boxed{V_{jh} = \gamma_{Rd} f_y A_{s1} - V_{col}} \tag{8.169}$$

and

$$V_{col} = \frac{(l_2/l_{2n})M_2^*}{(l_c + l_c')/2} \tag{8.170}$$

8.4.2.2 Joint shear strength according to the Paulay and Priestley method

According to this method, beam column joints resist shear through two mechanisms (Figure 8.64; Paulay and Priestley, 1992):

1. A diagonal concrete strut, which provides a shear resistance V_{ch}.
2. A truss mechanism comprising a diagonal compression field combined with horizontal (hoops) and vertical (longitudinal rebars) reinforcement. This mechanism provides a shear resistance V_{sh}.

The strut shear resistance V_{ch} is caused by the following forces on the faces of the joints:

- The compression forces in concrete C_{cb1}, C_{cc1}, C_{cc2}, C_{cb2}
- The bond forces ΔT_{cb1}, ΔT_{cc1}, ΔT_{cc2}, ΔT_{cb2} transferred by the reinforcement bars within the compression zone
- The beam and column shear forces V_{b1}, V_{col1}, V_{col2}, V_{b2}

These forces acting in two opposite groups at the top left and bottom right corners of the joint, respectively, cause a diagonal compressive force D_c. Therefore,

$$V_{ch} = D_c \cos(\alpha) \tag{8.171a}$$

$$V_{cv} = D_c \sin(\alpha) \tag{8.171b}$$

where
   α is the angle of the diagonal of the joint with respect to the horizontal.
   It should be noted that ΔT_c refers to only a part of the bond forces, while the rest of them is introduced along the longitudinal rebars of beams and columns in the core of the joint in the form of bond stresses. It is assumed that these bond stresses, which are introduced along the longitudinal bars, cause the compression forces D_s (Figure 8.64b) of the truss mechanism. Therefore,

$$V_{sh} = D_s \cos(\alpha) \tag{8.172a}$$

and

$$V_{sv} = D_s \sin(\alpha) \tag{8.172b}$$

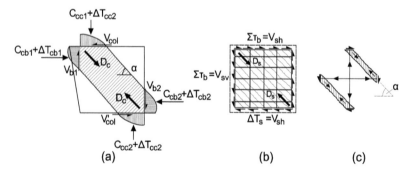

(a)   (b)   (c)

Figure 8.64 Shear transfer mechanism in a joint core: (a) diagonal concrete strut; (b) truss mechanism; (c) typical components of the truss.

The total shear resistance of the joint core can be expressed as the sum of the previously described mechanisms:

$$V_{jh} = V_{ch} + V_{sh} \tag{8.173a}$$

$$V_{jv} = V_{cv} + V_{sv} \tag{8.173b}$$

By making some reasonable assumptions and approximations (Paulay and Priestley, 1992), the following expressions may result for the factors of the above expressions 8.173a and 8.173b:

$$V_{ch} = \left(1.55 \frac{c}{b_c} + \beta - 0.55\right) T - V_{col} \tag{8.174}$$

$$V_{sh} = V_{jh} - V_{ch} = 1.55\left(1 - \frac{c}{b_c}\right) T \tag{8.175}$$

Taking into account that

$$\frac{c}{b_c} \cong 0.25 + 0.85 \frac{N_{Ed}}{f_{cd} A_c} \tag{8.176}$$

Equation 8.175 takes the form

$$V_{sh} \cong \left(1.15 - 1.30 \frac{N_{Ed}}{f'_c A_c}\right) T \tag{8.177}$$

Therefore, the required horizontal reinforcement in the form of hoops in the core of the joint may be defined as follows:

$$A_{jh} = V_{sh}/f_{ywd} \tag{8.178}$$

or

$$\boxed{A_{jh} = \left(1.15 - 1.30 \frac{N_{Ed}}{f_{cd} A_c}\right) \frac{\gamma_{Rd} f_{yd}}{f_{ywd}} A_{s1}} \tag{8.179}$$

In the above equations, the following notations have been used:

- c is the depth of the flexural compression zone of the column.
- β is the ratio of compression reinforcement content in beam sections to the corresponding tensile one $\left(0.5 \leq \beta = \dfrac{A_{s2}}{A_{s1}} \leq 1.0\right)$

- γ_{Rd} is the overstrength factor for steel ranging between $1.2 < \gamma_{Rd} < 1.4$.
- T is the tensile force of the steel at the upper flange of the beams $T = \gamma_{Rd}f_{yd}A_{s1}$.
- N_{Ed} is the minimum compression force action on the column.
- A_c is the area of the cross section of the column.
- f_{cd} is the design compressive strength of concrete.
- f_{yd} is the design yield strength of longitudinal steel bars.
- f_{yw} is the design yield strength of transverse steel hoops.

Keeping in mind expression 8.165, according to which

$$V_{jh} = \gamma_{Rd} \cdot f_{yd}(A_{s1} + A_{s2}) - V_{col} = (1+\beta)\gamma_{Rd} \cdot f_{yd} \cdot A_{s1} - V_{col} \tag{8.180}$$

and making the reasonable assumption that

$$V_{col} \approx 0.15(1+\beta)T \tag{8.181}$$

it is concluded that

$$\frac{V_{sh}}{V_{jh}} = \frac{1.15 - 1.30(N_{Ed}/f_{cd}A_c)}{0.85(1+\beta)} \tag{8.182}$$

So, for N_{Ed} min $\approx 0.10\, f_{cd}\, A_c$, the truss mechanism should resist 60–80% of the total horizontal joint shear force. From the above expression, it may easily be concluded that as N_{Ed} increases, the ratio V_{sh}/V_{jh} diminishes.

It is also worth noting that for $f_{ywd} = f_{yd}$, $N_{Ed} \cong 0.10\, f_{cd}\, A_c$, $\gamma_{Rd} = 1.25$, expression 8.179 takes the form

$$\boxed{A_{jh} \geq 1.28 A_{s1}} \tag{8.183}$$

This reinforcement should be placed in the space within the longitudinal steel (bars) of beams at their upper and lower flange and should comprise horizontal hoops and ties (Figure 8.63), placed normally to the column axis.

Finally, for the vertical joint shear reinforcement, the following expression is obtained:

$$\boxed{A_{jv} = \frac{1}{f_{yd}}[0.50(V_{jv} + V_b) - N_{Ed}]} \tag{8.184}$$

When Equation 8.184 becomes negative, obviously no vertical joint shear reinforcement will be required.

Equations 8.179 and 8.184 ensure the joints against cracking caused by diagonal tension. On the other hand, diagonal compression (Figure 8.64) may cause diagonal crushing. Therefore, the magnitude of the horizontal joint shear stress should be limited to acceptable levels. According to Paulay and Priestley (1992), this limit is given below:

$$\boxed{v_{jh} = \frac{V_{jh}}{b_j h_j} \leq 0.25 f_{cd} \leq 9 \ (\text{MPa})} \tag{8.185}$$

The above procedure has been adopted by the New Zealand Code (NZS 3101, 1995).
A similar model to that of Paulay and Priestley has been developed by A.G. Tsonos with very reliable results (Tsonos, 1999, 2001). A detailed presentation of this model is given in the first edition of this book (Penelis and Penelis, 2014).

8.4.2.3 Background for the determination of joint shear resistance according to ACI 318-2011 and EC8-1/2004

The collection and compilation of test results on interior joints (Kitayama et al., 1989, 1991) have caused serious reservations from a number of researchers, including the majority of those in North America, regarding the validity of the joint shear transfer models developed so far. In fact, according to the ACI special publication of Kitayama et al., the ultimate shear strength V_j of Equations 8.165 and 8.168 increases about linearly with the ratio of horizontal reinforcement ρ_{jh} (hoops) up to $\rho_{jh} = 0.4\%$ (Figure 8.65). Above this value and up to $\rho_{jh} = 2.4\%$, the ultimate joint strength remains independent of ρ_{jh}, while diagonal compressive failure mode always prevails. At the same time, the joint confinement via transverse beams and the slab at their top acting as a diaphragm increase the joint shear strength υ_{ju} significantly. Thus, the approach of the design of joints adopted by U.S. investigators and by ACI 318-2011 is based on a best-fitting procedure from a large number of available test results, allowing the adoption of maximum allowable nominal shear strength V_{jh} for each type of joint.

In view of the above, Eurocode EC8-1/2004 has adopted a homogeneous plane stress field in the core of the joint, depicted in Figure 8.66, for the calculation of joint horizontal reinforcement. This field includes

1. The shear stress $\upsilon_{jh} = \upsilon_{jv}$ from Equation 8.167
2. The vertical normal stress

$$\sigma_y = -N_{Ed}/A_c = -v_d \cdot f_{cd} \tag{8.186}$$

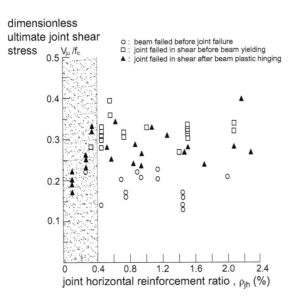

Figure 8.65 Effect of horizontal reinforcement ratio in interior joint ρ_{jh} on joint strength. (Adapted from Kitayama, K., Otani, S. and Aoyama, H. 1991. *Development of Design Criteria for R/C Interior Beam–Column Joints.* ACI Special Publication SP123, American Concrete Institute, Detroit, Michigan, 97–124.)

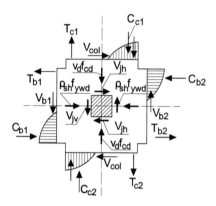

Figure 8.66 The homogeneous stress field in the joint. (Adapted from E.C.8-1/EN1998-1. 2004. *Design of Structures for Earthquake Resistance: General Rules, Seismic Actions and Rules for Buildings.* CEN, Brussels, Belgium.)

3. A smeared horizontal normal stress corresponding to the core reinforcement (transverse hoops)

$$\sigma_x = -\rho_{jh} f_{ywd} \tag{8.187}$$

which confines the concrete core in the horizontal direction at the yield stage of confinement steel

The principal stresses σ_I and σ_{II} of this plane stress field at failure stage should be equal to f_{ctd} (design tensile strength of concrete) and $-v_1 f_{cd}$ (design compressive strength of concrete), in inclined concrete struts. Therefore,

$$\left. \begin{array}{l} f_{ctd} = \sigma_I \\ -v_1 f_{cd} = \sigma_{II} \end{array} \right\} \frac{\sigma_x + \sigma_y}{2} \pm \sqrt{\left(\frac{\sigma_x - \sigma_y}{2}\right)^2 + \sigma_{xy}^2} \tag{8.188a}$$

or

$$\left. \begin{array}{l} f_{ctd} \\ -v_1 f_{cd} \end{array} \right\} \frac{v_d f_{cd} + \rho_{jh} f_{ywd}}{2} \pm \sqrt{\left(\frac{v_d f_{cd} - \rho_{jh} f_{ywd}}{2}\right)^2 + v_{jh}^2} \tag{8.188b}$$

From Equation 8.188b, the following two equations are obtained:

$$\frac{v_{jh}}{f_{cd}} = \sqrt{\left(\frac{f_{ctd}}{f_{cd}} + v_{top}\right)\left(\frac{f_{ctd}}{f_{cd}} + \rho_{jh}\frac{f_{ywd}}{f_{cd}}\right)}$$

or

$$\boxed{\rho_{jh} \cdot f_{ywd} \geq \dfrac{\upsilon_{jh}^2}{f_{ctd} + \nu_d f_{cd}} - f_{ctd}}$$ (8.189)

and

$$\boxed{\dfrac{\upsilon_{jh\,max}}{f_{cd}} \leq \sqrt{(\nu_d - \nu_1)\left(\rho_{jh}\dfrac{f_{ywd}}{f_{cd}} - \nu_1\right)}}$$ (8.190)

Equation 8.189 allows the determination of required reinforcement for shear in the core of the joint in the form of hoops and ties, so that diagonal tension cracks are avoided. The second equation (Equation 8.190) gives the upper limit for the nominal shear stress that may develop in the joint without any risk for failure due to diagonal compression.

At the same time, EC8-1/2004 has also adopted an alternative approach based mainly on the Paulay–Priestley model, which has been introduced with some simplifications.

Finally, it should be noted that the basic approach of EC8-1/2004 to the problem gives reliable results in relation to the statistical evaluation elaborated by Kitayama et al. (1991) *only for medium–high values of* ν_d (around 0.30). These values for columns are the most common (DCM: $\nu_d \geq 0.65$, DCH $\nu_d \geq 0.55$).

8.4.3 Code provisions for the design of joints under seismic action

From what has been presented so far, it is apparent that there are deviations from country to country in the Code specifications for the design of R/C joints against seismic action. Thus,

- The New Zealand Seismic Code NZS 3101 (1995) has adopted the Paulay and Priestley (1992) method as it was developed in Sections 8.4.2.1 and 8.4.2.2.
- The U.S. Code ACI 318-2011 has adopted a very simple approach based on statistical evaluations of test results. According to this,
  - Joints must be reinforced by hoops and ties of the same type as the critical regions of the adjacent columns.
  - Nominal shear υ_{jh} must be less than

$$\upsilon_{jh} \leq 1.7\sqrt{f_{cd}} \text{ (MPa) for internal joints (beams framing on four faces)}$$ (8.191)

$$\upsilon_{jh} \leq 1.2\sqrt{f_{cd}} \text{ (MPa) for external joints (beams framing on three faces)}$$ (8.192)

- The European Code EC8-1/2004 is based on the mechanical model presented in Section 8.4.2.3 and specifies in detail the following rules.

8.4.3.1 DCM R/C buildings under seismic loading according to EC 8-1/2004

The Code does not specify any analytical verification. The specified requirements may be summarised as follows:

- The horizontal confinement of the critical regions of the columns must also be extended in the joint.
- If beams frame into all four sides of the joint and their width is at least three-fourths of the parallel cross-sectional dimension of the column, the spacing of the horizontal confinement in the joint may be increased to twice that specified for the critical region of the columns but not greater than 150 mm.
- At least one intermediate vertical bar should be provided between corner bars at each side of the joint.

8.4.3.2 DCH R/C buildings under seismic loading according to EC 8-1/2004

1. Capacity design effects
   a. For the horizontal shear force V_{jhd} acting on the concrete core of the joints, expressions 8.165 and 8.169 may be used.
   b. The shear force of the column V_{col} above the joint corresponds to the most adverse value resulting from the analysis in the seismic design.
2. ULS verification and detailing
   a. The diagonal compression in the diagonal strut mechanism, taking into account in addition the transverse tensile strains, should not exceed the value $v_1 f_{cd}$ (see Equation 8.188b), where

$$v_1 = 0.6\left[1 - \frac{f_{ck}}{250}\right] (f_{ck} \text{ in MPa}) \tag{8.193}$$

The above requirement is deemed to be satisfied if the following rules are in effect:
   i. Interior beam–column joints
   Shear demand V_{jhd} should satisfy the following expression:

$$\boxed{V_{jhd} \le v_1 f_{cd} \sqrt{1 - \frac{v_d}{v_1}} \cdot b_j \cdot h_{jc}} \tag{8.194}$$

where

$$b_j = \min\{b_c; (b_w + 0.5h_c)\} \text{ if } b_c > b_w \tag{8.195}$$

$$b_j = \min\{b_w; (b_c + 0.5h_c)\} \text{ if } b_c < b_w \tag{8.196}$$

- h_{jc} is the distance between extreme layers of column reinforcement.
- v_d is the normalised axial force in the column above the joints.
   Expression 8.194 results from expression 8.190 with the simplification

$$\rho_{jh} = 0$$

which is on the safe side.

The above expression ensures that joints are safe against a diagonal compression failure mode.

    ii. Exterior beam–column joints

At the exterior beam–column joints, V_{jhd} should also satisfy expression 8.194 with its right-hand term reduced by 0.80.

  b. Shear reinforcement of the joint (horizontal reinforcement of hoops and vertical reinforcement of intermediate longitudinal bars) should satisfy the following expression:

$$\boxed{\frac{A_{sh} \cdot f_{ywd}}{b_j h_{jw}} \geq \frac{\left(\dfrac{V_{jhd}}{b_j \cdot h_{jc}}\right)^2}{f_{ctd} + v_d f_{cd}} - f_{ctd}} \tag{8.197}$$

where
- A_{sh} is the total area cross section of the horizontal hoops.
- V_{jhd} is as defined by expressions 8.165 and 8.169.
- h_{jw} is the distance between the top and the bottom reinforcement of the beam.
- b_{jc} is the distance between extreme layers of the column reinforcement.
- b_j is as defined above in expressions 8.195 and 8.196.
- v_d is the normalised design axial force of the column above ($v_d = N_{Ed}/A_c f_{cd}$).
- f_{ctd} is the design value of the tensile strength of concrete in accordance with EN 1992-1-1-2004.

Equation 8.197 results from Equation 8.189 presented in the previous paragraph.

  c. As an alternative to Equation 8.197, the following equations may be used:

    i. In interior joints:

$$\boxed{A_{sh} f_{ywd} \geq \gamma_{Rd}(A_{s1} + A_{s2})f_{yd}(1 - 0.8v_d)} \tag{8.198}$$

    ii. In exterior joints:

$$\boxed{A_{sh} f_{ywd} \geq \gamma_{Rd} A_{s2} f_{y1}(1 - 0.8v_d)} \tag{8.199}$$

where
- $\gamma_{Rd} = 1.20$.
- v_d refers to the column above the joint in the case of an interior column and to the column below the joint in the case of an exterior column.
- A_{s1} is the beam top reinforcement area.
- A_{s2} is the beam bottom reinforcement area.

Equations 8.198 and 8.199 result from Equation 8.179 after the introduction of some additional simplifications. It should be noted that the results of the basic expressions 8.194 and 8.197 differ significantly from those of the alternative expression of

Equations 8.198 and 8.199 presented above. It is the author's opinion that the whole issue of the design of column beam joints should be re-examined on the basis of the approach of ACI 318-2011, which, as was explained previously, is based mainly on statistical evaluation of test results, until more reliable expressions are established for the design.

d. Adequate vertical reinforcement of the column passing through the joint should be provided, so that

$$\boxed{A_{sv,1} \geq 2A_{sh}(h_{jc}/h_{jw})/3}$$

(8.200)

where
- A_{sh} is the required total area of the horizontal hoops in the joint.
- $A_{sv,1}$ denotes the total area of the intermediate bars placed in the relevant column sides between corner bars.

e. Rules referring to DCM buildings are in effect also for DCH buildings unless they are covered by the rules set forth above.

8.4.4 Non-conventional reinforcing in the joint core

Park and Paulay (1975) pointed out that the typical shear reinforcement of joint cores, which consists of hoops and vertical bars, may be replaced by cross-inclined bars resulting from bending part of the longitudinal reinforcement of the beam. It is clear that such a detailing causes construction difficulties, and it is infeasible to apply it in two orthogonal directions (when beams are framing into all four faces of a joint); hence, it has not been applied in practical situations.

On the other hand, the results of tests at the Aristotle University of Thessaloniki (Tsonos et al., 1995), where there was use of cross-inclined bars resulting from bending of column reinforcement, as shown in Figure 8.67, have shown superior performance of the non-conventionally reinforced beam–column sub-assemblages compared with similar specimens reinforced only with hoops and vertical straight bars. As shown in Figure 8.68, the hysteresis loops of specimens with cross-inclined column bars were considerably more stable and less pinched than the loops of similar conventionally reinforced specimens. The improvement was attributed mainly to the prevention of slippage of column bars within the joint core and to the increase in shear strength caused by the cross-inclined bars (see also Section 8.3.6.2).

It is also pointed out that for this arrangement of bars, the required hoop reinforcement is less than in conventionally reinforced joints. However, the use of cross-inclined column bars in two-way frames presents insurmountable difficulties. A possible way of resolving this problem might be the use of a pair of inclined bars placed along the diagonal of the column section; such a novel reinforcing pattern should, of course, first be studied experimentally to evaluate its efficiency as joint shear reinforcement.

8.5 MASONRY-INFILLED FRAMES

8.5.1 General

The masonry infills under consideration in this section are constructed after hardening of the concrete skeleton, in contact with it, but without special connection to it. In this context,

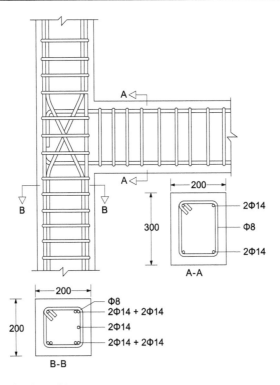

A-A

2Φ14
Φ8
2Φ14

B-B

Φ8
2Φ14 + 2Φ14
2Φ14
2Φ14 + 2Φ14

Figure 8.67 Arrangement of x-shaped bars in an exterior joint core.

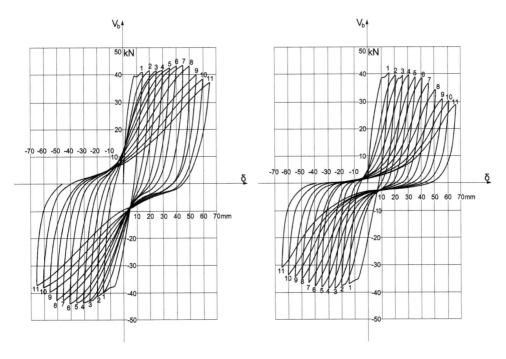

Figure 8.68 Hysteresis loops for exterior beam–column joints: (a) with x-shaped bars; (b) with conventional reinforcement.

they are considered in the first instance as non-structural elements and, therefore, as gravity loads on the underlying frame beams. This type of masonry infill is very common in southern Europe, where seismicity is very high.

Older versions of Codes (Vintzeleou, 1987) provide specific instructions for the design and construction of infilled structures, recommending two alternatives: either an effective isolation of the infills from the surrounding frames (Chapter 6.2.3, Figure 6.15) or a tight placing of the infills so that their interaction with the frames should be properly considered. In the first case, the structural system is clear and relatively reliable, but the infills are susceptible to overturning out of plane, in case of a strong earthquake. In the second case, the tight placing seems to improve the seismic behaviour of the building, a fact that counterbalances to a degree the additional inertial seismic forces corresponding to the significant masses of masonry infills.

Given that infills of this type have a considerable strength and stiffness, they provide a remarkable contribution to the seismic response of the structural system. In general, the presence of masonry infills affects the seismic behaviour of buildings in the following ways (Tassios, 1984b; Penelis and Kappos, 1997; Dowrick, 2005; Fardis, 2009; Stylianidis, 2012).

- The stiffness of the building increases, while the fundamental period decreases, and therefore the base shear due to seismic action also increases.
- The distribution of the lateral stiffness of the structure in plan and elevation is modified.
- Part of the seismic action is carried by the infills, thus relieving the structural system.
- The ability of the building to dissipate energy increases substantially.

The more flexible the structural system is, the more the above effects of the infills are observed in the structural system. Masonry infills have less deformability than the structural system, which means that they are brittle structural elements and, therefore, they fail first, presenting at first stage separations from the frame (see Chapter 11.1.7, Figure 11.30) and then x-shaped diagonal cracks or slide cracks at the mid-height of the panel (Figures 8.69 and 11.32). Thus, infills dissipate significant quantities of seismic energy, acting as a first line of seismic defence for the building. To this contribution of masonry infills must be attributed the limited damage in the structural system of R/C buildings with regular distribution of masonry in plan and elevation (residential buildings without stores at the ground level) in cities affected by strong earthquakes like Bucharest in 1977, Thessaloniki in 1978, Athens in 1981, etc.

However, there are many drawbacks in regard to the positive influence of masonry infills on the seismic behaviour of buildings. The early failure of infills under small inter-storey drifts due to their brittle character leads to a decrease in strength and stiffness of the building as a whole, and, therefore, to a simultaneous load transfer from the infill to the R/C structural system in the form of impulse loading. In this context, the load path in the case of a strong earthquake is transformed from one moment to the other with questionable reliability. Therefore, the various analytical models developed so far are complicated, due to the descending branch of the M–θ diagram; they are sensitive, and, in this respect, unreliable for the prediction of the inelastic response of the structure to the successive steps of infill failure under a strong seismic motion (Michailidis et al., 1995). On the other hand, the quality control of the masonry infills is also not reliable (brick and mortar strength, brick construction pattern, etc.). Finally, the impulse loading of R/C frames at their diagonally located corners by the infill masonry (Figure 8.70) often causes shear failure in the case of infill of high strength, as in the case of short columns (see Section 8.3.6).

The advantages and disadvantages above were presented in detail in the first edition of this book (Penelis and Penelis, 2014) on the basis of experimental evidence carried out for

Figure 8.69 Failure mode of the infilled frame F5. (Adapted from Stylianidis, K. 2012. Journal of O.C.B.T., 6, 194–212.)

Figure 8.70 Separation of the infill from the frame; local crushing of the infill corners of frame F4R. (Adapted from Stylianidis, K. 2012. Journal of O.C.B.T., 6, 194–212.)

many years in the lab of R/C structures of Aristotle University of Thessaloniki (Figures 8.69 through 8.72; Stylianidis, 1985, 2012; Valiasis, 1989; Sariyiannis, 1990; Valiasis et al., 1993). In the subsequent sections, detailed reference will be made to Code specifications for infills frames as far as modelling, analysis and design.

8.5.2 Code provisions for masonry-infilled frames under seismic action

8.5.2.1 Requirements and criteria

The effects of the infills in analysis and design of a building must be taken into account together with the high degree of uncertainty related to their behaviour, namely,

- The variability of their mechanical properties and, therefore, the low reliability in their strength, stiffness and energy dissipation capacities

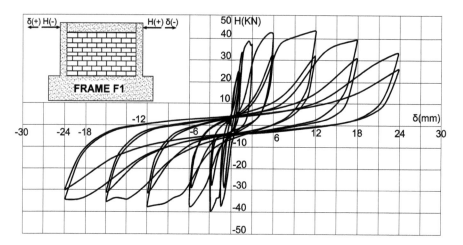

Figure 8.71 Lateral load–displacement loops. Infilled frame F1. (Adapted from Stylianidis, K. 2012. Journal of O.C.B.T., 6, 194–212.)

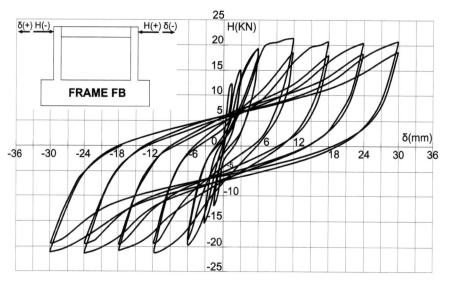

Figure 8.72 Lateral load–displacement loops. Bare frame FB. (Adapted from Stylianidis, K. 2012. Journal of O.C.B.T., 6, 194–212.)

- Their wedging condition, that is, how tightly they are connected to the surrounding frame
- The potential modification of their integrity during the life of the building
- Their irregular arrangement in plan and elevation in the building
- The possible adverse local effects due to the frame–infill interaction in critical regions of the frame (e.g. shear failure of columns under shear forces induced by the diagonal strut action of the infill)

Thus, according to EC 8-1/2004, *the safety of the structure cannot rely, not even partly, upon the infills and only their probable adverse influence is taken into account.*

Because of their high lateral stiffness, a large percentage of the seismic effects would have been transferred to them in the case that they were taken into account as structural elements. However, for increasing distortions, this initial distribution of seismic effects between infills and frames would change rapidly against the frame system. It is obvious that such an approach would be against the structural safety of the system.

So, according to EC 8-1/2004, the seismic analysis, in general, is carried out *on the bare frame system, and only additional measures are taken for an eventual adverse influence* of the masonry infills on the structural system. These measures are given in the following paragraphs and refer *only to frame and frame-equivalent dual systems* for DCH buildings. However, they provide criteria for good practice, which may also be useful for DCM and DCL R/C buildings. For all wall or wall-equivalent dual systems, the interaction between the concrete structural system and the infills may be neglected.

It is the author's opinion that the above approach of the Code to the problem is very conservative. Documented capacities in stiffness strength and energy dissipation of masonry infills could be exploited to a degree for low-rise buildings with regular distribution of infills in plan and elevation, combined with rules that would ensure the quality control of masonry construction. In this way, it is expected that a significant cost reduction could be obtained, particularly for low-cost residential blocks of apartments.

8.5.2.2 Irregularities due to masonry infills

1. Irregularities in plan
   a. Strongly irregular, non-symmetric or non-uniform arrangements of infills in plan should be avoided. Of course, this is not always possible due to the architectural constraints imposed on the design. For example, in the case of buildings located at the corner of a building block, with stores on the ground floor, there are infills along two consecutive sides of the building adjacent to the block and there are show windows on the other two. In such a case, the structural system should be analysed again, including in the structural model the infills simulated according to the recommendation presented in Section 8.5.2.4. It is apparent that the re-analysis should be carried out for a 3-D structural model. In this case, special attention should be paid to the verification of structural elements on the flexible sides of the plan against the effects of any torsional response caused by the infills.
   b. When masonry infills are not regularly distributed, but not in such a way as to constitute a severe irregularity as in the corner building above, these irregularities may be taken into consideration by increasing the effects of the accidental eccentricity imposed by Code with a factor λ equal to 2.0:

$$\boxed{\lambda = 2.0} \tag{8.201}$$

2. Irregularities in elevation
   a. If there are considerable irregularities in elevation (e.g. in the case of a pilotis-open ground storey imposed by the Building Code, or ground floors used as stores with open space and show windows at the facade), the internal seismic forces in the vertical elements of the respective storey should be increased (penalised).
   b. A simplified procedure for the determination of this magnifying factor η is specified by EC 8-1/2004 and is given below:

$$\boxed{\eta = \left(1 + \Delta V_{RW} \Big/ \sum V_{Ed}\right) \leq q} \tag{8.202}$$

where
- ΔV_{RW} is the total reduction of the resistance of masonry infills, in the storey concerned, compared to the more infilled storey above it.
- ΣV_{Ed} is the sum of the seismic shear forces acting on all vertical primary seismic members of the storey concerned.

c. If the magnification factor η, defined above, is lower than 1.10, there is no need for modification of the seismic action effects.
d. It should be remembered that the above penalty expressed by the magnifier η, at least for DCH buildings, should be added for the columns of the storey concerned to the penalty imposed on the whole structural system of a 20% reduction of the q-factor for DCM and DCH buildings with significant irregularities in elevation with abrupt changes.
e. It is important to note that for the determination of ΔV_{RW} in Equation 8.202, there is no need for an additional analysis, since V_{RW} refers to the strength capacity of the individual infill panels.

8.5.2.3 Linear modelling of masonry infills

For the linear analysis of structural systems, including masonry infills, particularly in the case of significant irregularities in plan or even in the case of strong irregularities in elevation (e.g. soft ground story), two procedures may be followed.

The most usual one is the simulation of the infill through a compression diagonal strut (Figure 8.73) with the following properties (Michailidis et al., 1995):

- Thickness t of the strut: the thickness of the wall t
- Width of the strut: $w \simeq 0.20d$, where d is the length of the diagonal
- Modulus of elasticity E_w of the strut: the modulus of elasticity of the masonry taken from EC 6

The second procedure, in use in the last decade, simulates the infill panels through shell finite elements (FEM). Taking into account that a rapid degradation of the infill occurs, it

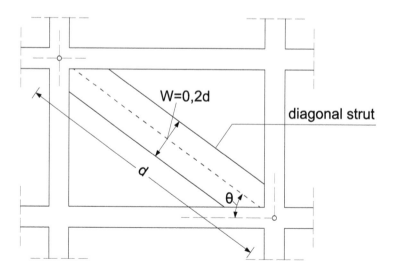

Figure 8.73 Compression diagonal model for the estimation of infill stiffness.

should not be forgotten that for the modulus of elasticity of the masonry, a reduction factor equal to 0.5 or less should be introduced.

In the case of perforated walls with more than one opening (e.g. door and window), the infill is disregarded in the model.

8.5.2.4 Design and detailing of masonry-infilled frames

8.5.2.4.1 General

Masonry-infilled frames may have two different forms:

1. Infill panels are separated from the structure and are fixed in the surrounding frame in such a way that they do not interfere with the structural deformation (Chapter 6, Figure 6.15). The main issues for these panels are their stabilisation against out-of-plane overturning in case of a strong earthquake and their heating and sound insulation problem at the gaps with the surrounding frame. This type of infill has no influence on the structural behaviour of the building in response to seismic actions.
2. Infill panels are integrated with the structural system. They are usually built in the form of masonry infills after concrete hardening with R/C tie belts and posts, they are edged at the surrounding frames and they are often connected with them. It is obvious that the presentation so far refers to this type of masonry-infilled structures (Figure 8.48). In any case, for both infill types belonging to all ductility classes DCL, M and H, appropriate measures are taken to avoid brittle failure and premature disintegration of the infill walls. The main measures refer to 'damage limitation' and particularly to 'the limitation of inter-storey drifts', which have been presented in detail in Chapter 6.2.3.

8.5.2.4.2 Masonry infills integrated into the structural system

For these masonry infills, in addition to the 'limitation of inter-storey drifts' specified by the Code, the following protective measures should be taken:

1. Particular attention should be paid to masonry panels with a slenderness ratio larger than 15 (ratio of the smaller of length or height to thickness). The same applies in case of openings.
2. The measures for the protection of the above masonry walls against out-of-plane overturning or shear failure x-shaped cracks are the following:
   a. Wall ties fixed to the columns and cast into the bedding planes of the masonry.
   b. Reinforced concrete posts and belts across the panels and through the full thickness of the wall.
   c. If there are large openings or perforations in any of the infill panels, their edges should be trimmed with belts and posts.

8.5.2.4.3 Adverse effects on columns adjacent to masonry infills

1. Taking into account the particular vulnerability of the infill walls of ground floors, it is reasonable that, due to partial or total failure of some of these walls, a seismically induced irregularity is to be expected. Appropriate measures should therefore be taken. The simplest approach to the problem, according to EC 8-1/2004, is the extension of the critical region at the columns of the ground floor to their entire length being confined accordingly.

2. If the height of the infills is smaller than the clear length of the adjacent columns, the following measures should be taken (Figure 8.48).
   - The entire length of the column is considered a critical region and should be reinforced with the number and pattern of stirrups for critical regions.
   - The consequences of the decrease in the shear span ratio of these columns should be appropriately confronted. In fact, in this case, the plastic hinges of the column are likely to be formed at their top, and at the captive point at the top of the masonry wall. Therefore, the capacity design shear force is increased, since the shear span ratio decreases. In this case, the clear length of the column l_{cl} is taken to be equal to the free part of the column, and $M_{i,d}$ at the column section at the top of the infill wall should be taken as being equal to $\gamma_{Rd} M_{Rc,i}$, with $\gamma_{Rd} = 1.1$ for DCM and 1.3 for DCH and $M_{Rc,i}$ the design resistance of the column.
   - The transverse reinforcement resulting above should cover the free part of the column plus a height equal to the dimension of the cross section of the column parallel to the infills in extension to the contact region of the column with the infill.
   - If the length of the free part of the column is less than $1.5\ h_c$, the shear force should be resisted *by diagonal* reinforcement (see Section 8.3.6.2).
3. When the masonry infill extends to the entire height of the columns, but only to one side of it while the other is free (e.g. corner columns), the entire length of the column should be considered a critical region and be reinforced with the number and pattern of stirrups required for critical regions.
   - The length l_c of the above columns, over which the diagonal strut force of the infill is applied, should be verified in shear for the smaller of the following shear forces:
     - The horizontal component of the strut force of the infill estimated to be equal to the horizontal *shear strength* of the panel, determined on the basis of the shear strength of bed joints.
     - The shear force estimated on the basis of shear capacity design for columns applied to the contact length l_c of the strut to column and with moment resistance at the ends of the contact length l_c, the overstrengthened moment resistance $\gamma_{Rd} M_{Rc,i}$ of the column. In most cases, this second value is much higher than the shear strength capacity of the infill along its horizontal joint bed.
4. All measures presented above for confronting the adverse effects of masonry infills on the columns hold for both ductility classes, more specifically DCM and DCH buildings.

8.5.3 General remarks on masonry-infilled frames

The approach of EC 8-1/2004 to masonry-infilled frames and masonry-infilled frame-equivalent dual systems of DCM and DCH seems to be rather prohibitive, for the reasons below:

- The increase in the accidental eccentricity for DCH by 2, even for buildings with limited irregularities in infill arrangement, increases the seismic effects.
- The increase in the calculated action effects on the bare structure for DCH by a factor η ranging up to 1.70 for irregularities in elevation leads to more reinforcement.
- The re-analysis of buildings with high irregularities in infill arrangement, taking into account their stiffness and redesigning of all structural elements (mainly columns) leads to an adverse influence on their design action effects due to infills.

The design rules specified by the Code for the columns in infilled frames lead the designer to avoid the use of masonry infills or to provide separation joints between masonry and the R/C structural system. However, extensive research (Valiasis and Stylianidis, 1989; Valiasis,

Stylianidis and Penelis, 1993; Michailidis et al., 1995; Stylianidis, 2012) has shown that masonry infills constitute a highly effective dissipation mechanism. In addition, extended statistical damage evaluation in areas affected by strong earthquakes (Penelis et al., 1987; Fardis, 2009) has shown, in general, a positive influence of masonry infills on the seismic behaviour of buildings. In this respect, infill masonry walls constitute a first line of defence of the building against earthquakes, acting as a type of damper that protects the R/C skeleton to a significant degree. Therefore, the subject should be reconsidered, in the future, by the Code so that methods could be developed that will allow, on one hand, local improvement of the structural elements suffering from the presence of masonry infills, while on the other favouring and promoting the extensive use of masonry infills in building construction.

8.6 EXAMPLE: DETAILED DESIGN OF AN INTERNAL FRAME

The internal frame along the y–y direction, grid axis B, of the eight-storey RC building studied in Chapter 5 has been selected for a detailed design (Figure 5.33). Since beams B8 and B19 are symmetrical to B57 and B68, the analytical calculations are performed only for B8 and B19, whereas beam B37 follows the design of B19. The same applies for columns C2, C8 and C14, which are symmetrical to C28, C22 and C18. In Figures 8.74 and 8.75, the moment and shear diagrams for the various loading cases are shown. (It should be recalled that the number of seismic load combinations is limited to 8. Details are provided in Chapter 5.9.1.)

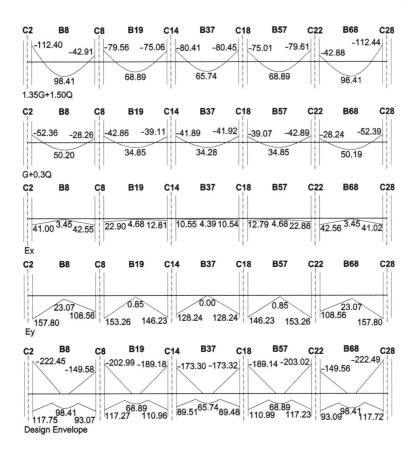

Figure 8.74 Moment diagrams for the various loading cases.

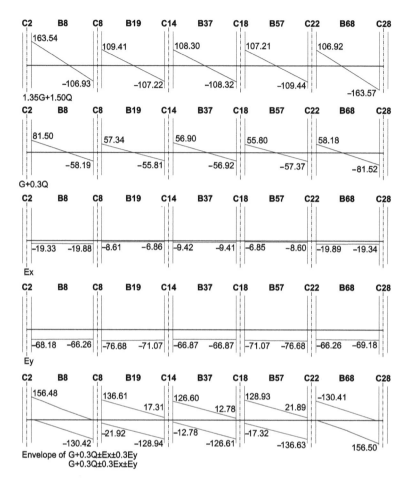

Figure 8.75 Shear diagrams for the various loading cases.

8.6.1 Beams: ultimate limit state in bending

Minimum percentage of longitudinal reinforcement (Equation 8.73)

$$\rho_{min} = 0.5 \frac{f_{ctm}}{f_{yk}} = 0.5 \frac{2.6}{500} = 0.0026 \rightarrow A_{s,min} = 4.66 \text{ cm}^2$$

8.6.1.1 External supports on C2 and C28 (beam B8 – left, B68 – right)

Hogging design moment for the top reinforcement: $M_d^- = -222.45$ kN m
Sagging design moment for the bottom reinforcement: $M_d^+ = 117.75$ kN m
Top longitudinal reinforcement:

$$\mu_{sd} = \frac{|M_d^-|}{b \cdot d^2 \cdot f_{cd}} = \frac{222.45 \cdot 10^6}{300 \cdot 597^2 \cdot \dfrac{25}{1.5}} = 0.125 \rightarrow \omega_{req} = 0.1368 \rightarrow$$

$$A_{s,req} = \omega_{req} \cdot b \cdot d \frac{f_{cd}}{f_{yd}} = 0.1368 \cdot 300 \cdot 597 \frac{25/1.5}{500/1.15} = 9.39 \text{ cm}^2 \rightarrow 5\varnothing16 \ (10.05 \text{ cm}^2)$$

Experimental evidence has shown that after flexural yielding of a beam in negative bending (tension in the upper reinforcement), part of the slab reinforcement up to a significant distance from the web of the beam is fully activated and contributes to the beam negative flexural capacity as tension reinforcement (Fardis et al., 2005). The slab width effective as the tension flange of a beam at the support of a column is specified in paragraph 5.4.3.1.1(3.b) of EC8-Part I (2004). The effective width of the external support is shown in Figure 8.76.

The slab reinforcement is $\varnothing8/150$. The reinforcement of the slab within the effective width is equal to $8\varnothing8$ (4.02 cm^2). The percentage of the provided tensile reinforcement, ρ_{prov}, is

$$\rho_{prov} = \frac{(4.02 + 10.05)}{300 \cdot 597} = 0.00786$$

Bottom longitudinal can be determined by

$$\mu_{sd} = \frac{|M_d^+|}{b_{eff} \cdot d^2 \cdot f_{cd}} = \frac{117.75 \cdot 10^6}{1700 \cdot 597^2 \frac{20}{1.5}} = 0.012 \rightarrow \omega_{req} = 0.0122 \rightarrow$$

$$A_{s,req} = \omega_{req} \cdot b_{eff} \cdot d \frac{f_{cd}}{f_{yd}} = 0.0122 \cdot 1700 \cdot 597 \frac{25/1.5}{500/1.15} = 4.76 \text{ cm}^2 \rightarrow 4\varnothing14 (6.16 \text{ cm}^2)$$

$$\rho'_{prov} = \frac{A_{s,req}}{b \cdot d} = \frac{6.16 \cdot 10^2}{300 \cdot 597} = 0.00344 < \frac{\rho_{prov}}{2} = 0.00393$$

Hence, the compressive longitudinal reinforcement has to be at least equal to $5\varnothing14$ (7.70 cm^2):

$$\rho'_{prov} = 0.00430 > \frac{\rho_{prov}}{2} = 0.00393$$

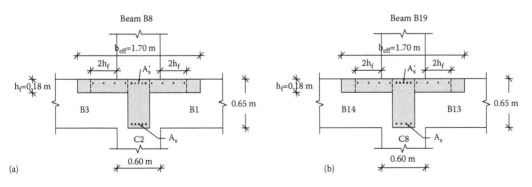

Figure 8.76 Slab effective width of the (a) external (beam 8) and (b) the internal support (beam 19).

The maximum reinforcement ratio of the tension zone, ρ_{lmax}, is determined by (Equation 8.69)

$$\rho_{l\max} = \rho_2 + \frac{0.0018}{\mu_{\varphi dem} \cdot \varepsilon_{sy,d}} \cdot \frac{f_{cd}}{f_{yd}}$$

For

$$T_1 = 0.82 \text{ s} \rightarrow T_1 > T_C = 0.6 \text{ s}: \mu_\varphi = 2q_o - 1 = 7.8$$

Hence,

$$\rho_{\max} = 0.00393 + \frac{0.0018}{7.8 \cdot \frac{(500/1.15)}{200000}} \cdot \frac{(25/1.5)}{(500/1.15)} = 0.00837 > \rho_{prov} = 0.00786$$

8.6.1.2 Internal supports on C8 and on C22 (beam B8 – right, B19 – left, B57 – right, B68 – left)

Hogging design moment for the top reinforcement: $M_d^- = -202.99 \text{ kN m}$
Sagging design moment for the bottom reinforcement: $M_d^+ = 117.27 \text{ kN m}$
Top longitudinal reinforcement:

$$\mu_{sd} = 0.114, \omega_{req} = 0.1237, A_{s,req} = 8.49 \text{ cm}^2, 5\emptyset16(10.05 \text{ cm}^2)$$

The percentage of the provided tensile reinforcement (ρ_{prov}) is (slab reinforcement is $8\emptyset8$): $\rho_{prov} = 0.00786$.
Bottom longitudinal reinforcement:

$$\mu_{sd} = 0.0116, \omega_{req} = 0.0122, A_{s,req} = 4.74 \text{ cm}^2 \rightarrow 4\emptyset14 \ (6.16 \text{ cm}^2)$$

$$\rho'_{prov} = 0.00344 < \rho_{prov}/2 = 0.00393$$

Hence, the compressive reinforcement is taken to be equal to $5\emptyset14$ (7.70 cm²):

$$\rho'_{prov} = 0.00430 > \rho_{prov}/2 = 0.00393$$

The maximum reinforcement ratio of the tension zone, $\rho_{\max}$ (Equation 8.69), for

$$T_1 = 0.82 \text{ s} \rightarrow T_1 > T_C = 0.6 \text{ s}: \mu_\varphi = 2q_o - 1 = 7.8, \rho_{\max} = 0.00837 > \rho_{prov} = 0.00786$$

8.6.1.3 Internal supports on C14 and C18
(beam B19 – right, B37 – left, B37 – right, B57 – left)

Hogging design moment for the top reinforcement: $M_d^- = -189.18$ kN m
  Sagging design moment for the bottom reinforcement: $M_d^+ = -110.96$ kN m
  Top longitudinal reinforcement:

$$\mu_{sd} = 0.1062, \omega_{req} = 0.1145, A_{s,req} = 7.86 \text{ cm}^2, 4\varnothing16 \ (8.04 \text{ cm}^2)$$

The percentage of the provided tensile reinforcement (ρ_{prov}) is (slab reinforcement is $8\varnothing8$):
$\rho_{prov} = 0.00674$.
  Bottom longitudinal reinforcement:

$$\mu_{sd} = 0.0110, \omega_{req} = 0.0115, A_{s,req} = 7.49 \text{ cm}^2 \rightarrow 3\varnothing14 \ (4.62 \text{ cm}^2)$$

$$\rho'_{prov} = 0.00258 < \rho_{prov}/2 = 0.00337$$

Hence, the compressive reinforcement is taken to be equal to $4\varnothing14$ (6.16 cm²):

$$\rho'_{prov} = 0.00334 \cong \rho_{prov}/2 = 0.00337$$

The maximum reinforcement ratio of the tension zone, ρ_{max} (Equation 8.72):
For

$$T_1 = 0.82 \rightarrow T_1 > T_C = 0.6 \text{ s}: \mu_\varphi = 2q_o - 1 = 7.8, \rho_{max} = 0.00751 > \rho_{prov} = 0.00674$$

8.6.1.4 Mid-span (beams B8, B68)

Sagging design moment for the bottom reinforcement: $M_d = 98.41$ kN m.
  Bottom longitudinal reinforcement:

$$\mu_{sd} = 0.0097, \omega_{req} = 0.0102, A_{s,req} = 4.33 \text{ cm}^2 < A_{s,min} = 4.66 \text{ cm}^2 \rightarrow 4\varnothing14 = 6.16 \text{ cm}^2$$

Top longitudinal reinforcement: In the case of DCH, one quarter of the maximum top reinforcement at the supports shall run along the entire beam length – $A_{s,sup} = 10.05 \text{ cm}^2 \ (5\varnothing16) A'_{s,mid} > 1/4 \cdot A_{s,sup} = 2.51 \text{ cm}^2 \ 2\varnothing16 \ (4.02 \text{ cm}^2)$ is provided as top reinforcement.

8.6.1.5 Mid-span (beams B19, B37, B57)

Sagging design moment for the bottom reinforcement: $M_d = 68.89$ kN m.
  Bottom longitudinal reinforcement:

$$\mu_{sd} = 0.0068, \omega_{req} = 0.0072, A_{s,req} = 3.03 \text{ cm}^2 < A_{s,min} = 4.66 \text{ cm}^2 \rightarrow 4\varnothing14 = 6.16 \text{ cm}^2$$

Top longitudinal reinforcement:

$A_{s,sup} = 10.05 \text{ cm}^2 (5\text{Ø}16) A'_{s,mid} > 1/4 \cdot A_{s,sup} = 2.51 \text{ cm}^2 \ 2\text{Ø}16$ is   provided   as   top reinforcement.

At this stage, the maximum allowable bar diameter of beam longitudinal bar passing through beam–column joints, d_{bL}, is defined:

Interior beam–column joints (Equation 8.81): for $N_{Ed} = 1739.99 \text{ kN} \rightarrow \nu_d = 0.289$,

$$\frac{d_{bL}}{h_c} \leq \frac{7.5 \cdot f_{ctm}}{\gamma_{Rd} \cdot f_{yd}} \cdot \frac{1+0.8 \cdot \nu_d}{1+0.75 \cdot K_D \cdot \rho'/\rho_{max}} = \frac{7.5 \cdot 2.6}{1.2 \cdot 434.78} \cdot \frac{1+0.8 \cdot 0.289}{1+0.75 \cdot 1 \cdot 0.00344/0.00751} \Leftrightarrow d_{bL} \leq 19.94 \text{ mm}$$

Exterior beam–column joints (Equation 8.82): for $N_{Ed} = 991.21 \text{ kN} \rightarrow \nu_d = 0.165$,

$$\frac{d_{bL}}{h_c} \leq \frac{7.5 \cdot f_{ctm}}{\gamma_{Rd} \cdot f_{yd}} \cdot (1+0.8 \cdot \nu_d) = \frac{7.5 \cdot 2.6}{1.2 \cdot 434.78} \cdot (1+0.8 \cdot 0.165) \Leftrightarrow d_{bL} \leq 25.4 \text{ mm}$$

8.6.2 Columns: ultimate limit state in bending and shear

The columns selected for design are columns C2 and C8.

8.6.2.1 Column C2 (exterior column)

8.6.2.1.1 Ultimate limit state in bending

For all the load combinations that appear in Table 8.4 at the top and base of column C2 of storey 1, $\omega_{req} = 0$. Thus, the minimum longitudinal reinforcement is placed. The number and bar diameters selected are 4Ø20 + 12Ø16 (36.69 cm² > $A_{s,min}$).

Minimum reinforcement: $A_{s,min} = \rho_{min} \cdot b \cdot h = 0.01 \cdot 600 \cdot 600 = 36.00 \text{ cm}^2$.

The minimum reinforcement is placed in all the columns of the building.

8.6.2.1.2 Calculation of the design values of the moments of resistance of beam 8 at the support on C2 (beam 8 – left)

→ E_y: Bottom reinforcement – 5Ø14, $\omega = 0.0198$, $\mu_{Rd} = 0.019$,

$$M_{ARd} = \mu_{Rd} \cdot b_{eff} \cdot d^2 \cdot f_{cd} = 0.019 \cdot 1.700 \cdot 0.597^2 \cdot \frac{25 \cdot 10^3}{1.5} = 190.27 \text{ kN m}$$

← E_y: Top reinforcement – 8Ø8 + 5Ø16, $\omega = 0.2050$, $\mu_{Rd} = 0.179$,

$$M'_{ARd} = \mu_{Rd} \cdot b \cdot d^2 \cdot f_{cd} = 0.0179 \cdot 0.600 \cdot 0.597^2 \cdot \frac{25 \cdot 10^3}{1.5} = 319.50 \text{ kN m}$$

8.6.2.1.3 Calculation of the design values of the moments of resistance at the ends of column C2

At the base of C2 at the second storey (Figure 8.77), using the corresponding earthquake forces from a table similar to Table 8.4

Table 8.4 Bending moments and axial forces at column C2 ends for column dimensioning – storey I

Load combinations		N	V_x	V_y	M_x	M_y
Top of column C2						
Lc1	$1.35G+1.5Q$	−2831.27	−1.67	−26.26	49.27	3.34
Lc2	$G+0.3Q+Ex+0.3Ey$	−1268.36	47.39	4.77	52.36	88.73
Lc3	$G+0.3Q+Ex−0.3Ey$	−1529.94	43.22	−15.65	21.26	81.21
Lc4	$G+0.3Q−Ex+0.3Ey$	−1675.1	−45.13	−9.67	25.82	−77.37
Lc5	$G+0.3Q−Ex−0.3Ey$	−1936.68	−49.3	−30.09	−5.27	−84.89
Lc6	$G+0.3Q+Ey+0.3Ex$	−1105.53	19.87	23.54	79.34	39.37
Lc7	$G+0.3Q+Ey−0.3Ex$	−1227.55	−7.88	19.2	71.38	−10.46
Lc8	$G+0.3Q−Ey+0.3Ex$	−1977.49	5.97	−44.52	−24.3	14.31
Lc9	$G+0.3Q−Ey−0.3Ex$	−2099.51	−21.78	−48.86	−32.26	−35.52
Base of column C2						
Lc1	$1.35G+1.5Q$	−2884.13	−1.67	−26.26	−64.98	−3.93
Lc2	$G+0.3Q+Ex+0.3Ey$	−1307.51	−49.3	−30.09	−78.56	−125.74
Lc3	$G+0.3Q+Ex−0.3Ey$	−1569.09	−45.13	−9.67	−20.82	−115.11
Lc4	$G+0.3Q−Ex+0.3Ey$	−1714.25	43.22	−15.65	−42.25	110.63
Lc5	$G+0.3Q−Ex−0.3Ey$	−1975.83	47.39	4.77	15.48	121.26
Lc6	$G+0.3Q+Ey+0.3Ex$	−1144.68	−21.78	−48.86	−133.21	−55.41
Lc7	$G+0.3Q+Ey−0.3Ex$	−1266.7	5.97	−44.52	−122.32	15.5
Lc8	$G+0.3Q−Ey+0.3Ex$	−2016.64	−7.88	19.2	59.24	−19.98
Lc9	$G+0.3Q−Ey−0.3Ex$	−2138.66	19.87	23.54	70.13	50.94

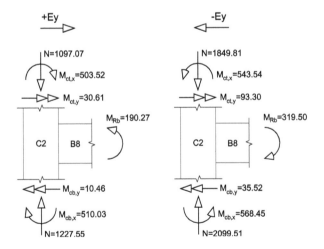

Figure 8.77 Capacity design moments for column C2.

$$\left.\begin{array}{l} \rightarrow E_y : G+0.3Q+E_y+0.3E_x, N=-991.21\,kN, v=-0.165 \\ M_y = -103.44\,kN, \mu_y = 0.035 \\ \omega = 0.2916 \end{array}\right\} \begin{array}{l} \mu_x = 0.166, \\ M_x = 498.24\,\text{kN m} \end{array}$$

$$\left.\begin{array}{l} \rightarrow E_y : G+0.3Q+E_y-0.3E_x, N=-1097.07\,\text{kN}, v=-0.183 \\ M_y = 30.61\,\text{kN}, \mu_y = 0.010 \\ \omega = 0.2916 \end{array}\right\} \begin{array}{l} \mu_x = 0.168, \\ M_x = 503.52\,\text{kN m} \end{array}$$

$$\left.\begin{array}{l} \rightarrow E_y : G+0.3Q-E_y+0.3E_x, N=-1743.95\,\text{kN}, v=-0.291 \\ M_y = -40.75\,\text{kN}, \mu_y = 0.014 \\ \omega = 0.2916 \end{array}\right\} \begin{array}{l} \mu_x = 0.179, \\ M_x = 535.78\,\text{kN m} \end{array}$$

$$\left.\begin{array}{l} \leftarrow E_y : G+0.3Q+E_y+0.3E_x, N=-1849.91\,\text{kN}, v=-0.308 \\ M_y = -93.30\,\text{kN}, \mu_y = 0.031 \\ \omega = 0.2916 \end{array}\right\} \begin{array}{l} \mu_x = 0.182, \\ M_x = 543.54\,\text{kN m} \end{array}$$

The results are summarised in Table 8.5.

The same procedure is followed for calculating the design values for the moments of resistance at the top and base of storey 1. The results appear in Tables 8.6 and 8.7, respectively.

Parameter κ_D (Equation 6.15), which accounts for the case of strong columns and weak beams, is calculated at the top of column C2 for both directions of seismic action.

Table 8.5 Determination of the design value of the moment of resistance M_x of column C2 at the base of storey 2

Load combination		N	M_y	μ_y	v	ω	μ_x	M_x
$\rightarrow E_y$	G+0.3Q+Ey+0.3Ex	−991.21	−103.44	0.035	−0.165	0.2916	0.166	498.24
	G+0.3Q+Ey−0.3Ex	−1097.07	30.61	0.010	−0.183	0.2916	0.168	503.52
$\leftarrow E_y$	G+0.3Q−Ey+0.3Ex	−1743.95	−40.75	0.014	−0.291	0.2916	0.179	535.78
	G+0.3Q−Ey−0.3Ex	−1849.81	93.3	0.031	−0.308	0.2916	0.182	543.54

Table 8.6 Determination of the design value of the moment of resistance M_x of column C2 at the top of storey 1

Load combination		N	M_y	μ_y	v	w	μ_x	M_x
$\rightarrow E_y$	G+0.3Q+Ey+0.3Ex	−1105.53	39.37	0.013	−0.184	0.2916	0.168	503.94
	G+0.3Q+Ey−0.3Ex	−1227.55	−10.46	0.003	−0.205	0.2916	0.170	510.03
$\leftarrow E_y$	G+0.3Q−Ey+0.3Ex	−1977.49	14.31	0.005	−0.330	0.2916	0.186	556.28
	G+0.3Q−Ey−0.3Ex	−2099.51	−35.52	0.012	−0.350	0.2916	0.190	568.45

Table 8.7 Determination of the design value of the moment of resistance M_x of column C2 at the base of storey I

Load combination		N	M_y	μ_y	v	ω	μ_x	M_x
→ E_y	G+0.3Q+Ey+0.3Ex	−1144.68	−55.41	0.019	−0.191	0.2916	0.169	505.90
	G+0.3Q+Ey−0.3Ex	−1266.7	15.5	0.005	−0.211	0.2916	0.171	511.98
← E_y	G+0.3Q−Ey+0.3Ex	−2016.64	−19.98	0.007	−0.336	0.2916	0.187	560.18
	G+0.3Q−Ey−0.3Ex	−2138.66	50.94	0.017	−0.356	0.2916	0.191	572.35

$$\rightarrow E_y : \sum M_{Rb} = 190.27 \text{ kN m}, \sum M_{Rc} = 1013.55 \text{ kN m}, \kappa_D = \min\left(1, \frac{190.27}{1013.55}\right) = 0.19$$

$$\rightarrow E_y : \sum M_{Rb} = 319.50 \text{ kN m}, \sum M_{Rc} = 1111.99 \text{ kN m}, \kappa_D = \min\left(1, \frac{319.50}{1111.99}\right) = 0.29$$

Parameter κ_C at the base of column C2 is taken to be equal to 1 in order to simplify the design procedure.

8.6.2.1.4 Capacity design seismic shear for column C2

The shear forces are determined according to the capacity design criterion as follows (Equation 6.11):

Clear height of the column: $l_c = h_{st} - h_b = 5 - 0.65 = 4.35$ m

$$\rightarrow E_y : V_{sd,CD} = \gamma_{Rd} \frac{\kappa_C M_{CRd} + \kappa_D M_{DRd}}{l_c} = 1.30 \frac{0.19 \cdot 510.03 + 1 \cdot 511.98}{4.35} = 181.62 \text{ kN}$$

$$\leftarrow E_y : V_{sd,CD} = \gamma_{Rd} \frac{\kappa_C M_{CRd} + \kappa_D M_{DRd}}{l_c} = 1.30 \frac{0.29 \cdot 568.45 + 1 \cdot 572.35}{4.35} = 219.86 \text{ kN}$$

The checks that follow are performed for $V_{sd,CD} = 219.86$ kN.

8.6.2.1.5 Shear resistance in the case of diagonal concrete crushing for δ = 21.8° (Equation 8.62)

$$V_{Rd,max} = \alpha_{cw} \cdot b_w \cdot 0.9d \cdot \frac{(v_1 \cdot f_{cd}) \cot \delta}{(1 + \cot^2 \delta)} = 1 \cdot 0.6 \cdot 0.9 \cdot 0.547 \cdot \frac{0.54 \cdot (25/1.5) \cdot 1000}{2.9}$$

$$= 916.70 \text{ kN}$$

$$V_{Rd,max} = 916.70 \text{ kN} > V_{sd,CD} = 219.86 \text{ kN}$$

8.6.2.1.6 Transverse reinforcement in the critical regions of column C2

The length of the critical region is

$$l_{cr} = \max\{1.5h_c; l_c/6; 0.6\} = \max\{1.5 \cdot 0.6; 4.35/6; 0.6\} = 1.35 \text{ m}$$

For the factor 1.50, see EC8-1/2004, Paragraph 5.5.3.2.2(13).
The spacing of hoops shall not exceed

$$s = \min\{b_o/3; 125; 6 \cdot d_{bL}\} = \min\{514/3; 125; 6 \cdot 16\} = 96 \text{ mm}$$

$$V_{wd} = V_{sd,CD} \Leftrightarrow s = \frac{A_{sw}}{V_{sd,CD}} 0.9d \cdot f_{ywd} = \frac{4 \cdot (\pi \cdot 8^2/4)}{219.86} \cdot 0.9 \cdot 0.547 \cdot \frac{500}{1.15} = 195.74 \text{ mm}$$

Hence, the shear reinforcement in the critical regions of column C2 is Ø8/95.

8.6.2.1.7 Transverse reinforcement outside the critical regions of column C2

The spacing of hoops shall not exceed

$$s = \min\{20\Phi_{L,\min}; \min(b_c; h_c); 400\} = \min\{20 \cdot 16; 600; 400\} = 320 \text{ mm}$$

Hence, the shear reinforcement outside the critical regions of column C2 is Ø8/180.

8.6.2.1.8 Detailing for local ductility – confinement reinforcement in the critical region at the base of column C2

The confinement reinforcement at the base of the column consists of a perimeter and an internal rectangular closed stirrup (four legs). The mechanical volumetric ratio of the required confining reinforcement, ω_{wd}, should satisfy Equation 8.127. Thus,

$$\omega_{wd,req} \geq \frac{1}{\alpha}\left(30 \cdot \mu_{\varphi} \cdot \nu_d \cdot \varepsilon_{syd} \frac{b_c}{b_o} - 0.035\right) = \frac{1}{0.641}\left(30 \cdot 7.8 \cdot 0.356 \cdot 0.0022 \frac{600}{520} - 0.035\right) = 0.27$$

This value is higher than 0.12, which is the lower limit set by the code: $\omega_{wd,req} = 0.27 > 0.12$.

The confinement effectiveness factor is Equation 8.129: $\alpha = \alpha_n \cdot \alpha_s = 0.776 \cdot 0.826 = 0.641$, where

Equation 8.130a:

$$\alpha_n = 1 - \sum_n \frac{b_i^2}{6 \cdot b_o \cdot h_o} = 1 - \frac{4 \cdot 247 + 8 \cdot 123.5}{6 \cdot 522 \cdot 522} = 0.776$$

Equation 8.131a:

$$\alpha_s = (1 - s/2b_o)(1 - s/2h_o) = (1 - 95/2 \cdot 522)(1 - 95/2 \cdot 522) = 0.826$$

The provided mechanical volumetric ratio for Ø8/95 is

$$\omega_{wd,prov} = \frac{V_o}{V_c} \cdot \frac{f_{yd}}{f_{cd}} = \frac{(4 \cdot 522 + 4 \cdot 522) \cdot \pi \cdot 8^2/4}{522 \cdot 522 \cdot 100} \cdot \frac{500/1.15}{20/1.5} = 0.212 < 0.27$$

Hence, hoops at the critical region are modified to Ø10/90

$$\omega_{wd,req} \geq \frac{1}{\alpha}\left(30 \cdot \mu_\varphi \cdot v_d \cdot \varepsilon_{syd}\frac{b_c}{b_o}-0.035\right)=\frac{1}{0.649}\left(30 \cdot 7.8 \cdot 0.356 \cdot 0.0022\frac{600}{520}-0.035\right)=0.268$$

The confinement effectiveness factor is Equation 8.129:

$$\alpha = \alpha_n \cdot \alpha_s = 0.778 \cdot 0.834 = 0.649, \text{ where}$$

Equation 8.130a:

$$\alpha_n = 1 - \sum_n \frac{b_i^2}{6 \cdot b_o \cdot h_o} = 1 - \frac{4 \cdot 245 + 8 \cdot 122.5}{6 \cdot 520 \cdot 520} = 0.778$$

Equation 8.131a:

$$\alpha_s = (1 - s/2b_o)(1 - s/2h_o) = (1 - 90/2 \cdot 520)(1 - 90/2 \cdot 520) = 0.834$$

$$\omega_{wd,prov} = \frac{V_o}{V_c} \cdot \frac{f_{yd}}{f_{cd}} = \frac{(4 \cdot 520 + 4 \cdot 520) \cdot \pi \cdot 10^2/4}{520 \cdot 520 \cdot 90} \cdot \frac{500/1.15}{20/1.5} = 0.350 > 0.268$$

8.6.2.2 Design of exterior beam–column joint

8.6.2.2.1 Ultimate limit state verification and design (Section 8.4.2)

The horizontal shear force acting on the concrete core of the exterior joint is (Equation 8.165)

$$\rightarrow E_y : V_{jh} = \gamma_{Rd} \cdot f_{yd} \cdot (0 + A_{s2}) - V_{col} = 1.2 \cdot \frac{500}{1.15}\left(0 + \frac{5 \cdot \pi \cdot 14^2}{4}\right) - \max\,(112.00; 100.32) = 289.58 \text{ kN}$$

$$\leftarrow E_y : V_{jh} = \gamma_{Rd} \cdot f_{yd} \cdot (A_{s1} + 0) - V_{col} = 1.2 \cdot \frac{500}{1.15}\left(\left(\frac{8 \cdot \pi \cdot 8^2}{4} + \frac{5 \cdot \pi \cdot 16^2}{4}\right) + 0\right) - \max\,(33.20; 44.8)$$

$$= 689.43 \text{ kN}$$

The shear force in the column above the joint, V_{col}, appears in Table 8.8 ($V_{col} = V_y$).

8.6.2.2.2 Diagonal compression induced in the joint by the diagonal strut mechanism

Equation 8.194 is applied in order to check the diagonal compression induced in the joint by the diagonal strut mechanism:

$$\rightarrow Ey : V_{jhd} \leq 0.8 \cdot v_1 \cdot f_{cd}\sqrt{1 - \frac{v_d}{v_1}} \cdot b_j \cdot h_{jc} = 0.8 \cdot 0.54 \cdot \frac{25 \cdot 1000}{1.5}\sqrt{1 - \frac{0.183}{0.54}} \cdot 0.60 \cdot 0.49 = 1735.57 \text{ kN}$$

Table 8.8 Bending moments and axial forces at the base of the second storey for column C2

Earthquake	Load combination	N (kN)	V_x (kN)	V_y (kN)	M_x (kN m)	M_y (kN m)
→ E_y	G+0.3Q+Ey+0.3Ex	−991.21	−67.26	−112	−171.83	−103.44
	G+0.3Q+Ey−0.3Ex	−1097.07	19.37	−100.32	−153.6	30.61
← E_y	G+0.3Q−Ey+0.3Ex	−1743.95	−26.62	33.2	52.13	−40.75
	G+0.3Q−Ey−0.3Ex	−1849.81	60.01	44.88	70.36	93.30

The normalised load above the joint is (Table 8.8): N_{Ed} = 1097.07 kN → v_d = 0.183

$$\leftarrow E_y : V_{jhd} \le 0.8 \cdot v_1 \cdot f_{cd}\sqrt{1-\frac{v_d}{v_1}} \cdot b_j \cdot h_{jc} = 0.8 \cdot 0.54 \cdot \frac{25 \cdot 1000}{1.5}\sqrt{1-\frac{0.308}{0.54}} \cdot 0.61 \cdot 0.49 = 1397.90 \text{ kN}$$

The normalised load above the joint is (Table 8.8): N_{Ed} = 1849.81 kN → v_d = 0.308.

8.6.2.2.3 Confinement of the joint

8.6.2.2.3.1 HORIZONTAL REINFORCEMENT

The horizontal and vertical reinforcement of the joint shall be such as to limit the maximum diagonal tensile stress of concrete to the design value of the tensile strength of concrete. For this purpose, Equation 8.197 needs to be satisfied:

$$\rightarrow E_y : A_{sh} = \frac{b_j \cdot h_{jw}}{f_{ywd}} \cdot \left[\frac{\left(\frac{V_{jhd}}{b_j \cdot h_{jc}}\right)}{f_{ctd} + v_d \cdot f_{cd}} - f_{ctd}\right] = \frac{0.60 \cdot 0.54 \cdot 10^6}{500/1.15} \cdot \left[\frac{\left(\frac{289.58}{0.60 \cdot 0.49 \cdot 10^3}\right)^2}{1.20 + 0.20 \cdot 25/1.5} - 1.20\right] < 0$$

$$\leftarrow E_y : A_{sh} = \frac{b_j \cdot h_{jw}}{f_{ywd}} \cdot \left[\frac{\left(\frac{V_{jhd}}{b_j \cdot h_{jc}}\right)^2}{f_{ctd} + v_d \cdot f_{cd}} - f_{ctd}\right] = \frac{0.60 \cdot 0.54 \cdot 10^6}{500/1.15} \cdot \left[\frac{\left(\frac{689.43}{0.60 \cdot 0.49 \cdot 10^3}\right)^2}{1.20 + 0.284 \cdot 25/1.5} - 1.20\right] < 0$$

Since $A_{sh} < 0$, it means that no hoops are required and that the concrete cross section may undertake the shear force at the joint. In this case, hoops and ties foreseen for the column critical region are arranged also in the joint. The alternative expression given by Equation 8.199 is also applied.

$$\rightarrow E_y : A_{sh} \ge \frac{\gamma_{Rd}}{\gamma_{ywd}} \cdot A_{s2} \cdot f_{yd} \cdot (1 - 0.8 \cdot v_d) = \frac{1.2}{500/1.15} \cdot \left(5 \cdot \frac{\pi \cdot 14^2}{4}\right) \cdot \frac{500}{1.15} \cdot (1 - 0.8 \cdot 0.205) = 7.72 \text{ cm}^2$$

The normalised load below the joint is (Table 8.9): N_{Ed} = 1227.55 kN → v_d = 0.205

$$\leftarrow E_y : A_{sh} \ge \frac{\gamma_{Rd}}{f_{ywd}} \cdot A_{s2} \cdot f_{yd} \cdot (1 - 0.8 \cdot v_d) = \frac{1.2}{500/1.15} \cdot \left(\frac{8 \cdot \pi \cdot 8^2}{4} + \frac{5 \cdot \pi \cdot 16^2}{4}\right) \cdot \frac{500}{1.15} \cdot (1 - 0.8 \cdot 0.350)$$

$$= 12.16 \text{ cm}$$

Table 8.9 Bending moments and axial forces at the top of storey I for column C2

Earthquake	Load combination	N (kN)	V_x (kN)	V_y (kN)	M_x (kN m)	M_y (kN m)
$\rightarrow E_y$	G+0.3Q+Ey+0.3Ex	−1105.53	19.87	23.54	79.34	39.37
	G+0.3Q+Ey−0.3Ex	−1227.55	−7.88	19.2	71.38	−10.46
$\leftarrow E_y$	G+0.3Q−Ey+0.3Ex	−1977.49	5.97	−44.52	−24.3	14.31
	G+0.3Q−Ey−0.3Ex	−2099.51	−21.78	−48.86	−32.26	−35.52

The normalised load below the joint is (Table 8.9): N_{Ed} = 2099.51 kN → v_d = 0.350.
Hoop layers provided in the joint: n = int(h_{jw}/s) + 1 = int(0.54/0.095) + 1 = 7 layers.
The total area of horizontal hoops is: $A_{sh,prov}$ = 7.4 · π · $10^2/4$ = 21.98 cm² > 12.16 cm².

8.6.2.2.3.2 VERTICAL REINFORCEMENT

The vertical reinforcement of the column passing through the joint is (Equation 8.200)

$$A_{sv,1} = \frac{2}{3} \cdot A_{sh} \cdot \frac{h_{jc}}{h_{jw}} = \frac{2}{3} \cdot 12.16 \cdot \frac{0.49}{0.54} = 7.36 \text{ cm}^2$$

The provided longitudinal bars at the face of the joint are 6Ø16 → 12.06 cm² > 7.36 cm².
The flexural and shear reinforcement placed in column C2 appear in Figure 8.78.

8.6.2.3 Column C8 (interior column)

8.6.2.3.1 Ultimate limit state in bending

The same design procedure followed in the case of column C2 applies in column C8 as well.
For all the load combinations that appear in Table 8.10 at the top and base of column C8,
ω_{req} = 0. Thus, the minimum longitudinal reinforcement is placed. The number and bar
diameters selected are 4Ø20 + 12Ø16 (36.69 cm² > $A_{s,min}$).
    Minimum reinforcement: $A_{s,min}$ = ρ_{min} b h = 0.01 600 600 = 36.00 cm².

8.6.2.3.2 Calculation of the design values of the moments of resistance of beams
8, 19 at the supports on C8 (beam 8 – right, beam 19 – left)

→ E_y: Top reinforcement beam 8:8Ø8 + 5Ø16, ω = 0.2050, μ_{Rd} = 0.179,

$$M_{ARd} = \mu_{Rd} \cdot b \cdot d^2 \cdot f_{cd} = 0.179 \cdot 0.600 \cdot 0.597^2 \cdot \frac{25 \cdot 10^3}{1.5} = 319.50 \text{ kNm}$$

Bottom reinforcement beam 19:5Ø14, ω = 0.0198, μ_{Rd} = 0.019,

$$M'_{BRd} = \mu_{Rd} \cdot b_{eff} \cdot d^2 \cdot f_{cd} = 0.019 \cdot 1.700 \cdot 0.597^2 \cdot \frac{25. \, 10^3}{1.5} = 190.27 \text{ kNm}$$

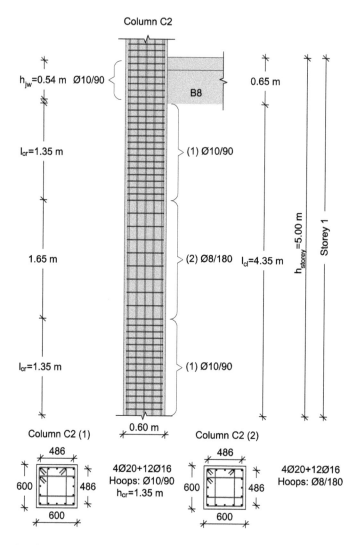

Figure 8.78 Flexural and shear reinforcement of column C2.

← E_y: Bottom reinforcement beam 8:5Ø14, ω = 0.0198, $μ_{Rd}$ = 0.019,

$$M'_{ARd} = μ_{Rd} \cdot b_{eff} \cdot d^2 \cdot f_{cd} = 0.019 \cdot 1.700 \cdot 0.597^2 \cdot \frac{25 \cdot 10^3}{1.5} = 190.27 \text{ kNm}$$

Top reinforcement beam 19:8Ø8 + 5Ø16, ω = 0.2050, $μ_{Rd}$ = 0.179,

$$M_{BRd} = μ_{Rd} \cdot b \cdot d^2 \cdot f_{cd} = 0.179 \cdot 0.600 \cdot 0.597^2 \cdot \frac{25 \cdot 10^3}{1.5} = 319.50 \text{ kNm}$$

Table 8.10 Bending moments and axial forces at column C8 ends for column dimensioning – storey I

Load combinations		N	V_x	V_y	M_x	M_y
Top of column C8						
Lc1	1.35G+1.5Q	−4266.45	−1.35	−4.73	9.67	3.66
Lc2	G+0.3Q+Ex+0.3Ey	−1998.25	41.1	20.88	42.38	73
Lc3	G+0.3Q+Ex−0.3Ey	−2156.99	38.73	−8.69	−9.1	68.57
Lc4	G+0.3Q−Ex+0.3Ey	−2605.81	−39.28	5.64	15.94	−66.08
Lc5	G+0.3Q−Ex−0.3Ey	−2764.55	−41.65	−23.93	−35.54	−70.5
Lc6	G+0.3Q+Ey+0.3Ex	−2025.71	15.73	50.04	93.19	29.48
Lc7	G+0.3Q+Ey−0.3Ex	−2207.97	−8.38	45.47	85.26	−12.24
Lc8	G+0.3Q−Ey+0.3Ex	−2554.83	7.83	−48.52	−78.41	14.74
Lc9	G+0.3Q−Ey−0.3Ex	−2737.09	−16.28	−53.09	−86.34	−26.99
Base of column C8						
Lc1	1.35G+1.5Q	−4319.3	−1.35	−4.73	−10.89	−2.18
Lc2	G+0.3Q+Ex+0.3Ey	−2037.4	41.1	−23.93	−61.7	108.3
Lc3	G+0.3Q+Ex−0.3Ey	−2196.14	38.73	5.64	15.45	102.36
Lc4	G+0.3Q−Ex+0.3Ey	−2644.96	−39.28	−8.69	−21.86	−102.26
Lc5	G+0.3Q−Ex−0.3Ey	−2803.7	−41.65	20.88	55.29	−108.2
Lc6	G+0.3Q+Ey+0.3Ex	−2064.86	15.73	−53.09	−137.77	41.53
Lc7	G+0.3Q+Ey−0.3Ex	−2247.12	−8.38	−48.52	−125.82	−21.64
Lc8	G+0.3Q−Ey+0.3Ex	−2593.97	7.83	45.47	119.4	21.74
Lc9	G+0.3Q−Ey−0.3Ex	−2776.24	−16.28	50.04	131.36	−41.43

8.6.2.3.3 Calculation of the design values of the moments of resistance at the ends of column C8 (Figure 8.79)

The design values of the moments of resistance M_x of column C8 at the base of storey 2 and at the top and base of storey 1 appear in Tables 8.11 through 8.13.

Parameter κ_D (Equation 6.15), which accounts for the case of strong columns and weak beams, is calculated at the top of column C8 for both directions of seismic action.

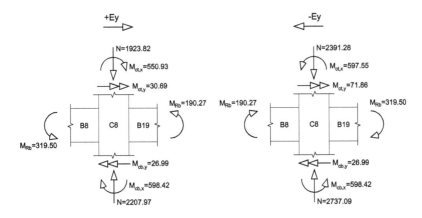

Figure 8.79 Capacity design moments for column C8.

Table 8.11 Determination of the design value of the moment of resistance M$_x$ of column C8 at the base of storey 2

Load combination		N	M$_y$	μ$_y$	v	ω	μ$_x$	M$_x$
→ E$_y$	G+0.3Q+Ey+0.3Ex	−1763.46	−82.91	0.028	−0.294	0.2916	0.179	536.75
	G+0.3Q+Ey−0.3Ex	−1923.82	30.69	0.010	−0.321	0.2916	0.184	550.93
← E$_y$	G+0.3Q−Ey+0.3Ex	−2230.92	−41.74	0.014	−0.372	0.2916	0.194	581.55
	G+0.3Q−Ey−0.3Ex	−2391.28	71.86	0.024	−0.399	0.2916	0.200	597.55

Table 8.12 Determination of the design value of the moment of resistance M$_x$ of column C8 at the top of storey 1

Load combination		N	M$_y$	μ$_y$	v	ω	μ$_x$	M$_x$
→ E$_y$	G+0.3Q+Ey+0.3Ex	−2025.71	29.48	0.010	−0.338	0.2916	0.188	561.09
	G+0.3Q+Ey−0.3Ex	−2207.97	−12.24	0.004	−0.368	0.2916	0.194	579.27
← E$_y$	G+0.3Q−Ey+0.3Ex	−2554.83	14.74	0.005	−0.426	0.2916	0.200	598.42
	G+0.3Q−Ey−0.3Ex	−2737.09	−26.99	0.009	−0.456	0.2916	0.200	598.42

Table 8.13 Determination of the design value of the moment of resistance M$_x$ of column C8 at the base of storey 1

Load combination		N	M$_y$	μ$_y$	v	ω	μ$_x$	M$_x$
→ E$_y$	G+0.3Q+Ey+0.3Ex	−2064.86	41.53	0.014	−0.344	0.2916	0.189	564.99
	G+0.3Q+Ey−0.3Ex	−2247.12	−21.64	0.007	−0.375	0.2916	0.195	583.17
← E$_y$	G+0.3Q−Ey+0.3Ex	−2593.97	21.74	0.007	−0.432	0.2916	0.200	598.42
	G+0.3Q−Ey−0.3Ex	−2776.24	−41.43	0.014	−0.463	0.2916	0.200	598.42

$$\rightarrow E_y : \sum M_{Rb} = 509.77 \text{ kN m}, \sum M_{Rc} = 1130.19 \text{ kN m}, \kappa_D = \min\left(1, \frac{509.77}{1130.19}\right) = 0.45$$

$$\leftarrow E_y : \sum M_{Rb} = 509.77 \text{ kN m}, \sum M_{Rc} = 1195.97 \text{ kN m}, \kappa_D = \min\left(1, \frac{509.77}{1195.97}\right) = 0.43$$

Parameter κ_C at the base of column C8 is taken to be equal to 1 in order to simplify the design procedure.

8.6.2.3.4 Capacity design seismic shear for column 8

The shear forces are determined according to the capacity design criterion as follows (Equation 6.11):

Clear height of the column: $l_c = h_{st} - h_b = 5 - 0.65 = 4.35$ m

$$\rightarrow E_y : V_{sd,CD} = \gamma_{RD} \frac{\kappa_C M_{CRd} + \kappa_D M_{DRd}}{l_c} = 1.30 \frac{0.45 \cdot 579.27 + 1 \cdot 583.17}{4.35} = 252.36 \text{ kN}$$

$$\leftarrow E_y : V_{sd,CD} = \gamma_{Rd} \frac{\kappa_C M_{CRd} + \kappa_D M_{DRd}}{l_c} = 1.30 \frac{0.43 \cdot 598.42 + 1 \cdot 598.42}{4.35} = 255.07 \text{ kN}$$

The checks that follow are performed for $V_{sd,CD} = 255.07$ kN.

8.6.2.3.5 Shear resistance in the case of diagonal concrete crushing for $\delta = 21.8°$ (Equation 8.62)

$$V_{Rd,max} = \alpha_{cw} \cdot b_w \cdot 0.9d \cdot \frac{(v_1 \cdot f_{cd}) \cot \delta}{(1 + \cot^2 \delta)} = 1 \cdot 0.6 \cdot 0.9 \cdot 0.547 \cdot \frac{0.540 \cdot (25/1.5) \cdot 1000}{2.9} = 916.70 \text{ kN}$$

$$V_{Rd,max} = 916.70 \text{ kN} > V_{sd,CD} = 255.07 \text{ kN}$$

8.6.2.3.6 Transverse reinforcement in the critical regions of column C8

The length of the critical region is

$$l_{cr} = \max\{1.5 h_c; l_c/6; 0.6\} = \max\{1.5 \cdot 0.6; 4.35/6; 0.6\} = 1.35 \text{ m}$$

The spacing of hoops shall not exceed

$$s = \min\{b_o/3; 125; 6 \cdot d_{bL}\} = \min\{514/3; 125; 6 \cdot 16\} = 96 \text{ mm}$$

$$V_{wd} = V_{sd,CD} \Leftrightarrow s = \frac{A_{sw}}{V_{sd,CD}} 0.9d \cdot f_{ywd} = \frac{4 \cdot (\pi \cdot 8^2/4)}{255.07} \cdot 0.9 \cdot 0.547 \cdot \frac{500}{1.15} = 168.73 \text{ mm}$$

Hence, the shear reinforcement in the critical regions of column C8 is Ø8/95.

8.6.2.3.7 Transverse reinforcement outside the critical regions of column C8

The spacing of hoops shall not exceed

$$s = \min\{20\Phi_{L,min}; \min(b_c; h_c); 400\} = \min\{20 \cdot 16; 600; 400\} = 320 \text{ mm}$$

Hence, the shear reinforcement outside the critical regions of column C8 is Ø8/165.

8.6.2.3.8 Detailing for local ductility – confinement reinforcement in the critical region at the base of column C8

The confinement reinforcement at the base of the column consists of a perimeter and an internal rectangular closed stirrup (four legs). The mechanical volumetric ratio of the required confining reinforcement, ω_{wd}, should satisfy Equation 8.127. Thus,

$$\omega_{wd,req} \geq \frac{1}{\alpha}\left(30 \cdot \mu_\varphi \cdot v_d \cdot \varepsilon_{syd} \frac{b_c}{b_o} - 0.035\right) = \frac{1}{0.641}\left(30 \cdot 7.8 \cdot 0467 \cdot 0.0022 \frac{600}{520} - 0.035\right) = 0.371$$

This value is higher than 0.12, which is the lower limit set by the code:

$$\omega_{wd,req} = 0.371 > 0.12$$

The confinement effectiveness factor is (Equation 8.129)

$$\alpha = \alpha_n \cdot \alpha_s = 0.776 \cdot 0.826 = 0.641$$

where
Equation 8.130a:

$$\alpha_n = 1 - \sum_n \frac{b_i^2}{6 \cdot b_o \cdot h_o} = 1 - \frac{4 \cdot 247 + 8 \cdot 123.5}{6 \cdot 522 \cdot 522} = 0.776$$

Equation 8.131a:

$$\alpha_s = (1 - s/2b_o)(1 - s/2h_o) = (1 - 95/2 \cdot 522)(1 - 95/2 \cdot 522) = 0.826$$

The provided mechanical volumetric ratio for Ø8/95 is

$$\omega_{wd,prov} = \frac{V_o}{V_c} \cdot \frac{f_{yd}}{f_{cd}} = \frac{(4 \cdot 522 + 4 \cdot 522) \cdot \pi \cdot 8^2/4}{522 \cdot 522 \cdot 100} \cdot \frac{500/1.15}{20/1.5} = 0.212 < 0.371$$

Hence, hoops at the critical region are modified to Ø12/90

$$\omega_{wd,req} \geq \frac{1}{\alpha}\left(30 \cdot \mu_\varphi \cdot v_d \cdot \varepsilon_{syd} \frac{b_c}{b_o} - 0.035\right) = \frac{1}{0.650}\left(30 \cdot 7.8 \cdot 0.467 \cdot 0.0022 \frac{600}{520} - 0.035\right) = 0.370$$

This value is higher than 0.12, which is the lower limit set by the code:

$$\omega_{wd,req} = 0.370 > 0.12$$

The confinement effectiveness factor is (Equation 8.129)

$$\alpha = \alpha_n \cdot \alpha_s = 0.780 \cdot 0.834 = 0.650$$

where
Equation 8.130a:

$$\alpha_n = 1 - \sum_n \frac{b_i^2}{6 \cdot b_o \cdot h_o} = 1 - \frac{4 \cdot 243 + 8 \cdot 121.5}{6 \cdot 518 \cdot 518} = 0.780$$

Equation 8.131a:

$$\alpha_s = (1 - s/2b_o)(1 - s/2h_o) = (1 - 90/2 \cdot 518)(1 - 90/2 \cdot 518) = 0.834$$

The provided mechanical volumetric ratio for Ø12/90 is

$$\omega_{wd,prov} = \frac{V_o}{V_c} \cdot \frac{f_{yd}}{f_{cd}} = \frac{(4 \cdot 518 + 4 \cdot 518) \cdot \pi \cdot 12^2/4}{518 \cdot 518 \cdot 90} \cdot \frac{500/1.15}{20/1.5} = 0.506 > 0.370$$

8.6.2.4 Design of interior beam–column joint

8.6.2.4.1 Ultimate limit state verification and design (Section 8.4.2)

The horizontal shear force acting on the concrete core of the interior joint is Equation 8.165:

$$\rightarrow E_y : V_{jh} = \gamma_{Rd} \cdot f_{yd} \cdot (A_{s1} + A_{s2}) - V_{col} \Leftrightarrow V_{jh}$$

$$= 1.2 \cdot \frac{500}{1.15} \left(\frac{8 \cdot \pi \cdot 8^2}{4} + \frac{5 \cdot \pi \cdot 16^2}{4} + \frac{5 \cdot \pi \cdot 14^2}{4} \right) - \max(143.02; 131.50) = 992.87 \text{ kN}$$

$$\leftarrow E_y : V_{jh} = \gamma_{Rd} \cdot f_{yd} \cdot (A_{s1} + A_{s2}) - A_{col}$$

$$V_{jh} = 1.2 \cdot \frac{500}{1.15} \left(\frac{8 \cdot \pi \cdot 8^2}{4} + \frac{5 \cdot \pi \cdot 16^2}{4} + \frac{5 \cdot \pi \cdot 14^2}{4} \right) - \max(116.86; 128.38) = 1007.51 \text{ kN}$$

The shear force in the column above the joint, V_{col}, appears in Table 8.14 ($V_{col} = V_y$).

8.6.2.4.2 Diagonal compression induced in the joint by the diagonal strut mechanism

Equation 8.194 is applied in order to check the diagonal compression induced in the joint by the diagonal strut mechanism:

$$\leftarrow E_y : V_{jhd} \leq v_1 \cdot f_{cd} \sqrt{1 - \frac{v_d}{v_1}} \cdot b_j \cdot h_{jc} = 0.54 \cdot \frac{25 \cdot 1000}{1.5} \sqrt{1 - \frac{0.321}{0.54}} \cdot 0.61 \cdot 4.49 = 1700.22 \text{ kN}$$

The normalised load above the joint is (Table 8.14): $N_{Ed} = 1923.82 \text{ kN} \rightarrow v_d = 0.321$

$$\leftarrow E_y : V_{jhd} \leq v_1 \cdot f_{cd} \sqrt{1 - \frac{v_d}{v_1}} \cdot b_j \cdot h_{jc} = 0.54 \cdot \frac{25 \cdot 1000}{1.5} \sqrt{1 - \frac{0.399}{0.54}} \cdot 0.60 \cdot 0.49 = 1365.31 \text{ kN}$$

The normalised load above the joint is (Table 8.8): $N_{Ed} = 2391.28 \text{ kN} \rightarrow v_d = 0.399$.

Table 8.14 Bending moments and axial forces at the base of the second storey for column C8

Earthquake	Load combination	N (kN)	V_x (kN)	V_y (kN)	M_x (kN m)	M_y (kN m)
$\rightarrow E_y$	G+0.3Q+Ey+0.3Ex	−1763.46	−54.47	−143.02	−218.04	−82.91
	G+0.3Q+Ey−0.3Ex	−1923.82	19.26	−131.5	−200.41	30.69
$\leftarrow E_y$	G+0.3Q−Ey+0.3Ex	−2230.92	−27.43	116.86	179.87	−41.74
	G+0.3Q−Ey−0.3Ex	−2391.28	46.3	128.38	197.51	71.86

8.6.2.4.3 Confinement of the joint

8.6.2.4.3.1 HORIZONTAL REINFORCEMENT

The horizontal and vertical reinforcement of the joint shall be such as to limit the maximum diagonal tensile stress of concrete to the design value of the tensile strength of concrete. For this purpose, Equation 8.197 needs to be satisfied:

$$\rightarrow E_y : A_{sh} = \frac{b_j \cdot h_{jw}}{f_{ywd}} \cdot \left[\sqrt{\frac{\left(\frac{V_{jhd}}{b_j \cdot h_{jc}}\right)^2}{f_{ctd} + v_d \cdot f_{cd}}} - f_{ctd} \right]$$

$$= \frac{0.60 \cdot 0.54 \cdot 10^6}{500/1.15} \cdot \left[\sqrt{\frac{\left(\frac{992.87}{0.60 \cdot 0.49 \cdot 10^3}\right)}{1.20 + 0.321 \cdot 25/1.5}} - 1.20 \right] = 3.86 \text{ cm}^2$$

$$\leftarrow E_y : A_{sh} = \frac{b_j \cdot h_{jw}}{f_{ywd}} \cdot \left[\sqrt{\frac{\left(\frac{V_{jhd}}{b_j \cdot h_{jc}}\right)^2}{f_{ctd} + v_d \cdot f_{cd}}} - f_{ctd} \right]$$

$$= \frac{0.60 \cdot 0.54 \cdot 10^6}{500/1.15} \cdot \left[\sqrt{\frac{\left(\frac{1007.51}{0.60 \cdot 0.49 \cdot 10^3}\right)^2}{1.20 + 0.399 \cdot 25/1.5}} - 1.20 \right] = 2.05 \text{ cm}^2$$

The alternative expression given by Equation 8.198 is also applied.

$$\rightarrow E_y : A_{sh} \geq \frac{\gamma_{Rd}}{f_{ywd}} \cdot (A_{s1} + A_{s2}) \cdot f_{yd} \cdot (1 - 0.8 \cdot v_d) \Leftrightarrow$$

$$A_{sh} \geq \frac{1.2}{500/1.15} \cdot \left(\frac{8 \cdot \pi \cdot 8^2}{4} + 5 \frac{4 \cdot \pi \cdot 16^2}{4} + \frac{5 \cdot \pi \cdot 16^2}{4} \right) \cdot \frac{500}{1.15} \cdot (1 - 0.8 \cdot 0.329)1 = 19.42 \text{ cm}^2$$

The normalised load below the joint is (Table 8.14): $N_{Ed} = 1923.82$ kN → $v_d = 0.321$

$$\leftarrow E_y : A_{sh} \geq \frac{\gamma_{Rd}}{f_{ywd}} \cdot (A_{s1} + A_{s2}) \cdot f_{yd} \cdot (1 - 0.8 \cdot v_d) \Leftrightarrow$$

$$A_{sh} \geq \frac{1.2}{500/1.15} \cdot \left(\frac{5 \cdot \pi \cdot 14^2}{4} + \frac{8 \cdot \pi \cdot 8^2}{4} + \frac{5 \cdot \pi \cdot 16^2}{4} \right) \cdot \frac{500}{1.15} \cdot (1 - 0.8 \cdot 0.329) = 17.80 \text{ cm}^2$$

The normalised load below the joint is (Table 8.14): $N_{Ed} = 2391.28$ kN → $v_d = 0.399$.
As may be observed, the two alternative expressions suggested by EC8-Part I (2004) for determining the required horizontal reinforcement of the joint yield very different results ($A_{sh} = 2.05$ cm² vs $A_{sh} = 17.80$ cm²), which is rather confusing for the designer. Fardis (2009) suggests that the least among the steel requirements of Equations 8.197 and 8.198 may be used with some confidence. Following this approach, the hoop layers provided in the joint are

$$n = \text{int}(h_{jw} / s) + 1 = \text{int}(0.54/0.095) + 1 = 7 \text{ layers}$$

The total area of a horizontal hoop is $A_{sh,prov} = 7 \cdot 4 \cdot \pi \cdot 12^2/4 = 31.64$ cm² > 19.42 cm².

8.6.2.4.3.2 VERTICAL REINFORCEMENT

The vertical reinforcement of the column passing through the joint is (Equation 8.200)

$$A_{sv,1} = \frac{2}{3} \cdot A_{sh} \cdot \frac{h_{jc}}{h_{jw}} = \frac{2}{3} \cdot 17.80 \cdot \frac{0.49}{0.54} = 11.76 \text{ cm}^2$$

The horizontal reinforcement of the joint is given below:
The longitudinal bars provided at the face of the joint are 6Ø16 → 12.06 cm² > 9.75 cm².
The flexural and shear reinforcement placed in column C8 appears in Figure 8.80.

Remark: Capacity design of columns to bending (Equation 6.8) has not been carried out, since the structural system has been classified as an *uncoupled wall* system (see Chapter 5.9.7).

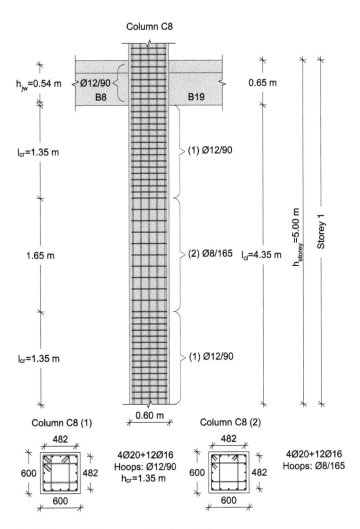

Figure 8.80 Flexural and shear reinforcement of column C8.

8.6.3 Beams: ultimate limit state in shear

8.6.3.1 Design shear forces

The design shear forces are determined in accordance with the capacity design rule (Chapter 6.1.3): $\rightarrow E_y$: Joint A:

Bottom reinforcement (beam 8 – left): 5Ø14, $\omega = 0.0198$, $\mu_{Rd} = 0.019$, $M_{AR} = 190.27$ kN m

$$\sum M_{Rb} = 190.27 \, \text{kN m}, \sum M_{Rc} = 1013.55 \, \text{kN m}, \, \kappa_A = \min(1, \Sigma M_{Rb}/\Sigma M_{Rc}) = 1$$
$$M_{A,d} = \gamma_{Rd} \cdot M_{AR} \cdot \kappa_A = 1.20 \cdot 190.27 \cdot 1 = 228.33 \, \text{kN m}$$

$\rightarrow E_y$: Joint B:

Top reinforcement (beam 8 – right): 8Ø8 + 5Ø16, $\omega = 0.2050$, $\mu_{Rd} = 0.179$, $M'_{BR,l} = 319.50$ kN m

Bottom reinforcement (beam 19 – left): 5Ø14, $\omega = 0.0198$, $\mu_{Rd} = 0.019$, $M_{BR,r} = 190.27$ kN m

$$\sum M_{Rb} = (319.50 + 190.27) = 509.77 \, \text{kN m},$$
$$\sum M_{Rc} = 1130.19 \, \text{kN m}, \, \kappa_B = \min\left(1, \sum M_{Rb} \Big/ \sum M_{Rc}\right) = 1$$

$$M'_{Bl,d} = \gamma_{Rd} \cdot M'_{BR} \cdot \kappa_A = 1.20 \cdot 319.50 \cdot 1 = 383.40 \, \text{kN m}$$

$$M_{Br,d} = \gamma_{Rd} \cdot M_{BR} \cdot \kappa_A = 1.20 \cdot 190.27 \cdot 1 = 228.33 \, \text{kN m}$$

$\rightarrow E_y$: Joint C:

Top reinforcement (beam 19 – right):

$$8\text{Ø}8 + 4\text{Ø}16, \omega = 0.1757, \mu_{Rd} = 0.157, M'_{CR,l} = 280.25 \, \text{kN m}$$

Bottom reinforcement (beam 37 – left): 4Ø14, $\omega = 0.0158$, $\mu_{Rd} = 0.015$, $M_{CR,r} = 152.22$ kN m

$$\sum M_{Rb} = (280.25 + 152.22) = 432.47 \, \text{kN m},$$
$$\sum M_{Rc} = 1157.83 \, \text{kN m}, \kappa_B = \min\left(1, \sum M_{Rb} \Big/ \sum M_{Rc}\right) = 1$$

$$M'_{Cl,d} = \gamma_{Rd} \cdot M'_{CR} \cdot \kappa_A = 1.20 \cdot 280.25 \cdot 1 = 336.30 \, \text{kN m}$$

$$M_{Cr,d} = \gamma_{Rd} \cdot M_{CR} \cdot \kappa_A = 1.20 \cdot 152.22 \cdot 1 = 182.66 \, \text{kN m}$$

Beam 8:

$$V_{sd,AB} = \frac{\left(M_{A,d} + M'_{Bl,d}\right)}{l} = \frac{-(228.33 + 383.40)}{4.40} = -139.03 \text{ kN}$$

$$V_{A,G+0.3Q} = 81.51 \text{ kN}, V_{B,G+0.3Q} = -58.19 \text{ kN}$$

$$V_{AS2} = V_{G+0.3Q} + V_{sd,AB} = 81.51 - 139.03 = -57.52 \text{ kN}$$
$$V_{BS1} = V_{G+0.3Q} + V_{sd,AB} = -58.19 - 139.03 = -197.22 \text{ kN}$$

Beam 19:

$$V_{sd,BC} = \frac{\left(M_{Br,d} + M'_{Cl,d}\right)}{l} = \frac{-(228.33 + 336.30)}{4.40} = -128.32 \text{ kN}$$

$$V_{B,G+0.3Q} = 57.35 \text{ kN}, V_{C,G+0.3Q} = -55.81 \text{ kN}$$
$$V_{BS2} = V_{G+0.3Q} + V_{sd,BC} = 57.35 - 128.32 = -70.98 \text{ kN}$$
$$V_{CS1} = V_{G+0.3Q} + V_{sd,BC} = -55.82 - 128.32 = -184.13 \text{ kN}$$

$\leftarrow E_y$: Joint A:
Top reinforcement (beam 8 – left):

$$8\varnothing 8 + 5\varnothing 16, \omega = 0.2050, \mu_{Rd} = 0.179, M'_{AR} = \mu_{Rd} \cdot b \cdot d^2 \cdot f_{cd} = 319.50 \text{ kNm}$$
$$\sum M_{Rb} = 319.50 \text{ kN m}, \sum M_{Rc} = 1111.99 \text{ kN m}, \kappa_A = \min\left(1, \sum M_{Rb} / \sum M_{Rc}\right) = 1$$

$$M'_{A,d} = \gamma_{Rd} \cdot M'_{A,R} \cdot \kappa_A = 1.20 \cdot 319.50 \cdot 1 = 383.40 \text{ kN m}$$

$\leftarrow E_y$, Joint B:
Bottom reinforcement (beam 8 – right):

$$5\varnothing 14, \omega = 0.0198, \ \mu_{Rd} = 0.019, M_{BR,l} = \mu_{Rd} \cdot b_{eff} \cdot d^2 \cdot f_{cd} = 190.27 \text{ kN m}$$

Top reinforcement (beam 19 – left):

$$8\varnothing 8 + 5\varnothing 16, \ \omega = 0.2050, \mu_{Rd} = 0.179, M'_{BR,r} = \mu_{Rd} \cdot b \cdot d^2 \cdot f_{cd} = 319.50 \text{ kN m}$$

$$\sum M_{Rb} = (190.27 + 319.50) = 509.77 \text{ kN m},$$
$$\sum M_{Rc} = 1195.97 \text{ kN m}, \kappa_B = \min\left(1, \sum M_{Rb} / M_{Rc}\right) = 1$$

$$M_{Bl,d}^{/} = \gamma_{Rd} \cdot M_{BR,l} \cdot \kappa_A = 1.20 \cdot 190.27 \cdot 1 = 228.33\,\text{kN m}$$

$$M_{Br,d}^{/} = \gamma_{Rd} \cdot M_{BR,r}^{/} \cdot \kappa_A = 1.20 \cdot 319.50 = 383.40\,\text{kN m}$$

← E_y, Joint C:
Bottom reinforcement (beam 19 – right):

$4\varnothing14, \omega = 0.0158, \mu_{Rd} = 0.015, M_{CR,l} = 152.22$ kN m

Top reinforcement (beam 37 – left):

$8\varnothing8 + 4\varnothing16, \omega = 0.0176, \mu_{Rd} = 0.157, M_{CR,r}^{/} = 280.25$ kN m

$$\sum M_{Rb} = (280.25 + 152.22) = 432.47\,\text{kN m},$$
$$\sum M_{Rc} = 1188.84\,\text{kN m}, \kappa_B = \min(1, \Sigma M_{Rb}/\Sigma M_{Rc}) = 1$$

$$M_{Cl,d} = \gamma_{Rd} \cdot M_{CR} \cdot \kappa_A = 1.20 \cdot 152.22 \cdot 1 = 182.66\,\text{kN m}$$
$$M_{Cr,d}^{/} = \gamma_{Rd} \cdot M_{CR}^{/} \cdot \kappa_A = 1.20 \cdot 280.25 \cdot 1 = 336.30\,\text{kN m}$$

Beam 8 (Figures 8.77 and 8.81):

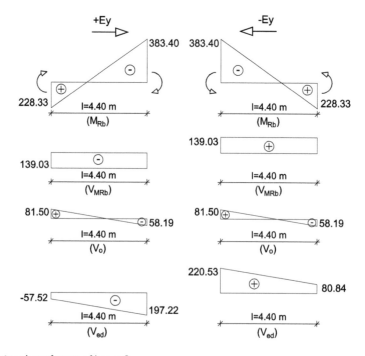

Figure 8.81 Design shear forces of beam 8.

$$V_{sd,AB} = \frac{\left(M'_{A,d} + M_{Bl,d}\right)}{l} = \frac{(383.40 + 228.33)}{4.40} = 139.03\,kN$$

$V_{A,G+0.3Q} = 81.51\ kN, V_{B,G+0.3Q} = -58.19\,kN$

$V_{AS1} = V_{G+0.3Q} + V_{sd,AB} = 81.51 + 139.03 = 220.53\,kN$

$V_{BS2} = V_{G+0.3Q} - V_{sd,AB} = -58.19 + 139.03 = 80.84\,kN$

Beam 19 (Figures 8.79 and 8.82):

$$V_{sd,BC} = \frac{\left(M'_{Br,d} + M_{Cl,d}\right)}{l} = \frac{(383.40 + 182.66)}{4.40} = 128.65\,kN$$

$V_{B,G+0.3Q} = 57.35\,kN, V_{C,G+0.3Q} = -55.82\,kN$

$V_{BS1} = V_{G+0.3Q} + V_{sd,BC} = 57.35 + 128.65 = 185.99\,kN$

$V_{CS2} = V_{G+0.3Q} + V_{sd,BC} = 55.82 + 128.65 = 72.84\,kN$

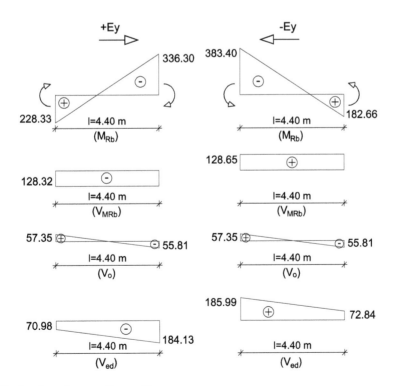

Figure 8.82 Design shear forces of beam 19.

Beam 8:

The algebraic value of the ratio between the minimum and maximum acting shear forces is (Equation 6.6)

$$\zeta = \frac{V_{AS2}}{V_{AS1}} = \frac{-57.52}{220.53} = -0.26 > -0.5; \zeta = \frac{V_{BS2}}{V_{BS1}} = \frac{80.84}{-197.22} = -0.41 > -0.5$$

Hence, shear resistance is provided by hoops only.

Beam 8 – left: $V_A = \max(|V_{AS1}|; |V_{AS2}|) = \max(220.53; 57.52) = 220.53$ kN
Beam 8 – right: $V_B = \max(|V_{BS1}|; |V_{BS2}|) = \max(197.22; 80.84) = 197.22$ kN

Beam 19:

The algebraic value of the ratio between the minimum and maximum acting shear forces is estimated for beam 19 as well (Equation 6.6):

$$\zeta = \frac{V_{BS2}}{V_{BS1}} = \frac{-70.98}{185.99} = -0.38 > -0.5; \zeta = \frac{V_{CS2}}{V_{CS1}} = \frac{72.84}{-184.13} = -0.40 > -0.5$$

Hence, shear resistance is provided by hoops only.

Beam 19 – left: $V_B = \max(|V_{BS1}|; |V_{BS2}|) = \max(185.99; 70.98) = 185.99$ kN
Beam 19 – right: $V_C = \max(|V_{CS1}|; |V_{CS2}|) = \max(184.13; 72.84) = 184.13$ kN

8.6.3.2 Shear reinforcement

8.6.3.2.1 Shear resistance in case of diagonal concrete crushing for δ = 21.8o (Equation 8.62):

Beam 8:

$$V_{Rd,max} = 1 \cdot 0.3 \cdot 0.9 \cdot 0.597 \cdot 0.54 \cdot (25/1.5) \cdot 1000/2.9 = 500.24 \text{ kN} > \max(V_A; V_B) = 220.53 \text{ kN}$$

Beam 19:

$$V_{Rd,max} = 1 \cdot 0.3 \cdot 0.9 \cdot 0.597 \cdot 0.54 \cdot (25/1.5) \cdot 1000/2.9 = 500.24 \text{ kN} > \max(V_B; V_C) = 185.99 \text{ kN}$$

8.6.3.2.2 Transverse reinforcement in the critical regions

The critical region is $l_{cr} = 1.5 \cdot h_w = 1.5 \cdot 65 = 97.5$ cm.

The maximum longitudinal spacing should not exceed s_{max}:

$$s_{max} = \min\left\{\frac{h_w}{4}; 24d_{bw}; 175; 6d_{bL}\right\} = \min\left\{\frac{650}{4}; 24 \cdot 8; 175; 6 \cdot 14\right\} = 84 \text{ mm}$$

Ø8/84 are the minimum allowed hoops.

Beam 8:

$$V_{wd} = \max(V_A; V_B) \Leftrightarrow s = \frac{A_{sw}}{V_A} 0.9d \cdot f_{ywd} = \frac{2 \cdot \pi \cdot 8^2/4}{220.53} \cdot 0.9 \cdot 0.597 \cdot \frac{500}{1.15} = 106.49 \text{ mm}$$

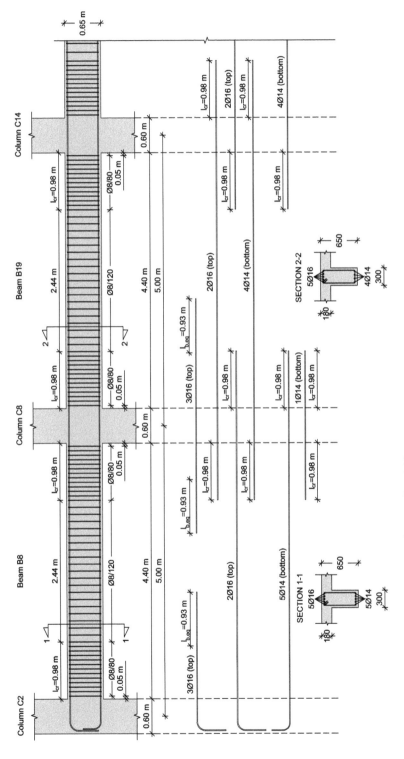

Figure 8.83 Flexural and shear reinforcement of beams 8 and 19.

Hence, the shear reinforcement in the critical regions of beam 8 is Ø8/80.
Beam 19:

$$V_{wd} = \max(V_B; V_C) \Leftrightarrow s = \frac{A_{sw}}{V_A} 0.9d \cdot f_{ywd} = \frac{2 \cdot \pi \cdot 8^2/4}{185.99} \cdot 0.9 \cdot 0.597 \cdot \frac{500 \cdot 10^3}{1.15} = 126.27 \, mm$$

Hence, the shear reinforcement in the critical regions of beam 19 is Ø8/80.

8.6.3.2.3 Transverse reinforcement outside the critical regions

The minimum shear reinforcement ratio is

$$\rho_{w,min} = \frac{0.08 \cdot \sqrt{f_{ck}}}{f_{yk}} = \frac{0.08 \cdot \sqrt{25}}{500} = 0.8?$$

The maximum longitudinal spacing between hoops should not exceed s_{max}:

$$s_{max} = 0.75 \cdot d = 075 \cdot 597 = 447.75 \, mm$$

In case of a two-leg 8 mm bar diameter hoops, stirrup spacing is defined as equal to

$$s = \frac{A_{sw}}{\rho_{w,min} \cdot b_w} = \frac{2 \cdot (\pi \cdot 8^2/4)}{0.0008 \cdot 300} \approx 419 \, mm$$

Ø8/419 are the minimum allowed hoops.
Beam 8:

$$V_{wd} = 189.58 \Leftrightarrow s = \frac{2 \cdot \pi \cdot 8^2/4}{189} \cdot 0.9 \cdot 0.597 \cdot \frac{500}{1.15} = 123.88 \, mm$$

Hence, the shear reinforcement outside the critical regions of beam 8 is Ø8/120.
Beam 19:

$$V_{wd} = 160.92 \Leftrightarrow s = \frac{2 \cdot \pi \cdot 8^2/4}{160.92} \cdot 0.9 \cdot 0.597 \cdot \frac{500}{1.15} = 145.95 \, mm$$

The shear reinforcement outside the critical regions of beam 19 is taken to be equal to Ø8/120 as in the case of B8 for simplicity in construction.
The flexural and shear reinforcement placed in beams B8–B19 appears in Figure 8.83.

Chapter 9

Seismic-resistant R/C walls and diaphragms

9.1 GENERAL

R/C walls together with frames and diaphragms constitute the main structural members in the planning and designing of the structural system of a seismic-resistant R/C building.

Thus far, many detailed references have been made to the structural behaviour of walls as wall structural systems (Chapter 4, Sections 4.5.3.3 and 4.5.3.4), to their behaviour factor q (Chapter 5.4), to their modelling for structural analysis (Chapter 5.6.3.1) and to their capacity design action effects (Chapter 6, Sections 6.1.5 to 6.1.7). All of the above refer to *demand*. In the following sections, detailed reference will be made to the *capacity* of R/C walls in terms of strength, ductility and energy dissipation under cyclic loading.

Before any further reference to 'capacity' issues, it will be useful to make a short summary of the main points on walls that have been presented so far.

1. 'R/C walls', according to modern Codes, are vertical structural members with an orthogonal cross section and a ratio of the sides of the cross section

$$\frac{l_w}{b} > 4.0 \tag{9.1}$$

2. R/C walls may constitute basic structural elements in the following building structural systems:
   a. Uncoupled wall systems (Figure 4.27)
   b. Dual systems (wall-equivalent of frame-equivalent systems; Figure 4.31)
   c. Coupled wall systems (Figure 4.29)
   d. Systems with large, lightly reinforced walls (Figure 9.2)
3. Apart from the structural system under consideration, the types of R/C walls in use may be classified into the following types:
   a. Slender ductile walls (Figure 4.27)
   b. Squat ductile walls (Figure 9.1)
   c. Coupled walls (Figure 4.29)
   d. Large, lightly reinforced walls (Figure 9.2)

The most common walls in use are the first (slender ductile walls). In the following sections, these four types of walls will be examined as far as their capacity is concerned.

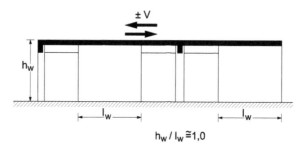

$$h_w / l_w \cong 1,0$$

Figure 9.1 A typical structural system with squat walls under seismic action ± V.

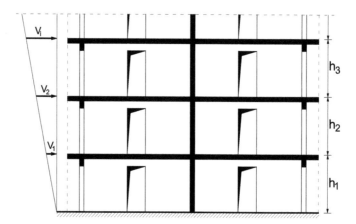

Figure 9.2 Large, lightly reinforced wall.

This chapter will also include design issues relating to diaphragms, for two reasons:

1. Diaphragms are planar elements, and, in this respect, their behaviour matches better that of walls.
2. The design of diaphragms in R/C buildings cast *in situ* in the form of slabs does not have any particular design concern; design issues arise only in special cases. Therefore, the presentation of the design of these special cases would not justify an independent chapter.

9.2 SLENDER DUCTILE WALLS

9.2.1 A summary on structural behaviour of slender ductile walls

1. The usual cross section of these members is the orthogonal one. However, barbell cross sections or T, U, L, Z and tubular cross sections are not uncommon (Figure 9.3a–g).
2. A recommended aspect ratio α_s of these walls with an orthogonal cross section lies between

$$7.0 \geq \alpha_s = \frac{h_w}{l_w} \geq 2.0 \qquad (9.2)$$

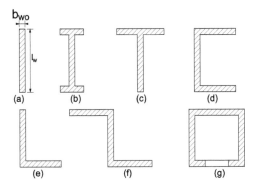

Figure 9.3 Various cross sections of ductile slender walls: (a) orthogonal; (b) barbell; (c) T-shaped; (d) C-shaped; (e) L-shaped; (f) Z-shaped; (g) tubular.

and

$$l_w \geq 2.00m$$

where
   l_w is the length of the cross section.
   h_w is the height of the wall.

3. The thickness b_{wo} of these walls should be (Figure 9.4)

$$b_{wo} \geq \max\{0.15, h_s/20\}\left[m\right]$$

(9.3)

where
   h_s is the clear storey height in meters.

   In addition, these walls should be strengthened with confined boundary elements with a minimum thickness of 200 mm. In case the thickness of the wall is greater than 200 mm, as we will see later, these confined boundary elements may be incorporated into the cross section of the wall. The depth l_c of the confined boundary element

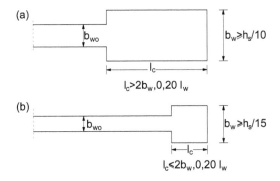

Figure 9.4 (a, b) Minimum thickness of confined boundary elements. (Adapted from E.C.8-1/EN1998-1. 2004. *Design of Structures for Earthquake Resistance: General Rules, Seismic Actions and Rules for Buildings.* CEN, Brussels, Belgium.)

determined in accordance with Section 9.2.4.2(3) should not be less than 0.15 l_w or less than 1.50 b_w, whichever is greater (Figure 9.4b). Moreover, the thickness b_w should not be less than $h_s/15$ (h_s: storey height).

If the depth of the confined part exceeds $2b_w$ or $0.20l_w$, whichever is greater, b_w should not be less than $h_s/10$ (Figure 9.4a). These requirements for confined boundary elements aim at ensuring a quantified *local ductility* and the protection of the edge of the walls from buckling (Figure 9.5).

4. The main structural characteristics of these walls are the following:
   • They are slender elements, and therefore their design for bending with axial force is separated from their design for shear, since the distribution of axial strains on a deformed cross section is planar (Bernoulli concept; Figure 9.9).
   • The capacity design for shear ensures the plastic behaviour to bending at the fixed end before the failure of the wall to shear. At the same time, the form of the envelope of the capacity design moment diagram (see Chapter 6.1.5) ensures that only one plastic hinge may be formed. In this respect, a robust backbone is formed in the structural system by means of the ductile walls, which minimises the hazard of a collapse in 'pancake' form.
   • In Subsection 9.2.1 the recommended values for the aspect ratio aim at the proper slenderness together with an adequate stiffness for the behaviour of the wall as a cantilever beam without change of the sign of curvature along the height of the walls.

In fact, according to what has been presented in Paragraph 4.5.3.3, the second part of the inequality (9.2) ensures that the wall is not a short one and, therefore, plastic formation at the fixed-end precedes the shear failure of the member (($M_{\text{fixed-end}}/V{\cdot}l_w) \geq 2.0$). At the same time, the first part of the inequality (9.2) ensures that the curvature of the wall does not change sign along its height, as may be concluded from engineering practice.

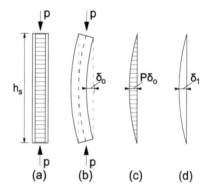

(a)   (b)   (c)   (d)

Figure 9.5 Deformations causing out-of-plane ductility: (a) compressive axial load *P*; (b) accidental displacement at the middle of wall height; (c) moment diagram generated by the axial compressive force; (d) additional displacement δ_1 due to the moment diagram P · δ_0 (second-order effects); if $\delta_1 \geq \delta 0$ then the wall passes to instability (out-of-plane buckling).

9.2.2 Behaviour of slender ductile walls under bending with axial load

9.2.2.1 General

Slender walls with an orthogonal cross-section are considered to fulfil the Bernoulli concept for linear distribution of axial strains on the cross-section like beams and columns if they have an aspect ratio a_s:

$$\boxed{a_s = \frac{M_{\text{fix-end}}}{V \cdot l_w} \geq 2}$$

(9.4)

where

$M_{\text{fix-end}}$ is the bending moment at the fixed end.
V is the relevant shear force.
l_w is the length of the cross section.

Having in mind that usually the axial compressive stresses of a wall due to the axial design forces are smaller than those of columns, it can be concluded that the structural behaviour of a wall lies between that of a beam and a column. Therefore, generally speaking, the same assumptions may be used for the design of ductile walls as for beams and columns, with some necessary changes, as we will see later. As in the case for beams and columns, the failure mode of walls may be of *flexural type* or *shear type,* depending on the reinforcement of the member.

As in the case of beams, the failure mechanism under flexure may have one of the following forms:

1. Yield of tensile steel reinforcement, wide flexural cracks in tension zone near the fixed end, large defections at the top and spalling of the compression zone, in other words ductile type of failure (Figure 9.6a,b).

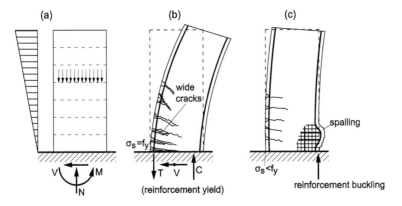

Figure 9.6 Failure mechanism of slender walls under flexure: (a) loading pattern; (b) ductile failure; (c) brittle failure.

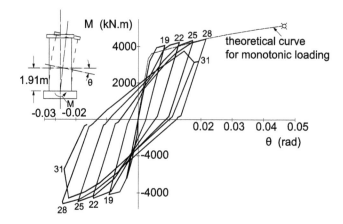

Figure 9.7 Moment–rotation hysteresis loops for slender R/C wall with barbell cross section subjected to cyclic loading. (From Oesterle, R.G., Fiorato, A.E., Aristizabal-Ochoa, J.D. and Corley, W.G. (1980). Hysteretic response of reinforced concrete structural walls. In ACI SP-63: Reinforced Concrete Structures Subjected to Wind and Earthquake Forces, American Concrete Institute, Detroit, pp. 243–273.)

2. Narrow cracks at the tension zone ($\sigma_s < f_y$), crushing of concrete at the compression zone near to the fixed end and buckling of reinforcement under compression (Figure 9.6a,c), that is, brittle failure due to over-reinforced tensile zone.

The $M-\varphi$ diagram for an orthogonal cross section is similar to that of a beam (Chapter 8, Figure 8.12). The length of the plastic branch depends to a high degree on the confinement of the boundary elements of the wall and on the degree of the axial compression. A more detailed approach will be made later. The same holds for the $M-\theta$ diagram under cyclic loading, as can be seen in the diagram of Figure 9.7, which is similar to that of a cantilever beam (Figure 8.21) and a column (Figure 8.46). Therefore, in the case that special care has been taken in the design of the wall for ductile behaviour, the $M-\delta$ diagram exhibits a stable form under reversed cycles of loading.

Interaction $M-N$ diagrams for symmetrically reinforced wall cross sections of orthogonal form are similar to those of columns (Figure 9.8). Additionally in the case of walls, the following remarks should be made:

- Moment carrying capacity is strongly influenced by the concentration of a large percentage of longitudinal reinforcement at the confined boundary elements.
- In the case of ductile walls, N_{Ed} almost always has values below N_{Rd} balanced. Therefore, ductile walls almost always enter the post-elastic region of $M-\varphi$ diagrams. This capacity is enhanced due to the proper confinement of the boundary elements.

9.2.2.2 Dimensioning of slender ductile walls with orthogonal cross section under bending with axial force

The dimensioning of walls under a bending moment with axial force is carried out under the same assumptions as for beams and columns (see Chapter 8, Sections 8.2.2.1 and 8.3.2.1). In case of walls, where the bending moment acts mainly about a main axis perpendicular to the length of the cross section (uniaxial bending with axial force), the dimensioning may

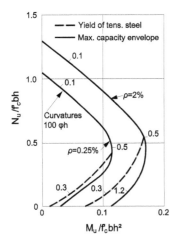

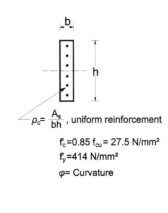

$\rho_c = \dfrac{A_s}{bh}$, uniform reinforcement

$f_c = 0.85\, f_{cu} = 27.5$ N/mm²

$f_y = 414$ N/mm²

φ = Curvature

Figure 9.8 Axial load–moment interaction curves for rectangular uniformly reinforced concrete walls. (After Salse, E.A.B. and Fintel, M. (1973). Strength, stiffness and ductility properties of slender shear walls. Proc. 5th World Conf. of Earthquake Eng., Rome 1: 919–928. Dowrick, D.: *Earthquake Risk Reduction*. 2005. Copyright Wiley-VCH Verlag GmbH & Co. KGaA. Reproduced with permission.)

be carried out either with the aid of design charts or simplified expressions of a closed form (Tassios, 1984b).

So, for example, in the case of a rectangular cross section reinforced with vertical grids corresponding to a reinforcement ratio $\rho_v = A_{sv}/b_w l_w$, where b_w is the width and l_w is the length of the cross section, and with concentrated reinforcement at the ends with an area $A_{s1} = A_{s2} = A_s$, the design moment M_{Rd} may be determined from the following expression (Figure 9.9a; Tassios, 1984b).

$$M_{Rd} = \left[\left(1 - \frac{\xi}{2}\right)\frac{A_s}{b_w l_w}f_{yd} + \frac{1}{2}(1-\xi)(\rho_v f_{yd} + \sigma_o)\right]b_w l_w^2 \tag{9.5}$$

where

$$\sigma_o = N_{Ed}/b_w l_w$$

is the average stress due to axial load alone (compression positive) and

$$\xi = \left(\frac{A_s f_y}{b_w l_w f_c} + \rho_v \frac{f_{yd}}{f_{cd}} + \frac{\sigma_o}{f_{cd}}\right)\frac{1}{1 + \rho_v f_{yd}/f_{cd}} \tag{9.6}$$

is the neutral depth ratio (x_u/l_w) at the ultimate limit state.

On the other hand, for walls with barbell section or with boundary-confined elements having a width l_c as shown in Figure 9.9b,c, with the main reinforcement concentrated at the boundary elements $(A_{s1} = A_{s2} = A_s)$, the design bending strength may be determined from the expression below (Tassios, 1984b):

$$M_{Rd} \cong \left(A_s f_{yd} + \frac{N}{2}\right)(l_w - l_c') \tag{9.7}$$

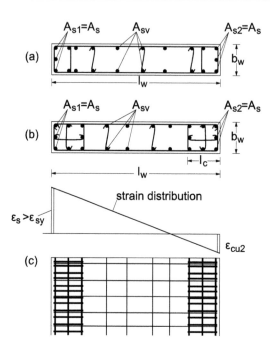

Figure 9.9 Dimensioning of slender ductile walls to bending with axial force: (a) main reinforcement A_s concentrated at the boundaries; (b) main reinforcement A_s concentrated in confined boundary elements; (c) concept of plane strain distribution.

However, the methods mentioned above should nowadays be considered obsolete, since various commercial computer platforms have been developed, which make possible the dimensioning of cross sections of any shape for biaxial bending with axial force (see Figure 9.10) like ECtools, CUBUS, NOUS and so on. These platforms may be used either for columns or for walls, since the design assumptions are the same. It should not be forgotten that for the cross sections of each vertical member and at each storey, at least 2 × 33 load cases should be examined at the top and bottom of the storey (see Chapter 5.8.3), a fact that makes almost impossible the use of manual calculations either with the aid of charts or simplified expressions.

9.2.2.3 *Dimensioning of slender ductile walls with a composite cross section under bending with axial force*

The cross section of the ductile walls usually has a composite form (e.g. L, T, U, I, Z hollow tubular form, etc.). These composite wall sections, according to EC8-1/2004, should be taken as integral units in analysis and design. The dimensioning of a cross section of this form may be carried out only with the aid of computational tools. Consider a cross section in the form shown in Figure 9.10, loaded by a vector group M_{Eyd}, M_{Ezd}, N_{Ed} at the geometrical centre S of the cross section. The ULS verification of bending resistance is expressed by the relation

$$\frac{M_{Ryd}}{M_{Eyd}} = \frac{M_{Rzd}}{M_{Ezd}} = \frac{N_{Rd}}{N_{Ed}} = \gamma \geq 1.0 \tag{9.8}$$

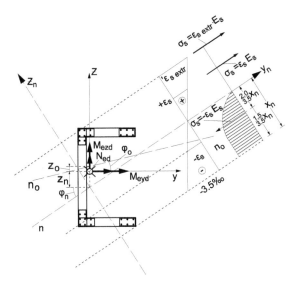

Figure 9.10 Dimensioning of a composite cross section bending with axial force.

or

$$
\left.
\begin{aligned}
M_{Ryd} &\geq \gamma M_{Eyd} \\
M_{Rzd} &\geq \gamma M_{Ezd} \\
N_{Rd} &\geq \gamma N_{Ed}
\end{aligned}
\right\}
\qquad (9.8a\text{–}c)
$$

If the neutral axis n–n has been determined (z_n, φ_n), the corresponding strain diagram at failure will be that given in Figure 9.10, together with the stress field of concrete and steel. Therefore, M_{Ryd}, M_{Rzd}, N_{Rd} can be determined and ULS verification of Equation 9.8a–c is easy. Starting from a first position of the neutral axis corresponding to the uncracked cross section (Stage I) and applying a Newton–Raphson procedure already presented in Chapter 8.2.2.4 for beams, the unknown parameters z_n, φ_n and γ can be defined very quickly via a proper commercial platform (NOUS/3P (2002) FAGUS/CUBUS-2011) or more sophisticated computer aids proper for academia purposes (Papanikolaou, 2012).

It is apparent that, at the same time, it is possible for the design of the cross section to be carried out automatically. Having an estimated steel reinforcement pattern in advance and introducing as unknown the change ΔA_s of the steel reinforcement in predefined positions (e.g. the corners and the ends of the composite cross section), in place of γ, which is predefined as $\gamma = 1$, the automated computer-aided design gives the required change of the reinforcement after all 33 load combinations have been checked, so that $[E_d] = [R_d]$.

9.2.2.4 Determination of M–φ diagram and ductility in terms of curvature under axial load for orthogonal cross sections

According to what has been presented so far, ductile walls are the main vertical members of the structural system at the base of which seismic energy will be dissipated, together with the ends of the beams and the columns at their bases (see Chapter 6.1.5, Figure 6.7). Therefore, a special concern must be given to the design of walls for *local ductility at their base*.

In the case that their cross section is orthogonal, it is self-evident that they resist bending on the long side of their cross section. In this respect, they behave like columns under uniaxial bending. Therefore, the procedure developed in Chapter 8.3.2.2 also holds for orthogonal ductile walls.

From the above-mentioned treatment for the ductility of columns, Equations 8.99b, 8.106 and 8.107 will be recalled and modified properly for the case of walls. According to Equation 8.106, κ should be introduced in Equation 8.107 with the value $\kappa = 1.44$. Therefore,

$$\mu_\varphi = \frac{\varepsilon_{cu}}{\varepsilon_y} \frac{0.72}{0.9\kappa(v+w_v)} \cong \frac{0.80}{1.44} \frac{\varepsilon_{cu}}{\varepsilon_y} \frac{1}{(v+w_v)}. \tag{9.9}$$

Equations 8.99b, 8.106 and 9.9 will be used a little bit later for the design of the local ductility of orthogonal walls according to EC8-1/2004.

For composite cross sections (L, T, U, I, Z, etc.), computer platforms have been developed that allow the determination of points A, B and C (Figure 8.12, Chapter 8.3.2.2; e.g. RCCOLA, FAGUS, NOUS/3P, etc.). Computer platform NOUS/3P (2002) has already been presented for beams and can also be used for composite wall sections and for columns.

9.2.3 Behaviour of slender ductile walls under prevailing shear

The failure modes of walls with orthogonal cross section under prevailing shear, as in the case of beams, may be classified in the following types:

- Diagonal tension (Figure 9.11a)
- Diagonal compression (Figure 9.11b)
- Sliding shear in the region of the plastic hinge (construction joint; Figure 9.11c)

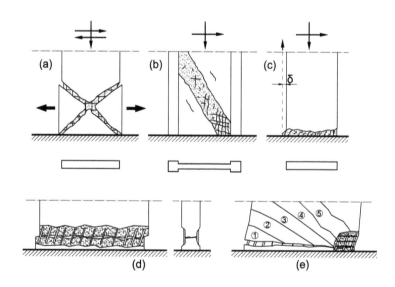

Figure 9.11 Failure modes of walls under prevailing shear: (a) diagonal tension; (b) diagonal compression; (c) sliding shear; (d) detail of sliding shear; (e) bending failure combined with sliding. (From Salonikios, T. 2007. Analytical prediction of the inelastic response of R/C walls with low aspect ratio. ASCE Journal of Structural Engineering, 133(6), 844–854. With permission of ASCE.)

Keeping in mind that the maximum normalised axial forces in ductile walls are limited by the current Codes, much lower than those in columns, as we will see a little later, it is apparent that the behaviour of ductile walls to shear lies between the behaviour of beams and columns. Therefore, the design of walls to shear follows in concept the methods developed for beams and columns with minor modifications, which will be presented in Section 9.2.4. Here, only two particular points should be presented:

1. From extended experimental work, it has been determined (Kowalsky and Priestley, 2000) that in the region of the plastic hinge at the base of the wall, due to the degradation of concrete compressive strength under loading reversals, the diagonal compression strength is also degraded. Therefore, a reduction factor should be introduced in the expressions used for the determination of V_{Rdmax}. This factor, as we will see later, is on the order of 0.40 for DCH buildings.
2. Sliding shear may appear in the region of a plastic hinge and particularly at the position of a construction joint. This failure at horizontal planes can be resisted by
   - Shear friction across the horizontal crack
   - Dowel action
   - Resistance of inclined steel bars arranged in the joint region

The main reason of this type of failure must be attributed to the low value of axial loading $(v_d \leq 0.35$ and, therefore, $v_k \leq (0.35/1.50) \cong 0.23$; see Section 9.2.4.3(2)).

As a consequence, the behaviour of a wall to cyclic loading after yield as far as the shear type of failure at the critical region is similar to that of beams under cyclic loading (Chapter 8.2.4.2, Figures 8.29 and 8.30).

Expressions for the strength calculation to sliding shear will be given in Section 9.2.4.

9.2.4 Code provisions for slender ductile walls

9.2.4.1 General

According to EC8-1/2004, ductile walls are defined as 'slender' if the aspect ratio h_w/l_w satisfies the known relation

$$\boxed{h_w / l_w \geq 2}$$

(9.10)

where
  h_w is the height of the wall.
  l_w is the length.

The design rules for slender ductile walls under seismic actions according to EC8-1/2004 are given below, together with the required justification in terms of the theory developed in previous subsections (Chapters 8 and 9). It should be noted that, as in the case of beams and columns, the design specifications are directly related to ductility demand in consideration (DCL, DCM and DCH). For a higher ductility class, these requirements tend to be stricter in order to obtain a *higher local ductility* in terms of curvature. Specifications for the design effects of ductile walls, taking into account capacity design procedure, have already been presented in Chapter 6. Material issues have also been presented in Chapter 7, Sections 7.2.4 and 7.3.5, for all ductility classes.

It should also be noted here that for the design of slender walls of class DCL, Code specifications for conventional R/C structures are applied (EC2-1-1/2004), except that concrete and steel reinforcement qualities are specified by EC8-1/2004 and have already been presented in Chapter 7. For this reason, the design of slender walls for DCL buildings does not require further treatment.

In conclusion, it should also be noted that all rules and specifications of the corresponding Code for conventional R/C structures (EC2-1-1/2004) continue to be in effect, unless they contradict what will be presented below.

9.2.4.2 Design of slender ductile walls for DCM buildings

1. *Geometrical constraints*
   These have been given already in Section 9.2.1.
2. *Resistance to flexure and shear*
   a. *Flexural* and *shear* resistances will be computed in accordance with EN1992-1-1/2004 using the axial force resulting from the analysis in the seismic design situation and the gravity load combination (basic).
   It is apparent that bending and shear demand, which are introduced in
   ULS verification, result from the *capacity design* procedure. In this respect, for each cross section under consideration, 33 load case combinations should be taken into account (1 basic +32 seismic combinations).
   b. The value of the normalised axial load v_d should be limited to 0.40, in other words,

$$\boxed{v_d \le 0.40}$$

(9.11)

   c. Vertical web reinforcement should be taken into account for the calculation of flexural resistance.
   d. Composite wall sections, more specifically wall sections of the form L, T, U, I, Z, should be taken as integral units. In any case, EC8-1/2004 allows the replacement of a composite cross section by webs connected with flanges and distribution of the bending moments and the axial forces to these T or I elements in a uniaxial bending load case. It is the author's opinion that this simplification cannot much help the design procedure, at least in flexure, since each load combination corresponds to a biaxial bending. Therefore, the T or I models should be formed for both main directions, the moment vectors in each main direction should be distributed to the elements acting in this direction and thus the number of verifications should be a multiple of 33 (Figure 9.12). Consequently, the easiest way is to design the composite section for each group of M_{Eyd}, M_{EZd}, N_{Ed} using a proper computer platform for biaxial bending (Section 9.2.2.3). Conversely, for shear design, the only realistic

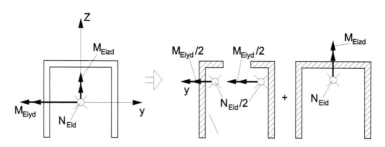

Figure 9.12 Biaxial bending of a composite cross section: simplified procedure replacing the composite cross section by orthogonal cross sections with flanges.

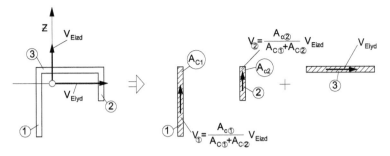

Figure 9.13 Biaxial shear of a composite cross section: simplified procedure replacing the composite cross section by orthogonal elements in each direction.

approach is the above simplification recommended by the Code. In this respect, shear forces V_{Eiyd} and V_{Eizd} of each load combination are distributed to the webs corresponding to their direction proportional to the area of each web. From this point on, these webs are dimensioned to shear as independent orthogonal cross sections (Figure 9.13).

3. *Detailing of slender ductile walls for local ductility*

   a.   The height of the critical region h_{cr} above the base of the wall is estimated as

$$h_{cr} = \max\{l_w, h_w/6\} \tag{9.12}$$

   but

$$h_{cr} \leq \begin{cases} 2l_w \\ h_s \quad \text{(for } n \leq 6 \text{ storeys)} \\ 2h_s \quad \text{(for } n > 6 \text{ storeys)} \end{cases} \tag{9.13}$$

where

   h_w is the total height of the wall from the base to the top.
   l_w is the length of the cross section of the wall.
   h_s is the clear storey height.
   base is defined as the level of the foundation (Figure 9.14) or the top of basement storeys with rigid diaphragms and perimeter walls (box-type foundation).

   b.   In the critical region *at the base* of the slender ductile walls, a value of the curvature ductility factor μ_ϕ should be provided, directly related to the basic value of the behaviour factor q_o of the building using Equations 5.52a and 5.52b.
        In order for the eventual overstrength at the fixed end of the wall to be taken into account, q_o should be introduced in the above expressions reduced by the factor M_{Ed}/M_{Rd} (counter-balance between ductility and overstrength), where M_{Ed} is the seismic design combination of bending moment from the analysis and M_{Rd} is the design flexural resistance. It is obvious that M_{Ed}/M_{Rd} is always less or equal to 1.0.
   c.   The above specified ductility μ_ϕ is ensured by means of a proper confinement of the edge regions of the cross section using steel hoops and ties (Figures 9.15 and 9.16).

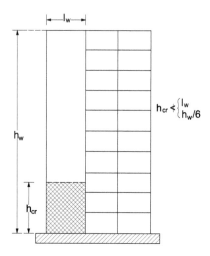

Figure 9.14 Determination of the wall critical region.

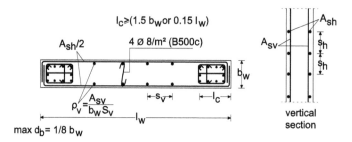

Figure 9.15 Arrangement of horizontal and vertical reinforcement in walls with a rectangular cross section.

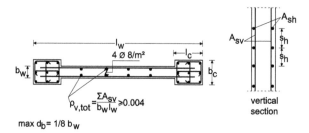

Figure 9.16 Arrangement of horizontal and vertical reinforcement in walls with a barbell cross section.

    d. For walls of a rectangular cross section, the mechanical volumetric ratio ω_{wd} of the required confining reinforcement in boundary elements should satisfy the following expression:

$$\boxed{\alpha\omega_{wd} \geq 30\mu_{\varphi}(v_d + \omega_{vd})\varepsilon_{sy,d}\frac{b_c}{b_o} - 0.035} \tag{9.14}$$

    The above equation is similar to Equation 8.127, which has been introduced in the case of columns. The only difference is that in the case of walls, the term ω_{vd}

is introduced in addition, which expresses the mechanical volumetric ratio of the longitudinal web reinforcement.

$$\omega_{vd} = \frac{A_{sv}}{b_c d_c} \cdot \frac{f_{yd}}{f_{cd}} \qquad (9.15)$$

The meaning of the above notation has been defined in Chapter 8.3.4.2.3 for columns. Equation 9.14 ensures a safety factor for local ductility of ductile walls on the order of

$$\gamma_{\mu\phi} \approx 2.30 \qquad (9.16)$$

See the first edition of this book (Penelis and Penelis, 2014).
e. The above-specified confinement must extend vertically over the height $_{hcr}$ of the critical region and horizontal along a length l_c (Figure 9.17) from the extreme compression fibre of the wall up to the point where compressive strain becomes less than

$$\varepsilon_{cu2} = 0.0035 \qquad (9.17)$$

Therefore, the confined boundary element for a wall of orthogonal section is limited to a distance (Figure 9.17)

$$l_c = x_u \left(1 - \frac{\varepsilon_{cu2}}{\varepsilon_{cu2,c}} \right) \qquad (9.18)$$

where

$$x_u = (v_d + \omega_{vd}) \frac{l_w \cdot b_c}{b_o} \qquad (9.19)$$

(see Equation 8.99b).

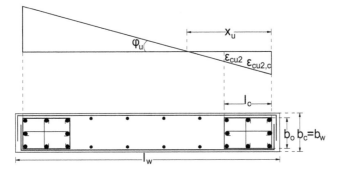

Figure 9.17 Confined boundary element of free-edge wall end (top: strains at ultimate curvature; bottom: wall cross section). (Adapted from E.C.8-1/EN1998-1. 2004. *Design of Structures for Earthquake Resistance: General Rules, Seismic Actions and Rules for Buildings.* CEN, Brussels, Belgium.)

The extreme compressive strain $\varepsilon_{cu2,c}$ is related to the degree of confinement $\alpha\omega_{wd}$ via the following expression:

$$\varepsilon_{cu2,c} = 0.0035 + 0.1\alpha\omega_{wd} \qquad (9.20)$$

(see Chapter 7.4.3.2; Equation 7.26).

   f. Therefore, for the determination of the required confinement in the critical regions of the wall, the following steps are followed:

Step 1: Determination of the required confinement $\alpha\omega_{wd}$ using Equation 9.14.

Step 2: Determination of the required $\varepsilon_{cu2,c}$ using Equation 9.20.

Step 3: Determination of the neutral axis depth x_u at ultimate curvature Equation 9.19.

Step 4: Determination of the depth l_c of the confined boundary using Equation 9.18.

Step 5: Formulation of the confinement detail and determination of $\alpha = \alpha_n \cdot \alpha_s$ Equation 8.129.

Step 6: Determination of $\omega_{wd} = \rho_w(f_{yd}/f_{cd})$ using the results of steps 1 and 5.

     It should be noted that the whole issue is handled by the American Code (ACI 318M-2011) in a very simplified manner, although the procedure followed is based on the same concept as EC8-1/2004, that is, the enhancing of the extreme compressive strain via concrete confinement with steel hoops and ties.

   g. The minimum dimensions of the confined boundary elements of free-edge wall ends have been specified in Section 9.2.1(3).

   h. The longitudinal reinforcement ratio in the boundary element should not be less than 0.005 over all the height of the element. It is obvious that the depth l_c of the boundary element may be decreased above the critical region to the lower limit that has been specified in Section 9.2.1(3).

   i. The provisions for the base of primary columns as far as the minimum value of ω_{wd} of 0.08 and the spacing of the hoops and ties should also be applied within the boundary elements of walls.

   j. In the height of the wall above the critical region, the rules of EN 1992-1-1/2004 apply basically regarding vertical horizontal and transverse reinforcement.

   k. The required area of transverse reinforcement A_{si} within the lap zone of the longitudinal reinforcement of boundary elements in walls, no matter if they are classified as DCM or DCH, is calculated as in the case of columns (Equation 8.141; Chapter 8.3.4.5).

   l. In the case of walls with barbells or with composite cross section (T, U, L, I, Z-shaped sections), the boundary elements should be formed in all boundaries and at the joints of the orthogonal parts (Figure 9.18). In this way, both local ductility and shear transfer at the joints between adjacent orthogonal parts is ensured. The

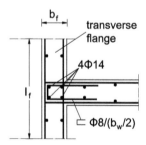

Figure 9.18 Detailing of the web-flange connection in flanged walls.

required confinement degree $\alpha\omega_{wd}$ and the confined area in this case are defined as follows:

i. The axial force N_{Ed} and the total area of the vertical reinforcement in the web A_{sv} are normalised to the area of the barbell or the flange $(h_c b_c)$ (Figures 9.19 and 9.20), that is,

$$v_d = \frac{N_{Ed}}{h_c b_c f_{cd}}, \omega_v = \frac{A_{sv} f_{yd}}{h_c b_c f_{cd}} \qquad (9.21a,b)$$

The neutral axis depth x_u at ultimate curvature φ_u after concrete spalling is given by expression 9.19. If the value of x_u does not exceed the depth of the barbell or the flange thickness after spalling of the cover concrete, then $\alpha\omega_{wd}$ may be defined using Equation 9.14. In this case, the flange or the barbell will be confined to their total area $b_c...h_c$ (Figures 9.19 and 9.20).

ii. In the case of a wall with barbells, if the value x_u exceeds the depth of the barbell, the simplest solution is to design a bigger barbell.

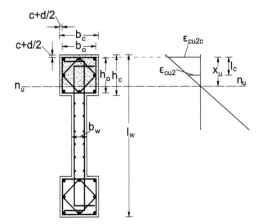

Figure 9.19 Determination of the confined boundary elements of a free-edge wall with barbells.

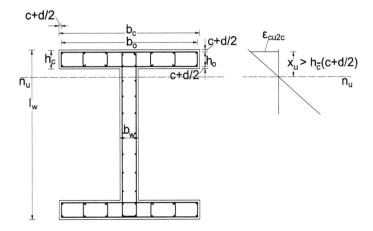

Figure 9.20 Determination of the confined region of a composite cross section.

iii. In the case of a composite wall, if the value x_u significantly exceeds the thickness of the flange, an approach to the problem would be an increase in the thickness of the flange. A second approach more suitable for an automatic computer-aided design is to sacrifice the flange and to confine the web as an orthogonal cross section with confined boundary elements (Fardis et al., 2005; Fardis, 2009).

Such an approach to the problem is compatible with the concept of the formation of confined boundary elements not only at the ends of the orthogonal components but also in all joints presented in a previous paragraph (Figure 9.21). However, this approach leads to very long confined boundary elements.

iv. In the case of a composite wall, if the value x_u exceeds the thickness of the flange, a computer-aided design could ensure an accurate determination of the required confinement $\alpha\omega_{wd}$ for a given local ductility μ_ϕ at the curvature level. This procedure is codified by EC8-1/2004 and is outlined below:

– The required value of μ_ϕ is determined using Equations 5.52a and 5.52b as a function of the basic value of the behaviour factor q_o.

– For successive values of $\alpha\omega_{wd}$ using Equation 9.20, the values of $\varepsilon_{cu2,c}$ are determined.

– For each of these values of $\varepsilon_{cu2,c}$, the values of ϕ_y and ϕ_u^* are determined using a computer-aided iterative procedure (e.g. platform NOUS; Figure 9.22). When the ratio reaches

$$\mu_\phi^{avail} = \frac{\phi_u^*}{\phi_y} \geq \mu_\phi^{required} \tag{9.22}$$

the procedure stops and the area with compressive strain greater than $\varepsilon_{cu2} = 3.5‰$ is confined with hoops and ties corresponding to $\alpha\omega_{wd}$, for which inequality 9.22 is fulfilled.

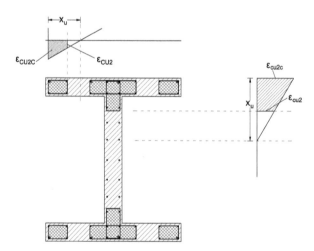

Figure 9.21 Determination of the confined regions of a composite cross section based on the assumption of sacrificed flanges.

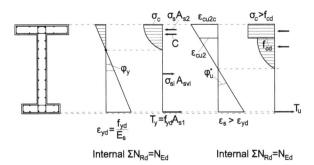

$$\varepsilon_{yd} = \frac{f_{yd}}{E_s}$$

$T_y = f_{yd} A_{s1}$

Internal $\Sigma N_{Rd} = N_{Ed}$

$\varepsilon_s > \varepsilon_{yd}$

Internal $\Sigma N_{Rd} = N_{Ed}$

Figure 9.22 Determination of the confinement of a composite cross section following an iterative step-by-step method.

9.2.4.3 Design of slender ductile walls for DCH buildings

1. *Geometrical constraints*
    In addition to what has been specified for ductile walls of DCM buildings, the following requirements are imposed:
    a. The minimum dimensions of barbells and flanges are given in the relevant paragraph for confined boundary elements.
    b. Random openings not regularly arranged to form coupled walls should be avoided, unless their influence is either insignificant or accounted for in the analysis dimensioning and design.
2. *Bending resistance*
    a. Flexural resistance is evaluated as for DCM walls.
    b. The value of the normalised axial load ν_d should be limited to 0.35:

$$\boxed{\nu_d \leq 0.35} \tag{9.23}$$

3. *Diagonal compression failure due to shear*
    The value of V_{Rdmax} should be calculated as follows:
    i. *Outside the critical region*
        The value of V_{Rdmax} should be calculated as in the case of columns with a length of the internal lever arm z equal to $0.8l_w$ and an inclination of the compression struts to the vertical equal to $\delta = 45°$ (Figure 9.23a).
    ii. *In the critical region, shear resistance is limited to 40% of the value outside the critical region* to confront the shear degradation of concrete in the critical region under cyclic loading. It should be noted that in the corresponding American Code (ACI 318M-2011), there is not any provision for this type of reduction.
4. *Diagonal tension failure of the web due to shear*
    a. The calculation of web reinforcement for the ULS verification in shear, in the case that

$$\alpha_s = \frac{M_{Ed}}{V_{Ed} l_w} \geq 2.0 \tag{9.24}$$

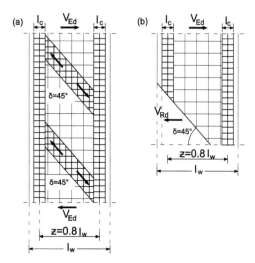

Figure 9.23 Diagonal shear resistance V_{Rd}: (a) compressive struts resistance; (b) tensile reinforcement resistance.

is carried out as in the case of columns, but with the values of z and $\tan \delta$ as follows (Figure 9.23b):

$$z = 0.80l_w \tag{9.25a}$$

$$\tan \delta = 1.0 \tag{9.25b}$$

The variable strut inclination concept is replaced by the Mörsh concept of 45° strut inclination, as in the case of beams of DCH buildings.

b. It is apparent that horizontal web bars must be properly anchored at the ends of the wall sections (Figures 9.15 and 9.16).

c. If the mechanical aspect ratio at the base of a storey is

$$\frac{M_{Ed}}{V_{Ed}l_w} \le 2.0 \tag{9.26}$$

even in case of slender walls, the design of diagonal tension should be carried out as in the case of squat walls (Section 9.4.4). This rule is meaningful basically for the upper storeys of dual systems. In these systems, while M_{Ed} is given by the diagram of Figure 6.8 in Chapter 6.1.5.2, which is linear over the entire height of the building, the relevant V_{Ed} diagram Figure 6.10, Chapter 6.1.5.3, due to capacity design, is overestimated at the upper storeys, and, therefore, very often, the aspect ratio $M_{Ed}/V_{Ed}l_w$ results in values less than 2. So, the web reinforcement at the upper storeys turns out to be unexpectedly high.

5. Sliding shear failure

a. At potential sliding shear planes (e.g. at construction joints) within the critical region, sliding shear failure may occur only in *slender walls of high ductility*. This has been properly explained in Section 9.2.3.2. As already noted there, the shear resistance against sliding, $V_{Rd,s}$, comprises three components:

$$\boxed{V_{Rd,s} = V_{dd} + V_{id} + V_{fd}} \tag{9.27}$$

where
V_{dd} is the dowel resistance of the vertical bars.
V_{id} is the shear resistance of inclined bars (at an angle φ to the potential sliding plane).
V_{fd} is the friction resistance.

The wall is deemed to be safe against sliding if the following condition is satisfied:

$$\boxed{V_{Ed} \leq V_{Rd,s}} \tag{9.28}$$

b. The above components V_{dd}, V_{id} and V_{fd} may be determined by means of the following expressions:

$$\boxed{V_{dd} = \min \begin{cases} 1.3 \sum A_{sj} \sqrt{f_{cd} \cdot f_{yd}} \\ 0.25 f_{yd} \cdot \sum A_{sj} \end{cases}} \tag{9.29a,b}$$

$$\boxed{V_{id} = \sum A_{si} \cdot f_{yd} \cdot \cos \varphi} \tag{9.30}$$

$$\boxed{V_{fd} = \min \begin{cases} \mu_f \left[\left(\sum A_{sj} f_{cd} + N_{Ed} \right) \xi + \dfrac{M_{Ed}}{z} \right] \\ 0.5 n \cdot f_{cd} \cdot \xi \cdot l_w \cdot b_{wo} \end{cases}} \tag{9.31a,b}$$

where
μ_f is the concrete-to-concrete friction coefficient under cyclic action, which may be assumed to be equal to 0.6 for smooth interfaces and to 0.7 for rough ones.
z is the internal lever arm with a value equal to $0.8 l_w$.
ξ is the normalised neutral axis depth at failure.
$\sum A_{sj}$ is the sum of the cross-section areas of the vertical bars of the web and of *additional* bars arranged in the boundary elements specifically for resistance against sliding.
$\sum A_{si}$ is the sum of the cross-section areas of all inclined bars in both directions; large-diameter bars are recommended for this purpose (Figure 9.24).
n: $0.6/(1 - f_{ck}[MP_a]/250)$.
N_{Ed} is introduced with a positive sign when compressive.

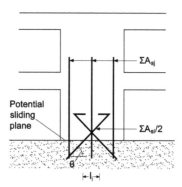

Figure 9.24 Bidiagonal reinforcement in structural walls.

c. The three mechanisms of shear resistance against sliding have been developed progressively in the past 50 years by various investigators.

The dowel effect mechanism V_{dd} (Equation 9.29a) is based on the Rasmussen (1963) reports in combination with an upper bound (Equation 9.29b) suggested by Paulay and Priestley (1992) as a reasonable limit for squat walls under cyclic loading.

The diagonal truss mechanism V_{id} is based on Mörsh and its 'truss analogy model', and has thus far been used many times (mainly for shear transfer in beams and short columns).

Finally, the friction mechanism V_{fd} (aggregate interlock) is based on the assumption that friction is developed along the compression zone only. Therefore, the axial forces contributing to friction are the following (Tassios, private communication, 1994):

i. The fraction of the wall axial force corresponding to the compression zone, ξN_{Ed}

ii. The clamping action of the vertical reinforcement intersecting the joint in the compression zone when sliding is activated $(\Sigma A_{sj})f_{yd} \cdot \xi$ (Figure 9.25; Park and Paulay, 1975)

iii. The compression force caused by the bending moment M_{Ed}/z

d. The arrangement of x-shaped reinforcement intersecting the sliding joint in the critical region is not obligatory; it is in the hands of the designer to decide if he uses additional vertical web reinforcement or x-shaped rebars. It should be noted that the arrangement of x-shaped bars in the walls is not an easy task from the constructability point of view.

e. Inclined bars should be fully anchored on both sides of the potential sliding joint. It should also be noted that inclined bars, if arranged after the general design of the wall as a result of a local design for the wall sliding at the base joint, lead to an increase in the bending resistance of the wall there and, therefore, to an increase in the base shear resulting from the capacity design procedure. This increase in the bending moment may be estimated as follows (Figure 9.26):

$$\Delta M_{Rd} = \frac{1}{2}\sum A_{si} \cdot f_{yd} \sin \varphi \cdot l_1 \qquad (9.32)$$

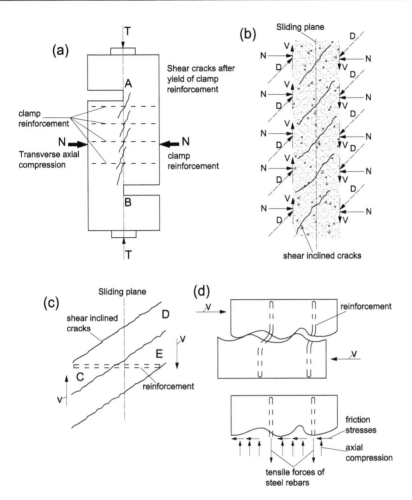

Figure 9.25 Shear transfer model at a sliding joint: (a) experimental set-up; (b) transfer model along an uncracked joint; (c) detail (b); (d) shear sliding. Friction stresses along a pre-cracked sliding plane.

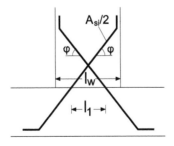

Figure 9.26 Increase in bending resistance at the base of the wall due to x-shaped rebar arrangement.

where
l_1 is the distance between the centrelines of the two sets of inclined bars at the x-shaped arrangement.

Alternatively, the shear resistance V_{id} of the inclined bars may be reduced instead of increasing V_{Ed} due to ΔM_{Rd}. In this context, V_{id} takes the form

$$V_{idred} = \sum A_{si} f_{yd} (\cos \varphi - 0.5 l_1 \ \sin \varphi / \alpha_s \cdot l_w) \qquad (9.33)$$

This expression results very easily from the following considerations. The increase ΔV_{Ed} of V_{Ed} due to the increase in base moments by ΔM_{Rd} may be derived from the expression

$$\alpha_s = \frac{\Delta M_{Ed}}{l_w \Delta V_{Ed}} \cong \frac{\Delta M_{Rd}}{l_w \Delta V_{Rd}} \qquad (9.34)$$

or

$$\Delta V_{Rd} \cong \frac{\Delta M_{Rd}}{l_w \alpha_s} \qquad (9.35)$$

Keeping in mind that

$$V_{idred} = V_{id} - \Delta V_{Rd} \qquad (9.36)$$

and introducing the values of V_{id} and ΔV_{Rd} from Equations 9.30 and 9.35, we get expression 9.33.

6. *Detailing of slender ductile walls for local ductility*
    In addition to the rules for DCM buildings presented in Section 9.2.4.2.3, EC8-1/2004 specifies some additional rules for DCH buildings, the most important of which are given below:
    a. Within the boundary elements in the critical region, a minimum value of ω_{wd} equal to

$$\omega_{wd\,min} = 0.12 \qquad (9.37)$$

is specified.
        At the same time, for the confining reinforcement of the boundary elements, that is, the hoop bar diameters, their spacing and the distance between consecutive longitudinal bars restrained by hoops, the rules specified for the base critical region of columns of DCH buildings are also specified in the case of DCH slender structural walls.
    b. The critical region is extended for one more storey than the critical region for DCM buildings, but only with half of the confining reinforcement.
    c. A minimum amount of web reinforcement,

$$\rho_{1,min} = \rho_{v,min} = 0.002 \qquad (9.38)$$

is specified to prevent premature web shear cracking. This reinforcement is provided in the form of two grids of bars, one in each face of the wall. The grids should be connected through cross-ties spaced at about 500 mm. Web reinforcement should have a diameter of those shown in Figures 9.15 and 9.16.

$$\boxed{8.0 \text{ mm} \leq d_{web} \leq b_{wo}/8} \tag{9.39}$$

and should be spaced at not more than 250 mm or 25 times the diameter of the bar, whichever is smaller.

d. For the protection of horizontal construction joints out of the critical region against cracking, a minimum amount of fully anchored vertical reinforcement should be provided across such joints.

The minimum ratio of this reinforcement ρ_{min} is given by the expression

$$\boxed{\rho_{min} \geq \begin{cases} \left[\left(1.3 \cdot f_{ctd} - \dfrac{N_{Ed}}{A_{sw}} \right) \Big/ \left(f_{yd} \left(1 + 1.5 \sqrt{f_{cd}/f_{yd}} \right) \right) \right] \\ 0.0025 \end{cases}} \tag{9.40}$$

where
 A_{sw} is the total horizontal cross-sectional area of the wall.
 N_{Ed} is the design axial load of the seismic load combinations.
 (N is considered positive when it is compressive.)

9.3 DUCTILE COUPLED WALLS

9.3.1 General

- As presented in Chapter 4.5.3.3, ductile coupled walls result when slender ductile walls are coupled with spandrels arranged at the level of the storeys above doors or windows. Moment diagrams of these systems are depicted in Figure 4.29.
- The main characteristic of these diagrams is that the fixed-end moment at the base of each of the coupled walls is smaller than half of the fixed-end moment that would be developed in the case that the stiffness of the spandrels was zero. This reduction of the fixed-end moment at the base is attributable to the function of the spandrels. This function has been explained in detail in Chapter 4.5.3.3.
- The analysis of coupled walls has passed through three stages related to the development of computers. In the late 1960s, the method of a continuous connecting medium was used, known also as 'laminar analysis' (Figure 9.27; Rosman, 1965; Penelis, 1969b). Frame models were introduced in the early 1980s, since their high redundancy could be dealt with by commercial computer platforms. Today, FEM discretisation is in favour, as it also may be used for any type of wall. Of course, this discretisation is supported by post-processing procedures for integration of resulting stresses to internal forces M_{Ed}, N_{Ed}, V_{Ed}, as we will see later (Section 9.6.1).

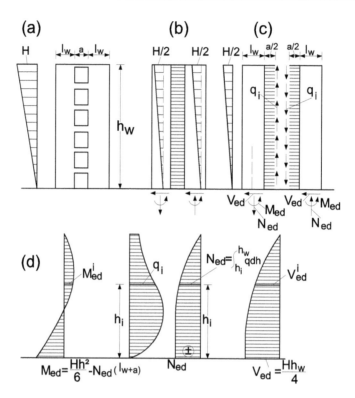

Figure 9.27 Structural behaviour of coupled walls: (a) structural system; (b) laminar analysis; (c) internal forces at the laminated spandrels; (d) diagrams of internal forces.

9.3.2 Inelastic behaviour of coupled walls

Apart from the analytical procedure for the determination of internal forces based on linear elastic procedure, it should be noted again (see Chapter 4.5.3.3) that the yield of spandrels under horizontal loading precedes the yield at the base of the walls. In fact, from analytical investigations and experimental evidence, the following conclusions have been drawn (Park and Paulay, 1975; Aristizabal-Ochoa, 1982; Shiu et al., 1984):

- Ductility demand of the spandrels in terms of rotations $\theta\rho/\theta_y$ reaches a value of about 3.5, while the coupled walls continue to be in elastic range.
- For a ductility demand of the walls at their base on the order of 4, the ductility demand of spandrels reaches a value of about 11. Therefore, special concern is needed for ensuring high-ductility capacities for the spandrels (Figure 9.28, position 1).
- Keeping in mind that the seismic horizontal loading is counterbalanced by fixed-end moments combined with axial loads of opposite sign, it is apparent that in every half-loading cycle, one of the walls is subjected to considerable tension in addition to flexure and shear. This load condition may adversely affect the diagonal tension capacity of the wall (Figure 9.28, position 2).
- Sliding of the walls at their base also constitutes an eventual failure mechanism (Figure 9.28, position 3).
- From all of the above, it can be seen that if the ductility of the spandrels is ensured and no pre-emptive shear failure appears at their web, coupled ductile walls provide an effective ductile system with an additional first line of defence at the spandrels

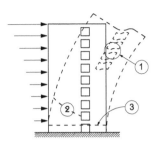

Figure 9.28 Critical areas of behaviour in coupled shear walls. (Park, R. and Paulay, T.: *Reinforced Concrete Structures.* 1975. Copyright Wiley-VCH Verlag GmbH & Co. KGaA. Reproduced with permission.)

against seismic action. In this respect, spandrels act like fuses (Abrams, 1991). For this reason, EC8-1/2004 classifies coupled walls in a higher 'behaviour' category than that for isolated (single) slender ductile walls (see Chapter 5.4.3). A basic condition for the characterisation of two or more ductile walls connected with beams is that their fixed-end moments are reduced by at least 25%, due to their coupling in relation to the fixed-end moments of the individual walls (see Chapter 4.5.3.3, Figure 4.29).

- It has been verified by extensive experimental evidence that spandrels reinforced with x-shaped reinforcement (Figure 9.29), well-anchored in the walls, exhibit a very high ductility in relation to spandrels reinforced conventionally (steel bars at the upper and bottom flanges and web reinforcement in the form of ties; Park and Paulay, 1975).

This type of reinforcement has also been adopted by the Code for DCH beams in the case of shear reversals (see Chapter 8, Sections 8.2.4 and 8.2.5). Even in the case of squat columns where axial load prevails, this type of reinforcing has been introduced as an alternative solution (Tegos and Penelis, 1988). For this reason, x-shaped reinforcing bars are arranged in the form of hidden inclined columns in the spandrel to avoid buckling. This arrangement has been adopted by all modern Seismic Codes.

Contribution to shear resistance V_{Rd} of x-shaped reinforcement is given by the following expression:

$$V_{Rd} = 2 \cdot A_{si} \cdot f_{yd} \cdot \sin \alpha \qquad (9.41)$$

where
V_{Rd} is the shear resistance in terms of design.
A_{si} is the total area of steel bars in each diagonal direction.
α is the angle between the diagonal bars and the axis of the beam.

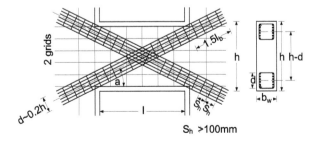

Figure 9.29 Arrangement of reinforcement in a coupling beam.

On the other hand, the contribution M_{Rdx} to moment resistance of the spandrel due to the x-shaped steel reinforcement is given by the expression

$$\boxed{M_{Rd} \cong (h - d) \cdot A_{si} \cdot f_{yd} \cdot \cos \alpha}$$
(9.42)

The notation in use in the above expressions is given in Figure 9.29.
As far as the coupled walls are concerned, two issues arise:

- The high tensile force at the base of the walls and its influence on their shear strength. From existing experimental evidence (Santhakumar, 1974; Park and Paulay, 1975; Abrams, 1991), it has been concluded that the shear design foreseen by modern Codes for slender ductile walls also provides adequate shear resistance for coupled walls.
- Design to flexure at the critical regions of the walls should take into account the sign of the axial force in relation to the sign of the corresponding moment (Figure 9.27c). This sign combination results in stronger reinforcement at the external flanges of the coupled walls than the internal ones. In the case that the analysis has been carried out using modal response spectrum analysis, this discrimination cannot be made easily, since all results have a positive sign.

9.3.3 Code provisions for coupled slender ductile walls

As already noted, two or more walls are considered *coupled* if the connecting beams reduce the fixed-end moments of the individual walls by at least 25%. Therefore, coupling of walls by means of slabs must not be taken into account, as it is not effective.

Rules in use for ductile beams of DCH may also apply for spandrels, only in the case that *at least one* of the following conditions is fulfilled:

- Cracking in both diagonal directions is unlikely, to occur, if

$$\boxed{V_{Ed} \leq f_{ctd} \cdot b_w (h - d)}$$
(9.43)

- A prevailing flexural mode of failure is ensured. An acceptable rule is

$$\boxed{l/h \geq 3.0}$$
(9.44)

- If neither of the above conditions is fulfilled, the resistance to seismic actions should be provided by reinforcement arranged along both diagonals of the beam (Figure 9.29). In this case,

$$\boxed{V_{Ed} \leq 2 \cdot A_{si} \cdot f_{yd} \cdot \sin \alpha}$$
(9.45)

$$\boxed{M_{Ed} \leq (h - d) \cdot A_{si} \cdot f_{yd} \cdot \cos \alpha}$$
(9.46)

(see Section 9.3.2, Equations 9.41 and 9.42)

where

$$V_{Ed} = 2M_{Ed}/l \qquad (9.47)$$

- The diagonal reinforcement is arranged in column-like elements with side lengths equal at least to $0.50 \cdot b_w$, its anchorage length greater by 50% than that required by the Code for conventional R/C structures. The hoops of these elements should fulfil the provisions for the critical region of columns in DCH buildings, to prevent buckling.
- Longitudinal and transverse reinforcement is provided on both lateral faces of the spandrel according to the Code for conventional R/C structures and particularly for deep beams. The longitudinal reinforcement should not be anchored in the coupled walls and should extend into them by 150 mm, so that a change in the structural behaviour of the spandrel can be avoided.
- Slender ductile coupled walls are designed according to the rules for individual (single) walls and in accordance with their ductility class (M or H), since there is no difference in their behaviour under seismic loading in the post-elastic stage.

9.4 SQUAT DUCTILE WALLS

9.4.1 General

Squat ductile walls with a ratio of height h_w to length l_w of less than 2,

$$\boxed{\alpha_s = \frac{h_w}{l_w} \leq 2.0}, \qquad (9.48)$$

find a wide application in earthquake-resistant low-rise buildings. They are also used in high-rise buildings as additional structural elements extending from their foundation to a number of lower storeys, contributing so significantly to the decreasing of stress and strain of the main structural system extending to the whole height of the building.

As already noted in previous sections, squat walls are nearly two-dimensional (planar) structural elements. Therefore, the design for bending cannot be separated from the design to shear, as also happens in the case of squat columns, spandrels and so forth. However, due to low load effects that develop in squat walls for the seismic combination, as these walls are used mainly in low-size buildings, approximate simplified methods are used for their design.

The main problem for these elements is that after reaching their shear-dependent flexural capacity, as in the case of short columns (see Chapter 8.3.6), they fail in shear at relatively low values of the chord rotation θ_p and, therefore, their available ductility $\mu_{\theta,avail}$ in terms of rotations is limited. This is the reason why EC8-1/2004 (see Chapter 5.4.3, Equation 5.7b) introduces for the determination of the q-factor a reduction factor K_w equal to

$$\boxed{0.5 \leq K_w = (1+\alpha_o)/3 \leq 1} \qquad (9.49)$$

for *wall* or *wall-equivalent* systems, where
α_o is the prevailing aspect ratio of the walls of the structural system.

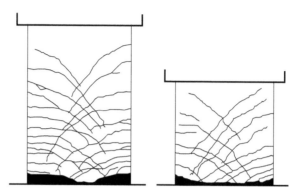

Figure 9.30 Crack patterns for specimens MSWI and LSW3 at the end of the experiment. (From Salonikios, T. 2007. Analytical prediction of the inelastic response of R/C walls with low aspect ratio. *ASCE Journal of Structural Engineering*, 133(6), 844–854. With permission of ASCE.)

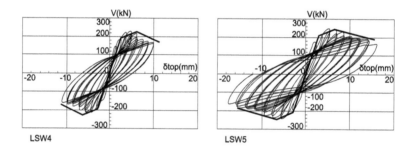

Figure 9.31 Comparative diagrams of the recorded hysteresis loops and the analytically calculated envelope curve, for specimens LSW4 and LSW5. (From Salonikios, T. 2007. Analytical prediction of the inelastic response of R/C walls with low aspect ratio. *ASCE Journal of Structural Engineering*, 133(6), 844–854. With permission of ASCE.)

For example, for $\alpha_o = 2.0$, $\kappa_w = 1.0$, while for $\alpha_o = 1.0$, $\kappa_w = 0.66$. That means there is a decrease in ductility demand of 33%.

However, from extensive experimental research, it has been concluded (Salonikios et al., 2000; Salonikios, 2007) that the ductility of properly designed squat ductile walls μ_θ in terms of rotation is rather high and may range from 5.0 to 7.0 (Figures 9.30 and 9.31).

9.4.2 Flexural response and reinforcement distribution

- Although Bernoulli's concept (linear strain distribution) is significantly violated in the case of short walls due to the strong interaction of normal and shear strains, the design in bending with axial force may be carried out as in the case of slender members (beams, columns, walls), since at ULS most of the steel reinforcement bars are at yield stage and, therefore, the strain distribution is insignificant. Therefore, the standard procedure used in the case of slender walls for bending may also be used for squat walls. The flexural strength at the base of the wall must be carefully evaluated, taking the contribution of all vertical bars into account, so that a realistic value of shear load effect resulting from the capacity design procedure can be derived (see Chapter 6.1.6).

The same process also holds for the analytical determination of ductility factors in terms of curvature (Section 9.2.2.4; Equation 9.9), although here the results usually depart more from experimental results.

- Various authors have adopted the view that evenly distributed vertical reinforcement at the base is preferable (Paulay and Priestley, 1992), because it results in an increased flexural compression zone and, therefore, in better conditions for inhibiting shear sliding, improving friction and dowel resistance.

However, Codes of Practice (EC8-1/2004 and ACI318M-2011) do not make any discrimination between slender and squat ductile walls. Therefore, the concept of strengthening their flanges *with confined boundary elements* in critical regions holds in practice also for squat ductile walls. In this regard, bending strength and ductility are mainly ensured by the flange chords, while shear resistance is ensured by the web.

9.4.3 Shear resistance

Shear failure modes of squat ductile walls are similar to those of slender ones. Those are

- Diagonal tension failure (Figure 9.11a)
- Diagonal compression failure (Figure 9.11b)
- Sliding shear failure (Figure 9.11c)

Keeping in mind that the normalised axial forces in squat ductile walls are usually very low (<0.20), their behaviour to shear is closer to the shear behaviour of spandrels.

9.4.4 Code provisions for squat ductile walls

Rules for DCM and DCH buildings will be presented together, as the differences of squat walls from slender ones are not significant.

It should also be noted that for the design of squat walls of DCL buildings, Code specifications for conventional R/C structures apply (EC2-1-1/2004), except for requirements for concrete and steel reinforcement qualities, as in the case of slender walls.

- *Geometrical constraints.* These are the same as those of slender walls (i.e. 9.2.4.2(1), 3(1)).
- *Bending resistance.* As was already explained in Section 9.4.2, the rules in force for a slender wall hold also for squat walls.
- *Detailing for local ductility.* Rules presented in previous paragraphs for slender walls are also in effect in the case of squat walls.
- Shear resistance.
    It should be recalled (Chapter 6.1.6) that capacity design rules for shear of squat walls require
- For DCM buildings

$$V_{Ed} = 1.50V'_{Ed}$$ (9.50)

where
  V'_{Ed} is the analysis load effect for shear force.

- For DCH buildings

$$\boxed{V_{Ed} = \gamma_{Rd}\left(\frac{M_{Rd}}{M_{Ed}}\right)V'_{Ed} \leq q V_{Ed}}$$

(9.51)

(see Chapter 6.1.6.1, Equation 6.23).
- From Equation 9.51, it can be seen that in the case of DCH buildings, special care should be taken that flexural resistance M_{rd} is kept near M_{Ed}; otherwise, high over-strength in bending would also lead to shear over-strength.
- Shear resistance for DCM buildings is computed in accordance with EC2-1-1/2004, as in the case of conventional R/C buildings.
- Shear resistance for DCH buildings should be computed in accordance with the following rules:

Design for *diagonal compression failure* of the web due to shear should be carried out as in the case of slender ductile walls (Section 9.2.4.3(3)).

Design for *diagonal tension failure* of the web due to shear should be carried out as follows:
  - Horizontal web bars should satisfy the expression

$$\boxed{V_{Ed} \leq V_{Rd,c} + 0.75\rho_h \cdot f_{yd,h} \cdot b_{wo}\alpha_s l_w}$$

(9.52)

where

ρ_h is the reinforcement ratio of horizontal web bars ($\rho_h = A_h/(b_{wo}\cdots s_h)$).
s_h is the distance between successive horizontal web bars.
b_{wo} is the web width.
$f_{yd,h}$ is the design yield strength of the horizontal reinforcement.
$V_{Rd,c}$ is the design shear resistance for members without shear reinforcement (see Chapter 8.2.4).

Equation 9.52 is in accordance with the concept for diagonal tension design followed by EC2-1-1/2004 for beams with an aspect ratio lower than 2 and by the American Code of Practice ACI M318-2011. Shear resistance of the horizontal reinforcement is based on the diagonal tension failure mode (Figure 9.32), where the strut inclination according to experimental evidence is defined by the corner-to-corner diagonal of the wall ($b_w = \alpha_s l_w$), and on the free-body equilibrium depicted in Figure 9.32. The coefficient 0.75 is introduced in accordance with EC2-1-1/2004 for beams with a short aspect ratio ≤ 2.0, where only the central 75% of the web reinforcement intersecting the inclined crack should be taken into account. In the critical region of the wall, $V_{Rd,c}$ should be taken *as zero* if the axial force N_{Ed} *is tensile*.
  - Vertical web bars should be provided to satisfy the condition

$$\boxed{\rho_h \cdot f_{yd,h} \cdot b_{wo}z \leq \rho_v \cdot f_{yd,v} \cdot b_{wo}z + \min N_{Ed}}$$

(9.53)

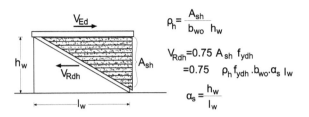

$$\rho_h = \frac{A_{sh}}{b_{wo}\ h_w}$$

$$V_{Rdh} = 0.75\ A_{sh}\ f_{ydh}$$
$$= 0.75\quad \rho_h f_{ydh} \cdot b_{wo} \cdot \alpha_s\ l_w$$

$$\alpha_s = \frac{h_w}{l_w}$$

Figure 9.32 Diagonal tension failure mode on squat walls.

where

ρ_v is the reinforcement ratio of vertical web bars ($\rho_v = A_v/b_{wo}...s_v$).
$f_{yd,v}$ is the design yield strength of the vertical web reinforcement.
$-z \cong 0.8\ l_w$.
N_{Ed} is positive when compressive.

Expression 9.53 results from the condition that the distributed vertical reinforcement force $\rho_v f_{yd}$, together with the distributed axial force $v_{Ed} \leq 0.40 = N_{Ed}/b_{wo}z$ composed with the horizontal $\rho_h f_{yd}$, should give a resultant with an angle of 45° (compressive strut inclination). This condition does not comply with the assumption of the corner-to-corner diagonal compressive strut inclination adopted in the previous paragraph for the determination of ρ_h (Equation 9.52). If the assumption of corner-to-corner diagonal strut were adopted, expression 9.53 would take the form

$$\rho_h \cdot f_{yd,h} \cdot b_{wo} z \alpha_s^2 \leq \rho_v \cdot f_{yd,v} \cdot b_{wo} z + \min N_{Ed} \tag{9.54}$$

The above design procedure, which refers mainly to squat walls, according to EC8-1/2004, should also be applied to slender DCH walls in the case that the ratio $M_{Ed}/V_{Ed} \times l_w$ at the base of a storey (shear span ratio) is lower than 2 (see Section 9.2.4.3(4)).

- Sliding shear failure
    Design for sliding shear failure at a base of the squat wall should be carried out in accordance with the procedure developed previously in Section 9.2.4.3(5), Equation 9.27 of this chapter. In the case of squat walls, half of the design base shear V_{Ed} should be carried by the contribution of diagonally arranged reinforcement, that is,

$$\boxed{\frac{V_{Ed}}{2} \leq V_{id}} \tag{9.55}$$

At higher levels, V_{id} should exceed only 25% of V_{Ed}.

9.5 LARGE LIGHTLY REINFORCED WALLS

9.5.1 General

This type of R/C structural system has already been defined in Chapter 4.4. Its capacity design has been presented in Chapter 6.1.7, while the corresponding behaviour factors and their justification have been outlined in Chapter 5.4.3. In this subsection, the design of these walls will be presented and discussed in detail.

First, a short overview will be made of the main structural characteristics and seismic behaviour of these structural systems.

- The main structural system consists of large R/C walls, which carry a large part of the gravity loads and the seismic action as well, while the fundamental period T_1 of the building, assuming fixed walls at the foundations, is less than 0.5 s.
- The energy dissipation mechanism of these buildings is not based on the formation of plastic hinges at the base of the walls, since it is not feasible due to the large length of the cross sections of the walls. Instead of a 'plastic hinge' dissipating mechanism, an 'uplift' mechanism of the walls from the soil or through opening and closing of horizontal cracks (Figure 9.33) is anticipated. In fact, such a wall is expected to transform seismic energy to potential energy through temporary uplift of structural masses and then to kinetic energy dissipated in the soil or in the flexural cracks through rigid body 'rocking'. In this respect, the dissipation zone is not concentrated at a plastic hinge in a critical zone at the base of the wall, but it is distributed to the soil joint and a number of flexural cracks over the entire height of the building.
- So, although this dissipation mechanism is not quantitatively controlled and justified as in the case of all other structural systems in effect, EC8-1/2004 has also adopted this type of structure, *but only for DCM buildings*. Additionally, it should be noted that, due to its low construction cost, this type of structural system is very popular in southern France and has been implemented for many years with very good results in response to seismic actions. So, the design rules used in these buildings thus far have also been mostly adopted by EC8-1/2004.
- The basic design concept includes the following principles:
  Design to bending is carried out using the moment diagram directly from the analysis, without any capacity design envelope, in contrast to all other types of wall systems.

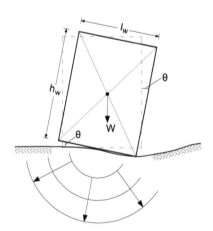

Figure 9.33 Rocking of large, lightly reinforced walls.

This moment diagram is combined with the gravity axial load fluctuating between $(1 \pm 0.5)N_{Ed}$ in order for the dynamic vibrating masses in the vertical direction to be taken into account.

- Design for shear is carried out for increased shear forces (see Chapter 6.1.7) in order to ensure that flexural failure precedes shear failure.

9.5.2 Design to bending with axial force

- The ULS in bending with axial force should be verified, assuming
  Horizontal cracking
  Plane strain distribution no matter what the aspect ratio h_w/l_w. Therefore, the design for bending is carried out in accordance with the relevant provisions of EN 1992-1-1/2004.
- Due to the small thickness of the wall in relation to its other two dimensions, normal stresses in concrete should be limited to prevent out-of-plane buckling. This requirement is satisfied if the rules of EN 1992-1-1/2004 for second-order effects are satisfied.
  These rules are stated in
  Paragraph 12.6.5.2/EC2-1-1/2004: the simplified design method for walls and columns of plain or lightly reinforced structures
  Paragraph 5.9 (Equation 5.40b)/EC2-1-1/2004: the lateral instability of slender beams
- In case that the dynamic axial force $(1 \pm 0.5)N_{Ed}$ must be taken into account, the limiting strain ε_{cu2} for concrete may be increased to

$$\boxed{\varepsilon_{cu2} = 0.005} \tag{9.56}$$

This increase allows a higher efficiency of the concrete compression zone and therefore diminishes the cases where steel in tension zones does not yield at ULS.

9.5.3 Design to shear

Due to the high safety margin for shear imposed by the capacity design rules (Chapter 6.1.7), the following requirements are specified by EC 8-1/2004:

- If

$$\boxed{V_{Ed} \leq V_{Rdc}} \tag{9.57}$$

the rules for minimum web reinforcements $\rho_{w,min}$ foreseen in the Code for conventional R/C walls are not obligatory.
- If

$$\boxed{V_{Rdc} \leq V_{Ed}} \tag{9.58}$$

the web reinforcement should be calculated in accordance with EN 1992-1-1/2004 on the basis of the *variable inclination truss model* or *a strut-and-tie model*, whichever is most appropriate for the particular geometry of the wall. If a strut-and-tie model

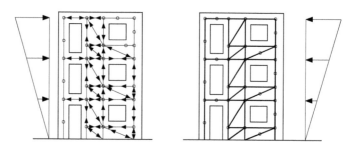

Figure 9.34 A strut-and-tie model for the shear design of a large, lightly reinforced wall.

is used, the *width of the compressive strut should take into account* the presence of openings and should not exceed 0.25 l_w or 4 b_{wo}, whichever is smaller (Figure 9.34).

The ULS against sliding shear at horizontal construction joints should be verified in accordance with EN 1992-1-1/2004 § 6.2.5, with 50% longer anchorage lengths.

9.5.4 Detailing for local ductility

- At the ends of the cross sections of the walls, boundary elements should be formed, reinforced with vertical steel rebars combined with hoops and cross ties. These elements should have a length c in the direction of the long side of the cross sections in the range of

$$\max \left\{ \begin{matrix} b_w \\ 3b_w \sigma_{cm}/f_{cd} \end{matrix} \right\} < c \tag{9.59}$$

where

σ_{cm} is the mean value of the concrete stress in the compression zone in the ULS of bending with axial force.

- The diameter of the vertical bars should *not be less than 12 mm* in the lower storey and upwards up to the storey where the length of the wall l_w is reduced by one-third. From that storey on, the diameter of the vertical rebars may be reduced to 10 mm.
- The diameter of the hoops or cross ties should *not be less than 6 mm* or one-third of the diameter of the vertical bars.
- In order for flexural failure to precede that of shear, the amount of vertical reinforcement should not exceed that which results from the design for bending with axial force at the successive storeys of the building.
- Continuous horizontal and vertical steel ties should be provided (Figure 9.35):
  Along all intersections of walls or connections with flanges
  At all floor levels
  Around openings in the wall

These ties should be designed in accordance with EN 1992-1-1/2004 9.10.

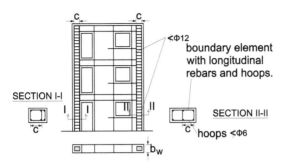

Figure 9.35 Reinforcement of a large, lightly reinforced wall.

9.6 SPECIAL ISSUES IN THE DESIGN OF WALLS

9.6.1 Analysis and design using FEM procedure

From the presentation thus far, it is apparent that walls either have an orthogonal cross section or a composite one (Γ T, I, Π, Z, etc.), and they are considered in concept for the analysis and design as linear members with an orthogonal or composite cross section developing a planar distribution of strains under bending with axial force.

However, it is well known that most of the modern computer platforms (SAP 2000, ETABS, SCIA, etc.) allow the introduction of the FEM procedure for planar members, which, of course, is not prohibited by modern Codes. In this respect, the modelling is of higher reliability than that of linear members. At the same time, modelling can proceed directly from the formwork drawings elaborated already in an AutoCAD or a similar platform.

Therefore, the outputs of the analysis are normal and shear-distributed forces in the middle plane of the walls combined with distributed bending moments, torsional moments and shears corresponding to their parallel behaviour *as plates* under bending. Together these two types of action constitute the *shell* function, as is well known from structural engineering (Figure 9.36).

This force pattern is proper for the design of steel walls or cores, but it is improper for the design of R/C walls for the following reasons:

1. The design of R/C members is based on the assumption of cracked tensile zones. So, the steel rebars under tension are arranged mainly at the extreme tensile chord of a cross section under consideration. As a result, the lever arm of the internal forces in the case of bending with axial force takes its maximum value and, therefore, the required reinforcement is minimised.

   Conversely, in the case that the internal normal tensile forces are covered by distributed reinforcement, as in the case of shell design due to the smaller internal lever arms, the amount of the required reinforcement would be much more than that resulting from conventional R/C design.

   For example, in case of a wall of orthogonal cross section loaded in its main direction (Figure 9.37a) with a lateral load P, if it is analysed as a linear member, it will be subjected at its base to a bending moment $MEd = h_w P$. Therefore, the required reinforcement in the tensile zone will be derived from the following expressions (Figure 9.37b):

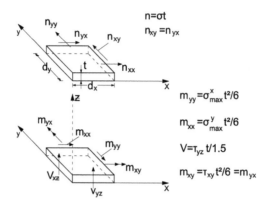

Figure 9.36 Notation for internal forces of a 'disk' and a 'shell'.

$$M_{Ed} \le M_{Rd} = z \cdot T \text{ or } T \approx \frac{M_{Ed}}{0.85 l_w} \tag{9.60}$$

Therefore,

$$A_s = \frac{T}{f_{yd}} \cong \frac{M_{Ed}}{0.85 f_{yd} l_w} = \frac{1.18 M_{Ed}}{f_{yd} l_w} \tag{9.61}$$

On the other hand, if this wall is analysed as a disc, the resulting axial stress distribution at its base is given in Figure 9.37c. Taking into account that the reinforcement is distributed in the tensile zone proportionally to the stress distribution, meaning that

$$dA_s = \sigma_m b_w dx / f_{yd} \tag{9.62}$$

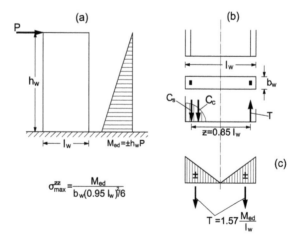

Figure 9.37 (a) Loading pattern of an orthogonal wall and bending moment diagram; (b) dimensioning of the wall as a linear member; (c) dimensioning of the wall using FEM method.

and that

$$\sigma_{max}^{zz} = \frac{M_{Ed}}{b_w l_w^2 / 6} \tag{9.63}$$

It follows that

$$T = \frac{1}{2} \frac{l_w}{2} b_w \sigma_{max}^{zz} = \frac{1}{2} \frac{l_w}{2} b_w \frac{M_{Ed}}{b_w l_w^2 / 6} = 1.50 \frac{M_{Ed}}{l_w} \tag{9.64}$$

Therefore,

$$A_s = \frac{T}{f_{yd}} = \frac{1.50 M_{Ed}}{f_{yd} \cdot l_w} \tag{9.65}$$

Consecutively, 27% more reinforcement is required where dimensioning follows the procedure used in the case of shells.

2. The above calculated over-reinforcement *must be considered the minimum* that can result, and it refers to systems where the 'lateral force method' is implemented. In case of 'the modal response spectral analysis', the stress distribution is a value resulting from the application of the 'SRSS' or 'CQC' method. In this respect, this output does not correspond to any consistent state of stress of the structure, but represents the distribution of the most probable values of stresses during the seismic action. At the same time, as these stresses are the output of a square root, they are values of undefined sign and, therefore, might be tensile or compressive. So, in the usual case where axial load effects also act together with bending if it were decided to cover the whole diagram by reinforcement in tension, the resulting amount of reinforcement seems to be even 2.0 times more than the reinforcement that would result in the case of implementation of the 'lateral force method' or the 'time history linear dynamic analysis'.

In order to overcome the above difficulties, many computer platforms for the seismic analysis of concrete or masonry structures (SAP 2000, ETABS/2011, SCIA—ECtools) have incorporated a post-processor that proceeds for each eigenvalue (i) to the integration of the stresses corresponding to this eigenvalue to a group of M_{Ed}^i N_{Ed}^i V_{Ed}^i load effects. From these load effects of the successive eigenvalues, using the SRSS or CQC procedure, the most probable values of M_{Ed}, N_{Ed} and V_{Ed} are computed, for each cross section of course again with an undefined sign. This procedure is implemented at the walls of each storey and at eventual spandrels where 'piers' and 'spandrels' are generated accordingly. From this point on, the design proceeds as in the case of linear members with a composite cross section.

9.6.2 Warping of open composite wall sections

9.6.2.1 General

The conventional approach to the torsional effects on cross sections of structural members is that they remain undeformed to their axial direction, and, in this respect, torsional loading is resisted only by shear stresses (torsion according to Saint-Venant theory). This is completely true only for members with a symmetric closed cross section. It is well known from the technical theory of elasticity that when structural members with composite cross sections that are also characterised as thin-walled sections are loaded by torsional external moments, they develop axial strains leading to warping of the cross section (Figure 9.38). Therefore, when this warping is prohibited by boundary constraints, it is reasonable for axial stresses to develop in addition to shear stresses due to torsion. Therefore, parallel to Saint-Venant uniform torsion, a *torsional bending* develops known as *torsion according to Vlasov theory* (Taranath, 2010).

This normal stress state *does not follow the assumption of plane distribution of stresses,* since it is the result of the prevention of the cross-section warping caused by the torsional moment. The resultants of these normal stresses are zero ($N_{Ed} = 0, M_{Ed} = 0$), since there is not such external loading on the structural member. This normal stress state changes along the axis of the structural member. Therefore, additional shear stresses are generated as the derivatives of normal stresses. Their sum over the section is also zero, but their torsional resistance contributes to the counterbalance of the external torsional moment together with the torsional resistance of the Saint-Venant shear stress state. In this respect, one part of the external torsional moment $T_{(z)}$ is resisted by the Saint-Venant shear $T_{v(z)}$ and the rest by the shear stresses due to warping behaviour (Vlasov stress state) $T_{w(z)}$

$$T_{(z)} = T_{v(z)} + T_{w(z)} \tag{9.66}$$

The scope of this book does not allow a detailed reference to the torsional theory of thin-walled cantilevers, since such an approach requires formation and integration of the relevant differential equations that relate the external loading torsional moment $T_{(z)}$ to the angle of twist $\theta_{(z)}$ and the geometrical and mechanical properties of the structural member and therefore the distribution of $T_{(z)}$ to $T_{v(z)}$ and $T_{w(z)}$.

However, it is useful to stay on this subject a little while longer for a thorough understanding of R/C wall behaviour. In fact, keeping in mind that R/C walls and cores in buildings

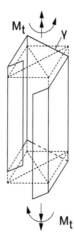

Figure 9.38 Warping of a thin-walled composite section under torsion.

usually constitute the main structural component resisting seismic action, and that analysis and design are carried out basically by means of computer-aided procedures, it is of paramount importance to know, at least qualitatively, the degree of approximation made by the conventional programming platforms and the way these results could be improved using post-processing procedures by hand or computationally so that the warping issue may be dealt with.

9.6.2.2 Saint-Venant uniform torsion

First, it is important to recall the torsional behaviour of cantilevers with circular, orthogonal and closed thin-wall cross sections (Figure 9.39a). According to Saint-Venant theory, the total angle of twist θ for a length z measured from the base of a cantilever is given by the expression (Anastasiadis, 1989; Taranath, 2010)

$$\theta_{(z)} = \frac{T}{GJ_p} z \qquad (9.67)$$

or

$$\frac{d\theta}{dz} = \frac{T}{GJ_p} \text{ and } T = GJ_p \frac{d\theta}{dz} \qquad (9.68)$$

where
  T is the twisting moment.
  z is the ordinate on the cantilever wherein to which twist $\theta_{(z)}$ is referred.
  G is the shear modulus of elasticity.
  J_ρ is the polar moment of inertia of the symmetric section (Figure 9.39a, b).

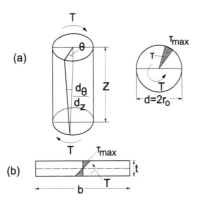

Figure 9.39 Torsion according to Saint-Venant theory (zero axial strain of the cross sections): (a) notation for uniform torsion, and τ distribution for circular cross section; (b) τ distribution for orthogonal cross section.

$$J_{pc} = \frac{\pi d^4}{32} = \frac{\pi r_o^4}{2} \quad \text{(circular cross section)} \tag{9.69a}$$

$$J_{po} = \frac{bt^3}{3} \quad \text{(long orthogonal cross section)} \tag{9.69b}$$

On the other hand, the maximum shear stress due to twisting moment is given by the expression

$$\boxed{\tau_{max} = \frac{T r_o}{J_{pc}} \quad \text{(for circular cross section)}} \tag{9.70a}$$

$$\boxed{\tau_{max} = \frac{2Tb}{J_{po}} \quad \text{(for orthogonal cross section)}} \tag{9.70b}$$

where
r_o is the radius of the circular cross section.
b, t are the dimensions of the orthogonal cross section.

In the case of a closed thin-wall cross section, the shear stress (τ) developing in the wall of the cross section is given by the expression (Figure 9.40)

$$\boxed{\tau = \frac{T}{2 A_c \cdot t}} \tag{9.71}$$

(first Bredt's formula)
where
t is the thickness of the wall.
A_c is the area enclosed by the central line of the cross section.

At the same time, the angle of twist θ for a cantilever at a height z from the base is given by the expression

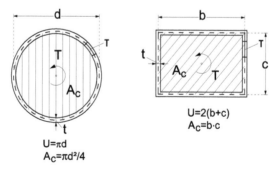

Figure 9.40 Torsional behaviour of members with symmetric tubular cross section.

$$\boxed{\theta = \frac{T}{GJ_{ph}} z}$$ (9.72)

or

$$\boxed{\frac{d\theta}{dz} = \frac{T}{GJ_{ph}} \quad \text{or} \quad T = GJ_{ph} \frac{d\theta}{dz}}$$ (9.73)

(second Bredt's formula)
where

$$J_{ph} = \frac{4A_c^2 t}{U}$$ (9.74)

U is the perimeter of the cross section.
The above equations are based on the following assumptions:

1. Plane cross sections in the undeformed state remain plane when torque is applied.
2. Cross sections remain undistorted in their own plane.

The first assumption is true for *circular sections, orthogonal sections* and *tubular sections*. As was already mentioned above, this approach to the problem is known as Saint-Venant's or *uniform torsion*. When the structural member is an open thin-walled structure, the above first assumption no longer remains true. This type of torsion is known as *warping torsion*, or as *constrained torsion* or *torsion bending*.

Being acquainted with the uniform torsion from their first contact with the classic strength of materials, designers generally encounter difficulties either with the concept of warping behaviour or with the methods of analysis and design referring to warping. Therefore, the task here is to introduce the concept of warping torsion and to give methods for taking into account the additional state of stress generated, or to ignore it in a justified way.

9.6.2.3 Concept of warping behaviour

We will try to describe the warping torsion on an I-shaped shear wall with unequal flanges as depicted in Figures 9.41 and 9.42 (Taranath, 2010). If the influence of the web is neglected, the centre of gravity (CG) of the cross section may be estimated with a close approximation by the relations

$$y_1 = \frac{A_2 L}{A_1 + A_2} \quad \text{and} \quad y_2 = \frac{A_1 L}{A_1 + A_2}$$ (9.75a,b)

It is well known that a horizontal load at the top of a cantilever with height h_w acting parallel to the flanges and passing through the CG of the cross section generates normal stresses at the base, accomplishing the plane stress distribution concept of Navier, given by the expression

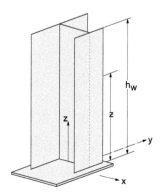

Figure 9.41 H-section core. (Adapted from Taranath, B.S. 2010. *Reinforced Concrete Design of Tall Buildings*, CRC Press, Taylor & Francis Group.)

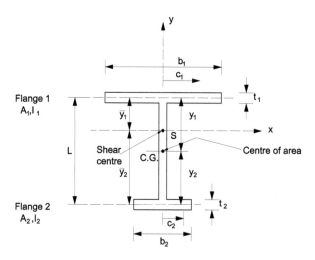

Figure 9.42 Core properties. (Adapted from Taranath, B.S. 2010. *Reinforced Concrete Design of Tall Buildings*, CRC Press, Taylor & Francis Group.)

$$\sigma_x = \frac{M}{J_y}x = \frac{Ph_w}{J_y}x \tag{9.76}$$

At the same time, the internal shear force V_x equal to P results in shear stresses at the flanges, given by the expression (Figure 9.43)

$$\tau = \frac{V_x S_{yi}}{J_y t_i} \tag{9.77}$$

where
   S_{yi} is the static moment on the part of the cross section to the left of point (i) in relation to the y–y axis (Figure 9.43).

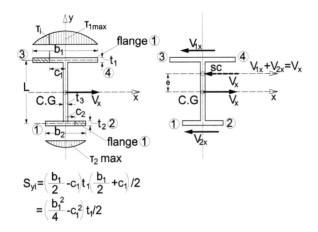

$$S_{yi} = \left(\frac{b_1}{2} - c_1\right) t_1 \left(\frac{b_1}{2} + c_1\right)/2$$

$$= \left(\frac{b_1^2}{4} - c_1^2\right) t_1/2$$

Figure 9.43 Shear diagrams due to V_x. The meaning of shear centre (SC).

t_i is the thickness of the section at the point of reference (i).
J_y is the moment of inertia of the cross sections in relation to the y–y axis.

The resultants of these stresses V_1 and V_2 of the flanges satisfy the expression

$$V_x = V_{1x} + V_{2x} \tag{9.78}$$

and they are equal to

$$V_1 = \frac{J_1}{J_1 + J_2} V \tag{9.79}$$

$$V_2 = \frac{J_2}{J_1 + J_2} V \tag{9.80}$$

In fact, taking into account that

$$\tau_{1max} = \frac{(b_1/2) \cdot (b_1/4)t_1}{J_y t_1} V_x = \frac{b_1^2}{8 J_y} V_x \tag{9.81a}$$

$$\tau_{2max} = \frac{(b_2/2) \cdot (b_2/4)t_1}{J_y t_2} V_x = \frac{b_2^2}{8 J_y} V_x \tag{9.81b}$$

It is concluded that V_{1x} and V_{2x} have the following values:

$$V_{1x} = \frac{2}{3} \tau_{1max} t_1 b_1 = \frac{b_1^3 t}{12 J_y} V_x \tag{9.82a}$$

$$V_{2x} = \frac{2}{3}\tau_{2\max}t_2 b_2 = \frac{b_2^3 t}{12 J_y} V_x \tag{9.82b}$$

or

$$V_{1x} = \frac{J_1}{J_y} V_x, \quad V_{2x} = \frac{J_2}{J_y} V_x \tag{9.83a,b}$$

The resultant of these internal forces passes through a point SC called the *shear centre* (SC), which is located at the position

$$\bar{y}_{1s} = \frac{J_2}{J_1 + J_2} L \quad \text{and} \quad \bar{y}_{2s} = \frac{J_1}{J_1 + J_2} L \tag{9.84a,b}$$

This means that in order for no torsional loading to exist, the horizontal load P should pass through the SC of the cross section. Otherwise, the structural member is subjected to a twisting moment T about this centre, equal to

$$T = P \cdot e \tag{9.85}$$

where
  e is the distance between GC and SC (Nitsiotas, 1960).

From Equations 9.84a and 9.84b, it may easily be concluded that in symmetric cross sections, the SC coincides with GC, and, therefore, for horizontal loading passing through the GC axis, no torsional effect is generated.

When a torque T is now applied at the top of the cantilever of Figure 9.44, this cantilever twists about the SC axis, causing the flanges to

1. Bend in opposite directions about the y–y (Figure 9.44a) axis
2. Twist about their vertical axes (Figure 9.44b), passing through their own SC, which coincides with their GC, since their cross section is orthogonal and, therefore, presents double symmetry

The effect of this flange bending, according to the Bernoulli assumption for planar distribution of strain, results in a rotation of the flange sections in an opposite direction and in this respect to a warping of the cross section. Diagonally opposite corners 1 and 4 in Figure 9.44 displace downwards, while 2 and 3 displace upwards. All along the axis of the cantilever at every ordinate (z), the torque $T = T_{(z)}$ is resisted internally by a couple $T_{w(z)}$, resulting from the shears in the flanges related to their bending in plane, and a couple $T_{v(z)}$, resulting from the twisting of the flanges and the web according to Saint-Venant torsion (Equation 9.67).

The angle of twist $\theta_{(z)}$ of the cantilever about the SC axis at a height (z) from the base results in a horizontal displacement of flange (1) at this level equal to

$$u_{x1}(z) = \bar{y}_1 \theta(z) \tag{9.86}$$

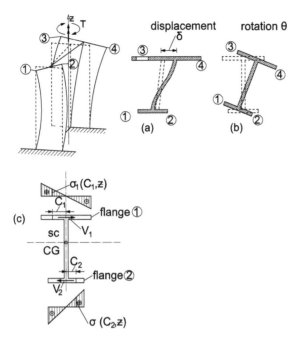

Figure 9.44 Combination of warping and uniform torsion; normal stresses due to warping: (a) the concept of warping; (b) displacements in-plane for bending in torsion; (c) normal stress distribution due to warping.

Therefore,

$$\frac{du_{x1}(z)}{dz} = \bar{y}_1 \frac{d\theta(z)}{dz} \tag{9.87}$$

$$\frac{d^2u_{x1}(z)}{dz^2} = \bar{y}_1 \frac{d^2\theta(z)}{dz^2} \tag{9.88}$$

$$\frac{d^3u_{x1}(z)}{dz^3} = \bar{y}_1 \frac{d^3\theta(z)}{dz^3} \tag{9.89}$$

Similar expressions are in effect for flange (2).

Shear related to the bending in flanges (1) and (2) may be expressed by the following equations:

$$V_1(z) = -EJ_1 \frac{d^3u_{x1}(z)}{dz^3} = -EJ_1\bar{y}_1 \frac{d^3\theta(z)}{dz^3} \tag{9.90}$$

$$V_2(z) = -EJ_2 \frac{d^3u_{x2}(z)}{dz^3} = -EJ_2\bar{y}_2 \frac{d^3\theta(z)}{dz^3} \tag{9.91}$$

The torsional moment of the above internal shear forces to the SC of the cross section is the warping moment:

$$T_{w(z)} = V_1 \bar{y}_1 + V_2 \bar{y}_2 = -\left(EJ_1 \bar{y}_1^2 + EJ_2 \bar{y}_2^2\right) \frac{d^3 \theta(z)}{dz^3} \tag{9.92}$$

or

$$T_{w(z)} = EJ_w \frac{d^3 \theta(z)}{dz^3} \tag{9.93}$$

where

$$J_w = J_1 \bar{y}_1^2 + J_2 \bar{y}_2^2 \tag{9.94}$$

Equation 9.94 is identical to Equation 2.87 of Chapter 2. From the development thus far, what becomes apparent is the close relationship between warping of composite open-wall sections and the torsional behaviour of a pseudospatial structure subjected to torsional action. In fact, in the second case, the main part of the external torsional loading is resisted by bending of the vertical structural elements and a smaller part by the torsional resistance of the vertical structural members. Therefore, the approach to the problem is identical to that of a pseudospatial structure, since the influence of the web of the I section has been neglected. Nevertheless, the following steps will allow the formulation of the warping differential equations in a comprehensible way. J_w, above, is a geometric property of the section similar to the moments of inertia J_1 and J_2 and is called the warping moment of inertia or warping constant. The general form of J_w is given in Equation 9.118.

This parameter expresses the capacity of the section to resist warping torsion, as J_1 and J_2 express the capacity of the section to resist bending moments. Neglecting the web and taking into account Equation 9.68, the torque resisted by the twisting of the cross section according to Saint-Venant theory becomes

$$T_v(z) = GJ_p \frac{d\theta(z)}{dz} \tag{9.95}$$

where
   J_p is the polar moment of inertia of the section, that is,

$$J_p = \frac{b_1 h_1^3}{3} + \frac{b_2 h_2^3}{3} \tag{9.96}$$

The sum of $T_w(z)$ and $T_v(z)$ results in the external torque $T(z)$, that is,

$$-EJ_w \frac{d^3 \theta(z)}{dz^3} + GJ_p \frac{d\theta(z)}{dz} = T(z) \tag{9.97}$$

The above differential equation 9.97 is the fundamental equation for restrained warping torsion and has been derived in a simplified inductive way, bypassing the technical theory of elasticity for torsion of open thin-wall members (Taranath, 2010). Of course, in the case of a cross section of general form, the geometric parameter J_w cannot be expressed by Equation 9.94. Its general form will be given in the next paragraph. The integration of the above equation gives θ, and from Equations 9.93 and 9.95, $T_w(z)$ and $T_v(z)$ are determined.

The normal stresses in the flanges due to bending (Figure 9.44; a, b, c) caused by V_1 and V_2 (warping action) are determined by the expression (see Equation 9.76)

$$\sigma_1(c_1, z) = \frac{M_1(z)c_1}{J_1} \tag{9.98}$$

and

$$\sigma_2(c_2, z) = \frac{M_2(z)c_2}{J_2} \tag{9.99}$$

Keeping in mind that

$$V_1(z) = -V_2(z) = V(z) \tag{9.100}$$

and

$$M_1(z) = -M_2(z) = M(z) \tag{9.101}$$

Equation 9.98 takes the form

$$\sigma_1(c_1, z) = \frac{M(z)c_1}{J_1} \tag{9.102}$$

or

$$\sigma_1(c_1, z) = \frac{M(z)c_1}{(J_1\bar{y}_1^2 + J_2\bar{y}_2^2)} \frac{J_1\bar{y}_1^2 + J_2\bar{y}_2^2}{J_1} \tag{9.103}$$

$$\sigma_1(c_1, z) = \frac{M(z)c_1}{J_w}\left(\bar{y}_1^2 + \bar{y}_2^2 \frac{J_2}{J_1}\right) \tag{9.104}$$

Taking into account from Equations 9.84a and 9.84b that

$$\frac{J_1 + J_2}{L} = \frac{J_2}{\bar{y}_1} = \frac{J_1}{\bar{y}_2} \tag{9.105a}$$

or

$$\frac{J_2}{J_1} = \frac{\bar{y}_1}{\bar{y}_2}$$ (9.105b)

and introducing Equation 9.105b into Equation 9.104, the following expression is obtained:

$$\sigma_1(c_1,z) = \frac{M_{(z)}c_1}{J_w}\bar{y}_1(\bar{y}_1 + \bar{y}_2)$$ (9.106)

or

$$\sigma_1(c_1,z) = \frac{M_{(z)}Lc_1\bar{y}_1}{J_w}$$ (9.107)

or

$$\boxed{\sigma_1(c_1,z) = \frac{B_{(z)}w_{(ci)}}{J_w}}$$ (9.108)

where

$$\boxed{B(z) = M(z)L}$$ (9.109)

is an action effect termed by Vlasov as *bimoment* and

$$\boxed{w(c) = \bar{y}_1 c_1}$$ (9.110)

is a coordinate termed as *principal sectorial area* or *principal sectorial ordinate* for the point of reference of the section. Of course, in the general case of an open composite cross section, the principal sectorial ordinate cannot be expressed by Equation 9.110, which relates to a special form of cross section. Its general form will be given in the next paragraph.

On the other hand, bimoment $B_{(z)}$, being in the special case of an I section, the product of the two opposite directed bending moments $M_{(z)}$ of the flanges with the flanges distance L. This magnitude corresponds to the bending moment in case of simple bending.

Returning to differential equation 9.97, it should be noted that this equation holds for any form of a composite open thin-walled member.

The integration of this equation takes the form (Anastasiadis, 1989)

$$\boxed{\theta = \kappa_1 \cos h(\lambda z) + \kappa_2 \sin h(\lambda z) + \kappa_3 + \frac{T(z)}{GJ_p}z}$$ (9.111)

Table 9.1 Analogy between bending and warping torsion

Bending	Warping torsion
$p_x(z)$: distributed load per m	$m(z)$ = distributed moment per m
$u_x(z)$: deflection	$\theta(z)$ = rotation
$u_x'(z)$: inclination of the deformed axis	$d\theta(z)/dz$ = twist
$J_y = \int y2\ dA$	$J_w = \int w^2\ dA$
$M_y(z) = EJ_y \cdot u_x''(z)$	$B_\kappa(z) = -EJ_w\ (d\theta^2(z)/dz^2)$
$Q_x(z) = -EJ_y \cdot u_x'''(z)$	$T_\kappa(z) = -EJ_w\ (d\theta^3(z)/dz^3)$
$\sigma = -x\ (M_y/J_z)$	$\sigma_\kappa(z) = -w_{(s)}B_\kappa(z)/J_w$

where
  κ_1, κ_2, κ_3 are constants determined by the boundary conditions and

$$\lambda = \sqrt{\dfrac{GJ_p}{EJ_w}} \qquad (9.112)$$

If θ and its derivatives are introduced in Equations 9.93 and 9.95, $T_w(z)$ and $T_v(z)$ may be determined. This procedure exceeds the scope of this subsection. However, it will be used next to draw useful conclusions for design.
  If J_p is small enough, as happens in the case of very thin walls, Equation 9.97 takes the form

$$-EJ_w \dfrac{d^3\theta(z)}{dz^3} = T(z) \qquad (9.113)$$

It is apparent that this differential equation, which expresses the warping torsion, exhibits a *complete formal analogy* with *simple bending*. This may easily be seen by the comparison of the corresponding values given in Table 9.1 (Anastasiadis, 1989).

9.6.2.4 Geometrical parameters for warping bending

From what has been presented so far, the following geometrical parameters of the cross section are needed for warping torsion:

  Coordinates of SC (x_s,y_s) (Figure 9.45)
  The *principal sectarian ordinate* $_{(w_{cs})}$ of the successive points of the centre line of the cross section with reference to the SC
  The *warping moment of inertia* or warping constant J_w

- Shear Centre (SC)
      As clarified in a previous paragraph, the SC of an open section is defined as the point through which a horizontal load should pass so that no torsional effect is developed in the cross section.

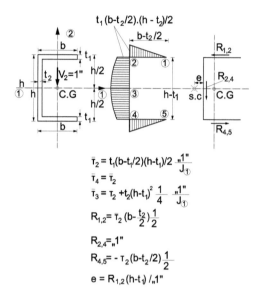

Figure 9.45 Determination of the SC of a U section.

For the determination of the coordinates of the SC, the following procedure is used (Figure 9.45):

Determine the CG of the cross section, the principal axes 1 and 2 and the principal moments of inertia J_1 and J_2 of the cross section.

For a vertical shear force equal to '1' parallel to axis 2 and passing through the CG, the shear flow diagram is determined by implementing the known expression

$$\bar{\tau}_{(2)s} = \tau_{(2)s}t = \frac{{}_n1''S_{(1)s}}{J_1} \tag{9.114}$$

where

$S_{(1)s}$ is the static moment at the successive points of the cross section with respect to axis 1-1, that is,

$$S_{(1)s} = \int_0^s s \cdot t \cdot ds \tag{9.115}$$

J_1 is the principal moment of inertia of the cross section with respect to axis 1-1.
    t is the thickness of the wall element.

Integrate the shear flow along each wall element:

$$R_{i(1)} = \int_i \bar{\tau}_{(2)s} \cdot ds \tag{9.116}$$

Find the axis of the resultant of successive $R_{i(2)}$. The SC of the cross section will lie on this axis.

Repeat the above steps 2–4 for a shear force equal to '1' parallel to axis 1 and passing through the CG. The new axis of the resultant of $R_{i(2)}$ will pass through the SC, and therefore its intersection with the axis defined at step 4 will determine the SC.

In the case that the cross section is symmetric to one principal axis, as in the example depicted (Figure 9.45), the SC will be located on the axis of symmetry. In case of cross sections of double symmetry, the SC coincides with the CG. In the case that all legs of the composite section pass through a common point, it is apparent that this point will be also the SC of the cross section. For usual forms of cross section, the position of the SC is given by closed expressions in conventional manuals (e.g. Table 9.2 where an extract from a relevant table is given; Taranath, 2010).

- Principal sectorial or warping ordinate $(w_{(s)})$
  The principal sectorial or warping coordinate at a point of the cross section of a thin-walled member is a parameter that expresses the warping response (i.e. displacement, strain and stress) at that point relative to the response at all other points of the section. Therefore, this value is similar to the ordinate x_i of a point i of a section under plain bending in the expression $\sigma_i = M_y x_i / J_y$, where σ_i is determined.
  The principal warping coordinate is defined with respect to the SC and a point of origin p_o on the section selected in such a way that the integration of the principal warping ordinates along the cross section gives a resultant equal to zero. This is determined by the fact that the resultants of axial stresses due to warping are zero. In this respect, in the case of simple symmetric sections, the point of origin p_o is selected *on*

Table 9.2 Torsion constants for open sections

Cross-section reference	Constants
Channel	$e = \dfrac{3b^2}{h+6b}$ $J_p = \dfrac{t^3}{3}(h+2b)$ $J_w = \dfrac{h^2 b^3 t}{12} \cdot \dfrac{2h+3b}{h+6b}$
Wide flanged beam with equal flanges	$J_p = \dfrac{1}{3}(2t^3 b + t_w^3 h)$ $J_w = \dfrac{h^2 t b^3}{24}$

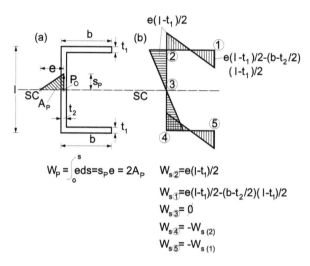

$$W_p = \int_0^s eds = s_p e = 2A_p \qquad W_{s\,2} = e(l\text{-}t_1)/2$$
$$W_{s\,1} = e(l\text{-}t_1)/2\text{-}(b\text{-}t_2/2)(l\text{-}t_1)/2$$
$$W_{s\,3} = 0$$
$$W_{s\,4} = \text{-}W_{s\,(2)}$$
$$W_{s\,5} = \text{-}W_{s\,(1)}$$

Figure 9.46 Determination of the principal warping ordinate w(s): (a) definition of w(s); (b) w(s) diagram of a C section.

the symmetry axis. The value of the principal sectorial coordinate at any point P on the profile is given by the area (Figure 9.46a)

$$w_c = \int_0^s eds \qquad (9.117)$$

where
   e is the perpendicular distance of the SC to the tangent of the profile at P.
   s is the distance between p_o and p measured on the central line of the profile.
   In Figure 9.46b, the principal warping ordinate of a C section is given as an example.
• *Warping moment of inertia* (J_w)

   This parameter expresses the warping torsional resistance of the cross section and is analogous to the moment of inertia in bending.
   The warping moment of inertia is defined by the expression

$$J_w = \int_0^A w_{(s)}^2 \, dA \qquad 9.118$$

and is easily determined if the warping ordinate $w_{(s)}$ has been defined. For usual cross sections, closed expressions are given for the determination of J_w, J_{pe} and e (see Table 9.2 where an extract of a relevant table is given; Taranath, 2010). Implementing Equation 9.118 in the case of the cross section of Figure 9.42, the expression given by Equation 9.94 (Figure 9.47) is obtained.

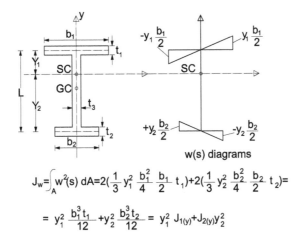

$$J_w = \int_A w^2(s)\, dA = 2(\frac{1}{3}\, y_1^2\, \frac{b_1^2}{4}\, \frac{b_1}{2}\, t_1) + 2(\frac{1}{3}\, y_2^2\, \frac{b_2^2}{4}\, \frac{b_2}{2}\, t_2) =$$

$$= y_1^2\, \frac{b_1^3 t_1}{12} + y_2^2\, \frac{b_2^3 t_2}{12} = y_1^2\, J_{1(y)} + J_{2(y)} y_2^2$$

Figure 9.47 Determination of the principal warping moment of inertia J_W of an I section with unequal flanges.

9.6.2.5 Implications of warping torsion in analysis and design to seismic action of R/C buildings

9.6.2.5.1 General

It has been noted many times so far that walls constitute the main backbone of the structural system of a building in resisting earthquakes. They are usually arranged in cores of a composite open cross section in the form of U, Γ, T, Z or similar sections. Therefore, in addition to their torsional rigidity due to bending and pure torsion, they are considered 3-D linear elements with their axes localised in the SC of their cross section, and they develop a warping torsional rigidity that should be taken into account. However, this stiffness parameter is ignored by almost all commercial computer programs for analysis and design. At the same time, EC8-1/2004 par. 5.4.3.4.1/(1) and (4) provides that the design of bending resistance of composite wall sections should be carried out, assuming that the cross sections are *integral units* and the strain distribution is planar, without any reference to warping torsion issues. In this respect, the analysis and design in almost all conventional commercial programs are in accordance with the Code.

However, the question remains as to how strong the influence of warping torsion on the results of the conventional approach would be in case this influence was taken into account.

This influence depends basically on the torsional stiffness of the structural system. In fact, keeping in mind that the torsional stiffness of a structural system has the form of the following expression (see Chapter 2.4.4(2): Equation 2.82):

$$T.R. = \sqrt{\sum J_{ix} y_i^2 + \sum J_{iy} x_i^2 + \sum J_p} \qquad (9.119)$$

where

J_{ix}, J_{iy} are the moments of inertia of the resisting vertical structural members (walls and columns).

J_p are their polar moments of inertia in reference to their shear centres.

x_i, y_i are the coordinates of the axes of these members passing through their local SCs with respect to the global stiffness centre.

It is clear, according to what has been presented in Section 9.6.2.3 that *the warping stiffness of each composite section to its local SC will be a small percentage of expression 9.119* if the global torsional stiffness results from many vertical elements distributed throughout the plan of the building. In this case, the warping torsion can be regarded as insignificant for the structural behaviour of the building. However, in the case that almost all walls are concentrated in a core while all other vertical elements are columns of low stiffness compared to the core stiffness, then warping torsion should be taken into account, since it constitutes the main part of the torsional stiffness of the building. The results included in Example 9.8.2 are indicative of the soundness of the above remark.

9.6.2.5.2 Analysis and design to warping torsion in the case of FEM modelling of the walls

In this case, warping torsion has already been taken into account in the determination of stresses, since the analysis is based on the theory of elasticity where all elastic compatibility considerations are taken into account, and not on the Bernoulli concept for planar distribution of strains. The results of the structural analysis of a U section at the level of stresses are given in Figure 9.48.

However, the axial stresses due to warping torsion cannot be taken into consideration in dimensioning and design for the following reasons. As noted in a previous paragraph, the normal stresses due to bimoment (warping) have a resultant axial force and bending moments with reference to the composite section equal to zero. In this respect, the integration of normal stresses on wall sections to a group of internal forces N, M_x and M_y (pier assumption) reduces to zero the influence of warping normal stresses and in this respect the influence of bimoment, too. It should be noted that a composite section is designed according to EC2-1-1/2004 as cracked, with a planar distribution of strains, and in this respect the existence of warping normal stresses contradicts the above assumptions of EC2-1-1/2004.

In the case that warping of open-wall R/C sections should be taken into account (small torsional stiffness of the structural system), the following approximate approach could be followed:

Step 1: From the static or dynamic analysis, the rotation angle θ_i at the level of the successive storeys is determined.

Step 2: From the diagram z–θ, the diagram of $d\theta_i/dz$ may be produced:

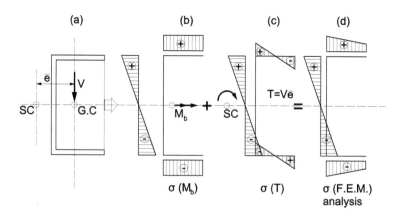

Figure 9.48 Normal stress state of a C section due to pure bending and warping torsion; (a) composite wall section loaded in the GC; (b) normal stress σ_l diagram according to Bernoulli concept; (c) warping normal stress σ_w diagram; (d) normal stress σ_t diagram resulted from the FEM analysis.

$$\theta_i' = \frac{d\theta_i}{dz} = \frac{\Delta\theta_i}{h_{storey}} \tag{9.120}$$

and from θ_i', the diagram of the second derivatives may also be derived:

$$\theta_i'' = \frac{d^2\theta_i}{(dz)^2} = \frac{\Delta\theta_i}{h_{storey}^2} \tag{9.121}$$

Step 3: Using the expression of Table 9.1, which relates $d^2\theta/dz^2$ to bimoment, meaning

$$\boxed{B_\kappa(z) = EJ_w \frac{d^2\theta(z)}{(dz)^2}} \tag{9.122}$$

the bimoment $B_\kappa(z)$ is determined at each storey level for every open thin-walled member. It should be noted that J_w refers to the local SC of the open composite wall section.

Step 4: Having defined the bimoments $B_\kappa(z)$ at each storey level of each composite section, the warping normal stresses are calculated using the relevant equation of Table 9.1,

$$\boxed{\sigma_\kappa(z) = -w_{(s)}B_\kappa(z)/J_w} \tag{9.123}$$

where
    $w(s)$ is the principal sectorial ordinate of the point of reference of the cross section.

Step 5: The above normal stresses are then integrated along each straight component of the composite section, and the tensile resultants are used to define the additional reinforcement at each leg due to warping torsion.

It is obvious that the above procedure overlooks cracking and violates the Bernoulli concept for planar distribution of strain over the section.

The above five steps may be programmed in a conventional program (e.g. ECtools) in the form of a post-processor and be activated only in case the designer decides that warping torsion should be taken into account.

9.6.2.5.3 Analysis and design to warping torsion in the case of linear 3-D modelling of the walls

In the case that the walls are modelled as 3-D linear elements with their axes passing through their local SC, warping torsion should be taken into account for the formation of the stiffness matrix of the element. In this respect, instead of a square symmetric matrix of 12 × 12 elements, a matrix of 14 × 14 should be introduced with two extra 'displacements' at the end of the member, due to the warping deformations (Taranath, 2010). This means that the main computer program for the analysis and design should be drastically modified in order for warping torsion to be taken into account.

9.7 SEISMIC DESIGN OF DIAPHRAGMS

9.7.I General

In addition to their main role in carrying gravity loads, the floor slabs of R/C buildings contribute to the formation of the structural system resisting seismic or wind actions. They act as horizontal diaphragms that transfer the inertial forces to the vertical structural system and ensure that these systems act together in carrying the seismic actions.

Very early in this book (see Chapter 2.4.4), a thorough reference was made to diaphragms and their significance in generating pseudospatial structural systems. For conventional R/C buildings, orthogonal in plan, with their dimensions l_x and l_y almost equal, and a smooth distribution of the stiffness of vertical elements over the plan of the building, there is no concern for the seismic analysis and design of the diaphragms, as they are formed automatically at the floor level of the storeys by the cast *in situ* slabs.

On the other hand, special attention should be given to the following cases:

1. Non-compact or very elongated in-plan shapes and in the case of large floor openings, particularly if these openings are located in the vicinity of the main vertical elements, prohibiting the effective connection between the vertical and horizontal elements, or at

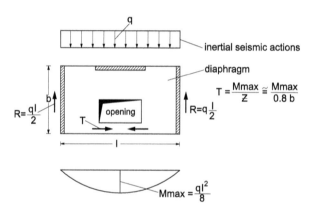

Figure 9.49 A diaphragmatic slab with an opening near the tension chord.

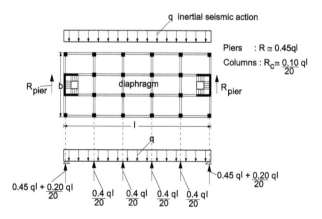

Figure 9.50 The structural behaviour of a long diaphragm with two cores at its ends.

the perimeter of the diaphragm, interrupting in this way the formulation of the tension chord of the diaphragm (Figures 9.49 and 9.50).
2. Complex and/or non-uniform layout of the lateral load resisting system, such as basements with R/C walls located only in parts of their perimeter.
3. Dual systems where structural systems of different stiffness characteristics are coupled together through the floor slab. Major problems may be expected only in the case where the distribution of walls is not appropriate (e.g. cores at the opposite sides of the building in plan; Figure 9.50).

Diaphragms must have sufficient in-plane stiffness for the distribution of horizontal inertial forces to the vertical structural system in accordance with the assumptions of the analysis (pseudospatial system). In this respect, plan shapes of L, C, H, I and X form should be carefully examined. According to EC8-1/2004, a solid reinforced concrete slab may be considered sufficient to serve as a diaphragm for conventional structures, if it has a thickness of not less than 70 mm and is reinforced in both directions with at least the minimum reinforcement specified in EC2-1-1/2004.

9.7.2 Analysis of diaphragms

9.7.2.1 Rigid diaphragms

- Floor diaphragms are commonly assumed and modelled as rigid in-plane bodies. In this respect, the analysis of the diaphragm is separated from the analysis of the global structural system, which in this case is considered as a pseudo-structural system.
- Lateral loading of the diaphragm at any storey may result as the difference of the shear forces due to seismic action at the vertical shear-resisting elements above and below the diaphragm (Figure 9.51). The resultant of these seismic actions in each main direction will be distributed in trapezoidal form along the dimension of the diaphragm, perpendicularly to the seismic action, to simulate the inertial forces of the masses of the diaphragm.

    In this respect, the resultant of the distributed inertial forces of the diaphragm will act on the same axis of the shear actions of the vertical elements on the diaphragm (Figure 9.51). Eurocode EC8-1/2004 makes no reference to lateral loading for the analysis of a diaphragm. American Code ASCE7-05 provides that this loading should satisfy design seismic force from the structural analysis as presented above, and additionally requires that this loading should not be less than that determined in accordance with the following equation:

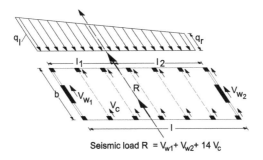

Seismic load $R = V_{w1} + V_{w2} + 14 V_c$

Figure 9.51 Determination of diaphragm loading due to seismic action.

$$F_{px} = \frac{\sum\limits_{i=x}^{n} F_i}{\sum\limits_{i=x}^{n} W_i} W_{px}$$

(9.124)

where

F_{px} is the diaphragm design force at level x.
F_i is the design force applied at level i (lateral force method of analysis; Chapter 5.6.4).
W_i is the weight tributary to level i.
W_{px} is the weight tributary to the diaphragm at level x.

- The action effects may then be estimated by modelling the diaphragm as a deep beam or a plane truss or strut-and-tie model, considered simply a supported beam with reactions at the supports almost equal to zero. (The simple supports are introduced only for avoiding the geometric redundancy of the system.) If the in-plane stiffness of the diaphragm is introduced, the deflections of the diaphragm may be calculated. If these deflections are less than 10% of the corresponding horizontal displacements of the pseudospatial structural system at that storey, the procedure is deemed to be correct. Otherwise, the diaphragm should be introduced in the overall structural system as a flexural planar member.

    For analysis of the diaphragm, Eurocode EC8-1/2004 proposes the introduction of elastic supports at the location where vertical members are connected with the diaphragms. This approach is not easily implemented, since the spring coefficients of the supports are not easily determined. A criterion for the correct choice of the spring coefficients would be the following: The spring reactions from the analysis of the diaphragm should be equal to the shear forces at those points from the analysis of the pseudospatial structure that have been taken into account from the beginning for the determination of the inertial forces of the diaphragm.
- The design values of the action effects should be derived by introducing an overstrength factor. The value of this overstrength factor recommended by EC8-1/2004 is

$$\gamma_d = 1.3$$

(9.125)

9.7.2.2 Flexible diaphragms

In the case that the diaphragm that has been modelled separately from the overall pseudospatial structural system exhibits displacements of more than 10% of the corresponding absolute horizontal displacements of the pseudospatial structural system due to seismic action, then the structural system must be reanalysed from the beginning by introducing the diaphragms in the model with their actual in-plane stiffness.

In this respect, the pseudospatial model of analysis is abandoned and diaphragms are introduced in the form of a finite element mesh (Chapter 4.2.8, Figure 4.14). It is obvious that in this case the analysis becomes very time consuming, since the number of the unknown displacements and rotations and the number of modes – if the model response spectrum analysis is followed – increase significantly. However, nowadays, with the rapid development of computer capacities, the procedure must not be considered prohibitive. It is apparent that in this case the output of the analysis is expressed in the form of stresses σ_x,

σ_y, τ_{xy} in the middle plane of the diaphragm and in the form of axial forces along the beams that act as chords and collectors (Chapter 4.2.8, Figure 4.15).

9.7.3 Design of diaphragms

The seismic design of diaphragms must include ULS verifications of reinforced concrete. The design resistances should be defined in accordance with EC2-1-1/2004.

In cases of core or wall structural systems, it should be verified that the transfer of the horizontal forces from the diaphragm to the cores or walls is ensured by the following provisions:

1. The design shear stress in the interface of the diaphragm and the core or the wall should be limited to $_{1.5fctd}$ for cracking control.
2. Shear sliding failure should be ensured (see Section 9.2.4(3) and (5)), assuming strut inclination of 45°. In this respect, properly anchored additional bars should be provided, contributing to the shear strength of the interface between diaphragms and cores or walls.

9.7.4 Code provisions for seismic design of diaphragms

The above procedure for analysis and design of diaphragms is provided by EC8-1/2004 only for DCH R/C buildings.

9.8 EXAMPLE: DIMENSIONING OF A SLENDER DUCTILE WALL WITH A COMPOSITE CROSS SECTION

The design of a wall of composite cross section, wall W2, is presented here (Figure 5.44). The C-shaped wall is considered an integral unit and is modelled with shell elements (Chapter 5.6.3.1). W2 wall comprises a web parallel to the $y-y$ axis and two flanges normal to it. The analysis in biaxial bending is conducted according to the computational platform NOUS (NOUS/3P, 2002). The procedure described in Section 9.2.2.3 is followed. The algorithm is based on the ultimate limit state design method, considering the following assumptions:

1. A parabolic-rectangular concrete stress diagram is used.
2. Ultimate compressive concrete strain is taken to be ε_{cu} = 3.5‰.
3. Ultimate tensile steel strain is taken to be ε_{cu} = 20‰.
4. Concrete tensile strength is taken to be equal to zero.

Moreover, the flexural capacity is estimated by taking into account only the vertical reinforcement of the boundary elements, whereas the shear capacity is estimated by considering only the web horizontal reinforcement and for a web length approximately equal to 0.8 l_w.

9.8.1 Ultimate limit state in bending and shear

The load combinations at the base of wall W2 appear in Table 9.3. The design of the composite wall W2 in bending and shear is performed by using the cross-section analysis program NOUS in combination with ECtools (2013). An automatic iterative procedure is followed for establishing cross-section equilibrium for the applied action effects and the considered length and percentage of longitudinal reinforcement of the confined boundary

Table 9.3 Bending moments and axial forces for Wall W2 dimensioning

Load	Combinations	N	V_x	V_y	M_x	M_y
Lc1	1.35G+1.5Q	−11378.01	−86.63	0.42	4.02	−205.69
Lc2	G+0.3Q+Ex+0.3Ey	−4348.01	−1442.03	1061.81	7140.18	−7146.97
Lc3	G+0.3Q+Ex-0.3Ey	−4348.28	−1441.97	−805.91	−5645.90	−7146.75
Lc4	G+0.3Q-Ex+0.3Ey	−8824.76	1368.35	806.41	5651.47	6966.00
Lc5	G+0.3Q-Ex-0.3Ey	−8825.04	1368.41	−1061.31	−7134.61	6966.22
Lc6	G+0.3Q+Ey+0.3Ex	−5914.55	−458.47	3151.42	21536.22	−2207.69
Lc7	G+0.3Q+Ey-0.3Ex	−7257.58	384.65	3074.80	21089.61	2026.20
Lc8	G+0.3Q-Ey+0.3Ex	−5915.47	−458.27	−3074.30	−21084.04	−2206.95
Lc9	G+0.3Q-Ey-0.3Ex	−7258.50	384.85	−3150.92	−21530.65	2026.94

elements (defined according to the provisions of EC8-Part I, 2004). The section at the base of the first storey is loaded at the CG, C, by $M_{Eyd} = -2207.69$ kN m, $M_{Exd} = 21536.22$ kN m and $N_{Ed} = -5914.55$ kN. This is considered, according to the checks performed by NOUS, to be the critical load combination, $G+0.3Q+E_y+0.3E_x$ (see Table 9.3); it is reminded that the effect of the accidental eccentricity is included in the response spectrum analysis, and thus the total number of load combinations, including the gravity load combination, is limited to 9.

Five iterations are performed before reaching cross-section equilibrium. In order for the safety factor γ to become equal to 1, the required percentage of longitudinal reinforcement of the entire wall cross section is 0.19%. Due to restrictions related to

1. The maximum distance between consecutive longitudinal bars restrained by hoops, which is defined as equal to $s_{min} = 150$ mm for high ductility walls.
2. The mechanical volumetric ratio of the required confining reinforcement in the boundary elements the longitudinal reinforcement percentage is increased to 0.55%.

The strain and stress profiles along with the position of the neutral axis (point N) are depicted for the elastic (non-cracked) and final iteration state in Figure 9.52a and b, respectively.

The ordinates of the geometrical centre (point C) are $x = 2.650$ m, $y = 2.184$ m, whereas the neutral axis position is modified from point N, in the first iteration (uncracked cross section; $x = 3.900$ m, $y = 1.656$ m, $r_z = 67.09$ deg), to point P, in the last (fifth) iteration ($x = 2.595$ m, $y = 2.943$ m, $r_z = 49.54°$).

The internal forces developed in the concrete and the reinforcing bars at the last iteration (fifth) step are equal to $N_{Rd} = -16184.80$, $M_{Rxd} = 30788.39$ kN m, $M_{Ryd} = -36479.92$ kN m relative to original coordinate system passing through point C.

The safety factor at the final iterative step is defined as equal to $γ = 2.23$. The length of the boundary elements is shown in Figure 9.53, while the longitudinal and transverse reinforcements of the wall are depicted in Figure 9.54.

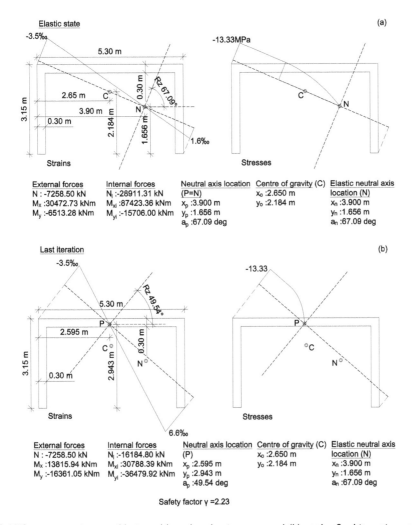

Figure 9.52 W2 cross-section equilibrium: (a) at the elastic state, and (b) at the final iteration step.

The mechanical volumetric ratio of the required confining reinforcement $\alpha\omega_{wd,req}$ within the critical height of the wall has been defined as equal to 0.193 for web element A and 1.04 for flange elements B and C (Figure 9.53). These volumetric ratios correspond to the hoop configuration shown in Figure 9.54 consisting of hoops of Φ8/70. The above computations have been carried out automatically using ECTools (2013) and are based on the assumption of sacrificed flange (see Section. 9.2.4.2(iii)).

The shear actions, for which the horizontal reinforcement of the web is defined, are $V_{Exd} = 1442.03$ kN (load combination $G + 0.3Q + E_x + 0.3E_y$, Table 9.3) and $V_{Eyd} = 3151.42$ kN (load combination $G + 0.3Q + E_y + 0.3E_x$, Table 9.3). The same amount of horizontal and vertical reinforcement is placed in the web of the wall being equal to 2#Ø8/160 (i.e. two legs of Ø8 at 160 mm).

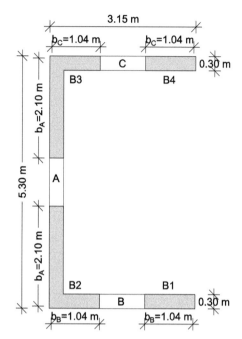

Figure 9.53 Definition of boundary elements.

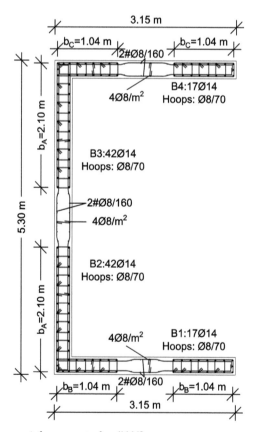

Figure 9.54 Flexural and shear reinforcement of wall W2.

9.8.2 Estimation of axial stresses due to warping torsion

The eight-storey RC building presented in the example of Chapter 5 has a sufficient number of vertical members distributed all over the plan of the building as well as two walls placed at the perimeter, *implying that warping torsion is insignificant and can be ignored in the structural design*. The objective of this example is to estimate the developed warping normal stresses for wall W2 (Figure 5.44) and compare them to pure bending normal stresses. Therefore, the procedure described in detail in Section 9.6.2.5.2 is implemented.

9.8.2.1 Estimation of the geometrical parameters for warping bending of an open composite C-shaped wall section

9.8.2.1.1 Shear centre

The SC lies on the horizontal symmetry axis at a distance h from the central line of the vertical component of the wall (Figure 9.55). The distance h is estimated as (Table 9.2)

$$h = \frac{3 \cdot b^2}{1 + 6 \cdot b} = \frac{3 \cdot 300^2}{500 + 6 \cdot 300} = 117.39 \text{cm}$$

The CG is estimated at a distance 81.61 cm from the central line of the vertical component.

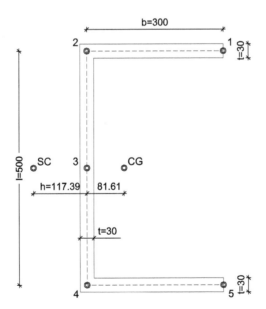

Figure 9.55 CG and SC position.

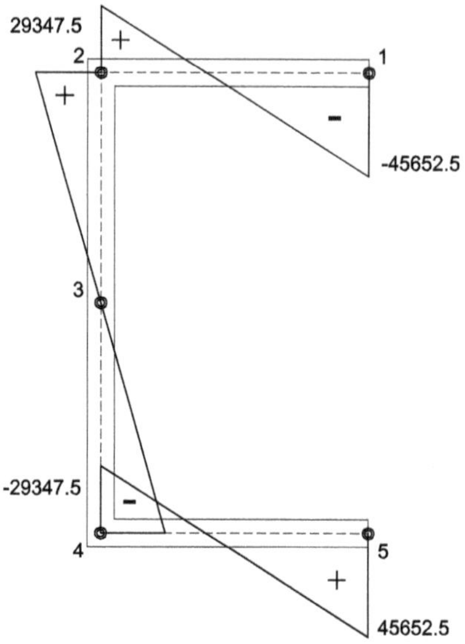

Figure 9.56 Principal sectorial ordinates, w_s, at the successive points of the central line of the cross section.

9.8.2.1.2 Principal sectorial ordinates, w_s

The determination of the principal warping ordinates of the successive points 1–5 of the central line of the cross section of the C-shaped wall with reference to the SC is given at each position by (Figure 9.56)

$$w_s(1) = h \cdot \frac{(l-t)}{2} - \left(b - \frac{t}{2}\right)\frac{(l-t)}{2} = 117.39 \cdot \frac{(530-30)}{2} - \left(315 - \frac{30}{2}\right)\frac{(530-30)}{2}$$

$$= -45652.5 cm^2$$

$$w_s(2) = h \cdot \frac{(l-t)}{2} = 117.39 \cdot \frac{(530-30)}{2} = +29347.5 cm^2$$

$$w_s(3) = 0$$

$$w_s(4) = -w_s(2) = -29347.5 cm^2$$

$$w_s(5) = -w_s(1) = 45652.5 cm^2$$

9.8.2.1.3 Warping moment of inertia, J_w

The warping moment of inertia, J_w, is estimated as follows (Table 9.2):

$$J_w = \frac{(l-t)^2 \cdot (b-(t/2))^3 \cdot t}{12} \cdot \frac{[2 \cdot (l-t) + 3 \cdot (b-(t/2))]}{(l-t)+6 \cdot (b-(t/2))} \Leftrightarrow$$

$$J_w = \frac{(530-30)^2 \cdot (315-(30/2))^3 \cdot 30}{12} \cdot \frac{[2 \cdot (530-30)+3 \cdot (315-(30/2))]}{(530-30)+6 \cdot (315-(30/2))}$$

$$= 1392 \cdot 10^{10} cm^6$$

9.8.2.2 Implementation of the proposed methodology for deriving the normal stresses due to warping

The normal stresses due to warping are estimated according to the following steps:

Step 1: The rotation angle θ_i at the centre of mass at each storey diaphragm for seismic loading in the x–x and y–y directions is presented in Tables 9.4 and 9.5, respectively. The rotation angle profile for each loading direction is presented in Figures 9.57 and 9.58.

Step 2: The profiles of the first derivative, $\theta_i'=d\theta_i/d_z=\Delta\theta_i/h_{st}$, and the second derivative, $\theta_i''=d^2\theta_i/(d_z)^2=\Delta\Delta\theta_i/h_{st}^2$, of the rotation angle are estimated for all the storeys in the two directions of loading (Tables 9.4 and 9.5, Figures 9.57 and 9.58).

Step 3: The bimoment $B_\kappa(z)$ is estimated at each storey via Equation 9.122 for the two loading directions, utilising the estimated warping moment of inertia, J_w, and the profile of the second derivative of the rotation angle, θ_i'' (Tables 9.4 and 9.5).

Step 4: Having estimated the principal sectorial ordinates, $w_s(i)$, and the warping moment of inertia, J_w, the axial stresses due to warping are estimated for the C-shaped wall section $\sigma_\kappa(z)$ in Equation 9.123 and presented in Tables 9.6 and 9.7 for seismic loading in the x–x and y–y directions, respectively.

In light of the above, it may be seen that the effect of warping torsion is insignificant, with the estimated normal stresses receiving almost zero values. This is expected, since the

Table 9.4 Estimation of the bimoment B_κ at each storey in the x–x direction

Storey	θ_i (rad)	$\theta_i'(1/mm)$	$\theta_i''(1/mm^2)$	B_κ (N mm²)
		Seismic action in the x–x direction		
8th	7.92E–04	2.93E–08	1.63E–15	+7.05E+08
7th	7.04E–04	1.47E–08	−1.63E–15	−7.05E+08
6th	6.60E–04	2.93E–08	0.00E+00	0.00E+08
5th	5.72E–04	2.93E–08	−1.63E–15	−7.05E+08
4th	4.84E–04	4.40E–08	1.63E–15	7.05E+08
3rd	3.52E–04	2.93E–08	−1.63E–15	−7.05E+08
2nd	2.64E–04	4.40E–08	1.96E–15	8.46E+08
1st	1.32E–04	2.64E–08	1.06E–15	4.56E+08

Table 9.5 Estimation of the bimoment b_κ at each storey in the y–y direction

Storey	θ_i (rad)	$\theta_i'(1/mm)$	$\theta_i''(1/mm^2)$	B_κ (Nmm²)
		Seismic action in the y–y direction		
8th	1.01E–03	2.93E–08	0.00E+00	0
7th	9.24E–04	2.93E–08	−4.04E–30	17.49E–07
6th	8.36E–04	2.93E–08	−1.63E–15	7.05E+08
5th	7.48E–04	4.40E–08	4.41E–30	19.04E–07
4th	6.16E–04	4.40E–08	−2.21E–30	9.54E–07
3rd	4.84E–04	4.40E–08	0.00E+00	0E+00
2nd	3.52E–04	4.40E–08	−7.35E–31	−3.16E–07
1st	2.20E–04	4.40E–08	1.76E–15	7.62E+08

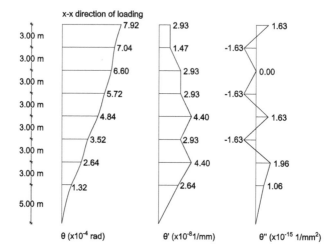

Figure 9.57 Rotation angle, θ_i, θ_i' and θ_i'' profile along building height for the x–x direction of loading.

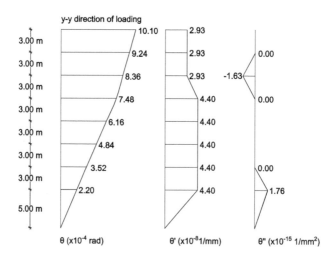

Figure 9.58 Rotation angle, θ_i, θ_i' and θ_i'' profile along building height for the y–y direction of loading.

Table 9.6 Estimation of the warping normal stresses (in MPa) for wall W2 at each storey in the x–x direction

	Seismic action in the x–x direction				
Storey	$\sigma_\kappa(1)$	$\sigma_\kappa(2)$	$\sigma_\kappa(3)$	$\sigma_\kappa(4)$	$\sigma_\kappa(5)$
8th	2.31E–03	−1.48E–03	0.00E+00	1.48E–03	−2.31E–03
7th	−2.31E–03	1.48E–03	0.00E+00	−1.48E–03	2.31E–03
6th	0.00E+00	0.00E+00	0.00E+00	0.00E+00	0.00E+00
5th	−2.31E–03	1.48E–03	0.00E+00	−1.48E–03	2.31E–03
4th	2.31E–03	−1.48E–03	0.00E+00	1.48E–03	−2.31E–03
3rd	−2.31E–03	1.48E–03	0.00E+00	−1.48E–03	2.31E–03
2nd	2.77E–03	−1.78E–03	0.00E+00	1.78E–03	−2.77E–03
1st	1.49E–03	−9.61E–04	0.00E+00	9.61E–04	−1.49E–03

Table 9.7 Estimation of the warping normal stresses (in MPa) for wall W2 at each storey in the y–y direction

	Seismic action in the y–y direction				
Storey	$\sigma_\kappa(1)$	$\sigma_\kappa(2)$	$\sigma_\kappa(3)$	$\sigma_\kappa(4)$	$\sigma_\kappa(5)$
8th	0.00E+00	0.00E+00	0.00E+00	0.00E+00	0.00E+00
7th	−5.72E−18	3.68E−18	0.00E+00	−3.68E−18	5.72E−18
6th	−2.31E−03	1.48E−03	0.00E+00	−1.48E−03	2.31E−03
5th	6.24E−18	−4.01E−18	0.00E+00	4.01E−18	−6.24E−18
4th	−3.12E−18	2.01E−18	0.00E+00	−2.01E−18	3.12E−18
3rd	0.00E+00	0.00E+00	0.00E+00	0.00E+00	0.00E+00
2nd	−1.04E−18	6.69E−19	0.00E+00	−6.69E−19	1.04E−18
1st	2.49E−03	−1.60E−03	0.00E+00	1.60E−03	−2.49E−03

Table 9.8 Estimation of the normal stresses due to bending (in MPa) for wall W2 at the first storey for both directions of loading

	Normal stresses due to bending				
1st storey	$\sigma(1)$	$\sigma(2)$	$\sigma(3)$	$\sigma(4)$	$\sigma(5)$
x–x dir.	3.84	2.53	1.60	2.53	3.84
y–y dir.	2.71	4.02	0.62	4.02	2.71

building is not torsionally flexible due to the existence of the walls W1 and W4 at the perimeter of the building (Figure 5.44) and the vertical elements distributed throughout its plan. Hence, for buildings with similar characteristics to the eight-storey building, warping has no significant effect and can be safely ignored.

For comparison purposes, the normal stresses due to bending in the first storey at successive points 1–5 of the central line of the cross section of the C-shaped wall (Figure 9.56) are presented in Table 9.8. Comparison with the corresponding values in Tables 9.6 and 9.7 confirms that the increase in the normal stresses due to warping is negligible.

Chapter 10

Seismic design of foundations

10.1 GENERAL

It is known that the foundation design of conventional buildings consists of two main components:

- The safety and service verification of the soil where the building is embedded
- The safety and service verification of the structural members of the foundation

It should be noted from the beginning that the building material of foundations has been reinforced concrete for almost the past 100 years, no matter what the building material of the superstructure has been.

While the analysis and design of the structural system has been gradually developed since the beginning of the nineteenth century, the study of the structural behaviour of the soil where the loads of the superstructure are transferred was delayed for almost one century. This delay must be attributed

- To the uncertainties of soil properties.
- To the soil composition of various strata with different properties.
- To the difficulty of generating constitutive laws for the mechanical properties of soil analogous to those of conventional structural materials. In fact, the soil composition in particles, the cohesion among them, the existing voids among the grains, the water content in the voids and so on generate a very complicated mixture of parameters that does not allow an easy and reliable interpretation of the behaviour of this mixture by mechanical laws for a continuous medium.
- Finally to the 3-D solid-state structural form of the foundation ground, since it is in fact a half 3-D continuous medium.

Therefore, a new engineering discipline was developed (soil mechanics) constituting the background of foundation design. The specialist in soil mechanics, after thorough *in situ* investigations and extended tests in the lab, gives all necessary data for strength, stiffness, damping and time-dependent properties for the soil under consideration in the form of a technical report.

The above difficulties, and particularly the 3-D solid state of the soil, had made an approach to soil–structure interaction issue, in other words, the relationship among (Penelis et al., 1995) 'subsoil–foundation–superstructure' as a structural continuum, almost impossible for many decades (Figure 10.1).

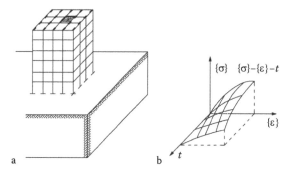

Figure 10.1 Soil–structure interaction analysis inelastic procedures: (a) structural model; (b) $\{\sigma\}-\{\varepsilon\}-t$ relation.

To the above difficulties common for all buildings, some extra parameters should be added for buildings under seismic actions. These parameters are the following:

1. While all other loadings are transferred from the building to the ground, seismic actions are transferred from the ground to the building in the form of seismic waves (time-dependent constraints). This means that the soil during the earthquake is under dynamic excitation, and, therefore, its mechanical properties should be re-examined.
2. The consequences of the capacity design concept established for the superstructure should be investigated in detail with regard to foundation behaviour.
3. The ductility concept, on which the analysis, design and detailing of the superstructure is based, should be linked with the relevant procedures for the foundation members.
4. Perspectives on taking into account inelastic soil cyclic deformations due to seismic actions as a damping mechanism for the structure as a whole should be investigated thoroughly.

In the following sections, after a short overview of the mechanical properties of soil under dynamic loading, the presentation will concentrate on the design of foundations under seismic action.

10.2 GROUND PROPERTIES

10.2.1 Strength properties

According to EC8-5/2004, the value of the soil strength parameters applicable under static undrained conditions may also generally be used for soil under seismic action, taking into account that water draining of the soil during the short duration of an earthquake is not possible. However, some additional information should be given.

10.2.1.1 Clays

For cohesive soils, undrained shear strength c_u should be used. However, the value of c_u is affected both by the loading rate and by the number of cycles of loading. Rate effects may increase c_u up to 25% compared with the static strength. Conversely, the number of cycles reduces c_u, particularly in case of over-consolidated clays. For example, a normally consolidated clay (OCR = 1) can sustain 10 cycles at 90% of the undrained static shear strain c_u.

10.2.1.2 Granular soils (sands and gravels)

For cohesionless soil, the appropriate strength parameter is the undrained shear strength $\tau_{cy,u}$. This parameter relies on the friction among grains and, therefore, on the angle of internal friction φ', which is not influenced by cyclic loading. However, the effective stress between grains may be reduced in saturated granular soils, and more particularly in sands, if pore water pressures increase during an earthquake. This reduction leads to shear strength reduction. Pore water pressure may increase during an earthquake in the case of loose sands, where granular material tends to increase in density under the shaking actions of the earthquake. This results in an increase in pore pressure, since pore water has not had time to drain away. In due time, pore water will find paths to leak out of the voids, and in this way, strength is restored. In the meantime, due to a significant temporary soil shear strength reduction, which may be drastically decreased, dramatic collapses of buildings may occur. This phenomenon is known as *soil liquefaction*. In a subsequent paragraph, more extended reference will be made to soil liquefaction, and more particularly to soils susceptible to liquefaction. EC8-5/2004 notes that this probability should be taken into consideration for the determination of $\tau_{cy,u}$.

10.2.1.3 Partial safety factors for soil

The partial safety factors for soil properties c_u, $\tau_{cy,u}$, q_u and $\tan \varphi'$, namely γ_{cu}, $\gamma_{\tau cy}$, γ_{qu} and $\gamma_{\varphi'}$, may be found according to EC8-5/2004 in the National Annex of each EU country. The recommended values of EC8-5/2004 are given below:

$$\begin{aligned}
\gamma_{cu} &= 1.4 \\
\gamma_{\tau cy,u} &= 1.25 \\
\gamma_{qu} &= 1.40 \\
\gamma_{\varphi'} &= 1.25
\end{aligned}$$

(10.1)

10.2.2 Stiffness and damping properties

The main stiffness parameter of the ground under seismic action is the shear modulus G given by the expression

$$G_s = \rho V_s^2$$

(10.2a)

where
  – ρ is the unit mass.
  – V_s is the shear wave propagation velocity of the ground.

Table 10.1 contains representative values of G_s and E_s of various soil categories. The value of G_s has been calculated by applying the well-known expression of the theory of elasticity:

$$G_s = \frac{E_s}{2(1+v_s)}$$

(10.2b)

Table 10.1 Typical values of E_s and G_s for soils and rocks

Soil type	E_s (MPa)	G_s (MPa)
Soft clay	Up to 15	5.3
Firm, stiff clay	10–15	3.0–18.0
Very stiff, hard clay	25–200	9.0–70.0
Silty sand	7–70	2.0–2.5
Loose sand	15–50	5.0–18.0
Dense sand	50–120	18.0–43.0
Dense sand and gravel	90–200	32.0–70.0
Sandstone	Up to 50,000	Up to 18,000
Chalk	5,000–20,000	1,800–7,000
Limestone	25,000–100,000	9,000–35,000
Basalt	15,000–100,000	5,000–35,000

Source: Dowrick, D.: *Earthquake Risk Reduction*. 2005. Copyright Wiley-VCH Verlag GmbH & Co. KGaA. Reproduced with permission.

where

v_s is Poisson's ratio taken to be equal to 0.40 (mean value).

Values of E_s have been taken from Dowrick (2005). Shear wave propagation velocity V_s and therefore G_s have been taken into account in EC8-1/2004 (see Chapter 3.4.2.2) for the ground classification of various types (A to S_1) and, consequently, for the determination of the parameter S of the elastic response spectra. It should also be noted that G_s is the basic parameter for the determination of the overall foundation spring stiffness. Detailed reference will be made to this issue in the following sections.

Damping, on the other hand, should be considered as an additional ground property in the cases where the effects of soil–structure interaction should be taken into account.

From extended laboratory and *in situ* tests, it has been concluded that stiffness and damping under cyclic loading are affected primarily by the shear strain amplitude (Figure 10.2). Keeping

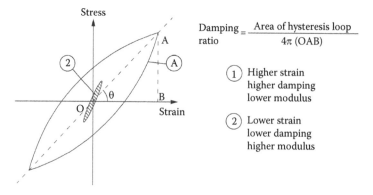

Figure 10.2 The effect of shear strain on damping and shear modulus of soils. (Dowrick, D.: *Earthquake Risk Reduction*. 2005. Copyright Wiley-VCH Verlag GmbH & Co. KGaA. Reproduced with permission.)

Table 10.2 Average soil damping ratios and average reduction factors (±1 standard deviation) for shear wave velocity V_s and shear modulus G within 20 m depth

Ground acceleration ratio, $a.S$	0.10	0.20	0.30
Damping ratio	0.03	0.06	0.10
$\dfrac{V_s}{V_{s,max}}$	0.90 (±0.07)	0.70 (±0.15)	0.60 (±0.15)
$\dfrac{G}{G_{max}}$	0.80 (±0.10)	0.50 (±0.20)	0.36 (±0.20)

Source: Adapted from EC8-5/2004. Design of Structures for Earthquake Resistance, Part 5: Foundations, Retaining Structures and Geotechnical Aspects. European Committee of Standardization, CEN, Brussels.

Note: $V_{s,max}$ is the average V_s value at small strain ($<10^{-5}$), not exceeding 360 m/s. G_{max} is the average shear modulus at small strain.

in mind that V_s measured by *in situ* tests refers to small strain values, it is apparent that for the strain values caused by the design earthquake, which are bigger, a reduction factor should be introduced. EUROCODE 8-5/2004 includes information about these reduction factors of

- Damping ratio
- Shear wave propagation velocity V_s
- Shear modulus G_s

in relation to $a_g.S$ (Table 10.2), where

- a_g is the ratio of the design ground acceleration to the acceleration of gravity.
- S is the soil factor.

For both properties, that is, shear modulus and damping, more information may be found in specialised books on soil mechanics and foundations.

10.2.3 Soil liquefaction

In Section 10.2.1.2, the soil liquefaction phenomenon was briefly outlined. Soil types that are susceptible to liquefaction basically have the following features:

1. They are soils that tend to be densified under induced cyclic shear strains (e.g. geologically young deposits).
2. There is water between the soil particles (e.g. the water table is not deeper than 15.0 m from foundation level; Youd, 1998).
3. A part of the shear strength of the soil is attributable to friction between soil particles (e.g. loose sands and even silts).
4. There are restrictions on the drainage of pore water from the soil (e.g. gravel is not susceptible to liquefaction).

Eurocode EC8-5/2004 makes special recommendations for the analytical assessment of liquefaction, which are beyond the scope of this book.

If soils are found to be susceptible to liquefaction to the degree that the safety of the foundation is endangered, special measures must be taken.

These measures are

- Ground improvement
- Piling

Ground improvement is accomplished either by soil compaction to increase soil resistance beyond the dangerous zone, or by the use of drainage.

Pile foundation makes possible the transfer of loads to layers not susceptible to liquefactions. The use of pile foundation alone should be considered with caution due to the large forces induced in the piles by the loss of soil support in the liquefiable layer or layers in combination with the uncertainties of the thickness of such layers.

10.2.4 Excessive settlements of sands under cyclic loading

Sand deposits, even if they are not under water level, are susceptible to densification and to excessive settlements under seismic-induced cyclic stresses. Therefore, when extended layers of loose unsaturated cohesionless materials exist in a small depth below the foundation, special care should be taken for the foundation and the building. The most effective method for limiting these settlements is soil improvement by compaction or grouting.

The densification and settlement potential of the previously mentioned soils should be evaluated properly using analytical models based on parameters justified by static or cyclic laboratory tests (Seed and Idriss, 1982; Kramer, 1996; Kramer and Elgamal, 2001).

The scope of this book does not allow further discussion of this issue.

10.2.5 Conclusions

From what has been presented thus far, it may be concluded that the structural engineer, responsible for the analysis and design of major R/C buildings in a seismic region, must have the support of a geotechnical engineer specialised in soil dynamics. This support must cover at least the following issues:

- *In situ* and laboratory tests necessary for ground evaluation
- Soil stratification
- Soil strength parameters, including the influence of the design seismic action
- Soil deformation and damping parameters
- Soil time-dependent deformations (consolidation of clays)
- Soil classification in one of the categories specified by EC8-1/2004 or other relevant Codes
- Susceptibility of soil to liquefaction
- Susceptibility of sandy soils to settlements under seismic excitation
- Recommendations for eventual soil improvement in relation to the recommended foundation system (shallow foundations, pile foundation, pier foundations)

10.3 GENERAL CONSIDERATIONS FOR FOUNDATION ANALYSIS AND DESIGN

10.3.1 General requirements and design rules

In addition to the conceptual design requirements and rules for foundations of conventional buildings, the following should be taken into account for buildings in seismic areas:

1. Gravity loads and seismic actions from superstructures must be transferred to the ground without substantial settlements. Permanent settlements must be compatible with the design requirements of the structure and of pre-existing buildings adjacent to the new one. Special care should be taken for the permanent differential settlements, which influence the service limit state of the superstructure.
2. The foundation must be stiff enough in relation to the superstructure so that the localised actions coming from the vertical members (columns–walls) to the ground are uniformly distributed. In this respect, permanent differential settlements could be minimised.
3. Usually, mixed foundations (e.g. piles with shallow foundations) should be avoided in the same structural system. In fact, the dynamic response of two different foundation parts belonging to the same structural system of the superstructure will differ, so the assumption of the analysis of the superstructure with a base that is uniformly excited is violated. Therefore, mixed foundation types may be used in dynamically independent units that are in parts of a building separated with structural joints into more than one unit. Only in special cases, where the need of mixed foundations is proved through specific studies, such a solution may be acceptable.
4. Vertical members (columns and walls) must be tied at the level of the foundation to a diaphragmatic disk using tie beams in the form of a grillage, foundation beams or a raft so that the uniform seismic vibrations of the base of the building is ensured (see Chapter 4.2.12).
5. Foundations at different levels should be avoided (see Chapter 4.2.12, Figure 4.18).
6. Special care should be taken of the strain dependence of the dynamic properties of soils and of effects related to the cyclic nature of seismic loading.

10.3.2 Design action effects on foundations in relation to ductility and capacity design

10.3.2.1 General

The design action effects on foundations, according to EC 8-5/2004 and EC 8-1/2004, are strongly related to the design considerations of the superstructure in relation to *ductility* level and *capacity* design. So,

- For *dissipative buildings*, the action effects for the foundations must be based on capacity design considerations. The basic principle is that the order of formation of yielding mechanisms must stop at the base of the superstructure above the foundations. *Foundation members and ground must remain in elastic range during the seismic excitation* (Figure 10.3a). Alternatively, the ductility mechanism may be extended even

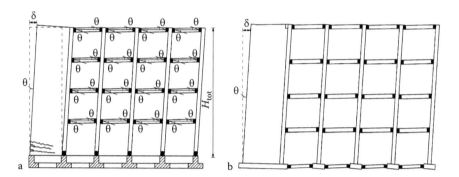

Figure 10.3 Dissipative mechanisms accepted by the Code: (a) dissipative superstructure–non-dissipative foundation mechanism (recommended); (b) dissipative superstructure–dissipative foundations mechanism. (With kind permission from Springer Science+Business Media: Fardis, M.N. 2009. *Seismic Design, Assessment and Retrofitting of Concrete Buildings.*)

to the foundation structural members (Figure 10.3b), but *never to the ground level,* which must always be in elastic range. The above extension of the ductile system to the foundation members should be used with due concern because damaged foundations are retrofitted with great difficultly. The New Zealand concrete code NZS3101/1995 follows a similar concept, while this approach is uncommon in US practice. As is well known, buildings of ductility class M and H (DCM, DCH) belong to the category of dissipative buildings.

- *For non-dissipative buildings* (DCL), the action effects will result from the analysis in the seismic design combination without any capacity design consideration ($q = 1.5$). The above procedure may also be followed for dissipative buildings (DCM, DCH) on the condition that a q-factor equal to 1.50 is introduced for the design of foundation members and ground. This is based on the reasonable consideration that capacity design effects cannot overcome those corresponding to $q = 1.50$.

10.3.2.2 Design action effects for various types of R/C foundation members

R/C foundation members, *footings, tie beams, foundation beams, foundation slabs, pile caps* and *piles,* as well as the joints between them and the vertical structural R/C elements of the buildings, are analysed and designed according to EC8-1/2004, following rules which result from the conceptual approach of the previous paragraph.

1. *Dissipative superstructure–non-dissipative foundation elements and foundation ground.*
   a. For dissipative buildings (DCM or DCH), foundations of structural walls or of columns of moment-resisting frames are analysed for the following design actions effects E_{Fd}:

$$\boxed{E_{Fd} = E_{F,G} + \gamma_{Rd}\,\Omega\,E_{F,E}}$$

(10.3)

where

- γ_{Rd} is the overstrength factor taken as being equal to

$$
\begin{aligned}
\gamma_{Rd} &= 1 && \text{for } q \le 3 \\
\gamma_{Rd} &= 1.2 && \text{for } q \ge 3
\end{aligned}
\tag{10.4}
$$

- $E_{F,G}$ is the action effect due to the non-seismic actions included in the seismic combinations (e.g. gravity loads).
- $E_{F,E}$ is the action effect from the analysis of the design seismic action.
- Ω is the value of $R_{di}/E_{di} \le q$ of the dissipative zone or element of the structure.
- R_{di} is the design resistance of the zone or element i.
- E_{di} is the design value of the action effect on the zone or element i in the seismic design situation.

b. For foundations of independent walls or columns,

$$
\Omega = \min \text{ of } \left\{ \frac{M_{Rd,x}}{M_{Ed,x}}, \frac{M_{Rd,y}}{M_{Ed,y}} \right\}
\tag{10.5}
$$

in the two orthogonal principle directions in the seismic design situation.

c. For common foundations of more than one vertical element (e.g. foundation beams, rafts, etc.), Ω is derived from the vertical element with the largest horizontal shear force for the design seismic situation, or, alternatively, a value for $\gamma_{Rd}\Omega$ is introduced in Equation 10.3 equal to

$$
\gamma_{Rd}\Omega = 1.40
\tag{10.6}
$$

It is apparent that in the case that

$$
\frac{q_{design}}{\gamma_{Rd}\Omega} \le q(DCL),
\tag{10.7a}
$$

which means that

$$
\frac{q_{design}}{1.40} \le 1.50
\tag{10.7b}
$$

or

$$
q_{design} \le 2.10
\tag{10.8}
$$

the design action effects for the seismic design situation, as was also previously noted, may be derived for a q-factor value equal to 1.50 for the foundation members and foundation ground.

d. The design of the above foundation elements, *except for concrete piles*, will be dimensioned and designed following the design rules for DCL buildings that don't have/follow any requirements for ductility or capacity design. In fact, according to what has been presented in the previous paragraph, no energy dissipation is expected in these elements.

e. The same also holds for the verification and dimensioning of the foundation ground. This is carried out under the assumption that the ground under the capacity design effects of item (1) (Equation 10.3) *responds linearly*.

2. *Dissipative superstructure and dissipative foundation elements–non-dissipative foundation ground.*

a. In the case that the alternative solution of a complete dissipative structural system (DCM or DCH) is decided on by the designer, as far as the superstructure and *the foundation elements* are concerned, the capacity coefficient $\gamma_{Rd}\Omega$ is taken in Equation 10.3 to be equal to

$$\boxed{\gamma_{Rd}\Omega = 1.0},\qquad\qquad (10.9)$$

which means that the design action effects of the foundation elements are derived on the basis of the analysis of the seismic design situation. In this case, the design and dimensioning of the foundation R/C members is carried out according to the ductility and capacity rules that have been followed for the superstructure.

b. Action effects for dimensioning and design of *foundation ground* must be increased to values given by Equations 10.4 through 10.6, since ground must be kept in linear range for these action effects so that dissipative mechanisms and, therefore, inelastic deformations are not derived in the foundation ground.

3. *Non-dissipative superstructure–non-dissipative foundation elements and foundation ground.*

This case refers to DCL buildings where the q-factor is assumed to be equal to 1.5 for an R/C structural system. For these buildings, the action effects, as previously mentioned (Section 10.3.2.1), will result from the analysis in the seismic design combinations without any capacity design consideration. In this case, the whole structural system will be dimensioned and designed with the procedure followed for DCL structural systems, while the ground should be considered also to respond elastically.

4. *Box-type basements of dissipative structures*

As noted in Chapter 4.2.12 on the conceptual design of R/C buildings to seismic actions, this type of basement comprises (Figure 10.4)

a. A concrete slab at the roof of the basement

b. A foundation slab (raft foundation) or a grillage of tie beams connecting foundation pads or pile caps or foundation beams at the foundation level

c. Perimetric and eventually interior walls, designed in the superstructure above the basement roof as dissipative structures

This box-type basement constitutes a very robust and stiff foundation system that minimises

- Differential settlements
- The risk of overturning failure, since the integral foundation acts as a box-like structure
- Sliding risk due to the additional activation of passive resistance of the surrounding ground in the case that a small slide is tolerated

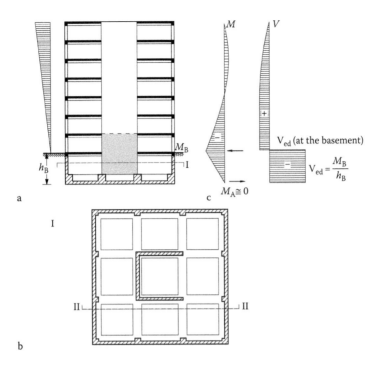

Figure 10.4 Box-type basement and its structural behaviour in relation to the superstructure (dual system): (a) elevation; rastered area is considered to be the critical region for ductile behaviour; (b) plan view at basement; (c) M_{Ed}, V_{Ed} diagrams of the structural wall core.

However, special care should be taken of the uplift risk in the case that the level of underground water is expected to be above the contact surface of the foundation slab and ground in case of a watertight basin.

In this system, the *columns* and *beams* (including those of the basement roof) are expected to remain elastic under the seismic design situation. Therefore, they may be designed in the framework of the box as DCL structures without special measures for ductile behaviour and capacity design procedure. Conversely, *structural walls* should be designed for plastic hinge development at the level of the basement roof slab with a critical region also extending downwards in the basement (Figure 10.4a).

Moreover, the full free height of these walls within the basement should be dimensioned for shear, assuming that the wall develops its flexural strength $\gamma_{Rd}M_{Rd}$ (with $\gamma_{Rd} = 1.1$ for DCM and $\gamma_{Rd} = 1.2$ for DCH) at the basement roof level and zero moment at the foundation level.

In conclusion, it should be noted that the analysis of the box-type basement must be carried out for seismic loadings magnified by $\gamma_{Rd}\Omega$ (Equation 10.6) so that its elastic behaviour is ensured.

10.4 ANALYSIS AND DESIGN OF FOUNDATION GROUND UNDER THE DESIGN ACTION EFFECTS

10.4.1 General requirements

The design action effects due to gravity loads at service and to seismic combination must be transferred to the ground without substantial permanent settlements (SLS); in other words, they must be compatible with the functionality requirements of the superstructure.

At the same time, the foundation ground must be ensured against *failure* caused by the basic combination of gravity loads introduced with the codified partial safety coefficients as well as the 'seismic combination' (ULS).

It should be noted that in addition to the ground resistance at the base of the foundation, friction shear developing between the side of the foundation body and the ground in the case that the foundation body is embedded in the ground, could be taken into account.

10.4.2　Transfer of action effects to the ground

For transferring the design action effects, that is, *shear forces*, on one hand, and *normal forces combined with bending moments* on the other, to the ground, the following mechanisms may be taken into consideration. For piles, additional mechanisms are taken into account, which will be presented in an independent paragraph.

10.4.2.1　Horizontal forces

The design horizontal shear effect V_{Ed} is transferred by the following mechanisms:

1. By means of a design shear resistance F_{Rd} developed between the horizontal base of the footing of the foundation slab of foundation raft and the ground (Figure 10.5a,b)
2. By means of the design lateral shear resistance E_{pd} arising from earth pressure on the vertical sides of the foundation parallel to V_{Ed} (Figure 10.5a,b)
3. By means of the design resisting earth pressure on the other sides of the foundation transversally to V_{Ed} (passive earth resistance minus earth pressures at rest)

It should be noted that a combination of the shear resistance with up to 30% of the resistance arising from fully developed passive earth pressure is allowed according to EC8-5/2004.

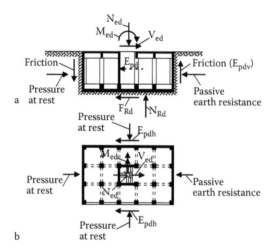

Figure 10.5 Mechanisms for transferring design action effects V_{Ed}, M_{Ed}, N_{Ed} to the ground: (a) elevation; (b) plan view.

10.4.2.2 Normal force and bending moment

Normal force N_{Ed} and bending moment M_{Ed} may be transferred to the ground by means of one or a combination of the following mechanisms (Figure 10.5a,b):

1. By the design value of resisting vertical forces acting at the base of the foundation (Figure 10.5a,b; N_{Rd})
2. By the design values of bending moments developed due to the design horizontal shear resistance between the sides of deep foundations like box-type basements and the ground (Figure 10.5a,b; E_{pdh})
3. By the design value of vertical shear resistance between the sides of embedded deep foundation elements (box-type basements) and the ground (Figure 10.5a,b; E_{pdv})

The last two contributions must be ensured with special measures relating to the formation of the backfill of the excavation, so that the development of shear resistance between backfill and the sides of the foundation may be ensured. The above last two contributions are taken into account only in special cases, like those of very high buildings with large aspect ratios h_w/l_w of total height h_w to horizontal length l_w, which cannot be counterbalanced easily by the reactions between the foundation base and the ground.

10.4.3 Verification and dimensioning of foundation ground at ULS of shallow or embedded foundations

10.4.3.1 Footings

Footings at ultimate limit state are verified and dimensioned at the level of the ground foundation against sliding and bearing capacity failure.

10.4.3.1.1 Failure by sliding

1. The design shear resistance F_{Rd} between the horizontal base of the footing or of the foundation slab and the ground above the water table may be calculated according to EC8-5/2004 from the following expression:

$$F_{Rd} = N_{Ed} \frac{\tan \delta}{\gamma_m} \qquad (10.10)$$

   where
   - N_{Ed} is the design normal force on the horizontal base.
   - δ is the structure–ground interface friction angle at the base of the footing.
   - γ_m is the partial factor for material property equal to $y'_\phi = 1.25$ (Equation 10.1).

   For footings below the water table, the design shear resistance is evaluated on the basis of undrained strength.
2. The design lateral resistance E_{pd} arising from earth pressure on the sides of the foundation and the ground (Figure 10.4a,b) E_{pd} may be calculated by the following expression:

$$E_{pd} = E_d \frac{\tan \delta}{\gamma_m} \qquad (10.11)$$

where

E_d is the earth pressure at rest soil state without static and hydrodynamic water force (see EC8-5/2004 ANNEX E).

It is apparent that appropriate measures must be taken on site, such as the compaction of backfill against the sides of the footing so that E_d develops.

From what has been presented above, it may be concluded that the safety verification criterion for sliding may be formed as follows:

$$V_{Ed} \le F_{Rd} + E_{pdh} \tag{10.12}$$

The mobilisation of sliding shear between foundation and ground supposes the development of a limited amount of sliding. This amount is tolerated provided those special measures are taken for the performance of lifelines connected with the building.

10.4.3.1.2 Bearing capacity of ground to failure

The bearing capacity verifications of the foundation ground under N_{Ed}, M_{Ed} and V_{Ed} for footings, raft foundations or a grillage of foundation beams are carried out by methods that have been developed for many years in soil mechanics (Hansen and Brinch, 1961; Meyerhof, 1965; Anagnostopoulos et al., 1994; Terzaghi et al., 1996). EC8-5/2004, in its Annex F, gives expressions for this verification, the detailed presentation of which exceeds the scope of this book.

It should be noted that N_{Ed}, M_{Ed} and V_{Ed}, introduced to the bearing capacity verifications of ground, are either the original design values derived as reactions of the foundation from the superstructure properly modified to take into account the capacity design procedure, or reduced values derived after the subtraction of the soil shear friction resistance at the sides of the foundation in the case that these restoring forces are taken into consideration. The same applies for the weight of removed soils in case of deep excavations.

10.4.3.2 Design effects on foundation horizontal connections between vertical structural elements

10.4.3.2.1 General

As already noted many times, analysis reasons and good performance to seismic action require the provision of horizontal connections at the base of vertical members (columns–walls) so that relative horizontal displacements at their bases are excluded.

The above conceptual requirement is fulfilled if the foundation ground for all vertical members is positioned at the same horizontal level, and a tie-beam grillage or a grillage of foundation beams or an adequate foundation slab is provided at the level of footings or pile caps (Figure 10.6). In this respect, and keeping in mind that from the analysis no action effects are derived at these ties, since the base vibrates in two main directions during the seismic action as a solid diaphragm, it is apparent that these design action effects must be defined explicitly by the Code.

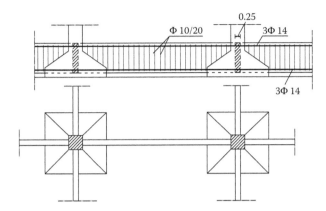

Figure 10.6 Independent footings with a tie-beam grillage.

10.4.3.2.2 Tie beams

According to EC8-5/2004, tie beams or a connecting slab may be omitted only in the following cases:

- Foundation ground Type A
- In low seismicity cases, for foundation ground Type B

In all other cases, the tie beams are designed to resist an axial force with reversed sign equal to

$\circ$For ground type B: $\pm 0.3 \cdot \alpha_g \cdot S \cdot N_{Ed}$

$\circ$ For ground type C: $\pm 0.4 \cdot \alpha_g \cdot S \cdot N_{Ed}$ (10.13a–c)

$\circ$ For ground type D: $\pm 0.6 \cdot \alpha_g \cdot S \cdot N_{Ed}$

where

N_{Ed} is the mean value of the design axial forces of the connected vertical members in the seismic design situation (see Chapter 4.2.12, Figure 4.17).

Values a_g and S have already been defined in Chapter 3.4.3.4.

10.4.3.2.3 Foundation slab

Tie zones should be designed to resist axial tensile or compressive forces equal to those of tie beams (Equation 10.13a–c). The width of these zones may be estimated at *ten times* their thickness.

10.4.3.3 Raft foundations

A raft foundation for the sliding or bearing capacity failure verification may be considered a solid footing loaded by V_{Edi}, N_{Edi}, M_{Edi} of all vertical members acting at their connection to the raft foundation. The same holds in case of a grillage of foundation beams.

It is apparent that, in this case, the slab of the raft foundation or the foundation beams also act as tie beams, and in this respect must be dimensioned additionally for axial load effects equal to those given by Equation 10.13a–c. The width of the connecting zone may be taken as *ten times* its thickness.

10.4.3.4 Box-type foundations

This foundation type is considered a solid body (footing) for the verification and dimensioning of the foundation ground. It must be remembered that seismic loading should be magnified by $\gamma_{Rd}\Omega$ (Equation 10.6), so that its elastic behaviour may be ensured. It should also be noted that this type of foundation is the most proper for shear friction forces developing on the sides of the box and resulting from the earth pressure to be taken into account, so that the load effects at the horizontal base may be significantly reduced.

10.4.4 Settlements of foundation ground of shallow or embedded foundations at SLS

10.4.4.1 General

It should be remembered that the analysis and design of foundation ground is carried out under the assumption that it remains in the elastic range. It is beyond the scope of this book to undertake a rigorous treatment of methods for estimating the expected settlements and rotations of the foundation ground under loading effects. However, an overview of this issue will be necessary, especially for the analysis and design of the structural members of the foundation.

10.4.4.2 Footings

Under the assumption that an orthogonal footing is a rigid body based on a homogeneous ground, the expected settlement and rotation are given by expressions that are functions of the shear modulus G, the Poisson's ratio v and the dimensions of the foundations. These expressions are based on the 'half-space' Boussinesq theory properly modified to comply with experimental results. So, for example, according to Whitman and Richart (1967),

$$\Delta_z = \frac{N_{Ed}}{K_z} = N_{Ed}\left(\frac{G}{1-v}\beta_z\sqrt{BL}\right)^{-1} \tag{10.14}$$

$$\Delta_x = \frac{V_{Ed,x}}{K_x} = N_{Ed,x}\left(2G(1+v)\beta_x\sqrt{BL}\right)^{-1} \tag{10.15}$$

$$\Delta_{\varphi y} = \frac{N_{Ed,y}}{K_{\varphi y}} = M_{Ed,y}\left(\frac{G\beta_\varphi BL^2}{1-v}\right)^{-1} \tag{10.16}$$

where
  - B, L are the dimensions of the footing.
  - β_z, β_x, β_φ are the factors related to the dimensions B and L of the footing and given by the diagram depicted in Figure 10.7b.
  - G, v are the modulus of shear and the Poisson's ratio, respectively.
  - K_z, K_x, $K_{\varphi y}$ are the spring constants (Figure 10.7a).
  - N_{Ed}, $V_{Ed,x}$, $M_{Ed,y}$ are the design load effects on the footing.

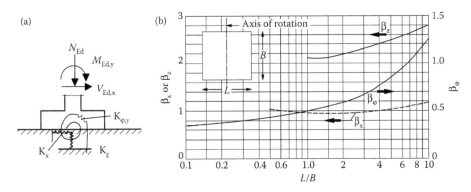

Figure 10.7 Load effects on a footing and stiffness simulation of the ground by elastic springs: (a) structural model; (b) coefficients β_x, β_z and β_φ, for estimating spring stiffness of rectangular footings. (Dowrick, D.: *Earthquake Risk Reduction*. 2005. Copyright Wiley-VCH Verlag GmbH & Co. KGaA. Reproduced with permission.)

Spring constants K_z, K_x, $K_{\varphi y}$ may be introduced as restraints in the analysis of the super-structure and the foundation, considered as an integrated system. Representative values of G for seismic conditions are given in Table 10.1. In the case that successive strata of soil underlying the foundation base have different mechanical properties, determined by lab tests, the simplest and most modern way to determine K_z, K_x, $K_{\varphi y}$ is to use a proper computer platform (e.g. PLAXIS-2D, 2012) that allows the determination of the settlement Δ_z, the sliding Δ_x and the rotation Δ_φ of the footing for a given group of loading effects N_{Ed}, M_{Ed}, V_{Ed} and then to generate the ratios N_{Ed}/Δ_z, $V_{Ed,x}/\Delta_x$, $M_{Ed,y}/\Delta_{\varphi y}$, which represent K_z, K_x, $K_{\varphi y}$. It should be noted that for many decades, this procedure was carried out through extended tiresome manual calculations (Anagnostopoulos et al., 1994; Terzaghi et al., 1996).

10.4.4.3 Foundation beams and rafts

For the determination of the mean value of settlements Δ_z of the ground below a foundation beam or a raft, equations in effect for footings may be used (Equation 10.14). Keeping in mind that the lab tests output result is usually E_s instead of G_s, Equation 10.14 takes the form

$$\Delta_z = \frac{N_{Ed}}{K_z} = N_{Ed} \left(\frac{E_s}{2(1-v^2)} \beta_z B \sqrt{\frac{L}{B}} \right)^{-1} \tag{10.17}$$

The above expression is used as a tool for the determination of *sub-grade modulus* or *Winkler constant*, which is necessary for the determination of the distribution of the soil reaction in the form of stresses on the foundation beam, so that internal forces M_{Ed} and V_{Ed} of the foundation beam may be determined (Figure 10.8).

It is well known that according to the Winkler approach, which is the most common assumption for the analysis of foundation beams or rafts, foundation ground is simulated with independent springs (Figure 10.9), the settlement Δ_z of which is given by the expression

$$q_o = \overline{\kappa}_s \Delta_z \tag{10.18}$$

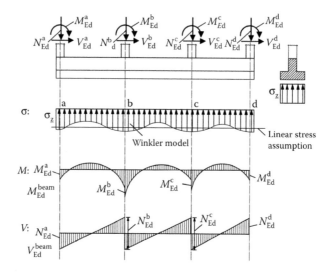

Figure 10.8 Ground reactions σ_z on a foundation beam and internal forces diagram.

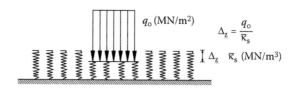

Figure 10.9 Ground simulation according to the Winkler model.

where
- q_o is the compression of the foundation ground (expressed in MPa).
- Δ_z is the settlement (in m) of the point where q_o is induced.
- $\overline{\kappa}_s$ is the *sub-grade modulus* (in MN/m³).

Using Equation 10.17 and taking into account that

$$N_{Ed} = q_o LB \tag{10.19}$$

it follows that

$$\Delta_z = q_o LB \left(\frac{E_s}{2(1-v^2)} \beta_z B \sqrt{\frac{L}{B}} \right)^{-1} \tag{10.20}$$

Therefore,

$$\boxed{\overline{\kappa}_s = \frac{E_s}{(1-v^2)B} \rho} \tag{10.21}$$

Table 10.3 ρ Values for various L/B ratios

L/B	1.0	1.50	2.0	3.0	5.0	10.0	20.0	30.0
ρ	1.05	0.86	0.78	0.66	0.56	0.44	0.39	0.33

where

$$\rho = \frac{0.5\beta_z}{\sqrt{L/B}} \qquad (10.22)$$

For various ratios of (L/B), ρ takes the values given in Table 10.3. A very common expression for $\bar{\kappa}_s$ is given below (Anagnostopoulos et al., 1994):

$$\boxed{\bar{\kappa}_s = \frac{E_s}{B}} \qquad (10.23)$$

where
   B is the shorter dimension of the orthogonal foundation base.

10.4.5 Bearing capacity and deformations of foundation ground in the case of a pile foundation

10.4.5.1 General

Piles and piers are usually designed to resist the following two types of action effects:

1. *Inertia forces* due to seismic action from the superstructure combined with gravity loads (N_{Ed}, V_{Ed}, M_{Ed}).
2. *Kinematic forces* caused by soil deformations during the passage of the seismic waves through the region of the foundation. This second type of forces is taken into account only in the case that all conditions mentioned below are in effect simultaneously:
   a. The ground profile is of type D, S_1 and S_2, including soil strata of sharply different stiffness.
   b. The building's zone is not of low seismicity ($a_g S$ is higher than 0.10 g).
   c. The building is of importance class III or IV.

   In this respect, this second type of forces may be considered an unusual load case that is taken into consideration in special cases. Therefore, in this short overview of pile foundations, it will not be analysed. The ultimate load resistance of the ground to vertical and horizontal loading is verified according to rules included in EC7-1/2004, which refers to foundation issues. Although pile foundations are beyond the scope of this book, a short overview will be given about the way in which the pile restraints are introduced to the model for the analysis of the foundation and the superstructure.

The design resistance of an R/C pile at the ULS is defined by the load capacity of the soil around the pile and by the load capacity of the body of the pile. Therefore, the action effects on the top of the pile must fulfil the following two groups of inequalities:

$$
\{N_{Ed}, M_{Ed}, V_{Ed}\} \leq \left\{ \begin{array}{lll} N_{Rd}^s, & M_{Rd}^s, & V_{Rd}^s \\ N_{Rd}^p, & M_{Rd}^p, & V_{Rd}^p \end{array} \right. \quad \begin{array}{l} 10.24a \\ 10.24b \end{array} \qquad (10.24)
$$

where symbol s refers to soil and p to piles.

10.4.5.2 Vertical load resistance and stiffness

The normal force resistance of a pile is determined analytically, and in most cases, there is a Code requirement for these load effects that should be verified by trials *in situ*. Computationally, this resistance includes two components (Figure 10.10):

1. The base resistance

$$
R_{bd} = R_{b\kappa}/\gamma_b \tag{10.25a}
$$

2. The shaft resistance

$$
R_{sd} = R_{s\kappa}/\gamma_s \tag{10.25b}
$$

where γ_b and γ_s are partial safety factors with recommended values given in EC7-1/2004.

For the determination of $R_{b\kappa}$ and $R_{s\kappa}$, various methods have been developed (Anagnostopoulos et al., 1994; Terzaghi et al., 1996; Dowrick, 2005).

Very useful information on this issue is included in DIN 4014/1990 (B.K. 1994), the German Code for bored piles. In this Code, detailed information is given for the elaboration of $R_{ck}-\Delta_{zk}$ diagrams, which relate the settlements of the pile to its base and shaft resistance (Figure 10.11).

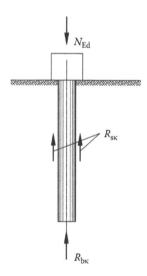

Figure 10.10 Base and shaft reactions $R_{b\kappa}$ and $R_{s\kappa}$ of a pile, respectively.

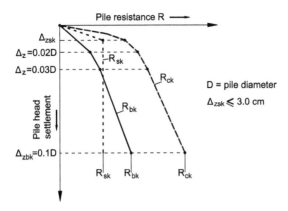

Figure 10.11 R_c–Δ_z curve according to DIN 4014/1990.

For the elaboration of these two diagrams of R_{sk}, R_{bk} and their resultant curve R_{ck}, codified tables are used to relate the soil mechanical properties to the base and shaft resistance of the pile for critical settlements of the pile head given by the Code. Thus, R_{cx} is the result of the sum of R_{bk} and R_{sk} for an ultimate settlement equal to 010D, where D is the pile diameter. Thus,

$$\boxed{R_{cx} = R_{bx} + R_{sx}}$$ (10.26)

It is obvious that after these calculations, partial safety factors must be introduced. It should be noted that this diagram must be verified by tests *in situ*.

The above outlined procedure also allows the determination of the spring stiffness of the pile κ_z^p for the vertical settlements.

In fact, keeping in mind that

$$R_c = \kappa_z^p \Delta_z$$ (10.27)

it follows that κ_z^p may be derived from the R_c–Δ_z diagram using the relation (Figure 10.12)

$$\kappa_z^p = \frac{R_{cd}}{\Delta_{zd}}$$ (10.28)

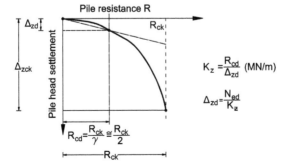

Figure 10.12 Determination of the axial spring stiffness of a pile from an R–Δ_z diagram.

where

 - R_{cd} is the design resistance of the pile.
 - Δ_{zd} is the settlement of the pile corresponding to R_{cd} and taken from the diagram of Figure 10.12 (secant modulus).

The above crude approach to the stiffness of piles in axial direction may be replaced by more refined computational methods using proper computer platforms (e.g. PLAXIS-2D, 2012b).

10.4.5.3 Transverse load resistance and stiffness

For the analysis of the ground surrounding pile and for the analysis and design of the body of the pile, proper models are elaborated for loading of the pile head with a bending moment M_{Ed} and a shear force V_{Ed}. The results of the analysis contain

 - Flexural stiffness of the pile–ground interaction
 - Soil reactions along the pile
 - Pile-to-pile dynamic interaction (dynamic pile group)

To verify that a pile is in position to carry the design transverse load and bending moment with adequate safety according to EC7-1/2004, one of the following failure mechanisms should be considered:

 - For short piles, rotation and displacements as a rigid body (Figure 10.13)
 - For long piles, bending failure of the pile accompanied by local yielding and displacement of the soil near the top of the pile (Figure 10.14)

For the determination of M_{Rd}^s, V_{Rd}^s at the pile head, simplified models have been developed, accompanied by relevant diagrams and tables (Anagnostopoulos et al., 1994; Dowrick, 2005), the presentation of which exceeds the scope of this book.

At the same time, computational methods have been developed allowing the determination of M_{Rd}^s, V_{Rd}^s, the corresponding horizontal displacements Δx, bending rotations $\Delta \varphi$, the horizontal soil reactions on the pile and its internal forces M_{Rd}^p and V_{Rd}^p. The most common method is that of a beam on elastic sub-grade based on the Winkler assumption already presented in the case of foundation beams. Keeping in mind that the Winkler method assumes linear soil behaviour, it is obvious that the reversal of Δx and $\Delta \varphi$ for M_{Ed} = '1' (kN m) and

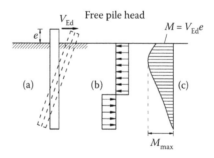

Figure 10.13 Short pile response to V_{Ed} loading in cohesive soil: (a) load pattern – deflections; (b) soil reactions; (c) pile bending moments.

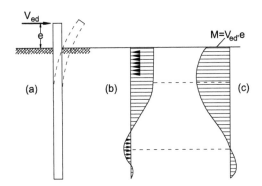

Figure 10.14 Long pile response to V_{Ed} loading in cohesive soil: (a) load pattern – deflections; (b) soil reactions; (c) pile bending moments.

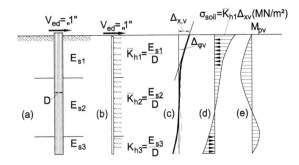

Figure 10.15 Response of a long pile to V_{Ed} = 'I' using the Winkler model: (a) soil layers and pile position; (b) 'Winkler' model; (c) pile-soil stiffness $\bar{K}_{vv} = ($'I'$/\Delta x_v)$, $K_{\varphi v} = ($'I'$/\Delta\varphi_v)$; (d) soil reactions $\sigma_{soil} = \bar{K}_h\Delta x_v$ (MN/m²); (e) pile bending moments.

V_{Ed} 'I' (kN) gives the head stiffness of the pile that may be introduced in the analysis of the foundation of the building (Figure 10.15).

The *horizontal sub-grade modulus* $\bar{\kappa}_{sh}$ (Figure 10.15) along the pile axis for each soil layer may be determined by the following relation (DIN 4014/1990):

$$\bar{\kappa}_h = \frac{E_{sh(z)}}{D} \tag{10.29}$$

where
- E_{sh} is the soil modulus of elasticity with a usual value equal to 50% of E_s in the vertical direction.
- D is the pile diameter.

According to DIN4014/90, the failure criterion of soil for the determination of M_{Rd}^s and V_{Rd}^s is defined by the condition that the soil reactions to the pile should not exceed the passive strength of the soil, that is,

$$\boxed{\sigma_u = \kappa_p \cdot z} \tag{10.30}$$

where
  − σ_u is the ultimate value of soil reaction on the pile:

$$\kappa_p = \tan^2\left(\frac{\varphi}{2} + 45°\right)$$

(10.31)

  − φ is the angle of soil internal friction.

In case of homogeneous soil, the soil modulus of elasticity may vary with the depth z from the soil surface to the pile base with one of the following expressions:

$$E_{s(z)} = E_s z/D \text{ (linear)}$$

(10.32a)

$$E_{s(z)} = E_s \text{ (constant)}$$

(10.32b)

$$E_{s(z)} = E_s \sqrt{z/D} \text{ (parabolic)}$$

(10.32c)

where
  − E_s is the modulus of elasticity of soil at a depth of D.
  − D is the pile diameter.
  − z is the depth of reference along the pile.

In this case, the *pile head stiffness* $K_{H,H}$, K_{MM} and $K_{HM} = K_{MH}$ may be found in a closed form in EC8-5/2005 Annex C.

It should be noted that in the above calculations for the determination of the bearing capacity of the foundation ground around the pile and the horizontal displacements and rotations of the pile head, the side resistance of the soil layers that are susceptible to liquefaction or to substantial strength degradation should be ignored.

10.5 ANALYSIS AND DESIGN OF FOUNDATION MEMBERS UNDER THE DESIGN ACTION EFFECTS

10.5.1 Analysis

10.5.1.1 Separated analysis of superstructure and foundation

Until recently and even nowadays, the most usual method for the analysis and design of foundations has been the following:

1. The superstructure is analysed and designed under the assumption of fixed-end columns and walls at the level of foundation.
2. The action effects (N_{Ed}, M_{Ed}, V_{Ed}) of the fixed ends of vertical members are used as loads for the foundation. For these loads, ULS and SLS verification is carried out:
   a. For the foundation ground
   b. For the structural members of the foundation

EC8-1/2004 permits the above procedure for modelling of the superstructure. However, in the case that the deformability of the foundation has an adverse overall influence on the structural response, the participation of the ground deformability in the formation of a model for dynamic analysis is obligatory.

For many decades, the basic assumption for the stress distribution of the ground reaction on the foundation members was the concept of *planar stress distribution*. It should be noted that this stress distribution is necessary for the determination of the internal forces (M_{Ed}, V_{Ed}) of foundation members so that their dimensioning and design are possible. It is apparent that using the information presented in Sections 10.4.3 and 10.4.4, the verification and design at ULS and SLS may be carried out for the ground and the foundation members.

After the introduction of computer techniques in analysis and design in the early 1950s, a simplified system of linear reaction springs was introduced for soil simulation, the well-known Winkler model, which has been discussed in detail in the preceding subsections.

In this respect, a more reliable model of soil reaction distribution has been introduced, which allows a more reliable determination of the internal forces of the foundation members and of the developing differential settlements than the 'planar stress distribution'.

The *sub-grade modulus* of the foundation ground, and the pile head stiffness or footing stiffness, are calculated separately using the procedures presented in Sections 10.4.4 and 10.4.5 and are introduced as restraints at the surface of the foundation. It should be noted that soil springs must have *unilateral compressive* reactions. In this respect, in the case that the analysis output includes tensile reactions in various positions, an interactive procedure should be adapted. This procedure is carried out manually from step to step or automatically if a proper computer platform is available. In any case, at the successive steps, the springs of tensile reactions are deactivated before the next step iteration is carried out.

In foundation modelling, the following cases may be met with:

1. *Footing tied with tie beams or a slab*
   In this case, the foundation is modelled as a linear space structural system with off-sets at the joints of the vertical elements to simulate the footing. In this system, spring constants K_z, K_x, K_y, $K_{\phi x}$, $K_{\phi y}$ are introduced at the joints, simulating the ground stiffness (see Section 10.4.4.2).
2. *Foundation beams or rafts*
   Foundation beams are usually simulated as foundation straps using the FEM procedure. They are supported on the ground, which is simulated by springs with a sub-grade modulus defined according to what has been presented in Section 10.4.4.3 (Equations 10.21 and 10.23). These straps are connected at the joints of the mesh with the beams of the foundation. It should be noted that the straps and the beam are considered to be interconnected at the centre line of the beam so that the developing bending moment of the beam is not divided in one part as bending moment and a second one as two equal internal axial forces with opposite sign, one at the centre of the beam and the other in the strap.
   Depending on the available computing platform, the *sub-grade modulus* $\bar{\kappa}_s$ is introduced either as the sub-grade modulus of a continuous support system (see Section 10.4.4.3) or as a concentrated spring at the joints of the FEM mesh. In this case, the concentrated spring constant $K_{(z)}$ is introduced with a value equal to

$$\boxed{K_{(z)} = \bar{\kappa}_s \cdot a \cdot b} \tag{10.33}$$

where

    a, b are the dimensions of the mesh of finite elements.

The same procedure is followed in the case of a raft foundation.

3. *Pile foundation tied with tie beams or a slab*

In the case of piled foundations, their structural system is simulated as a linear space structure with offsets at the joints of the vertical elements to simulate the pile cup. In this system, spring constraints K_{MM}, $K_{MH} = K_{HM}$, K_{HH} are introduced at the joints, simulating the interacting stiffness of ground and piles (see Sections 10.4.5.2 and 10.4.5.3) in the form of pile-head stiffness. In the case of groups of piles, it is advisable to replace the stiffness of each pile group by an integrated stiffness (EC7-1/2004) for each pile group.

10.5.1.2 Integrated analysis of superstructure and foundation (soil–structure interaction)

In the past decade, the rapid development of computer techniques has allowed the unification of the analysis and design of the superstructure and foundation of a building in an integrated model, while the foundation ground is simulated with elastic springs in the form they have been presented in the previous paragraph. In this respect, an approximate reliable simulation of the soil–structure interaction has been accomplished, since the influence of the differential settlements of the foundation at the joints with the vertical members of the superstructure is taken into account in the analysis and design of the integrated system 'subgrade foundation–superstructure'. At the same time, the integrated approach of the whole system has made the elaboration of input data simpler, since there is no need for using the output of the superstructure analysis as input for the foundation. EC8-1/2004 requires that the soil–structure interaction be taken into account in the case that this parameter leads to adverse results for the structural response. On the other hand, Code *allows* this procedure, even if this approach leads to beneficial results of the structural response.

In conclusion, it should be noted that in the procedure presented thus far, the influence of soil damping has not been taken into account, nor have the ground inertial forces. In fact, in the implementation of the modal response spectrum analysis, which is according to EC8-1/2004, the 'reference method', it is not possible to introduce different damping values for the various members of the structure since a common design response spectrum is used. Therefore, if the soil damping was going to be taken into account, a composite new viscous damping ratio $\bar{\zeta}$ should be introduced in the calculation of the design spectrum ordinates instead of the viscous damping ratio ζ of the structure (see Chapter 3.4.3). This procedure has been elaborated by Veletsos and Nair (1975) and Veletsos and Meek (1974) and is presented in detail by Dowrick (2005).

In any case, for the majority of common building structures, the effects of soil–structure interaction tend to be beneficial. Only in special cases of buildings, explicitly defined in EC 8-5/2004, the soil–structure interactions derive adverse results in structural response. Some of these categories of buildings are

1. Slender tall structures such as towers and chimneys
2. Structures supported on very soft soil
3. Structures susceptible to P–δ effects

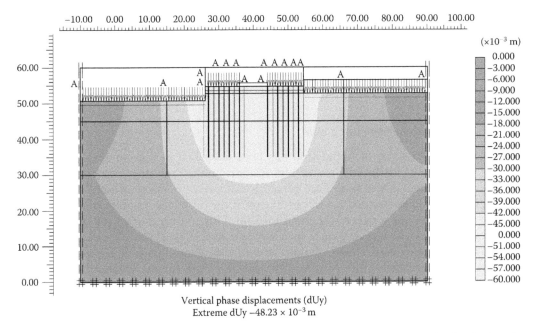

Vertical phase displacements (dUy)
Extreme dUy −48.23 × 10⁻³ m

Figure 10.16 Vertical displacements of a raft foundation combined with a network of piles for the foundation of a nine-storey building in Thessaloniki, Greece. (From Penelis, S.A. 2011. Structural and seismic design of a nine storey building with one basement in Vasileos Irakliou Str. Nr. 45 (Domotechniki real estate) Thessaloniki, Greece, documents for permit issue. Penelis S.A. archives.)

10.5.1.3 Integrated analysis of superstructure foundation and foundation soil

Recently, the introduction of the FEM procedure for soil simulation has allowed a more reliable approach to the problem in elastic or inelastic range. In this procedure, damping and vibrating masses (inertial forces) of the ground are introduced in the model. Of course, in this case, time–history dynamic analysis must be implemented. It should also be noted that serious problems continue to arise with the boundary conditions of the ground that should be introduced into the model.

In any case, this procedure is rather complicated and is used in special cases such as the retaining of deep excavations or construction of underground works, for example, metro tunnels and so forth, for which special computer platforms have been developed for elastic or inelastic range (e.g. PLAXIS-2D, 2012; Figure 10.16).

10.5.2 Design of foundation members

10.5.2.1 Dissipative superstructure–non-dissipative foundation elements and foundation ground

Design action effects for this design assumption, which is the most common, have been given already in Section 10.3.2.2(1) for all types of foundation members. It should be remembered that the basic concept for this alternative is that for foundations, the internal forces at the base of the vertical members (columns–walls) caused by seismic actions are multiplied by a magnifying factor $\gamma_{Rd}\Omega$ (Section 10.3.2.2) for them to be used as loads for the foundation. Of course, in the case that an integrated system of analysis for the superstructure and foundation has been used, the above-mentioned magnification factor $\gamma_{Rd}\Omega$ will be introduced for an additional group of load case combinations for the foundation design.

10.5.2.1.1 Footing with tie beams or a slab

The modelling of this type of foundation has been described in Section 10.5.1.1(1). The reactions N'_{Ed}, M'_{Ed}, V'_{Ed} of the stiffness springs result from the analysis of this system for the corresponding loading. These values are generally reduced in relation to input values N_{Ed}, M_{Ed}, V_{Ed} from the superstructure, as the tie beams or the foundation slab participates with their stiffness in carrying a part of the input loading. This contribution of the tie beams or foundation slab is usually ignored, since such an approach lies on the safety side. Of course, in the case of an integrated model of superstructure–foundation–ground, this reduction cannot be seen immediately since N_{Ed}, M_{Ed}, V_{Ed} are not loads for the foundation, but load effects at the base of columns and structural walls.

The verification of *foundation ground* and the *footing* itself is carried out for the combination of N'_{Ed}, M'_{Ed}, V'_{Ed}. This verification includes (Figure 10.17)

- Ground failure under the bending moment M'_{Ed} and axial force N'_{Ed}
- Sliding failure under V'_{Ed}
- Overturning according to EN1990/2003
- Footing failure to bending
- Footing failure to shear

For the determination of the internal forces in the footing, a planar stress distribution of soil reactions is assumed. Generally, for the dimensioning and design of the footings, the rules for the conventional design of R/C structures (EC2-1/2004) are implemented.

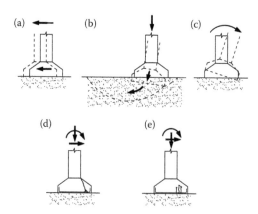

Figure 10.17 Modes of failure in pad foundations: (a) sliding failure; (b) soil-bearing capacity failure; (c) overturning; (d) shear failure in footing; (e) bending failure in footing. (Adapted from Booth, E. and Key, D. 2006. *Earthquake Design Practice for Buildings*. Thomas Telford Ltd.)

Tie beams must have minimum cross-sectional dimensions included in the National Annex of each EU member state. In EC8-1/2004, the following values are recommended:

$$b_{w\,min} = 0.25\,m \qquad\qquad\qquad (10.34a)$$

$$h_{w\,min} = 0.40\,m \qquad\qquad\qquad (10.34b)$$

for buildings up to three storeys, or $h_{w\,min} = 0.50$ m for those with four storeys and more above the basement. It is recommended by EC8-1/2004 that tie beams have along their length a longitudinal reinforcement of at least ρ_{min} both at the top and the bottom

$$\rho_{min} = 0.4\% \qquad\qquad\qquad (10.35)$$

of the area of the cross section of the tie beam.

On the other hand, foundation slabs should have a minimum thickness h_t. The recommended value of h_t is

$$h_t = 0.20\,m \qquad\qquad\qquad (10.36)$$

and the recommended longitudinal tie reinforcement should be ρ_{smin} at the top and the bottom, equal to

$$\rho_{smin} = 0.2\% \qquad\qquad\qquad (10.37)$$

Stub columns (Figure 10.18) between the top of a footing and the soffit of tie beams or foundation slabs *must be avoided*. In fact, the creation of a short column between the tie

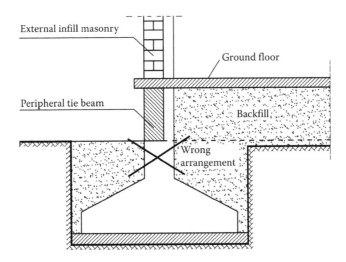

External infill masonry

Ground floor

Peripheral tie beam

Backfill

Wrong arrangement

Figure 10.18 Wrong arrangement of the peripheral tie beam due to the formation of a stub column.

beam or the slab and the footing very often leads to shear explosive cleavage failure (see Chapter 8.3.6).

The minimum axial forces for the design of the tie beams or the slab connecting zone have already been given in Section 10.4.3.2.2 (Equation 10.13a–c). However, it should be noted that tie beams or connecting slab zones should be dimensioned and designed for the combinations of the above axial forces with the bending moments and shear forces that derive from the analysis for the seismic load combinations. It should be noted that these bending moments and shear forces have high values in the case of structural walls where large overturning moments develop. Therefore, in the case of multi-storey buildings with dual structural systems, tie beams should be replaced by grid foundation beams with large cross sections (Figure 4.16, Chapter 4.2.12).

The dimensioning and design of tie beams and foundation slabs are carried out according to EC2-1/2004, as in the case of conventional R/C structures.

10.5.2.1.2 Foundation beams and rafts

These are the most common foundation systems in seismic areas. This type of foundation ensures, on one hand, the mutual connection of the vertical members of the superstructure at their base, and on the other, due to their high stiffness, the uniform distribution of the loads to the ground and the safe transfer of the high-level overturning moments of structural walls to the ground. The modelling of this type of foundation has been presented in Section 10.5.1.1(2). From the analysis of this system, the following output is obtained:

- The distribution of the soil reactions
- The slab moment and shear diagrams
- The beam moment shear and axial forces diagrams

Using these data, the dimensioning in bending and shear is carried out according to EC2-1/2004, that is, *without any extra provision for local ductility*.

10.5.2.1.3 Cast in place concrete piles and pile cups connected with tie beams

The modelling of this type of foundation has been described in Section 10.5.1.1(3). From the analysis of this system for loading described at the beginning of this section, the reactions N'_{Ed}, M'_{Ed}, V'_{Ed} of the stiffness springs simulating the piles result. These values are reduced in relation to input N_{Ed}, M_{Ed}, V_{Ed} from the superstructure, as the tie beams or the foundation slab participates with their stiffness in carrying the input loading.

Next, the interaction of the pile and the foundation ground must be analysed under the reactions N'_{Ed}, M'_{Ed}, V'_{Ed} of the pile head. This analysis will be carried out in accordance with the procedure described in Sections 10.4.5.2 and 10.4.5.3. From this analysis, the following output is obtained:

1. Stress distribution diagram on the surrounding ground (Figure 10.15)
2. Internal force diagrams N_{Ed}, M_{Ed}, V_{Ed} of the pile

From this point on, the safety verification may be easily carried out as follows:

1. *For the ground resistance*
   a. The design axial loading must be less than the ground design resistance (see Section 10.4.5.2)

b. The transverse soil reactions due to M'_{Ed}, V'_{Ed} must be less than the ground design resistance to horizontal loading (see Section 10.4.5.3, Equation 10.30)

2. *For pile dimensioning, EC2-1/2004 will be implemented, as in the case of columns of conventional buildings*

It should be noted that, although according to the design alternative under consideration, the structural members of the foundations and the foundation ground are considered to be in the elastic range, piles under certain conditions according to EC8-5/2004 are allowed to develop plastic hinges.

In this context, according to EC8-1/2004, the top of the pile up to a distance from the underside of the cup of $2d$, where d is the pile diameter, as well as the regions up to a distance of $2d$ on each side of an interface between two soil layers with strongly different shear stiffness (ratio of shear model greater than 6) will be detailed as a potential plastic hinge of a column, with proper confinement of reinforcement in the form of hoops and a minimum longitudinal reinforcement. The design of these critical regions will follow the rules for local ductility of columns at least of DCM (see Chapter 8.3.4(2)). What has been presented for these members in the case of footings also holds true for the tie beams or foundation slabs.

10.5.2.1.4 Joints of vertical elements with foundation beams

The joints of vertical elements (columns or walls) and foundation beams must be designed according to the rules in effect for joints of frame systems and according to the ductility class of the superstructure, although the foundation in this design alternative is considered to remain in non-dissipative condition. It should be remembered that joints at the foundation of DCM buildings are reinforced according to the rules described in Chapter 8.4.3 *without any analytical justification.*

Conversely, joints at the foundation of DCH buildings have to be dimensioned analytically. The shear action effect V_{jhd}, which is introduced for the ULS verifications of joints (see Chapter 8.4.3), is determined on the basis of analysis results that have been derived for the seismic action effects magnified by $\gamma_{Rd}\Omega$.

10.5.2.2 Dissipative superstructure–dissipative foundation elements–elastic foundation ground

Design action effects for this design alternative have been presented in Section 10.3.2.2(2). It should be remembered that the basic concept of this alternative is that superstructure and foundation are designed to dissipate energy, while the foundation ground is considered to remain in the elastic range during the seismic action. In this respect, all structural members of the superstructure and foundation are designed according to capacity design procedures in combination with local ductility requirements. Foundation ground, on the other hand, is designed at ULS for magnified seismic loading by a magnifying factor $\gamma_{Rd}\Omega$ (Section 10.3.2.2).

The structural members of various types of foundation are designed for the above seismic action effects as the members of the superstructure following all code specifications for capacity design and local ductility.

In particular, piles are designed and detailed for potential plastic hinging at the head as in Section 10.5.2.1.3. In this case, the confinement length at critical regions must be increased by 50% in comparison to Section 10.5.2.1.3.

In addition, the ULS verification of the pile to shear must be carried out for shear seismic action magnified by $\gamma_{Rd}\Omega$. Obviously, the piles in this case will be designed for local ductility for the same ductility class as the superstructure.

Table 10.4 An overview of the design procedures for the foundation and the ground

Parts of the Structure	Ductility classes				
Superstructure	DCL	DCM		DCH	
Foundation in general	DCL	DCL	DCM	DCL	DCH
		Seismic load magnification $\gamma_{Rd}\Omega$	$\gamma_{Rd}\Omega=1.0$	Seismic load magnification $\gamma_{Rd}\Omega$	$\gamma_{Rd}\Omega=1.0$
Box-type basement	DCL	DCL*		DCL*	
		Seismic load magnification $\gamma_{Rd}\Omega$		Seismic load magnification $\gamma_{Rd}\Omega$	
Connections	DCL	DCM		DCL	DCH
		Confinement of crit region extended in the connection		Seismic load magnification $\gamma_{Rd}\Omega$	$\gamma_{Rd}\Omega=1.0$
Piles and piers	DCL	DCL	DCM**	DCL	DCH**
		Seismic load magnification $\gamma_{Rd}\Omega$	$\gamma_{Rd}\Omega=1.0$	Seismic load magnification $\gamma_{Rd}\Omega$	$\gamma_{Rd}\Omega=1.0$
Ground	DCL	DCL		DCL	
		Seismic load magnification $\gamma_{Rd}\Omega$		Seismic load magnification $\gamma_{Rd}\Omega$	

* Structural walls in the basement are designed following the rules of dissipative structures (critical region).

** ULS shear verification must use V_{Ed} corresponding to DCL.

Finally, the horizontal shear force V_{jhd} that should be introduced for the ULS verification of joints (see Chapter 8.4.3) in the case of DCH buildings should be determined in accordance with the capacity design procedure (see Chapter 8.4.2.1).

In closing, it should be noted that the ULS verification of the foundation ground will be carried out following the same procedure as that presented in Section 10.5.2.1.

10.5.2.3 Non-dissipative superstructure–non-dissipative foundation elements and foundation ground

As noted in Section 10.3.2.2(3), this case refers to DCL buildings where the q-factor is equal to 1.50. In this case, superstructure, foundation and ground are designed for seismic load effects resulting from the analysis without any measure for capacity design of local ductility.

10.5.2.4 Concluding remarks

As may be seen from the previous presentation of this subsection, various design alternatives are given to the designer by the Code, which are summarised in Table 10.4.

10.6 EXAMPLE: DIMENSIONING OF FOUNDATION BEAMS

The design of the foundation beams B9, B20, B58 and B69 along grid axis C of the foundation shown in Figure 10.19 is presented here. The procedure implemented follows EC8-Part I, paragraphs 4.4.2.6(8) and 5.8.1(5).

The foundation is considered to be of box type, since it consists of the basement slab, the perimeter walls and the foundation beams. The perimeter walls are based on strip footings,

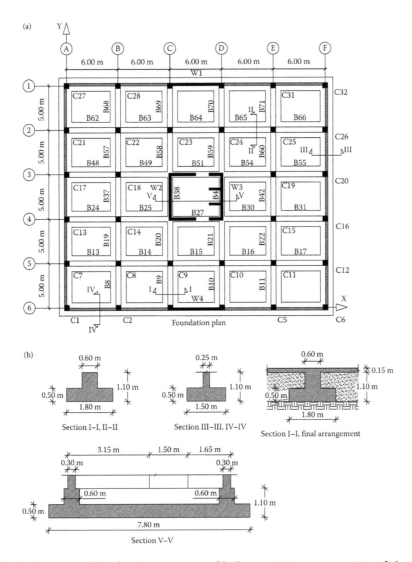

Figure 10.19 (a) Plan view of the foundation system; (b) characteristic cross sections of the foundation beams.

the columns on foundation beams, whereas walls W2 and W3 are based on a mat foundation. Line and shell elements are used to model the web and the flange of the foundation beams, respectively. In order to account for soil deformability, shell elements are considered to be on elastic support by assigning area springs at their vertical direction (Figure 10.20).

The mechanical properties of the soil are modulus of elasticity of soil E_s = 54 GPa; spring constant $K_s = E_s/b = 54{,}000/1.8 = 30{,}000$ kN/m³, where b = 1.8 m is the width of the foundation beam; angle of soil internal friction $\varphi = 42°$; soil weight density $\gamma = 18$ kN/m³; and cohesion intercept $c = 180$ MPa.

The minimum cross-section dimensions for the beams are

- Cross-sectional width: $b_{w,min}$ = 0.25 m (Equation 10.34a)
- Cross-sectional depth: $h_{w,min}$ = 0.50 m for buildings with four storeys or more above the basement (Equation 10.34a)

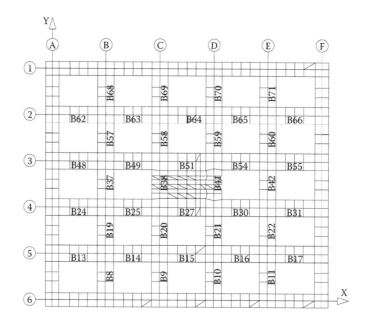

Figure 10.20 Model of the foundation by means of line and shell elements.

The minimum percentage of longitudinal reinforcement is (Equation 10.35)

$$\rho_{min} = 0.4\%_0 \rightarrow A_{s,min} = 0.40\%_0 \cdot 60 \cdot 103.2 = 24.77 \text{ cm}^2$$

Details regarding the geometry of the foundation beams appear in Figure 10.19. The design action effects considered in the analysis for the case of dissipative buildings (DCH for the building studied here) are estimated as follows (Section 10.3.2.2(1), Equation 10.3):

$$E_{Fd} = E_{F,G} + \gamma_{Rd} \cdot \Omega \cdot E_{F,E} = E_{F,G} + 1.40 \cdot E_{F,E}$$

The term $\gamma_{Rd} \Omega$ is equal to 1.40 according to the requirement of EC8-Part I (2004) for common foundations of more than one vertical element. The moment and shear envelopes that appear in Figures 10.21 and 10.22 are extracted after having been multiplied by 1.40.

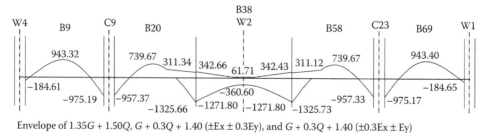

Envelope of 1.35G + 1.50Q, G + 0.3Q + 1.40 (±Ex ± 0.3Ey), and G + 0.3Q + 1.40 (±0.3Ex ± Ey)

Figure 10.21 Moment envelope for the foundation beams along grid axis C.

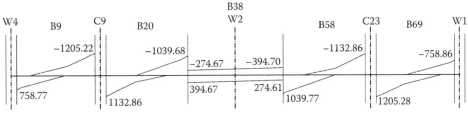

Envelope of $1.35G + 1.50Q$, $G + 0.3Q + 1.40\ (\pm Ex \pm 0.3Ey)$, and $G + 0.3Q + 1.40\ (\pm 0.3Ex \pm Ey)$

Figure 10.22 Shear envelope for the foundation beams along grid axis C.

10.6.1 Ultimate limit state in bending

Owing to the fact that beams B69 and B58 are symmetrical to B9 and B20, detailed design is presented only for beams B9 and B20.

External supports on the boundary element of W4 and on C9 (Beam B9 – left, B69 – right):
Hogging design moment for the bottom reinforcement: $M_d^- = 184.61$ kN m

- Bottom longitudinal reinforcement:

$$\mu_{sd} = \frac{|M_d^-|}{b \cdot d^2 \cdot f_{cd}} = \frac{184.61 \cdot 10^6}{600 \cdot 1032^2 \cdot (25/1.5)} = 0.0173 \rightarrow \omega_{req} = 0.0178 \rightarrow$$

$$A_{s,req} = \omega_{req} \cdot b \cdot d \frac{f_{cd}}{f_{yd}} = 0.0178 \cdot 600 \cdot 1032 \frac{25/1.5}{500/1.15} = 4.21\,\text{cm}^2 < A_{s,min} = 24.77\text{cm}^2 \rightarrow$$

$$\rightarrow 8\varnothing20(25.13\,\text{cm}^2)$$

- Top longitudinal reinforcement: The minimum amount of reinforcement placed at the top of the beam is 8Ø20.

Internal supports on C9 and on C23 (Beam B9 – right, B20 – left, B58 – right, B69 – left):
Hogging design moment for the top reinforcement: $M_d^- = -975.19$ kN m

- Bottom longitudinal reinforcement:

$$\mu_{sd} = 0.0916 \rightarrow \omega_{req} = 0.0979 \rightarrow A_{s,req} = 23.23\ \text{cm}^2 < A_{s,min} = 24.77\text{cm}^2 \rightarrow 8\varnothing20$$

- Top longitudinal reinforcement: The minimum amount of reinforcement placed at the top of the beam is 8Ø20.

Internal supports on the boundary elements of W2 (B20 – right, B58 – left):
Hogging design moment for the top reinforcement: $M_d^- = -1325.66$ kN m
Sagging design moment for the bottom reinforcement: $M_d^+ = 311.34$ kN m

- Bottom longitudinal reinforcement:

$$\mu_{sd} = 0.1245 \rightarrow \omega_{req} = 0.1364 \rightarrow A_{s,req} = 32.37\text{cm}^2 > A_{s,min} \rightarrow 11\varnothing20\ (34.56\ \text{cm}^2)$$

- Top longitudinal reinforcement:

$$\mu_{sd} = 0.0097 \rightarrow \omega_{req} = 0.0102 \rightarrow A_{s,req} = 7.29 \text{ cm}^2 < A_{s,min} = 24.77 \text{ cm}^2 \rightarrow 8\varnothing20$$

Mid-span (Beam B9, B69):
Sagging design moment for the bottom reinforcement: $M_d^+ = 943.32 \text{ kN m}$

- Top longitudinal reinforcement:

$$\mu_{sd} = \frac{|M_d^+|}{b_{eff} \cdot d^2 \cdot f_{cd}} = \frac{943.32 \cdot 10^6}{1800 \cdot 1032^2 (25/1.5)} = 0.0295 \rightarrow \omega_{req} = 0.0310 \rightarrow$$

$$A_{s,req} = \omega_{req} \cdot b_{eff} \cdot d \frac{f_{cd}}{f_{yd}} = 0.0310 \cdot 1800 \cdot 1032 \frac{25/1.5}{500/1.15} = 23.53 \text{ cm}^2 < A_{s,min}$$

$$= 24.77 \text{cm}^2 \rightarrow 8\varnothing20$$

- Bottom longitudinal reinforcement: 8Ø20 were placed according to the minimum requirements.

Mid-span (Beam B20, B58):
Sagging design moment for the bottom reinforcement: $M_d^+ = 739.67 \text{ kN m}$

- Top longitudinal reinforcement:

$$\mu_{sd} = 0.0232 \rightarrow \omega_{req} = 0.0243 \rightarrow A_{s,req} = 18.45 \text{ cm}^2 < A_{s,min} = 24.77 \text{ cm}^2 \rightarrow 8\varnothing20$$

- Bottom longitudinal reinforcement: 8Ø20 were placed according to the minimum requirements.

10.6.2 Ultimate limit state in shear

Shear resistance in case of diagonal concrete crushing for $\delta = 21.8°$ (8.62):
Beam 9 – left:

$$V_{Rd,max} = 1 \cdot 0.6 \cdot 0.9 \cdot 1.032 \cdot 0.54 \cdot (25/1.5) \cdot 1000/2.9 = 1729.49 \text{ kN} > 758.77 \text{ kN}$$

Beam 9 – right:

$$V_{Rd,max} = 1729.49 \text{ kN} > 1205.22 \text{ kN}$$

Beam 20 – left:

$$V_{Rd,max} = 1729.49 \text{ kN} > 1132.86 \text{ kN}$$

Beam 20 – right:

$$V_{Rd,max} = 1729.49\,kN > 1039.68\,kN$$

Transverse reinforcement in the critical regions:
The critical region is $l_{cr} = 1.5 \cdot h_w = 1.5 \cdot 110 = 165$ cm.
The maximum longitudinal spacing should not exceed s_{max}:

$$s_{max} = \min\left\{\frac{h_w}{4};24d_{bw};175;6d_{bL}\right\} = \min\left\{\frac{1100}{4};24\cdot10;175;6\cdot20\right\} = 120\,mm$$

Ø10/120 are the minimum allowed hoops (a bar diameter of 10 mm was considered as the minimum diameter for hoops).
Beam 9– left:

$$V_{wd} = 758.77 \Leftrightarrow s = \frac{A_{sw}}{V_A}0.9d \cdot f_{ywd} = \frac{4\cdot\pi\cdot10^2/4}{758.77}\cdot0.9\cdot1.032\cdot\frac{500}{1.15} = 167.20\,mm$$

Hence, the estimated shear reinforcement in the critical region at the left end of beam 9 is 4Ø10/120.
Beam 9 – right:

$$V_{wd} = 1205.22 \Leftrightarrow s = 105.26\,mm \rightarrow 4Ø10/100$$

Beam 20 – left:

$$V_{wd} = 1132.86 \Leftrightarrow s = 111.98\,mm \rightarrow 4Ø10/100$$

Beam 20 – right:

$$V_{wd} = 1039.68 \Leftrightarrow s = 122.02\,mm \rightarrow 4Ø10/120$$

For simplicity in construction, the same number of stirrups was considered for both beams, that is, 4Ø10/100 (Figure 10.23).
Transverse reinforcement outside the critical regions:
The length outside the critical regions for beam 9 is 1.1 m, whereas for beam 20, it is 1.25 m.
The minimum shear reinforcement ratio is

$$\rho_{w,min} = \frac{0.08\cdot\sqrt{f_{ck}}}{f_{yk}} = \frac{0.08\cdot\sqrt{25}}{500} = 0.8‰$$

The maximum longitudinal spacing between hoops should not exceed s_{max}:

$$s_{max} = 0.75\cdot d = 0.75\cdot1032 = 774\,mm$$

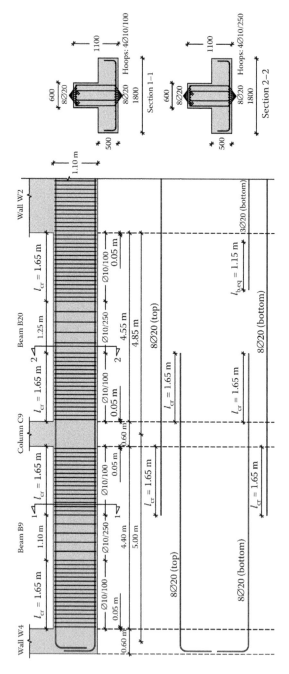

Figure 10.23 Flexural and shear reinforcement of foundation beams 9 and 20.

In the case of four-leg 10 mm bar diameter hoops, stirrup spacing is defined as equal to

$$s = \frac{A_{sw}}{\rho_{w,min} \cdot b_w} = \frac{4 \cdot (\pi \cdot 10^2/4)}{0.0008 \cdot 600} \approx 655\,\text{mm}$$

Ø10/655 are the minimum allowed hoops.
   Beam 9:

$$V_{wd} = 468.72 \Leftrightarrow s = \frac{4 \cdot \pi \cdot 10^2/4}{468.72} \cdot 0.9 \cdot 1.032 \cdot \frac{500}{1.15} = 270.66\,\text{mm}$$

The shear reinforcement outside the critical regions of beam 9 was taken to be 4Ø10/250.
Beam 20:

$$V_{wd} = 345.02 \Leftrightarrow s = \frac{4 \cdot \pi \cdot 10^2/4}{318.16} \cdot 0.9 \cdot 1.032 \cdot \frac{500}{1.15} = 367.71\,\text{mm}$$

4Ø10/250 were also considered for beam 20 for simplicity in construction.

Chapter 11

Seismic pathology

11.1 CLASSIFICATION OF DAMAGE TO R/C STRUCTURAL MEMBERS

11.1.1 Introduction

Seismic pathology of R/C buildings is of major importance for the assessment of their structural condition just after a strong earthquake. In fact, after a strong earthquake, a large number of structural engineers are mobilised for the emergency inspection and assessment of the buildings in the affected area. Therefore, it is necessary that they must be in a position

1. To focus their attention to the most frequent types of damage
2. To assess qualitatively the existing hazard for an eventual aftershock collapse from the damage under inspection and, therefore, to prohibit the use of certain buildings until more detailed evaluation or retrofitting actions take place
3. To proceed with temporary general or local propping of buildings in case the existing hazard of collapse is evident

At the same time, damages to the structural system constitute an excellent benchmark for the quantitative evaluation of the structural post-earthquake condition of a building, and, therefore, a reliable guide for an effective retrofitting.

A strong earthquake puts the whole structure through a hard test. As a result, all the weaknesses of the structure, due to either Code imperfections or analysis and design errors, or even bad construction, are readily apparent. It is not unusual that a strong earthquake leads to improvements or even drastic changes in design Codes and modifications in design methods. Similarly, it triggers liability and responsibility issues for the design and execution of construction projects.

It is difficult to classify the damage caused by an earthquake, and even more difficult to relate it in a quantitative manner to the cause of the damage. This is because the dynamic character of the seismic action and the inelastic response of the structure render questionable every attempt to explain the damages by means of a simplified structural model and elastic analytical procedures. Furthermore, the coincidence of more than one deficiency in the building makes the quantitative evaluation of the influence of each deficiency very difficult and sometimes even impossible.

Despite all the difficulties inherent in a damage classification scheme, an attempt will be made in this chapter to classify the damage into categories and to identify the cause of the damage in each case, according to current concepts of the behaviour of structural elements under cyclic inelastic loading, which sufficiently simulates the response of structural members to a strong earthquake.

In this section, damage classification will refer to individual structural elements (Tegos, 1979), while in the following section, reference will be made to the main causes of damage to R/C buildings as a whole. In both sections, the qualitative analysis will be supplemented with statistical data for the behaviour of structural elements and buildings during strong earthquakes.

The basic source of the statistical data is the research project, 'A statistical evaluation of the damage caused by the earthquake of June 20, 1978, to the buildings of Thessaloniki, Greece', which was carried out at the Laboratory of Reinforced Concrete of Aristotle University of Thessaloniki, in collaboration with the Ministry of Environment and Public Works (Penelis et al., 1987, 1988b), and also the report, 'The Earthquake of 19 September 1985 – Effects in Mexico City', by the Committee for Reconstruction of Mexico City (Rosenblueth and Meli, 1985). In addition, the technical report for the 'Evaluation of damages and their cause for the 103 near-collapse R/C buildings in Athens after the earthquake of September 07, 1999' prepared by an (OASP) committee (Kostikas et al., 2000) will be a valuable source of statistical information.

In the classification that follows, there is no reference to damage due to analysis errors, bad concrete quality, improper reinforcement detailing and so on, since the classification is based mainly on statistics. On the other hand, weaknesses of this type are always present in structures, and their frequency and severity depend on the level of technological development of a country. Of course, these weaknesses contribute to the degree of damage caused by an earthquake, and occasionally they become fatal for the stability of buildings.

Finally, it should be noted that the classification presented below is based mainly on the behaviour of R/C buildings designed and constructed according to a previous generation of Codes, since the main volume of the existing building stock was constructed before the introduction of modern Codes (before 1985). However, there are some first results for the behaviour of buildings designed and constructed recently according to modern Codes that have been affected by recent earthquakes, and, therefore, a first indicative reference will also be made to these cases. In any case, after future strong earthquakes, an extended survey and evaluation must be carried out on the stock of buildings designed according to the modern Code philosophy that will have been stocked in the meantime, so that justified conclusions can be drawn about the effectiveness of modern building Codes.

11.1.2 Damage to columns

Damage to columns caused by an earthquake is mainly of two types:

- Damage due to cyclic flexure but low shear under strong axial compression
- Damage due to cyclic shear but low flexure under strong axial compression

The first type of damage is displayed by failures at the top and bottom of the column (Figures 11.1, 11.2 and 11.3). It occurs in columns of moderate-to-high aspect ratio. For

$$\alpha = \frac{M}{Vh} = \frac{L}{2h} \geq 5.0\text{--}6.0, \tag{11.1}$$

this is the prevailing mode of failure. For

$$5.0\text{--}6.0 \geq \frac{L}{2h} \geq 2.0, \tag{11.2}$$

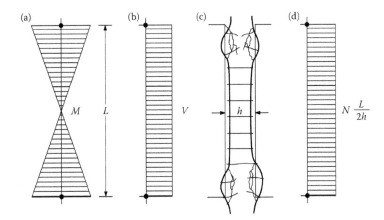

Figure 11.1 Column damage due to strong axial compression and cyclic bending moment: (a) bending moment diagram; (b) shear force diagram; (c) sketch of damage; (d) axial force diagram.

Figure 11.2 Column damage due to high axial compression and cyclic bending moment: Bucharest, Romania (1977).

Figure 11.3 Column damage due to high axial compression and cyclic bending moment: Kalamata, Greece (1986).

mode of failure depends on the degree of shear reinforcement (transverse ties; see Chapter 8, Sections 8.3.3 and 8.3.4). If shear reinforcement is high, the prevailing mode of failure continues to be of flexural type at the ends of the column; otherwise shear-compression failure prevails. It should be noted that, at least in Greece, for buildings designed before 1985, where no special concern for shear reinforcement in columns existed (capacity design to shear), for columns of moderate slenderness, bending failure is combined with shear compression failure at the ends (Figures 11.4d, 11.5 and 11.6).

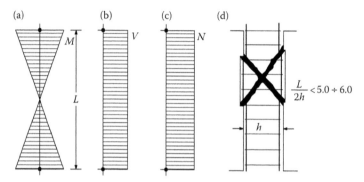

Figure 11.4 Column damage due to strong axial compression and shear: (a) bending diagram; (b) shear force diagram; (c) axial force diagram; (d) sketch damage.

Figure 11.5 Column damage due to high axial compression and cyclic shear: Kalamata, Greece (1986).

Figure 11.6 Column damage due to high axial compression and cyclic shear: Kalamata, Greece (1986).

The high bending moment at the ends of the column, combined with the axial force, leads to the crushing of the compression zone of concrete, successively on both faces of the column. The smaller the number of ties in these areas is, the higher their vulnerability to this type of damage will be. The crushing of the compression zone is displayed first by spalling of the concrete cover to the reinforcement. Subsequently, the concrete core expands and crushes. This phenomenon is usually accompanied by buckling of bars in compression and by hoop fracture. The fracture of the ties and the disintegration of concrete lead to shortening of the column under the action of the axial force. Therefore, this type of damage is very serious, because the column loses not only its stiffness but also its ability to carry vertical loads. As a result, there is a redistribution of stresses in the structure, since the column has shortened due to the disintegration of concrete in the above-mentioned areas.

This type of damage is very common; 23.3% of R/C buildings with damages in their structural systems by the Thessaloniki earthquake of 20 June 1978 displayed damage of this type (Penelis et al., 1987, 1988b). The great majority of failures of buildings with frame systems during the Mexico City earthquake of 19 September 1985 were caused by column damage (Rosenblueth and Meli, 1985). As the main causes for this brittle type of failure, one should consider the low quality of concrete, the inadequate number of ties in the critical areas and the presence of strong beams, which lead to columns failing first, and finally, of course, the strong seismic excitation inducing many loading cycles in the inelastic range.

The second type of damage is that of shear type and is displayed in the form of X-shaped cracks in the weakest zone of the column (Figures 11.4, 11.5 and 11.6). As explained in the previous paragraph, this type of failure occurs in columns of moderate slenderness ratio $L/2h$ between 5.0 or 6.0, and 2.0 in the case that transverse shear reinforcement is inadequate.

The ultimate mode of this type of damage is the explosive cleavage failure of short columns (Figures 11.7, 11.8 and 11.9), for a slenderness ratio

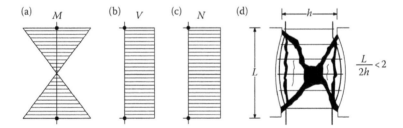

Figure 11.7 Explosive cleavage of a short column: (a) bending moment diagram; (b) shear force diagram; (c) axial force diagram; (d) sketch of damage.

Figure 11.8 Explosive cleavage failure of a short column: Kozani, Greece (1996).

Figure 11.9 Explosive cleavage failure of a short column: Kalamata, Greece (1986).

$$\alpha = \frac{M}{Vh} = \frac{L}{2h} \leq 2.0 \tag{11.3}$$

(see Chapter 8.3.6.1). This type of failure usually leads to a "spectacular" collapse of the building.

Generally speaking, the main reason for this type of damage is that the flexural capacity of columns with moderate to small slenderness ratio is higher than their shear capacity, and as a result shear failure prevails. The frequency of this type of damage is lower than the failure at the top and bottom of the column. It usually occurs in columns on the ground floor, where the slenderness ratio is low because of the large dimensions of the cross section of the columns. It also occurs in short columns, which have either been designed as short or have been reduced to short because of adjacent masonry construction that was not accounted for in the design (Chapter 4.2.5, Figure 4.8).

Finally, sometimes in the case of one-sided masonry-infilled frames, masonry failure is followed by shear failure of the adjacent columns (Figures 11.10 and 11.11; Stylianidis and Sariyiannis, 1992).

From the survey and evaluation of the 103 R/C buildings most affected by the earthquake of 7 September 1999 in Athens (Kostikas et al., 2000), it was found that 27% of the affected buildings presented shear damage at their short columns. It was also ascertained that

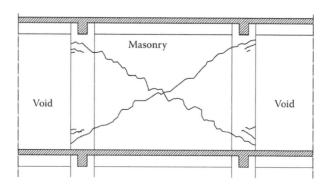

Figure 11.10 Damage in columns in contact with masonry on one side only.

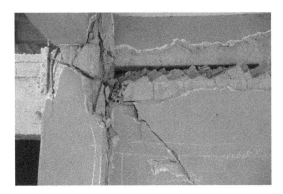

Figure 11.11 Damage of column in contact with masonry on one side only: Kalamata, Greece (1986).

89 out of 103 buildings presented damages in columns due to inadequate transverse hoop and tie reinforcement either in critical regions (confinement reinforcement) or along the column axis (shear reinforcement). Moreover, this was one of the main types of damage near collapse (Figure 11.51) for 55% of the cases under consideration.

In conclusion, it must be noted that column damages are very dangerous for the structure, because they change or even destroy the vertical elements of the structural system. Thus, when damage of this type is detected, emergency measures of temporary support should be provided immediately.

11.1.3 Damage to R/C walls

The damage caused to R/C walls by earthquakes is of the following types:

- X-shaped shear cracks
- Sliding at the construction joint
- Damage of flexural character (horizontal cracks – crushing of the compression zone)

During the Thessaloniki earthquake of 20 June 1978, 28.6% of the R/C buildings that suffered damage to their structural system displayed damage in the R/C walls (Penelis et al., 1987).

The most frequent type of damage is the appearance of cracks at the construction joint (Figures 11.12 and 11.13). Damage of this type occurred in 88% of the buildings with wall damage caused by the Thessaloniki earthquake of 20 June 1978 (Penelis et al., 1987, 1988).

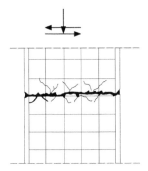

Figure 11.12 Shear wall damage at a construction joint.

Figure 11.13 Shear wall crack at the construction joint at the top of an R/C wall, Kalamata, Greece (1986).

This damage is mainly due to the fact that fresh concrete was not properly bonded with the hardened old one. All seismic codes in effect today require that extra care is taken when construction work is discontinued in order to ensure proper bonding of concrete (rough surface, cleaning, soaking, pouring of strong cement mortar first and then concrete). In addition, placement of connecting reinforcement is also required in the form of dowels. The introduction of these requirements is the result of the high frequency of this type of damage. However, it has to be mentioned that this type of damage does not pose a threat to the stability of the building, because, with the horizontal arrangement of the cracks, the wall can still carry vertical loads. Also, from the stiffness point of view, this type of damage has only a slight effect on the entire structural system.

The appearance of X-shaped cracks in R/C walls is the next most frequent damage (Figures 11.14–11.16). During the above-mentioned Thessaloniki earthquake, the frequency of this damage reached 30% of the buildings with wall damage. This is a shear type of brittle failure. Due to the arrangement of the cracks, the isosceles triangles that form on the two sides, under the action of vertical loads, tend to separate from the structure and therefore cause its collapse (Figure 11.14). In order to protect the structure from this type of failure, all current

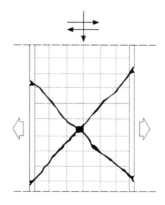

Figure 11.14 Shear wall damage due to shear (X-shaped cracks).

Figure 11.15 Shear wall failure: Kalamata, Greece (1986) failure to shear and simultaneously to bending.

Figure 11.16 Shear wall failure due to shear: Kalamata, Greece (1986).

codes require the formation of a column at each side of the wall that will carry the vertical loads after the shear failure of the web. These columns can either be thicker than the wall and visible (barbells), or they can be incorporated into the wall (Chapter 9.2.4).

Damage of flexural type occurs very rarely (Figures 11.15 and 11.17). It is the author's opinion that this is due to the fact that the bending moments developing at the base of the wall are much smaller than those calculated for the design, because the footing rotates as the soil deforms during the earthquake. On the other hand, this soil deformation does not significantly alter the shear force that is carried by the wall, and, as a result, shear failure

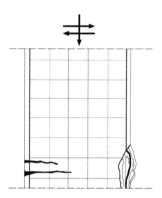

Figure 11.17   Shear wall damage due to flexure and compression.

prevails. It is hoped that for buildings designed according to the new generation of Codes, the ductile mode of failure will prevail.

11.1.4 Damage to beams

The damage that occurs to R/C beams due to an earthquake is as follows:

- Cracks perpendicular to the beam axis along the tension zone of the span
- Shear failure near the joints
- Flexural cracks on the upper or lower face of the beam at the joints
- Shear or flexural failure at the points where secondary beams or cut-off columns are supported by the beam under consideration
- X-shaped shear cracks in short beams that connect shear walls (coupled walls)

Although damage to beams does not jeopardise the safety of the structure, it is the most common type of damage in R/C buildings; 32.6% of the buildings that displayed structural system damages during the Thessaloniki earthquake of 20 June 1978 exhibited some type of beam damage.

Cracks in the tension zone of the span constitute the most common type of damage – 83% of the structures with beam damage in Thessaloniki due to the June 1978 earthquake had damage of this type. This type of damage (Figure 11.18) cannot be explained using analytical evidence, given the fact that the action of the seismic forces does not increase the bending moment in the span. However, the vertical component of the seismic action, due to its cyclic character, simply makes visible the micro cracks that result from bending of the tensile zone under tension, thus creating the impression of earthquake damage. This is the reason why the large majority of the cases of beams with this type of damage do not jeopardise the overall stability of the structure. It is also understood that the high frequency of damage of this

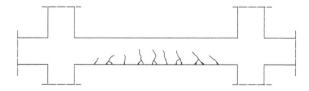

Figure 11.18 Flexural cracks at a beam span.

type is rather misleading, since in most cases it is just a display of already existing normal cracking rather than of earthquake damage.

The bending-shear failure near the joints (Figure 11.19) is the second most frequent type of damage (43%) in beams. Undoubtedly, it constitutes a more serious type of damage than the previous one, given its brittle character. However, only in a very few cases does it jeopardise the overall stability of the structure.

The flexural cracks on the upper and lower face of the beam at the joints (Figure 11.20) can be fully explained if the earthquake phenomenon is statically approximated by horizontal forces. From the frequency point of view, this type of damage is rarer than the shear type (28%). In most cases, cracking of the lower face is due to bad anchorage of the bottom reinforcement into the supports, in which case one or two wide cracks form close to the support.

The shear or flexural failure at the points where secondary beams or cut-off columns are supported (Figure 11.21) appears quite frequently. This is due to the vertical component of the earthquake, which amplifies the concentrated load.

X-shaped shear cracks in short beams coupling shear walls also appear quite often. This is a shear failure similar to that which occurs in short columns (Figure 11.22) but not dangerous for the stability of the building (see also Chapter 9.3.3).

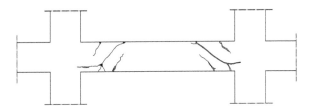

Figure 11.19 Bending-shear cracks near the joints of a beam.

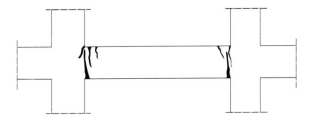

Figure 11.20 Flexural cracks at the lower face of the beam near the joint.

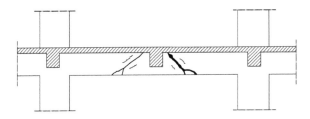

Figure 11.21 Shear failure at the location of an indirect support.

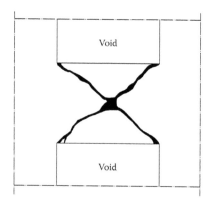

Figure 11.22 Shear failure of a shear wall coupling beam.

11.1.5 Damage to beam–column joints

Damage to beam–column joints, even at the early stages of cracking, must be considered extremely dangerous for the structure and be treated accordingly. Damage of this type reduces the stiffness of the structural element and leads to uncontrollable redistribution of load effects. Common failures of beam–column joints (corner joint, exterior joint of a multi-storey structure and interior joint) are shown in Figures 11.23 through 11.26.

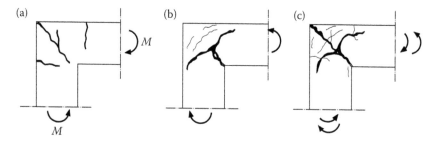

Figure 11.23 Failure of a corner joint: (a) moments subjecting the inner fibre to compression; (b) moments subjecting the inner fibre to tension; (c) cyclic bending moment loading.

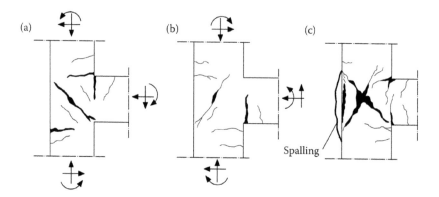

Figure 11.24 Failure of exterior joint in a multi-storey building: (a) moments inducing compression at the lower fibre of the beam; (b) moments inducing compression at the upper fibre of the beam; (c) cyclic bending moment loading.

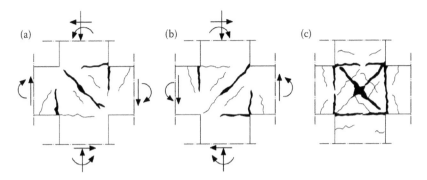

Figure 11.25 Failure of a cross-shaped interior joint: (a) seismic action in the right to left direction; (b) seismic action in the left to right direction; (c) cyclic seismic action.

Figure 11.26 Failure of an exterior joint in a multi-storey building: Kalamata, Greece (1986).

The flow of internal forces in the reinforcement and the concrete during the successive phases of cyclic loading has already been explained in Chapter 8.4.2 and will not be discussed here.

11.1.6 Damage to slabs

The most common types of damage that occur to slabs are the following:

- Cracks parallel or transverse to the reinforcement at random locations
- Cracks at critical sections of large spans or large cantilevers, transverse to the main reinforcement
- Cracks at locations of floor discontinuities, such as the corners of large openings accommodating internal stairways, light shafts and so on
- Cracks in areas of concentration of large seismic load effects, particularly in the connection zones of slabs to shear walls (diaphragmatic action) or to columns in flat slab systems

With the exception of the last type, damage to slabs cannot generally be considered dangerous for the stability of the structure. However, it creates serious aesthetic and functional problems, so it must be repaired. Moreover, the creation of such damage leads to the reduction of the available strength, stiffness, and energy dissipation capacity of the structure in the case of a future earthquake, and this is an additional reason for repair.

The first type of damage is the most frequent. Most cases are due to the widening of already existing micro cracks that are formed either because of bending action, temperature changes or shrinkage, and they become visible after dynamic seismic excitation. They are rarely caused by differential settlement of columns. In such cases, however, the phenomenon is accompanied by extensive cracking of the adjacent beams and masonry infills, and in this context it is detectable.

The second and third types of damage are typically due to the vertical component of the earthquake action (Figures 11.27 and 11.28).

The fourth type of damage is usually related to punching shear failure, aggravated by the cyclic bending caused by the earthquake (Figure 11.29). It has already been stressed in Chapter 4 that slabs on columns are seismically vulnerable structures, and they must be avoided, as they are not covered by the Codes in effect unless they are combined with other seismic-resistant systems (i.e. shear walls or ductile frames).

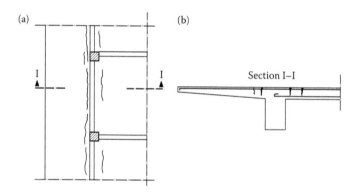

Figure 11.27 Slab damage at the critical area of a cantilever: (a) floor plan of the slab (upper side); (b) Section I–I.

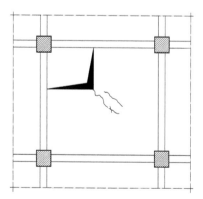

Figure 11.28 Slab damage at the corner of a large opening (down side of the slab).

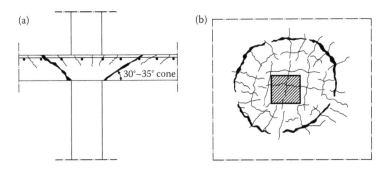

Figure 11.29 Damage at a slab-to-column connection: (a) section; (b) top side of the slab.

11.1.7 Damage to infill walls

As discussed earlier, almost all the infill walls in southern Europe were constructed in the past with masonry in contact with the surrounding structural members of the frame. Since these infills are constructed with materials (bricks, mortar and plaster) of lower strength and deformability than the structural members, they are the first to fail. Thus, the failure of the infills starts before damage to the frame occurs, and therefore, if not accompanied by damage to the structural members, the infills cannot be considered dangerous for the stability of the structure. However, the largest portion of the repair cost is usually attributable to damage to the infills, because they involve extensive repair of installations and of finishes, such as plastering, painting, tiling, plumbing, electric installations and so on.

The damage in the infills occurs in the following sequence: during the excitation of the structure due to an earthquake, the R/C frame starts to deform, and at this stage the first cracks appear on the plastering along the lines of contact of the masonry with the frame. As the deformation of the frame becomes larger, the cracks penetrate into the masonry. This is displayed by the detachment of the masonry from the frame (Figure 11.30a). Subsequently, X-shaped cracks appear, small at first and becoming larger later, in the masonry itself, in a stepwise pattern following the joint lines (Figures 11.30b, 11.31 and 11.32). When the cracks do not penetrate the whole thickness of the wall, the damage is characterised as 'light'; otherwise, it is 'serious' damage (X-shaped cracks).

From the above discussion, one can conclude that the damage in the infill panels must be the first in frequency of occurrence, since they usually precede damage in the R/C structural system (Tiedemann, 1980; Penelis et al., 1987, 1988b; Stylianidis, 2012).

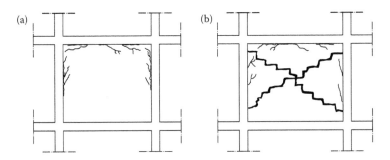

Figure 11.30 Damage in the infill panels: (a) detachment from the frame; (b) X-shaped through cracks.

Figure 11.31 Damage to masonry walls: Kalamata, Greece (1986).

Figure 11.32 Damage to masonry walls: Kozani, Greece (1996).

During the Thessaloniki earthquake of 20 June 1978, while damage of R/C buildings occurred in 7.4% of the beams, in 5.3% of the columns and in 6.5% of the shear walls, damage in infill panels occurred in 22.9% of the buildings with an R/C structural system. It should also be noted that out of the R/C buildings that suffered damage at the infill panels, 96% exhibited detachment from the surrounding frame, 79% exhibited X-shaped full-depth cracks and 12% exhibited out-of-plane collapse of the masonry wall.

11.1.8 Spatial distribution of damage in buildings

At this point, it would be useful to discuss the distribution of damage in buildings.

Along the vertical direction, the most serious damage occurs on the ground floor. The frequency and intensity of damage is gradually reduced in the upper floors. This distribution has been observed in most recent earthquakes: the Bucharest, Romania, earthquake of 1977; the Thessaloniki, Greece, earthquake of 1978; the Alkyonides, Greece, earthquake of 1981; the Montenegro, Former Yugoslavia, earthquake of 1980; the Kalamata, Greece, earthquake of 1986; and the Athens (Parnitha) earthquake of 1999. The Mexico City earthquake of 1985 constitutes an exception to the above observation and will be discussed later.

The methodology for analysis and design of earthquake-resistant structures cannot explain this phenomenon. In fact, the lower storeys, particularly the ground storey, due to the higher inertial forces, are subjected to larger seismic effects. At the same time, their structural elements are also designed for these higher seismic effects according to rules that apply to the whole building, and, therefore, they conform to common partial safety factors. Thus, damage is expected to be uniformly distributed throughout the building. Dynamic inelastic analysis of multi-storey buildings (Figure 11.33; Kappos and Penelis, 1987) also supports the notion of uniform distribution of damage. It is the author's opinion that the higher degree of damage in the ground floor is due to the fact that the infills contribute with the same amount of additional strength to all floors (given the fact that the masonry layout is the same on every floor), a fact that is not taken into account when analysing the structure. Indeed, if the masonry did not exist and the required strength for the earthquake was higher than the one available by a given percentage, the same for all storeys, there would be a uniform vertical distribution of damage. The addition of the strength of masonry to that of the R/C structural system exceeds the required strength in the upper storeys but not in the lower ones, and that is where damage occurs (Figure 11.34).

In the case of a flexible ground floor (shops with glass panels or the pilotis system with an open ground floor), the damage to this floor is much more severe and usually occurs only there (Figure 11.35). This subject will be discussed in detail in Section 11.2.5.

As far as the horizontal distribution is concerned, most of the damage occurs in areas that are far from the stiffness centre of the building and mainly on the perimeter of the building.

The Mexico City earthquake (1985) was the first seismic activity during which a large percentage of collapses and large-scale damage occurred in the upper floors of buildings (38%; Figure 11.36). This can be fully explained, considering the fact that the damaged buildings were very tall (with more than 12–15 storeys), with very flexible structural systems (flat slabs), wherein higher modes generate large seismic effects in the upper floors. Furthermore, the same types of damage occur in masts or tower-like structures (e.g. bell towers, minarets, chimneys).

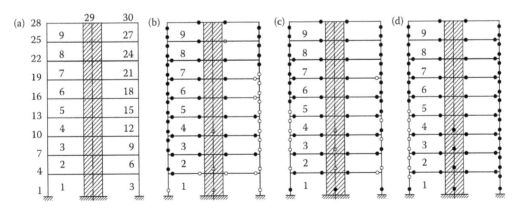

Figure 11.33 Distribution of plastic hinges in a dual nine-storey building, subjected to an El Centro (1940) excitation scaled by 0.75, 1.0 and 1.5, respectively; • = yielding of reinforcement on both ends (top and bottom); o = yielding of reinforcement on one end (top and bottom). (a) Geometry of the struct. system; (b) El Centro (1940) excitation scaled by 0.75; (c) scaled by 1.0; (d) scaled by 1.50.

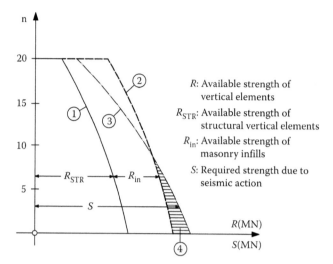

Figure 11.34 Explanation of the higher sensitivity of the lower storeys of a building: (1) shear-strength curve of the vertical structural elements of a building; (2) additional shear strength due to masonry, constant for all storeys; (3) required strength (seismic action) per storey, almost proportional to R_{STR} (e.g. 1.5 R_{STR}); (4) shaded area – storeys with damage.

Figure 11.35 Column damage of the open ground storey (pilotis system), Kalamata, Greece (1986).

11.1.9 Stiffness degradation

Strong earthquakes induce inelastic deformations to buildings accompanied in most cases by visual damage. As a result, the buildings sustain a stiffness degradation that is displayed by an increase in their fundamental period. Site investigations on actual buildings before and after an earthquake (Ogawa and Abe, 1980) have shown that there is a strong correlation between the extent of damage and the value of the ratio of the fundamental period of the building after the earthquake to that before the event. Shear cracks can be found by visual observation of buildings in which the value of the fundamental periods ratio is more than 1.3,

Figure 11.36 Large-scale damage at the upper floors of a flexible high building, Mexico City (1985).

$$\frac{T_2}{T_1} \geq 1.30 \qquad (11.4)$$

where
T_1 is the fundamental period before the earthquake.
T_2 is the fundamental period after the earthquake.

Taking into account the fact that

$$\frac{T_2}{T_1} = \left(\frac{K_1}{K_2} \right)^{1/2} \qquad (11.5)$$

where
K_1 is the equivalent stiffness of the building before the earthquake.
K_2 is the equivalent stiffness of the building after the earthquake.

it follows that where visual damage is observed in the building, the stiffness degradation is on the order of 40%.

$$\frac{K_1}{K_2} \geq \approx 0.60 \qquad (11.6)$$

11.2 FACTORS AFFECTING THE DEGREE OF DAMAGE TO BUILDINGS

11.2.1 Introduction

In the subsections that follow, there will be an attempt to present systematically the most important factors that seem to affect the degree of damage to buildings. The presence of one of these factors does not necessarily mean that it is the only reason for the damage. In most cases, there is more than one adverse factor in a structure; therefore, the determination of how much each of the factors contributes to the damage is not clear, even after a systematic statistical analysis of the damages.

11.2.2 Deviations between design and actual response spectrum

The first and most important reason for damage to structures is the deviation of the seismic actions that has been taken into account for the design at the time of construction from that of the seismic loading that caused the damage. In fact, there are still constructions and monuments, coexisting with the modern buildings that have been designed according to old Codes. Thus, a strong earthquake acts upon a variety of structures, some built with no structural design at all, others designed only for gravity loads, others designed for static earthquake horizontal loads with no consideration of ductility requirements and still others, the most recent ones, designed according to the current knowledge of seismic design. Therefore, it is reasonable to expect that this spectrum of structures, the great majority of which does not conform to design specifications based on the current state of knowledge, will experience some damage. Furthermore, it is possible for damage to occur in engineered structures designed according to the current Codes, mainly for the following three reasons:

1. Even though there has been significant progress in the design of earthquake-resistant structures during the last few decades, it does not mean that the seismic protection problem has been solved. Consequently, every generation believes that it has taken important steps towards the advancement of an area of interest; future developments, though, usually come to prove this belief wrong. Thus, it is not impossible that structures built today according to the most recent advances in earthquake engineering will not conform to the specifications in effect in a few years' time.
2. Contemporary structures are designed in such a way that when the design earthquake occurs, they should respond inelastically, which means that they are expected to sustain a controllable degree of damage, i.e. damage is not excluded.
3. Quite often, the design spectrum, scaled according to the behaviour coefficient and the safety factor, does not correlate with the actual response spectrum. The Mexico City (1985) earthquake, the Kalamata (1986) earthquake and the Athens (Parnitha, 1999) earthquake can be cited as examples (Figures 11.37–11.39). Therefore, before rushing into conclusions about the contribution of each damage factor, one should first study carefully the actual response spectrum of the earthquake that has caused the damage in relation to the provisions of the Code according to which most of the structures in the area were built (Rosenblueth and Meli, 1985; Anagnostopoulos et al., 1986; Penelis et al., 1986, 2000; Kostikas et al., 2000).

Apart from the characteristics of the seismic excitation, a number of the structure's own characteristics, which will be discussed subsequently, are factors that contribute to the vulnerability of the structures.

11.2.3 Brittle columns

In Section 11.1, which refers to the typology of damage in structural elements, the types of column failure have been discussed in detail. The vast majority of failures in buildings with R/C structural systems are due to column failure, caused by bending and axial load, or by shear under strong axial compression. There are clear indications that in buildings designed

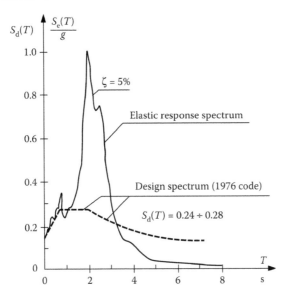

Figure 11.37 The 1985 Mexico City earthquake: comparison of the actual response spectrum with the design spectrum.

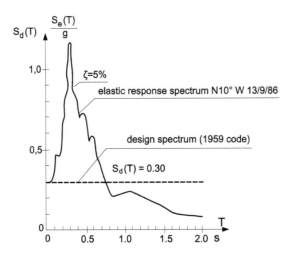

Figure 11.38 The Kalamata, Greece, earthquake of 13 September 1986: comparison of the elastic response spectrum with the design spectrum.

during recent decades, due to high axial loading level, most of the time the column reinforcement does not reach the yield point. In this respect, column failures must be attributed to the degradation of the mechanical properties of the concrete due to high inelastic strains under cyclic loading (low cycle fatigue). Quite often, the main reason for failure is the large spacing of ties at the critical regions of the column, which has a consequence of buckling of longitudinal rebars.

It should be noted that the main cause for 55% of the 103 most affected buildings in the Athens earthquake (1999) must be attributed to poor hoop and tie reinforcement of columns either at critical regions of confinement or along the axis of the column for shear (Figure 11.51).

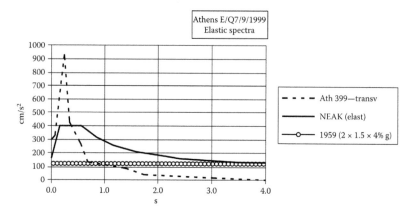

Figure 11.39 The Athens (Parnitha) 1999 earthquake: comparison of the elastic response spectrum with the design spectrum.

11.2.4 Asymmetric arrangement of stiffness elements in plan

It is well known that the core of staircases and elevators is the basic stiffness element in the structural system of a building; therefore, according to what has been discussed in the chapter on the analysis of structures, its central or eccentric position should be of major importance for the behaviour of the building in an earthquake (Figure 11.40). However, a statistical evaluation of the damage that the 1978 earthquake caused to the buildings of Thessaloniki (Penelis et al., 1987, 1988b) shows that this factor affected only 0.6% of the mean value of the percentage of the damaged structures (Figure 11.41). This phenomenon must be mainly attributed to the fact that the infills change the stiffness distribution in the building drastically, and, as a result, the effect of eccentricities due to asymmetric arrangement of R/C stiffness elements is reduced. In contrast, asymmetric arrangement of masonry causes remarkable inferior behaviour. This asymmetric arrangement of masonry is usually observed in the ground floors of structures located at the corners of building blocks, where the two sides of the perimeter are not filled with masonry because of their usage as shops (Figure 11.42). This is one of the topics that will be discussed next.

Figure 11.40 Torsional collapse of a building in Mexico City (1985).

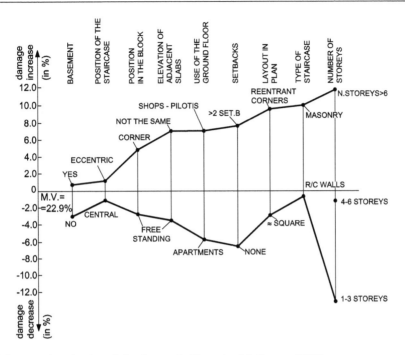

Figure 11.41 Statistical evaluation of the damage in Thessaloniki, Greece (1978).

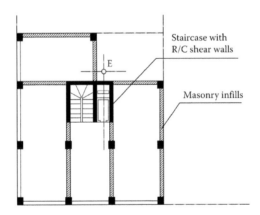

Figure 11.42 Stiffness centre location of a corner structure, when masonry infills are taken into account (approximately).

It should be noted that according to the survey and evaluation of damages for the 103 most affected buildings in Athens (1999), the asymmetric arrangement of structural stiffness elements in plan covered 15% of the cases, while the asymmetric arrangement of masonry infills in plan corresponded to 7% of cases (Figure 11.51).

11.2.5 Flexible ground floor

The sudden reduction of stiffness at a certain level of the building, typically at one of the bottom floors, results in a concentration of stresses in the structural elements of the flexible floor, which causes damage to those elements. An illustrative example of this fact is the distribution of shear forces that are developed on the R/C staircase core of a 20-storey building with no masonry at the four lower floors (Figure 11.43; Dowrick, 2005). The shear force distribution has been determined using dynamic inelastic analysis. This example makes obvious the fact that, for the floors with masonry, the shear force acting on the staircase core is much smaller if the infills are taken into account for the analysis, while for the four lower floors without masonry, the resulting shear force is much higher. For this reason, the Seismic Codes in effect today require an increase in the design shear for the storey with reduced stiffness compared to that of upper floors (see Chapter 8.5.2.2). They also require a high degree of confinement through closely spaced ties or in the form of spirals, throughout the height of the columns of the weak floor, in order to increase their ductility.

The most common case of a flexible floor is the open ground floor (pilotis system) or the ground floor used as a commercial area. In such a case, while the upper floors have high stiffness due to the presence of masonry infills, the ground floor has a drastically reduced stiffness because the vertical structural members contribute almost exclusively to it. In these buildings, almost all the damage occurs in the vertical structural elements of the ground floor, while the rest of the building remains mostly unaffected (Figure 11.44). In contrast, in buildings with masonry infills in the ground floor, the damage spreads throughout the

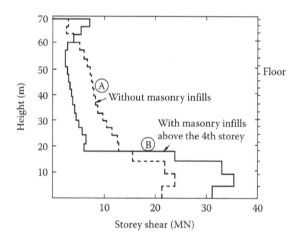

Figure 11.43 Shear force diagram of the staircase core of a 20-storey building, which is indicative of the effect of masonry infills on the storeys above the fourth floor.

Figure 11.44 Collapse of a building with a flexible ground floor: Bucharest, Romania (1977).

structure with usually decreasing intensity from the ground to the upper floors. The 1978 Thessaloniki earthquake caused damage to only 16.4% of the buildings with masonry infills in the ground floor, while damaged buildings having the pilotis system or shops on the ground floor reached 29.8% of the total number of this type of building. During the Mexico City earthquake of 1985, 8% of the buildings that collapsed or exhibited severe damage had a flexible ground floor.

Finally, during the Athens earthquake (1999), one of the main causes of damage for 22% of the 103 most affected R/C buildings was the 'pilotis effect' (Figure 11.51).

11.2.6 Short columns

It has already been noted (see also Chapter 8.3.6) that short columns can experience an explosive shear failure that can lead to a spectacular collapse of the building. This phenomenon, however, appears to be rarer than the failure of regular columns.

In the case of the Athens earthquake (1999), the failure of short columns was one of the main causes of damage near collapse for almost 16% of the 103 most affected R/C buildings (Figure 11.51).

11.2.7 Shape of the floor plan

Buildings with a square-shaped floor plan have the best behaviour during an earthquake, while buildings with divided shapes such as +, I, X or with re-entrant corners have the worst. During the 1978 Thessaloniki earthquake, among the damaged buildings (with damage in the R/C system), 19.5% had a square-shaped floor plan, while 32.5% had non-convex shapes of floor plan. This is the reason why EC8 and the CEB/MC-CD/85 do not allow simplified methods of analysis for earthquake actions when the building under consideration does not have a regularly shaped floor plan.

11.2.8 Shape of the building in elevation

Buildings with upper storeys in the form of setbacks have markedly inferior behaviour than buildings with regular form in elevation. During the 1978 Thessaloniki earthquake, among the total number of damaged buildings, 15.9% were buildings regular in elevation, while 29.9% were buildings with three or more successive setbacks (see also Figure 11.41).

11.2.9 Slabs supported by columns without beams (flat slab systems)

This type of failure has been discussed in the section on slab failures. In the seismically active regions of southern Europe, this type of structure is rather recent and, therefore, there are no statistical data regarding this failure mode. Experimental data, however, as well as statistical data from the 1985 Mexico City earthquake, suggest that this is a very vulnerable type of structure. Indeed, in Mexico City, where this kind of structural system is widely used, 41% of collapses or serious damage occurred in buildings of this type. Structures with such slabs are very flexible and with low ductility. Most of the failures in Mexico City occurred in columns. However, in more than 10% of the cases, the columns punched through the slab, under the action of a combination of both vertical and horizontal seismic loads. Moreover, the small thickness of the slab did not allow the development of the required bond stresses around the longitudinal reinforcement of the columns, and, therefore, after a few loading cycles, the joints failed due to the failure of bond mechanisms

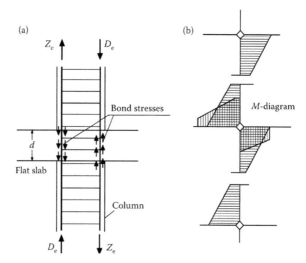

Figure 11.45 Degradation of bond between concrete and column reinforcement within the small thickness *d* of the slab: (a) sketch of the flow of forces; (b) corresponding bending moment diagram.

along the thickness of the slabs (Figure 11.45). For this reason, EC 8-1/2004 does not cover this type of structural system if it is not combined with other seismic-resistant systems (i.e. shear walls, ductile frames).

11.2.10 Damage from previous earthquakes

Buildings that have sustained damage during a previous earthquake and have been repaired usually exhibit the same type of damage in the next earthquake, to a larger extent. This phenomenon was observed in Bucharest during the 1977 earthquake, where many of the buildings repaired after the 1940 earthquake collapsed (Figure 11.46), and also more recently in Mexico City. The main reason for this phenomenon is that the repair work was not carried

Figure 11.46 Collapse of a building in Bucharest during the 1977 earthquake. The building, as depicted in the illustration, had been repaired after the 1940 earthquake.

out carefully enough, and also the fact that 60 to 70 years ago, the repair technology of earthquake damage was in the early stages of development.

In the case of the Athens earthquake (1999), only 3% of the 103 most affected buildings exhibited damages due to poor repair or strengthening from the previous earthquake of 1981 (Alkyonides earthquake).

11.2.11 R/C buildings with a frame structural system

Frame systems, not inferior to dual systems as far as strength is concerned, but superior with regard to available ductility, have lower stiffness than dual systems. As a result, during a seismic excitation, large inter-storey drifts develop, which cause extensive damage in the infill system. Given that the repair of this damage is a very costly procedure, it is understood why the 'frame system' constitutes a source of vulnerability for buildings. Therefore, although ductile behaviour can be achieved more easily with frame systems than with dual ones, and although this had led to extensive use of the frame systems in the 1960s, the idea that shear-wall systems are more suitable for R/C buildings has become more widely accepted since about 1975. Comparative studies of building behaviour during the earthquakes of Managua (1972), San Fernando (1971), Caracas (1967) and Skopje (1963) support the above opinion (Fintel, 1974b). From the Thessaloniki earthquake (1978), among the damaged buildings (damage in the structural system), 22% were buildings with shear walls and 32.9% were buildings without shear walls. Finally, one of the main observations of the research team from the University of Thessaloniki, which visited and studied the earthquake damage in Kalamata, Greece, in 1986, was the large extent of masonry damage in most multi-storey buildings with a frame structural system (Penelis et al., 1986). It is important here to note again that this type of structural system is susceptible to collapse in a pancake mode (Figure 11.51) in the case of poor design of columns or joints.

11.2.12 Number of storeys

The number of storeys is directly related to the fundamental period T of the structure, as discussed in previous chapters. Therefore, at least theoretically, the vulnerability of the structure to an earthquake depends on the ordinate of the acceleration spectrum of that specific earthquake corresponding to T, in relation to that of the design response spectrum of the building. In this context, the vulnerability of the building should be independent of the number of storeys in case the design spectrum and the spectrum of the seismic event were similar. However, the existing statistical data from earthquakes show that the vulnerability increases with the height of the buildings. As typical examples, one can cite Bucharest (1977), where damage and collapse were focused mainly in buildings with more than six storeys, Mexico City (Table 11.1; Rosenblueth and Meli, 1985) and Thessaloniki (1978), where among the damaged buildings, 10.9% were low-rise buildings (one to three storeys), while 34.9% were high-rise buildings (over six storeys). In Bucharest and Mexico City, the concentration of damage in high-rise buildings is compatible with the response spectrum of the corresponding earthquake, since large acceleration values correspond to high natural periods, corresponding to high-rise buildings. In the case of Thessaloniki, however, this correlation does not exist. The author's opinion on this issue is that the infill system drastically increases the stiffness as well as the strength of the structure. Given the fact that the masonry layout is more or less the same in every floor and independent of the height of the building, the percentage of additional stiffness and strength due to the presence of masonry is higher in low-rise buildings than in high-rise ones. As a result, the behaviour of low-rise buildings appears to be better (Figure 11.47).

Table 11.1 Percentages of collapses and serious
damage in Mexico City (1985)

Number of storeys	Percentages of collapses and serious damage (% of every building category)
1–2	0.9
3–5	1.3
6–8	8.4
9–12	13.6
>12	10.5
Total	1.4

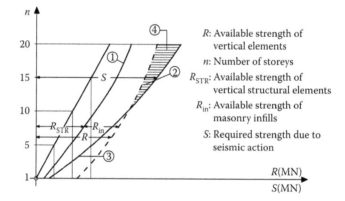

Figure 11.47 Explanation of the seismic vulnerability of high-rise buildings: (1) shear-strength curve of vertical structural elements; (2) additional strength due to masonry, constant for all floors; (3) required shear strength (seismic action) in case of a strong seismic event, almost proportional to R_{STR}; (4) shaded area – buildings with damage.

11.2.13 Type of foundations

The form of the foundation of the structure has two types of effects on the extent of damage in the building: direct and indirect.

Direct effects are displayed in the following characteristics:

• Failure of the foundation members (e.g. fracture of foundation beams)
• Fracture of the foundation soil
• Soil liquefaction
• Differential settlements of the ground
• Partial or general landslide of the foundation soil

The most usual form of the effects listed above is the differential settlements of the ground, especially in soft soils. Failures due to soil liquefaction are seldom seen, but they are spectacular (Figure 11.48).

The *indirect effects* are related to the out-of-phase motion of the bases of the individual columns, when their footings are not interconnected (see Chapter 4.2.12, Figure 4.17), or

Figure 11.48 Consequences of soil liquefaction on a high building in Mexico City (1985).

when the existing connection is too flexible. These differential displacements in both the horizontal and the vertical directions subject the structure to additional strains. As a result, buildings with isolated footings suffer more under seismic action than others. A characteristic example of the above is the fact that in the 1986 Kalamata, Greece, earthquake, the damage to the buildings on the sea-front avenue, where the foundations had great stiffness, was limited. However, the same did not happen in Thessaloniki, Greece. Although the foundations of the buildings in the coastal zone were either mat foundations or grids of foundation beams, the percentage of damage was high. The interference of other factors irrelevant to the type of the foundation, such as the amplification factor of the seismic excitation that is applied to soft soil deposits, and the shift of the prevailing period of the exciting force towards higher values when such soils are present, does not allow a clear statistical evaluation of how the presence of a good foundation indirectly affects the vulnerability of structures.

11.2.14 Location of adjacent buildings in the block

The location of adjacent buildings on the block has a great effect on the behaviour of the structure in response to an earthquake. More particularly, corner buildings are much more sensitive to earthquakes than free-standing ones. In the 1977 Bucharest earthquake, 35 out of the 37 buildings that collapsed were located at the corner of the block (Figure 11.44). In the 1985 Mexico City earthquake, 42% of the buildings that suffered serious damage or collapsed were corner structures (Rosenblueth and Meli, 1985; Table 11.2, Figure 11.44). In the 1978 Thessaloniki earthquake (Figure 11.49), among the damaged buildings (with damage to the R/C structural systems), only 19.9% were free-standing buildings, while 27.9% were corner ones. As sources of the higher vulnerability of corner structures, the following can be mentioned:

- Asymmetric distribution of stiffness elements on the floor plan due to the lack of masonry on two sides of the perimeter of the ground floor, where the space is usually occupied by stores.
- Transfer of kinetic energy to the corner buildings during the seismic interaction of adjacent buildings (Figure 11.50) through pounding. This transfer of energy causes a substantial increase in the inertial forces acting on the end structures (Anagnostopoulos, 1988; Athanassiadou et al., 1994).

Table 11.2 Causes of failure in Mexico City (1985)

Reason for failure	Percentage
Asymmetric stiffness	15
Corner structure	42
Weak ground floor	8
Short columns	3
Exceeded vertical design load	9
Pre-existing ground settlement	2
Pounding of adjacent structures	15
Damage from previous earthquakes	5
Punching failure of flat slabs	4
Failure of upper floors	38
Failure of lower floors	40

Figure 11.49 Collapse of an eight-storey corner building in Thessaloniki, Greece (1978).

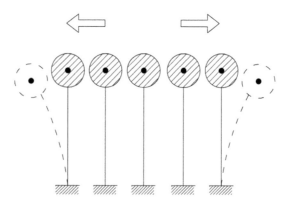

Figure 11.50 Transfer of energy at the end of SDOF oscillators in a series of adjacent systems.

11.2.15 Slab levels of adjacent structures

The impulse loading that a building receives from an adjacent structure during an earthquake is a major source of damage. The problem becomes even more serious when the floor slab levels of adjacent buildings do not coincide. In this case, the slabs of one structure during the oscillation pound on the columns of the adjacent building, and this results in fracture of the columns. In the 1978 Thessaloniki earthquake, among the damaged structures, the percentage of free-standing buildings or buildings that had the same floor slab levels as adjacent ones was only 19%, while the percentage of buildings with slab levels different from those of adjacent ones was 30.5%. In the 1985 Mexico City earthquake, in more than 40% of the structures that collapsed or suffered serious damage, pounding of adjacent structures took place (Table 11.2).

In the Athens earthquake of 1999, 21% of the most affected 103 R/C buildings had pounding from adjacent buildings as a main cause of damage.

11.2.16 Poor structural layout

This category includes buildings with asymmetric arrangement of stiffness elements in plan and elevation, irregularities of form in plan and elevation, cut-offs of columns, setbacks and so forth. Most of these categories have already been examined in previous items. However, as a whole, they reflect the result of poor collaboration of the architect and the structural engineer during the phase of conceptual design.

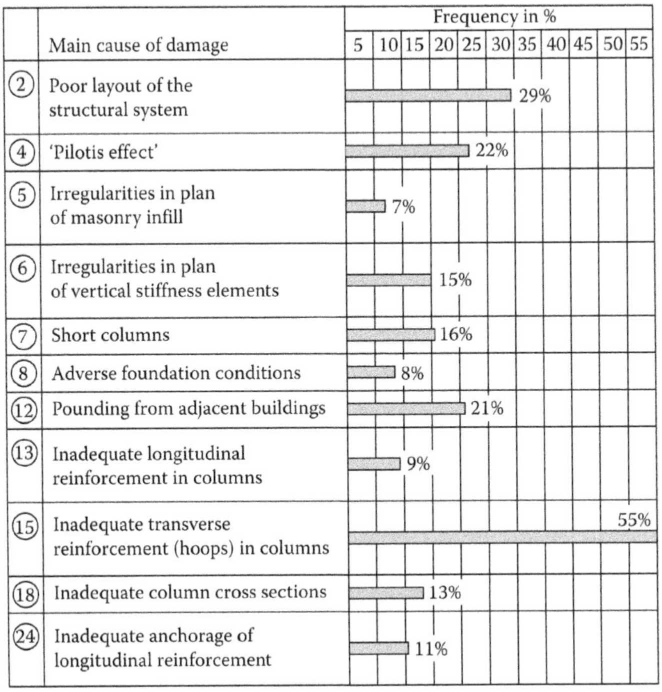

	Main cause of damage	Frequency in %
②	Poor layout of the structural system	29%
④	'Pilotis effect'	22%
⑤	Irregularities in plan of masonry infill	7%
⑥	Irregularities in plan of vertical stiffness elements	15%
⑦	Short columns	16%
⑧	Adverse foundation conditions	8%
⑫	Pounding from adjacent buildings	21%
⑬	Inadequate longitudinal reinforcement in columns	9%
⑮	Inadequate transverse reinforcement (hoops) in columns	55%
⑱	Inadequate column cross sections	13%
㉔	Inadequate anchorage of longitudinal reinforcement	11%

Figure 11.51 Frequency (in %) of appearance of various main causes of damage in the 103 most affected R/C buildings in Athens after the earthquake of 1999. (Adapted from Kostikas, Ghr. et al. 2000. Evaluation of damages and their cause for 103 near to collapse R/C buildings in Athens after the earthquake of September 7, 1999, Report of the OASP, Seismic Risk Management Agency of Greece.)

From the evaluation of the 103 most affected R/C buildings in Athens (1999), it has been concluded that almost 29% of them had poor structural layout as a main cause of failure (Figure 11.51). This fact, together with poor transverse column reinforcement, the 'pilotis effect' and the pounding of adjacent buildings on the block *were the four main reasons for collapse or damage near to collapse of the 103 most affected buildings in the Athens earthquake* (1999; Figure 11.51).

11.2.17 Main types of damage in buildings designed on the basis of modern codes

The earthquake of Athens, 1999, was a chance for a preliminary evaluation of damages displayed in R/C buildings constructed after 1985, which is the 'benchmark year' in Greece for new-generation Seismic Codes.

In fact, out of 103 of the most affected R/C buildings, 12 were designed and constructed according to these Codes. Although the number is limited and, therefore, statistically justified conclusions cannot be drawn, some initial remarks can be made:

1. The main causes of heavy damage continued to be the following:
   - Poor confinement and poor shear reinforcement of columns
   - 'Pilotis effect'
   - Short columns
   - Pounding of adjacent buildings
   - Poor structural layout
2. The ratio of the number of heavily affected R/C buildings of the period between 1985 and 1999 (14 years) to the total number of heavily affected buildings (12/103) did not display any improvement in relation to the ratio of the building stock of this period to the total stock (0.12–0.14). In the author's opinion, this must be attributed to the fact that there is always a lag between establishment and implementation of a new Code in practice.
3. The above explanation is based on the findings *in situ*, which means the poor structural layout, inadequate confinement in critical regions, inadequate joints between adjacent buildings and so on, although there was detailed reference in the Codes after 1985 for special concern on these points. In any case, the above remarks must be considered of limited credibility since they refer to a transient period of 14 years with a limited number of new buildings under consideration.

Chapter 12

Emergency post-earthquake damage inspection, assessment and human life protection measures

12.1 GENERAL

The aim of this chapter is to present a reliable procedure that should be followed for the emergency inspection of buildings after an earthquake so that all the structures of the affected area will be inspected in a credible way. The data collected by such inspections assist the State in achieving the following goals (UNIDO/UNDP, 1985, Penelis and Kappos, 1997):

- Reduce the number of casualties and injuries to occupants of damaged buildings, which might be caused by collapses due to subsequent aftershocks.
- Help people of the affected area gradually return to a normal way of life, which pre-supposes a reliable characterisation of the dangerous buildings and full knowledge of the extent of hazardous structures.
- Develop a database for a uniform assessment of risk in economic, social, political and other terms.
- Record and classify earthquake damage, so that the repair of damaged buildings will follow a priority order.
- Improve earthquake-resistant design, based on the recorded damage.

It should be obvious from this brief introduction that after a destructive earthquake, two levels of building inspection follow. The first level of inspection is performed by the State, during which there is a recording of damage, characterisation of hazardous structures, demolition of buildings close to collapse and propping for those that need it. This is an operation that needs to be carried out quickly, in order to gradually restore the normal way of life in the affected area.

During the second level of inspection, which will be discussed in detail in Chapter 13, the residual strength of every affected structure is estimated and the degree of intervention is decided. This is a laborious procedure that starts as soon as the first level of inspection is completed and the frequency and intensity of the aftershocks have diminished. It is also a procedure that is directly related to the decision about the repair and/or strengthening of the structures (EC8-3, CEN, 2005).

12.2 INSPECTIONS AND DAMAGE ASSESSMENT

12.2.1 Introductory remarks

The purpose of this section is a brief treatment of the problems related to the evaluation of damage to structures after an earthquake. A strong earthquake, like every other hazard, puts on trial not only the citizens but also the State. The authorities have to face chaotic situations due to lack of information, delays in locating the affected areas, eventual interruption of communications and multiple requests for assistance and for inspections of damaged buildings. In the first tragic hours, even days sometimes, the affected area stands almost alone, and it is during this initial period when good construction, efficient communications, quick decision-making and correct planning pay off in terms of lives and properties saved. The credibility of the State in its citizens' eyes depends on what the State can or cannot do during this early period. These are the views of government officials in charge of disaster relief, who have experienced the situations and problems that arise after a strong earthquake (Office of Emergency Preparedness, 1972; Penelis, 1984, 2008).

The foregoing remarks aim at depicting the environment in which the structural engineer is called upon to do an assessment; this should be the prevailing element for the design of the entire operation. Indeed, since damage evaluation sometimes refers to thousands of buildings, which have to be assessed in a short period of time in order for the affected area to return to a normal way of life, a special procedure has to be followed, completely different from that used for the assessment of the structural resistance of an individual building. In this context, procedures developed for 'seismic evaluation of buildings' for the pre-earthquake period (FEMA 154-158-178, FEMA 310; see Chapter 13) cannot be implemented in post-earthquake emergency conditions. After a strong earthquake, only *damage-oriented evaluation* may be implemented.

In the subsections that follow, there is systematic reference to matters of inspection, as well as to problems that the structural engineers face during evaluation of individual cases. At the same time, reference is made to propping measures that are taken in case of serious damage susceptible to collapse. This presentation is based on experience from the organisation and implementation of such an operation in the Thessaloniki (Greece) metropolitan area in 1978, for which one of the co-authors was the person in charge. Since then, this procedure, with minor modifications, has been implemented for various strong earthquakes that have affected Greece (Alkionides earthquake 1981, Kalamata earthquake 1986, Grevena earthquake 1989, Aigion earthquake 1990, Athens [Parnitha] earthquake 1999, etc.). It is also based on publications of several international organisations and national committees on the same subject (ATC, 1978; Greek National Report, 1982; Yugoslav National Report, 1982).

12.2.2 Purpose of the inspections

The main purpose of the inspection procedure for the structures after a destructive earthquake is to minimise the probability of casualties or injuries for the occupants. The hazard of such an event in buildings where damage has occurred in the main shock is serious enough, because it is possible for some of them to partly or completely collapse due to repeated aftershocks, as happened with the Alkyonides, Greece, earthquake (1981), and also with the Kalamata, Greece, earthquake (1986).

There are also other reasons for the inspections, beyond the above-mentioned, which are also of great importance. Thus, after the classification of those damaged buildings which are hazardous to use, life gradually returns to normal, given the fact that the rest of the buildings – as soon as the first psychological reactions begin to disappear – gradually return to their normal usage. Also, based on the first damage assessment, an approximate idea of the magnitude of the disaster in economic terms may be obtained. These data are needed by all levels of administration almost immediately for them to be able to start a proper planning of relief. Of course, the development of fragility curves in the pre-earthquake period in relation to the seismic action intensity of the seismic event allows the cost estimate of retrofitting activities that will follow (see Chapter 16 of the first edition of this book).

Finally, the statistical data from such an operation are very useful, not only for short-term decision-making regarding temporary housing but also in the long run on matters of evaluation of the construction procedures followed in the past and the factors that could affect them positively.

In closing, it should be noted that in the case that the affected area includes a large city, the whole procedure should be managed very carefully, as this becomes a large-scale operation with very high cost and organising requirements, accompanied by long-term implications. Indeed, the classification of a building as 'damaged' by the State leads to a long-term degradation of its market value, even though strengthening interventions after the earthquake could make it stronger than other non-damaged buildings.

12.2.3 Damage assessment

12.2.3.1 Introduction

It has already been mentioned that the main concern during damage inspection in structures after an earthquake is to minimise the probabilities of accidents to occupants, caused by partial or total post-earthquake collapses. Therefore, the problem that the structural engineer has to face in every case is to estimate the residual strength, ductility and stiffness of the structure, and decide whether or not they are sufficient to allow the use of the building at an acceptable level of risk.

It is understood that this evaluation, based on the existing evidence, is probably the most difficult problem for the structural engineer – much more difficult than the design of a new building. Extensive site inspections are required, first of all for a damage survey, and then to check the geometry of the structural system, the quality of the construction materials, the placement of reinforcement at critical structural elements compared to the original drawings, the vertical loads of the structure and the quality of the foundation soil. Subsequently, extensive calculations are needed, using the information collected from site observations, in order to determine the residual strength, stiffness and ductility of the structure. Finally, it has to be estimated whether or not the seismic excitation under consideration did not exceed the level of seismic hazard adopted by the code for the particular zone, as expressed by the design response spectrum.

Such a procedure is time-consuming and requires the full involvement of specialised personnel with a variety of technical means at their disposal. It is therefore obvious that such a procedure cannot be activated during the phase of emergency classification of structures as usable or not, since it is not feasible due to time and cost constraints.

Thus, engineers come face to face with buildings struck by the earthquake, without being able to use the scientific tools that they possess for the quantitative evaluation of the structure, which are the *in situ* measurements, the tests and the analysis. They are compelled by the circumstances to restrict themselves *to qualitative evaluations and make decisions based solely on visual observation of damage, using of course their knowledge, their experience and good engineering judgment.* This last statement is extremely important because it shows that these evaluations are very subjective. Notwithstanding its weaknesses, according to international practice, damage assessment is based on the above procedure (ATC 3-06, 1978; Greek National Report, 1982; Yugoslav National Report, 1982; UNDP/UNIDO, 1985).

12.2.3.2 General principles of damage assessment

Although damage assessment and decision making regarding the degree of usability of a structure are very subjective, there are some general principles that structural engineers must keep in mind when they have to make a decision concerning a building damaged by an earthquake. These are as follows:

1. They must have clearly in mind that their judgment must be limited to the evaluation of the risk of partial or total collapse in case of an aftershock, which is an earthquake of smaller magnitude than the main earthquake that comes from the same tectonic fault, therefore having characteristics similar to the main one (similar acceleration spectrum with smaller maxima). Thus, if the building does not exhibit damage to the structural system from the main shock, it means that it has not exceeded the elastic range; therefore the probability of damage, and, even more, the probability of collapse caused by aftershocks are statistically insignificant.

2. The risk of partial or total collapse of a structure damaged by the main earthquake comes from failure of vertical structural members (columns, structural walls, load-bearing masonry) under the action of vertical loads in combination with the horizontal seismic loads from aftershocks, which are expected to be smaller than the loads from the main event. The engineer should keep in mind that if damage appears in the structural system of a building, it means that the elastic range has been exceeded and, therefore, the resistance of the structure to seismic loading has been reduced by the main seismic event.

3. According to the above, engineers who perform the inspection first have to find out the layout of the structural system of the building, at least in the cases when it is damaged. If necessary, in the absence of drawings, they should use hammer and chisel in order to determine the location of the vertical structural elements of the building. No reliable damage assessment is possible without a clear understanding of the structural system of the building.

4. In order to estimate the residual strength, stiffness and ductility of a structure, the engineer has to trace out the damage to the structural system as well as to the infill panels. Particularly hazardous is damage to the vertical elements, especially at the ground floor. Crushing of concrete at the top or the bottom of a column accompanied by buckling of the longitudinal reinforcement, X-shaped cracks in shear walls with significant axial loading and X-shaped cracks in short columns are some of the types of damage that should seriously worry the engineer who performs the inspection. Extensive X-shaped cracks in the infills accompanied by permanent deviation of the structure from the vertical are also alarming indications. In contrast, cracks in horizontal structural elements, caused by either flexure or shear, are not particularly alarming. The same holds for limited spalling of vertical elements or flexural cracks. However, it has to be stressed that the familiarity of the engineer with the various types of earthquake damage is very useful, if not indispensable, for damage evaluation (see Chapter 11).

5. Before concluding the final evaluation, the engineer should pay particular attention to the following:
   a. The configuration of the structure. Buildings with symmetric or approximately symmetric floor plans have better seismic behaviour than asymmetric ones.
   b. The location of the vertical stiffness members on the floor plan. Symmetrically located stiffness members drastically reduce the consequences of eccentric loading.

   c. The existence of a flexible storey. The open ground floor (pilotis system) or ground floor occupied by stores makes the building vulnerable to seismic actions.

   d. The quality of the construction material. *In situ* tests with hammer and chisel, or, better, with special hammer testing equipment, constitute a very good relative indicator for the engineer, who can draw very useful conclusions after a few repetitions.

   e. The location of the structure in the block. It should not be forgotten that the vast majority of total collapses worldwide have occurred in corner structures.

6. The conclusion of the foregoing discussion is that the engineer can make one of the following decisions:

   a. Allow use of the building without any restriction, provided it does not exhibit any visible damage to the structural system (classification with, say, the colour *green*)

   b. Classify it as temporarily unusable and limit access to it (shifting the responsibility to the occupant) because of limited damage, until it is repaired (classification with, say, the colour *yellow*)

   c. Classify it as out of use because of extensive damage, until based on a detailed study, it is decided either to repair or demolish it (classification with, say, the colour *red*)

   d. Classify it as a *near-collapse building* and activate propping measures together with prohibition of any approach to the area around the building

It is obvious that buildings that fall into the first and last categories can easily be distinguished from buildings of the other two categories, for which capacity to resist an earthquake has been reduced because of damage, but they do not exhibit signs of a near-collapse condition. In contrast, the distinction between the 'yellow' and 'red' categories is not always easy; therefore, in case of doubt, the more conservative decision should be made.

12.3 ORGANISATIONAL SCHEME FOR INSPECTIONS

12.3.1 Introduction

The earthquake damage assessment – a job greatly affected by subjective judgment – requires hundreds, and sometimes even thousands, of engineers, each with a different level of knowledge, experience and engineering judgment. Therefore, prior to the earthquake, an appropriate organisational scheme should be developed, which would ensure the following:

- Immediate start of the inspections just after the earthquake and the main aftershocks, and the completion of the entire operation in a short time
- Damage assessment in a uniform and in as objective as possible way, so that mistakes are statistically minimised
- Detection of possible mistakes made during the first inspection and damage evaluation
- Timely notification of the authorities about buildings in need of propping or demolition.

The basic features of such a scheme are discussed in the sections that follow.

12.3.2 Usability classification–inspection forms

For an objective and uniform damage evaluation, the inspection must be performed by teams consisting of at least two engineers, so that a judgment that is as objective as possible is formulated. It is also necessary to prepare ahead of time special forms for damage description, strength evaluation and usability classification of structures, which should be based on the following principles:

- Easy completion of the data of the structure and the degree of damage, based on visual inspection as already mentioned.
- Assignment of the damage degree into a few clearly defined categories.
- Assessment of the usability of the building on the basis of clearly defined categories. International practice (ATC 3-06, 1978; Yugoslav National Report, 1982; UNIDO/UNDP, 1985; FEMA, 1986) has adopted the three levels of usability mentioned in the preceding section, each one of which is characterised by the colour of the sticker posted on the buildings, the already mentioned green, yellow and red stickers, which were also used after the 1978 Thessaloniki, Greece, earthquake.
- Codification of data for future statistical processing.

It should be mentioned that most countries currently have standard inspection forms (ATC 3-06, 1978; Yugoslav National Report, 1982; FEMA, 1986). In the framework of the UNIDO/UNDP-funded program 'Earthquake-Resistant Structures in the Balkan Region', such an inspection form was developed (Anagnostopoulos, 1984) as a result of cooperation of the Balkan countries supported by international experts.

In Greece, between 1978 and 1981, simple inspection forms were used, which partly fulfilled the requirements given above. An attempt was made to use a form similar to that proposed by UNIDO (Constantinea and Zisiadis, 1984) in Kalamata (1986). However, the attempt failed for the following reasons:

1. In most of the buildings, some apartments were locked, and hence detailed recording of damage was not possible.
2. The engineers were not trained to complete the forms in a uniform way.
3. The requirement for quick completion of the inspections did not allow a systematic and detailed description of the situation.

The author's opinion is that the forms for the preliminary (emergency) inspection must be as simple as possible, while for the statistical evaluation of the consequences of the earthquake, the engineer who has the responsibility for the repair should submit a detailed form to the authorities, along with the repair design, after performing a second detailed inspection aiming at designing the repair of the building.

12.3.3 Inspection levels

In order to locate probable mistakes made during the first (emergency) inspection, a second degree of inspection must be carried out, performed by two-member teams of engineers with high qualifications and experience. In both ATC 3-06 (1978) and the final manual of UNIDO/UNDP (1985), such an inspection is suggested for buildings that have been classified as red, as well as for buildings with owners or occupants that hindered the evaluation of the first inspection. This procedure was also followed in Thessaloniki in 1978.

The timely notification of the authorities about buildings in need of *immediate propping* or *demolition* is one of the first priorities of the inspection teams, during both the first- and second-degree inspection.

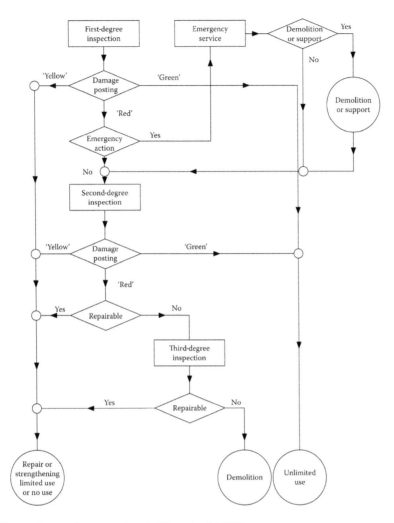

Figure 12.1 Chart of inspection procedure in Thessaloniki 1978.

Finally, it should be emphasised that for the inspection of buildings that are of vital importance for post-earthquake life in the affected area, committees of highly qualified engineers must be formed to inspect these buildings first. In Figure 12.1, the chart of the successive inspection levels of structures in the Thessaloniki earthquake is shown, the application of which did not exhibit any particular problem.

12.4 EMERGENCY MEASURES FOR TEMPORARY PROPPING

12.4.1 General

Immediate shoring (temporary propping) is recommended for buildings with serious damage in the vertical structural elements (columns, walls) or serious inclination from the vertical. By using shoring, the damaged elements are relieved of their loads by temporary additional structures, and, therefore, the danger of collapse, due to aftershocks, is mitigated.

Propping must take place *initially* at the floor where the damage of the vertical element occurred. It is necessary, however, to estimate the ability of adjacent beams to carry the vertical load of the damaged element, and if this is not adequate, support must be extended to other floors as well (Figure 12.2). The support system must be placed at a certain distance from the damaged element so that enough room is left for the repair work that will follow.

When there are problems of lateral instability in a structure, displayed mainly by vertical inclination, lateral support is provided either in the form of ribs or in the form of diagonal braces between the frames formed by beams and columns; even internal tension ties can be used for supporting buildings close to collapse.

The design of temporary supports must be done promptly, with the aid of approximate analysis and design, performed to determine only the order of magnitude of actions and action effects (stresses). The materials and techniques foreseen must be readily available, for instance, metal scaffolds, timber, steel profiles, timber grillage and so on.

Given the fact that the shoring of damaged structures is a very hazardous work for the people involved, the time these people spend in the building must be kept to a minimum. Therefore, it is recommended that the preparation of all the supporting elements is done away from the damaged structure (based on the dimensions measured on site), so that the work of shoring will be limited to the installation of these elements in the damaged building. It is recommended that extensiometers in the form of glass pieces bonded through gypsum plaster on critical cracks should be established before any propping work on site to safeguard the stability of the structure and, therefore, the life safety of the personnel involved in propping activities.

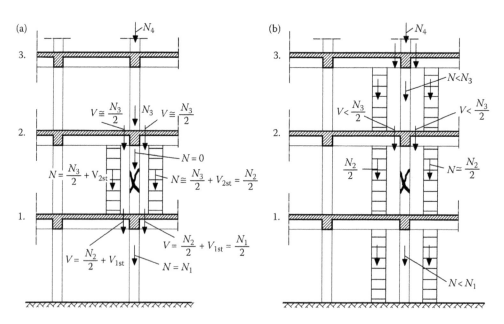

Figure 12.2 Shoring (temporary supporting) of a multi-storey building with a damaged column: (a) shoring of only one floor; (b) shoring of more floors.

12.4.2 Techniques for propping vertical loads

12.4.2.1 Industrial-type metal scaffolds

In the case of small loads, independent industrial-type metal tube shores are used (Figure 12.3), having a load-bearing capacity of about 20 kN and a height of about 3.00 m.

For the shoring of beams or slabs, prefabricated metal towers are used (Figure 12.4), which are wedged to the surface to be shored with the aid of the special screw-type bolts with which all industrial-type scaffolds are equipped. Usual SLS (Service load) capacity: 4× (20–35) kN for a free height of 3.0 m.

12.4.2.2 Timber

Timber elements can also be used for carrying vertical loads, either in the form of logs or telephone poles, or in the form of timber grillages. For every damaged column, at least one 250-mm-diameter log should be used on each side of the column. The allowable (service) load for this diameter and for floor heights about 3.00 m is estimated at 300 kN per pole for timber of good quality. If the height is greater than 3.00 m or the diameter smaller than 250 mm, the pole must be checked for buckling (EC 5/2004).

In case two or more supporting elements are used on each side, they must be connected to each other with X-shaped braces. If no logs are available, shoring can be achieved with timber grillage (Figure 12.5).

12.4.2.3 Steel profiles

Steel profiles can be used either in the same manner as timber or as an immediate strengthening means of the damaged column (Figure 12.6). In this case, they can be incorporated later into the concrete jacket. In the first case, there should always be a buckling verification.

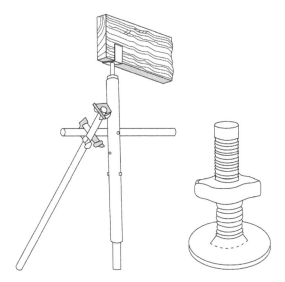

Figure 12.3 Independent industrial-type metal supports.

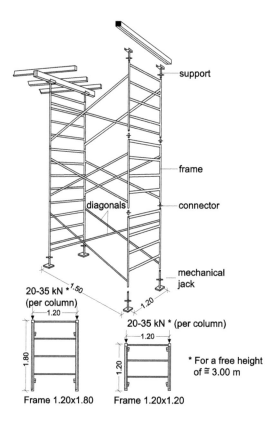

Figure 12.4 Industrial-type metal towers.

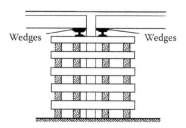

Figure 12.5 Shoring with timber grillage.

In the second case, the key to success is the tightening of the vertical steel angles to the column with transverse angles and prestressed ties before the transverse connecting straps are welded to the vertical angles.

12.4.3 Techniques for resisting lateral forces

12.4.3.1 Bracing with buttresses

Bracing with buttresses is the most common way of resisting lateral forces. These forces are due to the deviation of the building from the vertical axis either because of failure of vertical structural elements or because of settlement of the foundations (Figure 12.7). Some critical points of such a bracing system are the following:

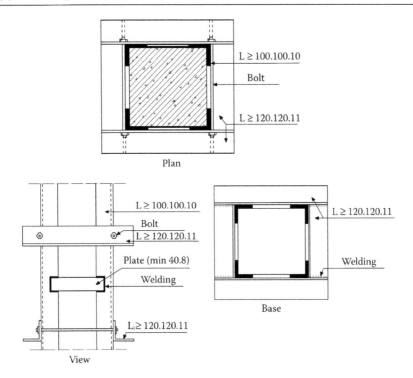

Figure 12.6 Immediate tying of a column with steel profiles.

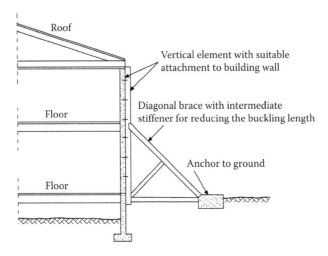

Figure 12.7 Bracing with buttresses.

- Anchoring of the bracings to the ground, so that they may resist horizontal thrusts
- Attachment of the vertical member to the building so that it prevents relative slipping
- Limiting the unbraced length of the inclined member of the lateral bracing to low values, to avoid in-plane or out-of-plane buckling

For this type of shoring, timber members are used more often than steel members. It should be noted that the horizontal forces that such a system is assumed to resist, for small

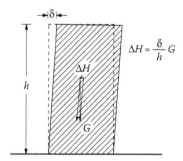

Figure 12.8 Estimate of the horizontal forces due to deviation from the vertical axis.

deviations from the vertical axis, are not very large and can easily be estimated approximately, using the relationship (Figure 12.8)

$$\Delta H = \frac{\delta}{h}G \qquad\qquad (12.1)$$

where δ/h is the deviation from the vertical axis and G is the total vertical load of the structure, which for normal buildings is estimated to be 10.0–12.0 kN/m² of floor area.

12.4.3.2 Bracing with diagonal X-braces

On the one hand, the use of diagonal timber or steel members in the plane of R/C frames allows the partial transfer of gravity loads to undamaged vertical elements, and on the other, prevents lateral deformation (Figure 12.9).

Frame bracing can consist of timber, tree logs or steel profiles of sufficient strength considering their tendency for buckling. This method is used when external bracing with buttresses cannot be easily installed.

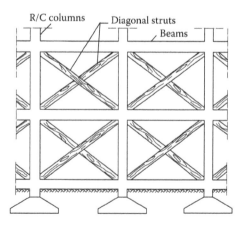

Figure 12.9 Bracing with diagonal struts.

12.4.3.3 Bracing with interior anchoring

In the case of hybrid structures consisting of R/C slabs supported by masonry, in order to retain external walls that have been detached and deviate from the vertical axis, metal tensioners are often used, which are prestressed with the aid of tensioner couplers (Figure 12.10).

12.4.3.4 Bracing with tension rods or rings

In the case of deviation from the vertical axis due to arch thrusts, prestressed metal rings or prestressed rods are used, depending on whether the structure is a dome or an arch.

12.4.4 Wedging techniques

The wedging procedure is a crucial part of every supporting or bracing procedure, because the transfer of the loads of a damaged element to the shoring or bracing system is accomplished through wedging. Wedging can be achieved by the following means:

- Wooden twin wedges (Figure 12.11)
- Mechanical jacks (screws; Figure 12.4)

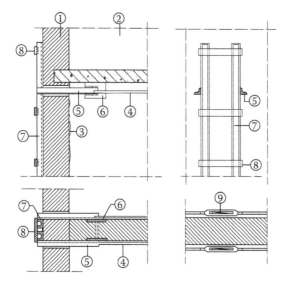

Figure 12.10 Bracing with internal tension ties: 1 = exterior wall; 2 = interior wall; 3 = crack; 4 = steel tensioner; 5 = angle 50 × 50 × 5 mm; 6 = steel plates; 7 = steel profiles; 8 = steel plates; 9 = tensioner coupler.

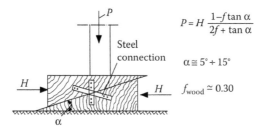

$$P = H \frac{1 - f \tan \alpha}{2f + \tan \alpha}$$

$$\alpha \cong 5° \div 15°$$

$$f_{\text{wood}} \cong 0.30$$

Figure 12.11 Wedging with twin wooden wedges (f is the friction coefficient).

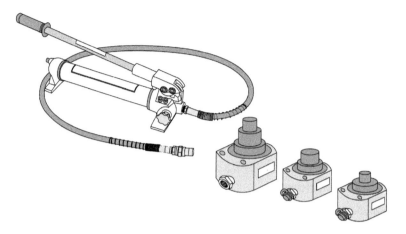

Figure 12.12 Wedging with hydraulic jacks.

Figure 12.13 Timber propping of a damaged column; Bucharest, Romania (1977).

- Hydraulic jacks (Figure 12.12)
- Couplers (Figure 12.10)

12.4.5 Case studies

In Figures 12.13 through 12.16, various propping case studies have been depicted from post-earthquake shoring measures.

12.5 FINAL REMARKS

The main conclusion of this chapter is that emergency damage inspection and assessment after an earthquake is a completely different procedure from that followed for the evaluation of residual seismic resistance of a structure in the pre- or post-earthquake period. This happens because in the case of an earthquake, very often thousands of buildings must be inspected and evaluated in a short period of time, so that the affected area can return to normal life.

Therefore, given the fact that hundreds and sometimes even thousands of engineers must be engaged in damage evaluation, each with a different level of knowledge, experience and

Figure 12.14 Timber propping of a damaged column; Kalamata, Greece (1986).

Figure 12.15 Steel profile props for the repair of a damaged column; Kalamata, Greece (1986).

Figure 12.16 Confinement of an R/C wall using external profiles, for temporary propping and in parallel final repair.

engineering judgment, the main task is the creation of the appropriate organising scheme before the earthquake, which will ensure the following:

- A damage evaluation that is uniform and as objective as possible, so that statistical mistakes will be minimised.
- Quick detection of any serious mistakes in evaluation from the first inspection.
- Timely notification about buildings in need of immediate support or demolition.

It is understood that such operations, which require a high degree of staff organisation and are directly associated with public security, should be undertaken by the central or regional government.

For the uniform and objective evaluation of the damage degree, it is necessary to prepare in advance special forms for damage description, residual seismic resistance evaluation and usability classification of structures, which must be based on the following principles:

- Easy completion with the data of the structure and the damage level, based on visual inspection
- Assignment of the damage degree into a few clearly defined categories
- Assessment of the usability of the building, on the basis of clearly defined categories
- Organisation of data for future statistical processing

For the effective completion of the engineers' mission, there is a need to organise short training courses, accompanied by visual aids (PowerPoint, DVDs, etc.) on the expected damage classified into categories, and on the procedure of completion of the inspection forms.

For the detection of serious errors and for inspection of special buildings, there is a need for a second-level inspection by experienced, highly qualified structural engineers.

For timely notification for emergency action on buildings needing propping or demolitions, the development of an appropriate coordination mechanism is required.

Chapter 13

Seismic assessment and retrofitting of R/C buildings

13.1 GENERAL

The seismic evaluation of an aggregate of existing buildings has as its main objective the decision-making about the need for rehabilitation of each individual building. On the other hand, any retrofitting activity aims at upgrading the seismic performance of the building to a level similar to that of new buildings designed and constructed in accordance with the Seismic Codes now in effect. Keeping in mind that the retrofitting cost (structural–non-structural) is very high, ranging around a mean value of 10–12% of the original cost of construction, it is apparent that an evaluation procedure must be established for screening the buildings under consideration in successive steps of accuracy, and until the number of buildings designated for eventual intervention has been minimised. At the same time, the reduction of the number of buildings from step to step also reduces the cost of evaluation of the whole procedure. For these buildings, a detailed quantitative evaluation is implemented and eventually a retrofitting design is applied.

National programmes for seismic risk mitigation, through seismic assessment and retrofitting, may differentiate between 'pre-earthquake' and 'post-earthquake' programmes. The first category is triggered by 'political decisions' and, in this respect, they may be characterised as 'active' programmes, while the second is triggered by a strong earthquake or, rather, the damages caused by the earthquake, and, in this respect, they may be characterised as 'passive' programmes (EC8-3/2005).

Buildings for which the owners have decided to intervene mainly for commercial reasons, such as a change in use that increases occupancy or importance class, remodelling above certain limits and so forth, must be classified in an independent category. In fact, in this case, the 'screening' procedure is omitted and the building is evaluated and retrofitted in detail directly, since the decision for intervention has already been taken for operational reasons.

It is reasonable that in the main two categories, since the trigger is different, the methods of seismic evaluation at the stage of screening display differences, while at the final stage of quantitative detailed seismic evaluation and retrofitting design, the procedure that is implemented is the same for both (Figure 13.1). In the following two sections, the above two cases will be presented:

- Pre-earthquake seismic evaluation of R/C buildings
- Post-earthquake seismic evaluation of R/C buildings

Subsequently, the quantitative detailed seismic evaluation and retrofitting design will be presented in detail.

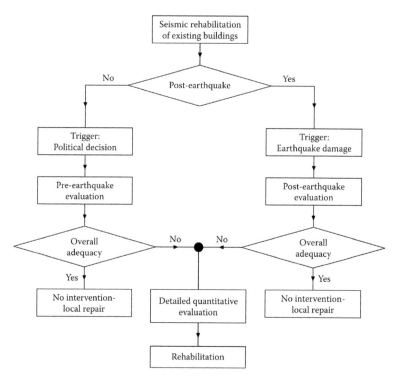

Figure 13.1 Flow chart for pre- and post-earthquake evaluation and rehabilitation.

13.2 PRE-EARTHQUAKE SEISMIC EVALUATION OF R/C BUILDINGS (TIERS)

For many decades, the main tool for seismic risk reduction was considered to be the design and construction of earthquake-resistant buildings based on modern Codes into which new knowledge on earthquake engineering and engineering seismology had been incorporated. However, as engineering seismology and earthquake engineering have been passing through a transitional period of development over the past 100 years, buildings designed and constructed in the past based on Codes then in effect are no longer compatible with modern Codes. These buildings constitute the majority of the building stock of a region and, therefore, they also constitute the main problem in case of a strong earthquake. For example, in the case of the Athens earthquake of 1999, the number of buildings constructed in the most affected area (Karabinis, 2002; Figure 13.2) during the 14 years after the benchmark year of establishment of a modern Seismic Code (1985) was only 17 out of 214 R/C buildings. Collapses of buildings constructed during this 14-year period were only 4 out of 93 in total.

In this context, it can be understood, particularly in the United States since the early 1980s, that a drastic reduction of seismic risk could be achieved only by an 'active' intervention in the existing building stock, where the most damages and collapses take place in case of a strong earthquake.

Therefore, systematic initiatives were undertaken for the development of a framework of procedures for seismic evaluation of buildings and retrofitting of those with critical deficiencies. These initiatives were mainly developed in the United States (ATC-3-06/1978, FEMA 154/ATC-21/1988, FEMA 178/1992, ICSSC PR-4/1994, FEMA 273-274/1997, ASCE 31-02/FEMA 310/1998, FEMA 356/2000, ASCE/SEI 41-06).

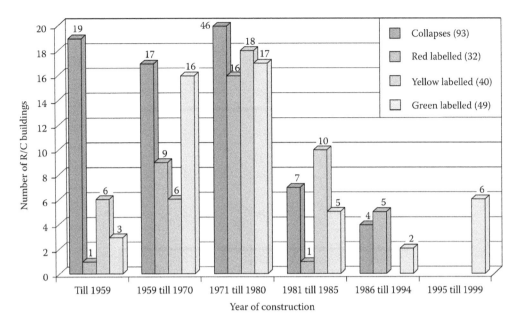

Figure 13.2 Distribution of vulnerability of R/C buildings in the most affected area of Athens in the 1999 earthquake, by year of construction. (Adapted from Karabinis, A. 2002. Validation of the pre-earthquake assessment procedure of R/C buildings in Greece, Technical Report (O.A.S.P.) Athens, Greece.)

At the same time, similar procedures have been developed to a degree in other countries (e.g. UNIDO/UNDP, 1985; Japan, 1990; New Zealand, 1996; Italy, 1998; Greece, 2000). In most cases, the procedure is articulated *in three successive steps (Tiers) of higher detail and accuracy, from the first step to the third.*

Nowadays, the U.S. framework of standards in effect for pre-earthquake seismic evaluation and rehabilitation of existing buildings comprises

- FEMA 154/ATC-21/1988. Rapid visual screening of buildings for potential seismic hazards (Tier 1).
- ASCE 31-02/FEMA 310/1998. 'Seismic evaluation of buildings' articulated in three successive steps of higher detail and accuracy (Tier 1 to 3, Figure 13.3).
- ASCE/SEI 41-06/FEMA 356/2000. Seismic rehabilitation of buildings is the third step of FEMA 310, together with detailed procedures for rehabilitation (Tier 3).
- Recently (2014), the above standards have been incorporated in ASCE/SEI 41-13 including some modifications of low importance.

In the first edition of this book (Penelis and Penelis, 2014), a brief reference has been made to the U.S. system of seismic evaluation of buildings, which may be considered the most integrated system in effect in the world.

Pre-earthquake rehabilitation programmes, due to their high cost, have been activated so far only for special categories of buildings (schools, hospitals, etc.) and in countries of high GNP. Among them, the pre-earthquake seismic evaluation programme of the United States (Presidential Order 12941, 1994) for government-owned or leased buildings may be considered the most extended and significant one.

The European framework of standards includes only EC8-3/2005, which is relevant to ASCE/SEI 41-06/FEMA 356 and refers to a detailed quantitative seismic assessment and

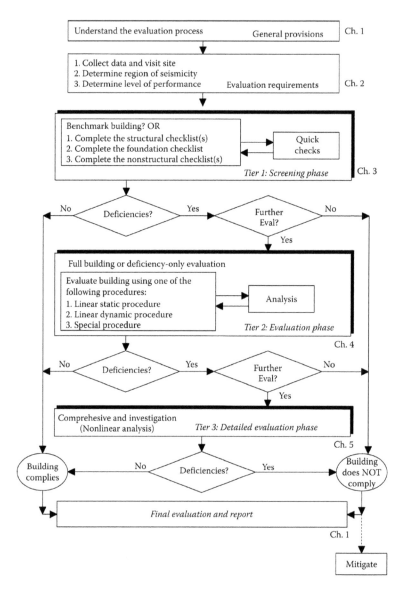

Figure 13.3 Evaluation process according to ASCE 31-02/FEMA 310/1998. (From FEMA-310/1998/ASCE 31-02. *Handbook for the Seismic Evaluation of Buildings* – A standard prepared by the ASCE for the Federal Emergency Management Agency, Washington, DC. With permission.)

retrofitting of buildings, leaving to the member states the initiative for development of standards for the preliminary seismic evaluation of buildings.

13.3 POST-EARTHQUAKE SEISMIC EVALUATION OF R/C BUILDINGS

13.3.1 Introduction

As noted in Section 13.1, strong earthquakes and the damage that they cause are the trigger for post-earthquake assessment and retrofitting actions. In this context, the main criterion for seismic evaluation and the degree of retrofitting should be the extent of damage and its seriousness.

However, most of the methods developed so far do not make any distinction in evaluation procedure between pre- or post-earthquake cases. In this respect, methods that have been mentioned in the previous section could also be implemented in the case of post-earthquake seismic evaluation. Keeping in mind that damage due to earthquakes may extend from plaster detachment to heavy damage of structural elements or overall structural degradation to a near-collapse situation or even collapse, it is obvious that the implementation of methods that are applied in pre-earthquake conditions needs to have a well-defined limit of the level of damage, above which the evaluation process is activated. For example, in the case of the Thessaloniki earthquake (1978), which is a typical case study, buildings with severe damage corresponded only to 4.5% of the building stock of the city, buildings with intermediate damage corresponded to 21% and buildings with damage only at partition walls corresponded to 22.9%. In this context, it is apparent that the cost of seismic evaluation and retrofitting increases rapidly in relation to the level above which evaluation is activated (Penelis, 2008).

In practice, in most cases, the degree and extent of seismic evaluation, and, therefore, the consecutive retrofitting actions, are strongly related to the seismic damage (Penelis and Kappos, 1997). *This approach radically decreases the rehabilitation cost of an affected region and the time required for the whole campaign.*

13.3.2 Objectives and principles of post-earthquake retrofitting

The main objectives of intervention in an earthquake-damaged building are to protect the structure from collapse in a future strong earthquake, to keep damage at tolerable levels in earthquakes of moderate intensity and to eliminate damage in earthquakes with a relatively short return period. In other words, the objectives of an intervention more or less coincide with those set for the design of a new structure.

In this sense, the only rehabilitation action that can guarantee the above objectives is strengthening, which means the repair of damaged structural elements and the increase in the seismic resistance of the structure up to the value of the required seismic resistance by the Code in effect, using additional strengthening measures. However, such an approach must be considered to be unrealistic.

In fact, taking into account that the buildings with severe seismic damage are limited to 4.5–5.0% of the building stock of a city (see Section 13.3.1), it is obvious that strengthening actions cannot be extended to all buildings with any degree of damage, even only to the partition walls, because in this case, the strengthening activities would be extended to almost 50% of the building stock, and, in this context, the intervention cost would be multiplied by almost 10. In other words, it is not reasonable to transform in practice the post-earthquake repair and strengthening intervention to a pre-earthquake rehabilitation activity having as trigger the seismic event and not a thorough cost-benefit analysis, which might activate such a rehabilitation action for a special category of buildings in case of a pre-earthquake intervention and of course without the panic of the consequences of an earthquake.

In view of the foregoing, the most realistic approach appears to be the repair of the damaged building to pre-earthquake condition. This is based on the notion that if the damaged elements, structural and non-structural, are repaired, the structure more or less regains its pre-earthquake seismic resistance and, therefore, will behave similarly in a future earthquake with the same characteristics. However, this approach has a weak point: the extent and the seriousness of damage caused by an earthquake constitute the most reliable criterion regarding the difference of the available seismic resistance V_C and the required one, V_B.

Therefore, in the case of serious damage with signs that the structure came near to collapse, we cannot consider repair to be enough. The structure should be strengthened to a degree so that in a future earthquake, it will behave like buildings with no or only light damage (Holmes, 1994).

Based on the above, the recommended approach to the intervention procedure may be stated as follows:

1. In buildings with light damage, of local nature, intervention should be limited to repair (see Section 13.10.1).
2. In buildings with extensive or heavy damage, of global type, intervention should include strengthening of the structure (see Section 13.10.1).

Extended discussion accompanied by divergent opinions on this issue may be found in the literature (Anagnostopoulos et al., 1989; Freeman, 1993).

In conclusion, it should be noted that in post-earthquake rehabilitation practice, the trigger for structural repair or strengthening is the type and the extent of damage to the structural system. Therefore, only buildings labelled as 'yellow' or 'red' by the post-earthquake inspecting of the affected region are further evaluated for *repair* or *strengthening* (see Chapter 12.2.3.2).

In the manual for *Post-Earthquake Damage Evaluation and Strength Assessment of Buildings under Seismic Conditions* of UNIDO/UNDP (1985), a detailed procedure has been developed, presented in short in the first edition of this book (Penelis and Penelis, 2014).

Greek practice for post-earthquake interventions since 1978, and up to the recent Athens earthquake (1999), has used the ratio of residual post-seismic resistance V_D to the pre-seismic capacity V_C of the building in terms of base shear (V_D/V_C) as an index for the seismic assessment (Chronopoulos, 1984; Penelis, 1979; Tassios, 1984c). The values of this index are estimated in practice by simplified procedures.

13.4 QUANTITATIVE DETAILED SEISMIC EVALUATION AND RETROFITTING DESIGN

As was clarified in Section 13.1, detailed seismic assessment and retrofitting of buildings is activated after a seismic evaluation procedure has been completed and buildings designated for detailed investigation and eventual rehabilitation have been identified. Whether the designation of these buildings is the result of a pre-earthquake or a post-earthquake procedure, the method for this detailed assessment or retrofitting is the same.

It should be noted in advance that detailed assessment and retrofitting is basically governed by the 'displacement-based design concept' in all modern guidelines and Codes. Therefore, the next section will be devoted to the presentation of this method, before any reference to Code specifications is made.

The core of this method is the elaboration of the *capacity curve* of the structure, which means the implementation of an inelastic static analysis, for a lateral force or displacement pattern (see Chapter 5.7.3), and the transformation of the MDOF structural system to an *equivalent SDOF* one, for which target displacement can easily be determined using relevant response spectra reduced to proper ductility or damping level (Chapter 5.7.5).

It should be noted that in the United States, detailed seismic evaluation and rehabilitation is specified by ASCE/SEI41-06/(FEMA 356/2000). In Europe, it is specified by EC8-3/2005.

In the sections that follow, a detailed analysis of this procedure will be presented based mainly on EC8-3/2005, as done thus far in this book. Additionally, an overview will be made of methods used in the United States.

13.5 OVERVIEW OF DISPLACEMENT-BASED DESIGN FOR SEISMIC ACTIONS

13.5.1 Introduction

As explained in Chapter 3.2.3, new buildings are analysed and designed according to all modern Seismic Codes, in effect following the *force-based design* procedure.

The basic concept of this procedure is the *reduction* of the elastic inertial seismic forces by a *q-factor* and counterbalancing this reduction by providing adequate *ductility* at all critical regions of the structural members ensuring in this way the overall *q*-factor for the structural system.

For seismic loads reduced by *q*, linear analysis is implemented. Additionally, capacity-design procedures are introduced to postpone any early brittle failure of various structural members or an early overall collapse due to the creation of a soft storey.

The above procedure, which is followed for new buildings, presents serious problems for implementation in the case of existing buildings, because the *q*-factor cannot be codified since the ductility of the structural members of existing buildings built in various periods in the past does not comply with the provisions of modern Codes for local ductility and capacity design requirements.

Thus, a reliable approach to the problem would be to calculate the capacity curve of the structural system, taking into account the inelastic deformation capacities (θ_p) at the ends of the structural members (Figure 3.22a and b), and, therefore, the implementation of the displacement-based design *in the form of an inelastic static analysis*, as presented in concept in Chapter 3.2.2 and in detail in Chapter 5.7.3.

In the next subsection, this method will be presented in detail together with various simplifications that also allow the use of elastic procedures for the analysis and design of existing R/C buildings.

13.5.2 Displacement-based design methods

The most rigorous displacement-based design is based on the complete non-linear dynamic time–history analysis, which even today is considered very complex and impractical for general use (see Chapters 3.2.2 and 5.7.2). So, simplified non-linear methods have been developed based on a non-linear static analysis procedure known as *'pushover analysis'* (see Chapters 3.2.2 and 5.7.3). This method, developed in the last 20 years, is implemented mainly for the assessment, verification and retrofitting of existing buildings.

Various approaches to the problem have been developed that have been adopted by various Codes of Practice. An overview of the following methods will be presented below:

- N2 method (EC8-1/2004)
- Capacity-spectrum method (ATC 40-1996)
- Coefficient method (ASCE/SEI 41-06; FEMA 356/2000)
- Direct displacement-based design (DDBD; Priestley et al., 2007)

13.5.2.1 N2 method (EC8-1/2004)

This method has been adopted by Eurocode EC8-1/2004 and is recommended in Annex B as an informative procedure (Fajfar, 1996).

The steps for the implementation of this method may be summarised as follows (see Chapter 3.2.2.2):

Step 1. Determination of the capacity curve of the MDOF model
Consider the structural system of Figure 13.4a and its eigen mode (Φ_i) corresponding to its fundamental period T_1. For a lateral load pattern F_i normal to Φ_i ($\Phi_n = 1$), i.e.

$$F_i = \Phi_i \cdot m_i \tag{13.1}$$

increasing from zero values stepwise to collapse or at least to 150% of the target displacement (see Section 13.5.2.1, step 5), the *capacity curve* V_b–δ_n of the structure

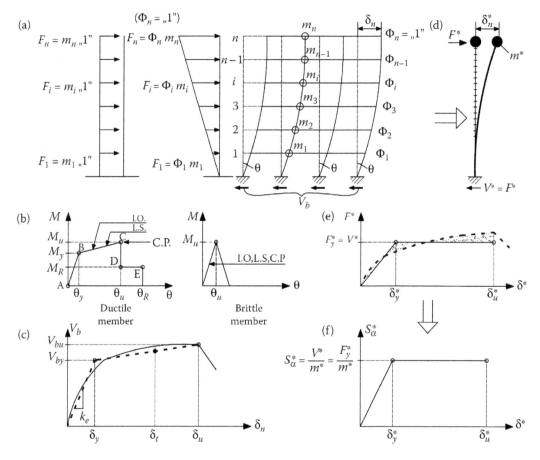

Figure 13.4 Procedure to obtain capacity curves of a structural system: (a) structural system (MDOF) loaded laterally by two different load patterns (eigen modes) for the implementation of a 'push-over' analysis; (b) input data (*M*–θ curves) for the member rotations at their ends; (c) capacity curve of the MDOF system; (d) equivalent SDOF model; (e) capacity curve of the equivalent SDOF model transformed to an EPP curve; (f) EPP capacity curve of the SDOF model normalised to a 'capacity spectrum'.

is derived, where V_b is the base shear and δ_n is the displacement of the top storey n (Figure 13.4c).

For the calculation of the coordinates of the above curve, it is necessary to know

- The geometry of the structure
- The quality of the construction materials (concrete–steel)
- The reinforcement of the structure in detail (bending-shear)
- Gravity loads of the system

This information makes possible the elaboration of $M-\theta$ diagrams for the ends of each structural member (beams, columns, shear walls; Figure 13.4b), which are input data required for the capacity curve calculation. It should be noted that these diagrams may have the form depicted in Figure 13.4b, depending on the member *ductility*. It should also be noted that the length of the plastic plateau $(\theta_u-\theta_y)$ depends on the axial load of the member (see Chapters 8 and 9). Therefore, an analysis of the structural system for gravity loads must precede any other calculation so that an approximate value of normal forces of the columns is determined, which is necessary for the elaboration of $M-\theta$ curves.

From the above, it may be easily concluded that the calculation of the capacity curve presumes that the structural system is known in detail. That is why inelastic static or dynamic analysis basically suits the assessment procedure of existing buildings. It may also be concluded that the formulation of the capacity curve is the core of the displacement-based design method.

It is important to note here that the adoption of the first eigen mode Φ_i of the elastic spectral dynamic analysis (e.g. inverted triangle) as a load pattern is not an indisputable assumption. It has been found through inelastic dynamic analysis that the results of the displacement δ_n (or building inter-storey drift) at various levels of deformation may be satisfied only if various load patterns are adopted (Elnashai and Di Sarno, 2008). Thus, in recent years, various methods have been developed for adapting the force distribution to the non-elastic state. These modified methods are known as 'adaptive pushover methods' (Bracci et al., 1997; Gupta and Kunnath, 2000; Elnashai, 2002). Research to refine adaptive pushover methods is still in progress, for example, modal pushover analysis (Antoniou and Pinho, 2004; Chopra and Chintanapakdee, 2004). Eurocode 8-1/2004 specifies two vertical distributions of the lateral loads (Figure 13.4a):

- A 'uniform' pattern
- A 'modal' pattern ('inverted triangle')

Step 2. Equivalent SDOF elastoplastic model

The capacity curve of the MDOF system is transformed, then, to the capacity curve of an equivalent SDOF model (Figure 13.4d,e; Fajfar, 1996; Fajfar and Dolšek, 2000).

As explained in Chapter 5.7.5, this transformation is achieved by introducing the following transformed quantities corresponding to the SDOF model:

$$m^* = \sum m_i \Phi_i = \sum F_i \tag{13.2}$$

$$V^* = V_b / \Gamma \tag{13.3}$$

$$F_i^* = \frac{F_i}{\Gamma}, V^* = \frac{\sum F_i}{\Gamma} \tag{13.4}$$

$$\delta_n^* = \frac{\delta_n}{\Gamma} \tag{13.5}$$

where

$$\Gamma = \frac{m^*}{\sum m_i \Phi_i^2} = \frac{\sum F_i}{\sum (F_i^2 / m_i)} \tag{13.6}$$

is the 'transformation factor' (see Chapter 5.7.5, Equation 5.92).

Thereafter, the corresponding capacity curve may be transformed to that (Figure 13.4e) of an idealised elasto-perfectly plastic (EPP) force–displacement diagram.

For the SDOF model in EPP form, the original stiffness K^* and the fundamental period T^* result from the following relations (Figure 13.4e):

$$K^* = F_y^* / \delta_y^* = V^* / \delta_y^* \tag{13.7}$$

and

$$T^* = 2\pi \sqrt{\frac{m^*}{K^*}} = 2\pi \sqrt{\frac{m^* \delta_y^*}{F_y^*}} \tag{13.8}$$

Thereafter, this EPP capacity curve of the SDOF model may be normalised in terms of acceleration,

$$S_a^* = \frac{V^*}{m^*} \tag{13.9}$$

This curve is known as the 'capacity spectrum' (Figure 13.4f).

Step 3. Demand spectra

For the next step of the procedure, the acceleration elastic response spectrum must be known (demand spectrum; Figure 13.5a). From this spectrum, the elastic acceleration–displacement response spectrum (ADRS) is determined using the relation (Figure 13.5b; see Chapter 2.2.4.5)

$$\boxed{S_{ae} = \left(\frac{4\pi^2}{T^2} \right) S_{de}} \tag{13.10}$$

From the above expression, for various values of T, and for the corresponding values of S_{ae}, the relevant values of S_{de} may be determined, and, therefore, the ADRS graph may be easily plotted.

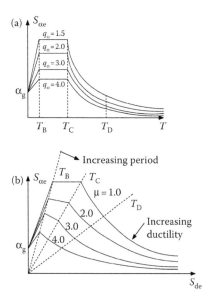

Figure 13.5 Demand spectra for various ductility μ values: (a) traditional acceleration response spectrum in relation to period T; (b) acceleration–displacement response spectra (ADRS) for various ductility μ values.

In this respect, radial lines correspond to constant periods T. From the family of the successive graphs, each corresponds to *a given ductility* μ and expresses for this ductility the corresponding inelastic response spectrum of the SDOF model.

The successive graphs may result from the elastic spectrum S_{ae}–S_{de} if the following expressions are used:

$$S_a = S_{ae}/q_0, \quad S_d = \left(\frac{\mu_\delta}{q_0}\right) S_{de} \tag{13.11}$$

where

$$q_0 = \mu_\delta \text{ for } T > T_c \tag{13.12}$$

$$q_0 = (\mu_\delta - 1)\left(\frac{T}{T_c}\right) + 1 \text{ for } T < T_c \tag{13.13}$$

(see Chapter 5.4.4.1, Equation 5.17). This form of design loading, as we will see next, can be compared directly to the non-linear capacity curve of the equivalent SDOF model.

Step 4. *Target displacement of the equivalent SDOF model*

Having calculated the capacity spectrum (Figure 13.4f), it may then be plotted upon the corresponding ADRS demand spectrum curve, the ordinates of which give the maximum acceleration of the elastic SDOF system in relation to the corresponding relative displacement S_e (Figure 13.6).

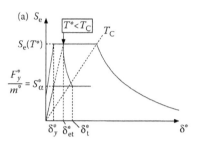

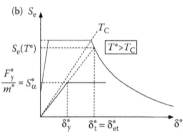

Figure 13.6 Determination of the target displacement for the equivalent SDOF system: (a) short period; (b) medium and long period.

The target displacement demand of the SDOF model with period T^* and *unlimited elastic behaviour* is given by the following expression:

$$\frac{\delta_{et}^*}{\delta_y^*} = \frac{S_e(T^*)}{S_\alpha^*} \tag{13.14a}$$

or

$$\delta_{et}^* = S_e(T^*)\frac{\delta_y^*}{S_\alpha^*} = S_e(T^*)\left(\frac{T^*}{2\pi}\right)^2 \tag{13.14b}$$

where

$S_e(T^*)$ is the elastic acceleration response spectrum at the period T^* (Figure 13.6).

Equation 13.14b is obtained by substituting Equation 13.10 into Equation 13.14a.

For the determination of the target displacement δ_t^* for inelastic behaviour, different expressions are used depending on the period T^* of the structure in relation to the corner period T_c of the plateau of the ADRS demand spectrum (Figure 13.6a,b).

- *Short period range* (Figure 13.6a)

$$T^* < T_c \tag{13.15}$$

In this case, if

$$S_a^* = \frac{F_y^*}{m^*} \geq S_e(T^*), \tag{13.16}$$

it means that the yield point of the capacity curve is higher than the corresponding elastic response spectrum, and therefore the response is elastic. Therefore,

$$\delta_t^* = \delta_{et}^* \tag{13.17}$$

Otherwise, if

$$S_a^* = \frac{F_y^*}{m^*} < S_e(T^*) \tag{13.18}$$

it means that the response is non-linear. Therefore (see Chapter 5, Equation 5.17),

$$\boxed{\delta_t^* = \frac{\delta_{et}^*}{q_0}\left(1 + (q_0 - 1)\frac{T_c}{T^*}\right) \geq \delta_{et}^*} \tag{13.19}$$

where

q_0 is the ratio between the acceleration of the structure with unlimited linear behaviour $S_e(T^*)$ and the structure with limited strength $F_y^*/m^* = S_a^*$,

$$\boxed{q_0 = \frac{S_e(T^*)}{S_a^*} = \frac{S_e(T^*)m^*}{F_y^*}} \tag{13.20}$$

- *Intermediate or long period*

$$T^* > T_c$$

In this case (Figure 13.6b),

$$\boxed{\delta_t^* = \delta_{et}^*} \tag{13.21}$$

It is recommended that δ_t^* should be limited to values less than $3\delta_y^*$.

Step 5. Global target displacement

Then, target displacement δ_t^* of the SDOF model is transformed to top target displacement of the MDOF model using the expression

$$\boxed{\delta_t = \Gamma\delta_t^*} \tag{13.22}$$

This value is compared with δ_u (near collapse global displacement),

$$\frac{\delta_u}{\delta_t} \geq \gamma_d \tag{13.23}$$

(see Chapter 3.2.2.2, Equation 3.13)

where

γ_d is the safety factor in terms of displacements.
This factor is specified in EC8-1/2004 to 1.50,

$$\boxed{\frac{\delta_u}{\delta_t} \geq 150\%} \qquad (13.24)$$

Step 6. Local seismic demands

For the defined target displacement δ_t, the local deformation quantities (e.g. θ_{dem}) of the ductile members, and the local strength demands (V_{dem}, M_{dem}) of the brittle elements as well, are determined by using the output of the pushover analysis of the system.

Step 7. Performance evaluation

Having the target displacement δ_t as the output of the 'pushover' analysis completed at a previous step, performance evaluation is carried out at the level of global and local performance as well, according to Code specifications. These specifications refer to

- *Global deformation performance* (e.g. $\delta_{target}/\delta_{collapse}$, global drift, distribution of the plastic hinges over the structure for the control of eventual premature collapse)
- *Local deformation performance for ductile members* (e.g. capacity rotation θ_u to rotation demand θ_p at the plastic hinges of the ductile members, inter-storey drifts, etc.)
- *Local strength performance for brittle members* (e.g. bending strength M_u to output moments for the target displacement or shear strength verification)

13.5.2.2 Capacity-spectrum method ATC 40-1996

This method was developed by the Applied Technology Council (ATC) in the framework of the 'Seismic Retrofit Practices Improvement Program' that was initiated by the 'Seismic Safety Commission of California'. It basically refers to R/C buildings. The steps for the implementation of this method maybe summarised as follows:

Step 1. Determination of the capacity curve of the MDOF model

This procedure is the same as that developed in Section 13.5.2.1.

Step 2. Equivalent SDOF model

The capacity curve of the MDOF system is transformed into the capacity curve of an equivalent SDOF model using the same procedure as in Section 13.5.2.1. The only difference from the previously presented method is that this capacity curve is not transformed further into an EPP diagram, but is used in the next step (step 4) in a *bilinear form (BI)*, as will be presented next.

Consider the capacity curve of an equivalent single-degree-of-freedom model transformed to a *capacity spectrum* in Figure 13.7. If the target displacement point were a_{pl}–d_{pl}, then the curve of the capacity spectrum could be transformed into a bilinear diagram of equivalent energy dissipation, as shown in Figure 13.7. The damping that occurs when seismic action drives the structure to point a_{pl}–d_{pl} may be given as a combination of viscous damping equal to 0.05 for R/C structures and hysteretic damping ζ_o,

$$\zeta_{eq} = 0.05 + \zeta_o \qquad (13.25)$$

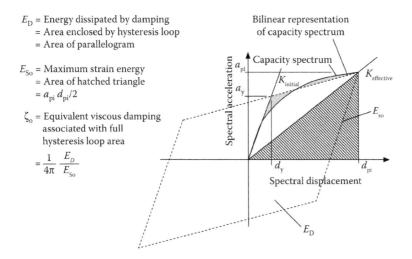

E_D = Energy dissipated by damping
  = Area enclosed by hysteresis loop
  = Area of parallelogram

E_{So} = Maximum strain energy
  = Area of hatched triangle
  = $a_{pi} d_{pi}/2$

ζ_o = Equivalent viscous damping
  associated with full
  hysteresis loop area

$= \dfrac{1}{4\pi} \dfrac{E_D}{E_{So}}$

Bilinear representation
of capacity spectrum

Figure 13.7 Derivation of damping for spectral reduction. (From ATC 40. 1996. *Seismic Evaluation and Retrofit of Concrete Buildings*. Applied Technology Council, Redwood City, California. With permission.)

Hysteretic damping ζ_o may be given by the relation (see Chapter 2.3.3, Equation 2.38)

$$\zeta_o = \frac{1}{4\pi} \frac{E_D}{E_{SO}} \tag{13.26}$$

Values of ζ_{eq} may be taken from Table 13.1 (ATC 40-96) for various slope ratios $((a_{pl}/a_y)-1)/(d_{pl}/d_y-1)$ and ductilities d_{pl}/d_y. The notation above is explained in Figure 13.7.

Step 3. Demand spectra

A family of successive demand spectra is plotted, each for a given *damping ratio* ζ (Figure 13.8), in standard S_{ea} versus T format, following the same procedure as in Section 13.5.2.1. Successive ARDS spectra may result, then, by combining the above spectra with a family of successive standard S_d (displacement) versus T spectra given by

Table 13.1 Effective damping, ζ_{eff}, in percent – structural behaviour type B

d_{pi}/d_y	Slope ratio: $[(a_{pi}/a_y) - 1]/[(d_{pi}/d_y) - 1]$						
	0.5	0.4	0.3	0.2	0.1	0.05	0
10	9	10	12	16	23	27	29
8	9	11	13	17	24	27	29
6	10	12	15	19	25	27	29
4	11	14	17	21	25	27	29
3	12	14	17	21	25	27	29
2	12	14	16	19	22	24	25
1.5	11	12	14	15	17	18	18
1.25	9	10	10	11	12	13	13

Source: ATC 40. 1996. *Seismic Evaluation and Retrofit of Concrete Buildings*. Applied Technology Council, Redwood City, California. With permission.

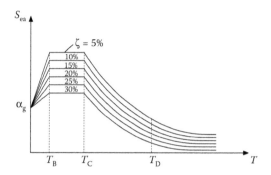

Figure 13.8 Family of demand spectra in standard S_{ea} versus T format.

Equations 3.39, 3.40 and 3.41 (Chapter 3.4.3.6; EC8-1/2004). The value of n in these equations is determined using the expression (see Chapter 3.4.3.4, Equation 3.33)

$$n = \sqrt{10/(5+\zeta)} \le 0.55 \qquad\qquad (13.27)$$

with ζ (%) values ranging from 5.0% to 30%. A family of such spectra is depicted in Figure 13.9.

Step 4. *Target displacement of the equivalent SDOF model*
Demand spectra and the capacity spectrum are plotted on the same diagram (Figure 13.10). The intersection points of the capacity spectrum with the demand

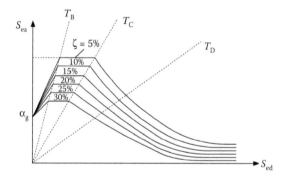

Figure 13.9 Family of demand spectra in ADRS format.

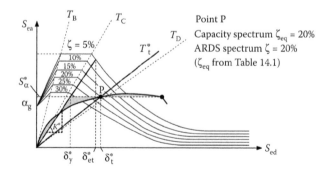

Figure 13.10 Determination of the target displacement implementing iterative procedure.

spectrum for which ζ_{eq} of the capacity spectrum is equal to the damping ζ of the demand spectrum presents the target displacement δ_t^* of the SDOF system. It is apparent that this point is determined by an iterative procedure.

Next steps for the determination of

- Global target displacement
- Local seismic demands
- Performance evaluation

follow the same procedure as in Section 13.5.2.1.

13.5.2.3 Coefficient method/ASCE/SEI 41-06 (FEMA 356/2000)

This method was originally developed by ATC in the framework of the FEMA 273-274/1997 report of 'Guidelines for the Seismic Rehabilitation of Buildings'. Soon thereafter, the American Society of Civil Engineers (ASCE) prepared the FEMA 356/2000 report, which was later issued as a national standard (ASCE/SEI 41-06). This is the most often used method in practice, at least in the United States.

This method allows the estimate of the target displacement of the system by applying a series of approximate inelastic displacement modifiers (i.e. coefficients) to the *elastic spectral displacement* of the system.

According to this method, the target displacement δ_t, which corresponds to the displacement at roof level, can be estimated by the expression

$$\delta_t = C_0 \cdot C_1 \cdot C_2 \cdot C_3 \cdot S_\alpha \frac{T_e^2}{4\pi^2} g \qquad (13.28)$$

where

C_0 is the modification factor to relate spectral displacement of an equivalent SDOF system to the roof displacement of the building MDOF system. It can be calculated as developed in Section 13.5.2.1, or it may be taken from Table 13.2 (FEMA 356).

C_1 is the modification factor to relate the expected maximum displacements of an *inelastic SDOF oscillator with EPP hysteretic properties to* displacements calculated for the linear elastic response.

Table 13.2 Values for modification factor C_0^a

| Number of stories | Shear buildings[b] | | Other buildings |
	Triangular load pattern (1.1, 1.2, 1.3)	Uniform load pattern (2.1)	Any load pattern
1	1.0	1.0	1.0
2	1.2	1.15	1.2
3	1.2	1.2	1.3
5	1.3	1.2	1.4
10+	1.3	1.2	1.5

Source: FEMA 356. 2000. *Prestandard and Commentary for the Seismic Rehabilitation of Buildings*, Washington, DC. With permission.

[a] Linear interpolation shall be used to calculate intermediate values.
[b] Buildings in which, for all stories, inter-storey drift decreases with increasing height.

$$C_1 = 1.0 \text{ for } T_e > T_c \tag{13.29a}$$

or

$$C_1 = \frac{1.0 + \left((R-1)/T_e\right)T_c}{R} \quad \text{for } T_e < T_c \tag{13.29b}$$

The upper limits of values of C_1 are given below:

$$C_1 = 1.50 \text{ for } T_e < 0.10 \text{ s} \tag{13.30}$$

$$C_1 = 1.0 \text{ for } T_e > T_c \text{ s} \tag{13.31}$$

In no case may C_1 be taken as less than 1.0.

T_c is the characteristic period corresponding to the end of the plateau of the elastic acceleration response spectrum (Chapter 3.4.3.3, Figure 3.16).
T_e is the effective period of the structural system (Figure 13.4c). This means that the capacity curve of the MDOF system must have been determined before the estimate of the target displacement.
R is the ratio of elastic strength demand to calculated yield strength coefficient (see below for additional information).
C_2 is the modification factor to represent the effect of hysteresis shape on the maximum displacement response. Values for C_2 are given in Table 13.3 (FEMA 356).
C_3 is the modification factor to represent increased displacements due to dynamic p–Δ effects. For buildings with positive post-yield stiffness, it takes the value

$$C_3 = 1.0 \tag{13.32}$$

Table 13.3 Values for modification factor C_2

Structural performance level	$T \leq 0.1 \ s^a$		$T \geq T_s \ s^a$	
	Framing Type I[b]	Framing Type 2[c]	Framing Type I[b]	Framing Type 2[c]
Immediate occupancy	1.0	1.0	1.0	1.0
Life safety	1.3	1.0	1.1	1.0
Collapse prevention	1.5	1.0	1.2	1.0

Source: FEMA 356. 2000. *Prestandard and Commentary for the Seismic Rehabilitation of Buildings*, Washington, DC. With permission.

[a]   Linear interpolation shall be used for intermediate values of T.
[b]   Structures in which more than 30% of the story shear at any level is resisted by any combination of the following components, elements or frames: ordinary moment-resisting frames, concentrically braced frames, frames with partially restrained connections, tension-only braces, unreinforced masonry walls, shear critical, piers and spandrels of reinforced concrete or masonry.
[c]   All frames not assigned to Framing Type I.

S_α is the elastic response spectrum acceleration at the effective fundamental period T_e and damping ratio ζ for concrete 5.0%.

The strength ratio R is equal to

$$R = \frac{S_\alpha}{V_{by}/W} \cdot \frac{1}{C_0} = \frac{q_0}{C_0} \tag{13.33}$$

where

V_{by} is the yield strength of the MDOF system (Figure 13.4c).

W is the total gravity loads that have been taken into account for the deformation of inertial seismic forces.

13.5.2.4 Direct displacement-based design (DDBD)

This method has been developed in the last 25 years (Moehle, 1987), and it is still in progress. Basically, it aims at mitigating the deficiencies in the current force-based design of *new buildings*. It is based on the substitution of an MDOF building by the equivalent SDOF model, already presented in Section 13.5.2.1, and the use of displacement design spectra developed for various levels of damping. Below, this method, which is presented in detail elsewhere (Priestley et al., 2007), will be presented in concept.

The method is illustrated in Figure 13.11a. A multi-storey building is replaced by an SDOF model (Figure 13.11a). The capacity curve of the SDOF model is assumed to be a bilinear elastic–plastic diagram for a predefined displacement (target displacement) δ_t (Figure 13.11b). Therefore, the ductility demand μ_D of the model is predefined. For this ductility, μ_D, the corresponding damping ratio ζ_{eq} is determined based on existing relations between the ductility μ_D and damping ratio ζ_{eq} (see Chapter 2.3.3, Equation 2.45; Figure 13.11c).

For the damping ratio determined above, and using displacement spectra directly (e.g. the displacement spectra of EC8-1/2004), the effective period T_e may be determined (Figure 13.11d). Thereafter, using the well-known equation

$$\boxed{K_e = 4\pi^2 m^* / T_e^2} \tag{13.34}$$

the effective stiffness K_e of the SDOF model may also be determined. Thus, the base shear of the SDOF model is given by the relation

$$\boxed{V_{base} = F_u = K_e \delta_t} \tag{13.35}$$

The base shear of the SDOF model is then transformed into the corresponding base shear of the MDOF structure. This value is distributed between the mass elements of the real structure as inertial forces, and then the structure is analysed for these forces to determine the design moments at locations of potential plastic hinges.

From what has been presented so far, the above method more or less follows the same steps as N2 and capacity-spectrum methods with some modifications relating, mainly, to the direct introduction of the displacement design spectra versus period T elaborated for various damping ratios (see Section 13.5.2.2).

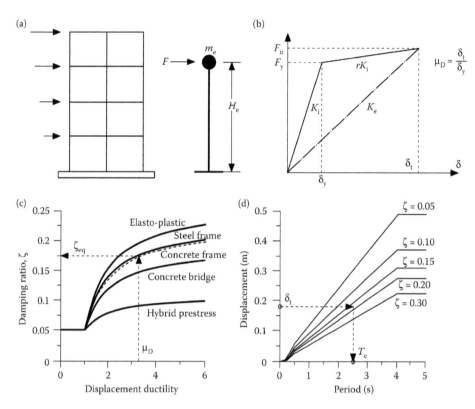

Figure 13.11 Fundamentals of direct displacement-based design. (a) SDOF simulation; (b) effective stiffness K_e; (c) equivalent damping vs. ductility; (d) design displacement spectra. (From Priestley, M.J., Calvi, G.M. and Kowalsky, M.J. 2007. *Displacement-Based Seismic Design of Structures*. IUSS Press, Pavia. With permission.)

The main question in the above procedure is how to define a series of structural information necessary for the assessment of the base shear V_b of the building before it has been designed and detailed. In particular, the following structural information is needed:

- The design displacement δ_t (target displacement)
- Displacement eigen modes
- Yield displacement δ_y
- Effective height H_e
- Distribution of the design-base shear

Thus, the main contribution of this method is the systematic elaboration of methodologies for the determination of the above-noted necessary information for various types of structures, taking into account their overall behaviour from the elastic range to collapse. More particularly, this information is given for

- Frame buildings
- Structural wall buildings
- Dual systems
- Masonry buildings
- Timber buildings
- Structures with isolation

The scope of this book does not allow further development of this issue, but a detailed presentation is made in the book *Displacement-Based Seismic Design of Structures* (Priestley et al., 2007).

13.5.2.5 Concluding remarks

From what has been presented thus far, the following conclusions may be drawn:

1. Methods N2 and capacity spectrum are quite similar.
   a. They use the capacity curve of the MDOF system.
   b. They transform this curve and all relevant mechanical properties of the system into those of an SDOF model.
   c. They both use ADRS demand spectra in the form of a family of curves; the N2 method for successive ductilities μ_B and the capacity spectrum for successive damping ratios ζ.
   d. The capacity-spectrum method makes use of the demand and the capacity spectra of the SDOF model plotted on the same diagram for the iterative estimate of the target displacement of the equivalent SDOF model.
   e. The N2 method, simplifying the capacity spectrum of the SDOF to an EPP-equivalent diagram, makes possible the direct determination of the target displacement. In this context, it must be considered a more simplified method for easy computer programming.
   f. From that point on (determination of target displacement of the SDOF model), they both follow the same procedure for the quantitative evaluation of the MDOF system.
2. The 'coefficient method' seems to be the most easily applicable in practice and computerised for the following reasons:
   a. After the elaboration or the capacity curve of the MDOF system, there is no need to transfer the procedure to an equivalent SDOF system.
   b. Target displacement is determined using a direct method of coefficients without any iterative procedure.
3. In the last 10 years, extended efforts have been made for the evaluation of the above methods, and for their improvement on the basis of extended inelastic dynamic analyses, which are used as a benchmark. It is worth referring, at this point, to the FEMA 440/2005 publication *Static Seismic Analysis Procedures* (FEMA 440, 2005) and to the National Institute of Standards and Technology (NIST) GCR 10-917-9/2010 publication on the *Applicability of MDOF Modeling for Design* (NIST GCR 10-917-9/2010, 2010).
4. Displacement-based design may also be used for new buildings. From what has been presented so far, the building must have been analysed, dimensioned and detailed using the traditional codified procedures, so that all necessary input for the non-elastic static analysis has been set down. In this context, the already designed new building may be evaluated for the following purposes:
   a. To verify or revise the overstrength ratio values α_u/α_1 (see Chapter 5.4.3).
   b. To estimate the expected plastic mechanisms and distribution of damage.
   c. As an alternative to the design based on linear-elastic analysis, which uses the behaviour factor q. In this case, the q-factor is verified using the target displacement and the capacity curve of the MDOF model (Figure 13.4c).
5. Finally, it should be noted that significant efforts have been recently made for the development of a DDBD of new buildings without a previous force-based design,

as specified by contemporary Codes (Priestley et al., 2007). It is the author's opinion that this is a very promising approach to the seismic design of structures, with high prospects for further development at the level of applications in codified form.

13.6 SCOPE OF THE DETAILED SEISMIC ASSESSMENT AND REHABILITATION OF R/C BUILDINGS

After the detailed presentation in Section 13.5 of the basic procedure for a quantitative seismic assessment of an existing building and for redesign of its rehabilitated form in case an intervention has been decided upon, it is necessary to give in concept the scope of seismic assessment and rehabilitation. This scope is described in EC8-3/2005 as follows:

- Provide criteria for the seismic performance of existing individual R/C buildings.
- Describe procedures for the selection of necessary corrective measures.
- Display methods and procedures for structural analysis and design, taking into account the corrective measures that have been chosen.
- Examine the degree of uncertainty of the partial safety factors of materials, taking into account that R/C buildings have been built in various periods in the past and probably contain hidden malfunctions. In this respect, the introduction of *confidence factors* (CFs) should be examined, in addition to the original partial safety factors.

In conclusion, it should be noted that in the case of listed (historical) R/C buildings, these may be assessed and retrofitted using the framework that will be presented below. However, any additional considerations have to be taken into account, related to the obligation for retaining the authenticity of the building and avoiding the falsifying of its original characteristics. All these rules are included in restoration charts (e.g. Chart of Venice) and are beyond the scope of this book.

13.7 PERFORMANCE REQUIREMENTS AND COMPLIANCE CRITERIA

13.7.1 Performance requirements

Performance levels introduced in EC8-3/2005 do not comply with those introduced for new R/C buildings in EC8-1/2004. On the contrary, they are much closer to performance levels in effect in US codes and guidelines (FEMA 273-274/1997, FEMA 356/2000). They are characterised as limit states and are listed below:

- *LS of near collapse (NC).* The structure is heavily damaged with low residual lateral strength and stiffness.
- *LS of significant damage (SD).* The structure is significantly damaged with some residual lateral strength and stiffness.
- *LS of damage limitation (DL).* The structural system is only lightly damaged, with structural elements kept below or at the limit of yielding.

Detailed relations between limit states and structural damage are given in Table 13.4 (Elnashai and Di Sarno, 2008).

Table 13.4 Correlation of engineering limit states and performance levels

	Performance levels		
Engineering limit states	*Damage limitation*	*Significant damage*	*Near collapse*
Cracking			
First yielding			
Spalling			
Plastification			
Local buckling			
Crushing			
Fracture/fatigue			
Global buckling			
Residual drift			

Source: Adapted from Elnashai, A.S. and Di Sarno, L. 2008. *Fundamentals of Earthquake Engineering*, Wiley, West Sussex, UK.

Note: Main symptoms at each limit state are in grey boxes.

The appropriate levels of protection are achieved by selecting for each of the limit states a return period for the seismic action. Values recommended by EC8-3/2005 are given below:

- *For NC*: 2475 years return period (2% probability of exceedance in 50 years); very rare earthquake, equal to 150% of design seismic action (DSA) earthquake
- *For SD*: 475 years return period (10% probability of exceedance in 50 years); rare earthquake DSA
- *For DL*: 225 years return period (20% probability of exceedance in 50 years); occasional earthquake, equal to 70% of DSA earthquake.

The above presented performance levels and the relevant seismic excitation levels are coupled in the matrix of Figure 13.12. The above return periods of the seismic action are given in the form of a note in EC8-3/2005 as 'recommended values'.

The authors agree with the position of Professor Fardis on this issue (Fardis, 2009), namely that this return period of DL is very high, near to DSA, and, therefore, may prevail in design. Therefore, this level should be decreased in National Annex to the level of 95 years, in other words, to a seismic action equal to 50% of DSA, and, in this respect, to coincide with the DL of new R/C buildings specified in EC8-1/2004.

If this change is introduced in the matrix of Figure 13.12, then the relevant matrices of Figures 3.24 (EC8-1) and 13.10 (EC8-3) take the same form.

In addition, it should be noted that the limit state of *no (local) collapse* requirement of EC 8-1 corresponds to the limit state of *SD*, and both correspond to *Life Safety* level of FEMA 273-274/1997 (FEMA 356/2000).

The limit states of EC8-3/2005 correspond to the following levels of the diagram of Figure 3.22:

- *LS of NC*: Level 4
- *LS of SD*: Level 3
- *LS of DL*: Level 2 or Level 1

depending on the adopted return period.

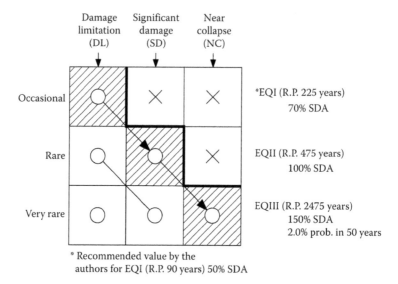

Figure 13.12 Relationship between performance level and earthquake design level (according to EC8-3/2005).

In closing, it should be noted that EC8-3/2005 allows the national authorities of each member state to decide whether all three limit states should be checked, or two of them or just one of them. It is the author's opinion that the limit states of DL and of SD should be the two obligatory limit states, so that the procedure of assessment and rehabilitation complies with that of the design of new R/C buildings.

It should be noted that in the new version of EC8-3, which is under elaboration, the performance levels are modified to comply with those of EC8-1 in its new approach (see Chapter 3.5.2).

13.7.2 Compliance criteria

13.7.2.1 Seismic actions

1. For all methods of analysis except that of the q-factor method, *elastic response spectra* are introduced, that is, spectra not reduced by q-factor.
2. For the q-factor method, the design spectra in use for new buildings, which are elastic acceleration response spectra reduced by the q-factor, are also used for existing buildings.

13.7.2.2 Safety verification of structural members

1. A distinction is made between 'ductile' and 'brittle' elements (see Section 13.5.2.1, Figure 13.4b).
2. For all methods of analysis, except the q-factor method, the safety verifications are carried out according to the following procedure:
   a. For 'ductile' elements, safety is verified by checking that demands do not exceed the corresponding capacities in terms of deformation (Figure 13.4b).
   b. For 'brittle' elements, the above verifications are carried out *in terms of strength* (Figure 13.4b).

3. For the *q*-factor method, the safety verification is carried out for all types of elements *in terms of strength*, as in the case of new buildings, with some additional measures that will be examined later.

13.7.2.3 'Primary' and 'secondary' seismic elements

Some of the existing structural elements may be designated as 'secondary seismic elements', like those of new buildings. These elements are verified with the same compliance criteria as primary ones, but using less conservative estimates of their capacity than for the 'primary' ones.

The introduction of the concept of 'secondary' seismic elements is particularly significant for rehabilitation procedures, because it allows the designer to limit the intervention only to a number of vertical elements, and, therefore, to decrease the extent of the rehabilitation.

13.7.2.4 Limit state of near collapse (NC)

1. Analysis may be carried out using any elastic or non-elastic method of analysis, except the *q*-factor method, which is considered to be unsuitable for this limit state. The reason is obvious, since the *q*-factor method has been developed for new buildings (EC8-1/2004) for the ultimate limit state compliance criterion, which for existing buildings corresponds to the SD limit state (Figure 13.12) and not to the NC limit state.
2. Seismic action for which the analysis of this level must be carried out is *150% of the seismic actions for DSA of EC8-1/2004* corresponding to the ultimate limit state.
3. Demands will be based on the results of the analysis for the above seismic actions.
4. Capacities will be based on proper ultimate deformations for ductile elements and on ultimate strengths for brittle ones.

13.7.2.5 Limit state of significant damage (SD)

1. As explained previously, this limit state corresponds to the ultimate limit state of new buildings. Therefore, the seismic action for which the analysis of this level will be carried out is that of EC 8-1/2004 for the ultimate limit state of new buildings.
2. This limit state is the most important; as it corresponds to the ultimate limit state of new buildings and to the 'Life Safety' of the U.S. codes. *Therefore, verifications based on this limit state must be considered obligatory.*
3. All methods of analysis may be used, including the *q*-factor method approach. However, a critical point is the value of the *q*-factor that is introduced in the design, since the design and detailing of existing buildings usually do not comply with the specifications of modern seismic Codes.
4. Except when using the *q*-factor approach, capacities will be based on damage-related deformations for ductile elements and on strengths for brittle ones.
5. In the *q*-factor method, demands shall be based on the seismic actions reduced by *q*. Capacity verifications in this case will be made in *strength terms*, no matter if the members are ductile or brittle.

13.7.2.6 Limit state of damage limitation (DL)

1. Analysis may be carried out by using any one of the accepted methods (elastic, inelastic, *q*-factor approach).

2. Seismic actions for this limit state must be considered at 50% of DSA of EC8-1/2004 corresponding to ultimate limit state. This seismic action corresponds to the level of 95 years of return period.

3. Except when using q-factor capacities, it should be based *on yield strengths* either for ductile or brittle members. Capacities of infills must be based on mean inter-storey drift capacity of the infills.

4. In the case of the q-factor approach, demands and capacities will refer to mean inter-storey drift.

13.8 INFORMATION FOR STRUCTURAL ASSESSMENT

13.8.1 General

The selection and evaluation of input data necessary for quantitative seismic assessment of an R/C building may be considered the most significant part of the assessment procedure. For this reason, all modern codes of quantitative evaluation of existing buildings pay special attention to this issue, since all subsequent steps of evaluation are based on this information. Besides keeping in mind that the level of knowledge of this information cannot always be equally reliable, Codes usually introduce in codified form three *knowledge levels* for the building under examination and thereafter relevant *CFs*, increasing the partial safety factors of materials.

At the same time, the selection and evaluation of input data constitutes the bigger part of the assessment cost, as will be clear from the presentation below.

Eurocode 8-3/2005 specifies in detail the procedure for the selection of information and the validation of the knowledge level. The input data according to EC8-3/2005 will be collected from a variety of sources, including

- Available documentation of the building (e.g. structural drawings, structural analyses, technical reports referring to the structure and the soil, etc.)
- Relevant generic data sources (e.g. Codes and standards contemporary with the time of the building design)
- Field investigations (e.g. survey of the geometry of the structural system, arrangement of reinforcement detailing, etc.)
- *In situ* and laboratory measurements and tests (destructive and non-destructive tests of concrete and steel, soil investigations, etc.)

13.8.2 Required input data

The information for structural evaluation must basically cover the following issues.

13.8.2.1 Geometry of the structural system

This information is collected from the existing structural drawings of the building under evaluation (formwork drawings). This information must be cross-checked with *in situ* survey of the main critical general dimensions and of the crucial cross sections as well (e.g. large spans, column cross sections, main beams, shear walls, etc.).

In the case that the structural drawings cannot be found, which is a very usual case for buildings of the past, a detailed survey must be carried out so that new formwork drawings of the structural system of the building can be elaborated on, accompanied by proper detail drawings. It is obvious that in order for such an effort to be reliable, scaffoldings must be erected, and plasters and finishes at various critical positions must be removed so that the structural engineer in charge is in a position to identify the structural system and the credibility of the survey.

Special concern must be given to the foundation of the building. In the case that the structural drawings are available, limited local excavations should be carried out just for a cross-check on the existing drawings. Otherwise, these excavations should be extended to the degree that using a back analysis (see below) the foundation layout may be at least estimated. Infill masonry should also be surveyed.

In the case of a post-earthquake quantitative assessment, a detailed survey of damage at structural elements and infill masonry should be carried out, accompanied by sketches and illustrations. Damage pattern in post-earthquake cases constitutes a high-level tool for the structural system identification, which means that the assessment that will follow must be in a position to justify the damage pattern in order to be considered a reliable result.

13.8.2.2 Detailing

This includes the number of bars in the cross section and their arrangement in the structural members (longitudinal and stirrup reinforcement, rebar overlapping) and reinforcement detailing in the joints.

This information is collected basically from the existing structural drawings (i.e. general reinforcement drawings, detailing drawings, barbending schedules, etc.). This information must be cross-checked with *in situ* investigation using destructive (DT) or non-destructive (NDT) tests at selected critical positions. If the *in situ* findings do not comply with the existing drawing information, then a statistical evaluation of the total information must be carried out based on the information collected by the *in situ* investigations.

In the case that the reinforcement drawings cannot be found, an extended investigation procedure is necessary, destructive and non-destructive, at a statistically reliable number of critical positions.

Of course, these data cannot cover all of the structural system, for which reinforcement detail is necessary for a quantitative assessment. Thus, a back analysis of the structural system must be carried out according to the Codes in effect at the time of original design, and the reinforcement output must be cross-checked against the *in situ* findings. This is a time-consuming job and must be accompanied by a statistical elaboration of the back-analysis results based on the *in situ* findings.

This procedure has been implemented extensively by the authors at the complex of buildings of the Army Pension Fund in Athens, Greece, a complex of nine structurally independent eight-storey buildings with a total useful surface of 80,000 m² (Figure 13.13). For this complex, designed and constructed between 1930 and 1939, fewer than 35% of the structural drawings were found, and, therefore, an extended investigation program was needed for the quantitative evaluation and retrofitting of the complex (Penelis and Penelis, 2001).

13.8.2.3 Materials

Points of reference for the materials (concrete–steel) that were used for the construction of a building are the structural drawings – if they exist – because the specifications for the materials in use have been clearly defined on them for many decades past. However, an extended investigation on the construction materials (steel–concrete) must be carried out

Figure 13.13 Complex of buildings of the army pension fund in Athens, Greece.

using *in situ* non-destructive tests (hammer tests, ultrasonic tests, electromagnetic detectors, etc.; Figures 13.14 and 13.15) and cross-checked with destructive ones (Figure 13.16; core-taking, rebar extraction for tests, etc.) executed in the lab.

It should be noted that on the core samples, carbonation tests should be also carried out for checking the degree of rebar protection against corrosion, particularly in older buildings. This information is very important in the case that rehabilitation is planned, which will

Figure 13.14 Hammer tests at MELISSARI Complex (Thessaloniki).

Figure 13.15 Electromagnetic steel detecting (MELISSARI Complex, Thessaloniki).

Figure 13.16 Core taking (Papanikolaou Hospital, Thessaloniki).

undoubtedly prolong the life of the structure and will upgrade its operational status. It should also be noted that the hammer testing results on old concrete surfaces may mislead the engineer in charge to wrong conclusions, since carbonation hardens the concrete and, in this respect, increases the strength indication of the equipment. Therefore, in this case, hammer testing must always be accompanied by a core taking test so that hammer testing results may be reduced properly.

In order to give a sense of the extent of an investigation program for materials, we note that in the case study of the Pension Army Fund Complex in Athens, to which reference was made above, the material tests included (Penelis and Penelis, 2001)

- NDT
  - Schmidt hammer testing: 320 test groups
  - Steel detecting: 300 positions
- DT
- Core taking of 400 cores
- Dismantling of rebars at R/C members: 600 positions
- Bore holes for soil investigation: 3

13.8.2.4 Other input data not related to the structural system

In addition to the above three categories of input data that are related exclusively to the structural system, a series of input data must be clarified. Those are

- Ground conditions.
  - Geotechnical soil properties based on bore holes and laboratory or *in situ* tests.
  - Classification of the ground conditions as categorised in EC 8-1/2004.
  - Information on the seismic design criteria used in the original design. This information is necessary for an eventual back-analysis as mentioned above.
- Description of the present and the planned use of the building. This information will include
  - Changes in dead loads (partition walls, floors, etc.).
  - Changes in live loads (change of the use of the building, for example, from residence to department store).
  - Changes in the importance class.

13.8.3 Knowledge levels and CFs

For the purpose of choosing the *admissible type of analysis* and the appropriate *CF values*, EC8-3/2005, like the relevant U.S. Code, classifies knowledge about the structural system of the existing building into *three levels*:

- *KL1*: Limited knowledge
- *KL2*: Normal knowledge
- *KL3*: Full knowledge

The factors determining the appropriate knowledge level focus on

- Geometry
- Details
- Materials

which are the input data relating to the structural system of the building. EC8-3/2005 specifies rules in a descriptive form for this classification. At the same time, it recommends the extent of inspections, details and testing of materials for each type of primary element (beams, columns, walls) that should be carried out for each knowledge level (see Table 13.5). The knowledge level adopted determines

- The admissible method of analysis
- The CFs

Table 13.5 Knowledge levels and corresponding methods of analysis and CF

Knowledge level	Geometry	Details	Materials	Analysis	CF
KL1	From original outline construction drawings with sample *visual* survey *or from full survey*	Simulated design in accordance with relevant practice *and from limited in situ* inspection	Default values in accordance with standards of the time of construction *and from limited in situ testing*	LF-MRS[(2)]	CF_{KL1}[l]
KL2		From incomplete original detailed construction drawings with *limited in situ* inspection *or from extended in situ inspection*	From original design specifications with *limited in situ testing or from extended in situ testing*	All	CF_{KL2}[l]
KL3		From original detailed construction drawings with *limited in situ* inspection *or from comprehensive in situ inspection*	From original test reports with *limited in situ testing or from comprehensive in situ testing*	All	CF_{KL3}[l]

Source: EC8-3/2005. *Design of Structures for Earthquake Resistance. Part 3: Assessment and Retrofitting of Buildings*. European Committee for Standardization, BSI, CEN, Brussels. With permission.

Abbreviations: LF: lateral force procedure, MRS: modal response spectrum analysis.

(1) The values ascribed to the CFs to be used in a country may be found in its National Annex. The recommended values are

$CF_{KL1} = 1.35$, $CF_{KL2} = 1.20$, $CF_{KL3} = 1.00$.

(2) LF-MRS: lateral force – multimodal response spectrum analysis

The above relation, along with the values recommended by EC8-3/2005, is given in Table 13.5. CFs are used as *denominators* for the determination of the material properties to be used in the calculation of the force capacity (strength), when capacity is to be compared with demand for safety verification. As *nominators* in the above relations, the mean values obtained from *in situ* or laboratory tests will be used. In other words, in the case of existing buildings, *the mean values of in situ*–defined strengths play the role of characteristic strength, and the CFs play the role of partial safety factors of materials for new buildings.

13.9 QUANTITATIVE ASSESSMENT OF SEISMIC CAPACITY

13.9.1 General

Assessment is a quantitative procedure for checking whether or not an existing undamaged or damaged R/C building will satisfy the required limit state appropriate to the seismic actions under consideration (see Sections 13.7.1 and 13.7.2).

This procedure includes

1. Selection of seismic actions
2. Structural modelling
3. Structural analysis
4. Safety verification

These issues will be examined in detail below.

13.9.2 Seismic actions

For all admissible methods of analysis – except that of the *q*-factor procedure – no matter if they are based on elastic or non-elastic models, *the elastic response spectra* of EC8-1/2004 are used *without any reduction by a q*-factor.

In the case of the *q-factor approach*, the *design factors* of EC8-1/2004 are used with a q_o-factor for R/C buildings equal to 1.50:

$$q_o = 1.50 \tag{13.36}$$

However, higher values of *q* may be adopted if this could be justified in reference with the local and global available ductility. This may be accomplished with a sufficient accuracy if a pushover analysis of the structural system is carried out. Alternatively, as will be explained next, simplified approximate methods may be used for the determination of the *q*-factor of the structural system (Tassios, 2009).

What is specified by EC8-1/2004 for new buildings also holds for load combinations in the case of retrofitting.

13.9.3 Structural modelling

Structural modelling is based on the input data information presented in a previous subsection. For the material properties, the *mean values* of the data that have been determined are used. All provisions of EC 8-1/2004 regarding modelling and accidental torsional effects may be applied here, also without modification.

The strength and stiffness of secondary seismic elements against lateral actions may, in general, be neglected in the analysis. However, in no case could the selection of secondary elements be made in such a way that the irregularities of the structure disappear.

In the case of post-earthquake assessment, special care should be paid to the evaluation of the residual stiffness and strength of damaged structural members (Tassios, 2009).

13.9.4 Methods of analysis

13.9.4.1 General

Seismic action effects may be calculated using one of the following methods:

- Lateral force (LF) analysis (linear)
- Multimodal response spectrum (MRS) analysis (linear)
- Non-linear static (pushover) analysis
- Non-linear time–history dynamic analysis
- q-factor approach

It should be noted that the most easily applicable of the above methods are the non-linear static (pushover) analysis and the q-factor approach, since both may be implemented directly without the risk that by the end of the analysis, the method that has been implemented is prohibited for the particular structural system (see the following paragraphs).

13.9.4.2 Lateral force elastic analysis

Conditions for the applicability of this method according to EC 8-1/2004 have been presented in Chapter 5.6.4.

These conditions aim at ensuring a smooth fundamental eigen mode and a limited influence of the higher eigen modes, so that an inverted triangular load distribution results in reliable seismic load effects. In this context, regularity in elevation and limited value of the fundamental period T constitute the conditions for applicability of this method. In addition to the above conditions, a further one is specified for the applicability of the method to existing buildings. The aim of this new *basic condition* is ensuring that the inelastic deformations at the plastic joints are distributed smoothly throughout the structural system, so that local or storey chord plastic deformations (inter-storey drifts) do not present serious irregularities in relation to the overall structural deformation (Figure 13.17).

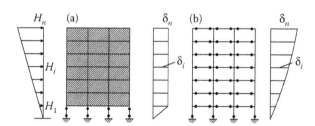

Figure 13.17 Storey displacements under a triangular lateral loading: (a) irregular distribution of inter-storey drifts due to masonry infills (generation of a soft storey); (b) regular distribution of inter-storey drifts in the case of a frame with strong columns–weak beams.

The above additional condition is quantified as follows:

1. The structural system is analysed using the lateral force method according to what has been presented in Chapter 5.6.4 for a base shear V_b, resulting from the codified *elastic response* acceleration spectrum (without any reduction due to *q*-factor).
2. For each member *i* of the system under bending, the *demand* D_i in terms of strength is determined (M_{iD}, N_{iD}) from the preceding analysis. At the same time, the capacity C_i of the member is determined, that is, the bending moment M_{iR} for an axial load equal to N_{iD}.
3. For the ends of each member, the magnitudes

$$\rho_i = \frac{D_i}{C_i} \qquad (13.37)$$

are computed. The above value ρ_i indicates approximately the *ductility demand* μ_D at each critical region (see Figure 13.18). Usually μ_D resulting from the above procedure is higher than 1.0, since the analysis has been carried out for the elastic base shear, which is 3.0–4.0 times higher than the design base shear, for which the *q*-factor has been taken into account at the derivation of the design acceleration response spectrum. If, for example, this demand is very high at a soft storey in relation to the other storeys (Figure 13.17), it is obvious that the fundamental eigen mode, at the inelastic stage, stops being a smooth inverted triangle, and so the lateral load distribution assumed in the analysis is no longer valid. Therefore, the method is not permitted to be used for the analysis.
4. From the above members, those for which

$$\rho_i < 1.0 \qquad (13.38)$$

it is anticipated that even for the target displacement expressed by ρ_{imax} (Figure 13.18), they remain in the elastic region and therefore do not contribute to the plastic deformation of the system.

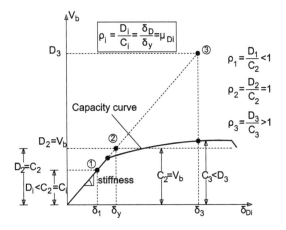

Figure 13.18 Concept for the lateral force elastic analysis.

For those for which

$$\rho_i > 1.0 \qquad\qquad (13.39)$$

it is anticipated that they enter in the plastic region. Among them, ρ_{imax} and ρ_{imin} (>1.0) are selected, and the ratio ρ_{imax}/ρ_{imin} is determined. If

$$\boxed{\dfrac{\rho_{imax}}{\rho_{imin}} \leq 2.5 \quad \text{(recommended value)}} \qquad\qquad (13.40)$$

the use of the method is permitted. It should be noted that the ratio $\rho_i = D_i/C_i$ may be replaced by the ratio of the reinforcement demand A_{siD} in the tensile zone to the available reinforcement A_{siav} at the same zone of the cross section:

$$D_i/C_i \cong A_{siD}/A_{siav} \qquad\qquad (13.41)$$

The above acceptability criterion requires that the structural system will be first analysed, the ratios ρ_i will be determined and then a decision will be made if the method is acceptable or not.

In closing, it should be clarified that the above procedure is a criterion for the acceptance of the method and has nothing to do with the *safety verification procedure* to which reference will be made later.

13.9.4.3 Multimodal response spectrum analysis

The conditions of the applicability of this method have been given in Chapter 5.6.5. In addition, the condition above (Equation 13.40) for the ratio of ρ_{imax}/ρ_{imin} must also be in effect.

It must also be remembered here that the seismic input for this method is the *elastic* acceleration response spectrum and not the design spectrum, which, as known, is reduced by the *q*-factor.

13.9.4.4 Non-linear static analysis

13.9.4.4.1 General

Non-linear static (pushover) analysis has already been presented in Chapter 5.7.3, while the use of this method for the seismic assessment of the structural response of a building has been set forth in detail in Section 13.5.2.1.

As already known, it is a non-linear static analysis under constant gravity loads and monotonically increasing loads.

13.9.4.4.2 Lateral loads

As was mentioned and explained in Section 13.5.2, at least two vertical distributions of lateral loads should be applied (Figure 13.4a):

- A 'uniform' pattern, based on lateral forces that are proportional to mass distribution at height
- A 'triangular' pattern corresponding to the fundamental eigen mode of the structural system

These two patterns of lateral forces are applied at the centres of masses of the storeys.

At the same time, accidental eccentricities should be taken into account, as presented for new buildings ($e_{xi}, e_{yi} = 5\% L_x, L_y$). These patterns of lateral loads are applied successively in two main directions perpendicular to each other, as in the case of new buildings.

13.9.4.4.3 Capacity curve–target displacement–torsional effects

As explained in Section 13.5.2, the relation between base shear and the displacement at a control point at the roof results in the capacity curve of the system. The control displacement is taken *at the centre of mass at the top storey* of the building. It is obvious that for each lateral load pattern, two such capacity curves are generated, one for each main direction of the building. The target displacement is defined following one of the procedures presented in Section 13.5.2.

Pushover analysis performed with the force patterns presented above may significantly underestimate deformations at the stiff/strong side of a *torsionally flexible structure*. In fact, the static approach to the eccentricities between the centre of mass and the centre of rigidity (see Chapters 2.4.4 and 5.3) underestimates the torsional deformations φ_i of the space structural system, *even in the case of elastic static analysis*, in comparison with the corresponding multimodal response spectrum analysis (dynamic amplification of torsional deformations).

For such structures, EC 8-1/2004 specifies that displacements at the stiff/strong side should be increased by applying an *amplification factor* to the displacements of the stiff/strong side based on the results of an elastic modal analysis of the spatial model, meaning that the amplification factor for the inelastic static response would result in the ratio $\varphi_{idynamic}/\varphi_{istatic}$ of the elastic system.

It is apparent that this procedure requires

Non-linear static analysis
Linear static analysis
Multimodal response spectrum analysis

for the determination of the amplification factors for each direction at the stiff/strong side of the structural system. During the past 10 years, efforts have been made for a reliable simulation of the torsional effects on spatial systems designed by means of non-linear static analysis. One of them, developed by the authors, has already been presented in Chapter 5.7.5.2 (Penelis and Kappos, 2005; Penelis and Papanikolaou, 2010).

13.9.4.5 Non-linear time–history analysis

The procedure developed for new buildings in Chapter 5.7.6 may also be implemented for the assessment of existing buildings. It should be remembered, however, that this procedure is proper only for the evaluation of all the others and the results are used as benchmarks for such evaluations.

13.9.4.6 The q-factor approach

The *q*-factor approach is the 'reference method' of the design of new buildings adopted by EC 8-1/2004 and the relevant U.S. standards. This has been presented in detail in Chapters 2 through 10 of this book, and in the same way, it is also implemented for the seismic assessment of existing buildings.

Taking into account that the global and the local ductility of an existing building is not controlled by rules that are taken into account in the design of new buildings, for existing R/C buildings, EC8-3/2005 specifies a low value of *q*-factor equal to

$$q = 1.50 \tag{13.42}$$

It has already been explained in Chapter 5.4.3 that this value corresponds to R/C buildings designed according to Codes for conventional R/C structures.

However, as noted above, higher values of *q* may be adopted if they were suitably justified with reference to the local and global available ductility. This justification may be based either on approximate procedures (Tassios, 2009) or on the implementation of a pushover procedure up to failure (δ_u) for the generation of the capacity curve of the system.

From this curve, the upper limit of available ductility μ_{av} is easily determined (Figure 13.19) by the expression

$$\mu_{av} \cong \frac{\delta_{u,av}}{1.50\delta_y} \tag{13.43}$$

Keeping in mind that there is a direct relation between μ_{av} and q_{av} (e.g. for $T_1 > T_C$, $q = \mu$), it follows that the *q*-factor can then be determined very easily (Figure 13.19).

From this point on, the *q*-factor approach is implemented as in the case of new buildings.

13.9.4.7 Additional issues common to all methods of analysis

For the following issues:

- Combination of the components of the seismic actions
- Additional measures for masonry infills
- Combination coefficients for variable actions
- Importance classes

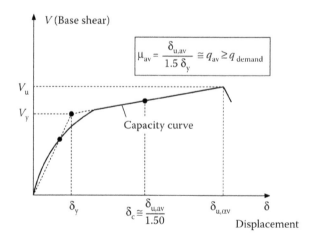

Figure 13.19 Determination of the q-factor value using the pushover capacity curve.

the rules in effect for new buildings included in EC8-1/2004 are also adopted by EC8-3/2005 for existing buildings.

13.9.5 Safety verifications

13.9.5.1 General

As already known, structural elements are classified into two main categories:

- Ductile ones
- Brittle ones

In the first category, failure is caused in the case of bending moments with low axial force presenting plastic deformations, so that the element is *deformation-controlled*.

In the second category, failure is caused either by low bending moment with high axial force or by prevailing shears presenting brittle failures, so the element is *force-controlled*.

13.9.5.1.1 Ductile elements

(Moment–rotation) M–θ diagrams at critical regions of ductile members have the form of Figure 13.4b or Figure 13.21 (left) and are defined by the coordinates of the following points:

Point B: yield stage, $M_y\ \theta_y$
Point C: failure stage, $M_u\ \theta_u$
Point D: residual strength stage, $M_R\ \theta_R$

They constitute the capacity curves of the critical regions of ductile members and are introduced as input data for capacities in inelastic methods. It should be noted that the residual resistance from D to E may be non-zero in some cases and zero in others. Many computer programs for the non-linear static procedure can have only a bilinear capacity curve as input (points A, B, C).

As thoroughly explained in Chapters 5, 8 and 9, end rotations θ at plastic hinges may be determined by the relevant cross-section curvatures φ_u and the length L_{pl} of the plastic hinge. Both magnitudes have been examined in detail in the above chapters, and closed expressions have been formulated for *beams*, *columns* and *ductile walls*. Therefore, curves M–θ may be generated easily up to point C (Figure 13.4b). EC 8-3/2005 includes in Annex A (Concrete Structures) in the form of information, closed expressions for θ_y and θ_u. In ATC 40-1986 and FEMA 356-2000, the coordinates of points B, C and D of Figure 13.4b are given in a normalised form in tables, which can be easily used as an input in computer programs for non-elastic analysis.

The plateau length between B and C depends basically on the axial force and on the reinforcement in the tension and compression zones of the member. It should be noted that the slope of branch B–C according to FEMA 356 and ATC 40 ranges between 5% and 10% of the initial slope (effective stiffness of the member).

13.9.5.1.2 Brittle members

M–θ diagrams (moment–rotation) at the critical regions of brittle members have the form of Figure 13.4b (right). The critical point B of this diagram is defined either by the bending strength at the end of the brittle member or by its shear strength. In both cases, as explained in Chapters 8 and 9, in the diagrams of beams, columns and walls, there is no ductility plateau, so the element is *force-controlled*. Ultimate moment or shear force at failure may be calculated using procedures presented in Chapters 8 and 9. At the same time, EC 8-3/2005 in Annex A includes information for the evaluation of the above strength for beams, columns and shear walls in the form of closed expressions similar to those presented in Chapters 8 and 9.

13.9.5.2 Linear methods of analysis

For the lateral force method or for modal response spectrum analysis, the following principles and rules for safety verifications are adopted by EC 8-3/2005.

13.9.5.2.1 Ductile members

- These are members under bending with low axial force (beams, shear walls, columns with low axial force). They are verified in terms of deformations, meaning chord rotation θ_E at the ends of the members.
- Demand θ_E results from the linear analysis.
- Capacities θ_u result according to what has been presented in a previous paragraph for *mean values* of the material properties divided by the *CF* of the structural system.

13.9.5.2.2 Brittle members

- Safety verification of these members is carried out in *terms of strength*.
- Demands V_E result either from the analysis or from a *capacity design* as presented in Chapter 6. More particularly (Figure 13.20)

If the ratio

$$\rho_i = D_i/C_i = V_{iE}/V_{iRm} \leq 1.0 \tag{13.44}$$

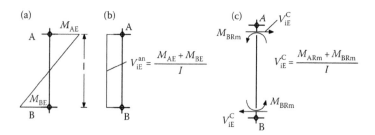

If $V_{iE}^{an}/V_{iRm} \leq 1.0$: for safety verification $V_{iE} = V_{iE}^{an}$

If $V_{iE}^{an}/V_{iRm} \geq 1.0$: for safety verification $V_{iE} = V_{iE}^{C}$

V_{iRm}: capacity of the column to shear

M_{ARm}, M_{BRm}: capacity of the ends of the column to bending

Figure 13.20 Determination of V_{iE} for the safety verification of a column to shear (brittle failure): (a and b) moment and shear diagrams from the analysis for seismic lateral loads; (c) shear force V_{iE}^{C} from the capacity design procedure.

demand D_i (e.g. V_{iE}) resulting from the analysis is introduced in the safety verification inequities.

Capacity C_i (e.g. V_{iRm}) of the ductile component (to shear) is evaluated using mean values of the material properties and introducing the relevant CF.

If the ratio

$$\rho_i = D_i/C_i = V_{iE}/V_{iRm} \geq 1.0 \tag{13.45}$$

the capacity design value V_{iE}^C is introduced in the safety verification inequities.

- Capacity design values V_{iE}^C introduced in the above equations are calculated for overstrengths:

$$\gamma_{Rd} = 1.0 \tag{13.46}$$

- In the above equations, bending moments M_{Rm} at the ends of the columns, which are introduced for the determination of capacity design shear $D_i = V_{iE}^C$, may be calculated for an axial load due to the gravity loads only.

13.9.5.3 Non-linear methods of analysis (static or dynamic)

- In this case, the demands for both 'ductile' or 'brittle' components result from the analysis, performed in accordance with Sections 13.9.4.4 and 13.9.4.5, using mean value properties of the material.
- The values of the capacity of ductile or brittle components to be compared to demand in safety verifications are determined in accordance with Sections 13.9.5.1.1 and 13.9.5.1.2 for *mean values* of the material properties divided by the *CF* of the structural system.

13.9.5.4 The q-factor approach

The values of both the demand and capacity of ductile and brittle members should be in accordance with what has been set forth in Section 13.7.2.2 for the *q*-factor procedure.

13.9.5.5 Acceptance criteria

- It is obvious that the acceptance criteria vary according to
  - The limit state under consideration
  - The type of control mechanism of the member ('ductile' or 'brittle')
  - The structural contribution of the member (primary or secondary) to the seismic resistance
- It should be remembered that for each limit state from DL to NC limit, the seismic action for which the structure is examined has been gradually increased (see Section 13.7.2), and, therefore, demand is also enhanced.
- Chord rotation capacities θ_{um} at the end of ductile members, according to EC 8-3/2005 recommendations at its Annex A (concrete structures), must be reduced by one standard deviation σ for *primary members*. This reduction is estimated at

$$\theta_{u,m-\sigma} = \frac{\theta_{um}}{1.50} \tag{13.47}$$

For the secondary members, the denominator of expression 13.47 is taken to be equal to 1.0. Therefore, rotation chord capacity of secondary members is taken to be equal to θ_{um}.

- In ATC 40-1996 and FEMA 356-2000, these capacities are tabulated in a normalised form and, therefore, they may be easily incorporated as general type input in a computer program. However, the concept of these tables is the same as that of the procedure followed by EC 8-3/2005.
- For all methods of analysis (elastic or non-elastic) – except for the q-method procedure – the following safety verifications are specified by EC 8-3/2005 in its informative Annex A (concrete structures).
  1. Primary ductile members (deformation controlled; Figure 13.21a)

$$\text{Limit states:} \begin{cases} \text{damage limitation (DL)} \\ M_E \leq M_y, \theta_E \leq \theta_y \\ \text{significant damage (SD)} \\ \theta_E \leq 0.75\theta_{u,m-\sigma} \\ \text{near collapse(NC)} \\ \theta_E \leq \theta_{u,m-\sigma} \end{cases} \qquad (13.48a\text{--}c)$$

  2. Primary brittle members (force-controlled; Figure 13.21b)

$$\text{Limit states:} \begin{cases} \text{damage limitation(DL)} \\ V_E \leq V_{Rm,EC2} \\ \text{significant damage (SD)} \\ V_E \leq V_{Rm,EC8/1.15} \\ \text{near collapse(NC)} \\ \text{joints:} V_{CD,j} \leq V_{Rm,jEC8} \end{cases} \qquad (13.49\ a\text{--}c)$$

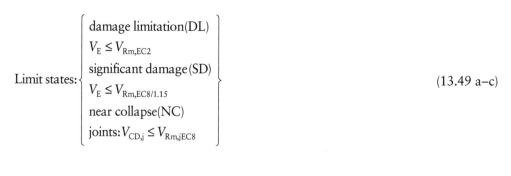

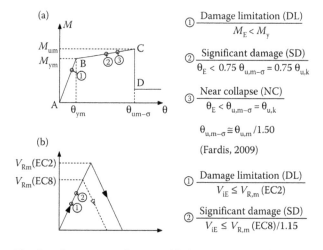

Figure 13.21 Safety verifications for primary elements: (a) ductile elements; (b) brittle elements.

3. Secondary ductile members

$$\text{Limit states:}\begin{cases} \text{damage limitation (DL)} \\ M_E \leq M_y, \text{or}\, \theta_E \leq \theta_{y,ef} \\ \text{significant damage(SD)} \\ \theta_E \leq 0.75\theta_{um} \\ \text{near collapse(NC)} \\ \theta_E \leq \theta_{um} \end{cases} \qquad (13.50a\text{--}c)$$

4. Secondary brittle members

$$\text{Limit states:}\begin{cases} \text{damage limitation (DL)} \\ V_E \leq V_{Rm,EC2} \\ \text{significant damage(SD)} \\ V_E \leq V_{Rm,EC8/1.15} \\ \text{near collapse(NC)} \\ \text{joints:}\, V_{CD,j} \leq V_{Rm,jEC8} \end{cases} \qquad (13.51a\text{--}c)$$

- For the *q-factor approach*, safety verification checks are carried out as in the case of new structures *in terms of strength*, taking into account in addition the following remarks that have been already noted.
  - For the three limit states (DL, SD, NC), relevant design acceleration spectra should be introduced increasing accordingly from DL to NC (see Section 13.7.1) for the determination of relevant seismic effects.
  - For the NC limit state, *q*-factor may be increased by one-third of its value for SD (corresponding to the no [local] collapse requirement for new buildings).
- In Table 5.5 of Chapter 5 (EC 8-3/2005), the following requirements are summarised:
  - The values of the material properties to be adopted in evaluating both the demand and capacities of the elements for all types of analysis
  - The criteria that should be followed for the safety verification of both ductile and brittle elements for all types of analysis

13.10 DECISIONS FOR STRUCTURAL RETROFITTING OF R/C BUILDINGS

13.10.1 General

As presented in Section 13.1, a preliminary seismic evaluation comes first, both in the case of post- or pre-earthquake decisions for evaluation and eventual intervention. A detailed quantitative evaluation follows only in the case that serious damages have been identified or serious deficiencies have been revealed during the preliminary evaluation.

Therefore, after the quantitative seismic evaluation of an R/C building, one of the following two types of intervention may be decided:

- Repair
- Strengthening

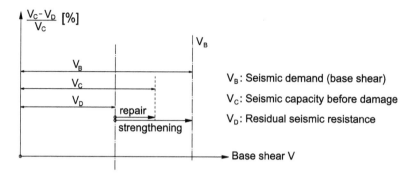

Figure 13.22 Schematic presentation of V_B, V_C, V_D.

The term 'repair' means that the damaged structural or non-structural members reach again the minimum strength, stiffness and ductility they ought to have before the earthquake. This means that 'repair' is limited only to the damaged elements, and in this sense 'repair' must be considered as a *local intervention*. Residual seismic resistance V_D is increased with the repair at least up to the value of seismic capacity V_c before damage (Figure 13.22).

The term 'strengthening' is used to express the increase in the seismic resistance of the structure with interventions beyond repair, so that the available seismic resistance becomes equal to seismic demand V_B, or to a predefined percentage of it (Figure 13.22). This means that in addition to the local interventions to the damaged elements, interventions of *global type* will be carried out, so that the overall structural behaviour of the building is improved.

The effects of rehabilitation on *stiffness*, *strength* and *deformability* must be taken into account in an analytical model of the rehabilitated structure.

Having in mind the above qualitative definitions of 'repair' and 'strengthening' and taking into account the quantitative approach to seismic evaluation at the global (general) as well as local level (Figure 13.4b,c), the above definitions may be quantified as follows:

- *Seismic repair refers* to cases where the global stability is satisfied (the target displacement δ_t for SD limit does not exceed $\delta_u/1.50$)

$$1.50\delta_t(SD) \le \delta_u(NC) \tag{13.52}$$

However, some members exhibit deficiencies or damages (i.e. local demand in terms of deformation $[\theta_E]$ or strength $[V_E]$ exceeds the limits defined by the safety verification). Therefore, in these cases, intervention is limited only to the *local retrofitting* of these members, so that their stiffness, strength and ductility cover the safety verification checks.

- Conversely, *seismic strengthening* refers to cases where the global stability mentioned above is not satisfied. In this case, a *general type intervention* is required (strengthening). This intervention may aim at one or more of the following improvements of the structural behaviour. Those are
    - Increasing the overall structural ductility (e.g. confinements of columns in frame systems) (Figure 13.23)
    - Increasing strength and stiffness (e.g. by strengthening existing structural members and particularly vertical ones, or by adding shear walls in the spans of existing frames)
    - Increasing stiffness, strength and ductility by strengthening and confining existing vertical members

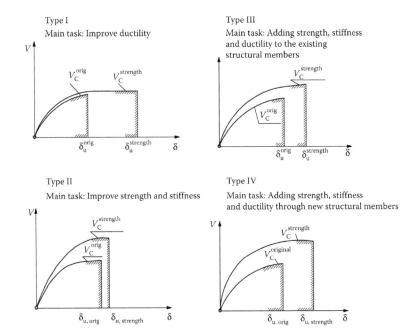

Figure 13.23 Diagrams illustrating the four strengthening types.

- Drastically increasing stiffness, strength and ductility by adding new seismic-resisting members like R/C cores, new structural walls, jacketing successive frame columns and filling their span with structural walls (Figure 13.23)

The general type intervention may often be limited to one or two storeys to improve 'soft storey' vulnerability (irregularities in elevation) or to a region in the plan view of the building to improve its irregularities in plan.

Seismic risk is radically reduced in the fourth case above, particularly when new R/C cores or structural walls are added. In this case, if a significant part of the seismic action is resisted, say 70%, by these new members, then the intervention presents a high degree of reliability, since there are no uncertainties about the effectiveness of the intervention. These new elements can be designed, constructed and supervised as for a new structure, and, therefore, the professional who is responsible for the intervention feels safe enough.

This type of intervention is extended to the foundation, where special concern should be given to resistance to the overturn moments newly appearing there, due to the high stiffness of the new elements that concentrate a large percentage of the base shear. The above type of intervention is proper for pre-earthquake interventions in buildings with a usage that is changing and, therefore, where there is high flexibility due to the high degree of architectural modifications. In case of post-earthquake interventions of a general type (strengthening) where broad constraints exist due to building operation, particularly in the case of residential buildings, and where the intervention cost plays a considerable role, these types of interventions are questionable.

13.10.2 Criteria governing structural interventions

Criteria governing structural interventions in R/C buildings may be classified into two categories, namely, *general* and *technical* ones, according to generally accepted Codes of Practice (e.g. EC 8-3/2005 FEMA 396/ASCE/SEI 41-06).

13.10.2.1 General criteria

The retrofitting scheme should consider the following points:

- Costs, both initial and future. This parameter is particularly crucial after a destructive earthquake when the intervention is obligatory by undisputable needs based on damage exhibited.
- Durability of original and new elements, particularly their *compatibility* (e.g. position of new R/C cores ensuring the diaphragmatic action of the floors), existing carbonation or corrosion of steel reinforcement below new concrete jacketing and so on.
- Available skilled personnel, equipment and materials. For example, in the case of the Thessaloniki earthquake (1978), issuing of permits for new buildings was postponed for one year to ensure the availability of skilled personnel for retrofitting works.
- Existing means for quality control (e.g. proper labs for testing materials and techniques).
- Occupancy (impact on the building) both during and after the work.
- Aesthetics of the building.
- Preservation of the historical and architectural identity of historical buildings.
- Duration of the work in relation to the needs of the occupants (occupancy disruption).

13.10.2.2 Technical criteria

The following technical aspects should be taken into account:

- All identified deficiencies or damages should be appropriately repaired, implementing local modification of the damaged members.
- High irregularities in stiffness and strength should be improved as much as possible, both in plan and elevation.
- Increase in the local ductility should be implemented, where required.
- Overall strength increase in the structure should not lead to a reduction of the overall ductility of the structural system.
- All strength requirements of the relevant Code should be fulfilled after the intervention, taking into account the provisions of Chapter 14 for the redesign of elements.
- Spreading the areas of potential inelastic behaviour as much as possible across the entire structure should be one of the tasks of the intervention.

13.10.2.3 Types of intervention

Bearing in mind the above general and technical criteria, an intervention may be selected from the following representative types individually or in combination:

- Restriction to the use of the building or modification of its use
- Local or global modification (repair or strengthening) of damaged or undamaged elements
- Possible upgrading of existing non-structural elements into structural ones
- Modification of the structural system aiming at stiffness regularity, elimination of vulnerable elements or a beneficial change of the natural period of the structure
- Mass reduction
- Addition of new structural elements (e.g. bracings, infill walls)
- Full replacement of inadequate or heavily damaged elements
- Redistribution of action effects, for example, by means of re-levelling (bringing columns back to their original position) of supports, or by adding external pre-stressing

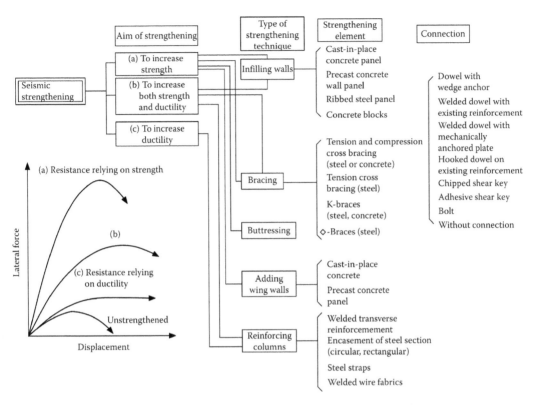

Figure 13.24 Typical strengthening methods used in Japan.

- Addition of a new structural system to carry the seismic action
- Addition of damping devices at appropriate parts of the structure
- Partial demolition
- Seismic isolation

In Figure 13.24, typical strengthening methods used in Japan are given in a schematic form (Sugano, 1981; Rodriguez and Park, 1991).

13.10.2.4 Examples of repair and strengthening techniques

Finally, to give a clearer picture of the frequency of implementation of the various techniques, some statistics on R/C building repair and strengthening techniques in Japan and Mexico are given.

In the case of the 1966 Tokachi-Oki earthquake, the strengthening methods used for the rehabilitation of 157 R/C buildings are listed in Figure 13.25 (Endo et al., 1984; Rodriguez and Park, 1991). In general, more than one method was used for a building, and the most common method of strengthening (in 85% of cases) was the addition of shear walls cast into existing frames. Column jacketing was used in 35% of cases.

In the case of the 1985 Mexico City earthquake, the various strengthening methods used for 114 R/C buildings are listed in Table 13.6, in relation to the number of floors of the structures. According to these data, jacketing of columns (designated as concrete JC in Table 13.6) was the most commonly used technique for buildings with 12 storeys or fewer (Aguilar et al., 1989; Rodriguez and Park, 1991).

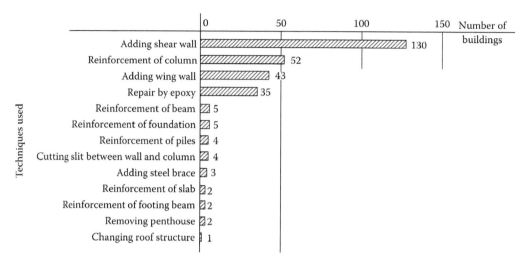

Figure 13.25 Repair and strengthening techniques used for 157 buildings in Japan.

Table 13.6 Repair and strengthening techniques for 114 reinforced concrete buildings in Mexico versus number of floors

Repair and strengthening techniques	Number of floors			
	<5	6–8	9–12	>12
Sealing	1	1	0	0
Resins	2	2	3	2
Replacement	7	8	5	6
Hydraulic jacks	1	1	1	0
Concrete JC[a]	11	18	26	5
Steel JC	2	7	10	2
Concrete JB[b]	4	7	14	2
Steel JB	1	0	3	1
Shear wall	8	12	16	9
Infill wall	4	9	2	2
Steel diagonals	0	7	7	2
Concrete frames	1	3	3	3
Additional elements	3	3	4	2
Straightening	0	1	2	2
New piles	2	4	8	3

[a] JC = Column jacketing.
[b] JB = Beam jacketing.

13.11 DESIGN OF STRUCTURAL REHABILITATION

13.11.1 General

The retrofit design procedure includes the following steps:

- Conceptual design
- Analysis
- Verifications

- Drawings
- Technical reports

13.11.2 Conceptual design

Conceptual design usually covers the following issues:

1. The type and configuration of the retrofit scheme
2. Selection of intervention techniques and materials
3. Preliminary estimation of dimensions of additional structural components
4. Preliminary estimation of the modified stiffness of the retrofitted elements

13.11.3 Analysis

All methods accepted for use at the stage of assessment (see Section 13.9.4) may also be used at the stage of analysis for retrofitting. Of course, modified or repaired structural members will be modelled, taking into account their modified characteristics. It is apparent that in both stages, the same method is selected, since in this case the remodelling at the stage of retrofitting usually needs minor modification relative to that used for assessment.

It should be remembered that of all methods permitted by EC 8-3/2005, the non-linear static (pushover) analysis and the q-factor approach are the most easily applied in practice. For both of them, the computational determination of the corresponding $M-\theta$ diagrams for all new or repaired elements is a prerequisite. For new elements, the procedure is well known and is similar to that followed at the assessment stage. For the elements for which a repair is foreseen, the $M-\theta$ diagram depends on the method of repair that will be followed (e.g. R/C jacketing, FRPs, etc.). This issue will be presented in Chapter 14.

13.11.4 Safety verifications

According to EC 8-3/2005, safety verifications are carried out in general as in the stage of assessment (see Section 13.9) for all types of members, modified or repaired as well as new ones. For existing members, mean values from *in situ* tests and any additional sources of information will be used for the determination of local capacities, as explained in detail in the case of assessment. These values will also be reduced by the CF, as in the case of assessment. For new elements or added materials for repair or strengthening of existing members, nominal design values are used without reduction by CF. The remarks in Section 13.11.3 in regard to the capacity diagrams $M-\theta$ of repaired members also hold in the case of safety verifications.

A more detailed approach to safety verifications for the non-linear static analysis method and for the q-factor approach will be made below.

13.11.4.1 Verifications for non-linear static analysis method

In this case, the following verifications must be carried out:

1. *Global safety verification*
   Target displacement (δ_t) for loading at SD limit state (e.g. ULS for new buildings) should not be bigger than NC displacement (δ_u) reduced by 1.50 (Figure 13.4),

$$\boxed{\delta_t = \frac{\delta_u}{1.50}} \tag{13.53}$$

2. Local safety verification of ductile members

This category includes members under prevailing bending with or without axial force, at the end of which plastic joints may be formed (beams, columns, walls).

In rehabilitation stage, three cases may appear:

- Existing members without any intervention
- Existing members modified through repair or strengthening
- New members

Safety verification in this case is carried out in terms of deformation, as in the case of assessment.

In the case that some of the members under consideration do not comply with the safety verification requirements presented in Section 13.9.5, one of the following alternatives may be implemented.

- For new elements, the most convenient way is enhancing the ductility of the new member.
- For existing or modified members, the easiest way is to increase their strength. In this case, if the required end rotation θ_E exceeds θ_{ud} (Figure 13.26), based on the concept of 'energy balance' (see Chapter 3.2), M_{ud} should be increased by an overstrength factor γ_{ov} (Figure 13.26) equal to

$$\gamma_{ov} = \frac{M_{ud}^{ov}}{M_{ud}} \cong \frac{\mu_E - 1}{\mu_{ud} - 1} \qquad (13.54)$$

In this respect, the required extra energy capacity E_1 due to θ_E, which is greater than θ_{ud}, is balanced by extra energy capacity E_2 due to overstrength γ_{ov}. In fact, referring to the diagram of Figure 13.26, it may be seen that

$$E_1 = (\theta_E - \theta_{ud})M_{ud} \qquad (13.55)$$

$$E_2 = (M_{ud}^{ov} - M_{ud})(\theta_{ud} - \theta_{yd}) \qquad (13.56)$$

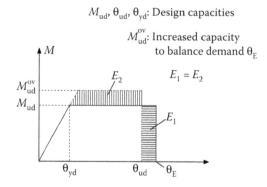

$M_{ud}, \theta_{ud}, \theta_{yd}$: Design capacities

M_{ud}^{ov}: Increased capacity to balance demand θ_E

Figure 13.26 Strength of a member increased to balance the deformation θ_{ud} inefficiency to cover the deformation demand θ_E.

Expression 13.54 results from the condition that $E_1 = E_2$ where
- μ_E is the ductility demand θ_E/θ_{yd}.
- μ_{ud} is the design ductility θ_{ud}/θ_{yd}.

It should be noted that θ_{ud} refers to the specified limit for the limit state for which the design is carried out.

Another way to approach the problem is to characterise these elements as 'secondary', and in this respect to increase the specified deformation limit θ_{ud}.

3. *Local safety verification of brittle members or mechanisms*

This category basically includes members under prevailing shear that fail in a brittle way before plastic hinges can be formed at their ends due to bending. It may also include columns with high axial load, which can develop a very low plastification at their ends and, therefore, they function as brittle elements.

In the rehabilitation stage, the same cases of characteristic members may appear, as in the case of ductile ones, which are new members, existing unmodified members and strengthened or repaired ones.

Safety verification in this case is carried out in terms of *strength*, as in the case of assessment.

In the case that some of the members under consideration do not comply with the safety verification parameters presented in Section 13.9.5.5, strengthening of the member is the only way to improve its response.

13.11.4.2 Verifications for the q-factor approach

It is obvious that similar safety verifications with the non-linear static method must also be carried out in the q-factor approach.

1. *Global safety verification*

This safety verification has the objective of verifying that

$$q_{Dem} \leq q_{cap} \tag{13.57}$$

where
- q_{Dem} is the q-factor introduced in the design acceleration response spectrum for the force-based design of the system.
- q_{cap} is the available global q-factor of the system. Of course, this value should by no means exceed values of the q-factor codified for new buildings.

The verification of q_{Dem}-factor that is introduced in the design spectrum of new R/C buildings is ensured explicitly by design Code requirements for local ductility of the members of the new building and by the capacity design requirements. In the case of the rehabilitation of an existing R/C building, it comprises new members, existing unmodified members and those under repair or strengthening. Therefore, for the determination of the available q-factor, a reliable method must be mobilised.

The safest procedure is to determine the capacity curve of the system using a pushover static analysis procedure. Having this curve, the q-factor can easily be determined. A reduction factor of about 1.50 should be introduced to this value. The resulting final q_{Dem}-factor should by no means be bigger than q-factors recommended by Code for new R/C buildings.

In the case that the main structural system carries the seismic actions by means of new structural elements (e.g. R/C cores, R/C walls, R/C jacketed columns), the existing structural elements might be designated as 'secondary'. In this case, the building is designed as a new one, with values of q-factor, those recommended by the Code.

An alternative approach could be the design of the new structural system using the q-factor recommend by Code, while for an existing system for which the R/C members do not comply with the rules of the Code for local ductility, a value of $q = 1.50$ should be implemented. In this case, neither local ductility rules nor capacity design rules would be obligatory for the existing members. It should be noted that in this case, the analysis of the overall structural system is carried out for a q-factor corresponding to a new building. In this respect, the seismic effects of the existing structural system must be increased by $q/1.50$, and as a consequence they will exhibit very high values.

2. *Local safety verifications*

Local safety verifications at ULS or DLS are carried out for all structural members, new or existing, using the force-based design that has been presented for new buildings.

In the case that the existing elements are considered 'secondary', or in the case that a q-factor equal to 1.50 is introduced for these members, nothing new should be added here in relation to the procedure specified by EC 8-1/2004 for new buildings.

In the case that a common q-factor is determined by means of the capacity curve of the building, overload factors γ_{ov} for the existing elements should be introduced to the seismic effects resulting from the common q-factor given by expression 13.54.

It should be noted that the mobilisation of the capacity curve as an additional tool for the q-factor approach is a very effective means for the reduction of the degree of strengthening of existing R/C members.

13.11.5 Drawings

Drawings for rehabilitation may be classified in three groups, namely:

- Demolitions
- Temporary propping
- Rehabilitation

1. *Demolitions*

Demolition drawings display
   a. The regions of the structural system that should be demolished
   b. The order in which the various members should be demolished so that local or total collapse is avoided
   c. The equipment that might be used for the demolition
   d. Life protection measures that should be taken during the demolition
2. *Temporary propping*

These drawings must comply with those of demolitions. They must include information
   a. About supporting systems of dead loads
   b. About bracing systems to protect the building from lateral inclination and, therefore, from significant second-order effects that might lead to 'pancake' collapse
   c. About stabilising systems of weakened diaphragms due to demolition of broader regions for passing new R/C shafts or staircase cores and so forth
   d. About the order of installation of the supporting systems
   e. About details of connections of these systems to the structural system

It should be noted that temporary propping is one of the most significant actions of rehabilitation procedures, because during the demolition, the structural system is weakened or sometimes loses its robustness and integrity. If it is not well propped, the building may collapse. The most common accidents with fatalities happen during demolition activities.

In closing, it should be noted that propping drawings should be accompanied by the relevant technical report, including necessary structural design of the various parts of the propping system.

3. *Rehabilitation drawings*

Rehabilitation drawings include, as do the drawings of new buildings, the following:

a. General layout drawings in plan, also including characteristic cross sections in elevation. In these drawings, the following must be displayed:
b. All new members
c. All repaired or modified members with detailed information about the material and the degree of repair (e.g. dimensions of R/C jackets with the new reinforcement, etc.)
d. General information about
   i. Dead loads
   ii. Live loads
   iii. Seismic loads
   iv. Ground conditions
   v. Construction materials of the existing system
   vi. Construction materials of new parts or repaired members and so on
e. Detailing drawings of all new elements, repaired elements, connections of new to old parts of the building and so on.

It should be noted that for all drawings of the above three groups (demolitions, temporary propping and rehabilitation), various colours should be used to differentiate the existing parts from the new ones and the interventions to existing elements.

4. *Technical report*

Technical reports should accompany the design of the rehabilitation scheme. These reports should include

a. Justification of the structural intervention
b. Technical description of the intervention
c. Quantity estimates
d. Bill of quantities
e. Cost estimates

13.12 FINAL REMARKS

From what has been presented so far, it can be concluded that very few structural problems are as challenging for a professional as confronting the consequences of an earthquake.

From the scientific point of view, the main tool available to the engineer, the analysis, has often been proved to be inadequate to explain the damage patterns, possibly because the assumptions on which it is based are over-simplified (static loading, elastic response of the system, not taking into account the infill system, etc.). Thus, there is always doubt regarding the effectiveness of whatever intervention has been decided upon. This doubt is much higher in case of a pre-earthquake rehabilitation, since in this case there is no damage pattern for system identification.

From a practical point of view, the determination of the 'available' and the 'residual' seismic resistance of existing members, damaged or not, involves a high degree of uncertainty because of the subjectivity involved in the determination of the seismic resistance of the structural elements.

Referring to conceptual design, the various types of intervention that are decided upon are not always feasible. For example, structures that were built without respecting the provisions of modern Codes (most existing structures fall into this category) cannot meet ductility DC 'M' requirements, and possibly not even ductility DC 'L' requirements. On the other hand, a large increase in base shear V_B creates the need for additional strength and stiffness elements, which leads to foundation problems, as well as to functional problems when the structure is in use again.

Based on the above, the legal framework that is set every time after a destructive earthquake for the rehabilitation of damage cannot withstand strictly scientific criticism. This happens because this framework attempts to establish a balance between the desirable and the feasible. In other words, it is a political decision within the broader meaning of the term, which tries to optimise the combination of scientific knowledge with the technological and financial capabilities in order to face the acute social problem of the safe retrofitting of damaged structures.

Of course, in the case of pre-earthquake assessment and rehabilitation, the above controversial parameters, cost and safety, are not so acute, since the rehabilitation procedure is not as urgent as in the case of post-earthquake rehabilitation activities.

Independently of the previous general remarks, in summarising, reference should be made to the following special points:

1. The rehabilitation of a seismically damaged building is a much more difficult task than the original design and construction of the building. The same holds for an intervention in the pre-earthquake period.
2. The difficulties arise during the design, as well as during the supervision and execution of the intervention works.
3. A basic factor for the successful outcome of the whole operation is the correct diagnosis of the causes of damage or the deficiencies, in the case of a pre-earthquake intervention. The level of intervention depends on this diagnosis, meaning the repair or strengthening of the structure.
4. The design of the rehabilitation must aim at
   a. Providing the structure with the stiffness, strength and ductility that it had before the earthquake by means of repair in the case of local damage.
   b. Providing the structure with the strength, stiffness and ductility required by the current Codes in the case of damage of global character (strengthening). The same holds for structures in pre-earthquake conditions, for which a global instability is verified.
5. Independent of the local or global character of the damage, the structural elements must be repaired in such a way that they regain the strength and ductility required by the current Codes.
6. For the choice of the repair technique, the market conditions and the feasibility of application of the chosen techniques in every particular case must be taken into account.
7. The rehabilitation is usually accompanied by the removal of many structural members, and, therefore, special care should be taken of the temporary support of the structure.
8. The outcome of the intervention depends to a large extent on the quality control of the design and construction. Therefore, very careful supervision is necessary during the execution of the rehabilitation works.
9. Intervention in heavily damaged infills is very important for the structure, and for this reason, appropriate care should be taken.
10. Finally, it has to be stressed once again that structural rehabilitation must have as its point of reference the proper combination of strength, stiffness and ductility.

Chapter 14

Technology of repair and strengthening

14.1 GENERAL

The purpose of this chapter is to present in brief the technological problems associated with interventions in structures damaged by earthquakes or at pre-earthquake stage in the case of an active seismic retrofitting.

In the preceding chapters, detailed reference has been made to the procedure followed for decision-making about the extent and the type of interventions. At the same time, the successive steps for the design of the interventions were discussed in detail. In this chapter, reference will be made to the materials and techniques of interventions and to the dimensioning of the structural elements for various types of intervention. However, given the fact that several manuals, specifications and Codes have been published to date (UNDP, 1977; AUT, 1978, 1979; GMPW, 1978; NTU, 1978; UNIDO/UNDP, 1983; Penelis and Kappos, 1997; OASP, 2001; Dritsos, 2004), where numerous technical details are given, the focus here will be mainly on some typical repair and strengthening techniques and on the dimensioning of the relevant structural elements.

Particularly on the subject of dimensioning, there are many concerns with regard to the reliability of the proposed methods, for the following reasons:

- There is no adequate experimental verification of these methods.
- Most of them are based on rough and/or simplified models, since analytical models based on experimental and theoretical knowledge have not yet been developed to a degree suitable for practical use.
- The quality of execution of the repair and strengthening works on site influences the results significantly.
- The evaluation of the redistribution of stresses from the old element to its strengthening presents reliability problems.

It should be stressed here that the main issues concerning repair and strengthening, which are materials, techniques and redesign considerations, exhibit different degrees of development in regard to research, implementation and codification level. Table 14.1 gives a qualitative picture of this development (Zavliaris, 1994).

Before the individual topics of this chapter are addressed, it will be useful to summarise the intervention procedure as given in the previous chapters.

After a destructive earthquake, an inspection operation is usually organised by the State in order to locate the buildings that are unsuitable for use. At the same time, all necessary demolitions and shoring are carried out. Next to this initial phase and once the aftershocks have attenuated, the procedure for the design of the intervention in every individual damaged building to be retrofitted begins.

Table 14.1 Development in materials, techniques and redesign considerations

	Materials	Techniques	Redesign considerations
Research and development (R&D)	◯	◯	◯
Implementation	○	○	○
Codification	∘	∘	∘

Note: The diameter of the circles represents the degree of development (qualitatively).

A similar procedure is followed in the case of a pre-earthquake active rehabilitation. After successive tiers of evaluation and screening of the buildings that are susceptible to damage or even collapse in the case of a strong earthquake, a detailed quantitative assessment is carried out, which, most of the time, is followed by a rehabilitation procedure. This second phase, that is, the quantitative assessment and retrofitting, is much more systematic than the first, more laborious and more effective, and it requires much time and expenses. The preceding chapter, as well as the present one, cover the approach to problems associated with this second phase.

In closing, in this brief introduction, it must be mentioned that the main reference for the intervention techniques for individual structural members presented here was the UNIDO/UNDP manual, *Repair and Strengthening of Reinforced Concrete, Stone and Brick-Masonry Buildings* (UNIDO/UNDP, 1983), which represents a synthesis of experience and expert knowledge at an international level. By this choice, it is felt that some contribution is being made to the realisation of one of the UNIDO/UNDP goals, that is, the dissemination of this widely accepted up-to-date knowledge on intervention techniques to the international scientific community with special interest in the subject.

14.2 MATERIALS AND INTERVENTION TECHNIQUES

In this section, reference will be made to the materials and intervention techniques that are frequently encountered in the repair or strengthening of structures after an earthquake. Given the fact that these special materials, as well as their application techniques, are governed by detailed specifications that are typically related to the know-how that accompanies them, the designer, before including any of these materials in a rehabilitation project, must be fully informed about them. In the following, a general presentation of the materials and techniques is given, and some critical points related to their advantages, disadvantages and successful application are discussed.

14.2.1 Conventional cast-in-place concrete

Conventional concrete is very often used in repairs as a cast-in-place material. In many cases, the results are not satisfactory because of the shrinkage of conventional cement, which causes reduced bonding between old and new concrete. In order to improve bonding conditions and cover additional uncertainties in construction operations, the use of concrete having a strength higher than that of the element to be repaired is recommended $(f_{c_{rep}} \geq f_{c_{exist}} + 5 \text{ MPa})$, as well as a low slump and water/cement ratio. Such a choice, however, renders compaction very difficult, especially when thin jackets are foreseen, thus making

Figure 14.1 Dedusting and wetting of the old concrete and reinforcement at the complex of the Army Pension Organization (Athens).

necessary the use of super-plasticisers to increase slump up to 200 mm with the standard method of Abram's cone. The maximum size of aggregates should not exceed 20 mm in order for the mix to pour through the narrow space between the old concrete and the forms.

The procedure of casting the concrete is critical for the success of the intervention. Old surfaces should be made as rough as possible and cleaned in order to increase the adhesion between old and new concrete. After the placement of the reinforcement, the forms are placed, which have special lateral openings for casting of concrete. Before concreting, there should be a final dedusting of the surfaces with compressed air, as well as extensive wetting of the old concrete and the forms. Concrete should be thoroughly vibrated to ensure a high degree of compaction (Figure 14.1).

14.2.2 High-strength concrete using shrinkage compensating admixtures

For the construction of cast-in-place concrete jackets, special dry-packed mortar is used very often, which is available in the market under several commercial names. This mortar consists of cement, fine sand (up to 2.0 mm), super-plasticisers and expansive admixtures in the appropriate proportions, so that mixing with water of about 15% of weight produces fluid mortar that attains high strength in a very short time (e.g. 30 MPa in a 24-h period, 70 MPa in 28 days), while at the same time it does not shrink. The attainment of high strength in a short period of time is due to the formation of a special silica calcium hydrate from the reaction between the expansive admixture and the cement. Therefore, very satisfactory repairs are accomplished, without voids and shrinkage cracks, using very thin jackets, for example, 40 mm. In order for these products to be used, they must be accompanied by a quality control certificate. As far as the rest of the procedure is concerned, it is the same as for conventional concrete.

14.2.3 Shotcrete (gunite)

If the appropriate equipment and trained personnel are available, shotcrete is considered to be a very good repair solution. Indeed, due to the fact that forms are not needed, it can be applied on surfaces of any inclination, even on ceilings. Its use is more common on extended surfaces such as R/C and masonry walls, but it can also be used for the construction of jackets around columns or beams.

As far as strength is concerned, a strength higher than that of the repaired element is always specified ($f_{c_{rep}} \geq f_{c_{exist}} + 5$ MPa).

The main advantages of the method are the absence of forms, the very good adhesion between old and fresh concrete due to the high degree of compaction energy during shotcreting and high strength due to the low water/cement ratio.

Two different processes have been developed so far for gunite application, namely

- Dry process
- Wet process

14.2.3.1 Dry process

This process requires the following equipment for production and application (NTU, 1978):

- Concrete mixer for dry mixing
- Water tank
- Centrifugal water pump
- High-capacity compressor
- Gun with one or two chambers
- High-pressure hoses
- Nozzle

The production procedure is as follows (Figure 14.2):

1. A mixture of 0.5 kN of cement and about 2.0 kN of aggregates with maximum grain size of 7, 12 or 16 mm, depending on the case, is dry-mixed in the concrete mixer.
2. The mixture is fed into the gun and still in dry form, in suspension, reaches the nozzle through a hose with the aid of compressed air.
3. At the nozzle, water is injected into the material. From there, the mixture is forcefully shot onto the surface to be repaired, which has been previously roughened, wetted and appropriately reinforced. Every layer has a maximum thickness of 30–40 mm. If larger thickness is required, a second layer should be applied.
4. The resulting surface is very rough; therefore, after hardening, it must be covered with plain plaster or mortar.

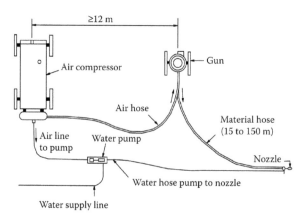

Figure 14.2 Typical arrangement of equipment for shotcreting.

14.2.3.2 Wet process

This process was introduced in the late 1970s. The required equipment includes the following:

- Concrete mixer for wet mixing ready for use. This mixer may be avoided in case of the use of ready-mixed wet concrete.
- High-capacity compressor.
- Gun of continuous supply (open).
- High-pressure hoses.
- Nozzle.

The production procedure is as follows:

Ready-mixed concrete of aggregates with maximum grain size of 7–12 mm, depending on the case, is fed into the gun (Figure 14.3); it reaches the nozzle through a hose with the aid of compressed air and is forcefully shot onto the surface to be repaired.

14.2.3.3 Final remarks

- The following may be considered as disadvantages of the dry procedure: (1) the fact that the water/cement ratio cannot be quantitatively controlled, given the fact that the fluidity of the mix is controlled only visually by the operator; (2) the waste of a large fraction of the material due to reflection on the surface of application; and (3) the cement dust sprayed in the space, which makes application in close spaces difficult.
- Flexibility and fluidity must be considered the main advantage of the material, as it is controlled visually by the operator who attends the shooting of the material onto the old surface.
- The above disadvantages are eliminated in the case of the wet process. However, the advantage of flexibility in controlling fluidity does not exist in this case.
- In Tables 14.2 and 14.3 (ACI 506, 1995), representative values of experimental results for the dry and wet processes are displayed. From the comparative evaluation, the following conclusions may be drawn:
  - The scattering of compression strength in the dry process is higher than that of the wet one.
  - The bond strength of the dry process is almost twice as high as that of the wet process.

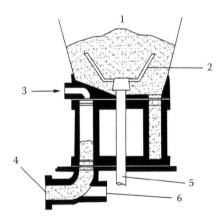

Figure 14.3 The principle of functioning of a shotcreting machine: 1 = wet material supply; 2 = mixer; 3 = compressed air; 4 = material exit under compression; 5 = rotor; 6 = compressed air.

Table 14.2 Dry-mix shotcrete on old concrete

Sample no.	Compressive strength of shotcrete cores, psi (MPa)		Bond strength in shear, psi (MPa)	
1	5850	(40.3)	720	(5.0)
2	7140	(49.2)	598	(4.1)
3	5900	(40.7)	422	(2.9)
4	5410	(37.3)	520	(3.6)
5	7060	(48.7)	874	(6.0)
6	4620	(31.9)	411	(2.8)
7	4580	(31.6)	508	(3.5)

Source: ACI 506 R-90. 1995. Guide to shotcrete. ACI Manual of Concrete Practice, Part 5. Farmington Hills, MI, USA. With permission from ACI.

Data are from a single project. It is presented for illustrative purposes only.
All tests on 6 in (150 mm) diameter cores.
Shotcrete placed by dry method.
Shear test conducted by 'guillotine' method where load is applied parallel to the bonded surface.

Table 14.3 Wet-mix shotcrete on old wet-mix shotcrete

Sample no.	Compressive strength of shotcrete core, psi (MPa)	Bond strength in shear, psi (MPa)
11	4810 (33.2)	131 (0.9)
12		181 (1.3)
13	4420 (30.5)	243 (1.7)
14		220 (1.5)
15	4860 (33.5)	336 (2.3)

Source: ACI 506 R-90. 1995. Guide to shotcrete. ACI Manual of Concrete Practice, Part 5. Farmington Hills, MI, USA. With permission from ACI.

Data are from a single project. It is presented for illustrative purposes only.
All tests on 6 in (150 mm) diameter cores.
Shotcrete placed by wet method.
Shear test conducted by 'guillotine' method where the load is applied parallel to the bonded surface.

- The above disadvantages are eliminated in the case of the wet process. However, the advantage of flexibility in controlling fluidity does not exist in this case.
- In Tables 14.2 and 14.3 (ACI 506, 1995), representative values of experimental results for the dry and wet processes are displayed. From the comparative evaluation, the following conclusions may be drawn:
  - The scattering of compression strength in the dry process is higher than that of the wet one.
  - The bond strength of the dry process is almost twice as high as that of the wet process.

14.2.4 Polymer concrete

Polymer-modified concrete is produced by replacing part of the conventional cement with certain polymers that are used as cementitious modifiers. The polymers, which are normally supplied as water dispersants, act in several ways. They function as water-reducing plasticisers, they improve the bond between old and new elements, they improve the strength of the hardened concrete and so on. However, it should be noted that polymer concrete also has several disadvantages. It is vulnerable to fire conditions and, due to its lower alkalinity, presents inferior resistance against carbonation compared to conventional concrete.

14.2.5 Resins

Resins are usually used for grouting injections into cracks in order to glue together cracked concrete or for bonding thin metal or fibre-reinforced plastic (FRP) sheets onto concrete surfaces. These are materials made up of two components that react and harden after they are mixed together. More specifically, one component is the resin in fluid form (epoxy, polyester polyurethane, acrylic, etc.), while the second is the hardener (AUT, 1978, 1979; NTU, 1978). There is a great variety of such products with different properties depending on the chemical composition of the components, the mixing ratios and the possible additives such as fillers or sand. Therefore, the engineer must have good knowledge of the properties of such a material before selecting the proper one for a specific use.

Epoxy resins are the most common type of these materials in use today. Resins must have an adequate *pot life* so that a usual dosage can be used before it hardens. Curing requirements should be compatible with the temperature and moisture conditions of the structure. The resin must have excellent bonding and adhesion to concrete and steel and must present small to negligible shrinkage. Also, its modulus of elasticity must be generally compatible with that of the concrete to be glued. Resins lose their strength at temperatures higher than 100°C, and, therefore, such repairs are not fireproof without fire protection (e.g. plaster). Resins used in the form of injections must have a viscosity appropriate for the crack width to which the injection is applied. Resins used for bonding metal or FRP sheets usually have high viscosity. Table 14.4 shows comparative data for strength and deformability of conventional concrete and of epoxy resins (AUT, 1978).

There are several techniques for the application of resin injections. In the simplest case, the resin is mixed with the hardener in a separate receptacle and a gun with an injection nozzle is filled with the mixture (Figure 14.4). Sometimes the mixing is done within the gun with separately controlled supply of the two components. The injection, in case it is done by hand, is applied with low pressure (up to 1 MPa). In case the injection is done by pump, a pressure up to 20 MPa may be applied. The gun is equipped with a pressure gauge. Since epoxy resins are materials that cause irritation to skin, eyes and lungs, the appropriate means of personnel protection are required when working with them (gloves, protective eyeglasses, masks). When the crack width is small (0.3–0.5 mm), pure resin is used. In the case of wider cracks, it is useful to mix the resin with filler, having a grain diameter not larger than 50% of the crack width or 1.0 mm, whichever is smaller. The ratio of resin to filler is usually about 1:1 in weight.

Before the application of resin injections, the crack is cleaned with compressed air. Then holes of 5–10 mm in diameter are opened with a drill at certain distances along

Table 14.4 Comparison between mechanical properties of concrete and epoxy resins

Property	Concrete	Epoxy resin
Compressive strength (MPa)	20–90	Up to 250
Tensile strength (MPa)	2–6	3.5–35
Flexural strength (MPa)	3.5–9.0	10–35
Elongation (%)	0.01	0.2–50

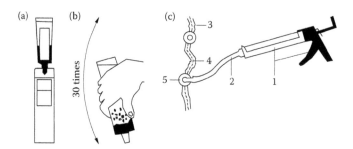

Figure 14.4 Procedure for the application of resin injections: (a) mixing of resin with the hardening agent; (b) shaking of the mixture for it to become homogeneous; (c) application of the resin injection: 1 = injection gun, 2 = plastic hose, 3 = crack, 4 = sealer, 5 = nipples.

the length of the crack, and nipples or ports of the appropriate diameter are placed on the mouths of the holes to facilitate the execution of the resin injections. The crack is then sealed on the surface with a quick-hardening resin paste and the injections are applied. On vertical surfaces, the procedure starts from the lowest nipple or port, and as soon as the resin leaks from the mouth of the next nipple, the procedure is discontinued, the mouth is sealed and the same process is repeated for the next nipple. The next day, when the epoxy resin hardens, the resin paste is removed from the surface with an emery wheel.

14.2.6 Resin concretes

Resin concretes are concretes in which the cement has been replaced by resin. They are mainly used for replacing pieces of concrete that have been cut off. In order to make sure that there will be enough bonding between the old and the new parts, it is recommended that the old concrete be well cleaned and its surface coated with pure resin before the new resin concrete is cast in the place of the cut-off piece. Resin concretes require not only a special aggregate mix to produce the desired properties, but also special working conditions, since all two-component systems are sensitive to humidity and temperature.

14.2.7 Grouts

Grouts are often used for the filling of voids or cracks with large openings on masonry or concrete. The usual grouts consist of cement, water, sand, plasticisers and expansive admixtures in order to obtain high strength and minimum shrinkage during hardening. Details on the composition of conventional grouts can be found in all prestressed concrete manuals where they are used for bonding of post-tensioned tendons. Grouts are mainly used for the repair of structural masonry. In the case of traditional or monumental buildings, the grouts that are used must be compatible with the original construction materials as far as strength and deformability are concerned. Therefore, a large percentage of the cement is replaced in this case by pozzolans or fly ash and calcium hydroxide (UNIDO/UNDP, 1984; Penelis et al., 1984).

For application, the same procedure is followed, as in the case of resin injections. Figure 14.5 shows the general set-up for the application of grouts.

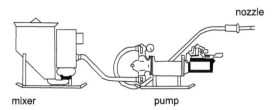

Figure 14.5 Arrangement for application of cement grouts.

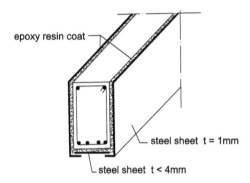

Figure 14.6 Strengthening of a beam with resin-bonded metal sheets.

14.2.8 Epoxy resin-bonded metal sheets on concrete

In this method of intervention (Figure 14.6), the bonding is carried out with epoxy resin spread on the lower face of beams, on the vertical faces of beams or on the joints. The sheets are made of stainless steel (usually 1.00–1.50 mm thick) so that they can be fitted well and bonded on the surface of the element to be strengthened (NTU, 1978; AUT, 1979).

The intervention procedure includes the following phases: careful smoothing of the concrete surface with an emery wheel or emery paper; washing and drying of the concrete surface; roughening up the sheet surface using the process of sandblasting; coating of the concrete surface with an epoxy resin of high viscosity; covering the steel sheet with an epoxy resin layer; and putting it up and keeping it in place with tightening screws for 24 h, so that it will be glued onto the concrete. This procedure should be repeated if a second sheet is necessary. Finally, the sheets should be covered up with wire mesh and cement plaster or shotcrete. The introduction of FRPs has minimised the use of the above method.

14.2.9 Welding of new reinforcement

The most usual way to strengthen regions under tension is the use of new reinforcement. The force transfer from the old reinforcement to the new is accomplished through welding (Figure 14.7). New bars are welded onto the old ones with the aid of connecting bars (bar pieces of the same diameter, but not smaller than 16 mm, and of at least 5ø length, spaced about 500 mm apart).

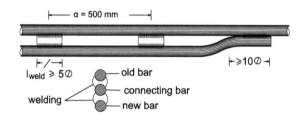

Figure 14.7 Welding of a new reinforcement bar.

14.2.10 FRP laminates and sheets bonded on concrete with epoxy resin

14.2.10.1 General

The strengthening of R/C structural members (e.g. beams, slabs, walls, columns, joints) with externally bonded FRP systems is a relatively new method of intervention that has replaced to a large extent the use of externally bonded steel sheets. FRPs exhibit the following advantages compared to steel sheets:

- They are light in weight.
- They do not corrode.
- They are available in large dimensions.
- They have very high strength accompanied by a linear elastic behaviour up to failure (Figure 14.8).

On the other hand, they exhibit a series of disadvantages:

- They have a brittle type of failure (Figure 14.8). Therefore, they must be considered materials of low ductility. However, they may be used in the form of sheets for external confinement of concrete, which, as is well known, positively influences the ductility of columns (see Section 14.5.4).
- They are susceptible to fire.
- They lose a large percentage of their initial strength under permanent loading, ranging from 15% to 60% (Table 14.5).

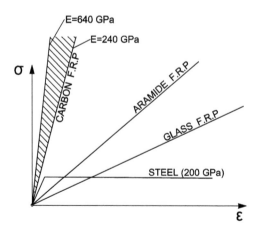

Figure 14.8 Constitutive laws of FRPs.

Table 14.5 Mechanical properties of FRPs

Material	Elastic modulus (GPa)	Tensile strength (MPa)	Failure deformation (%)	Loss of strength under permanent load (%)
Glass-FRP	50	1700–2100	3	60
Aramid-FRP	65–120	1700–2100	2–3	50
Carbon-FRP	165–600	1400–3000	0.5–1.7	15
Steel	200	220–400	0.2[a]	

[a] Yield deformation.

FRPs are formed by embedding continuous fibres into a resin matrix that binds the fibres together.

The common fibres are

- Carbon fibres (CFRP)
- Glass fibres (GFRP)
- Aramid fibres (AFRP)

The most usual resins in practice are the epoxy ones.

14.2.10.2 Technical properties of FRPs

All three types of FRPs, CFRP, GFRP and AFRP, used for strengthening of R/C structures exhibit a wide variety of tensile strength and stiffness, as depicted in Table 14.5.

On the other hand, the comparison of their σ–ε diagrams to that of mild steel shows that some of them, and particularly the CFRPs, have a high E-modulus equal to or two to three times higher than that of steel (Figure 14.8).

At the same time, it can easily be concluded that their strength is much higher – 5 to 10 times – than the strength of steel. However, their brittle behaviour and debonding problems, which will be examined later, do not allow an exploitation of this strength to a high degree.

For the choice of the proper type of FRP for an intervention, various parameters should be taken into account. In Table 14.6, a series of properties is evaluated for each of the three main types of FRPs in use (Meier and Winistorfer, 1995), taking into consideration the integrated behaviour of the type of FRP and the resin matrix.

Table 14.6 Holistic evaluation of FRPs

Criteria	Behaviour		
	CFRP	GFRS	AFRP
Tensile strength	Very good	Very good	Very good
Compressive strength	Very good	Poor	Very good
E-module	Very high	Medium	Low
Fatigue	Excellent	Good	Acceptable
Creep	Good	Acceptable	Poor
Resistance to alkaline environment	Very good	Good	Poor
Durability	Very good	Good	Acceptable
Resistance to fire	Poor	Poor	Poor

14.2.10.3 Types of FRP composites

Two common methods of forming FRP composites have been used so far in strengthening R/C members:

- The wet lay-up method
- The use of prefabricated laminates

The first one involves the in situ application of resin to a woven fabric of a unidirectional woven sheet applied on the R/C surface (Teng et al., 2002). These sheets are applied in successive layers until the proper resistance is ensured. The wet lay-up method is more versatile for in situ applications in case of bonding to curved surfaces and wrapping around corners. It is basically used for the confinement of R/C columns, for shear strengthening of beams and for shear strengthening of R/C or masonry walls.

In the second method, prefabricated laminates are industrially produced. In this procedure, fibres are embedded in an epoxy resin matrix, and then laminates are thermically cured until hardening. With this procedure, laminates may contain a percentage of up to 70% of fibres. Laminates are produced with various cross sections, varying from 50 × 1.2 to 120 × 1.4 mm. These laminates are bonded on-site with epoxy resin on a concrete surface, which has been properly prepared. Prefabricated laminates are used basically for strengthening R/C members against bending.

The most common FRP material in practice for seismic repair and strengthening of R/C members is CFRP. It may be found in markets either in the form of laminates or in the form of sheets.

Laminates are produced in three different degrees of E-modulus, namely, soft, medium and high. The mechanical properties of these three types are given in Table 14.7.

Type H is not used very often due to the inability of effective exploitation of its high stiffness.

CFRP sheets are basically used for

- Confinement of columns
- Shear strengthening of beams
- Shear strengthening of walls

They are usually found in two main categories, namely, C sheets with unidirectional fibres and *low* or *high* modulus of elasticity. The mechanical properties of these two types are displayed in Table 14.8.

Table 14.7 Mechanical characteristics of CFRP laminates (indicative values)

Properties	Type S	Type M	Type H
Modulus of elasticity	>165.000 N/mm²	>210.000 N/mm²	>300.000 N/mm²
Tensile strength	>2.800 N/mm²	>2.400 N/mm²	>1.350 N/mm²
Mean tensile strength at failure	>3.050 N/mm²	>2.900 N/mm²	>1.450 N/mm²
Failure deformation/design	>1.7%	>1.2%	>0.45%
Strain	0.6–0.8%	0.6–0.8%	0.2–0.3%
Colour		Black	
Fibre content (volumetric)		≈68–70%	

Table 14.8 Mechanical characteristics of CFRP sheets (indicative values)

Properties	Low E-modulus	High E-modulus
Modulus of elasticity	>240.000 N/mm²	>640.000 N/mm²
Tensile strength	>3.500 N/mm²	>3.500 N/mm²
Failure deformation	1.55%	0.4%
Weight	200 g/m²	400 g/m²
Mean thickness	0.117 mm	0.190 mm
Design strain	0.4–0.6%	0.2%
Safety coefficient S	1.2	1.2

CFRP sheets of low modulus of elasticity are used for column confinement, while sheets of high E-modulus are used basically for shear strengthening.

Figures 14.9 through 14.11a,b exhibit the application of CFRPs for R/C member strengthening. Producers' manuals give in detail all necessary information on technical properties of the available materials, their application techniques and the specifications with which they comply.

In closing, it should be noted that C laminates or C sheets may also be used in successive layers.

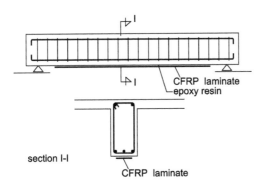

Figure 14.9 Strengthening of a T beam against bending by applying CFRP laminates.

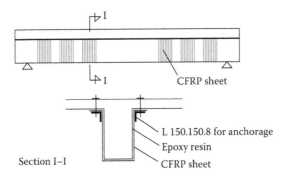

Figure 14.10 Strengthening of a T beam against shear by applying CFRP sheets.

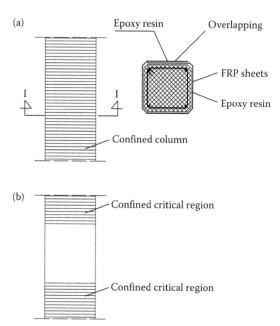

Figure 14.11 Confinement of an orthogonal R/C column by applying CFPR sheets: (a) wrapping with FRP sheets for axial strength enhancement; (b) wrapping with FRP sheets at the plastic joints for increasing ductility capacity of the plastic joints.

14.3 REDIMENSIONING AND SAFETY VERIFICATION OF STRUCTURAL ELEMENTS

14.3.1 General

Repair and strengthening are related mainly to several interface issues, which are due to the damage itself or are created by the intervention. New materials are added to the existing structural elements, for example, concrete to concrete, epoxy resin to concrete, steel to concrete, steel acting through welding and so on. Consequently, load transfer from the original element to the additional 'reinforcing' materials is carried out through discontinuities, by means of unconventional mechanisms like friction, dowel action, large pull-out action, adhesion and so on. The systematic study of these mechanisms constituting a kind of new mechanics of the non-continuum appears to be a fundamental prerequisite for the rational design of repaired and strengthened structural elements (Tassios, 1983; Tassios and Vintzeleou, 1987; CEB, 1991b). However, besides the independent study of these force transfer mechanisms, the proper combination of several of them in integrated physical and mathematical models is needed for the safety verification of the structural elements, since the various repair or strengthening techniques may activate several force transfer mechanisms simultaneously. In this context, extensive research is needed to bridge the existing gaps in knowledge in this area until this process is applicable to practical problems.

Therefore, at present, redimensioning and safety verification follow in interventions practice a semi-empirical procedure based on practical rules supported by experimental evidence. In subsequent subsections, these two methods will be presented in detail.

14.3.2 Revised γ_m-factors

No matter which one of the two methods mentioned above is followed for the redimensioning and safety verification, it should be stressed that special attention should be given to the γ_m-factors introduced in the calculation.

Original materials will be factored by the confidence factor (see Chapter 13) applied to the mean values of their mechanical properties, as specified in EC 8-3/2005. The strengths of additional materials bonded to the original structural elements must be divided by increased γ_m-factors in recognition of the additional uncertainties in reconstruction operations. Keeping in mind the above considerations, particularly in the case of cast-in-place new concrete, the use of concrete with a strength of 5 MPa higher than that of the original elements has already been recommended (Section 14.2). Thus, the designer may retain the same γ_m-factor for both the original and the new element, on the condition that the strength introduced in the redesign calculations will be that of the original concrete.

14.3.3 Load transfer mechanisms through interfaces

In the following paragraphs, the most common transfer mechanisms along the several discontinuities or interfaces between existing and additional material will be presented as they were grouped in EC8/Part 1.4/Draft (CEN, 1993). A separate section will be devoted next (Section 14.5) to load transfer mechanisms between FRPs and concrete due to the significant importance of this mechanism for repair and strengthening of R/C members using FRPs.

14.3.3.1 Compression against pre-cracked interfaces

During reloading after cracking due to tension, compressive forces may be developed prior to full recovery of the previous extensional deformation, since the protruding elements constituting the rough surface at both faces of a crack may come into earlier contact due to their transversal microdisplacement (uneven bearing). Consequently, it is allowed to account for this phenomenon by means of an appropriate model (Figure 14.12; Tassios, 1983; Gylltoft, 1984). The quantitative evaluation of such a model needs extensive experimental support.

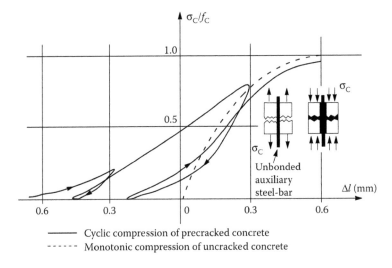

Figure 14.12 Monotonic and cyclic compression of cracked concrete.

14.3.3.2 Adhesion between non-metallic materials

Local adhesion versus local slip between old and new materials may be accounted for by means of appropriate models, while taking into account their sensitivity to curing conditions and the characteristics of possible bonding agents. Taking into consideration that the value of the slip needed to mobilise adhesion is very low, it is permissible to consider that the entire adhesion resistance is developed under almost zero displacement (Figure 14.15; Hanson, 1960; Ladner and Weber, 1981; Tassios, 1983).

14.3.3.3 Friction between non-metallic materials

In several cases, friction resistance may be accounted for as a function of relative displacement (slip) along the discontinuity or along the interface. A constitutive law must be formulated for this purpose based on experimental data (Figure 14.13).

In some cases, when the slip needed to activate the maximum friction resistance (τ_u) is relatively low, the concept of a 'friction coefficient',

$$\tau_u = \mu\sigma_u \tag{14.1}$$

may be used. However, for relatively low σ-values, the strong relation between 'μ' and 'σ' values must be taken into account (Figure 14.14; Tassios, 1983, 2009).

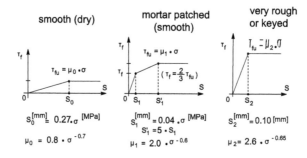

Figure 14.13 Formalistic models for concrete-to-concrete friction as a function of normal compressive stress σ.

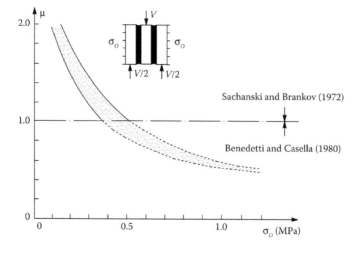

Figure 14.14 Friction coefficients for masonry as a function of the average normal stress.

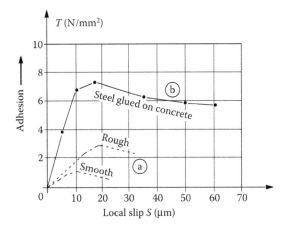

Figure 14.15 Constitutive law of adhesion: (a) concrete-to-concrete through bonding agent (Hanson, 1960); (b) steel sheets glued to concrete by means of epoxy resin (Ladner and Weber, 1981).

14.3.3.4 Load transfer through resin layers

The tensile strength of the contact interface between a resin layer and a given material (e.g. concrete) may be taken as equal to the tensile strength of the weaker of the two. Therefore, in the case of concrete, its tensile strength f_{ctm} must be introduced in all calculations related to the load transfer through this interface. Of course, this value must be divided by a γ_m at least equal to 1.5. The local shear resistance generated along such an interface is a function of the local slip and the normal stress acting on the area under consideration. Figure 14.15 gives the constitutive law of the shear resistance as a function of the slip for σ equal to zero (adhesion).

14.3.3.5 Clamping effect of steel across interfaces

The friction generated across a sheared interface transversely reinforced by well-anchored steel bars may be evaluated as follows (Figure 14.16; Chung and Lui, 1978):

1. In the case of an expected large relative displacement along the interface, the ultimate friction resistance may be estimated as

$$\tau_R = \mu\sigma_{tot} \not> \tau_{um} \tag{14.2}$$

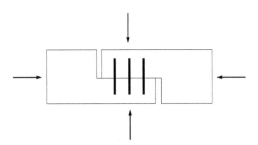

Figure 14.16 Clamping effect of steel across interfaces.

where μ denotes the friction coefficient available under normal stress (Figure 14.13) and

$$\sigma_{tot} = \rho f_y + \sigma_o \tag{14.3}$$

- f_y is the yield strength of steel.
- σ_o is the external normal stress across the interface.
- ρ is the effective steel ratio along the interface.
- $\tau_{u,m}$ is the shear resistance of the material itself.

2. If large slips along the interface are not tolerated, the generated friction resistance is evaluated, taking into account the displacement compatibility on both faces of the interface.

14.3.3.6 Dowel action

The design value of the maximum shear force that may be transferred by a bar crossing an interface may be calculated, taking into account the strength and deformability of the dowel and the connected material as well as the distance of the dowel from the edges. According to Rasmussen (1963), for the plastic compressive stage,

$$D_u \cong 1.3 d_b^2 \sqrt{f_c f_y} \tag{14.4}$$

while according to Vintzeleou and Tassios (1986), for the post-cracking stage,

$$\frac{D}{D_u} \cong 0.7 \sqrt[4]{s} \, (mm) \text{ for } d_b \cong 12 - 22 \text{ mm} \tag{14.5}$$

where
D_u is the ultimate capacity of a dowel embedded in uncracked concrete.
$-f_c$ is the unconfined strength of concrete.
f_y is the yield strength of steel.
d_b is the dowel diameter.
s is the local slip at the interface (in mm).
D is the dowel action for slip equal to s.

14.3.3.7 Anchoring of new reinforcement

1. Anchorage lengths of steel bars in new concrete must follow the criteria of relevant codes such as EC2. In the case of bar anchorages in holes bored in old concrete where special grouts are used (e.g. high-strength concrete with shrinkage compensating admixtures, resin concretes, etc.), shorter anchorage lengths are needed. These are specified in the manuals of the material used and must be verified by pull-out tests performed by an authorised laboratory.
2. In most cases, the anchoring of additional steel bars is accomplished by welding them onto the existing bars directly or by means of additional welded spacers (Figure 14.7); such force transfers may be considered rigid. In such cases, it is necessary to verify that the bond ensured by the existing bar is sufficient to anchor the total force acting on both bars.

14.3.3.8 Welding of steel elements

In designing steel-to-steel connections by means of welding, in addition to the checks of welding resistance, the following mechanical behaviour should be considered, since the activation of force transfer depends on the concept of the connections:

- Direct welding of additional bars or steel profiles on existing ones ensures a complete generation of force transfer with almost zero slip.
- Intermediate deformable steel elements necessitate the introduction of proper models so that compatibility of deformations may be ensured (Tassios, 1983).

14.3.3.9 Final remarks

From the preceding presentation, the following conclusions may be drawn:

a. The constitutive laws of the transfer mechanisms need to be supported by additional experimental evidence covering several parameters related to the intervention techniques.
b. It should be stressed that in designing the repair or strengthening of a structural element, several force transfer mechanisms are generated, so that only an integrated model based on the finite element method (FEM) may take all of them into account, the interrelations among them and the level at which each of them is activated during loading, as happens with the analysis of original R/C elements or masonry walls (Ignatakis et al., 1989, 1990).
c. Furthermore, even if such models based on the FEM were available, they would have to be verified through experimental evidence on repaired or strengthened structural sub-assemblages.
d. From the foregoing follows that, at present, the formation of integrated analytical models cannot yet lead to dimensioning or safety verification methods for general use, suitable for practical applications. However, it is hoped that, in the near future, this procedure will lead to the derivation of reliable models.
e. For the time being, the approach to the problem is based on a simplified estimation of resistances originating from practical rules that are verified by laboratory tests. Sometimes this approach is combined with simplified models of force transfer mechanisms, as we will see later. In the next subsection, the basic concept of this semi-empirical method used in practice will be given in detail.

14.3.4 Simplified estimation of the resistance of structural elements

1. The basic concept in developing any repair or strengthening technique is to ensure that failure of the repaired structural element as a monolithic unit will precede any failure at the interfaces between old and new material. This is verified by tests, and where failure occurs at the interfaces first, extra connecting means are provided on an empirical basis (e.g. closer-spaced dowels, a resin layer between old and new concrete). In order for this basic concept to be accomplished, the specifications referring to each intervention technique should be rigorously followed during the execution of the work.
2. With the above concept as a prerequisite, specimens of the repaired or strengthened structural elements are tested in the laboratory under monotonic or cyclic loading to

failure, and relevant displacement versus resistance diagrams are plotted. From these diagrams, the basic values of ultimate strength $R_{u,rep}$, stiffness K_{rep} and energy dissipation $E_{u,rep}$ are determined.

3. At the same time, the above values are calculated based on the assumption that the structural element under consideration was constructed as a monolithic unit, including the initial element and its additional elements in the form of repair (e.g. R/C jackets). It is obvious that the respective values of $R_{u,monol}$, K_{monol} and $E_{u,monol}$ will be greater than or at least equal to those of the repaired element due to the fact that interface deficiencies are not taken into account in the calculation of $R_{u,monol}$, K_{monol} and $E_{u,monol}$. Therefore, 'model reduction factors' are introduced (EC8 Part 1.4/ENV draft, EC8-3/2005 Annex A [informative]),

$$\varphi_R = \frac{R_{u,rep}}{R_{u,monol}} \quad \varphi_K \frac{K_{rep}}{K_{monol}} \quad \varphi_E = \frac{E_{u,rep}}{E_{u,monol}} \tag{14.6}$$

The index 'monol' refers to a monolithic element consisting of the initial element and the repair. These factors allow the redimensioning and safety verification of the repaired element to be carried out as if it were a monolithic unit. In fact, the results of the calculation of the resistance of a repaired or strengthened element that is based on monolithic considerations are multiplied by the model reduction factors, in order to comply with the capacity expected for the repaired or strengthened element. These reduced results of the resistance are introduced in the design verification at ULS. The whole procedure is accomplished with some additional simplified force transfer checks at the critical interfaces, as will be discussed later.

4. From the preceding presentation, it may be concluded that 'model reduction factors' have reliable values only for the special cases for which laboratory tests have been performed. If the geometrical data of the original and the added sections are different, or the span or the height of the structural element changes, there is no evidence that these values will still be valid. Therefore, it is clear that additional experimental and analytical research is urgently required to provide information about the seismic behaviour of structures repaired or strengthened using different techniques (Rodriguez and Park, 1991).

14.4 REPAIR AND STRENGTHENING OF STRUCTURAL ELEMENTS USING CONVENTIONAL MEANS

14.4.1 General

Structural elements, depending on the desirable seismic resistance, the damage level and the type of their joints, may be repaired or strengthened with resin injections, replacement of broken-off parts, R/C jackets, metal cages or FRPs.

As mentioned in Section 14.3.3, the key to the success of the repair or strengthening procedure is to attain a high degree of bonding between the old and the new concrete. This can be accomplished as follows:

- By roughening the surface of the old concrete
- Coating the surface with epoxy or another type of resin before concreting
- Welding reinforcement bars
- Using steel dowels

The ductility of the repaired element is improved by proper confinement with closely spaced hoops, steel jackets, composite material (FRP) jackets and so on.

It should be kept in mind that changes in the sectional area of the structural elements lead to a redistribution of stress due to resulting changes in the stiffness of the various structural elements.

Metal cages made of steel angles and straps are used exclusively for column repair. However, the repair of the joint between column and beam is not possible.

The bonding of metal plates or FRP laminates on concrete is generally a technique easy to apply, whereby zones under tension can be strengthened without altering the stiffness.

The last two methods require special means of fire protection, which is not necessary in case of R/C jackets.

Repair and strengthening of R/C members using FRP sheets or laminates will be examined separately in Section 14.5 because their safety verification requires a detailed examination of the load transfer mechanisms between FRPs and concrete, and because in the last 20 years, this technique has been gaining more ground every day.

14.4.2 Columns

Damage to columns appears at different levels, such as

- Fine cracks (horizontal or diagonal) without crushing of concrete or failure of reinforcement
- Surface spalling of concrete without damage to the reinforcement
- Crushing of concrete, breaking of the ties and buckling of the reinforcement

Depending on the degree of damage, different techniques may be applied, such as resin injections, removal and replacement of parts or jacketing.

14.4.2.1 Local interventions

Resin injections and resin mortars are applied only for the repair of columns with small cracks or peelings, without crushing of concrete or damage to the reinforcement. The degree of retrofit can be checked by comparing the force–displacement (H–δ) diagrams of the original column and the repaired one with epoxy resins (Figure 14.17; Sariyiannis, 1990; Sariyiannis and Stylianidis, 1990). The results from such comparisons are very encouraging, with regard to the effectiveness of the repair.

Removal and replacement are applied in columns with a high degree of damage, meaning crushing of concrete, breaking of ties and buckling of longitudinal reinforcement. Of course, before carrying out such work, a temporary support system is always provided to carry the column loads. Then, if concrete failure is only superficial, partial removal and repair are carried out (Figure 14.18). Otherwise, in the case of a total failure, there is a complete removal of the material, placement of new longitudinal reinforcement with welding, placement of new closely spaced ties and concreting (Figure 14.19). It should be mentioned that, in the first case, good bonding between old and new concrete is absolutely necessary. In the second case, the construction of an R/C jacket usually follows the replacement.

14.4.2.2 R/C jackets

R/C jackets are applied in the case of serious damage or inadequate seismic resistance of the column. Jackets are applied basically at all sides of the perimeter of the column,

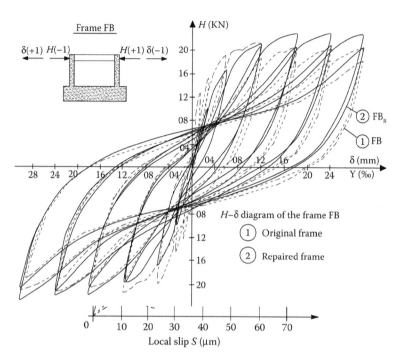

Figure 14.17 H–δ diagram of the original frame and then when repaired with epoxy resin injections.

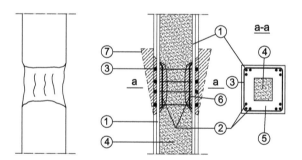

Figure 14.18 Column repair in the case of superficial damage: 1 = existing reinforcement; 2 = added new reinforcement; 3 = added new ties; 4 = existing concrete; 5 = new concrete; 6 = welding; 7 = temporary cast form.

which is the ideal case. However, sometimes, depending on the existing local conditions, jackets are applied on three or less sides (Figure 14.20). In cases where the jacket is limited to the storey height, an increase in the axial and shear strength of the column is achieved with no increase in flexural capacity at the joints. Therefore, it is recommended that the jackets extend beyond the ceiling and the floor slabs of the storey where column repair is necessary (Figure 14.21).

In the case of one-sided jackets, special care should be taken to connect the old with the new part of the section; this can be accomplished by welding closely spaced ties to the old reinforcement (Figure 14.22).

In the usual case of full jackets, the composite action of the old and new concrete is sometimes left solely to the natural bonding of the two materials, which can be strengthened with roughening of the old surface. It is also sometimes strengthened by welding some bent-up

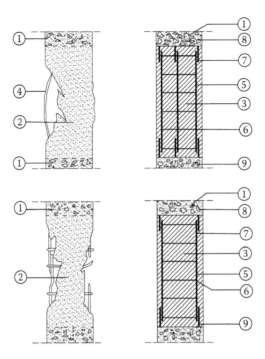

Figure 14.19 Repair of a seriously damaged column: 1 = existing undamaged concrete; 2 = existing damaged concrete; 3 = new concrete; 4 = buckled reinforcement; 5 = added new reinforcement; 6 = added new ties; 7 = welding; 8 = existing ties; 9 = existing reinforcement.

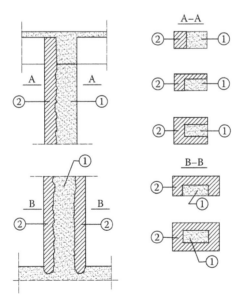

Figure 14.20 R/C column jacketing arrangement: 1 = existing column; 2 = jacketing concrete.

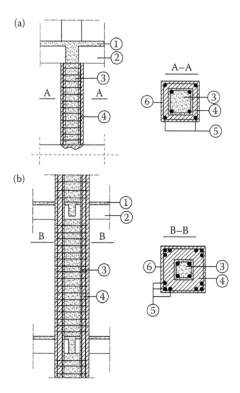

Figure 14.21 Column jackets: (a) jacket along the height of one storey; (b) jacket extended to the upper and lower storey; 1 = slab, 2 = beam, 3 = existing column, 4 = jacket, 5 = added longitudinal reinforcement, 6 = added ties.

bars between the old and new longitudinal reinforcement (Figure 14.23). This connection is necessary when the column has completely deteriorated or when its height is too great, in which case there is a danger of buckling of the new longitudinal reinforcement. However, laboratory tests have shown that, in general, the degree of composite action obtained is very satisfactory even without the strengthening of force transfer by welding the longitudinal reinforcement (Zografos, 1987).

14.4.2.3 Steel profile cages

In general, this is a technique not widely used. The cage consists of four steel angles of minimum dimensions L 50.50.5, which are connected to each other with welded blades of minimum dimensions 25.4 mm (see Chapter 12.4.2.3, Figure 12.6). Prior to welding, the angles are held tight on the column with the aid of transverse angles and pre-stressed ties. The voids between the angles and the concrete are filled with non-shrinking mortar (EMACO, EMPECO, etc.) or resin grout, and then the column is covered with gunite or cast-in-place concrete reinforced with welded wire fabric. It is obvious that, with this arrangement, increase in the flexural capacity of the column at the joints with the top and bottom is impossible, due to the fact that the cage is not extended into the floors above and below.

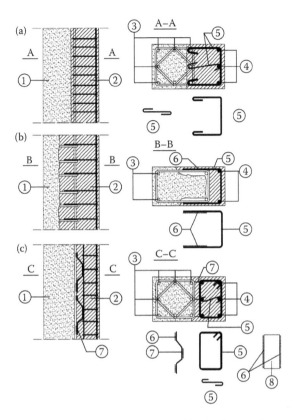

Figure 14.22 One-sided strengthening of a column: (a) use of hooks for reinforcement connection; (b) use of welding of ties for reinforcement connection; (c) use of welded bent bars for reinforcement connection. 1 = existing column; 2 = jacket; 3 = existing reinforcement; 4 = added longitudinal reinforcement; 5 = added ties; 6 = welding; 7 = bent bars; 8 = metal plate.

14.4.2.4 Steel or FRP encasement

Steel or FRP encasement is the complete covering of an existing column with thin steel or FRP sheets. This type of intervention offers the possibility of only a small increase in column size. Steel sheets (with 4–6 mm thickness) are welded together throughout their length and located at a distance from the existing column. The voids between the encasement and the column are filled with non-shrinking cement grout.

The strengthening with FRP can be accomplished basically with encasement using FRP sheets bonded on the concrete surface. Unidirectional fibres of the sheet are arranged transversally to the column axis.

In this context, strengthening may aim at improving

- Ductility through confinement
- Shear resistance
- Lap splice resistance of the original reinforcement

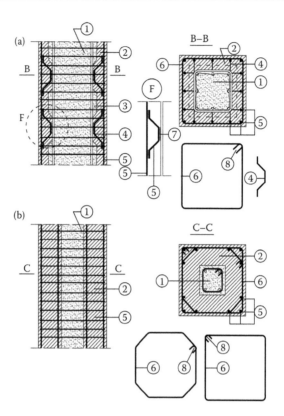

Figure 14.23 Connection of the old to the new reinforcement of the jacket: (a) protection of new bar against buckling with welding; (b) protection of new bars against welding with octagonal ties. 1 = existing column; 2 = jacket; 3 = key; 4 = bent bars; 5 = added reinforcement; 6 = ties; 7 = welding; 8 = alternating corners.

Bending enhancement of the column may also be achieved by bonding of FRP laminates parallel to the column axis. However, the flexural strength of the frame structure cannot be improved because it is impossible to pass the encasement through the floors.

14.4.2.5 Redimensioning and safety verifications

Experimental results (French et al., 1990; Sariyiannis, 1990; Sariyiannis and Stylianidis, 1990; Stylianidis, 2012) regarding the dimensioning of repaired columns have shown the following:

1. In the case of repair with resin injections, the ratio of the strength of the repaired element to that of the original one is about 1. In general, the epoxy-repaired cracks do not reopen in tests; new cracks tend to develop adjacent to the repaired ones. The stiffness of the repaired column appears to exceed 85% of the original one, and the same happens with the energy dissipation capacity,

$$\frac{R_{d,rep}}{R_{d,orig}} \cong 1 \qquad \frac{K_{rep}}{K_{orig}} \cong 0.85 \qquad \frac{E_{rep}}{E_{orig}} \cong 0.85 \qquad (14.7a)$$

The bond between reinforcement and concrete also appears to be restored, even for high inter-storey drifts exceeding 4%.

Similar results have been incorporated in FIB state-of-the-art-report Bulletin 24 Seismic assessment and retrofit of R/C buildings (2003). According to this document, the statistical evaluation of 33 experimental tests of repaired columns and R/C walls injected with epoxy resins has given the following results:

$$\frac{M_{y,exp}}{R_{y,pred}} = 1.04, \quad \frac{K_{ef,exp}}{K_{ef,pred}} \cong 0.89, \quad \frac{\theta_{u,exp}}{\theta_{u,pred}} \cong 0.95 \qquad (14.7b)$$

Predicted capacity values have resulted from the design of the original member.

2. In the case of repair with reinforced cast-in-place jackets, the experimental results have shown (Zografos, 1987; Bett et al., 1988; Bush et al., 1990) that the lateral capacity of the strengthened column can be reliably predicted, assuming complete compatibility between the jacket and the original column. For jackets with gunite concrete, despite all the contrary estimation (NTU, 1978), the results fall slightly below those of conventional R/C jackets cast in forms. However, given the fact that field conditions are not as ideal as those of a laboratory, the author's opinion is that, on the one hand, the new concrete must have a strength 5 MPa greater than that of the original element, and on the other, a model correction factor $\varphi \cong 0.90$ for the strength and the stiffness of the repaired element should be introduced.

$$\frac{R_{d,rep}}{R_{d,monol}} \cong 0.90, \quad \frac{K_{rep}}{K_{monol}} \cong 0.90 \qquad (14.8a)$$

The index 'monol' refers to a monolithic element consisting of the initial element and the jacket.

Similar results have been presented in the above-mentioned FIB Bulletin 24 (2003) document as a result of the statistical evaluation of 15 experimental tests of jacketed columns, as follows:

$$\frac{M_{y,exp}}{M_{y,monol}} = 0.97, \quad \frac{K_{ef,exp}}{K_{ef,monol}} \cong 1.02, \quad \frac{\theta_{u,exp}}{\theta_{u,monol}} \cong 1.08 \qquad (14.8b)$$

3. In the case of repair with metal cages of straps and angles (Arakawa, 1980; Tassios, 1983), redimensioning may be carried out according to what was suggested in (2).
4. In the case of repair with bonded steel sheets, the additional shear resistance V_{fc} of the column may be estimated by the following expression:

$$V_{fc} = 2tf_y h \cot \delta \qquad (14.9)$$

where
- f_y is the yield strength of the steel sheet.
- t is the thickness of the plate.
- h is the dimension of the column cross section parallel to V_{fc}.
- δ is the angle between the column axis and diagonal cracks. It may be considered that $\delta = 30°$ (Priestley and Seible, 1991).

The coefficient 2 has been introduced to take into account that the plates are bonded on both sides over the shear crack. In both cases, a load transfer verification control is necessary on the bond interface between concrete and sheet (Section 14.3.3).

5. Safety verification of strengthening with FRPs will be examined in Section 14.5 collectively for all cases of strengthening with FRPs.

14.4.2.6 Code (EC 8-3/2005) provisions

Eurocode EC 8-3/2005 gives the following recommendations in its Annex A in the form of information:

1. *Concrete jacketing*
   a. For the purpose of evaluating strength and deformation capacities of jacketed members, the following approximate assumptions may be made.
      i. The jacketed member behaves monolithically.
      ii. The fact that axial dead load is originally applied to the original column alone is disregarded.
      iii. The concrete properties of the jacket may be assumed to apply over the full section.
   b. The following relations may be assumed to hold between the values V_R, M_y, θ_y and θ_u calculated under the assumptions above and the values V_R^*, M_y^*, θ_y^*, and θ_u^* to be adopted in the capacity verifications:

$$\frac{V_R^*}{V_R} = 0.9, \frac{M_y^*}{M_y} = 1.0, \frac{\theta_y^*}{\theta_y} = 1.05, \frac{\theta_u^*}{\theta_u} = 1.0 \qquad (14.8c)$$

2. *Steel jacketing*
   a. Steel jackets are mainly applied to columns for the purpose of
      i. Increasing shear strength
      ii. Increasing ductility through confinement
      iii. Improving the strength of deficient lap splices
   b. Shear strength

The contribution of the jacket to shear strength given by Equation 14.9 is recommended by EC 8-3/2005 to be reduced to 50% of its value so that the jacket remains in elastic stage and, in this respect, is able to control the width of internal cracks.

14.4.3 Beams

As in the case of columns, depending on the degree of damage to the beams, several techniques are applied, such as resin injections, bonded metal or FRP sheets, FRP laminates and removal and replacement of concrete and R/C jackets.

14.4.3.1 Local interventions

Resin injections are applied only for the repair of beams with light cracks without crushing of concrete.

Removal and replacement are applied to beams with a high degree of damage such as crushing of concrete or failure of reinforcement, loss of bonding and spalling due to dowel action. Propping with temporary supports always precedes repair work of this type. The procedure that is then followed is similar to that described for column repair. However, at this point, it has to be stressed that difficulties may arise regarding the compaction of concrete if it is not possible for casting to be carried out from the upper side of the beam with special openings in the slab.

14.4.3.2 R/C jackets

Reinforced concrete jackets can be applied by adding new concrete to three or four sides of the beam. In the same technique, one should also include the strengthening of the tension or compression zone of a beam through concrete overlays. In order to accomplish force transfer between old and new concrete, roughening of the surface of the old concrete is required, as well as welding of connecting bars to the existing bars and new reinforcement.

Reinforced overlays on the lower face of the beam (Figure 14.24) can only increase its flexural capacity. Existing reinforcement is connected to the new reinforcement by welding. Jacketing on all four sides of the beam is the most effective solution. The thickness of the concrete that is added to the upper face is such that it can be accommodated within the floor thickness (maximum: 50–70 mm). The placement of the ties is achieved through holes, which are opened in the slab at closely spaced distances and are also used for pouring the concrete. The longitudinal reinforcement bars of the jacket are welded to those of the old concrete (Figure 14.25).

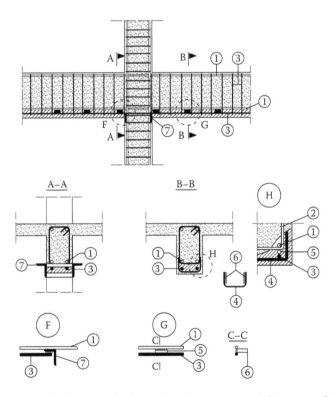

Figure 14.24 Strengthening of a beam on the lower face: 1 = existing reinforcement; 2 = existing stirrups; 3 = added longitudinal reinforcement; 4 = added stirrups; 5 = welded connecting bar; 6 = welding; 7 = collar of angle profiles.

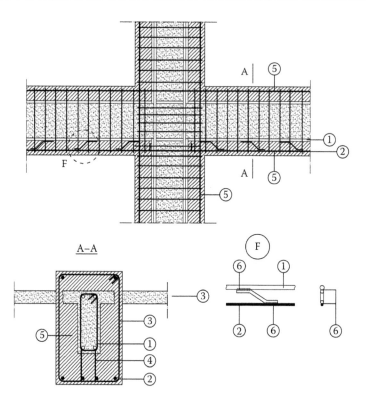

Figure 14.25 Jacket on four sides of a beam: 1 = existing reinforcement; 2 = added longitudinal reinforcement; 3 = added stirrups; 4 = welded connecting bar; 5 = concrete jacket; 6 = welding.

Jackets on three sides of the beam are used to increase the flexural and shear capacity of the beam for vertical loading, but not for seismic actions, given that strengthening of the load-bearing capacity of the section near the supports is impossible. The key to the success of such an intervention is the appropriate anchorage of the stirrups at the top of the sides of the jacket (Figure 14.26). Due to the fact that using forms and pouring the concrete from the top is not possible, the only feasible solution is gunite concrete.

14.4.3.3 Bonded metal sheets

The technique for bonding metal sheets onto concrete was described in detail in a previous section. These sheets are bonded either on the lower face of the beam under repair, for strengthening of the tension zone, or on the vertical sides of the beam near the supports, for shear strengthening. This procedure should be preceded by crack repair with epoxy resin. The bonded plates must be protected by welded wire mesh and cement plaster or shotcrete.

14.4.3.4 Redimensioning and safety verification

1. *Resin injections:* Extensive laboratory tests (Popov and Bertero, 1975; French et al., 1990; Economou et al., 1994) have shown that if there is no concrete degradation, epoxy resin injections are very effective. The repaired beam is capable of resisting several loading cycles, the initial strength is completely restored, while stiffness and

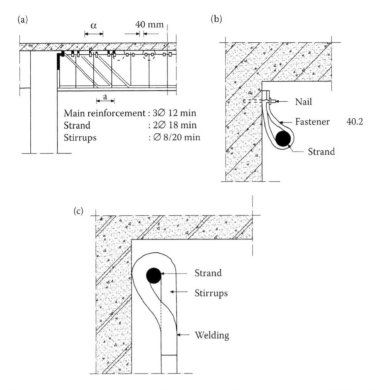

Figure 14.26 Jacket on three sides of a beam: (a) general reinforcement pattern; (b) detail of fixing of the strand; (c) detail of anchoring the ties on the strand.

energy dissipation appear to be somewhat lower than those of the original beam. Consequently, 'the model correction factor' φ may be considered equal to 1 in this case:

$$\frac{R_{d,rep}}{R_{d,orig}} \cong 1.0, \quad \frac{K_{rep}}{K_{orig}} \cong 1.0, \quad \frac{E_{rep}}{E_{orig}} \cong 1.0 \tag{14.10}$$

2. *R/C overlays or jacketing:* Extensive experimental results have shown (Vassiliou, 1975; Tassios, 1983; Abdel-Halim and Schorn, 1989; Saiidi et al., 1990) that concrete overlays and jacketing are an effective technique for repair or strengthening. The additional layers and the parent concrete remain bonded throughout loading until failure, provided that construction specifications given in the previous paragraphs are met. The reduction in strength of the repaired beam varies between 8% and 15% of the strength of the monolithic beam (initial + jacket). The reduction in stiffness of the repaired beam is somewhat higher (10–20%) with respect to the stiffness of the monolithic beam. Consequently, 'the model correction factor' φ may be considered as follows:

$$\frac{R_{d,rep}}{R_{d,monol}} \cong 0.85, \quad \frac{K_{rep}}{K_{monol}} \cong 0.80 \tag{14.11}$$

EC 8-3/2005 makes no reference to jacketed beams, nor does FIB Bulletin 24 (2003). It is anticipated that the recommendations for the reduction factor of jacketed R/C columns and walls might also be adopted for jacketed beams on all their sides.

In addition to the general strength and stiffness verifications described previously, specific verifications for the force transfer mechanisms along the several interfaces between existing and additional material should be performed. In the case where adhesion between old and new concrete is proved to be inadequate, the transfer mechanism should be ensured with extra connectors on the interface. Two such cases can be identified:

a. *Interface of connection in the tension zone* (Figure 14.27a): The shear stresses developing on the interface between old and new concrete are given, according to the theory of strength of materials applied to reinforced concrete, by the approximate relationship (Tassios, 1984a)

$$\tau_{02} = V_d / \left[b_w z \left(1 + \frac{A_{s1} (d_1 - x) z_1}{A_{s2} (d_2 - x) z_2} \right) \right]$$

(14.12)

where

$$z = \frac{A_{s1} z_1 + A_{s2} z_2}{A_{s1} + A_{s2}}$$

(14.13)

Bearing in mind that special care is taken in ensuring the adhesion of the new to the old concrete through resin coats of higher strength than that of concrete, the value

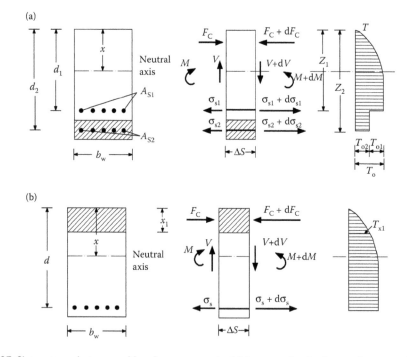

Figure 14.27 Shear stress between old and new concrete: (a) intervention in the tension zone; (b) intervention in the compression zone.

resulting from the above relationship at the interface must be compared with the basic concrete shear strength (see Section 14.3.3.4).

Therefore, if τ_{02} is greater than τ_{Rdc},

$$\tau_{02} \geq \tau_{Rdc} \qquad (14.14)$$

where

$$\tau_{Rdc} = c_{Ed,c}k(100\rho_1 f_{ck})^{1/3} \qquad (14.15)$$

(see Equation 8.53)
then the total shear flow $(T = \tau_{02}b)$ must be carried by welding of the new reinforcement to the old one.

Therefore, for a distance α between successive weldings, welding thickness t and number η of new bars, the welding length l_{wel} must be equal to (Figure 14.7)

$$l_{weld} = \frac{\tau_{02} \cdot b_w \cdot \alpha}{t \cdot n \cdot 0.8 f_{yd}} \qquad (14.16)$$

where $f_{yd} = f_{yk}/1.15$ is the yield stress of the welding steel divided by the safety factor γ_s of the material (design strength). It is understood that a 'model correction' factor equal to 0.8 has been introduced in formula 14.16.

b. *Interface of connection in the compression zone* (Figure 14.27b): The shear stresses developing at the interface between the old and the new concrete are again given according to the classic theory of strength of materials by the approximate relationship

$$\tau_{x1} = \frac{V_d}{b_w z} \frac{x_1}{x} \left(2 - \frac{x_1}{x} \right) \qquad (14.17)$$

If the resulting value of τ_{x1} is greater than τ_{Rdc}, as defined above, the total shear flow $(T = \tau_{x1}b)$ must be carried by shear connectors (Figure 14.28). The ultimate shear carried by the two legs of such a connector is equal to (Tassios, 1984a)

$$D_u \cong 2d^2 \sqrt{f_{cd}f_{yd}} \qquad (14.18)$$

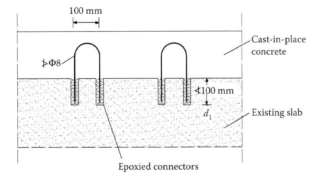

Figure 14.28 Shear connectors between old and new concrete in the compression zone.

where d is the diameter of the connector, f_{cd} is the design strength of concrete and f_{yd} is the design strength of the connector's steel. In Equation 14.18, a model correction factor $\varphi = 1.30$ has been introduced (Section 14.3.3.6).

3. *Bonded metal sheets.* The required section b_t of a sheet in a flexural area is expressed by the relationship

$$\Delta M_d \leq (bt)zf_{yd} \tag{14.19}$$

and hence

$$(bt)_{req} \geq \frac{\Delta M_d}{zf_{yd}} \tag{14.20}$$

(model correction factor equal to 1) where ΔM_d is the additional moment (strengthening) beyond the ultimate M_{du} carried by the original section (ΔM_d should not be greater than $0.5M_{du}$ for construction reasons), z is the lever arm of the internal forces and f_{yd} is the design strength of the sheet.

The required anchorage length of the sheet is given by the relationship

$$l_a = \varphi \left(\frac{f_{yd}}{r \cdot \tau_u} t \right) \tag{14.21}$$

where φ is the model correction factor ($\varphi \cong 1.3$), f_{yd} is the design strength of the sheet, t is the thickness of the sheet and τ_u is the maximum local adhesion strength between concrete and the steel sheet. For sheet thickness $t < 1$ mm, the recommended value for $\tau_u \cong 2f_{ctd}$, while for $t = 3$ mm, the recommended value for $\tau_u \cong f_{ctd}$ (Figure 14.15). Note that f_{ctd} is the tensile design strength of concrete and r is the correction factor to take into account the non-uniform distribution of τ_u over the bonding area due to the different slippage from point to point, from the crack to the end of the sheet. Recommended value: $r = 0.40$ (Tassios, 1983).

The required thickness t of the sheets that are bonded on both sides of a beam over shear cracks to carry additional shear forces may be given by the relationship (Tassios, 1984a)

$$\Delta V_d \leq 2t \cdot z \cdot f_{yd} \cot \delta$$

$$t_{req} \geq \frac{1}{2} \frac{\Delta V_d}{z \cdot f_{yd} \cot \delta} \tag{14.22}$$

For the meaning of the symbols included in the above relationships, see Section 14.4.2.5, item (4).

The safety verification of the force transfer through the bonded interface may be carried out with the aid of the following expressions (Figure 14.29; Tassios, 1983):

$$\frac{\Delta V_d}{2} \leq F_s = \left(\frac{1+\xi}{2} \right) d \cdot l_o \cdot r \cdot \tau_u \tag{14.23}$$

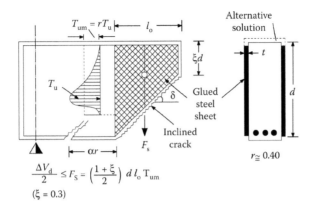

Figure 14.29 Shear force transfer through epoxy resin glued steel sheet.

taking into account that

$$l_o = (1 - \xi)d \cot \delta$$

Equation 14.23 takes the form

$$\frac{\Delta V_d}{2} = \frac{1 - \xi^2}{2} d^2 \cdot r \cdot \tau_u \cdot \cot \delta \tag{14.24}$$

For $\xi \cong 0.30$ and $\delta \cong 30°$, Equation 14.24 takes the form

$$\Delta V_d \le 0.7 \tau_u d^2 \tag{14.25}$$

14.4.4 Beam–column joints

Depending on the degree of damage, the following techniques are applied for the repair of beam to column joints:

- Resin injection
- X-shaped prestressed collars
- Bonded steel plates or FRP sheets
- R/C jackets

A very interesting report on Repair and Strengthening of Beam-Column R/C Joints is the state-of-the-art report No. 04-4 of the Georgia Institute of Technology (Engindeniz et al., 2004).

14.4.4.1 Local repairs

Resin injections are applied in the case of fine and moderate cracks, without degradation of concrete or buckling of the reinforcement bars. However, restoration of bonds between steel and concrete with the aid of epoxy resin is questionable, since contradictory results appear in the international literature (Popov and Bertero, 1975; French et al., 1990; Karayannis et al., 1998).

Therefore, the joint should be strengthened at the same time with one of the techniques that will be presented next, especially in the case of frame structural systems without R/C walls.

14.4.4.2 X-shaped prestressed collars

After the cracks have been filled in with resin injections, or after the decomposed concrete is removed and the voids are filled with epoxy or non-shrinking mortar, the joint is strengthened with external ties (collars), which are prestressed with tensioner couplers (Figure 14.30). Then the joint is covered with welded wire fabric and a jacket of gunite concrete. When four beams are framing into the joint, the application of this technique is not feasible because the X-shaped collars cannot pass through the joint (NTU, 1978).

14.4.4.3 R/C jackets

The construction of R/C jackets to a damaged joint is the safest method for strengthening. This is generally a difficult technique, given the fact that a jacket must usually be constructed for every structural element framing into the joint. It is obvious that roughening of the surfaces is required, as well as punching of the slabs, in order for the ties to go through; injecting of the damaged joint area with resins must precede the construction of the R/C jackets (Figure 14.31; Alcocer and Jirsa, 1993; Tsonos, 1999, 2002).

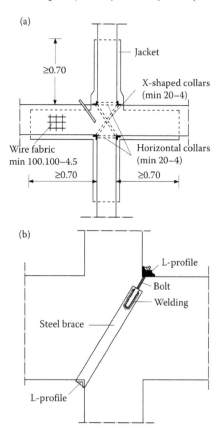

Figure 14.30 Strengthening of a joint with prestressed collars: (a) general arrangement of the strengthening; (b) detail of the prestressed collar.

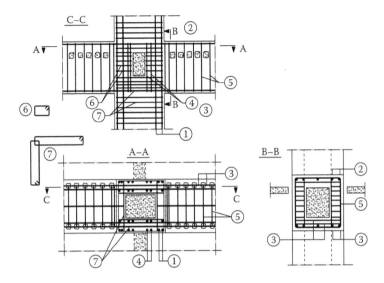

Figure 14.31 Strengthening of a joint with a jacket: 1 = column reinforcement; 2 = beam top reinforcement; 3 = beam bottom reinforcement; 4 = joint vertical stirrups; 5 = beam stirrups; 6 = column ties; 7 = column ties in joint.

14.4.4.4 Bonded metal plates

Bonded metal plates can only be applied to plane joints, as in the case of X-shaped collars. This is a technique that provides strengthening to the joint without altering its dimensions. Local repair precedes the bonding of the plates, then the plates are tied with prestressed bolts (Figure 14.32). The thickness of the plates in this case must be at least 4.0 mm, which does not create any problems with the bonding process, since the plates are kept tight to the concrete surface with the aid of prestressed bolts (Corazao and Durruni, 1989; Beres et al., 1992; Hoffschild et al., 1995).

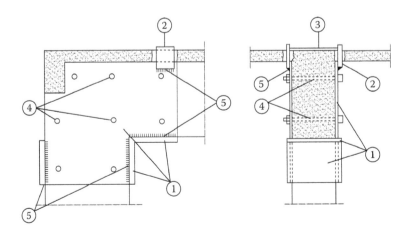

Figure 14.32 Bonded metal plates on a joint: 1 = steel plate; 2 = steel plate; 3 = steel strap; 4 = prestressed bolts; 5 = welding.

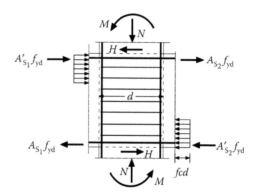

Figure 14.33 Schematic representation of the internal forces in a joint.

14.4.4.5 Redimensioning and safety verification

The redimensioning of the joint is carried out under the assumption that complete compatibility has been achieved between the original element and the added material, as happens with columns. The internal force distribution is given in Figure 14.33. However, given the fact that the field conditions are not as ideal as those in a laboratory, the author's opinion is that a model correction factor φ should be introduced, equal to

$$\frac{R_{d,rep}}{R_{d,monol}} \cong 0.80, \quad \frac{K_{rep}}{K_{monol}} \cong 0.80 \tag{14.26}$$

14.4.5 R/C walls

It is well known that R/C walls, due to their high stiffness and strength, are the most effective seismic-resistant elements of a structure. Therefore, the repair and strengthening of a damaged R/C wall can drastically improve the seismic resistance of a building.

14.4.5.1 Local repairs

If a properly reinforced wall exhibits cracks of small width, without bond deterioration or concrete crushing, it can be repaired with epoxy resins. Laboratory tests have shown that such an intervention fully restores the strength of the wall, but not its stiffness and energy dissipation capacity, due to the fact that resin cannot penetrate into the capillary cracks that accompany cracks with larger openings (Tassios, 1983; Lefas and Kotsovos, 1990; Lefas et al., 1990).

It should be mentioned here that most of the walls in older buildings have inadequate reinforcement due to the Code insufficiencies of earlier years. Thus, a simple repair with resin injections is very often not enough. It needs to be combined with R/C jackets to strengthen the wall.

14.4.5.2 R/C jackets

R/C jackets can have one of the forms shown in Figure 14.34. In the case of a jacket on both sides of the wall, the connection of the two layers with through-thickness ties is necessary (at least 3 bars $d = 14$ mm/m²).

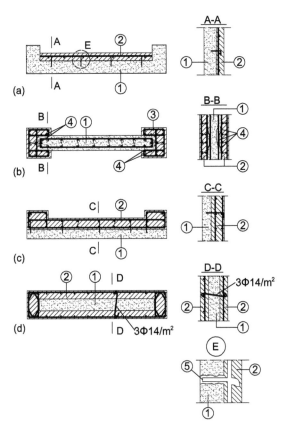

Figure 14.34 Strengthening of a wall with a jacket: (a) one-sided jacket; (b) thickenings at the ends of the existing wall; (c) one-sided jacket with end thickenings; (d) jackets, on both sides of the existing wall. 1 = existing wall; 2 = added wall; 3 = added columns; 4 = welding; 5 = epoxied bar.

At the points where the wall passes from one storey to the other, it is necessary to punch holes in the slab and place diagonal reinforcements through them (Figure 14.35).

In the construction of the R/C jackets, the following rules apply:

- The strength of the new concrete must be at least 5 MPa greater than that of the old concrete.
- The minimum thickness of the jacket should be 50 mm on each side.
- The minimum horizontal and vertical reinforcement should be 0.25% of the section of the jacket.
- The minimum reinforcement of the strengthening ends of the wall should be 0.25% of the section of the jacket end.
- The diameter of the ties at the wall ends should not be less than 8 mm, with a maximum spacing not exceeding 150 mm.
- The jacket must be anchored to the old concrete, with dowels spaced at no more than 600 mm in either direction (AUT, 1978; NTU, 1978).

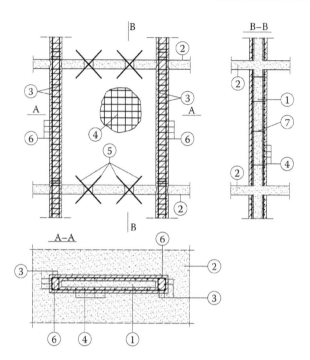

Figure 14.35 General arrangement for the strengthening of a wall: 1 = existing wall; 2 = existing slab; 3 = added longitudinal reinforcement; 4 = added wire fabric; 5 = diagonal connecting bars; 6 = added ties, 7 = connecting ties.

14.4.5.3 Redimensioning and safety verification

1. In the case of repair with resin injections, the ratio of the strength of the repaired element to the strength of the original may be taken to be equal to 1, as discussed earlier, while the ratio of the stiffness and energy dissipation capacity may be taken to be equal to 0.85,

$$\frac{R_{d,rep}}{R_{d,orig}} \cong 1, \quad \frac{K_{rep}}{K_{orig}} \cong 0.85 \quad \frac{E_{rep}}{E_{orig}} \cong 0.85 \tag{14.27}$$

2. In the case of repair with jackets, provided that the damaged wall was repaired earlier either with resins or resin mortars or non-shrinking cement mortars, the behaviour of the repaired element does not differ from that of the monolithic one (original + jacket), as far as both strength and stiffness are concerned. Therefore, as in the case of columns, walls are dimensioned based on the relationships

$$\frac{V_{R\,rep}}{R_{R\,monol}} = 0.9, \quad \frac{M_{y\,rep}}{M_{y\,monol}} = 1.0, \quad \frac{\theta_{y\,rep}}{\theta_{y\,monol}} = 1.05, \quad \frac{\theta_{u\,rep}}{\theta_{u\,monol}} = 1.0 \tag{14.28}$$

The required number of dowels between the original wall and the jacket can be estimated by the relationship (Tassios, 1984c)

$$n_d = \frac{\left[(V_d - V_{R,orig}) - l_w h_w \tau_{adh} \right]}{D_u} \tag{14.29}$$

where

V_d is the shear strength of the repaired wall (wall + jacket; MN).

$V_{R,orig}$ is the shear strength of the original wall after it is repaired, estimated to be 0.80 of the strength of the original undamaged wall (MN).

$l_w b_w$ are the dimensions of the wall under repair (m).

τ_{adh} is the average adhesion design strength of the new to the old concrete estimated to be equal to τ_{Rdc} (Section 14.4.3.4, item 2).

D_u is the dowel strength equal to

$$D_u = d^2 \sqrt{f_{cd} f_{yd}} \, (MN) \tag{14.30}$$

where

d is the diameter of the dowel (m).

f_{yd} is the design strength of the dowel (MPa).

f_{cd} is the design strength of concrete (MPa).

14.4.6 R/C slabs

It was stated in Chapter 11 that slab damage mainly appears in the form of cracks in the middle of large spans, above their supports, near discontinuities such as corners of large openings, at the connections of stairs to the slabs and so on. Depending on the extent and the type of damage, a different degree of intervention can be applied.

14.4.6.1 Local repair

If a properly reinforced slab exhibits cracks of small width without crushing of the concrete or bond deterioration, it can be repaired with epoxy resins. In the case of local failure accompanied by crushing or degradation of concrete, there can be a local repair for the full thickness of the slab (Figure 14.36). However, the need for such a repair is typically accompanied by the need to increase the slab thickness or to add new reinforcement.

14.4.6.2 Increase in the thickness or the reinforcement of a slab

Where the computational verification indicates that the slab resistance is insufficient, the slab can be strengthened either by increasing its thickness from the upper side with cast-in-place

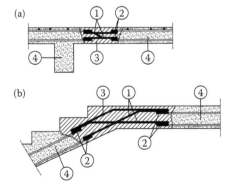

Figure 14.36 Local repair through the thickness of a slab: (a) repair in the span; (b) repair on the connection of a stair to the slab; 1 = added reinforcement; 2 = welding; 3 = added concrete; 4 = existing slab.

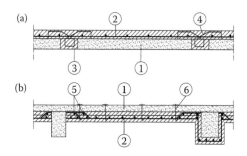

Figure 14.37 Increase in the thickness of the slab and addition of new reinforcement: (a) increase in the thickness on the upper face; (b) increase in the thickness on the lower face with the addition of new reinforcement; 1 = existing slab; 2 = added reinforcement; 3 = dowel; 4 = anchoring bent bars; 5 = welded connecting bars; 6 = hanging ties.

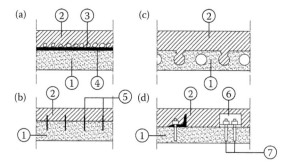

Figure 14.38 Details of connection of a new layer to the old concrete on a slab: (a) connection of new to old concrete using epoxy resin; (b) connection using epoxy bolts; (c) connection using existing voids of a voided slab; (d) connection using angle profiles; 1 = existing slab; 2 = new slab; 3 = sand corner; 4 = epoxy resin; 5 = epoxied bolts; 6 = angle profile; 7 = anchor bolts or shoot nails.

concrete or by increasing its thickness and placing additional reinforcement on its lower side with gunite concrete (Figure 14.37). The force transfer between the old and the new concrete is the key to the success of the intervention. This can be accomplished by other means in addition to roughening the old surface or resin coatings on the interface, such as anchors, dowels and so on (Figure 14.38).

14.4.6.3 Redimensioning and safety verifications

The dimensioning of slabs that have been strengthened with additional reinforcement and increase in thickness is carried out based on the assumption of a monolithic section (original + additional layer). The results are multiplied by the model *correction factor* φ, which is taken to be equal to 1.0 if the thickness of the new layer h is less than $h_o/3$ and $\varphi = 0.65$ if h is equal to or larger than $h_o/3$ (Tassios, 1983, 1984c),

$$\frac{M_{rep}}{M_{monol}} = 1.0 \quad \text{for } h < \frac{h_o}{3} \tag{14.31}$$

$$\frac{M_{rep}}{M_{monol}} = 0.65 \quad \text{for } h > \frac{h_o}{3} \tag{14.32}$$

The proposed values for the stiffness ratio are

$$\frac{K_{rep}}{K_{monol}} = 0.90 \quad \text{for } h < \frac{h_o}{3} \tag{14.33}$$

$$\frac{K_{rep}}{K_{monol}} = 0.40 \quad \text{for } h > \frac{h_o}{3} \tag{14.34}$$

However, in addition to the general safety verifications, there should be specific considerations for the force transfer mechanism through adhesion between the old and the new concrete, similar to those that were explained for the dimensioning of beams.

14.4.7 Foundations

The methods of repair or strengthening of foundations fall beyond the scope of this book, given the fact that they are related to interventions that belong to the field of foundation engineering. Indeed, when damage related to foundations occurs, it is not unusual for the need to arise for construction of retaining walls with anchorages to resist landslides, for construction of piles, for strengthening the soil with cement groutings and so on. Therefore, only the technique of connecting the column jacket to the footing will be dealt with here, as well as eventual strengthening of the footing itself.

14.4.7.1 Connection of column jacket to footing

Given the fact that the critical area of a column to flexure is at its top and bottom, the column jacket must continue beyond the point where the column frames into the footing, so that reinforcement bars will have the required anchorage length. This can be accomplished either with the arrangement of Figure 14.39 or with that of Figure 14.40.

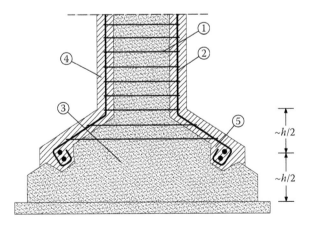

Figure 14.39 The end of a column jacket in the footing: 1 = new ties Φ12/100 mm; 2 = longitudinal reinforcing bars; 3 = existing concrete; 4 = added concrete; 5 = dowel in old concrete.

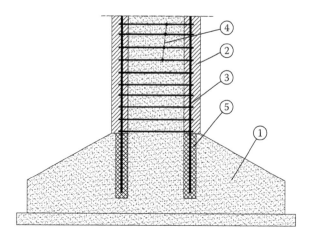

Figure 14.40 Anchorage of the column jacket reinforcement in the footing: 1 = old concrete; 2 = jacket; 3 = long reinforcement; 4 = new ties; 5 = epoxied connections.

14.4.7.2 Strengthening of footings

An increase in the area of a footing is decided on either because of inadequate bearing surface due to poor original estimation of the soil-bearing capacity, or because larger axial forces are transferred to the foundation due to the addition of new structural elements. In these cases, the increase in the area of the footing is carried out according to the arrangements shown in Figures 14.41 and 14.42.

The first arrangement, which is simpler than the second, is applied when the strengthening of the footing is extended to the column in the form of a jacket. In this case, the inclined

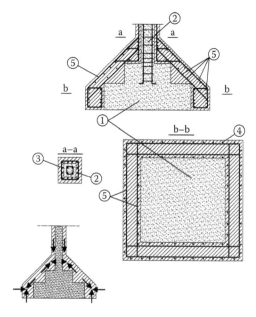

Figure 14.41 Strengthening of footing – column: 1 = existing foundation; 2 = existing column; 3 = reinforced jacket; 4 = added concrete; 5 = added reinforcement.

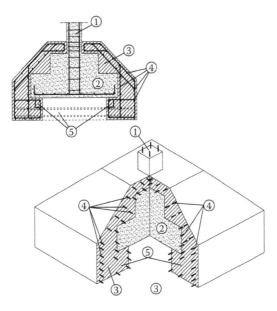

Figure 14.42 Strengthening of a footing without strengthening of the column: 1 = existing column; 2 = existing foundation; 3 = added concrete; 4 = added reinforcement; 5 = steel profile.

forces for the transmission of the soil pressure to the column jacket (Figure 14.41) are carried by rectangular closed reinforcement rings, which are formed either with large overlaps or welding.

The second arrangement is much more difficult because excavation under the existing footing is required. In this case, a temporary support is usually required, and special attention should be paid to avoiding settlement due to undermining.

14.4.8 Infill masonry walls

In previous chapters, there was systematic reference to the significance of the infill system to the seismic behaviour of structures, and the importance of its repair was explained (Bertero and Brokken, 1983; Sarigiannis et al., 1990; Stylianidis, 2012).

14.4.8.1 Light damage

Cracks that do not go through the thickness of the wall but appear only on the plaster have already been characterised as 'light damage' (Chapter 11.1.7). To repair this kind of damage, a band of plaster of a width equal to 100–150 mm on each side of the crack is removed, and it is replaced by a new plaster after the wall is moistened with water. A band of light wire mesh or a suitable FRP sheet is very often used as reinforcement underneath the new plaster.

14.4.8.2 Serious damage

This term refers to open (full-thickness) cracks in the infill wall, independently of the crack width. In this case, the strength, the stiffness as well as the ability of the infill to dissipate energy have obviously been reduced, and, therefore, an intervention more extended than the previous one is required. Therefore, if the crack is only a few millimetres wide, after the plaster is removed in a band of 100–150 mm on each side, the crack is widened on the surface of the wall, washed using a water jet, and filled with cement mortar of high cement content, pushing the mortar as deep as possible inside the crack with a thin trowel and smoothing the surface. Then a wire mesh or an FRP band is nailed on the area where the plaster has been removed and new plaster is applied (Figure 14.43).

If the cracks are wider, two solutions are possible: either the wall is removed and recon-structed, or the plaster on the whole surface of the wall is removed and the procedure of the previous paragraph is followed. The wire mesh or the FRP sheet in this case is placed on the whole surface of the wall and a plaster consisting of cement mortar of 20 mm thickness or a thin layer (about 30–40 mm) of gunite concrete is shotcreted. It is understood that inter-ventions of this type lead to strengths and stiffnesses of the masonry wall higher than the original ones (Figure 14.44). Therefore, there should be a verification of the relative strength and stiffness of the adjacent columns in order to avoid shear failure in the columns due to a new earthquake (Sariyiannis, 1990), in case the repaired masonry is not extended into the next span.

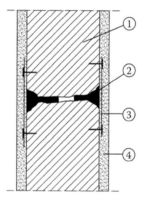

Figure 14.43 Repair of a through-thickness crack in an infill wall: 1 = existing masonry walls; 2 = sealing of the crack with cement mortar; 3 = wire mesh; 4 = plaster.

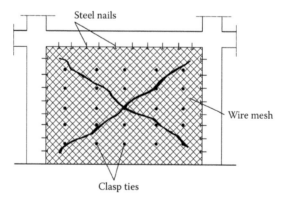

Figure 14.44 Repair of a seriously damaged infill masonry wall.

14.5 REPAIR AND STRENGTHENING OF STRUCTURAL ELEMENTS USING FRPs

14.5.1 General considerations

Summarising the main properties of FRP sheets and laminates that have been presented in Section 14.2.10, we should recall the following key points:

1. Carbon FRPs (CFRPs) in the form of sheets or laminates are the most suitable for the repair and strengthening of R/C members than all other FRP categories.
2. The E-modulus of CFRPs is almost equal to or higher than that of steel ($E_{cf} \cong 165$–640 GPa compared to $E_{steel} = 200$ GPa).
3. Tensile strength at failure is very high, ranging between 1.450 and 3.500 MPa.
4. Deformation at failure ranges between 0.45% and 1.7%.
5. Constitutive law σ–ε is linear without a plastic plateau at all (brittle behaviour).
6. Load transfer from an R/C member to an FRP sheet or laminate is accomplished through adhesive epoxy resins.
7. Taking into consideration remarks 5 and 6 above, it follows that the design strength of FRP is defined by two limits:
   a. The first is FRP tensile strength reduced by a rather high safety factor on the order of 2.0 due to the brittle failure character of FRP.
   b. The second is shear (bond) stress developing at the contact surface by which *delamination* of the FRP from the concrete may be caused. This value is related to the strain that develops on FRPs at the delamination stage.
   The second limit is usually the parameter defining the design strength of FRPs.
8. The fire resistance of an FRP intervention is very low. Therefore,
   a. The original member must be in a position to carry all dead loads with its original steel reinforcement at least with the safety factor specified for fire resistance. This implies that the strengthening degree with FRPs cannot exceed the original strength of the member by 100%.
   b. Protection measures against fire should be taken (e.g. mortar protective coatings).
9. Interventions based on FRPs cannot substantially change the stiffness of the member since they are flexible materials.
10. FRPs may be used for the following types of intervention:
    a. Strengthening to bending
    b. Strengthening to shear
    c. Strengthening to axial compression through confinement
    d. Ductility increase in columns through confinement at their ends
    e. Clamping of lap splices
    f. Strengthening of joints

14.5.2 Bending

Main failure modes of an R/C beam additionally reinforced by CFRP laminates to flexure are the following (Teng et al., 2002):

1. Intermediate flexural crack-induced debonding (Figure 14.45a)
2. Crushing of concrete under compression (Figure 14.45b)
3. Plate-end debonding (Figure 14.45c)
4. Shear failure (Figure 14.45d)

FRP design must cover all of the above failure modes.

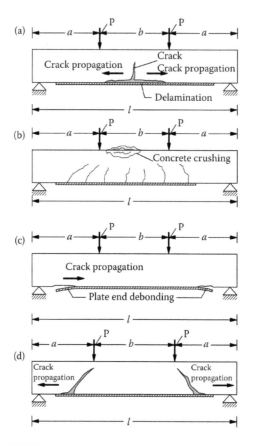

Figure 14.45 Failure modes of FRP-plated R/C beams: (a) intermediate flexural crack-induced delamination; (b) crushing of compressive zone; (c) plate-end debonding; (d) intermediate shear crack-induced debonding.

14.5.2.1 Intermediate flexural crack-induced debonding

In order for the strengthening of the beam to be ensured against crack-induced debonding, FRP elongation must be limited to a specified value defined as 'effective strain'. If this value is exceeded, cover delamination or FRP debonding may occur, since the force in the FRP cannot be sustained by the concrete substrate. For that value of the 'effective strain', there are strong deviations among various standards and specifications.

- According to the Swiss prenorm SIA 166/2004, design elongation is specified to

$$\varepsilon_{fe} = 0.6\%$$

- Therefore, design strength f_{fd} may by derived from the expression

$$f_{fd} = \frac{1}{\gamma_{fd}} E_f \cdot \varepsilon_{fe} = \frac{1}{\gamma_{fd}} 0.6\% \cdot E_f \qquad (14.35)$$

where
γ_{fd} is the partial safety factor for FRP laminates or sheets. The usually recommended value for $\gamma_{fd} = 1.2$.

- According to the Japanese standards (JBDPA, 1999), effective strain is specified to

$$\varepsilon_{fe} \leq 0.7\%$$

- According to FIB (2003)

$$\varepsilon_{fe} \leq 0.6\%$$

- According to ACI 440.2R-02/2002: (SI units)

$$\varepsilon_{fe} \leq \kappa_m \cdot \varepsilon_{fu} \qquad (14.36)$$

where
ε_{fu} is strain at failure, and

$$\kappa_m = \begin{cases} \dfrac{1}{60\varepsilon_{fu}}\left(1 - \dfrac{nE_f \cdot t_f}{360.000}\right) \leq 0.9, & \text{for } nE_{FRP} \cdot t_f \leq 180.000 \\[3mm] \dfrac{1}{60\varepsilon_{fu}}\left(\dfrac{90,000}{nE_f \cdot t_f}\right) \leq 0.9, & \text{for } nE_{FRP} \cdot t_f > 180.000 \end{cases} \qquad (14.37a,b)$$

The meaning of the above notation is the following:
- E_f is the modulus of elasticity of the FRP.
- n is the number of plies.
- t_f is the thickness of each CFRP laminate.

Application of Equations 14.37a and 14.37b results in a value for ε_{fe} in the range (35%–60%) ε_{fu}.
- In closing, it should be noted that according to EC 8-3/2005 Annex A, the following value is recommended for ε_{fe}:

$$\varepsilon_{fe} = \sqrt{\frac{0.6 \cdot f_{ctm} \cdot \kappa_b}{E_f \cdot t_f}} \text{ and } \sigma_{fe} = \sqrt{\frac{0.6 \cdot f_{ctm} \cdot \kappa_b \cdot E_f}{t_f}} \qquad (14.38)$$

where
- f_{ctm} is the concrete mean tensile strength.

$$\kappa_b = \sqrt{1.50(2 - w_f/s_f)/(1 + w_f/100 \text{ mm})} \qquad (14.39)$$

(known as the 'covering coefficient').
- w_f is the width of the FRP laminate.
- s_f is the spacing of the FRP laminates.
- E_f is the modulus of elasticity of the FRP.
- t_f is the thickness of the FRP.

EC 8-3/2005 recommends a partial safety factor for debonding equal to $\gamma_{fd} = 1.50$, since this mode of failure depends on the concrete tensile strength. Therefore,

$$\boxed{f_{td} = \frac{1}{\gamma_{fd}} \sqrt{0.6 \frac{E_f \cdot f_{ctm} \cdot \kappa_b}{t_f}}}$$

$\qquad$ (14.40)

This basic value is reduced further for the various types of applications (full warping, U-shaped [i.e. open] jackets and side-bonded sheets/strips).

It should be noted that although the provisions of EC 8-3/2005 Annex A (informational) are based on sound theoretical, experimental and statistical research (Biskinis, 2007), they are very conservative in relation to other standards and specifications in effect and, therefore, in the opinion of the authors, they should be reconsidered, because they almost overturn the capability of using FRPs in practice.

From what has been presented above, it may be concluded that the design strength to flexure is determined by the relation

$$\boxed{f_{fd} = \frac{1}{\gamma_{fd}} \varepsilon_{fe} \cdot E_f}$$

$\qquad$ (14.41)

since this value is always, at least for CFRPs, lower than

$$\boxed{f_{fu}/\gamma_{fm} = f_{fu}/2.0}$$

$\qquad$ (14.42)

The design may be carried out using existing computer platforms for the case. An approximate estimate is given below.

Consider the R/C cross section depicted in Figure 14.46, loaded by a bending moment

$$M_{Ed} = M_{0E} + M_{1E}$$

$\qquad$ (14.43)

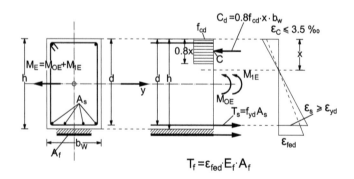

Figure 14.46 Strengthening of a beam under flexure using FRP laminates.

where

M_{0E} is the original bending moment for which the cross section was designed and reinforced through steel rebars.

M_{1E} is the additional bending moment for which FRP reinforcement will be used.

The equilibrium condition of the moments of the internal forces with reference to point C balancing M_E results in

$$M_{Rd} \cong 0.9d \cdot T_s + 0.9h \cdot T_f \geq M_E \tag{14.44}$$

or

$$0.9d \cdot f_{yd} \cdot A_s + 0.9h \cdot \varepsilon_{fed} \cdot E_f \cdot A_f \geq M_{Ed} \tag{14.45}$$

or

$$\boxed{A_f \geq \dfrac{M_{Ed} - 0.9d \cdot f_{yd} \cdot A_s}{0.9h \cdot \varepsilon_{fed} \cdot E_f}} \tag{14.46}$$

14.5.2.2 Crushing of concrete under compression before tension zone failure

The condition that the depth x of the compressive zone of the beam should fulfil in order for concrete crushing to precede steel yielding is given by the following expression:

$$\xi = \frac{x}{h} = \frac{\varepsilon_c}{\varepsilon_{fe} + \varepsilon_c} \geq \frac{3.5\%_0}{f_{fd}/E_f + 3.5\%_0} = \xi_{bal} \tag{14.47}$$

(balanced section, see Chapter 8.2.2.2)

Bearing in mind that

$$C_d \cong 0.8 f_{cd} x b_\omega \tag{14.48}$$

and that

$$C_d = T_d \tag{14.49a}$$

or

$$0.8 f_{cd} x b_\omega = f_{yd} \cdot A_s + \varepsilon_{fed} \cdot E_f \cdot A_f \tag{14.49b}$$

it may be concluded that

$$\boxed{x = \dfrac{f_{yd} \cdot A_s + \varepsilon_{fed} \cdot E_f \cdot A_f}{0.8 \cdot f_{cd} \cdot b_w}} \tag{14.50}$$

The value of A_f in Equation 14.50 has already been determined by Equation 14.46, and therefore, the right term of Equation 14.50 is completely determined. So, if

$$\xi = \frac{x}{h} \leq \xi_{bal} = \frac{3.5\%}{f_{fd}/E_f + 3.5\%},$$

(14.47a)

the condition for failure of the tensile zone at yield is fulfilled. Otherwise, the design must be revised, because crushing of the compressive zone prevails.

14.5.2.3 Plate-end debonding

- In order for a laminate to be ensured against plate-end debonding, the laminate should extend a distance l_b past the point along the span corresponding to the cracking moment M_{cr} given by the following expression (Figure 14.47):

$$l_e = l_b = \sqrt{\frac{E_f \cdot t_f}{2.5 \cdot f_{ctm}}} \text{ (mm)}$$

(14.51)

l_e is defined as the 'effective length', the meaning of which will be explained later (Triantafillou, 1998; Tassios, 2009).
- For this value, EC 8-3/2005 recommends the following expression:

$$l_e = l_b = \sqrt{\frac{E_f \cdot t_f}{\sqrt{4 \cdot \tau_{max}}}} \text{ (units N, mm)}$$

(14.52)

where

$$\tau_{max} = 1.8 \cdot f_{ctm} \cdot \kappa_b$$

(14.53)

is the maximum bond strength influenced basically by the concrete average tensile strength f_{ctm} and by the covering coefficient κ_b, already defined in Section 14.5.2.1, Equation 14.39.

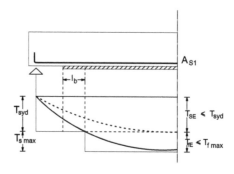

Figure 14.47 Graphic representation of the plate-end extension to avoid plate-end debonding.

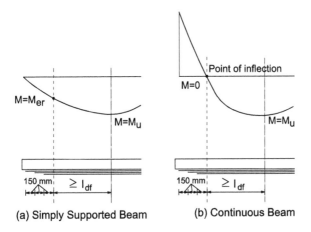

Figure 14.48 Graphic representation of the guidelines for allowable termination points of a three-ply FRP laminate. (a) Simply supported beam; (b) continuous beam. (From ACI 440-2R-02. With permission.)

- ACI 440.R2-02 specifies for l_b a value

$$l_b = \max(d, 150 \text{ mm})$$

where
d is the effective height of the cross section (Figure 14.48).

In the case of successive plies, l_b refers to each ply from the end of the one above.

14.5.2.4 Theoretical justification of debonding length l_b and strain ε_{fe}

Debonding strain and length has been the subject of extended analytical, experimental and
statistical investigations in the last 20 years, since they constitute the main parameters for
the design of FRPs in R/C structures (Hollaway and Leeming, 1999; Concrete Society,
2000; Teng et al., 2002; Tassios, 2009).
  A simplified conceptual approach to this issue will be made below so that an in-depth
understanding of the relations presented above may be achieved.
  Consider in Figure 14.49 a laminate bonded externally to a beam under bending. From
two successive intermediate hairline cracks of the concrete, debonding starts to propagate
in both directions as the loading and, therefore, the developing bending moment at the
crack positions along the beam increases. The debonding mechanical model, along with
the stresses and strains developing on the laminate and the bonding stresses developing at
the interface, are given conceptually in Figure 14.50a,b,c. Stresses and strains of the laminate are considered as a linear function along the effective length le, while bonding stresses
are considered to be uniform along the same length.
  Under the above assumptions at 'debonding stage', the following relations are in effect:

$$\sigma_{f(x)} = \frac{x}{l_e}\sigma_{f\max}, \quad \varepsilon_{f(x)} = \frac{x}{l_e}\varepsilon_{f\max} \qquad (14.54a,b)$$

$$\varepsilon_{f(x)} = \frac{\sigma_{f(x)}}{E_f} = \frac{x}{l_e}\frac{\sigma_{f\max}}{E_f} \qquad (14.55)$$

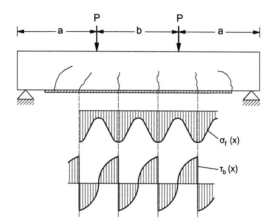

Figure 14.49 Load transfer from the beam to the bonded laminate.

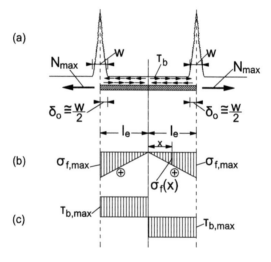

Figure 14.50 Intermediate crack-induced debonding: (a) stress–displacement pattern; (b) assumption for axial stress–strain distribution along the FRP laminate axis; (c) assumption for bond stress distribution along the contact surface.

The laminate elongation at the boundary of the crack will be equal to

$$\delta_o = \int_0^{l_e} \varepsilon_{f(x)} \, dx \tag{14.56}$$

or

$$\delta_o = \int_0^{l_e} \frac{x}{l_e} \varepsilon_{f\,max} \, dx = \frac{l_e}{2} \varepsilon_{f\,max} \tag{14.57}$$

Therefore,

$$\varepsilon_{max} = \frac{2\delta_o}{l_e}$$

(14.58)

and

$$N_{max} = \varepsilon_{fmax} E_f \cdot t_f \cdot b_f = \frac{2E_f \cdot t_f \cdot b_f}{l_e} \delta_o$$

(14.59)

On the other hand, according to the above assumption, the maximum bonding action balancing N_{max} is equal to

$$T_{max} = \tau_{bmax} b_f l_e$$

(14.60)

Bearing in mind that τb max is a function of the mean tensile strength of concrete fctm, since this strength is usually much lower than the shear strength of the adhesive bonding epoxy resin and of the covering coefficient κb, which means that

$$\tau_{bmax} = \lambda \cdot \kappa_b \cdot f_{ctm}$$

(14.61)

it follows that

$$T_{max} = \lambda \cdot \kappa_b \cdot f_{ctm} \cdot b_f \cdot l_e$$

(14.62)

At the debonding stage

$$N_{max} = T_{max}$$

(14.63)

Therefore, by introducing in Equation 14.63 the values of N_{max} and T_{max} given by Equations 14.59 and 14.62, the result is

$$\frac{2E_f \cdot t_f \cdot b_f}{l_b} \delta_e = \lambda \cdot \kappa_b \cdot f_{ctm} \cdot b_f \cdot l_e$$

(14.64)

or

$$l_e = \sqrt{\frac{2E_f \cdot t_f \cdot \delta_o}{\lambda \cdot \kappa_b \cdot f_{ctm}}}$$

(14.65)

$$\varepsilon_{fmax} = \frac{2\delta_o}{l_e} = \sqrt{\frac{2\lambda \cdot \kappa_b \cdot f_{ctm} \cdot \delta_o}{E_f \cdot t_f}}$$

(14.66)

For a reasonable crack width w equal to 0.4 mm, it follows that

$$\delta_o = \frac{w}{2} \cong 0.2 \text{ mm}$$

Consequently, Equations 14.65 and 14.66 take the following form:

$$l_e = \sqrt{\frac{0.4 E_f \cdot t_f}{\lambda \cdot \kappa_b \cdot f_{ctm}}}$$

(14.67)

and

$$\varepsilon_{f \max} = \varepsilon_{fe} = \sqrt{\frac{0.4\lambda \cdot \kappa_b \cdot f_{ctm}}{E_f \cdot t_f}}$$

(14.68)

It is obvious that le also represents the debonding length lb beyond the final crack before the end of the laminate.

It should be noted also that κb and λ may take various values as the result of a best-fit procedure with experimental results. At the same time, the original assumptions of linear distribution of strains and uniform distribution of bond stresses on the conduct surface may be modified, leading to a similar expression.

14.5.3 Shear

Additional shear capacity due to FRP sheets or laminates may be estimated using similar expressions, like those for steel stirrup rebars.

- So, according to EC 8-3/2005 recommendations, $V_{Rd,f}$ is given by the following expressions (Figure 14.51):
  - For side FRP stirrups or sheets (e.g. T beams):

$$V_{Rd,f} = 0.9d \cdot f_{fd,e} \cdot 2t_f \frac{\sin\beta}{\sin\delta} \cdot \frac{w_f}{s_f}$$

(14.69)

  - For full warping with FRP or for U-shaped FRP strips or sheets:

$$V_{Rd,f} = 0.9d \cdot f_{fd,e} \cdot 2t_f \left(\frac{w_f}{s_f}\right)^2 (\cot\delta + \cot\beta)\sin\beta$$

(14.70)

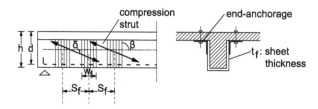

Figure 14.51 Additional shear capacity due to FRP wrapping strips.

where

d is the effective depth.

δ is the strut inclination angle.

β is the angle between the strong fibre direction in the FRP sheet or fabric and the axis of the member, usually $\beta = 90°$.

w_f is the width of the FRP strip or sheet orthogonally in the main direction of the fibres (for sheets $w_f = \min(0.9d, b_w)\sin(\theta + \beta)/\sin\theta$).

s_f is the spacing of the FRP strips ($=w_f$ for sheets).

$f_{fd,e}$ is the design FRP effective debonding strength, which depends on the strengthening configuration (fully wrapped FRP, U-shaped FRP, side-bonded FRP). The main parameter in these expressions for $f_{fd,e}$ is the effective maximum strain and the effective debonding, to which extended reference has been made in the previous subsection.

- According to ACI 440.2R-02, effective strain in FRP laminates or sheets is specified in the following simple form.
  - For completely wrapped members:

$$\varepsilon_{fr} = 0.004 \le 0.75\, \varepsilon_{fu} \tag{14.71}$$

The same value for bonded U wraps or bonded face plies is limited to

$$\varepsilon_{fr} = \kappa_r \cdot \varepsilon_{fu} \le 0.004 \tag{14.72}$$

where

κ_r is computed from Equations 14.73 to 14.76 (Khalifa et al., 1998).

$$\kappa_r = \frac{\kappa_1 \cdot \kappa_2 \cdot l_e}{11.9\varepsilon_{fu}} \le 0.75 \tag{14.73}$$

$$l_e = \frac{23.3}{(n \cdot t_f \cdot E_f)^{0.58}} \text{(SI units)} \tag{14.74}$$

n is the number of successive plies.

$$\kappa_1 = \left(\frac{f_d}{27}\right)^{2/3} \text{(SI units)} \tag{14.75}$$

$$\kappa_2 = \begin{cases} \dfrac{d_f - l_e}{d_f} & \text{for U - wraps} \\[2ex] \dfrac{d_f - 2l_e}{d_f} & \text{for two sides bonded} \end{cases} \tag{14.76}$$

d_f is the effective depth of the FRP sheet or stirrup.

14.5.4 Axial compression and ductility enhancement

14.5.4.1 Axial compression

- FRP systems can be used to increase the axial compression strength of a concrete member by providing confinement with FRP jackets (Teng et al., 2002; Fardis, 2009; Tassios, 2009). Apparently, in this case, the fibres are oriented transversally to the axis of the R/C member (column) in the form of hoops (Figure 14.11a,b). The concept of concrete confinement with FRPs is similar to that of the confinement with continuous steel spiral rebars (see Chapters 7.4.3 and 8.3.4). However, due to the different constitutive law of steel and FRPs (Figure 14.8) and to the different bond action of embedded steel rebars and externally bonded FRP sheets or laminates through adhesive epoxy resins, extended experimental and analytical research has been made for the establishment of design rules in the case of FRP concrete confinement (Pantazopoulou, 1995; Teng et al., 2002; Biskinis, 2007; Fardis, 2009; Tassios, 2009).
- The model of Lam and Teng (2003a,b) has been proven to be the most reliable among many others for the determination of the apparent confined strength f_c^* of concrete and its corresponding ultimate strain ε_{cu}^* (Vintzeleou and Panagiotidou, 2007). These two values for full jacketing of a rectangular column with FRP sheets are given in the following expressions (Fardis, 2009):

$$\frac{f_c^*}{f_c} = 1.0 + 3.3 \left(\frac{b}{h}\right)^2 \alpha_n \frac{\rho_f f_{fe}}{f_c} \tag{14.77}$$

$$\frac{\varepsilon_{cu}^*}{\varepsilon_{c2}} = 1.75 + 12 \sqrt{\frac{h}{b}} \alpha_n \frac{\rho_f f_{fe}}{f_c} \left(\frac{\varepsilon_{su}}{\varepsilon_{co}}\right)^{0.45} \tag{14.78}$$

where
b and h are the shorter and longer of the two sides of a rectangular section. In the case of a circular section, $b = h$ = diameter of the section.
α_n is the confinement effectiveness factor equal to $\alpha_n = 1$, for circular sections, and

$$\alpha_n = 1 - \frac{(b - 2R)^2 + (h - 2R)^2}{3b \cdot h} \tag{14.79}$$

for rectangular sections, where
R is the radius of rounded corners (Figure 14.52).

$$f_{fe} = E_f \varepsilon_{se} \tag{14.80}$$

$\varepsilon_{fe} \cong 0.60 \varepsilon_{fu}$ (for CFRP)
$\varepsilon_{fe} \cong 0.85 \varepsilon_{fu}$ (for GFRP, AFRP)
$\varepsilon_{c2} = 2‰$ (EC2-1-1/2004)

ρ_f is the geometric ratio of FRP to the wrapped concrete:

$$\rho_f = \frac{\pi D t_f}{\pi D^2 / 4} = \frac{4 t_f}{D} \quad \text{(circular cross section)} \tag{14.81}$$

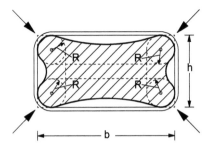

Figure 14.52 Confinement effectiveness factor for orthogonal cross section depending on rounded corner radius R.

or

$$\boxed{\rho_f = \frac{2(b+h)t_f}{bh}}\ \text{(orthogonal cross section)} \tag{14.82}$$

t_f is the sheet thickness.

It should be noted that in the case of a circular section (Figure 14.53), the value

$$\boxed{f_1 = \frac{1}{2}\rho_f f_{de} = \frac{1}{2}\rho_f \varepsilon_{de} E_f} \tag{14.83}$$

expresses the lateral pressure applied by the FRP sheet to the concrete core at the stage of bonding failure of the FRP. In fact, at this stage

$$f_1 D = 2t_f \cdot f_{de} \tag{14.84}$$

or

$$\boxed{f_1 = \frac{2t_f f_{de}}{D}} \tag{14.85}$$

Taking into account that the volumetric content of the FRP sheet to concrete is

$$\rho_f = \frac{\pi D t_f}{\pi D^2/4} = \frac{4t_f}{D} \tag{14.86}$$

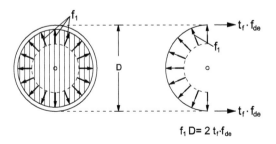

$$f_1 D = 2\ t_f \cdot f_{de}$$

Figure 14.53 Confinement of a circular cross section by wrapping with FRP sheets.

it follows that

$$t_f = \frac{\rho_f D}{4} \tag{14.87}$$

Introducing Equation 14.87 into Equation 14.85 results in

$$\boxed{f_1 = \frac{1}{2}\rho_f f_{de} = \frac{1}{2}\rho_f \varepsilon_{de} E_f} \tag{14.88}$$

Equation 14.77 enables the design of an existing R/C column for an axial load for which dimensions and reinforcement of the column are inadequate. In fact, the design capacity Ned of an R/C column with a cross section Ac and longitudinal reinforcement As in the case of confinement is given by the expression

$$\boxed{N_{Rd} = \frac{f_c^*}{\gamma_c} A_c + \frac{f_{y\kappa}}{\gamma_s} A_s} \tag{14.89a}$$

where
    f_c^* is the apparent strength of confined concrete given by Equation 14.77.
    γ_c is the partial safety factor for concrete ($\gamma_c = 1.5$).
    -$f_{y\kappa}$ is the characteristic yield strength of steel.
    γ_s is the partial safety factor for steel ($\gamma_s = 1.15$).

Capacity NRd must be bigger than demand NEd. Therefore, Equation 14.89a takes the form

$$\boxed{N_{Ed} \le N_{Rd} = \frac{f_c^*}{\gamma_c} A_c + \frac{f_{y\kappa}}{\gamma_s} A_s} \tag{14.89b}$$

Equation 14.89b results in f_c^* and then Equation 14.77 results in ρ_f. Finally, from Equation 14.81 or 14.82, the thickness of the wrapping sheet or the number of the successive sheets of a sheet of a given thickness may be calculated.

- EC8-3/2005 does not make any reference to strengthening of columns under axial loading using FRPs, since such an action does not refer directly to seismic retrofitting. However, the case of existing column inadequacy to axial loading in the framework of safety verifications for seismic loading is not infrequently encountered.
- ACI 440.2R-02 guidelines for 'the design and construction of externally bonded FRP systems for strengthening concrete structures' provide the following for the apparent strength of concrete f_c^* confined with FRPs:

$$\boxed{\frac{f_c^*}{f_c} = 2.25\sqrt{1+7.9\frac{f_1}{f_c}} - 2\frac{f_1}{f_c} - 1.25} \tag{14.90}$$

where

$$f_1 = \frac{\kappa_\alpha \rho_f}{2} f_{f\max} = \frac{\kappa_\alpha \rho_f \varepsilon_{fe} E_f}{2} \tag{14.91}$$

$$\varepsilon_{fe} = 0.004 \le 0.75 \varepsilon_{fu} \text{ (effective strain)} \tag{14.92}$$

κ_α is the 'efficiency factor' equal to $\kappa_\alpha = 1.0$ (for circular sections)

$$\rho_f = \frac{4nt_f}{D} \text{ (for circular sections)} \tag{14.93}$$

and
  n is the number of jacketing plies.
  The above approach is based on the model of Spoelstra and Monti (1999).

14.5.4.2 Ductility enhancement

- Ductility enhancement of an R/C member may be achieved by wrapping the critical regions where plastic hinges are potentially developed with FRP sheets. Obviously, these sheets are arranged with their fibres transversal to the axis of the member in the form of hoops (Figure 14.11b). This type of strengthening causes an increase in the strain capacity ε_u^* of the member due to confinement (Equation 14.78).
- According to EC8-3/2005, the necessary amount of confinement pressure f_1 to be applied depends on the ratio

$$I_x = \mu_{\varphi,tar} / \mu_{\varphi,avail} \tag{14.94}$$

of the target curvature ductility $\mu_{\varphi,tar}$ to the available curvature ductility $\mu_{\varphi,ava}$ and may be evaluated as

$$\boxed{f_1 = 0.4 I_x^2 \frac{f_c \varepsilon_{cu}^2}{\varepsilon_{fe}^{1.5}}} \tag{14.95}$$

where
- f_1 is the confinement pressure.
- f_c is the concrete strength.
- ε_{cu} is the concrete ultimate strain.
- ε_{fe} is the adopted FRP jacket effective strain, which is lower than the ultimate strain of FRP ε_{fu}.

  In the case of a circular cross section, f_1 is given by Equation 14.88, from which ρ_f results if the value of f_1 derived from Equation 14.85 is introduced into Equation 14.88.
  In the case of rectangular cross sections in which the corners have been rounded to allow wrapping, the confinement pressure demand f_1' is given by Equation 14.95 after it has been multiplied by $\kappa_s = 2R/D$, where D is the larger section width.

- ACI 440.2R-2 specifies that the maximum usable compressive strain in concrete for FRP-confined circular R/C members may be evaluated by the following equation (Mander et al., 1988):

$$\boxed{\varepsilon_{cu}^{*} \cong \frac{1.71\left(5f_{cu}^{*} - 4f_{c}\right)}{E_{c}}} \tag{14.96}$$

However, this specification does not make any reference to orthogonal cross sections.

14.5.4.3 Clamping of lap splices

In its Annex A, EC8-3/2005 also makes recommendations for the strengthening design of existing lap splices by means of clamping, using FRP wrapping of the region in question.

14.5.5 Strengthening of R/C beam–column joints using FRP sheets and laminates

- Since 1998, efforts for upgrading existing beam–column joints have been focused on the use of FRPs in the form of epoxy-bonded sheets and laminates, since they are attractive for their flexibility. The fibre orientation in each ply can be adjusted so that specific strengthening objectives can be achieved (Engindeniz et al., 2004).

  The literature on FRP-strengthened joints mainly consists of simplified two-dimensional tests, while three-dimensional tests are rather rare due to the difficulties of application. Below, two successful proposals are presented, the first for a two-dimensional joint arrangement (Figure 14.54; Clyde and Pantelides, 2002) and the second for a three-dimensional one (Tsonos, 2008; Figure 14.55).

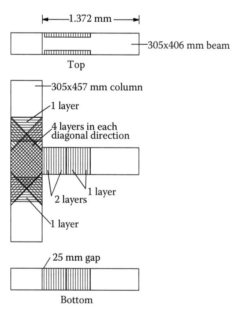

Figure 14.54 CFRP-strengthened specimen tested by Clyde and Pantelides. (Adapted from Clyde, C. and Pantelides, C.P. 2002. Seismic evaluation and rehabilitation of R/C exterior building joints. Proceedings of the Seventh U.S. National Conference on Earthquake Engineering, Georgia Institute of Technology, Boston, USA.)

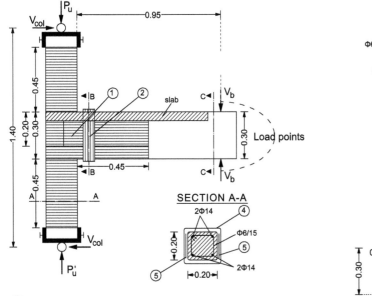

① 10 layers of CFRPs for increasing the shear strength of the joint
② Strips of CFRPs to secure the anchorage length of the joint layers
③ Drilled holes in the slabs of specimens FRPF₁ and FRPS₁
④ 7 layers of CFRPs for increasing the shear strength of the columns
⑤ 9 layers of CFRPs for increasing the flexural strength of the columns

Figure 14.55 Jacketing of column and beam–column connection of sub-assemblages FRPFI and FRPSI (dimensions in m). (Reprinted from *Engineering Structures*, 30, Tsonos, A., Effectiveness of CFRP-jackets in post-earthquake and pre-earthquake retrofitting of beam-column subassemblages, 777–793, Copyright 2008, with permission from Elsevier.)

• The additional shear capacities of the strengthened columns and beam–column joints due to FRPs may be evaluated by the following equation (Tsonos, 2002):

$$V_f = 0.9\,\varepsilon_{fe}\,E_f\,\rho_f\,b_w\,d \qquad (14.97)$$

where
d is the effective depth of the cross section.
b_w is the minimum width of the cross section over the effective depth.
ρ_f is the volumetric ratio of the FRP equal to $2t_f/b_w$ for continuously bonded shear reinforcement of thickness t_f.
E_f is the elastic modulus of elasticity of the FRP.
ε_{fe} is the effective FRP strain given by the following expression for fully wrapped or properly anchored FRPs (FIB, 2003):

$$\varepsilon_{f,e} = \left[0.17\varepsilon_{fu} \left(\frac{f_{cm}^{\frac{2}{3}}}{E_f \cdot \rho_f} \right) \cdot 0.006 \right] \qquad (14.98)$$

where
f_{cm} is the mean value of the concrete *compressive* strength.

14.6 ADDITION OF NEW STRUCTURAL ELEMENTS

The seismic resistance of a structure is drastically improved by the addition of new structural elements of great stiffness able to carry horizontal forces.

The new structural elements could be (Sugano, 1981; Bertero and Brokken, 1983; Bush et al., 1991; Rodriguez and Park, 1991; Penelis, 2001)

- R/C walls inside the frames that are formed by beams and columns (Figure 14.56). In this case, usually the columns at the ends of the wall are jacketed with R/C jackets.
- Additional R/C walls outside the frames (Figure 14.56d).
- New frames.
- Truss systems (made of metal or R/C) in the R/C frame (Figure 14.57).

The choice of type, number and size of the new elements depends on the characteristics of each structure. The most common type is the addition of R/C walls. Since interventions of this type alter the stiffness of the structure as well as its dynamic characteristics, they must be introduced into the original structure with special care. In this case, re-evaluation

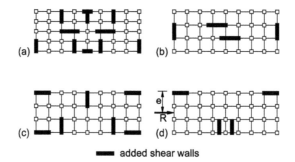

Figure 14.56 Addition of new R/C walls inside a frame or skeleton structure: (a), (b), (c) favourable layout (symmetric); (d) unfavourable layout (eccentric walls).

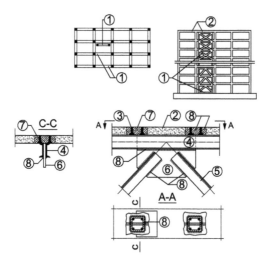

Figure 14.57 Addition of truss systems inside R/C frames: 1 = added steel truss; 2 = existing structure; 3 = steel dowel; 4 = horizontal steel rod; 5 = diagonal steel rod; 6 = steel joint plate; 7 = added concrete; 8 = welding.

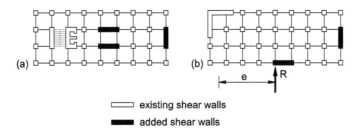

existing shear walls
added shear walls

Figure 14.58 Improvement of stiffness eccentricities with the addition of new R/C walls. (a) Initial eccentricity along y axis (b) Initial eccentricity along x and y axis.

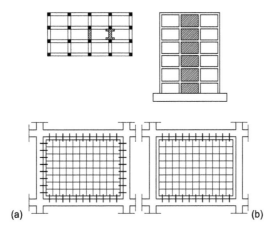

Figure 14.59 Addition of new R/C walls inside a frame: (a) connection along the four sides; (b) connection only with the beams.

of the structural response of the building is necessary based on a rigorous new analysis and design. This re-evaluation must also be extended to the foundations, since the addition of new stiffness elements (e.g. multi-storey walls) leads to a concentration of large shear forces and moments at the base of these elements, which requires an appropriate strengthening of their foundation with widened footings or additional connecting beams, new foundation beams or even micropiling. Finally, it has to be stressed that very often the addition of new elements is carried out not only to increase the stiffness or the strength of the structure but also mainly to alleviate some of the existing stiffness eccentricities that led to damage (Figure 14.58).

In the case of the addition of new elements, special care should be taken to ensure their connection with the existing elements so that force transfer is secured. Especially in the case of the addition of new R/C walls, the connection with the frame is made with dowels anchored with epoxy mortar or with reinforcement bars welded to the existing bars (Figure 14.59; Bertero and Brokken, 1983; Bush et al., 1991).

14.7 QUALITY ASSURANCE OF INTERVENTIONS

14.7.1 General

For a successful structural intervention, additional measures are needed in order to ensure quality of design and construction. Quality assurance of the design includes a thorough review by an independent reviewer so that it is made sure that the design criteria and solutions are the proper ones and that the computational work and structural details have been properly prepared. Quality assurance of construction includes inspection and testing of materials, elaboration of method statements for construction and assurance that the design is properly implemented during construction. While quality control is important for all constructions affecting the safety of the occupants, it is particularly important for seismic repair or strengthening due to the fact that these activities require a high degree of engineering judgment and careful attention to detail.

14.7.2 Quality plan of design

Seismic repair and strengthening projects require an appropriate scheme of counter-checking of design documents. When the design has been completed and the project manager has thoroughly reviewed the work, an additional check should be performed by an independent engineer. This may be a governmental or private agency responsible for verifying the criteria and checking the calculations and drawings to make sure that they conform to the criteria and regulations of the building codes.

14.7.3 Quality plan of construction

Construction inspection is carried out by an individual agency or firm, similar to conventional construction. However, an experienced engineer, with an extensive knowledge of repair materials and techniques, should be appointed as the construction inspector of repair and strengthening projects. The design engineer should continue to be involved with the inspection process and provide answers to questions arising during the implementation of the design details in the construction. This is extremely important for such projects, as many unexpected situations will be encountered during the construction, related mainly to hidden damage discovered after the finishes have been removed.

The quality of materials is verified by sampling and testing as in a conventional project. The differences involve only the verification of existing conditions and the testing of special materials such as resins, non-shrinking mortars, shotcrete and so forth.

The design documents should include a detailed description of the work schedule related to the repair and strengthening, as well as detailed specifications for the materials and construction techniques.

14.8 FINAL REMARKS

From the preceding presentation, the following final remarks can be made:

1. During the process of repairing damage caused by an earthquake or reforming the structural system in a pre-earthquake period, due to demolition works on the structure, a strong temporary shoring system is required to avoid collapse.
2. During the repair process, additional materials and techniques are used that are rarely applicable to new structures. Therefore, a detailed study of their characteristics is required, as well as very careful supervision during their application.
3. The form and extent of repairs cannot be completely foreseen during the design phase. The engineer is often compelled to improvise in order to adjust the materials and techniques to the needs of the existing special conditions.

4. Those interventions that drastically alter the original dynamic characteristics of the building must be applied with extreme care.

5. The redimensioning and safety verification of the repaired elements is achieved by more or less approximate procedures, firstly because no reliable analytical models based on laboratory tests have yet been developed for the variety of cases met with in a damaged structure, and secondly because there is a high degree of uncertainty with regard to the achieved degree of composite action of the old element and the new materials.

6. Finally, the repair cost of an element is much higher than the cost of its original construction, due to the fact that, on the one hand, repair involves complicated works such as demolition, supports, welding, injections and so on, and on the other, the original construction inhibits the unobstructed use of mechanical equipment. Therefore, the cost estimate of such an operation is to an extent unpredictable.

Chapter 15

Seismic isolation and energy dissipation systems

15.1 FUNDAMENTAL CONCEPTS

As has been explained in the previous chapters, the earthquake effects are felt on buildings as a direct result of the ground shaking. During the evolution of earthquake engineering, these effects have been controlled by the concepts of strength, ductility and damping, which allow the dissipation of energy through the accumulation of damage in appropriately engineered structural members.

In a different approach, the concept of disengaging the building from the ground shaking has been introduced and it is called *seismic isolation* or *base isolation*. Supplementary to that concept, or individually, the use of *increased damping*, though specially designed devices, has also been developed so that a typical steel building or reinforced concrete has a damping ratio of 20–25% of critical instead of the 2% or 5%, respectively (see Figures 3.7 and 3.8, Chapter 3.2.4).

The concepts of *base isolation* and *supplemental damping* are presented in the following sections. The pros and cons of each option are outlined, and the theoretical background for selecting the appropriate conceptual design is provided. Additionally, the state-of-the-art technological advances in such materials are presented, while the codified application of such technologies is presented according to the clauses of Eurocode 8 and the relevant European Norms for seismic isolation; and ASCESEI-41/13 for supplemental damping.

15.1.1 Seismic isolation

The principle of base isolation may easily be grasped even by the non-expert, when one refers to the case of a toy truck on top of a table, sketched in Figure 15.1. When one shakes the table abruptly, the toy truck remains stable in relation to the person shaking the table, as the wheels of the truck isolate it from the movement of the table. Obviously, the shaking of an earthquake takes place in two orthogonal directions (neglecting the vertical component is the standard practice for seismic isolation) so since inception of the isolation concept, a building was not put on top of wheels, but instead rested on spheres or balls.

Indeed, the concept of the seismic isolation has been introduced at the end of the nineteenth century by John Milne, a professor of mining engineering in Tokyo. This initial application included the placement of buildings on top of piles with steel balls on top of them. The actual configuration included a sandwich of two steel plates with 'saucer-like' edges and a steel ball in the middle; the top plate, slightly concave, was attached to the building and the bottom was attached to the pile (Naeim and Kelly, 1999).

Since the first application to date, all seismic isolation concepts have evolved around this simple concept, which results in the ground moving while the building remains stable in relation to an independent-from-the-ground-coordinate system. It is interesting to note that

Figure 15.1 Table and toy truck concept. The person moving abruptly the table sees the toy truck at rest (still) as it is isolated through its wheels.

the initial idea of Professor John Milne was later abandoned and replaced by *rubber* or *steel rubber bearings* with controlled mechanical properties only to resurface recently modified as the *inverted pendulum sliders*. The very first application of isolating a reinforced concrete building was in Skopje (Former Yugoslavia) in 1969 with a three-storey school being placed on rubber bearings (Naeim and Kelly, 1999).

The theoretical background of the concept is that the introduction of the isolation actually modifies the modal characteristics of the building through the introduction of the very flexible (soft) level, that of the seismic isolation. This results in a period elongation of the whole building, as is graphically explained in Figure 15.2, where the acceleration and displacement spectra of Eurocode 8 are shown as well as the spectral response values for a fixed base R/C building with eigen-period of 0.50 s and the same response for a base isolated R/C building with an elongated eigen-period of 2.5 s. From this figure, the following are clear:

- The increase in the fundamental eigen-period results in *significant decrease* in inertial forces (S_e).
- The increase in the fundamental eigen-period results in *significant increase* in displacements (S_d).
- The effectiveness of the system is for buildings with fundamental eigen-period, when fixed at base, less than 1.0 s (period within the plateau of the spectra).

Based on the above, if frictionless materials existed for the realization of sliders, a building could even have zero inertial forces from an earthquake, which in turn would also mean that the relative displacement would be equal to the peak ground displacement (see Chapter 3.4.3.6). The minimum static friction of any sliding isolation system is of the order of 4–5%, which means that at least this level of lateral forces is transmitted to the building. Additionally, the sliding friction also provides some damping, which in turn helps reduce the relative displacement. Similarly, in rubber bearings, there is a specific stiffness and damping coefficient that controls the movement. In both cases, the goal is to fine-tune these properties (stiffness and damping) for the optimal results.

15.1.2 Buildings with supplemental damping devices

Damping of reinforced concrete buildings is known to be 5% of critical damping. As is known in many international codes, the dissipative capacity of the structure through nonlinear behaviour is correctly considered as equivalent additional damping. This reduction is

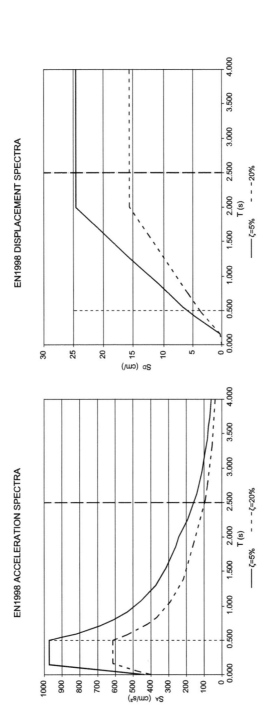

Figure 15.2 EN1998 type 1, acceleration and displacement spectra (5% and 20% damping) showing response values for a building with a fixed base period of 0.50 s elongated through isolation to 2.50 s.

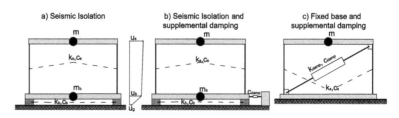

Figure 15.3 Portal frame with (a) seismic isolation, (b) seismic isolation and supplemental damping and (c) fixed base and supplemental stiffness and damping.

introduced in Eurocode 8 Part 3 as an inelastic spectra for a given ductility while in the US codes is presented, probably more accurately, as an overdamped spectra (see Figures 13.5 and 13.8, Chapter 13.5.2).

As has already been explained in seismic isolated buildings, a significant increase in the displacements takes place, which obviously would easily be controlled by the introduction of horizontal dampers to reduce this dynamic displacement. Figure 15.3b explains the application of damping to a seismically isolated two-degree-of-freedom system and Figure 15.3c to the fixed base equivalent single-degree-of-freedom system.

In the former case, the application of the damping is applied at the isolation interface (surface that separates the substructure and the superstructure, where the isolation system is located), thus controlling the total seismic force, which may easily extrapolated to actual buildings. In the latter case, the damping is applied, usually as bracing, at the single bay of the frame, thus controlling the seismic force. The problem in fixed base actual buildings is it to determine the appropriate locations in plan and over the height of the building, that the additional damping shall be introduced, and to determine the resulting total damping of the system (existing building plus dampers). The goal is to avoid making the building irregular and increasing the participation of higher modes of vibration.

15.2 CONCEPT DESIGN OF SEISMICALLY ISOLATED BUILDINGS

15.2.1 Main requirements of concept design

During the structural concept design of a seismically isolated building, there are decisions that have to be made and communicated to the other design disciplines, such as MEP engineers and architects, so that they may be taken into account into the other disciplines' concept design.

These decisions are mainly the following:

- The horizontal level at which the seismic isolation will be applied
- Accessibility to the isolators for inspection and maintenance
- The distribution of the isolator devices in plan
- The target fundamental period of vibration
- The target damping
- The estimated expected maximum displacement
- Selection of isolator type

All the above choices seriously affect other disciplines regarding the void required around the isolated building so that it may move during the earthquake, the design of pipes and systems over/through the isolator interface considering the displacement during the earthquake, etc. It is important to point out that the main cost of the seismic isolation of a building is not the purchase cost of the devices themselves, which is usually balanced by the reduction in the superstructure cost, but the indirect costs resulting from architectural detailing, MEP, special items to account for the foreseen seismic movement as for gas pipes and other lifelines and the provision of space for inspection, maintenance and local uplifting in case of eventual replacement of an isolator (Eurocode 8 par.10.4.(8))

In order to provide some insight in these issues, the following paragraphs lay out the basic requirement for the conceptual design of a seismic isolation system of a building.

15.2.1.1 Seismic isolation horizontal level

The seismic isolation should be placed at a continuous horizontal level and should not be arranged into two different levels. This is desired in order to avoid creating severe stiffness abnormality over the height, which is not desired in the seismic design of buildings and causing control issues regarding differential seismic ground motion (Eurocode 8 par.10.5.3). Although not explicitly, Eurocode 8 enforces this requirement by requiring that the following two conditions are both satisfied:

- A rigid diaphragm is provided above and under the isolation system, consisting of a reinforced concrete slab or a grid of tie beams.
- The devices constituting the isolation system are fixed at both ends to the rigid diaphragms either directly or, if not practicable, by means of vertical elements (RC plinths) with relative horizontal displacement, in the seismic design situation lower than 1/20 of the relative displacement of the isolation system.

Based on the above, the level that the seismic isolation is going to be placed is critical for many decisions that will affect the overall design of the building. There are two possible horizontal levels that one may apply the seismic isolation:

- Directly above the foundation
- Below the ground floor

In buildings without basement, obviously these two options are condensed to one, as is shown in Figure 15.4a. However, even in this case, there is the problem of the elevator shaft. As is well known to all structural engineers, the elevator shaft requires a pit of around 1.50–2.00 m below the lowest stop of the elevator, which means that there is always a part of the building going lower. However, considering the requirements of having the seismic isolation at one level, the most common solution is to hang the pit shaft from the top diaphragm of the seismic isolation as shown in Figure 15.4a. Having the shafts separated by the isolation interface is not a very common solution as it requires to take into account the seismic movement for all the elevator guides, which is complex if not impossible from an operational safety point of view.

One of the most extreme cases of such an application is found in the Athens Opera House (SNFCC), which has the seismic isolation interface at the ground floor level, yet required the stage pit to drop 15 m below, thus creating a hanging reinforced concrete structure of 3000 tons as shown in Figure 15.5.

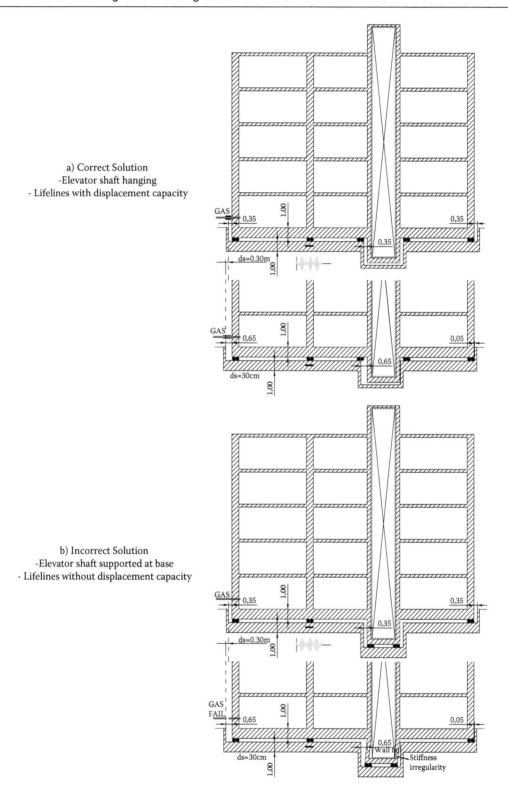

a) Correct Solution
-Elevator shaft hanging
- Lifelines with displacement capacity

b) Incorrect Solution
-Elevator shaft supported at base
- Lifelines without displacement capacity

Figure 15.4 Building with seismic isolation and (a) elevator shaft hanging and (b) elevator shaft incorrectly supported.

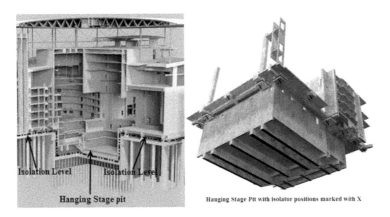

Figure 15.5 Athens Opera House (SNFCC) with seismic isolation level at GF and hanging 3000 tons stage pit.

15.2.1.2 In-plan distribution of isolator devices

The in-plan distribution of the isolator devices should ideally be matching the columns distribution of the superstructure. Obviously in buildings with shear walls or even to MRF with architectural constraints, this simple rule cannot be always followed.
So Eurocode 8 (par.10.5.2) sets the following requirements:

- To minimize torsional effects, the effective stiffness centre and the centre of damping of the isolation system should be as close as possible to the projection of the centre of mass on the isolation interface.
- The compressive stress induced in the isolator units by the permanent actions should be as uniform as possible.

The first condition, in the case of linear analysis, is quantified as 7.5% total eccentricity. In each of the two principal horizontal directions, the total eccentricity (including the accidental eccentricity) between the stiffness centre of the isolation system and the vertical projection of the centre of mass of the superstructure does not exceed 7.5% of the length of the superstructure transverse to the horizontal direction considered.

It should be noted that in the case of the seismically isolated Acropolis Museum in Athens, the superstructure columns do not correspond with the foundation piles below, which were arranged based on the archaeological excavations. Therefore, isolators are not, in all cases, directly below the vertical elements, while the rules of Eurocode 8 have been followed. Due to the irregular distribution of the isolators, this building was verified using nonlinear THA.

15.2.1.3 Theoretical background

The equations of motion for the two-DOF system of Figure 15.3a are given below, for the relative displacements $v_b = u_b - u_g$ and $v_s = u_s - u_b$:

$$(m + m_b)\ddot{v}_b + m\ddot{v}_s + c_b\dot{v}_b + k_b v_b = -(m + m_b)\ddot{u}_g \tag{15.1a}$$

$$m\ddot{v}_b + m\ddot{v}_s + c_s\dot{v}_s + k_s v = -m\ddot{u}_g \tag{15.1b}$$

where

v_b : relative displacement of base to ground
v_s : relative displacement of superstructure to base
m : the mass of the superstructure
m_b : the mass of the diaphragm above the isolation
k_s : the stiffness of the superstructure
c_s : the damping of the superstructure
k_b : the stiffness of the isolation
c_b : the damping of the isolation

The nominal frequencies are

$$\omega_b^2 = \frac{k_b}{m+m_b} \qquad \omega_s^2 = \frac{k_s}{m} \tag{15.2}$$

Given the fact that the stiffness of the seismic isolation k_b is much less than the stiffness of the superstructure k_s, we introduce the parameter ε:

$$\varepsilon = \frac{\omega_b^2}{\omega_s^2} \tag{15.3}$$

and we assume a value for this of the order (10^{-2}).

Using these assumptions, it is easily proven, that the following approximations apply (Naeim and Kelly, 1999):

$$\left.v_s\right|_{max} = \varepsilon S_D(\omega_b, \zeta_b) \tag{15.4}$$

$$\left.v_b\right|_{max} = S_D(\omega_b, \zeta_b) \tag{15.5}$$

$$C_s = S_e(\omega_b, \zeta_b) \tag{15.6}$$

where

C_s : the base shear coefficient
$S_D(\omega_b, \beta_b)$: the displacement response spectrum for the ground motion $\ddot{u}_g$ at frequency ω_b and damping factor ζ_b
$S_e(\omega_b, \beta_b)$: the acceleration response spectrum for the ground motion $\ddot{u}_g$ at frequency ω_b and damping factor ζ_b

with

$$S_D = S_e/\omega^2 \tag{15.7}$$

Based on the above, for the conceptual design phase, the isolation system design may be approximated by a relative base displacement based on the effective period and damping of the isolation system ($T_{b,eff}$, $\zeta_{b,eff}$), while the seismic force transmitted to the building is based on the total mass of the building ($m + m_b$) and the effective period and damping of the isolation system ($T_{b,eff}$, $\zeta_{b,eff}$). These assumptions have been incorporated in the simplified linear analysis procedure presented in par. 10.9.3 of Eurocode 8 and are elaborated in the following subsection.

15.2.1.4 Target fundamental period, damping and expected displacements

At the conceptual design stage, the following parameters need to be defined, so that the design process may continue:

- Target fundamental period with isolation
- Required damping of the isolation system
- Expected displacements for the design earthquake

As already mentioned, the period elongation achieved through the small lateral stiffness of the seismic isolation (k_b) is useful for buildings that have a fundamental eigen-period between 0.1 and 1.0 s, *while the target period is usually 2.5 to 3.0 s.*
The effective fundamental period of the seismically isolated building is calculated as follows:

$$T_{eff} = 2\pi\sqrt{\frac{M}{K_{eff}}} \tag{15.8}$$

where
$\quad M\ :\ $ the total mass ($m_b + m$)
$\quad K_{eff}:\ $ the effective horizontal stiffness of the isolation system, approximated as the sum of the effective stiffnesses of the isolator units (k_{eff}).

The stiffness of each isolator unit is defined by the manufacturer, although some useful indications are provided in the following sections.
The displacement of the stiffness centre due to the seismic action (design displacement) may be calculated as follows (Eurocode 8 par.10.9.3):

$$d_{dc} = \frac{M\,S_e(T_{eff},\,\zeta_{eff})}{K_{eff,\,min}} \tag{15.9}$$

where $S_e\left(T_{eff},\zeta_{eff}\right)$ is the spectral acceleration at $T = T_{eff}$, for ζ_{eff} damping.
The effective damping (ζ_{eff}) is an equivalent viscus damping of the isolator units and *is usually around 25%.* The energy dissipation should be expressed from the measured energy dissipated in cycles with frequency in the range of the natural frequencies of the modes considered. For higher modes outside this range, the modal damping ratio of the complete structure should be that of a fixed base superstructure, i.e. 5% for R/C buildings.

15.2.2 Isolation devices

In the previous paragraphs, there have been several references to different types of seismic isolation devices, which in general all called bearings. The previous references mainly focused on the historical evolution of such devices, while in this section, the state of the art of such devices shall be presented.

There are two main categories of such devices: *the inverted pendulum pot bearings*, which are essentially sliders, and *the elastomeric bearings*.

There are pros and cons for both of these devices; however, the main considerations are the following:

- The elastomeric bearings can be designed to provide higher damping than the inverted pendulum ones.
- The elastomeric bearings are *significantly cheaper* than the steel inverted pendulum ones.
- The steel inverted pendulum bearings are *more durable* than the elastomeric as they consist of *stainless steel and PTFE* surfaces.
- The steel inverted pendulum bearings are not subject to vertical deformations as are the elastomeric bearings.
- The inverted pendulum bearings *are not sensitive to temperature or fire.*

It should be mentioned that the production and installation of bearings within the European Union is governed by EN15129, which prescribes specific prototype and production tests for such devices. These tests are required to be executed at a third-party accredited facility, while in the United States similar tests are foreseen, though they may be executed in-house by the manufacturer. A critical difference between the European and US practice in the production of inverted pendulum isolators is the containment ring foreseen in the United States, which is essentially a stopper that does not allow the building to hop off the isolator in case the seismic excitation displacement exceeds the design provisions and the safety factors. The relevant European Norms explicitly forbid this containment ring following the notion that pounding on the containment ring may cause a shock effect on the building, which has not been taken into account by the design. It is the authors' view that such differences have mainly been introduced for commercial purposes.

Finally, it is interesting to note that the bearings (isolators) are mechanisms and, in this respect, they have shorter lifetime than the building itself. Therefore, there must be a provision for the necessary space for inspection, maintenance and eventual replacement. The last action, although very rare, requires the capacity of local lifting of each isolator. It should be noted that the guaranteed lifetime is a very serious parameter for the decision-making regarding the type of isolators that will be chosen.

15.2.2.1 Inverted pendulum bearings

The inverted pendulum bearings were invented by EPS (http://www.earthquakeprotection.com) in 1985. The simplicity of the device is leading back to the roots of the seismic isolation, as explained in the first paragraph of this chapter and is shown in Figure 15.6.

Since then, the system has been developed worldwide by several different manufacturers, while it has undergone improvements and evolved into the double and triple curvature inverted pendulums. Since the scope of this paragraph is to present the basic ideas, and not to be a brochure of available products, the basic equations for designing a single curvature inverted pendulum are presented.

The characteristic hysteretic loop of single curvature inverted pendulum is shown in Figure 15.7.

From Figure 15.7 the horizontal resisting force (F) is given by the following equation:

$$F = \frac{W}{R}d + \mu W(\text{sgn}\dot{d}) \tag{15.10}$$

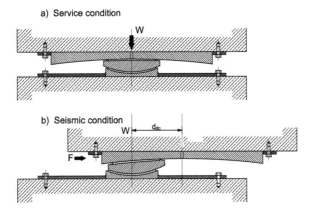

Figure 15.6 Single curvature sliding isolator concept. (Courtesy of mageba, http://www.mageba-group.com)

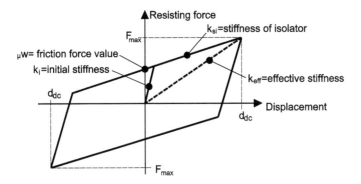

Figure 15.7 Single curvature sliding isolator theoretical hysteresis loop.

where

μ	:	the dynamic friction coefficient with values varying from μ_{low} to μ_{up} as per EN15129
W	:	the weight on the isolator $M = W/g$
R	:	radius of curvature of the device
d	:	horizontal displacement
sgnd	:	sign of sliding velocity vector; +1 or –1.

The horizontal resisting force (F) becomes independent of velocity for speeds above 51 mm/s and pressures more than 15 MPa. In this case, the horizontal force F takes a maximum value

$$F_{max} = \frac{W}{R} d_{dc} + \mu W \qquad (15.11)$$

where

d_{dc} is the design displacement of the device.

The effective stiffness of the device (k_{eff}) is calculated below:

$$k_{eff} = \frac{F_{max}}{d_{dc}} = \frac{\frac{W}{R}d_{dc} + \mu W}{d_{dc}} = \frac{W}{R} + \frac{\mu W}{d_{dc}} \qquad (15.12)$$

The effective period is calculated as

$$T_{eff} = 2\pi\sqrt{\frac{M}{k_{eff}}} = 2\pi\sqrt{\frac{\frac{W}{g}}{\frac{W}{R}d_{dc} + \mu W}} = 2\pi\sqrt{\frac{d_{dc}\frac{W}{g}}{\frac{W}{R}d_{dc} + \mu W}} = 2\pi\sqrt{\frac{R\,d_{dc}\frac{W}{g}}{Wd_{dc} + \mu WR}}$$

$$T_{eff} = 2\pi\sqrt{\frac{Rd_{dc}}{g(d_{dc} + \mu R)}} \qquad (15.13)$$

The effective damping ζ_{eff}, provided through friction of the stainless steel and the sliding material, is given from the following equation, as detailed in Equation 2.38 in Chapter 2, Sections 2.3.2 and 2.3.3:

$$\zeta_{eff} = \frac{1}{4\pi}\frac{\text{Area of hysterisis loop}}{\text{Area of circumscribed triange}} \qquad (15.14)$$

where

Area of hysteresis loop = $4\,\mu Wd_{dc}$
Area of circumscribed triangle = $1/2\,k_{eff}\,d_{dc}^2$

$$\zeta_{eff} = \frac{2}{\pi}\frac{\mu R}{(d_{dc} + \mu R)} \qquad (15.15)$$

These equations are used in order to determine the isolator properties based on the concept design values of T_{eff}, k_{eff} and d_{dc}. Such an application is shown in Table 15.1 from the case of the Athens Opera House with a target $T_{eff} = 2.60$ s.

Finally, the concept design should ensure that the selected isolator positions do not result into axial loads that are more than the capacity of the isolator devices. Indicatively it should be mentioned that the largest isolator installed with CE marking is the 70,000 kN isolator for the Athens Opera House. Table 15.2 includes indicative dimensions of isolators per axial load order of magnitude. The exact verifications required for the isolation devices include the following items, and are the responsibility of the manufacturer:

- Sliding disk stresses at the centre and at eccentric position
- Separation of sliding surface
- Stress in top and bottom anchor plates
- Deformation of backing plates
- Concrete stresses
- Anchoring/sliding verification
- Dynamic recentring

Table 15.1 Calculation of inverted pendulum isolator properties for the case of the Athens Opera House

Property	Values	Units
R	2700	mm
d_{dc}	0.234	m
μ_{up}	0.054	
μ_{low}	0.036	
T_{eff}	2.587	s
$(\zeta_{eff})_{up}$	0.244	
$(\zeta_{eff})_{low}$	0.187	
$(\zeta_{eff})_{aver}$	0.216	
n_{max}	0.650	
$Sa_{el}\,[n=1]$	1.407	m/s²
ω	2.428	1/s
q	1.5	
γ_I	1.4	
γ_X	1.5	
S_e	0.853	m/s²
S_d	0.234	m
$S_d \times \gamma_X$	0.351	m
S_v	0.568	m/s

Table 15.2 Indicative inverted pendulum in-plan dimensions (diameter) per axial load from the Athens Opera House

Load [kN]	D [mm]
3500	280
5500	320
6750	380
8000	420
10,000	450
12,500	500
15,000	540
21,500	650
26,500	800
32,500	930
70,000	1555

Finally, it is interesting to note that for this type of isolators, the centre of mass of the superstructure coincides with the stiffness centre of the isolators (zero eccentricity). In fact, having in mind that all isolators of a building have the same values of R, d_{dc} and μ, it is concluded that Equation 15.12 takes the form

$$k_{eff} = W\overline{k} \tag{15.12a}$$

where

$$\bar{k} = \left(\frac{1}{R} + \frac{\mu}{d_{dc}} \right) : \text{constant}$$

(15.12b)

meaning that k_{eff} for all isolators is proportional to their axial load.

15.2.2.2 Rubber bearings

Two types of rubber bearings are used worldwide for seismic isolation (Figure 15.8):

- Reinforced elastomeric bearings with steel connection plates
- Reinforced elastomeric bearings with steel connection plates and a lead core

The simple case of the rubber and steel laminated bearing is designed to support the weight of the structure and to provide post-yield elasticity. Rubber provides the isolation and the recentring of the bearing after a seismic event, while reaching a damping of up to 16%. At the same time, the steel plates ensure the confinement of the rubber so that it can carry considerable axial forces (loads) with very small axial deformations.

The more complex case of the rubber and steel laminated bearing with the lead core, additionally to the previous case, has the lead core that deforms plastically under shear deformations, and reaches a damping of up to 30%.

For the elastomeric bearing, the main property of interest is the horizontal stiffness $k_{H,eff}$ calculated by the following equation:

$$k_{eff} = \frac{GA}{h_s}$$

(15.16a)

where
- G: the shear modulus of the elastomer
- A: the area of the elastomer
- h_s: the total thickness of rubber

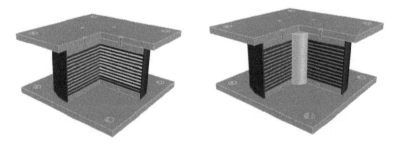

Figure 15.8 Reinforced elastomeric bearings with steel connection plates (left) without and (right) with lead core.

The vertical stiffness of the elastomer bearing is given by the following formula (Naeim and Kelly, 1999):

$$k_v = \frac{E_c A}{h_s} \tag{15.16b}$$

where
  E_c is the instantaneous compression modulus of the steel–rubber compound.
  For a circular pad with diameter D and a single layer thickness H_b, the value of E_c is

$$E_c = 6GS^2 \tag{15.17}$$

where S is the shape factor, defined as

$$S = \frac{D}{4H_b} \tag{15.18}$$

In the case of lead plug elastomeric bearings, the properties are mostly defined experimentally. Referring to the bilinear idealization of Figure 15.9, the effective stiffness is defined as follows:

$$k_{eff} = k_{p.y} + \frac{F_1}{d_{dc}} \tag{15.20}$$

where
  $k_{p.y}$: the post yield stiffness
  F_1 : the intercept of the hysteresis loop and the force axis, usually estimated from the lead core yield stress (10.3 MPa)

The effective damping ζ_{eff} is defined by the following equation, based on Equation 2.39 of Chapter 2.3.3:

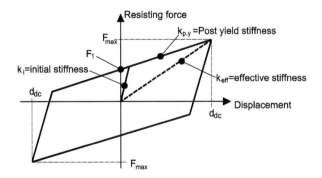

Figure 15.9 Idealized bilinear stress–strain diagram for lead plug elastomeric bearings.

Table 15.3 Indicative properties of circular elastomeric bearings with lead core

			Elastomeric Bearing with lead core d_{dc} = 400 mm							
D [mm]	h_s [mm]	H_b [mm]	W [kN]	N_E [kN]	F_y [kN]	F_{max} [kN]	$k_{p.y}$ [kN/mm]	k_{eff} [kN/mm]	k_v [kN/mm]	ζ [%]
500	160	326	3,600	1,250	315	755	1.1	1.89	814	29
600	176	350	5,950	2,150	420	990	1.45	2.49	1,346	28
700	192	374	8,750	3,450	515	1230	1.8	3.09	1,991	28
800	208	398	10,950	5,100	620	1500	2.17	3.73	2,725	26
900	216	410	16,250	6,750	690	1750	2.65	4.38	3,658	26
1000	224	422	18,750	10,100	760	2030	3.16	5.07	4,693	25

Source: Courtesy of mageba.

$$\zeta_{eff} = \frac{\text{Area of hysterisis loop}}{2\pi k_{eff}\, d_{dc}^2} \qquad (15.21)$$

The area of the hysteresis loop is $4F_1(d_u - d_y)$. So, for $k_i = 10k_{p.y}$, ζ_{eff} takes the following form:

$$\zeta_{eff} = \frac{4F_1\left(d_{dc} - \dfrac{F_1}{9k_{p.y}}\right)}{2\pi d_{dc}(d_{dc}k_{p.y} + F_1)} \qquad (15.22)$$

As already mentioned, these types of bearings have properties that are defined mainly experimentally, so in the initial design phase, it is useful to have reference values such as the ones provided in Table 15.3.

15.3 CONCEPT DESIGN OF BUILDINGS WITH SUPPLEMENTAL DAMPING

15.3.1 Concept design

The current version of Eurocode 8 does not foresee any provisions for the explicit design of buildings with supplemental damping. The new generation of Eurocode 8, not yet disseminated or approved for use, includes a very complex set of equations for the design of supplemental damping. To the author's opinion, this approach is not yet practical; therefore, the concept design, using elastic analysis approaches, is based on ASCESEI-41/13.

There are two main types of passive energy dissipation devices (dampers), displacement-dependent and velocity-dependent, although there are also other types, as outlined in Table 15.4.

As an initial guideline, four or more dampers should be provided in a given storey of a building in one principal direction of the building, with a minimum of two devices located on each side of the centre of stiffness of the storey in the direction under consideration. This condition being satisfied ensures the smallest demand (lower safety factors) for the dampers design.

Table 15.4 Passive energy dissipation systems (MCEER, Constantinou et al., 1998)

Types	Principle	Materials/Technology	Intervention
Displacement based	Hysteretic	Steel or lead	Energy dissipation
	Friction	Metal to metal	Stiffness increase
			Strength increase
Velocity based	Viscoelastic solids	Viscoelastic polymers	Energy dissipation
	Deformation of viscoelastic fluids	Highly viscous fluids	Stiffness increase
	Fluid orificing	Fluids passing through orifices and special seals	

In order to make a preliminary selection of the positioning of the devices and estimate their properties, in the concept design phase, the linear static procedure is utilized, while the fundamental period of the building is estimated either approximately or through a modal analysis. It is noted that the linear static analysis is not always permitted for such applications; however, even in the cases that this is true, this analysis should be the first step in the design.

The equivalent damping (ζ_{eff}) of a structure with added dampers is given by the following equation, based on Equation 2.39 of Chapter 2.3.3:

$$\zeta_{eff} = \zeta + \frac{1}{4\pi} \frac{\sum_j W_j}{W_k} \qquad (15.23)$$

where

ζ :   the damping of the framing plan equal to 0.05 for R/C buildings

$\sum_j W_j$:   the work done by device j in one complete cycle corresponding to floor displacements δ_j; the summation extends over all devices j

W_k :   the maximum strain energy in the frame defined below:

$$W_k = \frac{1}{2}\sum_i F_i \delta_i = \frac{1}{2}\sum_i k_{eff,i} \delta_i^2 \qquad (15.24)$$

and

F_i: the inertia force at floor level i, the summation extending over all floor levels.

Obviously in order to calculate W_j, for any damper device, the stress–strain curve and characteristic properties of this device are required. In the following paragraphs, guidance for determining or approximating these properties is given. It should, however, be noted that even if the selection is made and a more complex analysis is not required in the final design stages, these properties should be determined eventually by testing, and the final values be used to validate the preliminary design.

Based on the above, the basic design steps are the following.

Step 1: Linear static analysis of the existing building

- Calculation of fundamental period (T).
- Calculation of spectra acceleration (S_e) and spectral velocity (S_d) for (T) for $\zeta_{eff} = 5\%$.

- Calculation of storey displacements.
- Calculation of storey forces.

Step 2: Linear static analysis of the building with dampers

- Determination of required damping ζ_{eff} (20–30%).
- Calculation of fundamental period (T') (see Chapter 2.2.2, Equation 2.9).
- Calculation of spectral acceleration (S'_e) and spectral velocity (S'_d), for (T') for ζ_{eff} = 20–30%.
- Calculation of reduced storey displacements.
- Calculation of reduced storey forces.

Step 3: Calculation of actual ζ_{eff}

- From the analysis, the damper stroke (d_{dc}) and stroke velocity $(\dot{d})$ are calculated.
- Given that the building will mainly vibrate to the fundamental period (T'), the stroke velocity $(\dot{d})$ may be approximated as

$$\dot{d} = \frac{2\pi}{T'} d_{dc} \tag{15.25}$$

This is not the same as the peak ground velocity as the spectral acceleration (S_e) is not the same as the peak ground acceleration.

- The forces on the dampers are calculated and the dampers are selected.
- Using Equation 15.23 and the properties of the selected damper devices, the actual ζ_{eff} is calculated.

Step 4: Go to Step 2
A new loop is required in any of the following cases:

- If the ζ_{eff} calculated from Step 3 is different from the ζ_{eff} assumed
- If the damper forces are larger than available devices
- If the damper strokes are larger than available devices

15.3.2 Displacement-depended dampers

A displacement-dependent device has a force–displacement relationship that is a function of the relative displacement between each end of the device while being independent of the relative velocity between each end of the device and frequency of excitation.

Such devices are usually custom-designed steel elements and are not provided by a manufacturer. Although there are many different geometric configurations for the metallic devices, the basic principle is that energy dissipation takes place through the inelastic deformation of a metal. Usually this metal is mild steel, although it may also be lead.

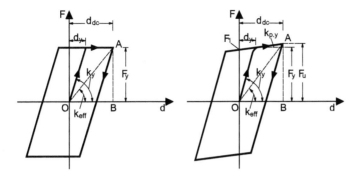

Figure 15.10 Idealized bilinear stress–strain diagrams for hysteretic dampers. (a) EPP. (b) Elastoplastic with strain hardening.

Such devices may be modelled in sufficient detail to capture their force–displacement response and their dependence, if any, on axial–shear–flexure interaction or bilateral deformation response. Through this analytical procedure, the static properties of the dampers are defined, i.e. initial stiffness k_1, the yield force F_y, the post yield stiffness $k_{p.y}$ and the force–displacement diagram of Figure 15.10 are calculated. Obviously prior to final design, these should be assessed experimentally.

The calculation of the equivalent damping (ζ_{eff}) of the building, defined in Section 15.3.2, requires the calculation of the work done by the dampers $\sum_j W_j$. Considering the force–displacement diagrams of a hysteretic steel damper in Figure 15.10, the definition of the work done by one device is as follows:

- For elastic perfectly plastic (EPP)

$$W_j = 4F_y(d_{dc} - d_y) \tag{15.26}$$

- For elastoplastic with strain hardening

$$W_j = 4F_1(d_{dc} - d_y) \tag{15.27}$$

where
d_y : the yield displacement of the damper
d_{dc}: the maximum seismic displacement (stroke) of the damper estimated from analysis of the building
F_y : the yield force of the damper

F_1 : the intercept of the hysteresis loop and the force axis, which for EPP is equal to F_y

15.3.3 Velocity-dependent dampers

15.3.3.1 Solid viscoelastic devices

Viscoelastic solid materials are usually polymers or glassy substances dissipating energy through shear deformation. A typical viscoelastic damper is shown in Figure 15.11.

The cyclic response of viscoelastic solids is generally dependent on the frequency and amplitude of the motion and the operating temperature (including temperature rise caused by excitation). Solid viscoelastic devices are modelled using a spring and dashpot in parallel (Kelvin model, see Chapter 2.3.2 and Figure 15.12). The spring and dashpot constants selected shall capture the frequency and temperature dependence of the device consistent with fundamental period of the building and the operating temperature range (ASCESEI-41/13).

The relevant hysteretic loop is given by the ellipse of Figure 15.13 (see also Figure 2.18). The force of the solid viscoelastic device is calculated as follows:

$$F = k_{eff}d + C\dot{d} \qquad (15.28)$$

where
  d : the relative displacement between each end of the device
  $\dot{d}$: the relative velocity between each end of the device
  k_{eff}: the effective stiffness of the device calculated below

$$k_{eff} = \frac{|F^+| + |F^-|}{|d_{dc}^+| + |d_{dc}^-|} = \frac{F^{max}}{d_{dc}} \qquad (15.29)$$

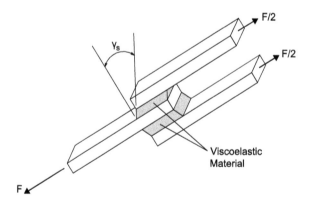

Figure 15.11 Typical viscoelastic solid damper. (Adapted from Christopoulos, C. and Filiatrut, A., 2006, Principles of Passive Supplemental Damping and Seismic Isolation, IUSS, Pavia.)

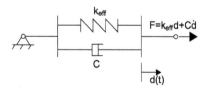

Figure 15.12 Spring and dashpot in parallel (Kelvin model).

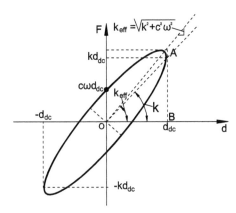

Figure 15.13 Hysteretic loop of viscoelastic damper.

for symmetric diagram

C :   the damping coefficient of the device calculated below (see Chapter 2.3.2, Equations 2.34 and 2.35):

$$C = \frac{W_{loop}}{\pi \omega d_{dc}^2} \qquad (15.30)$$

where

W_{loop} :   the area enclosed by one complete cycle of the force–displacement response of the device

ω :   the angular frequency of the excitation

The resulting equivalent damping of the device (ζ_{eff}) is derived from the following equation (Christopoulos and Fillatrault, 2006):

$$2\zeta_{eff} = \frac{G_c \omega}{G} \qquad (15.31)$$

where

G :   the elastic shear modulus

G_c :   the shear viscous damping constant (Christopoulos and Fillatrault, 2006)

while

$$k_{eff} = \frac{GA}{h_s} \qquad (15.32)$$

$$C = \frac{G_c A}{h_s} \qquad (15.33)$$

Table 15.5 Indicative solid viscoelastic dampers geometry

Damper type	Shear Area [mm²]	Thickness [mm]	Volume [mm³]
A	968	5.08	4917
B	1936	7.62	14752
C	11613	3.81	44246

Source: Christopoulos and Filiatrault, 2006.

Table 15.6 Indicative solid viscoelastic dampers properties (Christopoulos and Filiatrault, 2006) for 5% shear strain and angular frequency 3.5 Hz

Damper type	Temperature [°C]	G [MPa]	G_c ω [MPa]	ζ_{eff}
A	21	2.78	3.01	0.540
	24	2.10	2.38	0.564
	28	1.57	1.9	0.605
	32	1.17	1.37	0.585
	36	0.83	0.90	0.540
	40	0.63	0.63	0.50
B	25	1.73	2.08	0.600
	30	1.29	1.54	0.595
	34	0.94	1.11	0.590
	38	0.76	0.84	0.550
	42	0.62	0.65	0.525
C	25	0.19	0.17	0.450
	30	0.16	0.12	0.375
	34	0.14	0.10	0.355
	38	0.12	0.08	0.335
	42	0.11	0.07	0.320

where
  h_s :   the shear thickness
  A_s: the shear area

Tables 15.5 and 15.6 provide indicative values of these constants to be used in a concept design. As expected, the properties are a function of the frequency, the temperature and the allowed strain, which means that also the above equations for k_{eff} and C are a function of these properties.

15.3.3.2 Fluid viscoelastic devices

Only in the recent years (after 1992) have fluid dampers been migrating from the military and aerospace industry to the earthquake engineering practice (Taylor, 1999). Such devices are based on the high-velocity fluid flow through orifices. These devices are considered the state of the art for passive energy dissipation devices.

Fluid viscoelastic devices are modelled using a combination of springs and dashpot in series to represent the constitutive relation of the device.

The axial force of a fluid viscous device is computed according to the following equation (ASCESEI-41/13 par. 14.3.3.2.3) and is independent for the displacement (i.e. has practically zero stiffness):

$$\boxed{F = C \, |\dot{d}|^\alpha} \tag{15.34}$$

where, as before,

    C: the damping coefficient for the device 100–400 [kN s/m$^\alpha$]
    α: the velocity exponent of the device, with values 0.2 to 1.0
    $\dot{d}$:   the stroke velocity of the device

The force–displacement curves of this type of dampers are presented in Figure 15.14.

Force–displacement curves on fluid viscoelastic dampers, from tests on dampers for the Athens Opera House (SNFCC) executed at AUTh under the authors' supervision (Manos and Katakalos, 2016), are shown in Figure 15.15, where the axial force is constant over the displacement (stroke) and is only affected by the stroke speed.

Also Table 15.7 provides indicative damper properties for usual axial loads in buildings.

So, considering a usual moment resisting frame building (H = 3.50 m, L = 7.00 m) with a fundamental period of 1.00 s and an inter-storey displacement of 0.035 m, the resulting diagonal damper stroke would be 0.031 m and the resulting stroke velocity would be calculated as

$$\dot{d} = \frac{2\pi}{T} d_{dc} = \frac{2\pi}{1} 0.031 = 0.20 \ \mathrm{m/s} \tag{15.35}$$

Figure 15.16 shows viscous dampers with different selections of C and a, which for the stroke velocity foreseen (0.20 m/s) develop almost equal damper axial force. It is obvious, however, that for higher earthquakes than the foreseen, for increased stroke speed, the dampers with lower α values control the developed axial force, thus limit damage to the support frame.

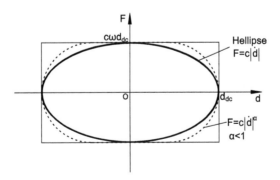

Figure 15.14 Force–displacement curve for fluid viscoelastic dampers (effective stiffness $k_{eff} = 0$).

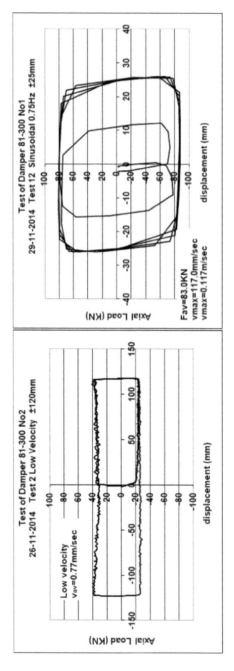

Figure 15.15 Test results on dampers for the Athens Opera House (SNFCC) for different strokes and velocities. (From Manos, G. and Katakalos K., 2016, Experimental investigation of the behaviour of structural components of the SNFCC canopy in Athens (in Greek), 17th National Conference of Concrete Structures, Thessaloniki.)

Table 15.7 Indicative damper dimensions and properties

Axial Force [kN]	Displacement capacity [mm]												
	50		100		150		200		250		300		400	
[mm]	Φ	L	Φ	L	Φ	L	Φ	L	Φ	L	Φ	L	Φ	L
50	110	720	110	1020	110	1320	110	1620	110	1920	110	2220	110	2820
100	120	750	120	1050	120	1350	120	1650	120	1950	120	2250	120	2850
200	180	780	180	1080	180	1380	180	1680	180	1980	180	2280	180	2880
500	195	820	195	1120	195	1420	195	1720	195	2020	195	2320	195	2920
750	215	835	215	1135	215	1435	215	1735	215	2035	215	2335	215	2935
1000	235	855	235	1155	235	1455	235	1755	235	2055	235	2355	235	2955
1250	280	920	280	1220	280	1520	280	1820	280	2120	280	2420	280	3020
1500	295	990	295	1290	295	1590	295	1890	295	2190	295	2490	295	3090
1750	325	1045	325	1345	325	1645	325	1945	325	2245	325	2545	325	3145
2000	365	1190	365	1490	365	1790	365	2090	365	2390	365	2690	365	3290
2500	405	1270	405	1570	405	1870	405	2170	405	2470	405	2770	405	3370
3000	455	1385	455	1685	455	1985	455	2285	455	2585	455	2885	455	3485
4000	505	1505	505	1805	505	2105	505	2405	505	2705	505	3005	505	3605

Source: Courtesy of mageba.

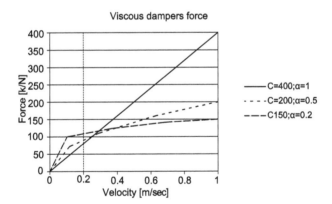

Figure 15.16 Viscous dampers options regarding C and α (from the Athens Opera House (SNFCC) design).

15.4 FINAL DESIGN OF BUILDINGS WITH SEISMIC ISOLATION AND/OR SUPPLEMENTAL DAMPING

15.4.1 Analysis methods

In the previous paragraphs, all the requirements and procedures for the concept design of buildings with seismic isolation or supplemental damping have been elaborated; however, it is necessary to provide some guidelines regarding the final structural design of such buildings. To that effect, we shall refer to the analysis methods elaborated in Chapter 5 and define their applicability, and at the same time provide the specific modelling guidelines for the isolator devices modelling per analysis type.

When designing\ a seismically isolated building or a building with supplemental damping, it is essential that the initial elastic approach of the previous section is used as a guideline. Further to that, the following options are available for the final design of the structure:

- Elastic spectral dynamic analysis (modal linear analysis)
- Linear time–history analysis (THA)
- Nonlinear THA

As mentioned in Paragraph 10.4 of Eurocode 8, only *full isolation is accepted*, that is, the superstructure remains in the elastic range. Although Paragraph 10.10 (5) allows the use of a *q*-factor equal to 1.50, for all analysis purposes, the superstructure is considered to remain elastic for all considered limit states.

15.4.2 Modal linear analysis for buildings with seismic isolation

According to Eurocode 8 for seismically isolated buildings, modal analysis is required when the behaviour of the devices may be considered as equivalent linear and any one of the following conditions is not met:

- The total eccentricity (including the accidental eccentricity) between the stiffness centre of the isolation system and the vertical projection of the centre of mass of the superstructure does not exceed 7.5% of the length of the superstructure transverse to the horizontal direction considered.
- The building does not conform to any of the following conditions:
  a. The distance from the site to the nearest potentially active fault with a magnitude $Ms \geq 6.5$ is greater than 15 km.
  b. The largest dimension of the superstructure in plan is not greater than 50 m.
  c. The substructure is sufficiently rigid to minimize the effects of differential displacements of the ground.
  d. All devices are located above elements of the substructure, which support vertical loads.
  e. The effective period T_{eff} satisfies the following condition:

$$3T_f \leq T_{eff} \leq 3 \text{ s}$$

where
  T_f: the fundamental period of the superstructure assuming a fixed base
  f. The lateral-load resisting system of the superstructure is regularly and symmetrically arranged along the two main axes of the structure in plan.
  g. The rocking rotation at the base of the substructure is be negligible.
  h. The ratio between the vertical (k_v) and the horizontal stiffness (k_{eff}) of the isolation system satisfies the following expression:

$$\frac{k_v}{k_{eff}} \geq 150$$

  i. The fundamental period in the vertical direction, T_V, is not greater than 0.1 s.
In practical terms, the building should be modelled using the requirements of Chapter 5 for the modal analysis, while the isolators should be introduced using linear elements with

stiffness corresponding to the effective stiffness (k_{eff}) of the isolator units. The design spectra should be calculated according to the effective damping of the system (ζ_{eff}) for the periods around the fundamental period of the isolation system, while if the software allows different damping per mode, the first three modes should be set to (ζ_{eff}) while the rest to 5% for R/C buildings. The behaviour factor (q) should have a maximum value of 1.50.

It should be noted that in the case that the following conditions apply:

- The centre of stiffness (K_{eff}) of the group of isolators coincides with the centre of mass (M) of the structure.
- The centre of the polar mass moment of inertia (J_d) of the superstructure coincides with the centre of torsional stiffness of the group of isolators (J_{TC}).
- The ratios M/K_{eff} and J_d/J_{TC} are equal.

it can be proven that the fundamental periods for the three DOF are equal (Chapter 2.4.4.3, Equation 2.109).

15.4.3 Modal linear analysis for buildings with supplemental damping

According to ASCESEI-41/13, modal linear analysis is required for a building with supplemental damping if any of the following conditions are met:

a. The ratio of the maximum resistance in each storey, in the direction under consideration, to the storey shear demand ranges beyond 80% to 120% of the average value of the ratio for all stories. *The maximum storey resistance shall include the contributions from all components, elements and energy dissipation devices.*
b. The maximum resistance of all energy dissipation devices in a storey, in the direction under consideration, exceeds 50% of the resistance of the remainder of the framing, where said resistance is calculated at the displacements anticipated for the selected excitation.

In practical terms, the building should be modelled using the requirements of Chapter 5 for the modal analysis, while the dampers should be introduced using linear elements with stiffness and damping corresponding to the effective values (k_{eff}, C) of the dampers. The design spectra should be calculated according to the effective damping of the system ($\zeta_{eff} \leq$ 30%), which is to be limited to 30%.

15.4.4 Time–history linear analysis

Linear THA is not essentially different from the spectral analysis. The same assumptions should be followed in the spectral analysis, while the selection of accelerograms should follow the requirements set in Chapter 5. The only difference is that it allows the definition of the damping per mode, so that the first three modes are set to ζ_{eff}, while the rest to 5% for R/C buildings.

15.4.5 Time–history nonlinear analysis for seismically isolated buildings

In case that the seismic isolation system may not be considered linear, nonlinear THA of the building should be executed. According to Eurocode 8, par. 10.9.2, the seismic isolation system requires nonlinear THA if any one of the following conditions is true:

a. The effective stiffness of the isolation system at a displacement d_{dc} is less than 50% of the effective stiffness at a displacement of $0.2d_{dc}$.
b. The effective damping ratio of the isolation system exceeds 30%.
c. The force–displacement characteristics of the isolation system vary by more than 10% due to the rate of loading or due to the vertical loads.

The nonlinear THA should be performed according to the requirements of Chapter 5. However, as the nonlinearity is only limited to the seismic isolation devices, the correct approach is to model the superstructure elastically and choose appropriate nonlinear springs (or elements) to model the horizontal behaviour of the seismic isolation.

15.4.6 Time–history nonlinear analysis for buildings with supplemental damping

In case that the seismic behaviour may not be considered linear, nonlinear THA of the building should be executed. According to ASCESEI-4/13, if any of the following conditions apply, the nonlinear THA should be selected:

a. The framing system exclusive of the energy dissipation devices does not remain essentially elastic for the selected excitation after the effects of added damping are considered.
b. The effective damping ratio of the building exceeds 30%.
c. The secant stiffness of each energy dissipation device, calculated at the maximum displacement in the device, is not included in the mathematical model of the rehabilitated building, but instead a mean k_{eff} has been introduced for groups of devices.
d. The building regularity checks are not fulfilled when the energy dissipation devices are considered.

The nonlinear THA should be performed according to the requirements of Chapter 5. The energy dissipation devices shall be modelled as nonlinear elements; however, if the members of the structure remain essentially elastic, they may be modelled with linear elements. Substitution of viscous effects in energy dissipation devices by global structural damping for nonlinear THA is not permitted.

If the properties of the energy dissipation devices are dependent on excitation frequency, operating temperature (including viscous heating), deformation (or strain), velocity, sustained loads and bilateral loads, such dependence shall be accounted for in the analysis by assuming upper- and lower-bound properties, i.e. the envelope of several such analyses should be utilized. This requirement actually renders the fluid viscous dampers the most appropriate solution, to the authors opinion, as they are usually unaffected by such variations.

The viscous forces in velocity-dependent energy dissipation devices shall be included in the calculation of the design actions and deformations.

References

Abdel-Halim, M.A.H. and Schorn, H. (1989). Strength evaluation of shotcrete-repaired beams. *ACI Structural Journal*, **86**(3), 272–6.

Abrams, D.P. (1991). Laboratory definitions of behaviour for structural components and building systems, in ACI SP-127: *Earthquake-resistant Concrete Structures-Inelastic Response and Design*, ACI, Detroit, Michigan, 91–152.

ACI (2011). *Building Code Requirements for Reinforced Concrete (ACI 318-11) and Commentary*. Detroit, Michigan.

ACI 318 M (2011). *Building Code Requirements for Structural Concrete*. American Concrete Institute, Farmington Hills, Michigan.

ACI 318-11 (2011). *Building Code Requirements for Structural Concrete and Commentary*. ACI Committee 318.

ACI 440.2R-02 (2002). *Guide for the Design and Construction of Externally Bonded FRP Systems for Strengthening Concrete Structures*. ACI, Detroit, Michigan.

ACI 506 R-90 (1995). *Guide to Shotcrete*. ACI Manual of Concrete Practice, Part 5. Farmington Hills, Michigan.

ACI Committee 408 (1991). *Splice and Development Length of High Relative Rib Area Reinforcing Bars*. American Concrete Institute, Farmington Hills, Michigan.

ADAPTIC (2012). *User Manual*, Department of Civil and Environmental Engineering, Imperial College, London SW7 2BU.

Aguilar, J., Juarez, H., Ortega, R. and Iglesias, J. (1989). The Mexico earthquake of September 19, 1985: Statistics of damage and retrofitting techniques in R/C buildings affected by the 1985 earthquake, *Earthquake Spectra, EERI*, **5**(1), 145–52.

AIJ (1994). *Structural Design Guidelines for Reinforced Concrete Buildings*. Architectural Institute of Japan, Tokyo.

Alcocer, S.M. and Jirsa, J.O. (1993). Strength of reinforced concrete frame connections rehabilitated by jacketing. *ACI Structural Journal*, **90**(3), 249–61.

Ambraseys, N.N. and Bommer, J.J. (1991). The attenuation of ground accelerations in Europe. *Earthquake Engineering and Structural Dynamics*, **20**(12), 1179–202.

Ambraseys, N.N., Douglas, I., Sarma, S.K. and Smit, P.M. (2005). Equations for the estimation of strong ground motions from shallow crustal earthquakes using data from Europe and the Middle East. Horizontal peak ground acceleration and spectral acceleration, *Bulletin of Earthquake Engineering*, **3**(1), 1–53.

Ambraseys, N.N., Smit, P., Berardi, D., Cotton, F. and Berge, C. (2000). *Dissemination of European Strong Motion Data*. CD ROM edition. European Council, Environment and Climate Research Programme.

Ambraseys, N.N., Simpson, K.A. and Bommer, J.J. (1996). Prediction of horizontal response spectra in Europe. *Earthquake Engineering and Structural Dynamics*, **25**(4), 371–400.

Anagnostopoulos, C., Georgiadis, M. and Pitilakis, K. (1994). *Foundation-Retaining Works* (in Greek). Civil Engineering Department, Aristotle University of Thessaloniki.

Anagnostopoulos, S. (1988). Pounding of buildings in series during an earthquake. *Earthquake Engineering Structural Dynamics*, **16**, 443–56.

Anagnostopoulos, S. et al. (1986). *The Kalamata earthquake of September 1986.* Report of the Institute of Engineering Seismology and Earthquake Engineering, Thessaloniki, Greece (in Greek).

Anagnostopoulos, S.A. (1984). Listing of earthquake damage of buildings and building characterization in emergency situations, *Proceedings of the Conference on Earthquakes and Structures,* EPPO, Athens, **I**, pp. 550–61 (in Greek).

Anagnostopoulos, S.A., Petrovski, J. and Bouwkamp, J.G. (1989). Emergency earthquake damage and usability assessment of buildings, *Earthquake Spectra,* 5(3), 461–76.

Anastasiadis, K.K. (1983). *Structural Dynamics* (in Greek). Ziti, Thessaloniki, Greece.

Anastasiadis, K.K. (1989). *Earthquake Resistant Structures* (in Greek). Computer Technics, Thessaloniki, Greece.

Anastasiadis, K.K. and Athanatopoulou, A. (1996). Torsional flexibility of buildings and seismic Codes (in Greek). *12th Greek Conference for Concrete Structures,* Technical Chamber of Greece, Limasol, Cyprus.

Anastasiadis, K., 2001. *Aseismic Structures.* Ziti Editions. Thessaloniki, Greece, p. 397 (in Greek).

Antoniou, S. and Pinho, R. (2004). Advantages and limitations of adaptive and non-adaptive force-based pushover procedures. *Journal of Earthquake Engineering,* 8(4), 497–522.

Antoniou, S., Rovithakis, A. and Pinho, R. (2002). Development and verification of a fully adaptive pushover procedure. *Proceedings of the 12th European Conference on Earthquake Engineering,* London. Paper No. 882.

Aoyama, H. and Noguchi, H. (1979). Mechanical properties of concrete under load cycles idealizing seismic actions, *Bulletin d'Information, CEB,* **131**, 29–63.

Arakawa, T. (1980). Effect of welded bond plates on seismic characteristics of R/C columns. *Proceed. 7th World Conf. on Earthquake Engng.* Istanbul, 7, pp. 233–40.

Aristizabal-Ochoa, J.D. (1982). Dynamic response of coupled wall systems. *ASCE Journal of the Structural Division,* **108**(ST8), 1846–57.

ASCE 7-05 (2005). *Minimum Design Loads for Buildings and Other Structures.* American Society of Civil Engineers, Reston, Virginia.

ASCE/SEI 41-13 (2014). *Seismic Evaluation and Retrofit of Existing Buildings.* American Society of Civil Engineers, Reston, Virginia.

ASCE/SE17-10 (2010). *Minimum Design Loads for Buildings and Other Structures.* American Society of Civil Engineers, Reston, Virginia.

ATC 3-06 (1978). *Tentative Provisions for the Development of Seismic Regulations for Buildings* (Chapter 13). US Government Printing Office, Washington DC.

ATC 40 (1996). *Seismic Evaluation and Retrofit of Concrete Buildings.* Applied Technology Council, Redwood City, California.

Athanassiadou, C.J., Penelis, G.G. and Kappos, A.J. (1994). Seismic response of adjacent buildings with similar or different dynamic characteristics. *Earthquake Spectra,* **10**(2), 293–317.

AUT (Aristotle University of Thessaloniki) (1978). *Earthquake Damage Repair in Buildings.* Thessaloniki (in Greek).

AUT (Aristotle University of Thessaloniki) (1979). *A Seminar for Earthquake Damage Repair in Buildings.* Thessaloniki (in Greek).

Baden Württemberg Innenministerium (1985). *Erdbebensicher Bauen.* Stuttgart.

Balázs, G.L. (1989). Bond softening under reversed load cycles. *Studie Ricerche, Politecnico di Milano,* (11), 503–24, Milano.

Bett, B.J., Klingner, R.E. and Jirsa, J.O. (1988). Lateral load response of strengthened and repaired concrete columns. *ACI Structural Journal,* **85**(5), 499–508.

Beres, A., El-Borgi, S., White, R.N. and Gergely, P. (1992). Experimental results of repaired and retrofitted beam-to-column joint tests in lightly reinforced concrete buildings. Technical report NCEER-92-0025, SUNY/Buffalo, USA.

Bertero, V. and Brokken, S.I. (1983). Infills in seismic resistant buildings. *Journal of Structural. Engineering ASCE,* **109**(6), 1337–61.

Bertero, V.V. (1979). Seismic behaviour of structural concrete linear elements (beams, columns) and their connections. CEB *Bulletin d'Information*, **131**, Paris.

Bertero, V.V. and Popov, E.P. (1977). Seismic behaviour of ductile moment resisting reinforced concrete frames. In *ACI SP 53: Reinforced Concrete Structures in Seismic Zones*. ACI, Detroit Michigan, pp. 247–91.

Biggs, J.M. (1964). *Introduction to Structural Dynamics*. McGraw-Hill, New York.

Biskinis, D.E. (2007). Resistance and deformation capacity of concrete members with or without retrofitting (in Greek). Doctoral Thesis, Civil Engineering Department, University of Patras, Patras, Greece.

Blakeley, R.W.G. (1971). Ductility of prestressed concrete frames under seismic loading, Ph.D. thesis, University of Canterbury, Christchurch.

Blume, J.A. (1960). Structural dynamics in earthquake resistant design. *Transactions of ASKE*, **125**, 1088–139.

Bommer, J.J. and Elnashai, A.S. (1999). Displacement spectra for seismic design. *Journal of Earthquake Engineering*, 3(4), 1–32.

Boore, D.M. (2005). Erratum: Equations for estimating horizontal response spectra and peak acceleration from western North American earthquakes. A summary of recent work. *Seismological Research Letters*, 76(3), 368–9.

Booth, E. and Key, D. (2006). *Earthquake Design Practice for Buildings*. Thomas Telford Ltd London, UK.

Borzi, B. and Elnashai, A.S. (2000). Refined force reduction factors for seismic design. *Engineering Structures*, 22(10), 1244–60.

Bousias, S.N., Panagiotakos, T.B. and Fardis, M. (2002). Modelling of R/C members under cyclic biaxial flexure and axial force, *Journal of Earthquake Engineering*, 6(3), 213–38.

Bracci, J.M., Kunnath, S.K. and Reinborn A.M. (1997). Seismic performance and retrofit evaluation of R/C structures, ASCE *Journal of Structural Engineering*, **123**(1), 3–10.

BSSC (2003). NEHRP. *Recommended Provisions for Seismic Regulations for New Buildings and Other Structures*. Building Seismic Safety Council (FEMA Rep. 368, 369). Washington, DC.

Bush, T.D., Tallon, R. and Jirsa, J.O. (1990). Behaviour of a structure strengthened using reinforced concrete piers. *ACI Structural Journal*, 87(5), 557–63.

Bush, T.D., Wyllie, L.A. and Jirsa, J.O. (1991). Observations on two seismic strengthening schemes for concrete frames. *Earthquake Spectra, EERI Journal*, 7(4), 511–27.

Cairns, J. (2006). *Proposals of Fib TG 4.5 for Bond Anchorage in the New Fib Model Code*. Fib, Lausanne.

CEB (1970) CEB-FIP. *International Recommendations for the Design and Construction of Concrete Structures: 1 Principles and Recommendations*. Bulletin No. 72. Comité Euro-international du Beton (CEB), Paris.

CEB (1985). *Model Code for Seismic Design of Concrete Structures*. Bulletin d'Information CEB, **165**, Lausanne.

CEB (1991a). *CEB-FIP Model Code 1990, Bulletin d'Information*, Nos. 203/204/205, Comité Euro-International du Béton, Lausanne.

CEB (1991b). *Behaviour and Analysis of Reinforced Concrete Structures under Alternate Actions Including Inelastic Response*, Vol. 1. CEB, Bul. d'Inf. 210.

CEB (1993). CEB-FIP Model Code 1990, *Bulletin d'Information*, CEB, **213/214**, Lausanne.

CEN Techn. Comm. 250/SC8 (1993). *Eurocode 8: Earthquake Resistant Design of Structures, Part 1 4 (Draft): Repair and Strengthening*, CEN, Berlin.

CEN Technical Committee 250/SC2 (1991). *Eurocode 2: Design of Concrete Structures – Part 1: General Rule and Rules for Buildings (ENV 1992-1-1)* CEN, Berlin.

CEN Technical Comm. 250/SC8 (1994). *Eurocode 8: Earthquake Resistant Design of Structures – Part 1: General Rules and Rules for Buildings* (ENV 1198-1-1), CEN, Berlin.

Chopra, A.K. and Goel R.K. (2002). A modal pushover analysis procedure for estimating seismic demands for buildings. *Earthquake Engineering and Structural Dynamics*, **31**, 561–82.

Christopoulos, C. and Filiatrault, A. (2006). *Principles of Passive Supplemental Damping and Seismic Isolation*, IUSS, Pavia.

Chopra, A.K. (1995). *Dynamics of Structures*. Prentice-Hall Inc, USA.

Chopra, A.K. (2001). *Dynamics of Structures: Theory and Application in Earthquake Engineering*, 2nd Edition. Prentice-Hall, New Jersey.

Chopra, A.K. and Chintanapakdee, C. (2004). Evaluation of Modal and FEMA pushover analyses. Vertically 'regular' and irregular generic frames. *Earthquake Spectra*, 20(1), 255–271.

Chronopoulos, M. (1984). Damage and rehabilitation cost, *Proceedings of the Conference on Earthquakes and Structures*, EPPO, Athens, I, pp. 459–469 (in Greek).

Chung, H.W. and Lui, L.M. (1978). Epoxy-repaired concrete joints under dynamic loads. *Journal of ACI*, 75(7), 313–316.

Clough, R.W. (1970). *Earthquake Response of Structures, Earthquake Engineering*. Prentice-Hall, New Jersey.

Clough, R.W. and Johnston, S.B. (1966). Effects of stiffness degradation on earthquake ductility requirements. *Proceedings of Japan Earthquake Engineering Symposium*, Tokyo, pp. 227–232.

Clough, R.W. and Penzien, J. (1975). *Dynamics of Structures*. McGraw-Hill, New York.

Clough, R.W. and Penzien, J. (1993). *Dynamics of Structures*. McGraw-Hill, New York.

Clyde, C. and Pantelides, C.P. (2002). Seismic evaluation and rehabilitation of R/C exterior building joints. *Proceedings of the Seventh U.S. National Conference on Earthquake Engineering*, Boston, USA.

Collier, C.J. and Elnashai A.S. (2001). A procedure for combining horizontal and vertical seismic action effects. *Journal of Earthquake Engineering*, 5(4), 521–539.

Concrete Society (2000). *Design Guidance for Strengthening Concrete Structures Using FRP Materials*. Concrete Society Technical Report No. 55, Crownthorne, Berkshire, UK.

Constantinea, A.M. and Zisiadis, N.A. (1984). A proposal for inspection procedure after an earthquake. *Proceedings of the Conference on Earthquakes and Structures*, EPPO, Athens. I, 574–594 (in Greek).

Constantinou, M., Soong, T. and Dargush, G. (1998). *Passive Energy Dissipation Systems for Structural Design and Retrofit*. Monograph, MCEER, Buffalo.

Corazao, M. and Durrani, A.J. (1989). *Repair and Strengthening of Beam-to-Column Connections Subjected to Earthquake Loading*. Technical Report NCEER-89-0013, SUNY/Buffalo, USA.

Cosenza, E., Munfredi, G. and Realfonzo, R. (2000). Torsional effects and regularity conditions in R/C buildings, *Proceedings of the 12th World Conference on Earthquake Engineering*, Auckland.

Darwin, D., McCabe, S.L., Browning, J.-A., Mutumoros, A. and Zuo, J. (2002b). Evaluation of development length design expressions. In Balázs, G.L. et al. (eds.) *3rd International symposium: Bond in Concrete – From Research to Standards*, Budapest, 747–754.

Darwin, D., Zuo, J., McCabe S.L. (2002a). Descriptive equations for development and splice strength of straight reinforcing bars. In Balázs, G.L. et al. (eds.) *3rd International Symposium: Bond in Concrete – From Research to Standards*, Budapest, 501–508.

Der Kiureghian, A. (1981). A response spectrum method for random vibration analysis of MDOF Systems, *Earthquake Engineering and Structural Dynamics*, 9, 419–435.

DIN 4014 (1990). Bohrpfähle, Herstellung, Bemessung, Tragverhalten, Beton Kalender, Band II, 1994, Ernst & Sohn, Berlin.

Dowrick, D. (2005). *Earthquake Risk Reduction*. John Wiley & Sons, Ltd, Chichester, West Sussex, England.

Drakopoulos, J. and Makropoulos, K. (1983). *Seismicity and Hazard Analysis Studies in the Area of Greece*. National University of Athens, Athens, Greece.

Dritsos, S. (2004). *Repair and Strengthening of R/C Structures* (in Greek). University of Patras, Patras, Greece.

Dwairi, H., Kowalsky, M.J. and Nau, J.M. (2007). Equivalent viscous damping in support of direct displacement-based design. *Journal of Earthquake Engineering*, 11(4), 512–530.

E.A.K. (Greek Seismic Code) (2000). *Seismic Risk Management and Protection Organisation (ΟΑΣΠ)*. Athens, Greece.

E.C.8-1/EN1998-1 (2004). *Design of Structures for Earthquake Resistance: General Rules, Seismic Actions and Rules for Buildings*. CEN, Brussels, Belgium.

EC2-1-1/EN1992-1-1 (2004). *Design of Concrete Structures - Part 1-1: General Rules and Rules for Buildings*. CEN, Brussels, Belgium.

EC7-1/2004. *Geotechnical Design. Part 1: General Rules*. The European Committee for Standardization, CEN, Brussels.

EC8-3/2005. *Design of Structures for Earthquake Resistance. Part 3: Assessment and Retrofitting of Buildings*. European Committee for Standardization, CEN, Brussels.

EC8-5/2004: *Design of Structures for Earthquake Resistance, Part 5: Foundations, Retaining Structures and Geotechnical Aspects*. European Committee of Standardization, CEN, Brussels.

Economou, C., Karayiannis, C. and Sideris, K. (1994). Epoxy repaired beams under cyclic loading. *Proceedings of the 11th Hellenic Conf. on Concrete*, Techn. Chamber of Greece. Corfu, 3, pp. 26–37 (in Greek).

ECtools (2013). User manual, 3Pi Software ltd. Thessaloniki, Greece, http://www.ectools.eu.

EERI Committee on Seismic Risk (1984). Glossary of terms for probabilistic seismic risk and hazard analysis. *Earthquake Spectra*, **1**, 33–40.

Eibl, J. and Keintzel, E. (1989). Seismic shear forces in R/C cantilever shear walls. *Proceedings of the 9th World Conference on Earthquake Engineering*, Tokyo–Kyoto, Aug. 1988, Maruzen, **VI**, 5–10.

Eligehausen, R., Lettow, S. (2007). *Formulation of Application Rules for Lap Splices in the New Fib Model Code*. University of California, Berkeley. Stuttgart.

Eligehausen, R., Popov, E.P. and Bertero, V.V. (1983). Local bond stress–slip relationships of deformed bars under generalized excitations. *Report EERC-83/23*, University of California, Berkeley.

Elnashai, A.S. (1998). *Use of Real Earthquake Time–History in Analytical Seismic Assessment of Structures*. HM Nuclear Installations Directorate (NH) Report.

Elnashai, A.S. (2000). Advanced inelastic static analysis for seismic design and assessment. G. Penelis International Symposium on Concrete and Masonry Structures, Aristotle University of Thessaloniki, Greece.

Elnashai, A.S. (2002). Do we really need inelastic dynamic analysis? *Journal of Earthquake Engineering*, **6**(1), 123–130.

Elnashai, A.S. and Di Sarno, L. (2008). *Fundamentals of Earthquake Engineering*. Wiley, West Sussex, U.K.

Elnashai, A.S. and Mwafy, A.M. (2000). Static pushover versus dynamic to collapse analysis of RC buildings. ESEE Research Report, No 00-1, Imperial College, January.

Elnashai, A.S., Papanikolaou, V.K. and Lee, D.H. (2002–2005). *Zeus NL, A Program for Inelastic Dynamic Analysis of Structures*. Mid-America Earthquake Center, University of Illinois at Urbana-Champaign, 2005.

Elnashai, A.S. and Papazoglou, A.J. (1997). Procedure and spectra for analysis of R.C. structures subjected to strong vertical earthquake loads, *Journal of Earthquake Engineering*, **1**(1), 121–155.

EN1990 (2002). *Eurocode – Basis of Structural Design*. CEN, Brussels, Belgium.

Endo, T., Okifugi, A., Sugano S., Hayashi, T., Shimizu, T., Takahara, K., Saito, H. and Yoneyama, Y. (1984). Practices of seismic retrofit of existing concrete structures in Japan. *Proceedings of the 8th World Conference on Earthquake Engineering*, San Francisco, U.S.A., **I**, pp. 469–476.

Engindeniz, M., Kahn, L. and Zureick, A. (2004). Repair and strengthening of non-seismically designed R/C beam-column joints. Georgia Institute of Technology: State-of-the-art Report 04-4.

Erdik, M. and Aydinoğlu, N. (2000). *Earthquake Vulnerability of Buildings in Turkey*. Internal Report, Boğaziçi University, Istanbul.

FAGUS/Cubus (2011). *R/C Cross Sections Definition and Analyses*. Zurich.

Fajfar, P. (1996). *Towards a New Seismic Design Methodology for Buildings*. University of Ljubljana, Faculty of Civil Engineering and Geodesy Ljubljana, Slovenia.

Fajfar, P. and Dolsek, M. (2000). A transparent nonlinear method for seismic performance evaluation, *3rd Workshop of the Japan-UK Seismic Risk Forum, Proceedings*. Imperial College Press.

Fajfar, P. and Dolšek, M. (2000). A transparent nonlinear method for seismic performance evaluation. *Collective edition entitled 'Implications of Recent Earthquakes on Seismic Risk*. A.S. Elnashai and S. Antoniou (eds.), Imperial College Press, London.

Fajfar, P., Marusic, D. and Perus, I. The extension of the N2 method to asymmetric buildings. *Proceedings of the 4th European Workshop on the Seismic Behaviour of Irregular and Complex Structures.* CD ROM, Thessaloniki, 2005.

Fardis, M., Carvalho, E., Elnashai, A., Faccioli, E., Pinto, P. and Plumier, A. (2005). *Designer's Guide to EN 1998-1 and EN 1998-5.* Thomas Telford Ltd, London.

Fardis, M.N. (2009). *Seismic Design, Assessment and Retrofitting of Concrete Buildings.* Springer, Heidelberg.

FEMA 172/NEHRP. (1992). *Handbook of Techniques for Seismic Rehabilitation of Existing Buildings.* Federal Emergency Management Agency, Washington, D.C.

FEMA (Federal Emergency Management Agency) (1986). *NEHRP: Recommended Provisions for the Development of Seismic Regulations for New Building.* Part III (App: Existing Buildings), Washington, D.C.

FEMA 154 (1988). *Rapid Visual Screening of Buildings for Potential Seismic Hazards: A Handbook.* Federal Emergency Management Agency, Washington, D.C.

FEMA 179/NEHRP (1992). *Handbook for the Seismic Evaluation of Existing Buildings.* Federal Emergency Management Agency, Washington, D.C.

FEMA 273, 274 (1997). *NEHPR Guidelines for the Seismic Rehabilitation of Buildings.* FEMA, Washington D.C.

FEMA 279 (1997). NEHPR. Guidelines for the seismic rehabilitation of buildings prepared by the Building Seismic Safety Council for the Federal Emergency Management Agency, Washington, D.C.

FEMA 274 (1997). NEHRP. Commentary on the guidelines for seismic rehabilitation of buildings prepared by the Building Seismic Safety Council for the Federal Emergency Management Agency, Washington, D.C.

FEMA 310 (1998). ASCE 31-02. *Handbook for the seismic evaluation of buildings.* A standard prepared by the ASCE for the Federal Emergency Management Agency, Washington, D.C.

FEMA 356 (2000). *Prestandard and Commentary for the Seismic Rehabilitation of Buildings.* Washington, D.C.

FEMA 356 (2000). Prestandard and Commentary for the seismic rehabilitation of buildings, prepared by the SAC Joint Venture for the Federal Emergency Management Agency, Washington, D.C.

FEMA 356 (2000). ASCE SEI 41-06. Standard and commentary for the seismic rehabilitation of buildings, Federal Emergency Management Agency, Washington, D.C.

FEMA 440 (2005). *Static Seismic Analysis Procedures.* Report prepared by the A.T.C. for the Federal Emergency Management Agency, Washington, D.C.

fib (2003). *Seismic Assessment and Retrofit of Reinforced Concrete Buildings, Bulletin 24: State-of-the-Art Report.* International Federation for Structural Concrete (fib), Lausanne, Switzerland.

fib (2012). *Model Code 2010, vol 1.* International Federation for Structural Concrete (fib), Lausanne, Switzerland.

Filippou, F.C., Popov, E.P. and Bertero, V.V. (1983). Modeling of R/C joints under cyclic excitations. *Journal of Structural Engineering,* ASCE, **109**(11), 2666–2684.

Fintel, M. (1974a). Multistorey structures. Chapter 10 of *Handbook of Concrete Engineering.* Van Nostrand Reinhold Co., New York.

Fintel, M. (1974b). Ductile shear walls in earthquake-resistant multistory buildings. *Journal of the ACI,* **71**(6), 296–304.

Fintel, M. and Derecho, A.R. (1974). Earthquake-resistant structures. In *Handbook of Concrete Engineering,* Van Nostrand Reinhold Co., New York, pp. 356–432.

Fischinger, M. and Fajfar, P. (1994). Seismic force reduction factors. *Proceedings of the 17th Regional European Seminar on Earthquake Engineering.* Haifa, Israel, September 1993, (Ruttenber A. ed.) Balkema, Rotterdam, pp. 279–296.

Freeman, S.A. (1993). Assessment of structural damage and criteria for repair. *Proceedings of the 3rd US/Japan Workshop on Urban Earthquake Hazard Reduction,* Publication No.93-B, EERI, Oakland, California.

French, C.W., Thorp, G.A., and Tsai, W.J. (1990). Epoxy repair techniques for moderate earthquake damage. *ACI Structural Journal,* **87**(4), 416–424.

"Gasparini, D.A. and Vanmarcke, E.H. (1976). *Simulated Earthquake Motion Compatible with Preselected Response Spectra.* Department of Civil Engineering Research Report R76-4, M I T, Cambridge, MA.

Georganopoulou, Am. (1982). *Structural Dynamics,* P. Ziti, Thessaloniki (in Greek).

GMPW (Greek Ministry of Public Works) (1978). *Recommendations for the Repair of Buildings with Earthquake Damage.* Thessaloniki (in Greek).

Grant, D.N., Blandon, C.A. and Priestley, M.J. (2005). *Modelling Inelastic Response in Direct Displacement-Based Design.* Report 0.3, IUSS Press, Pavia.

Greek National Report (1982). *On Damage Evaluation and Assessment of Earthquake Resistance of Existing Buildings.* Pr. Rep. 13/05 UNIDO/UNDP, Thessaloniki.

Gupta, A.K. (1990). *Response Spectrum Method in Seismic Analysis and Design of Structures.* Blackwell Scientific Publishing, Cambridge, MA.

Gupta, A.K. (1991). *Response Spectrum Method, in Seismic Analysis and Design of Structures.* Blackwell Scientific Publishing, Cambridge, MA.

Gupta, A.K. and Chu, S.L. (1977). Probable simultaneous response by the response spectrum method of analysis. *Nuclear Engineering and Design,* **44**, 93–97.

Gupta, B. and Kunnath, S.K. (2000). Adaptive spectra-based pushover procedure for seismic evaluation of structures. *Earthquake Spectra* **16**(2), 367–391.

Gupta, A.K. and Singh, M.P. (1977). Design of column sections subjected to three components of earthquakes. *Nuclear Engineering and Design,* **41**, 129–133.

Gutenberg, B. and Richter, C.F. (1942), (1956). Earthquake magnitude, intensity, energy and acceleration. *Bulletin of the Seismological Society of America* **32**, 163 and **46**, 105.

Gylltoft, K. (1984). Fracture mechanics model for fatigue in concrete. *Materials and Structures,* RILEM, **17**(97), 55–58.

Hansen, J. Brinch (1961). A general formula for bearing capacity. *Ingeniören,* **5**, 38–46.

Hanson, N.W. (1960). Precast-prestressed concrete bridges. 2. Horizontal shear connections. *JPCA Research and Development Laboratories,* **2**(2), 38–58.

Hassoun, M.N. and Al-Manaseer, A. (2008). *Structural Concrete.* John Wiley & Sons. Inc., Hoboken, New Jersey.

Hoffschild, T.E., Prion, H.G.L. and Gherry, S. (1995). Seismic retrofit of beam-to-column joints with grouted steel tubes. *Thomas Paulay Symposium, Recent Developments in Lateral Force Transfer in Buildings, SP-157.* pp. 397–425. ACI, Detroit, Michigan, USA.

Hollaway, L.C. and Leeming, M.B. (1999B). *Strengthening of Reinforced Concrete Structures Using Externally-Bonded FRP Composites in Structural and Civil Engineering.* Woodhead Publishing, Cambridge, UK.

Holmes, T.W. (1994). Policies and standards for reoccupancy repair of earthquake-damaged buildings. *Earthquake Spectra,* **10**(1), 197–208.

Housner, G.W., Martel, R.R. and Alford, J.L. (1953). Spectrum analysis of strong-motion earthquakes, *Bulletin of the Seismological Society of America,* **43**(2).

I.E.S.E.E. (Institute of Engineering Seismology and Earthquake Engineering, Greece) (2003). *Strong Motion Data Base of Greece 1978–2003.* Thessaloniki, Greece.

IBC (2012). International Building Code, International Code Consortium.

ICONS Project (2012). Center for International Development and Conflict Management, Department of Government & Politics, University of Maryland.

ICSSC PR-4 (1994). Standards of seismic safety for existing federally owned or leased buildings. National Institute of Standards and Technology (NISTIR 5382), Gaithersburg, Maryland.

Ignatakis, C. (2011). Reinforcement detailing for R/C structures according to EC2-1-1/2004 and EC8-1/2004 (in Greek). *Seminar of Hellenic Steel Industry on European Codes for R/C Structures.* Herakleion, Crete.

Ignatakis, C., Stavrakakis, E. and Penelis, G. (1989). Parametric analysis of R/C columns under axial und shear loading using the finite element method. *ACI Structural Journal,* **86**(4), 413–18.

Ignatakis, C., Stavrakakis, E. and Penelis, G. (1990). Analytical model for masonry using the finite element method. *Software for Engineering Workstations, International Journal,* **6**(2), 90–96.

Jacobsen, L.S. (1960). Damping in composite structures. *Proceedings of the 2nd World Conference on Earthquake Engineering*, Tokyo and Kyoto, 1029–1044.

JBDPA (1999). *Seismic Retrofit Design and Construction Guidelines for Existing Reinforced Concrete and Steel Encased R/C Buildings Using Continuous Fibre Reinforcing Materials*. Japan Buildings Disaster Prevention Association, Tokyo.

Jirsa, J.O. (1974). Factors influencing behaviour of reinforced concrete members under cyclic overloads. *Proceedings of 5th World Conf. on Earthq. Eng.* June 1973, Rome, Italy, 2, pp. 1198–1204.

Jirsa, J.O. (1994). Divergent issues in rehabilitation of existing buildings. *Earthquake Spectra*, 10(1), 95–112.

Jirsa, J.O., Maruyama, K. and Ramirez, H. (1980). The influence of load history on the shear behaviour of short RG columns, *Proceedings of the 7th World Conference on Earthquake Engineering*, Istanbul, 6, 339–146.

Kaku, T. and Asakusa, H. (1991). *Bond and Anchorage of Bars in Reinforced Concrete Beam-Column Joints*. ACI Special Publication SP123, American Concrete Institute, Detroit, Michigan, 401–124.

Kanaan, A.E. and Powell, G.H. (1973). DRAIN-2D: *A General Purpose Computer Program for Dynamic Analysis of Inelastic Plane Structures*, Rep. EERC-73/6 and EERC-73/22, Univ. of California, Berkeley.

Kappos, A. and Penelis, G.G. (1987). Investigation of the inelastic behaviour of existing R/C buildings in Greece. *Technica Chronica, TCG*, 7(3), 53–86.

Kappos, A.J. (1990). Sensitivity of calculated inelastic seismic response to input motion characteristics. *Proceedings of 4th U.S National Conference on Earthquake Engineering, Palm Springs, Calif.*, 2, pp. 25–34.

Kappos, A.J. (1991a). Analytical prediction of the collapse earthquake for R/C buildings: Suggested methodology. *Earthquake Engineering and Structural Dynamics*, 20(2), 167–176.

Kappos, A.J. (1991b). Analytical prediction of the collapse earthquake for R/C buildings: Case studies. *Earthquake Engineering and Structural Dynamics*, 20(2), 177–190.

Kappos, A.J. and Penelis, G.G. (1986). Discussion of influence of concrete and steel properties on calculated inelastic seismic response of reinforced concrete frames. *ACI Journal*, 83(1), 167–169.

Kappos, A.J. and Penelis, G.G. (1989). Evaluation of the inelastic seismic behaviour of R/C buildings designed by CEB model code. *Proceedings of the 9th World Conference on Earthquake Engineering*, Tokyo–Kyoto, 1988, Maruzen, V, 1137–1142.

Kappos, A.J., Stylianidis, K.C. and Penelis, G.G. (1991). Analytical prediction of the response of structures to future earthquakes. *Proceedings of European Earthquake Engineering*, 5(1), 10–21.

Kappos, A., Panagopoulos, G., Penelis, Gr. (2008). Development of a seismic damage and loss scenario for contemporary and historical buildings in Thessaloniki. *Soil Dynamics and Earthquake Engineering*, 28, pp. 836–850, Elsevier, Germany.

Kappos, A.J. and Kyriakakis, P. (2000). A re-evaluation of scaling techniques for natural records. *Soil Dynamics and Earthquake Engineering*, 20, 111–123.

Karabinis, A. (2002). Validation of the pre-earthquake assessment procedure of R/C buildings in Greece, Technical Report (O.A.S.P.) Athens, Greece.

Karakostas, Ch., Lekidis, V., Kappos, A., Panagopoulos, G., Kontoes, C. and Keramitsoglou, I. (2012). Evaluation of seismic vulnerability of buildings in Athens and L'Aquila in the framework of the MASSIVE seismic mitigation system. *Proceedings of 15th World Conference of Earthquake Engineering (WCEE)*, Lisbon.

Karayannis, C.G., Chalioris, C.E. and Sideris, K.K. (1998). Effectiveness of R/C beam-column connection repair using epoxy resin injections. *Journal of Earthquake Engineering*, 2(2), 217–240.

Karsan, I.D. and Jirsa, J.O. (1969). Behaviour of concrete under compressive loading. *Journal of the Structural. Division*, ASCE 95(ST 12), 2543–2563.

Kent, D.C. and Park, R. (1971). Flexural members with confined concrete. *Journal of the American Concrete Institute, Proceedings*, 67(3), 243–248.

Khalifa, A., Gold, W., Nanni, A., Abel-Aziz, M. (1998). Contribution of externally bonded FRP to the shear capacity of R/C flexural members. *Journal of Composites in Construction*, **2**(4), 195–203.

Kitayama, K., Otani, S. and Aoyama, H. (1989). Behaviour of reinforced concrete beam-column-slab subassemblies subjected to bi-directional load reversals. *Proceedings of the 9th World Conference on Earthquakes*, Tokyo–Kyoto, 1988, Maruzen, **VIII**, 581–586.

Kitayama, K., Otani, S., Aoyama, H. (1991). *Development of Design Criteria for R/C Interior Beam–Column Joints*. ACI Special Publication SP123, American Concrete Institute, Detroit, Michigan, 97–124.

König, G. and Liphardt, S. (1990). *Hochhäuser aus Stahlbeton*, Beton Kalender, Teil II, 457–539. Ernst & Shon Verlag, Berlin.

Kostikas, Ghr. et al. (2000). *Evaluation of Damages and Their Cause for 103 Near to Collapse R/C Buildings in Athens after the Earthquake of September 7, 1999*. Report of the O.A.S.P. (Seismic Risk Management Agency of Greece).

Kowalsky, M.J. and Priestley, M.J.N. (2000). Improved analytical model for shear strength of circular reinforced concrete columns in seismic regions. *ACI Structural Journal*, **97**(3), 388–396.

Kramer, S.L. (1996). *Geotechnical Earthquake Engineering*. Prentice Hall, Upper Saddle River, New Jersey.

Kramer, S.L. and Elgamal, A.W. (2001). Modelling soil liquefaction hazards for performance-based earthquake engineering. *Pacific Center for Earthquake Engineering*. Report PEER 2001/13, Berkeley, California.

Krawinkler, H. and Nassar, A.A. (1992). Seismic design based on ductility and cumulative damage demand and capacities. In Fajfar, P. and Krawinkler, H. (eds.), *Nonlinear Seismic Analysis and Design of R.C. Buildings*. Elsevier Applied Science, New York, NY.

Ladner, M. and Weber, C. (1981). *Geklebte Bewehrung im Stahletondau*. EMPA, Dübendorf.

Lam, L. and Teng, J.G. (2003a). Design oriented stress-strain model for FRP confined concrete. *Construction and Building Materials*, **17**(6–7), 471–489.

Lam, L. and Teng, J.G. (2003b). Design oriented stress-strain model for FRP-confined concrete in rectangular columns. *Journal of Reinforced Plastic Composites*, **22**(13), 1149–1186.

Lefas, I., Tsoukis, D. and Kotsovos, M. (1990). Behaviour of repaired R/C walls to monotonic and cyclic loading. *Proceedings of the 9th Greek Conf. on Concrete*, TCG, Kalamata, **2**. pp. 231–238 (in Greek).

Leonhardt, F. (1973). *Vorlesungen über Massivbau, 1. teil*. Springer Verlag, Berlin.

Leonhardt, F. (1977). *Vorlesungen uber massivbau, Vierter teil*. Springer-Verlag, Berlin.

Leonhardt, F. and Mönnig, E. (1973). *Vorlesungen über Massivbau, Erster Teil*. Springer-Verlag, Berlin.

Leonhardt, F. and Walther, R. (1962). *Schubversuche an einfeldrigen Stahlbetonbalken mit und ohne Schubbewehrung*. DA für Stahlbeton Heft 151, W. Ernst und Sohn, Berlin.

Lolas, I.D. and Kotsovos, M.D. (1990). Strength and deformation characteristics of reinforced concrete walls under load reversals. *ACI Structural Journal*, **87**(6), 716–726.

Luft, R. (1989). Comparisons among earthquake codes. *Earthquake Spectra*, **5**(4), 767–789.

Mahaney, J.A., Paret, T.F., Kehoe, B.E. and Freeman, S.A. (1993). The capacity spectrum method for evaluating structural response during the Loma Prieta Earthquake. *Proceedings of the National Earthquake Conference*, Memphis, Tennessee.

Makarios, T. and Anastasiadis, K. (1997). The real and the 'plasmatic' axis of centres of stiffness (in Greek). *Journal of Engineering*, **17**(3), 97–120. Technical Chamber of Greece, Athens, Greece.

Mander, J.B., Priestley, M.J.N. and Park, R. (1988). Theoretical stress-strain model for confined concrete. *ASCE Journal of Structural Engineering*, **114**(8), 1804–1826.

Manos, G., Kourtidis, B., Katakalos, K., Kotoulas, L., Nalmpantidou, A., Koidis, G. and Kiprioti, A. (2016). Experimental investigation of the behaviour of structural components of the SNFCC canopy in Athens (in Greek). *17th National Conference of Concrete Structures*, Thessaloniki.

Martinez-Rueda, J.E. (1997). Energy Dissipation Devices for Seismic Upgrading of R.C. Structures. Ph.D. thesis, University of London.

Meier, U. and Winistorfer, A. (1995). Retrofitting of structures through external bonding of CFRP sheets. Non-metallic (FRP) reinforcement for concrete structures. *Proceedings of the 2nd International RILEM Symposium*, Ghent, Belgium, ed. L. Tuerwe, pp. 509–516, E&FN Spon London, UK.

Meyerhof, G.G. (1965). Shallow foundations. *ASCE Journal of Soil Mechanics*, **91**(SM2), 21–31.

Michailidis, C.N., Stylianidis, K.C. and Kappos, A.J. (1995). Analytical modeling of masonry infilled R/C frames subjected to seismic loading. *Proceedings of the 10th European Conference on Earthquake Engineering*, Vienna, 1994, Balkama, **3**, 1519–1524.

Milutinovic, Z. and Trendafiloski, G. (2003). RISK-UE, WP4 – *Vulnerability of Current Buildings*. Contract: EVK4-CT-2000-00014, European Commission, Brussels.

Minami, K. and Wakabayashi, M. (1980). Seismic resistance of diagonally reinforced concrete columns, *Proceedings of the 7th World Conference on Earthquake Engineering*, Istanbul.

Ministry for Environment and Public Works (1992). *Greek Code for Earthquake Resistant Structures*, Athens.

Moehle, J.P. (1987). Displacement-based design of R/C structures subjected to earthquakes. *Earthquake Spectra*, **4**, 403–428.

Moore, G.E. (1965). Cramming more components onto integrated circuits (PDF). *Electronics Magazine*, **4**.

Mutsumura, K. (1992). On the intensity measure of strong-motions related to structural failures. *Proceedings of the 10th World Conference on Earthquake Engineering*, **1**, 375–380.

Naeim, F. and Kelly, J. (1999). *Design of Seismic Isolated Structures: From Theory to Practice*. John Willey & Sons, New York

National Annex of Greece. (2010). *National choice in EN 1998-1:2004 Athens*.

Nau, J.M. and Hall W.J. (1984). Scaling methods for earthquake response spectra. *Journal of Earthquake Engineering ASCE*, **110**(7), 1533–1548.

Newman, K. and Newman, J.B. (1971). Failure theories and design criteria for plain concrete. *Solid Mechanics and Engineering Design*. John Wiley Interscience, New York.

Newmark, N.M. and Hall, W.J. (1982). Earthquake spectra and design, *EERI Monograph Series*, EERI, Oakland, California.

NIST GCR 10-917-9/2010. *Applicability of MDOF Modeling for Design*. Prepared by the National Institute of Standards and Technology, Gaithersburg, Maryland.

Nitsiotas, G. (1960). *Technical Theory of Elasticity* (in Greek). Konstandopoulos Pb., Thessaloniki, Greece.

NOUS/3P (2002). *RC Section Designer*. 3P Penelis Software Ltd., Thessaloniki, Greece.

NTU (National Technical University) (1978). *Recommendations on the Repair of Earthquake Damaged Buildings*. Athens (in Greek).

NZS 3101 (1995). *The Design of Concrete Structures*. New Zealand Standards Parts 1 and 2. Wellington.

OASP (Seismic Risk Management and Protection Agency) (2000). *Causes of Damage to the 103 Most Affected Buildings (Near Collapse) in Athens after the Earthquake of Parnitha (7-9-1999)*. Report (Coordinator Kostikas C.), OASP, Athens.

OASP (2001). Recommendations for pre and post earthquake retrofitting of buildings (in Greek). Seismic Risk Management Agency of Greece, Athens, Greece.

Office of Emergency Preparedness (1972). *Disaster Preparedness*. Report to the Congress, Vols. 1–3, Washington, D.C.

Oesterle, R.G., Fiorato, A.E., Aristizabal-Ochoa, J.D. and Corley, W.G. (1980). Hysteretic response of reinforced concrete structural walls. In ACI SP-63: Reinforced Concrete Structures Subjected to Wind and Earthquake Forces, American Concrete Institute, Detroit, pp. 243–273.

Ogawa, J. and Abe, Y. (1980). Structural damage and stiffness degradation of buildings caused by severe earthquakes. *Proceedings of the 7th World Conf. on Earthquake Engineering*, Istanbul, **VII**, pp. 527–534.

Otani, S. and Sozen, M.A. (1972). *Behaviour of Multistorey Reinforced Concrete Frames during Earthquakes*. Civil Engineering Studies, Structural Research Series No. 392, Univ. of Illinois, Urbana.

Panetsos, P. and Anastasiadis, K. (1994). Design of R/C elements under seismic action. *Proceedings of the 11th Greek Conf. on Concrete*, Technical Chamber of Greece, Corfu, 2, pp. 267–81 (in Greek).

Pantazopoulou, S. (1995). Role of expansion on mechanical behaviour of concrete. *ASCE Journal of Structural Engineering*, **121**(12), 1795–1805.

Papaioannou, C.A., Kiratzi, A.A., Papazachos, B.C. and Theoduiidis, N.P. (1994). Scaling of normal faulting earthquake response spectra in Greece. *Proceedings of the 7th Congress of the Hellenic Geological Society*, Thessaloniki, May 1994.

Papapetrou, A. (1934). Dynamics in earthquake engineering (in Greek). *Journal of Engineering. Technical Chamber of Greece*, **49**, 21–26.

Papanikolaou V.K. (2012) Analysis of arbitrary composite sections in biaxial bending and axial load. *Computers and Structures*, **98–99**, 33–54.

Papazachos, B.C. (1986). Active tectonics in the Aegean and surrounding area. *Proceed. Summer School on Seismic Hazard in the Mediterranean Region*, Strasbourg, France, 21–30 July. Kluwer Academic Publishers, pp. 301–331.

Papazoglou, A.J. and Elnashai, A.S. (1996). Analytical and field evidence of the damaging effect of vertical earthquake ground motion, *Earthquake Engineering and Structural Dynamics*, **25**(10), 1109–1137.

Park, R. (1972). Theorization of structural behavior with a view to defining resistance and ultimate deformability. *Symposium on Resistance and Ultimate Deformability of Structures Acted on by Well-Defined Repeated Loads*. IABSE, Lisbon, 1973.

Park, R. (1986). Ductile design approach for reinforced concrete frames. *Earthquake Spectra, Earthquake Engineering. Research Institute*, **2**(3), 565–619.

Park, R. and Paulay, T. (1975). *Reinforced Concrete Structures*. John Wiley & Sons, New York.

Park, R., Priestley, M.J.N. and Gill, W.D. (1982). Ductility of square confined concrete columns, *Journal of the Structural. Division, ASCE*, **108**(4), 929–950.

Park, Y.-J. and Ang, A.H.-S. (1985). Mechanistic seismic damage model for reinforced concrete. *Journal of Engineering, ASCE*, **111**(4), 722–739.

Paulay, T. (1986a). A critique of the special provisions for seismic design of the Building Code Requirements for Reinforced Concrete (ACI 318-83). *Journal of the ACI*, **83**(2), 274–283.

Paulay, T. (1986b). The design of ductile reinforced concrete structural walls for earthquake resistance. *Earthquake Spectra, Earthquake Engineering. Research Institute*, **2**(4), 783–823.

Paulay, T. (1996). Seismic design for torsional response of ductile buildings. *Bulletin of the New Zealand National Society for Earthquake Engineering*, **29**(3), 21–26.

Paulay, T. (1997). Seismic torsional effects on ductile structural wall systems. *Journal of Earthquake Engineering*, **1**(4), 721–745.

Paulay, T., Bachmann, H. and Moser, K. (1990). *Erdbebenbemessung von Stahlbeton Hochbauten*. Birkhäuser, Berlin.

Paulay, T. and Bull, I.N. (1979). Shear effects on plastic hinges of earthquake resisting reinforced concrete frames. *CEB Bulletin d'Information*, Paris, **132**, 165–172.

Paulay, T., Park, R. and Priestley, M.J.N. (1978). Reinforced concrete beam-column joints under seismic actions. *Journal of the A.C.I.*, **75**(11), 585–593, also Discussion by D.F. Meinheit and Authors' closure, *Journal of the ACI*, **76**(5), May 1979, 662–7.

Paulay, T. and Priestley, M.J.N. (1992). *Seismic Design of Reinforced Concrete and Masonry Buildings*, John Wiley & Sons, New York.

Penelis, G.G. (1969a). Design of R/C cross sections under biaxial bending combined with axial force using inelastic stress-stain laws for concrete and steel. Habilitation Thesis (in Greek), Thessaloniki, Greece.

Penelis, G.G. (1969b). Eine Verbesserung der R. Rosman – H. Beck Methode, Der Bauingenieur H. 12.

Penelis, G.G. (1971). Die Knickung Raümlicher Mehrstockiger Rahmenträger. Der Bauingenieur, 11, 393–9.

Penelis, G.G. (1979). *Rehabilitation of Buildings Damaged by Earthquakes*. Ministry of Public Works, Thessaloniki (in Greek).

Penelis, G.G. (1984). Problems after earthquake inspections, expert appraisals. *Proceedings of the EPPO Conference on Earthquakes and Structures*, OASP, Athens, **I**, 521–537 (in Greek).

Penelis, G.G. (1996). An overview of seismic risk management, *Proceedings of the 1st Japan–U.K Seismic Risk Forum*. London.

Penelis, G.G. (2001). Pre-earthquake assessment of public buildings in Greece. *Greek-Turkish Conference on Seismic Assessment*. Istanbul, Technical Chambers of Greece and Turkey.

Penelis, G.G. (2008). The earthquake of Thessaloniki, in 1978: Turning point for seismic risk management in Greece (in Greek). *Special Publication of the Engineering Faculty of Aristotle University of Thessaloniki on the Occasion of the 30th Anniversary of the Earthquake of Thessaloniki in 1978.*

Penelis, G.G. and Kappos, A.J. (1997). *Earthquake-Resistant Concrete Structures*. SPON E&FN (Chapman and Hall), London.

Penelis, G.G. and Penelis, Gr. (2001). *Structural Design for Rehabilitation of the Buildings of the Army Pension Organization, Athens*. Technical report attached to the design permit, Athens, Greece.

Penelis, G. and Penelis, Gr. (2014). *Concrete Buildings in Seismic Regions*. CRC Press (A SPON BOOK), London.

Penelis, G.G., Karaveziroglu, M., Stylianidis, K. and Leontaridis, D. (1984). Case study: The Rotunda, Thessaloniki. In *Repair and Strengthening of Historical Monuments and Buildings in Urban Nuclei*, 6, UNIDO/UNDP, Vienna, pp.165–188.

Penelis, G.G. et al. (1986). *Technical Report on the September 13 and 15, 1986 Kalamata Earthquake*, Lab of R/C, Dept. of Structural Engineering, University of Thessaloniki, Thessaloniki (in Greek).

Penelis, G.G. et al. (1987). *A Statistical Evaluation of Damage Caused by the 1978 Earthquake to the Buildings of Thessaloniki*, Technical Report of the Laboratory of R/C, Aristotle University of Thessaloniki (in Greek).

Penelis, G.G., Sarigiannis, D., Stavrakakis, E. and Stylianidis, K. (1988a). A statistical evaluation of damage to buildings in the Thessaloniki, Greece, earthquake of June 20, 1978. *Proceedings of the 9th World Conference on Earthquake Engineering*. Tokyo–Kyoto, **VII**, pp. 187–192.

Penelis, G.G., Papagiannidou, C., Penelis Gr. (2000). Response of buildings designed according to the old and new Greek Seismic Code to 7.9.99 Athens earthquake. *Proceedings of the Japan–U.K. Seismic Risk Forum, 3rd Workshop, Imperial College*, London, Imperial College Press.

Penelis, G.G., Stylianidis, K., Kappos, A. (1988b). *Analytical Prediction of the Response of the Building Stock of Thessaloniki to Future Earthquakes*, invited lecture at the 1st Conference on Engineering Seismology in Greece, Department of Geophysics, Aristotle University of Thessaloniki, Greece.

Penelis, Gr.G., Stylianidis, K., Kappos, A. and Ignatakis, Ch. (1995). Reinforced concrete structures (university notes), Aristotle University of Thessaloniki, Thessaloniki (in Greek).

Penelis, Gr. and Kappos, A. (2005). Inelastic torsional effects in 3D pushover analysis of buildings. *Proceedings of the 4th European Workshop on the Seismic Behaviour of Irregular and Complex Structures*. CD ROM, Thessaloniki.

Penelis, Gr., Papanikolaou, V. (2010). *Nonlinear Static and Dynamic Behaviour of a 16-Storey Torsionally Sensitive Building Designed According to Eurocodes*. 14(5), pp. 706–725. Taylor & Francis Group, London, UK.

Penelis, S.A. (2011). Structural and seismic design of a nine storey building with one basement in Vasileos Irakliou Str. Nr. 45 (Domotechniki real estate), Thessaloniki, Greece, documents for permit issue. Penelis S.A. archives.

Penzien, J. and Watabe, M. (1974). Simulation of 3-dimensional earthquake ground motion. *Bull. of Int, Inst. of Seism. and Earrthq. Engng*, (Tsukuba, Japan), **12**.

PLAXIS-2D (2012a). User Manual, Plaxis bv P.O. Box 572, 2600 AN DELFT, Netherlands

PLAXIS-2D (2012b). *Finite Element Package for Geotechnical Engineering*. Plaxis bv, Delft.

Polyakov, S. (1974). *Design of Earthquake-Resistant Structures*. Mir Publications, Moscow.

Popov, E.P. (1977). Mechanical characteristics and bond of reinforcing steel under seismic conditions. *Proceedings of Workshop on Earthquake Resistant Reinforced Concrete Buildings Construction*, Univ. of California, Berkeley, II, 658–662.

Popov, E.P. and Bertero, V.V. (1975). Repaired R/C members under cyclic loading. *Earthquake Engineering and Structural Dynamics*, 4(2), 129–144.

Presidential Order 12.941 (1994). *Seismic Safety of Existing Federally Owned or Leased Buildings.* Executive Order of December 1, 1994, The White House, Washington, D.C.

Priestley, M.J., Calvi, G.M., Kowalsky, M.J. (2007). *Displacement-Based Seismic Design of Structures.* IUSS Press, Pavia.

Priestley, M.J.N. and Calvi, E.M. (1991). Towards a capacity-design assessment procedure of reinforced concrete frames. *Earthquake Spectra*, 7(3), 413–438.

Priestley, M.J.N. and Park, R. (1987). Strength and ductility of concrete bridge columns under seismic loading. *ACI Structural Journal*, 84(1), 61–76.

Priestley, M.J.N. and Seible, F. (1991). Design of retrofit measures for concrete bridges. In *Seismic Assessment and Retrofit of Bridges*, Struct. Syst. Res. Rep., No. SSRP 91/03, Univ. of California, San Diego.

Priestley, M.J.N. and Wood, J.H. (1977). Behaviour of a complex prototype box girder bridge. *Proceedings of the RILEM International Symposium on Testing In-Situ of Concrete Structures*, Budapest, vol. I, 140–153.

Rasmussen, B.H. (1963). *Betonindstobte Tvaer Belastede Boltes og Dornes Baereevne.* Bygningstatiske Meddelser. Copenhagen.

RCCOLA 90. *A Computer Program for R/C Column Analysis.* NISEE, Pacific Earthquake Engineering Research (PEEP) Center, University of California, Berkeley, California.

Reid, H.F. (1911). The elastic-rebound theory of earthquakes. *Bulletin of Geology*, 6, 413.

Richart, F.E., Woods, R.D. and Hall, J.R. (1970). *Vibrations of Soils and Foundations.* Prentice-Hall, New Jersey.

Richter, C.F. (1958). *Elementary Seismology.* W.H. Freeman, San Francisco.

Riddell, R. and Newmark, N.M. (1979). Force-deformation models for nonlinear analysis (technical note). *Journal of the Structural. Division, ASCE*, 105(12), 2773–2778.

RISK-UE (2001–2004). *An Advanced Approach to Earthquake Risk Scenarios with Applications to Different European Towns.* Contract: EVK4-CT-2000-00014, European Commission, Brussels.

Rodriguez, M. and Park, R. (1991). Repair and strengthening of reinforced concrete buildings for seismic resistance. *Earthquake Spectra, Journal of EERI*, 7(3), 439–460.

Rosenblueth, E. and Contreras, H. (1977). Approximate design for multicomponent earthquakes. *Journal of Engineering Mechanics, ASCE*, 103, EM5.

Rosenblueth, E. and Meli, R. (1985). The 1985 earthquake: Causes and effects in Mexico City. *Concrete International (ACI)*, 8(5), 23–36.

Rosman, R. (1965). *Zahlentafeln für die Schnittkräfte von Windscheiben mit Öffnungsreihen.* Bauhingenier-Praxis, Heft 65, W. Ernst and Sohn, Berlin.

Roussopoulos, A. (1956). *Earthquake Resistant Structures.* NTU, Athens (in Greek).

Rowe, R.E. (1970). Current European views on structural safety. *Journal of Structural Division, ASCE*, 96(ST3), 461–467.

Saiidi, M. and Sozen, M.A. (1981). Simple nonlinear seismic analysis of R/C structures. *Proceedings of ASCE*, 107(ST5), May, 937–952.

Saiidi, M., Vrontinos, S. and Douglas, B. (1990). Model for the response of reinforced concrete beams strengthened by concrete overlays. *ACI Structural Journal*, 87(6), 687–695.

Sakai, K. and Sheikh, S.A. (1989). What do we know about confinement in reinforced concrete columns? *ACI Structural Journal*, 86(2), 192–207.

Salonikios, T. (2007). Analytical prediction of the inelastic response of R/C walls with low aspect ratio. *ASCE Journal of Structural Engineering*, 133(6), 844–854.

Salonikios, T., Kappos, A.J., Tegos, I.A. and Penelis, G.G. (2000). Cyclic load behaviour of low slenderness R/C walls. Failure modes, strength and deformation analysis and design implications. *ACI Structural. Journal*, 97(1), 132–141.

Salse, E.A.B. and Fintel, M. (1973). Strength, stiffness and ductility properties of slender shear walls. *Proc. 5th World Conf. of Earthquake Eng., Rome*, 1, 919–928.

Santhakumar, A.R. (1974). The ductility of coupled shear walls. Ph.D. thesis, University of Canterbury, Christchurch.

SAP (2000). Three dimensional static and dynamic finite element analysis and design of structures, Computers and Structures Inc. (1999), Berkeley, California, 1999.

SAP V-14 (2000). Integrated software for structural analysis and design. Computers and Structures, Inc.

Sariyiannis, D. (1990). *Seismic Behaviour of R/C Frames Filled with Masonry After Repair*. Reports of the R/C Lab, AUT, Thessaloniki (in Greek).

Sariyiannis, D. and Stylianidis, K. (1990). Experimental investigation of R/C one storey frames repaired with resins under cyclic shear. *Proceedings of the 9th Hellenic Conf. on Concrete. TCG, Kalamata*, 2, pp. 223–230 (in Greek).

SCEEP/HAZUS-MH/FEMA. (2008). *Overview of Hazus-MH: The Earthquake Module*, South Carolina Earthquake Education & Preparedness Program.

Scia Engineer (2013). User Manual, Nemetschek Scia nv, Industrieweg 1007, 3540, Herk-de-Stad, Belgium, 2013.

Scott, B.D., Park, R. and Priestley, M.J.N. (1982). Stress–strain behaviour of concrete confined by overlapping hoops at low and high strain rates. *ACI Structural Journal*, **79**(1), 13–27.

Scribner, C.F. and Wight, J.K. (1980). Strength decay in R/C beams under load reversals. *ASCE Journal of the Structural Division*, **106**(ST4), 861–876.

SEAOC (1978). *Recommended Lateral Force Requirements and Commentary*. Structural Engineers Associations of California.

SEAOC (1995). *Performance Based Seismic Engineering of Buildings: Vision 2000*, Structural Engineering Association of California, Sacramento, California.

SEAOC (1999). *Recommended Lateral Force Requirements and Commentary*. Seismology Committee, Structural Engineers Association of California, Sacramento, California.

SEAOC (2009). *IBC Structural/Seismic Design Manual*. SEAOC.

Seed, H.B. and Idriss, I.M. (1982). *Ground Motions and Soil Liquefaction during Earthquakes*. Earthquake Engineering Research Institute, Oakland, California.

Seismostruct (2016). User Manual, Seismosoft Ltd, Via Boezio 10, 27100 Pavia, Italy, 2016.

Sextos, A., Pitilakis, K. and Kappos, A. (2003). Inelastic dynamic analysis of R/C bridges accounting for spatial variability of ground motion, site effects and soil-structure interaction phenomena. Part 1: Methodology and Analytical tools. *Earthquake Engineering and Structural Dynamics*, V, **32**(4), 607–627.

Sheikh, S.A. and Uzumeri, S.M. (1982). Analytical model for concrete confinement in tied columns. *Journal of the Structural Division*, ASCE, **108**(ST12), 2703–2722.

Shiu, K.N., Takayanagi, T. and Corley, W.G. (1984). Seismic behaviour of coupled wall systems. ASCE *Journal of the Structural Division*, **110**(5), 1051–1066.

Shohara, R. and Kato, B. (1981). Ultimate strength of reinforced concrete members under combined loading. IABSE Colloquium, Advanced mechanics of R/C, Delft, 701–716.

SIA 166 (2004). Schweizerische Vornorm Klebebewehrungen. Schweizerische Normenvereinigung, Zürich, Schweiz.

Sinha, B.P., Gerstle, K.H. and Tulin, L.C. (1964). Response of singly reinforced beam to cyclic loading. *Journal of the American Concrete Institute, Proceedings*, **61**(8), 1021–1038.

Soroushian, P., Obasaki, K. and Marikunte, S. (1991). Analytical modelling of bonded bars under cyclic loads. *Journal of Structural. Engineering ASCE*, **117**(1), 48–60.

Spoelstra, M.R. and Monti, G. (1999). FRP-confined Concrete Model. *Journal of Composites for Construction*, 3(3), 143–150.

Strobach, K.L. and Heck, H. (1980). Von Wegeners Kontinental Verschiebung zur modernen Plattentektonik. *Bild der Wissenschaft*, **11**, 99–109.

Stylianidis, C.K. (1985). Experimental Investigation of the behaviour of single-storey infilled R/C frames under cyclic quasi-static horizontal loading – parametric analysis. Ph.D. thesis (in Greek), Aristotle University of Thessaloniki, Greece.

Stylianidis, K. (2012). Experimental investigation of masonry infilled R/C frames. *The Open Construction and Building Technology Journal*, 6, 194–212.

Stylianidis, K. and Sariyiannis, D. (1992). Design criterion to avoid column shear failure in infilled frames due to seismic action. *Proceedings of the 1st National Conference on Engineering Seismology and Earthquake Engineering.* Athens, Greece, **2**, 220–230 (in Greek).

Sugano, S. (1981). Seismic strengthening of existing concrete buildings in Japan. *Bulletin of the New Zealand Society for Earthquake Engineering*, **14**(4), 209–222.

Takahashi, T., Kobayashi, S., Fukushima, Y., Zhao, J.X., Nakamura, H. and Somerville, P.G. (2000). A spectral attenuation model for Japan using strong motion database, *Proceedings of the 6th International Conference on Seismic Zonation; Managing Earthquake risk in the 21st century*, Earthquake Engineering Research Institute, Oakland, California.

Takeda, T., Sozen, M.A. and Nielsen, N.N. (1970). Reinforced concrete response to simulated earthquakes. *Journal of the Structural. Division*, ASCE, **96**(12), 2557–2573.

Taranath, B.S. (2010). *Reinforced Concrete Design of Tall Buildings.* CRC Press (Taylor & Francis Group), Boca Raton, FL.

Tassios, T.P. (1979). Properties of bond between concrete and steel under loads idealizing seismic actions. *CEB Bulletin d'Information*, **131**, Paris, 67–122.

Tassios, T.P. (1983). Physical and mathematical models for redesign of damaged structures. *Introductory Report, IABSE Symposium*, Venice, pp. 29–77.

Tassios, T.P. (1984a). Repairs after an earthquake. *Proceedings of the Conf. on Earthquake and Struct.* EPPO, Athens, 1, pp. 595–636 (in Greek).

Tassios, T.P. (1984b). Masonry infill and R/C walls under cyclic actions. *CIB Symposium on Wall Structures*, Warsaw, 1984.

Tassios, T.P. (1984c). Post-earthquake interventions. *Proceedings of the Conference on Earthquakes and Structures*, EPPO, Athens, I, 595–636 (in Greek).

Tassios, T.P. (1989). Specific rules for concrete structures-justification note no. 6: Required confinement for columns. In *Background Document for Eurocode 8-Part 1, Vol. 2 – Design Rules*, CEC DG III/8076/89, 23–49.

Tassios, T.P. and Vintzeleou, E.N. (1987). Concrete-to-concrete friction. *Journal of Structural Engineering*, ASCE, **113**(4), 832–849.

Tassios, Th. (2009). *Theory for the Design of Repair and Strengthening.* Symmetry Publishers, Athens (in Greek).

Taylor, D. (1999). *Buildings: Design for Damping.* Taylor Devices Inc, North Tonawanda.

Tegos, I. and Penelis, G.G. (1988). Seismic resistance of short columns and coupling beams reinforced with inclined bars. *ACI Structural Journal*, **85**(1), 82–88.

Tegos, I.A. (1979). *A Systematic Analysis of Damage in R/C Buildings.* A Seminar for the Repair of Buildings with Earthquake Damage, Ministry of Public Works, Thessaloniki (in Greek).

Tegos, I. (1984). Contribution to the study and improvement of earthquake-resistant mechanical properties of low slenderness structural elements. Ph.D. thesis, Aristotle University of Thessaloniki (in Greek).

Teng, J.G., Chen, J.F., Smith, S.T. and Lam, L. (2002). *FRP Strengthened R/C Structures.* Wiley, Chichester, West Sussex, UK.

Terzaghi, K., Peck, B.R. and Mesri, Gh. (1996). *Soil Mechanics in Engineering Practice* 3rd edition), John Wiley & Sons Inc., New York.

Tiedemann, H. (1980). Statistical evaluation of the importance of non-structural damage to buildings. *Proceedings of the 7th World Conference on Earthquake Engineering*, Istanbul, VI, pp. 617–624.

Tolis, S.V. and Faccioli, E. (1999). Displacement design spectra. *Journal of Earthquake Engineering*, **3**(1), 107–125.

Triantafillou, T.C. (1998). Strengthening of structures with advanced FRPs. *Progress in Structural Engineering and Materials*, **1**(2), 126–134.

Tsamkirani, A. (1982). *Structural Dynamics* (in Greek). Ziti, Thessaloniki, Greece.

Tsionos, A.G., Tegos, I.A. and Penelis, G.G. (1992). Seismic resistance of type 2 exterior beam-column joints reinforced with inclined bars. *ACI Structural Journal*, **89**(1), 3–12.

Tsonos, A. (1999). Lateral load response of strengthened reinforced concrete beam-to-column joints. *ACI Structural Journal Proceedings*, **96**(1), 46–56.

Tsonos, A. (2001). Seismic retrofit of R/C beam-to-column joints using local three-sided jackets. *Journal of European Association for Earthquake Engineering*, I, 48–64.

Tsonos, A. (2008). Effectiveness of CFRP-jackets in post-earthquake and pre-earthquake retrofitting of beam-column subassemblages. *Engineering Structures*, 30, 777–793.

Tsonos, A.G. (2002). Seismic repair of exterior R/C beam-to-column joints using two-sided and three-sided jackets. *Structural Engineering and Mechanics*, 13(1), 17–34.

Tsonos, A.G., Tegos, I.A. and Penelis, G.G. (1995). Influence of axial force variations on the seismic behaviour of exterior beam-column joints. *European Earthquake Engineering*, IX(3), 51–63.

Uang, C.-M. and Bertero, V. (1991). UBC seismic serviceability regulations: Critical review. *Journal of Structural Engineering, ASCE*, 117(7), 2055–2068.

Umehara, H. and Jirsa, J.O. (1984). Short rectangular R/C columns under bidirectional loading. *Journal of Structural. Engineering ASCE*, 110(3), 605–618.

UNDP (1977). *Repair of Buildings Damaged by Earthquakes*, United Nations Development Programme, New York.

UNIDO/UNDP (1975). *Post-Earthquake Damage Evaluation and Strength Assessment of Buildings under Seismic Conditions*. Vol. 4 of Building construction under seismic conditions in the Balkan Region. United Nations Development Programme, Vienna.

UNIDO/UNDP (1983). *Repair and Strengthening of Reinforced Concrete, Stone and Brick Masonry Buildings*, 5, Vienna.

UNIDO/UNDP (1984). *Repair and Strengthening of Historical Monuments and Buildings in Urban Nuclei*, 6, Vienna.

UNIDO/UNDP PR. RER/79/015 (1985). *Post-Earthquake Damage Evaluation and Strength Assessment of Buildings under Seismic Conditions*, Vol. 4, Chap. 2, Vienna.

Valiasis, N.T. (1989). Experimental investigation of the behaviour of R/C frames infilled with masonry panels subjected cyclic horizontal load-analytical modeling of the masonry panel. Ph.D. thesis (in Greek), Aristotle University of Thessaloniki, Greece.

Valiasis, N.T. and Stylianidis, C.K. (1989). Masonry infilled R/C frames under horizontal loading. *Experimental Results. European Earthquake Engineering*, 3, 10–20.

Valiasis, N.T., Stylianidis, C.K. and Penelis, Gr. G. (1993). Hysteresis model for weak brick masonry infills in R/C frames under lateral reversals. *European Earthquake Engineering*, 1, 3–9.

Vamvatsikos, D. and Cornell, C.A. (2002). Incremental dynamic analysis. *Earthquake Engineering and Structural Dynamics*, 31(2), 491–514.

Vassiliou, G. (1975). Behaviour of repaired R/C elements, Ph.D. Thesis, NTU, Athens (in Greek).

Veletsos, A.S. and Meek, J.W. (1974). Dynamic behaviour of building – foundation systems. *Earthquake Engineering Structural Dynamics*, 3(2), 121–138.

Veletsos, A.S. and Nair, V.V.D. (1975). Seismic interaction of structures on hysteretic foundations. *ASCE, Journal of Structural Division*, 101(ST1): 109–129.

Vidic, T., Fajfar, P. and Fischinger, M. (1994). Consistent inelastic design spectra. Strength and displacement. *Earthquake Engineering and Structural Dynamics*, 23(5), 507–521.

Vintzeleou, E. (1987). *Behaviour of Infilled Frames Subjected to Lateral Actions. A State-of-the-Art Report*. EC8 Editing Panel, Brussels.

Vintzeleou, E. and Panagiotidou, E. (2007). An empirical model for predicting the mechanical properties of FRP-confined concrete. In Triantafillou T.C. (ed.) *8th International Symposium on FRP Reinforcement of Concrete Structures*, Paper 6-4, Patras, Greece.

Vintzeleou, E.N. and Tassios, T.P. (1986). Mathematical models for dowel action under monotonic and cyclic conditions. *Magazine of Concrete Research*, 38(134), 13–22.

Walraven, J.C. (2002). Delft, University of Technology, Background document for EN1992-1-1. Eurocode 2: Design of Concrete Structures-Chapter 6.2: Shear, Delft.

Warburton, G.B. (1976). *The Dynamical Behaviour of Structures*, 2nd ed., Pergamon Press, Oxford.

Whitman, R.V. and Richart, F.E. (1967). Design procedures for dynamically loaded foundations. *Journal of Soil Mechanics and Foundation Division*, 93(SM6), 169–191.

Wiegel, L.R., ed. (1970). *Earthquake Engineering*, Prentice-Hall, New Jersey.

Wight, J.K. and Sozen, M.A. (1975). Strength decay of R/C columns under shear reversals. *Journal of the Structural Division, ASCE*, 101(STS), Proc. Paper 11311, 1053–1065.

Wilson, E.L. (1985). A new method of dynamic analysis for linear and nonlinear systems. *Finite Elements in Analysis and Design*, **1**(1), April, 21–23.

Wilson, E.L. (2002). *Three Dimensional Static and Dynamic Analysis of Structures*. Computers and Structures Inc. (CSI), Berkeley, California.

Wilson, E. and Habibullah, A. (1987). Static and dynamic analysis of multistory buildings including P-δ effects. *Earthquake Spectra*, **3**(2), 289–298.

Wilson, E.L. and Button, M.R. (1982). Three-dimensional dynamic analysis for multicomponent earthquake spectra. *Earthquake Engineering and Structural Dynamics*, **10**, 471–476.

Youd, T.L. (1998). *Screening Guide for Rapid Assessment of Liquefaction Hazard at Highway Bridge Sites*. National Center for Earthquake Engineering Research, Buffalo, New York, Technical Report MCEER-98-0005.

Yugoslav National Report (1982). *On Damage Evaluation and Assessment of Earthquake Resistance of Existing Buildings*, Pr. Rep. 13/05 UNIDO/UNDP, Skopje.

Zararis, P. (2002). *Design Methods of Reinforced Concrete* (in Greek). Kyriakides Bros. Ltd, Thessaloniki.

Zavliarii, K. (1994). Repairs in R/C structures. General report. *Proceedings of the 11th Hellenic Conf. on Concrete*, Techn. Chamber of Greece, Corfu (in Greek).

Zografos, P. (1987). Repair of columns with R/C jackets. Reports of the R/C Lab, AUT, Thessaloniki (internal report).

Index